Microdevices: Physics and Fabrication Technologies

Series Editors
Arden Sher
SRI International
Menlo Park, CA
USA

Marcy A. Berding
Applied Optics Laboratory
SRI International
Menlo Park, CA
USA

For other titles published in this series, go to
http://www.springer.com/series/6269

Junhao Chu • Arden Sher

Device Physics of Narrow Gap Semiconductors

Springer

Junhao Chu
National Laboratory for Infrared Physics
Shanghai Institute of Technical Physics, CAS
Shanghai 200083, China
jhchu@mail.sitp.ac.cn

and

Laboratory of Polar Materials and Devices
East China Normal University
Shanghai 200241
jhchu@sist.ecnu.edu.cn

Arden Sher
SRI International (Retired)
333 Ravenswood Ave.
Menlo Park
CA 94025-3493, USA
arden-sher@comcast.net

ISBN 978-1-4614-2480-2 e-ISBN 978-1-4419-1040-0
DOI 10.1007/978-1-4419-1040-0
Springer New York Dordrecht Heidelberg London

Softcover reprint of the hardcover 1st edition 2010

Printed on acid-free paper

Springer is part of Springer Science+Business Media (www.springer.com)

Foreword

The subject of semiconductor science and technology gained importance in the latter half of the twentieth century with the first stages studies of research on group IV elements and III–V group binary compound semiconductors. Much of the early device work concentrated on PbSnTe alloys, but this material had several problems related to its high-dielectric constant. The III–V compound semiconductor with the narrowest gap was InSb (energy gap is 0.18 eV at room temperature) and was not suited to the mid-wave or long-wave infrared ranges. To satisfy the demand of the infrared detection technology, a narrow gap semiconductor alloy appeared in 1959, the ternary alloy $Hg_{1-x}Cd_xTe$ based on HgTe and CdTe, from which narrower bandgaps can be reached. By changing the composition value of x, it was possible to meet the demands of infrared detectors in several important wave bands. Meanwhile, some interesting physical properties appeared due to the narrower energy bandgap. It immediately attracted intensive research attention. Professional discussions related to this material appeared around 1976. In 1976, an international conference named "Narrow Gap Semiconductors" was initiated. From then on, several other conferences related to "II–VI Group Semiconductors" or "Narrow Gap Semiconductors" have been held, for example, the conferences sponsored by NATO. The scale of every conference was large so that their proceedings formed the basis of an important branch of semiconductor science. During this rapid developmental process, several proceedings and review papers related to this type of semiconductor were published. To my knowledge, this book is the first international treatise which fully reviews the research results of narrow gap semiconductors.

One of the authors of this book, Professor Junhao Chu, has made many contributions to the field of narrow gap semiconductors. He is considered one of the most suitable authors for a comprehensive book on narrow gap semiconductor physics. He was invited to update the database of Hg-based semiconductors for the new versions of "Landolt-Bornstein: Numerical Data and Functional Relationships in Science and Technology" published in 1999 and 2008.

It has been my pleasure to read this book and the authors' previous book ("Physics and Properties of Narrow Gap Semiconductors," Springer 2008) in detail. They explore many research fields of HgCdTe, including basic crystallographic properties and growth methods from early studies to present research, principles and measurements of physical phenomena, and so on. Therefore, this book is not only a necessary reference for those who undertake research, teaching, and industrial work on narrow gap materials, but also of value for those undertaking work related to all semiconductors.

Prof. Ding-Yuan Tang
Member of Chinese Academy of Sciences
Shanghai Institute of Technical Physics, CAS

Preface

The physics of narrow gap semiconductors is an important branch of semiconductor science. Research into this branch focuses on a specific class of semiconductor materials which have narrow forbidden band gaps. Past studies on this specific class of semiconductor materials have revealed not only general physical principles applicable to all semiconductors, but also those unique characteristics originating from the narrow band gaps, and therefore have significantly contributed to science and technology. Historically, developments of narrow gap semiconductor physics have been closely related to the development of the science and technology of infrared optical electronics in which narrow gap semiconductors have played a vital role in detectors and emitters, and other high speed devices. This book is dedicated to the study of narrow gap semiconductors and their applications. It is expected that the present volume will be valuable not only in the understanding of the fundamental science of these materials but also the technology of infrared optical electronics.

There have been several books published in this field in recent decades. In 1977, a British scientist, D. R. Lovett, published "Semimetals and Narrow-Band Gap Semiconductors" (Pion Limited, London). Later, German scientists R. Dornhaus and G. Nimtz published a comprehensive review article in 1978, whose second edition, "The Properties and Applications of the HgCdTe Alloy System, in Narrow Gap Semiconductors," was reprinted by Springer in 1983 (Springer Tracts in Modern Physics, Vol. 98, p.119). These two documents included systematic discussions of the physical properties of narrow-gap semiconductors and are still important references in the field. In 1980, the 18th volume of the series "Semiconductors and Semimetals" (edited by R.K.Willardson and Albert C. Beer) in which very useful reviews were collected, was dedicated to HgCdTe semiconductor alloys and devices. In 1991, a Chinese scientist, Prof. D. Y. Tang, published an important article, "Infrared Detectors of Narrow Gap Semiconductors," in the book "Research and Progress of Semiconductor Devices" (edited by S. W. Wang, Science Publishers, Beijing, p.1–107), in which the fundamental principles driving HgCdTe-based infrared radiation detector technology were comprehensively discussed. In addition, a handbook, "Properties of Narrow Gap Cadmium-based Compounds" (edited by P. Capper), was published in the United Kingdom in 1994. This handbook contains a number of research articles about the physical and chemical properties of HgCdTe narrow-gap semiconductors and presents various data and references about Cd-based semiconductors.

This book, "Device Physics of Narrow Gap Semiconductors," is the second in a two–volume sequence on narrow gap semiconductors. The first volume, "Physics and Properties of Narrow Gap Semiconductors," (Springer 2008), focuses mainly on materials physics and fundamental properties. Both volumes describe a variety of narrow gap semiconductor materials and revealing the intrinsic physical principles that govern their behavior. In this book, narrow gap semiconductors are presented within the larger framework of semiconductor physics. In particular, a unique property of this book is its extensive collection of results of research deduced by Chinese scientists, including one of the authors of this book, although the results are integrated into the larger body of knowledge on narrow gap semiconductor materials and devices. In organizing the book, special attention was paid to bridging the gap between basic physical principles and frontier research. This is achieved through extensive discussions of various aspects of the frontier theoretical and experimental scientific issues and by connecting them to device-related technology. It is expected that both students and researchers working in relevant fields will benefit from this book.

The effort was encouraged by Prof. D. Y. Tang. J. Chu is most grateful to Prof. D. Y. Tang's critical reading of the manuscript and invaluable suggestions and comments. A. Sher is indebted to Prof. A-B Chen for invaluable suggestions. The authors are also grateful to numerous students and colleagues who over the years have offered valuable support during the writing of this book. They are Drs: Y. Chang, K. Liu, Y. S. Gui, X. C. Zhang, J. Shao, L. Chu, Y. Cai, B. Li, Z. M. Huang, X. Lü, L. He, M. A. Berding, and S. Krishnamurthy. We are indebted to Professor M. W. Müller for his careful reading of the English manuscript of Vol. I. The electronic files of the manuscripts were edited by Dr. H. Shen and Dr. X. Lü.

The research of J. Chu's group as is presented in this book was supported by the National Natural Science Foundation of China, the Ministry of Science and Technology of the People's Republic of China, the Chinese Academy of Science, and the Science and Technology Commission of the Shanghai Municipality.

March 2009

Junhao Chu
Arden Sher

Contents

Chapter 1
Introduction

This is the second volume of a two-volume set describing the properties of narrow gap semiconductors and devices made from them. The first volume entitled "Physics and Properties of Narrow Gap Semiconductors" sets the stage for the present volume. In the first volume most of the fundamental properties of the narrow gap semiconductors, both theoretical and experimental, are developed, while in this volume these fundamental properties are extended to those more specific to device physics. This volume contains extensive reviews of defects in semiconductors, recombination mechanisms, two-dimensional surface gases formed in inversion layers, superlattices and quantum wells, and a variety of devices.

Chapter 2 begins with a theoretical study of defects, shallow and deep impurity states as well as native point defects, and continues with an extensive treatment of experimental methods to characterize these states. The methods deal with defect concentrations resulting from different fabrication strategies, and their characteristic states in the band gap and those that resonate in the conduction and valence bands. The theory ranges from Green function methods based on a tight binding model to more accurate ones based on density functional theory with full potentials in the linearized muffin tin orbital approximation. Experimental methods that are treated to determine defect concentrations and their energy states include neutron, electron, and X-ray scattering; chemical techniques; and capacitance voltage, deep level transient, admittance, transmission, photoluminescence, and frequency swept conductance spectroscopes. These methods yield not only defect densities and their states, but also their state band widths, and their impact on carrier populations and lifetimes.

Chapter 3 focuses on recombination mechanisms. The mechanisms treated include both those that are intrinsic (Auger and radiative), and extrinsic (trap assisted). Auger recombination arises from two principal mechanisms, denoted Auger 1 and 7. Auger 1 is a mechanism in which an electron and a hole recombine with the excess energy taken up by a second electron in the conduction band being excited into a higher state. Auger 7 is a process in which the recombination energy excites an electron from the light hole band into an empty state in the heavy hole band. This process is more important in p-type material, while Auger 1 is more important in n-type material. Radiative recombination is one in which an electron and hole recombine with

J. Chu and A. Sher, *Device Physics of Narrow Gap Semiconductors*, Microdevices,
DOI 10.1007/978-1-4419-1040-0_1, © Springer Science+Business Media, LLC 2010

the excess energy taken up by the emission of a phonon. If the material has a direct band gap the process can proceed by a vertical process since photons have almost no momentum, but for an indirect band gap material the process must be accomplished with the assistance of a phonon to conserve both energy and momentum.

Trap-assisted recombination is a theory treated by several people but the one most quoted is Shockley-Read recombination theory. The simplest version of this theory deals with a situation where there is a trap with a state in the band gap that is neutral when it is empty and can accept a single electron so it is negative when full. The theory starts from two continuity equations, one for the electrons and the other for the holes. This provides two equations in three unknowns, the electron, the hole, and the trap populations. In the Shockley-Read treatment they make the so-called steady state approximation in which they assume that the rate at which the electron and hole populations change is the same. This provides the third equation. With this approximation, the problem is reduced to one that has a single relaxation time that depends on the net trap population, the trap energy, the Fermi level, and the temperature. This theory has been used extensively by many authors to treat a variety of problems, so while it is an unjustified approximation it appears repeatedly throughout the reminder of the book.

The third equation in a proper treatment, done here for the first time, is the charge neutrality condition. It along with the two continuity equations forms a set of nonlinear first-order differential equations that can be reformulated as a second-order nonlinear differential equation. The solutions to this equation have at least two exponentials with different relaxation times. Special cases are treated that display details clearly differing from those of the Shockley-Read approximate solutions. In view of this development, while it is not done in this book, much of the literature of trap-assisted recombination needs to be reformulated.

This chapter goes on to prescribe many experimental methods to measure the various relaxation processes along with results. The measurement methods include: optical modulation and infrared absorption, microwave reflection following pulsed irradiation, decay of the photoconductivity, scanning photoconductivity to measure the spatial variation of the decay, and the temperature dependence of these methods to separate out the Auger, radiative and trap-assisted contributions to the decay.

Chapter 3 deals with two-dimensional electron gases (2DEG) systems. These 2DEG systems occur when the bias voltages in MIS structures are such that they are in inversion, or even sometimes when they are in depletion. Then the potential adjacent to the insulator has a sharp variation that allows quantum states to form in the resulting potential well. Fixed charge in the insulator as well as slow and fast defect states can effect these quantum states. All these possibilities are captured in a theory of these quantum states. The theory of these 2DEG's reveals their energy levels and populations as functions of the spatial variation of the well, the effective Fermi level, and the temperature. Experimental methods to measure the responses of these states on both n- and p-type samples include: *C-V* spectroscopy, SdH oscillations as a function of magnetic field orientation and also the Fourier transform of the oscillations, and cyclotron resonance. These measurements reveal the ground state and some excited state energies of the 2DEG, as well as their populations, and broadening mechanisms. Finally there are experiments to explore the impact these effects have on various devices, particularly those that depend on conductivity or magnetoresistance.

Chapter 4 delves into effects arising from superlattices and quantum wells. The band structures of superlattices composed of multiple layers of low-dimensional materials depend on boundary conditions derived from envelope functions. The theory treats interface states among materials that do not have a lattice match as well as those that do. It also deals with superlattices in which at least one material has an inverted band structure, e.g. HgTe/CdTe. The influence of the presence of a magnetic field is also treated. Once again a collection of experiments to explore the properties of these materials are presented with an emphasis on the HgTe/CdTe superlattice. Finally there is a section devoted to metal/semiconductor interfaces with blocking contacts. The quantum Hall effect is also discussed in this section.

Finally, Chapter 5 begins by treating a collection of homo- and/or heterojunction semiconductor alloy-based infrared detector devices ranging from photoconductors to various photodiodes. Expressions for the responsivity (the ratio of the signal voltage to the incident radiative power), and the detectivity (the ratio of the signal to noise voltages divided by the incident optical power, times a normalization factor consisting of the square root of the product of the device area and the noise band width) are deduced. Several parameters important to device performance are explored in some detail. These include the quantum efficiency (the ratio of the incident photon flux to that actually absorbed by the device), all the different noise sources in the various operating modes, and the operating temperature. Once again the results of numerous experiments to explore these properties are presented.

Then a section is devoted to metal-insulator-semiconductor (MIS) structured devices. Now the metal/semiconductor blocking contact Schottky barrier replaces the p-n junction of the diode. This introduces some modifications to the factors influencing the operating mode analysis, but they are minor. Once again both theory and experiments are presented.

The next sections treats quantum well infrared photodetector (QWIP) based devices as well as other QW structures. These devices consist of a periodic set of QWs each imbedded in barrier layers. The quantum wells contain states trapped in them. There are numerous operating modes that have been tried. Some function by absorbing photons that excite electrons from a bound state into the continuum where they are collected because of an externally applied field. Others function by photon absorption exciting electrons from one bound state to another bound state from which they progressively tunnel under the influence of an external field through a succession of excited bound states until they are collected. There are two classes of QWIP materials, those with a small conduction band offset, e.g., GaAs/AlGaAs and InGaAs/AlGaAs, and those with a large conduction band offset, e.g., InAs/InGaSb and InAs/InAsSb. But for both classes because of the selection rules for the absorption among two-dimensional quantum wells, only the component of the incident radiation that is parallel to the interface is absorbed. This restricts the incident optical system, and along with a requirement for low-temperature operation severely limits the quantum efficiency of QWIPs. To some extent this failing is mitigated by quantum dots (QD) based detectors. Because QDs have three dimensional wells, selections rules allow normal incident photons to be absorbed, and while they are harder to fabricate they function better than QWIPs.

A subsequent section deals with quantum cascade lasers (QCL). They constitute the first semiconductor lasers that are tunable over a wide range of wavelengths from the near to the far infrared. The QCLs consist of a sequence of layers each containing an injector and a QW. The large tunable range is accomplished by changing the QW widths and/or their barrier heights. The materials used in QCLs are III-V compound semiconductor alloys grown on GaAs or InP substrates. A number of different operating modes have been devised and tested that feature different alloys and structures. A detailed evaluation of the benefits and disadvantages of these modes is presented.

The final section treats single-photon detectors ranging from traditional photomultiplier tubes (PMT) that operate in the visible at room temperature, to QD single-photon transistor (SPT) devices that operate in the FIR (175–210 μm) at $\sim$0.04 K. The general designation of this class of devices is avalanche photodiodes (APD). The fundamental mechanism involved is avalanche gain produced from the primary photoexcited electro/hole pair by further exciting them into hot electrons or holes in a high electric field. Not only the signal but also the noise mechanisms are treated. There are many modes of APDs that have been built from several materials, including Si, Ge, and InGaAs/InP, and tested in the NIR. Some have been demonstrated to operate at room temperature but function better at 77 K. Single-photon avalanche diodes (SPAD) devices are called out for special attention. Coupled with quenching circuits these devices detect single photon events with quenching times of a few nanoseconds. Thus, they are suitable for digital communications applications. Finally, a new class of single-electron transistor (SET) APD devices made from QDs in a high magnetic field has been demonstrated. The absorption is due to transitions among Landau levels. They operate at very low temperatures ($\sim$0.04 K), and function in the FIR (175–210 μm) in circuits that respond at frequencies above 10 MHz.

In sum, we hope that this detailed discussion and review of the extensive literature on the properties and device physics of narrow gap semiconductors fills an important need. The literature on this subject as reviewed here is extensive, and we believe the addition of some papers, previously published only in Chinese, adds important aspects.

Chapter 2
Impurities and Defects

Investigating impurities and defects for any semiconductor material is an important topic. Much research has been devoted to impurity and defect states in wide-gap semiconductor materials. For the pseudobinary semiconductor alloy HgCdTe (mercury cadmium tellurium (MCT)), which is a good material for preparing infrared detectors, the investigation of its defects has a special significance. The behavior of impurities and defects in HgCdTe has been discussed in many papers in recent years. However, the research on impurities and defects of HgCdTe has encountered considerably more complexity and difficulty than that encountered in other semiconductors because of its narrow band gap, the low conduction band effective mass, the ease with which Hg vacancies are formed, and complex with other native point defects and impurities. Despite these difficulties, research in recent years have provided a basic description of impurities and defects, and their diffusion and photoelectric behavior in HgCdTe.

For the narrow gap semiconductor material HgCdTe, we need to know which kind of impurities and defects exist in the material, their chemical composition and electrical activity, if they are p-type or n-type, the magnitude of the impurity concentration, the ionization energies of these defects, their impact on electrical and optical properties, how to experimentally observe their properties, and how to theoretically analyze their properties.

2.1 Conductivity and Ionization Energies of Impurities and Native Point Defects

2.1.1 Defects

One of the major differences between real crystals and ideal crystals is that, in a real crystal, there are many defective areas where the regular arrangement of atoms has been destroyed. Defects result from many factors such as impurity atoms, growth aberrations, e.g., dislocations or planar defects, point defects resulting in an excess or a deficiency of some elements in the compound. The primary defects in an HgCdTe (Swink and Brau 1970; Yu 1976; Bye 1979; Mirsky and Shechtman 1980;

J. Chu and A. Sher, *Device Physics of Narrow Gap Semiconductors*, Microdevices,
DOI 10.1007/978-1-4419-1040-0_2,

Wang et al. 1984; Cheung 1985; Cole et al. 1985; Bubulac et al. 1985; Cai 1986; Kurilo and Kuchma 1982; Datsenko et al. 1985; Rosemeier 1983; Petrov and Gareeva 1988; Schaake 1988; Yu et al. 1990; Chen 1990; Dean et al. 1991; Shin et al. 1991; Wang et al. 1992; Yang 1988) mainly derive from (1) intrinsic point defects, e.g., vacancies, interstitial atoms, antisites, and complexes of these defects; (2) impurities; (3) multidimensional structural defects, such as dislocation, grain boundaries, and strains. Thus the geometry of lattice defects includes point, line, planar and bulk defects.

Because of a Maxwell distribution of atomic kinetic energies during crystal growth, some atoms always have a sufficient kinetic energy to leave lattice sites and be excited into "interstitial" positions to create a "vacancy–interstitial" pair. This pair can either be bound or unbound. When it is unbound it can be treated as the "evaporation" of an atom from a lattice site into a remote interstitial position. This special defect type is called a "Frenkel" defect.

Now we turn to another kind of defect. During growth or annealing "Holes" may form on the surface of crystal and then diffuse into the interior of crystal. This kind of defect is called "Schottky" defect. It is the absence of an atom on a lattice site with the extra atom in the vapor phase. Thus for example, on the surface of a HgCdTe crystal, a Hg atom may evaporate leaving a V_{Hg} on the surface which then diffuses into the interior of crystal, to form a "Schottky" defect. There is also the possibility that the extra atom remains bound on the surface to cause the effective thermal expansion coefficient to be larger than otherwise expected. Both "Frenkel" and "Schottky" defects are caused by thermal motion within the crystal lattice. They are called thermally induced defects. There are many other thermally induced native point defects besides Frenkel and Schottky defects but these two are given special attention because they are so common.

The densities of thermally induced defects are decided by the temperature of crystal and the formation energies of defects. Let ΔU_{F} denote the formation energy of a Frenkel defect, the work required to excite the atom from a lattice site into a remote interstitial position. In general there are a number of nonequivalent interstitial sites in the lattice, but usually one type of site has a much lower value of ΔU_{F} than the others. This is the site one identifies with the Frenkel defect. Let N and N' be respectively the number of lattice sites and interstitials. Suppose at a temperature T, n atoms have been excited from lattice sites to interstitial position to form a vacancy interstitial pair. If under this condition the system is in equilibrium there will be a increase in the entropy. The system entropy S can be written as:

$$S = k_{\mathrm{B}}(\ln P' + \ln P) \tag{2.1}$$

where k_{B} is the Boltzmann constant, P' is the number of n atoms distributed among the N' interstitial positions, and P is the number of n vacancies distributed among the N lattice sites:

$$P' = \frac{N'!}{(N'-n)!n!}, \tag{2.2}$$

$$P = \frac{N!}{(N-n)!n!}. \tag{2.3}$$

Substituting (2.2) and (2.3) into (2.1), and using Stirling's formula (for a large value of x, then $\ln x! \cong x(\ln x - 1)$), we get:

$$S = k_B\{[N \ln N - (N-n)\ln(N-n) - n \ln n] \\ + N' \ln N' - (N'-n)\ln(N'-n) - n \ln n]\} \tag{2.4}$$

Also atoms jumping into interstitial sites leads to an increase of the internal energy of crystal lattice:

$$\Delta W = n\Delta U_F \tag{2.5}$$

The thermal equilibrium condition occurs when the free energy $F = W - TS$ is a minimum as a function of n:

$$\frac{\partial F}{\partial n} = \Delta U - k_B T \ln \frac{(N-n)(N'-n)}{n^2} = 0. \tag{2.6}$$

This gives the result:

$$n = \sqrt{(N-n)(N'-n)} \exp\left(-\frac{\Delta U_F}{2k_B T}\right). \tag{2.7}$$

Because both $N \gg n$ and $N' \gg n$, (2.7) can be written as:

$$n = \sqrt{NN'} \exp\left(-\frac{\Delta U_F}{2k_B T}\right). \tag{2.8}$$

The factor of 1/2 in the exponent in (2.8) is due to the existence simultaneously of two kinds of defects in the crystal, interstitials and vacancies.

Similarly, we can analyze of the number of Schottky defects, with the extra atom in the vapor phase, obtaining:

$$n = N \exp\left(-\frac{\Delta U_S}{k_B T}\right), \tag{2.9}$$

where ΔU_s is the formation energy of a Schottky defect.

Generally speaking, the formation energy of defect is temperature dependent. Assuming the formation energies depend linearly on the temperature, then we find:

$$\Delta U_F = \Delta U_{F0} - T\frac{\partial(\Delta U_F)}{\partial T} = \Delta U_{F0} - \alpha T, \tag{2.10}$$

and

$$\Delta U_S = \Delta U_{S0} - T\frac{\partial(\Delta U_S)}{\partial T} = \Delta U_{S0} - \beta T, \tag{2.11}$$

where ΔU_{F0} and ΔU_{S0} are the formation energies of defects at absolute zero, and α and β are constants.

Substituting (2.10) and (2.11) into the expression for n, leads to the results:

$$n = B_F \sqrt{NN'} \exp\left(-\frac{\Delta U_F}{2k_B T}\right), \tag{2.12}$$

and

$$n = B_S N \exp\left(-\frac{\Delta U_S}{k_B T}\right). \tag{2.13}$$

The values of B_F and B_S can be approximated by fits to experiments. For all crystals, both B_F and B_S lie in the range 2–50. For HgCdTe crystals, F defects and/or S defects form under different annealing conditions.

Besides thermal defects, there is another class of defects, irradiation-induced defects. When a high-speed particle such as a neutron, an α particle, deuteron, fragments of nuclear fission, γ radiation, a high-energy electron, a high-energy ion, or high-energy laser irradiates a crystal, it induces structural failures mostly in the form of "Frenkel" defects. When for example, γ radiation irradiates the crystal, secondary photoelectrons and Compton electrons are produced which in turn induce structural defects. These kinds of crystal defects generated by irradiation are named irradiation defects. Irradiation defects are different from thermal defects. After irradiation, irradiation-induced defects are unstable and the crystal is not in thermodynamic equilibrium. When a crystal is annealed after irradiation, many irradiation-induced defects will diffuse quickly to recombine.

When a massive neutral particle or charged high-speed particle irradiates a crystal, elastic collisions will take place between the high-speed incident particle and the atomic nuclei of the crystal. High-speed incident particles will cause the electrons bound in atoms of the crystal to be excited and to ionize. Also, an incident high-speed particle can induce a nuclear reaction leading to the activation of some atoms within crystal. These atoms will become impurity centers. However at lower incident energies in semi-conductor crystals, the excitation and ionization of the valance electrons are most prevalent.

When an elastic collision between a high-speed particle and the atoms of a crystal takes place, elastic waves will form in the crystal to transform energy into the thermal motion energy of the atoms. Ultimately this results in a structural failure of the crystal. If the energy of the atoms at a lattice point exceeds a critical value U_d of the formation energy of a defect in the crystal, it will result in the structural failure of crystal lattice. Generally speaking, U_d needs to be two or three times that required for atoms go from the lattice to an interstitial site. Because the binding energy of most crystal atoms is $\sim$10 eV, the value of U_d is about 25 eV. So for an irradiated crystal, only those atoms with the energy $U \geq U_d$ will form a Frenkel defect.

Now we introduce a kinetic energy parameter ε related to the kinetic energy of a moving particle:

$$\varepsilon = \frac{m}{M_1} E, \tag{2.14}$$

where m is the electron mass, M_1 the mass of the moving particle, E the energy of the moving particle, and ε is the energy of an electron that has the same speed as the moving particle. When $\varepsilon >> \varepsilon_i$ where ε_i is the excitation energy of a valence electron, the majority of the energy of the incident moving particle will be expend in the processes of excitation and increased ionization of the lattice ions, with little of the energy expend in elastic collisions. If the values of the parameter ε approaches ε_i, the processes of excitation and ionization become difficult and only elastic collisions can take place, still resulting in the creation of elastic waves and structural defects.

In an elastic impact process, there is approximately a proportional relation between the incident energy of the high-speed particle that results in a structure defect, and the energy consumed in the crystal. This proportional relation depends on the magnitude of the incident particle energy and properties of crystal. When an elastic impact between an incident particle occurs in a crystal, the energy loss per unit distance can be written as (Jones 1934; Mott and Jones 1936):

$$-\left(\frac{\mathrm{d}E}{\mathrm{d}x}\right)_c = \frac{2\pi Z_1^2 Z_2^2 e^4 N_0}{M_2 v^2} \ln \frac{E}{E*}, \tag{2.15}$$

where Z_1 and Z_2 are the atomic number of the moving particle and a static lattice particle, respectively, N_0 is the density of atoms in the crystal, v the velocity of the moving particle, E the energy of the moving particle, e the electron charge, M_2 the mass of the static particle, and E^* is given by the following expression:

$$E^* = 0.618 \left(Z_1^{2/3} + Z_2^{1/3}\right)^2 \frac{mM_1}{4\mu^2} R, \tag{2.16}$$

where R is the Rydberg constant which is equal to 13.6 eV and μ the reduced mass between the moving particle and the static particle. Similarly, the energy expended by the high-speed particle to form structural defects per unit distance can be written as:

$$-\left(\frac{\mathrm{d}E}{\mathrm{d}x}\right)_{\text{defect}} = \frac{2\pi Z_1^2 Z_2^2 e^4 N_0}{M_2 v^2} \ln \frac{E 4\mu^2}{U_\mathrm{d} M_1 M_2}. \tag{2.17}$$

Then the ratio of the energy to form structural defects to the total energy lost is:

$$\frac{\left(\frac{\mathrm{d}E}{\mathrm{d}x}\right)_{\text{defect}}}{\left(\frac{\mathrm{d}E}{\mathrm{d}x}\right)_c} = \frac{\ln\left(\frac{E}{U_\mathrm{d}} \frac{4\mu^2}{M_1 M_2}\right)}{\ln \frac{E}{E*}}. \tag{2.18}$$

This ratio is about 0.5 for most crystals. According to (2.15), (2.17), and (2.18), we can determine approximately the factors related to forming irradiation-induced defects in crystals. These factors differ from each other for different crystals, and are related to the atomic mass of the crystal. Jones (1934) deduced the energy threshold for forming irradiation-induced defects in crystals with different atomic masses (see Table 2.1).

Table 2.1 The energy threshold of a high-speed particle for forming irradiation defects in crystals with different atomic masses

	Crystal atom			
Atomic mass	10	50	100	200
Neutron-atom (eV)	75	325	638	1,263
Electron-γ ray (MeV)	0.10	0.41	0.65	1.10
α particle (eV)	31	91	169	325
Nuclear fission chip	85	30	25	27

The density N of defects induced by high-speed incident particles irradiating a crystal can be written as:

$$N = \frac{1}{2U_\mathrm{d}}\frac{M_1}{m}\left(\varepsilon_t + \frac{m}{M_1}E \times 10^{-3}\right)\ln\left(\frac{E}{U_\mathrm{d}}\frac{4\mu^2}{M_1 M_2}\right), \tag{2.19}$$

where the physical meaning of every parameter has been given above. Although the above equations are based on a relatively simplified model, they still can be employed as an approximate reference for analyzing the formation of irradiation defects in narrow gap semiconductor materials.

In the discussion above, we have emphasized the thermal defects and irradiation defects, but in real semiconductors, the most common defect is an impurity. Impurities derived from elements differing from those on the crystal also modify the periodicity of crystal lattice and change the physical properties of crystal. Impurities can have a great influence on the electrical and optical properties of semiconductors. Impurity atoms can become donors or acceptors which influence the conductivity of the crystal. They can form trapping centers or recombination centers which will influence the lifetime of minority carriers and so influence the electrical and optical properties of a crystal.

2.1.2 Chemical Analysis of Impurity Defects and their Conductivity Modifications

The technology involved in material growth and device fabrication often introduces impurity defects into the material. The states of these impurities always determine device characteristics. So, it is essential to understand the behavior of impurity defects. Accordingly, impurity defect states in a crystal lattice can be divided into shallow energy level, deep energy level, and resonant impurity defect states. Also they can be divided into donors, acceptors, and traps. Each kind of defect will serve as a scattering and/or a recombination center during transport to decrease the mobility and lifetime of carriers. So, device performance will be influenced.

To prepare p-type, n-type, or p–n junctions in HgCdTe material, it is usually doped intentionally. We can add the impurities into the material directly in the growth process, make impurities diffuse into the material in a heat treatment process, or employ ion implantation. In the case where the material is unintentionally doped, the defects are either native point defects or residual impurities. People generally employ "$7N$" material (impurity densities $<10^7$ cm^{-3}) to prepare HgCdTe devices. However, despite growers best efforts impurities still exist in the grown crystals in sufficient numbers to influence the properties of HgCdTe crystals (Pratt et al. 1986; Shen et al. 1980). Analyzing the relation between the impurity content, after purification, in the elemental materials used to prepare HgCdTe bulk crystal, and the ones in the grown bulk crystals indicates that the impurity density is almost the same. Generally speaking, the impurities found in HgCdTe bulk crystal are mostly derived from the elemental Te, Cd, and Hg materials. However, when HgCdTe bulk material is grown, the type and density of impurities also will be influenced by the quartz tube in which it is grown because it contains Al, Te, Ca, Mg, Ti, Cu, and B impurities.

Most impurities in HgCdTe crystals are electrically active. Some impurities that are not electrically active still induce deep levels in the band gap and as a consequence will reduce the lifetime of minority carriers. Therefore, in the case of unintentionally doping, the content must be limited not just for certain specified impurities but for all of them.

Theoretically, we can judge if an impurity is an acceptor or a donor in a HgCdTe semiconductor sample based on the atomic number of the impurity, its group number in the periodic table of elements and the sublattice on which it resides. Table 2.2 gives the results.

In fact, not all of the impurity elements listed in Table 2.2 can be activated in HgCdTe. To observe the electrical activity of a particular impurity element in HgCdTe, it can be intentionally doped into the material. The carrier concentration of sample then can be determined by measuring the Hall coefficient of the sample. Then the doping concentration can be obtained from atomic absorption spectrometry. Knowing carrier concentration and doping concentration, we can find the degree of electrical activity as a function of temperature of this kind of impurity.

Many authors have carefully investigated the degree of electrical activity for many kinds of impurity elements. The results indicate that the elements In, I, Cl, Al, and Si are definitely donors, and the elements P, As, and Sb, are amphoteric being donors if substituted on the cation sublattice and acceptors on the anion sublattice, while the electrical properties of Li, Cu, and Ag are definitely acceptors.

Now we will discuss issues related to donor impurities. For example, in $Hg_{1-x}Cd_xTe$ ($x = 0.20$) crystals doped with In are n-type. The carrier concentration

Table 2.2 Impurity elements their possible roles on different sublattices and in interstitial positions

group	I	II	III	IV	V	VI	VII	VIII
Possible role	acceptor		donor		acceptor		donor	
Element	Li	Cd	B	C	N		F	Fe
	Na	Hg	Al	Si	P	Te	Cl	Ni
	K		Ga	Ge	As		Br	
	Cu		In	Sn	Sb		I	
	Ag		Tl	Pb	Bi		Mn	
	Au			Ti	V			

increases with an increase of the In concentration and the Hg partial pressure in a heat treatment process. The carrier concentration is proportional to the square root of the In doping concentration, and to the square root of the Hg pressure (Vydyanath 1991). Only a fraction of the In atoms doped into HgCdTe crystals behave as single donors when In atoms are heavily doped because In can combine with Te to form In_2Te_3 (Vydyanath et al. 1981). However, In atom impurities are fast diffusers, their diffusion coefficient is about $5 \times 10^{14}\,cm^2/Vs$ at 300 K (the activation energy is ~11 eV) (Destefanis 1985). This limits the stability of p–n junctions formed from In doping. A p–n junction can be formed when In atoms are introduced into a p-type HgCdTe substrate by ion implantation. The electrically active n-type defects caused by irradiation damage can be eliminated in an annealing process. In this kind of process the junction depth is easy to control (Destefanis 1985, 1988; Gorshkov et al. 1984).

As stated before, because the Hg–Te bond is weak, it is easy to form Hg vacancies. Therefore, people have been motivated to study the interactivity between Hg vacancies and other impurities. The In atom is a common impurity in HgCdTe. The In atom is trivalent, and when substituted for Hg and Cd it is donor. But in higher concentrations it becomes compensated. One of the potential reasons is that In combines with Te to form In_2Te_3. The other reason is that In forms In-vacancy pair. Hughes et al. have analyzed theoretically the possibility of forming an In-vacancy pair (Hughes et al. 1991). They conclude an inactive In-vacancy pair will form when the material is quenched from above 350 K down to room temperature. These In-vacancy pairs can be eliminated if the material is annealed at the temperature of 160 K (Hughes et al. 1994) in an atmosphere with a high Hg partial pressure.

In is an element widely used to obtain an n-type-doped HgCdTe material. It is a suitable element to dope bulk materials, as well as liquid phase epitaxy (LPE), molecular beam epitaxy (MBE), and MOCVD film materials. However, because it is a fast diffuser, the so-called memory effect in the process of doping in which the raw In material having adsorption on the wall of growth tube or reactor, diffuses into the prepared p-on-n junction to cause it to be graded (Gough et al. 1991; Easton et al. 1991). To mitigate this problem Maxey et al. (1991), Easton et al. (1991), and Murakami et al. (1993) turned to the element I as a dopant. The element I is a group VII element which when substituted on the Te sublattice is a donor. Because I has a relatively small diffusion coefficient very steep p-on-n junctions can be prepared. I has and very weak "memory" effect. The element I is an adequate dopant not only for inter-diffusion of multilayer films but also for direct growth of alloys.

When the electron concentration of an I-doped $Hg_{1-x}Cd_xTe$ ($x = 0.23$) sample is between 5×10^{15} and $2\times10^{18}\,cm^{-3}$ at a temperature of 77 K about 20–100% of the I is activated as donors. The carrier mobility of an I-doped $Hg_{1-x}Cd_xTe$ ($x = 0.23$) sample is greater than that of the same sample doped with In at the temperature of 20 K. The activated I doping concentration in a HgCdTe crystal is related to the Hg pressure when diffusion doping is done. The higher the Hg pressure, the larger is the I doping concentration. The element I forms compounds of $(Hg,Cd)I_2$ in a HgCdTe crystal, to limit its electrical activity (Vydyanath et al. 1981; Vydyanath and Kröger 1982; Vydyanath 1991).

Elemental O is very active, and it is easily doped into HgCdTe crystals in the process of crystal growth. Because element O is a donor, the carrier concentration of an n-type HgCdTe crystal will increase when the concentration of O atoms is high. The O atoms derived from the walls of a quartz tube diffusing into a HgCdTe crystal can be obstructed when C atoms are deposited on the wall of quartz tube. Consequently, the carrier concentration of an n-type HgCdTe crystal will be decreased. Given this de-oxidation process the carrier concentration can be accurately controlled in the range of $\pm 1 \times 10^{14}\,\text{cm}^{-3}$ (Yoshikawa et al. 1985). Some electrical parameters of an n-type $Hg_{0.7}Cd_{0.3}Te$ crystal at 77 K obtained by this method are as the follows: electron concentration $\leqslant 5 \times 10^{13}\,\text{cm}^{-3}$, electron mobility $\sim 4 \times 10^4\,\text{cm}^2/\text{Vs}$, a maximum minority carrier lifetime of $\sim 74\,\mu\text{s}$. These results indicate that the O atom is an important background impurity donor for n-type HgCdTe crystals.

Group V elemental impurities are amphoteric, acting as a donor when occupying the Hg and Cd cation sublattice, and acting as an acceptor when located on the Te anion sublattice. For example, the As atom can be doped into a HgCdTe crystal by high-temperature diffusion (Capper 1982) and ion implantation (Destefanis 1988; Baars et al. 1988; Wang 1989; Ryssel et al. 1980). In an ion implantation experiment, people found that As ions forms nonelectrically active charge compensation complexes that do not migrate easily in HgCdTe crystals. However, some inactive As atoms will be activated to be donor impurities when the material is annealed at 300 K, and to be acceptors when annealed above 400 K. In HgCdTe crystals, the electrical behavior of Sb atoms is similar to that of As atoms. The Sb atoms will be acceptors when occupying the Te sublattice and be donors when located on the Hg sublattice (Wang 1989). The electrical behavior of P atoms is relatively complex. The P atoms, when the sample is held at high Hg partial pressure, are located on interstitial positions or on the Te sublattice and become single acceptors. However, when the Hg partial pressure is low, the P atoms are located at on the cation sublattice and are single donors (Vydyanath 1990, 1991; Gorshkov et al. 1984).

Next we discuss acceptor impurities. Cu atoms located on the cation sublattice are single acceptors in HgCdTe (Vydyanath et al. 1981; Gorshkov et al. 1984). Cu deposits in a cooling process when its concentration becomes high enough to reach saturation. Since Cu deposits always collect around the positions of extended defects, one can observe these defects in HgCdTe as they are Cu decorated. This method is destructive to materials. Later we will see that the Cu atoms are derived from the walls of the quartz tube during growth. When the Cu content is relatively high, it may become the most important background acceptor impurity.

The elements P, As, Ag, and Sb are all p-type impurities in materials grown from the Hg-rich side of the existence curve, and their diffusion rates are relatively slow (Vydyanath et al. 1981, Vydyanath 1991; Gorshkov et al. 1984; Wang 1989; Ryssel et al. 1980). The elements P, As, and Sb have been used widely as the doping agents to manufacture HgCdTe photovoltaic devices. Previous research indicates that Sb-doped HgCdTe samples are more stable than the As-doped ones (Wang 1989; Capper et al. 1985).

In HgCdTe crystals, Au atoms exhibit a low electrical activity, and almost do not change the carrier concentration or the minority carrier lifetime. Often only 3% of the Au are electrically active acceptors. But Au can be used to fabricate electrodes at temperatures below 150°C (Capper 1982; Jones et al. 1983). Chu's experimental results indicate that the concentration of Au acceptors can reach $10^{18}\,\mathrm{cm}^{-3}$ (Chu et al. 1992). They evaporated Au onto the surface of samples, then used a YAG laser to irradiate the sample, followed by annealing at a temperature of 250°C for 20 h. After the Au impurities diffused into the samples, a CV measurement was done on the resulting metal–insulator–semiconductor (MIS) structure. They found that the concentration of acceptors increases as diffusion progresses, finally reaching levels of about $10^{18}\,\mathrm{cm}^{-3}$. This indicates that Au exists in HgCdTe as an acceptor.

For most of the elements, the doping concentration N and carrier concentration n lie close to the straight line $N = n$. However, the electrical activity of both Fe and Au are very low. For Fe and Au atoms with the concentrations of about $10^{17}\,\mathrm{cm}^{-3}$, their corresponding excited carrier concentrations are only about $10^{15}\,\mathrm{cm}^{-3}$. Furthermore, even when the concentrations of Fe and Au increase up to $10^{18}\,\mathrm{cm}^{-3}$, the carrier concentrations are still $\sim 10^{15}\,\mathrm{cm}^{-3}$, having saturated. The transition metal elements, such as Fe, though they show little electrical activity in a HgCdTe crystal, still can reduce the lifetime of the minority carriers markedly, and so are called "lifetime killers" (Capper 1982, 1991). In the preparation of the raw materials and subsequent growth of HgCdTe, extensive efforts are needed to control their impurity content.

The electrical activity is different in $Hg_{1-x}Cd_xTe$ crystals with different compositions. For example, the electrical activity of the Cl atom is only several percent when the concentration is $x < 0.3$, while it is almost a 100% when $x > 0.3$. The As and Sb atoms behave similar to that of Cl atoms (Capper 1991).

Approximately, the main group and the subgroup elements of group I and the main group elements of group V are acceptors; and the subgroup elements of group II, the main group elements of group III and VII are donors. Neutral impurities are relatively complex. For example, for elements of group IV, the element Si behaves as donor, while Ge, Sn, and Pb are neutral impurities, having a different behavior than that anticipated. The anticipated conductivity of impurity elements is almost consistent with those encountered in practical cases, but there are still some deviations. Table 2.3 lists the electrical behavior of 26 elements in HgCdTe crystals (Capper 1991).

2.1.3 *Theoretical Estimation Method for Impurity Levels*

It is important to observe the ionization energy of impurity defects based on an understanding of their impact on the conductivity. There are some rough empirical expressions available to characterize the ionization energy of shallow impurities. For an n-type HgCdTe material, measurements of the Hall coefficient indicate that the ionization energy of a univalent donor is almost zero. For a p-type HgCdTe material, the measurements indicate that an acceptors ionization energy increases

Table 2.3 Electrical activity of impurities in HgCdTe crystals

			Practical case						
Element	Group	Anticipated behavior	SSR (Te)	II (Te)	LPE (Hg)	LPE (Te)	MOVPE (Hg)	MOVPE (Te)	MBE (Te)
H	IA	A(m)				A			
Li	IA	A(m)	A			A			A
Cu	IB	A(m)	A	A	A	A			A
Ag	IB	A(m)	A	A	A			A	A
Au	IB	A(m)	A		A				
Zn	IIB	I(m)D(i)		D	I	I			
Hg	IIB	I(m)D(i)		D					
B	IIIA	D(m)	D	D		D			
Al	IIIA	D(m)	D	D	D	D		I/D	D
Ga	IIIA	D(m)	D		D	D		D	
In	IIIA	D(m)	D	D	D	D	D	D	D
Si	IVA	D(m)A(t)	D			D?		D	D
Ge	IVA	D(m)A(t)	I		D				
Sn	IVA	D(m)A(t)	I		D?			D	
Pb	IVA	D(m)A(t)	I						
P	VA	A(t)	I/A + A		A	I(A)	A	I	
As	VA	A(t)	I/A + A		A	I(A)	A	I	D/A
Sb	VA	A(t)	I			I(A)	A	I/D	
Bi	VA	A(t)				I			D
O	VIA	D(i)I(t)	D			D			
Cr	VIB	I(t)	I						
F	VIIA	D(t)		D		D?			
Cl	VIIA	D(t)	D						(D)
Br	VIIA	D(t)	D						(D)
I	VIIA	D(t)	D			D	D		(D)
Fe	VIII	I(m)D(t)			I				
Ni	VIII	I(m)D(t)	I						

A acceptor, donor, *I* inactive, (m) = at the lattice point of metal, (t) = at the lattice point of Te, (i) = at interstitial position, I(A) = impurity that needs to be activated by annealing at high-temperature, I/D and D/A = different results obtained from that impurities are activated under different conditions, (D) = donor in CdTe buffer layer, II = ion implantation.

proportional to the composition x, and decreases proportional to the cube root of the acceptor concentration, $N_a^{1/3}$. In the range, $0.2 < x < 0.24$, the ionization energy can be written as the following empirical formulas (Capper 1991):

A. When N_a is in the range of $2.5\text{–}3.5 \times 10^{17}$ cm^{-3}

$$E_a = 91x + 2.66 - 1.42 \times 10^{-5} N_a^{1/3} \text{(unintentionally doped)}, \qquad (2.20)$$

B. When N_a is in the range of $0.8\text{–}2.5 \times 10^{17}$ cm^{-3}

$$E_a = 42x + 1.36 - 1.40 \times 10^{-5} N_a^{1/3} \text{(intentionally doped)}. \qquad (2.21)$$

The energy level of impurity defects also can be estimated theoretically. For the group II–VI compound semiconductor HgCdTe, the anion Te has valance 6, and the cations Hg and Cd have valance 2. Thus according the simple rules, if group VII elements are substituted on the anion sublattice they are donors, and group V elements are acceptors. Similarly, if group III elements are substituted on the cation sublattice they are donors, and group I elements are acceptors. Generally, group V elements substituted on the cation sublattice are also donors and group III elements substituted on the anion sublattice are acceptors.

The energy released to the conduction band by a donor atom is called the ionization energy E_{i}. The electron concentration n of the conduction band can be obtained from Hall coefficient measurements. The electron concentration of the conduction band is proportional to $\exp(-E_{\mathrm{i}}/kT)$. So E_{i} can be obtained from the slope of the measured curve, $\ln(n) \sim 1/T$. However, for narrow gap semiconductors, the donor ionization energy E_{i} to the conduction band is very small. So the error is relatively large when E_{i} is determined by this method.

The ionization energy of any impurity in a semiconductor with shallow ionization energy can be calculated in a simple effective mass theory. Generally, for a semiconductor with a large dielectric constant and a small effective mass, the binding energy of a shallow donor impurity can be calculated by means of the effective mass approximation (EMA) method. The physical picture behind this method is that, a crystal is looked upon as continuous dielectric medium, a carrier moves like free particle in this dielectric, the effective mass related to the energy band replaces the free particle mass, and the binding potential of the impurity is controlled by the dielectric constant of the host lattice. Assuming that the distance between an impurity ion and a bound electron is r, and $\varepsilon(0)$ is the static dielectric constant of crystal, we get the interaction potential between single positive charge and the electron located at the displacement $\mathbf{r}$:

$$V(\mathbf{r}) = -\frac{e^2}{4\pi\varepsilon(0)r}. \tag{2.22}$$

The electron can be regarded as one having an effective mass m^* that is moving in a potential $V(\mathbf{r})$. This hydrogen atom-like model can then be solved to deduce the ground state energy or equivalently the ionization energy. The following equation is obtained in the EMA:

$$\left[-\frac{\hbar^2}{2m^*}\nabla_{\mathbf{r}}^2 - \frac{e^2}{4\pi\varepsilon(0)r}\right]\psi(\mathbf{r}) = E\psi(\mathbf{r}). \tag{2.23}$$

Because the potential is spherically symmetrical, $\nabla_{\mathbf{r}}^2 \Rightarrow \frac{\partial^2}{\partial r^2}$. The eigenvalues of (2.23) are:

$$E_n = -\frac{R^*}{n^2}, \tag{2.24}$$

where

$$R^* = \frac{m^* e^4}{2\hbar^2[4\pi\varepsilon(0)]^2} = \frac{m^*}{m_0}\frac{1}{\varepsilon^2(0)}R^{\mathrm{H}}, \tag{2.25}$$

and R^* is called the equivalent Rydberg energy, with R^{H} being the Rydberg energy of a hydrogen atom equal to 13.6 meV. n is an integer, so (2.24) gives a series of binding energy levels with the ground state being $-R^*$. Therefore the ionization energy is, $E_{\mathrm{i}} = R^*$. For silicon, the effective mass is $m^* = 0.2m_0$ without considering the anisotropy of mass tensor, and the value of ε is 12, so one gets $E_n = -0.0181/n^2(\mathrm{eV})$. Then the binding energy of the ground state $n = 1$ is $E_1 = -0.0181\,\mathrm{eV} = -E_{\mathrm{i}}$. For $\mathrm{Hg}_{1-x}\mathrm{Cd}_x\mathrm{Te}(x = 0.4)$, with $m^* = 0.04m_0$, $\varepsilon(0) = 15$, then $E_n = -0.0024/n^2(\mathrm{eV})$ or $E_{\mathrm{i}} = 2.4\,\mathrm{meV}$. For the heavy hole band, $m^* = 0.55m_0$, then $E_n = -0.0033/n^2(\mathrm{eV})$ or $E_{\mathrm{ihh}} = 3.3\,\mathrm{meV}$. For $\mathrm{Hg}_{1-x}\mathrm{Cd}_x\mathrm{Te}(x = 0.2)$, with $m^* = 0.01m_0$, $\varepsilon(0) = 17.5$, then we find $E_{\mathrm{i}} = 0.5\,\mathrm{meV}$.

For an electron bound to an impurity center, the extent of spatial wave functions radius, a_{n}, corresponding to "the nth Bohr orbit radius" of a hydrogen-like wave function is:

$$a_n = \frac{n^2\varepsilon(0)}{(m^*/m_0)}a_0, \tag{2.26}$$

where a_0 is the first Bohr orbit radius of hydrogen, $a_0 = 0.53 \times 10^{-8}\,\mathrm{cm}$. For a group V donor in silicon, $a_1 = 30\,\mathrm{Å}$, and for a donor in germanium, $a_1 = 80\,\mathrm{Å}$. For donors in HgCdTe, if $m^*/m_0 = 0.04$, $\varepsilon = 15$, then $a_1 = 198\,\mathrm{Å}$. Therefore, when an impurity concentration is very high, the electron wave functions bound to impurities will overlap.

If an impurity is located in an interstitial rather than a substitutional position, its energy levels may not follow the simple hydrogenic effective mass model of a shallow impurity. Its energy levels may involve an impurity complex with its wave functions overlapping neighboring atoms. Strictly speaking, even for a substitutional impurity, its ground state wave function may be localized, so its binding energy can not be calculated simply by using the effective mass model. In fact strictly speaking, the electric potential of an impurities core can not be regarded as that of point charge, and more complete approximation divides the region close to impurity into two parts, an interior and an exterior region. In the exterior region, $r > r_c$, where r_c is the same order of magnitude as the distance to a nearest neighbor, an effective mass model is approximately correct and the potential energy can be regarded as $-e^2/\varepsilon r$. In the interior region where, $r < r_c$, the potential energy of the impurities atomic core can be regarded as a δ-function potential well. In this approximation the depth of the δ-function potential well is adjusted to fit the empirically determined binding energy (Lucovsky 1965). Other theories have been used to explain the deviation of the measured impurity binding energy from that calculated using the effective mass model. For example, see the Green's function method of Chen and Sher which also applies to deep levels (Chen and Sher 1985).

Glodeanu (1967) proposed a theoretical method that can be used to calculate the deep levels of impurities directly. He studied the bivalent substitutional impurities that can forms two localized levels in band gap. He adopted a helium-like model to approximate a single electronic energy band, and then calculated the deep donor and acceptor levels of many impurities in GaAs and Si. Chen and Sher have warned that the deep levels are quite sensitive to the quality of the underlying band structures of the host material, so this method can only be expected to produce approximate results (Chen and Sher 1985).

For a donor with two valence electrons, the Hamiltonian operator can be written as follows:

$$H = -\frac{\hbar^2}{2m^*}(\nabla_1^2 + \nabla_2^2) + V(\mathbf{r}_1) + V(\mathbf{r}_2) + U_{\text{eff}}(\mathbf{r}_1, \mathbf{r}_2), \tag{2.27}$$

and the corresponding Schrödinger equation is:

$$H\Psi(\mathbf{r}_1, \mathbf{r}_2) = E\Psi(\mathbf{r}_1, \mathbf{r}_2), \tag{2.28}$$

where $V(r_1)$ is a periodic potential which denotes the interaction between one of the two electrons and the effective crystal field, similarly, $V(r_2)$ is the periodic potential of the other electron. The last term in (2.27) is an approximation to the difference between the two electron Hamiltonian and its periodic part:

$$U_{\text{eff}}(r_1, r_2) = -\frac{Z_{\text{e}}e^2}{\varepsilon r_1} - \frac{Z_{\text{e}}e^2}{\varepsilon r_2} + \frac{e^2}{\varepsilon\,|\mathbf{r}_1 - \mathbf{r}_2|}. \tag{2.29}$$

The first two terms in (2.29) denote the interaction between the two electrons and the screened positive charge Z_{e} which is introduced to keep the crystal electrically neutral. The last term is the electron–electron interaction that gives rise to correlations and exchange.

In (2.28), the wave function takes the form:

$$\Psi(\mathbf{r}_1, \mathbf{r}_2) = \frac{1}{N\Omega}\sum_{\mathbf{k}_1,\mathbf{k}_2} c(\mathbf{k}_1, \mathbf{k}_2) U_{\text{c},0}(\mathbf{r}_1) U_{\text{c},0}(\mathbf{r}_2) \times \exp(\mathrm{i}\mathbf{k}_1 \cdot \mathbf{r}_1) \exp(i\mathbf{k}_2 \cdot \mathbf{r}_2), \tag{2.30}$$

where $U_{\text{c},0}$ is a Bloch function evaluated at the location of conduction band minimum i.e., at $k_1 = 0$. In general, $U_{\text{c},0}$ satisfies the following equation:

$$\left[-\frac{\hbar^2}{2m^*}\nabla_1^2 + V(\mathbf{r}_1)\right] U_{\text{c},0}(\mathbf{r}_1) = E_{\text{c},0} U_{\text{c},0}(\mathbf{r}_1), \tag{2.31}$$

where Ω is the volume of the primitive cell, and N is the number of primitive cells.

If Ψ from (2.30) is substituted into (2.28), it reduces to a helium-like relation:

$$\left[-\frac{\hbar^2}{2m^*}(\nabla_1^2 + \nabla_2^2) - \frac{Ze^2}{\varepsilon r_1} - \frac{Ze^2}{\varepsilon r_2} + \frac{e^2}{\varepsilon\,|r_1 - r_2|}\right] F_n(r_1, r_2) = EF_n(r_1, r_2). \tag{2.32}$$

Glodeanu (1967) determined the energy using variational calculus to solve (2.32). For $F_n(r_1, r_2)$, a trial function with the following form was chosen:

$$F_n(r_1, r_2) = \frac{Z'^3}{\pi r_0^3} \exp\left[-\frac{Z'}{r_0}(r_1 + r_2)\right], \quad \text{with } r_0 = \frac{\hbar^2 \varepsilon}{m^* e^2}. \tag{2.33}$$

Then two expressions for the ionization energy are obtained:

$$E_1 = a\left[Z_{\mathrm{eff}}^2 - \frac{5}{4}Z_{\mathrm{eff}} + \frac{25}{128}\right], \text{ and } E_2 = aZ_{\mathrm{eff}}^2, \quad (2.34)$$

where a is a parameter, $a \sim m^*/\varepsilon$, and

$$Z' = Z_{eff} - \frac{5}{16}. \quad (2.35)$$

For the impurities Cr, Cd, and Zn, the effective mass used is $m^* = 0.34m_0$, A is an acceptor, D is a donor.

Because it is difficult to determine the differences among impurity atomic cores correctly, Z_{eff} in (2.34) is regarded as an undetermined parameter chosen to minimize the quadratic sum of the relative error between the theoretical and the experimental results. Table 2.4 shows the predicted theoretical ionization energy and the experimental results in GaAs, Ge, and Si. The Z_{eff} is also listed. The reference value of a donor level is the minimum of the conduction band, and the reference value of an acceptor level is the maximum of valence band.

This is a relatively rough early method. It only provides an approximate estimate.

Swarts et al. (1982) calculated the energy state of vacancies in HgCdTe using a Green function method. They used an empirical tight binding model truncated at second neighbors. Hass et al. (1983), in a paper published at nearly the same time, also used an ETB model truncated at second neighbors but they chose a different parameter set. Both added a spin-orbit Hamiltonian and used impurity potentials derived from the differences between atom term values of the impurity and the

Table 2.4 Ionization energies of deep levels of impurities in GaAs, Si, and Ge taken from theory and experiments

				E_1		E_2	
Crystals	Impurities	Types	Z_{eff}	Theoretical value	Experimental value	Theoretical value	Experimental value
GaAs	Cu	A	2	0.171	0.15	0.407	0.47
Si	S	D	2.145	0.206	0.18	0.447	0.52
	Ni	A	2.380	0.280	0.23	0.550	0.70
	Co	A	2.515	0.328	0.35	0.614	0.58
	Zn	A	2.428	0.297	0.31	0.573	0.55
Ge	Se	D	2.275	0.138	0.14	0.285	0.28
	Te	D	2.195	0.124	0.11	0.262	0.30
	Mn	A	2.475	0.177	0.16	0.322	0.37
	Co	A	2.815	0.249	0.25	0.431	0.43
	Fe	A	3.060	0.312	0.34	0.509	0.47
	Ni	A	2.753	0.234	0.22	0.413	0.44
	Cr	A	2.580	0.068	0.07	0.123	0.12
	Cd	A	2.735	0.079	0.06	0.138	0.20
	Zn	A	2.050	0.034	0.03	0.078	0.09

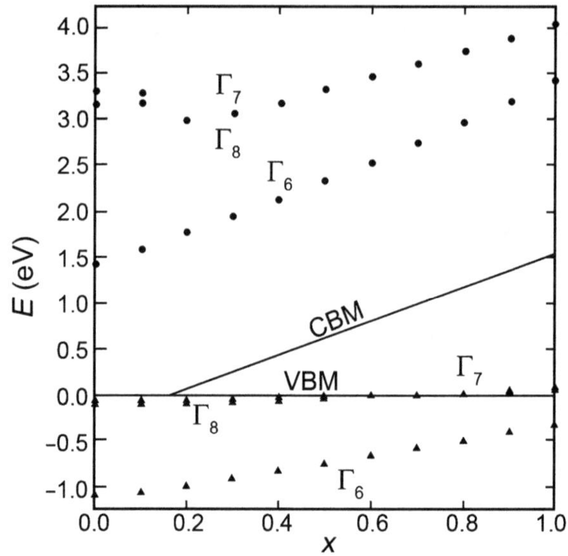

Fig. 2.1 The energy levels of vacancies as a function of the composition x, circular dots denote anion vacancies, and triangular dots denote cation vacancies

lattice atom for which it was substituted. They used the coherent potential approximation to get the host alloy band structures. The defect energies were calculated using a Green function method. The two calculations produced similar results. The Swarts calculation indicated that the Te vacancy has an energy level in the band gap, and one far above the conduction band edge. The energy level of a Hg or a Cd cation vacancy lies very close to valence band edge in an ideal crystal model. If a Coulomb interaction and lattice distortion corrections are taken into account, few energy levels lie in the band gap. Figure 2.1 shows the calculated energy levels of vacancies as a function of the composition x. This calculation does not predict the observed behavior. A likely cause of the error lies in the starting host band structures and the way the impurity potential is assigned. Unless they are close to correct the predicted defect levels will not be correct (Chen and Sher 1985).

Kobayashi et al. (1982) also calculated the energy band structure of CdTe, HgTe, and $Hg_{0.84}Cd_{0.16}Te$ with an empirical tight-binding approximation Hamiltonian truncated at first neighbors but with an extra s state added. They also used a Green function method. They have determined scaling relations for each of the tight binding parameters of the host lattice to generate their band structures, and provided rules for deciding on the defect potentials. Then, they calculated the energy levels of sp^3 bound defect states. In their calculations they have ignored the spin–orbit interaction, the long-range Coulomb interaction outside of the central primitive cell, the long-range part of the defect potential, and any lattice relaxation in the defects surroundings. So, the problem was simplified with the nondiagonal terms of matrix elements of defect potentials all set to zero. Also in the calculation no account is taken of the electron–electron interactions. When the spin is ignored, the group point symmetry of the lattice is T_d, and the energy levels of the bound states which take sp^3 as basis vectors are a onefold degenerate A_1 energy level (S-like state) and a threefold degenerate T_2 energy level (P–like state). If the spin is taken into account but the spin–orbit interaction is excluded, all the four of the above energy levels become twofold degenerate levels. Furthermore, if the spin–orbit interaction is taken into account, the above T_2 energy level with sixfold degeneracy becomes a Γ_7 energy level with twofold degeneracy ($P_{1/2}$-like state) and a Γ_8 energy level with fourfold degeneracy ($P_{3/2}$-like state). On the other hand, the A_1 energy level with twofold degeneracy is still twofold degenerate when the spin–orbit interaction is included, namely, it becomes a Γ_6 energy level ($S_{1/2}$-like state) in the two-group expression. These symmetry considerations change the secular equation for the defect energy levels E to (Chen and Sher 1985):

$$1 - v_\alpha g_\alpha(E), \tag{2.36}$$

where α designates the symmetry of a local state, e.g., Γ_6, Γ_7, and Γ_8 on an atomic site of the zinc-blend structure, and $g_\alpha(E)$ is the diagonal matrix element of the host-crystal green function. $g_\alpha(E)$ can be calculated from the partial density of states (PDOS) by:

$$g_\alpha(E) = \int \frac{\rho_\alpha(\varepsilon)}{(E-\varepsilon)} \mathrm{d}\varepsilon. \tag{2.37}$$

The PDOS is given by:

$$\rho_\alpha(\varepsilon) = \sum_{n\mathbf{k}} |a_n^\alpha(\mathbf{k})|^2 \, \delta\Big(\varepsilon - \varepsilon_n(\mathbf{k})\Big), \tag{2.38}$$

where $\varepsilon_n(\mathbf{k})$ are the band energies of the host crystal and $a_n^\alpha(\mathbf{k})$ are the probability amplitudes of the band states in the Bloch basis. The integrals in (2.37) over the Brillouin-zone must be done numerically. Solutions to (2.36) give the energy levels found by Kobayashi et al. for the substitutional impurities in HgCdTe, shown in Figs. 2.2 and 2.4.

Figure 2.2 shows the deep levels of $S_{1/2}$-like states with Γ_6-symmetry as a function of the alloy composition x when impurity atoms occupy cation-sites. The two thick lines in Fig. 2.2 denote the energies of the conduction band minimum and the valence band maximum, respectively. The energy levels of $S_{1/2}$-like states for all the impurities that occupy cation-sites in CdTe are shown on the right vertical axis where many element symbols are labeled. The thin lines denote the energy levels of the impurities in the band gap of $Hg_{1-x}Cd_xTe(x < 1)$. The trends with alloy composition x are shown only for selected impurities. The dashed curves indicate resonant levels in the conduction band or the valence band. Figure 2.3 shows $P_{1/2}$-like state and $P_{3/2}$-like state deep levels as a function of the alloy composition x when substitutional impurities occupy cation-sites. Figure 2.4 shows $P_{1/2}$-like state and $P_{3/2}$-like state deep levels as a function of the alloy composition x when substitutional impurities occupy anion-sites (Te-site).

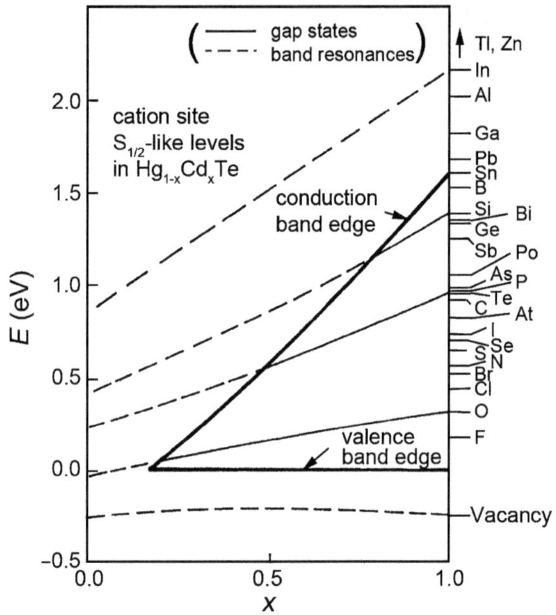

Fig. 2.2 The Γ_6-symmetric ($S_{1/2}$-like state) deep levels of substitutional impurities on cation-sites as a function of the alloy composition x

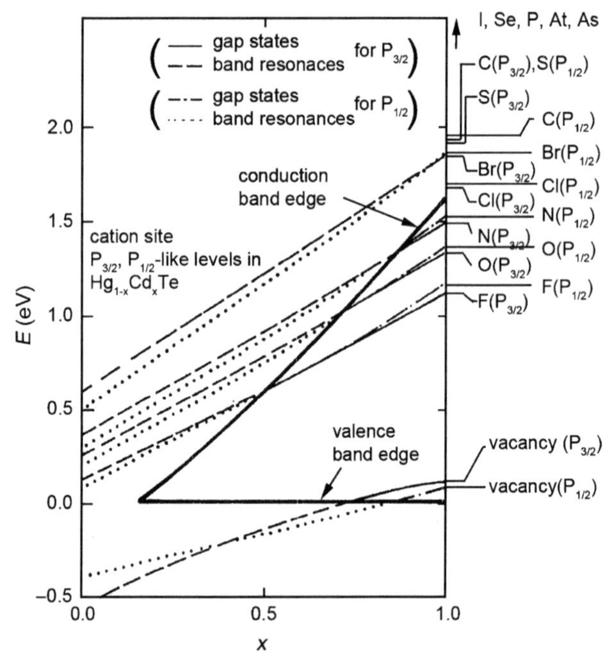

Fig. 2.3 The Γ_7-symmetric ($P_{1/2}$-like state) and Γ_8-symmetric ($P_{3/2}$-like state) deep levels of substitutional impurities on cation-sites as a function of the alloy composition x

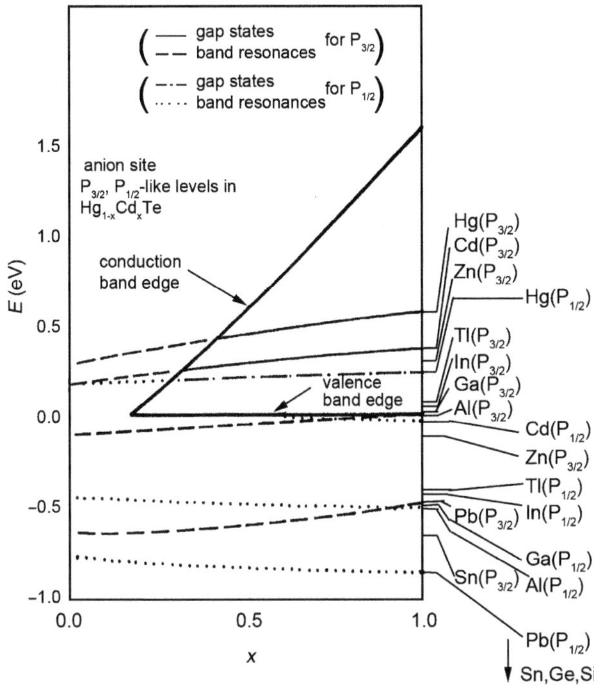

Fig. 2.4 The Γ_7-symmetric ($P_{1/2}$-like state) and Γ_8-symmetric ($P_{3/2}$-like state) substitutional impurity on anion-sites deep levels as a function of the alloy composition x

From these calculated results, we find that the deep levels have slopes $\mathrm{d}E/\mathrm{d}x$ somewhat smaller than $\mathrm{d}E_g/\mathrm{d}x$. Thus, a defect on a cation site with level at $E_g/2$ in CdTe would tend to remain near the center of the gap for a considerable range of alloy compositions. Anion site defects have $\mathrm{d}E/\mathrm{d}x$ values that are very small, and so appear to be "attached" to the valence band. The results of Jones et al. (see in Sect. 2.1.5) suggested that, for $0.2 \leqslant x \leqslant 0.4$, the deep levels at $E_g/2$ and $3E_g/4$ may be due to a cation vacancy (presumably singly and doubly charged, respectively) (Jones et al. 1982). This identification is inconsistent with this work where the cation vacancies only exhibit levels in the gap that lie close to the valence band edge at higher concentrations.

Because the conclusions deduced in the next paper to be discussed (Chen and Sher 1985), points out inaccuracies of the Swarts, Kobayashi, and Hass results, we will not elaborate further on the conclusions drawn form these papers. The sources of the errors are the choice of host band structures and the method of choosing the effective potentials to assign to the impurities.

The Hamiltonians used by Swarts, Kobayashi, and Hass are all ETB with short range cut-offs. Swarts and Hass add a spin orbit Hamiltonian and Kobayashi an additional s-state to the normal sp^3 hybrid basis. These Hamiltonians produce flawed band structures that have band states that are among other things too broad, and in some cases have secondary band gaps. The PDOS to which the impurity levels are sensitive are imperfect. To correct these defects, Chen and Sher (1985) employed their orthonormal orbital (ONO) band structures. Their method of generating the Hamiltonian consists of four steps.

1. It starts with four Gaussian orbitals per atom and empirical pseudopotentials (Chen 1977), and computes the Hamiltonian matrix $H(\mathbf{k})$ and the overlap $S(\mathbf{k})$ as was done by Kane (1976) and Chadi (1977).
2. The Gaussian orbitals are transformed into ONO's, so $H(\mathbf{k})$ is transformed into $H_0(\mathbf{k})$ and S into the identity matrix. The band structures calculated from $H_0(\mathbf{k})$ are accurate to within 5% of those done by more sophisticated means using the same potentials.
3. A spin–orbit Hamiltonian in the ONO basis is incorporated.
4. To compensate for the effects of the truncated basis and the nonlocal potentials, a perturbation Hamiltonian H_1 is added. H_1 has the form of a truncated ETB Hamiltonian with parameters adjusted to fine tune the important band energies and effective masses (Chen and Sher 1980, 1981a, b).

There are two major differences between this method and more traditional ETB methods. (1) This method includes all the long ranged interactions so the high Fourier components are present to properly reproduce sharp band curvature features. (2) Proper wave functions are produced so predictions of other quantities, like transport properties and lifetimes, are accurate.

The second ingredient to the impurity deep state calculation are the potentials, v_α. These potentials are set by the difference between the term values derived from free atom atomic pseudopotentials of the impurity and the host atom that is replaced. The potentials used by the different authors depend on the particular pseudopotentials they used. However, none of them are completely trustworthy, because the effective pseudopotentials of the atoms in the crystal differ from those of the free atom. Because of this uncertainty Chen and Sher did the deep level calculations for a range of potentials around the free atom result. For CdTe they used potentials varying from −0.5 to 2.0 eV to test the sensitivity to these choices for different impurities. The results are collected in Figs. 2.5–2.8 and Table 2.5 for the four papers, (Chen and Sher 1985) labeled a, (Hass et al. 1983) labeled b, (Swarts et al. 1982) labeled c and (Kobayashi et al. 1982) labeled d.

These figures are plots of E vs. v for the Γ_6, and Γ_7 symmetry states for a variety of impurities substituted on the Cd and Te sublattices. The vertical lines identify the locations of the potentials v derived from the term value differences using the Chen and Sher assignments. It is evident that there are substantial differences among the deep state assignments from the four papers. Because the E vs. v curves depend sensitively on the partial densities of states, which differ greatly among the four approaches, these assignments also differ. To emphasize the impact of the choice of v on these differences a list of E values is given in Table 2.5 for v and $v + 1$ eV. The E values deduced from $v + 1$ eV are denoted ∇E.

This paper emphasizes the uncertainties in the assignments of deep state energies in various approximations so it is a bit disingenuous to discuss results. To illustrate this point, consider the following examples: Table 2.5 has Li on a Te site with an

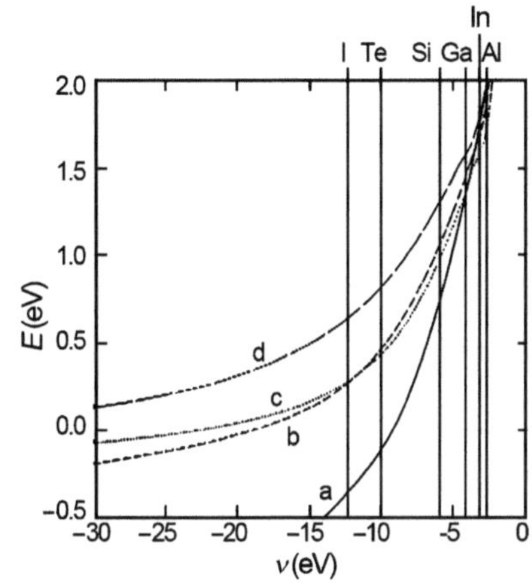

Fig. 2.5 The E vs. v curves for the Γ_6 states on a Cd site

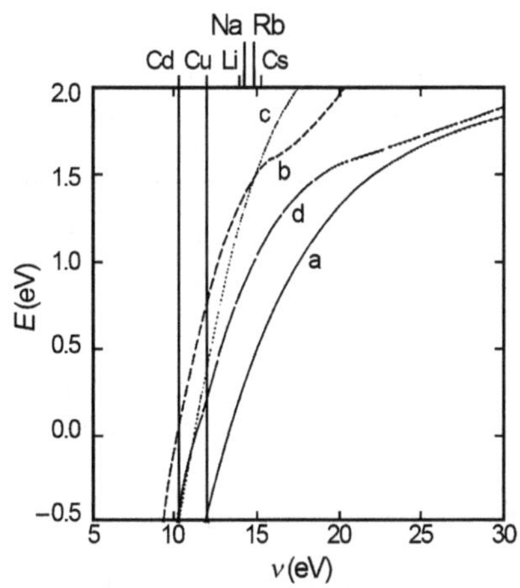

Fig. 2.6 The E vs. v curves for the Γ_6 states on a Te site

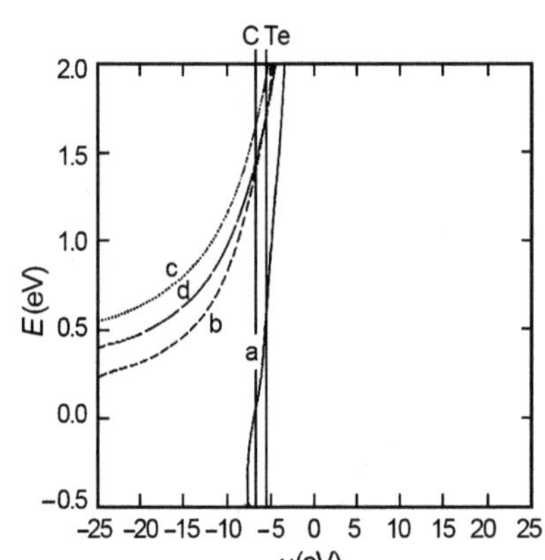

Fig. 2.7 The E vs. v curves for the Γ_7 states on a Cd site

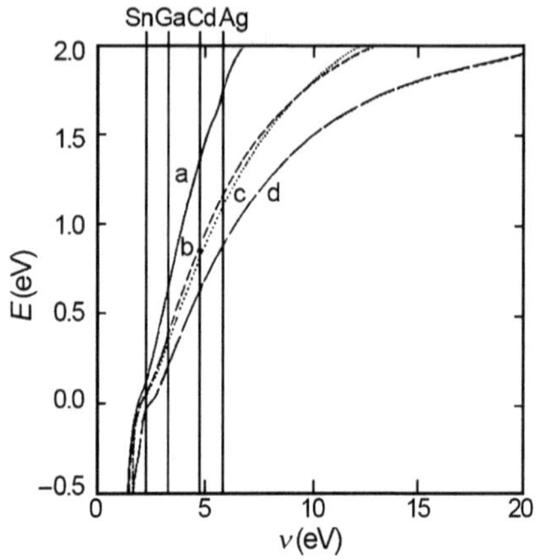

Fig. 2.8 The E vs. v curves for the Γ_7 states on a Te site

Table 2.5 Defect energy levels E and changes ∇E due to a 1 eV change in the impurity potential parameter. All energies are in units of eV. V_0 stands for an ideal vacancy

	Model a		Model b		Model c		Model d	
Defect	E	ΔE	E	ΔE	E	ΔE	E	ΔE
Γ_6 on Cd site								
Ga	1.29	0.39	1.42	0.24	1.33	0.23	1.57	0.18
C	−0.21	0.09	0.38	0.09	0.36	0.13	0.74	0.08
Si	0.67	0.30	1.02	0.10	0.93	0.19	1.27	0.15
P	−0.19	0.11	0.39	0.09	0.38	0.08	0.75	0.08
O	<-0.5		−0.02	0.02	0.04	0.01	0.32	0.02
Te	−0.13	0.13	0.44	0.10	0.42	0.08	0.79	0.09
Cl	<-0.5		0.06	0.03	0.10	0.02	0.41	0.04
V_0	<-0.5		<-0.5		−0.30		−0.20	
Γ_7 on Cd site								
C	−0.02	0.37	1.32	0.22	1.59	0.20	1.39	0.19
Si	1.57	0.65	>2.0		>2.0		>2.0	
P	0.16	0.38	1.48	0.26	1.73	0.23	1.52	0.21
O	<-0.5		0.89	0.14	1.22	0.13	1.03	0.12
Te	0.48	0.55	1.60	0.29	1.88	0.23	1.66	0.24
Cl	<-0.5		0.96	0.17	1.29	0.14	1.09	0.14
V_0	<-0.5		0.00		0.21		0.06	
Γ_6 on Te site								
Li	0.14	0.29	1.28	0.22	1.15	0.35	0.76	0.25
Cu	<-0.5		0.54	0.42	0.12	0.52	0.03	0.32
Γ_7 on Te site								
Ag	1.89	0.32	1.26	0.22	1.21	0.23	0.99	0.20
Cd	1.66	0.34	1.11	0.26	1.05	0.26	0.85	0.22
Ga	0.98	0.49	0.61	0.33	0.55	0.32	0.40	0.30
Si	−0.07	0.40	−0.11	0.36	−0.13	0.38	−0.38	0.72
Sn	0.28	0.47	0.15	0.31	0.13	0.24	0.02	0.28

s level of 0.14 eV in Model a, so one may be tempted to relate it to the acceptor state identified experimentally (Zanio 1978). However, this is not the hydrogenic acceptor stare on a Cd site that one might expect. One might also want to assign the $1/3$ and $2/3$ gap states for the Te antisite p levels on the Cd site found from Model a as those seen in experiments (Jones et al. 1982; Collins and McGill 1983). Because of the large uncertainties in the calculation, these results should be regarded as suspicious surprises rather than confirmations.

2.1.4 Doping Behavior

Berding et al. (1997, 1998c), Berding and Sher (1998a) were well aware of the uncertainties in the deep state predictions of the earlier work. They set out to eliminate the approximations responsible for these uncertainties and to advance the work from the one that focused on the 0 K predictions to one that could deal with finite temperatures and thus with realistic situations. In addition of their excellent band structures, they calculated energies for impurities that are situated on relaxed sites.

Berding et al. (1997) have discussed the behavior of the elements of group IB and IVA in periodic table as acceptor impurities in HgCdTe. By means of the full-potential linearized muffin-tin orbital method (FP-LMTO method), they have calculated approximately the electron total energy and localized energy level in the band gap, and have calculated the temperature dependence of the concentration of impurities and grown-in defects. Details of these theoretical calculations can be found in the articles of Sher's group, for reference (See Berding et al. 1998c; Berding and Sher 1998a).

The calculated results indicate that Cu, Ag, and Au occupy the cation vacancy and are p-type dopants. The calculated results of Berding are shown in Fig. 2.9, which gives the acceptor concentrations for a $10^{17}\,\mathrm{cm}^{-3}$ doping level of group I impurities occupy Hg sites in HgCdTe at 500°C as a function of the Hg partial pressure. 500°C is near to a LPE growth temperature. Figure 2.9 also shows the concentration of V_{Hg} as a function of Hg partial pressure in the range between 0.1 and 10 atm. The densities of Cu, Ag, Au, and Te occupying Hg sites, and, Cu or Ag occupying interstitial positions are also presented. When $10^{17}\,\mathrm{cm}^{-3}$, Cu is present $\sim 10^{13}\,\mathrm{cm}^{-3}$ occupy interstitial positions, and for Ag and Au, the interstitial densities are smaller and are $\sim 10^{11}$ and $\sim 10^{10}\,\mathrm{cm}^{-3}$, respectively. The LPE growth temperature of HgCdTe is about 500°C. Following growth by LPE, materials are often subjected to a low-temperature (200–250°C) anneal under Hg saturated conditions to remove the as-grown Hg vacancies. For material equilibrated under these conditions, 100% activation as acceptors is found for Cu, Ag, and Au, with the densities of interstitial atoms being less than $10^{10}\,\mathrm{cm}^{-3}$ for all three. However, it is impossible to prolong the annealing times infinitely so the slower diffusing species may not reach their equilibrium.

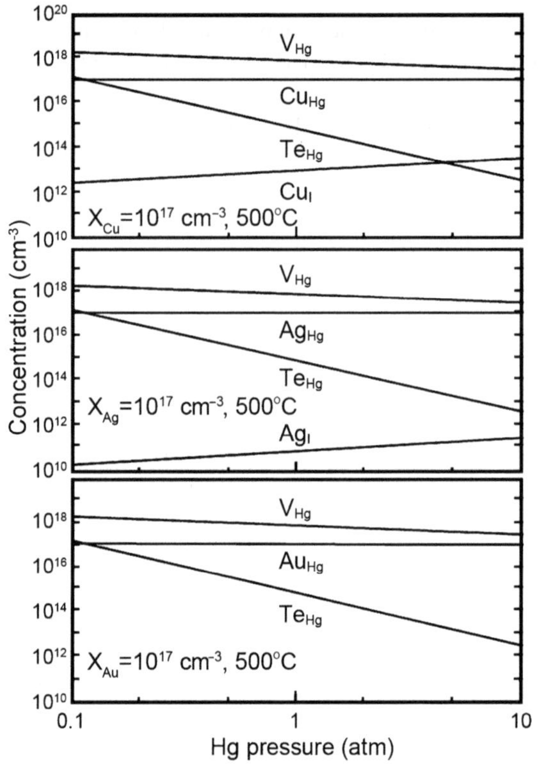

Fig. 2.9 Acceptor concentrations as a function of Hg partial pressure in HgCdTe at 500°C, near to typical liquid phase epitaxy (LPE) growth temperatures, when the elemental group I impurities occupy Hg vacancies

Figure 2.10 shows the results of concentration of the group V elements P, As, and Sb, including as a function of the Hg partial pressure when they are substituted on the anion and cation sublattices. From Fig. 2.10, we see that except at very high Hg partial pressures the concentration of As_{Hg} is greater than that of As_{Te}, indicating that As substitutes on Hg sites where it is a donor, more easily than on Te sites where it is an acceptor. Sb behaves the same as As. Only when Hg partial pressure approaches to 10 atm or more, does the concentration of As_{Te} and Sb_{Te} become greater than that of As_{Hg} and Sb_{Hg}. But for P doping, the concentration of P_{Te} is greater than that of P_{Hg} when the Hg partial pressure reaches only 1 atm. The group V dopants are exhibit compensation at lower Hg partial pressures. After annealing at a temperature of 240°C, the concentrations of P, As, and Sb occupying the anion sublattice do increase, while those occupying the cation sublattice decrease, but there is still quite a large concentration range were the materials are compensated semiconductors, as shown in Fig. 2.11.

From above analysis, we see that, for the group V dopants in HgCdTe with LPE growth on the Hg-rich side of the existence region, then the elements P, As, and Sb

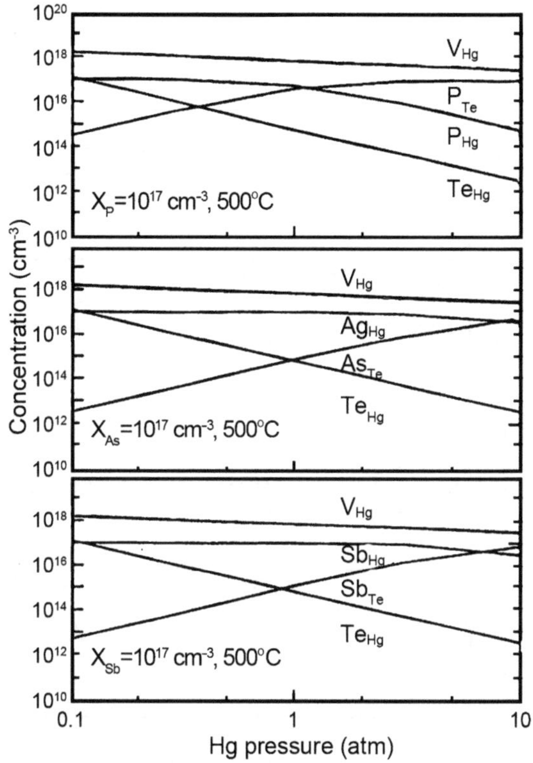

Fig. 2.10 Defect concentrations as a function of the Hg partial pressure in HgCdTe equilibrated at 500° C, near the LPE growth temperature, when some Te and Hg sites are occupied by P, As, and Sb respectively

will occupy the Te sublattice and be acceptors. Then elements of P, As, and Sb will occupy the Hg sublattice to become donors if the material is grown from Hg-rich solutions. They are some what compensated at the extremes of the partial pressure regimes. In order to get p-type dopants, it is best to grow HgCdTe by LPE from a Te-rich solution and then anneal the samples at a temperature of 240°C. Besides, when the materials are annealed for a long time at the temperature of 240°C, the self-compensation behavior can be eliminated effectively.

In the following, we will discuss As-doped HgCdTe grown by MBE. The growth temperature is about 180°C when HgCdTe is grown by MBE. Berding et al. developed a similar theoretical calculation for MBE growth (Berding et al. 1998c; Berding and Sher 1999a, b), and discussed the amphoteric behavior of impurity As in HgCdTe. The detail of theoretical calculation is found in Berding et al. (1998c) and Berding and Sher (1998a). Though MBE growth is not an equilibrium process, it can be regarded as equilibrium growth approximately, so the defect distribution is near-equilibrium. The optimal temperature for MBE growth is 185–190°C, but can be slightly lowered to 175°C to incorporate the As into the HgCdTe more effectively.

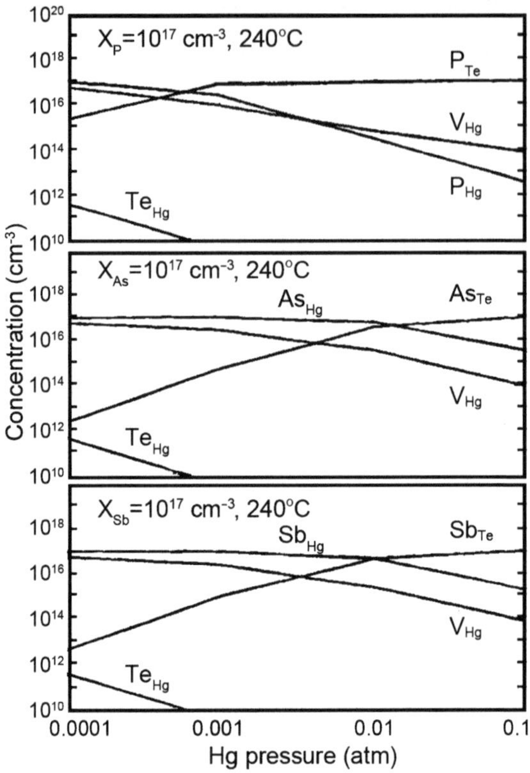

Fig. 2.11 P, As and Sb defect concentrations on the anion and cation sublattice sites as a function of Hg partial pressure for HgCdTe samples equilibrated at 240° C. The Hg vacancy and Te antisite concentrations are also shown

Under the condition of Te saturated, that is to say, $P_{Hg} = 10^{-5}$ atm, much less than 1% of the As will be incorporated on the Te sublattice. On the contrary, almost all the As occupy the Hg sublattice, and as shown in Fig. 2.12, 30–50% of the As_{Hg} forms a neutral complex bound to V_{Hg}. From Fig. 2.12 we can see that the concentration of As_{Hg}–V_{Hg} is only slightly less than that of As_{Hg}. The remaining As_{Hg} are donor impurities. Therefore, when the concentration of As is relatively low, the materials are p-type semiconductors because of the V_{Hg} acceptors, then as the concentration of As increases it becomes n-type with some compensating V_{Hg} acceptors.

Specifying the Cd concentration to be $x = 0.3$ and the total As concentration to be 10^{16} cm^{-3}, in their calculations Berding et al. presented the defect concentrations as a function of Hg partial pressure at the annealing temperature of 220°C shown in Fig. 2.13. From Fig. 2.13, we see that under a Hg saturated condition at a Hg partial pressure is 0.1 atm almost 99% of As impurities occupy Te sites and therefore are acceptors. Arsenic almost entirely occupies Hg sites when the sample is grown at 175°C by MBE, and can transfer to Te sublattice sites through a sequence of annealing steps ranging from 350 to 220°C.

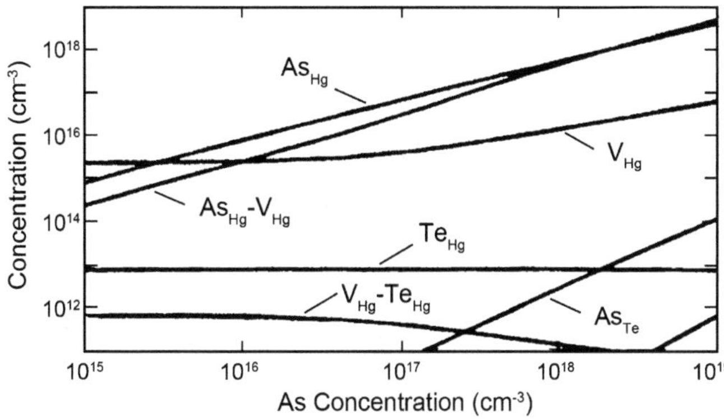

Fig. 2.12 Defect concentrations in HgCdTe as a function of the As concentration at 175° C in samples grown by molecular beam epitaxy (MBE) under Te saturated conditions

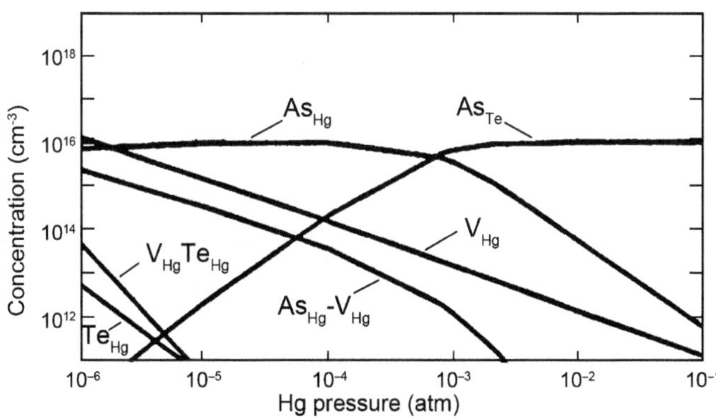

Fig. 2.13 Defect concentrations with As-doped HgCdTe as a function of the Hg partial pressure at an annealing temperature of 220° C for a sample grown by MBE under Te saturated conditions

Berding et al. (1998c) have proposed a model for the arsenic transfer from the cation to the anion sublattice. According to this model, in as-grown materials the As atoms occupy cation sites and some of these As_{Hg} atoms are bound to Hg vacancies in the initial state to form neutral complexes, As_{Hg}–V_{Hg}. This As_{Hg}–V_{Hg} defect complex will be present at high density in the as-grown material. The first step of the p-type activation process is, $Te \rightarrow V_{Hg} = Te_{Hg} + V_{Te}$, i.e., a Te atom transfers from a Te lattice site into a cation vacancy site, creating a Te antisite, Te_{Hg}. The second step permits, the V_{Te} to interact with the As_{Hg}–V_{Hg} complex and the As_{Hg} transfers into the vacated Te site, forming a As_{Te} leaving behind a Hg vacancy, or $(V_{Hg} - As_{Hg}) + (V_{Te} - Te_{Hg}) = As_{Te} + (V_{Hg} - Te_{Hg})$. In the final step, the V_{Hg} vacancy and Te_{Hg} antisite form a bound neutral pair (Berding et al. 1995), which diffuses to the surface where in picks up a Hg atom from the high partial pressure Hg atmosphere to become a new HgTe molecule or the Te_{Hg} evaporates. Figure 2.14 shows the schematic process of the As atom transfer in the proposed model.

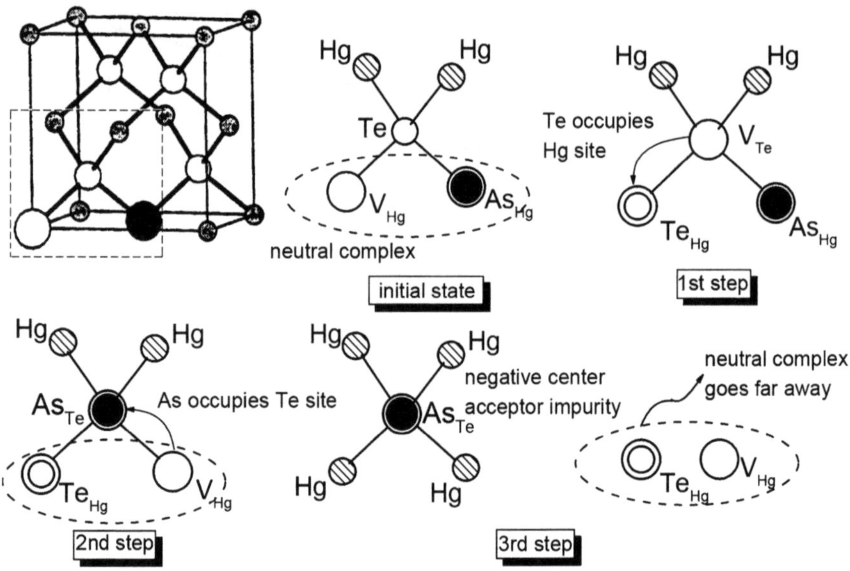

Fig. 2.14 Schematic of the arsenic transfer process from the Hg to the Te sublattice in the proposed model

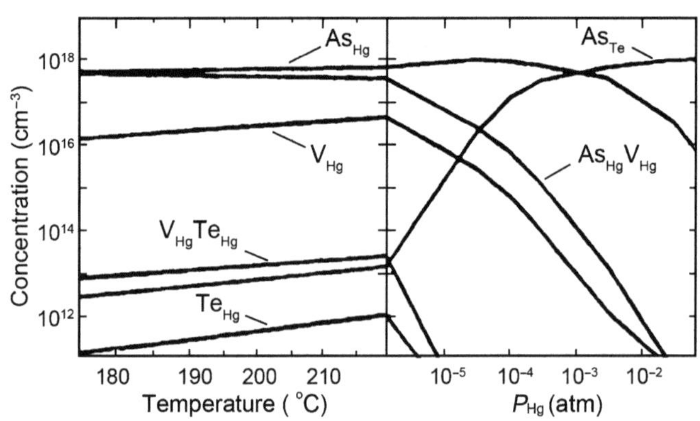

Fig. 2.15 Defect concentrations as a function of temperature and Hg partial pressure in a possible annealing process for $Hg_{0.7}Cd_{0.3}Te$

The above model describes the process through which an As atom transfers onto Te site after annealing when it initially occupied a Hg site in the as-grown material. Following this model, one can calculate the concentration of different kinds of defects under different annealing condition, and can give advice about a proper annealing process. Figures 2.15 and 2.16 show the calculated results. In Fig. 2.15, the material starts from the MBE growth condition at 175°C, is heated to 220°C under Te-saturated conditions, and is then subjected to increasing Hg partial pressures from 10^{-6} to 10^{-1} atm while the temperature is held at 220°C. It can be seen

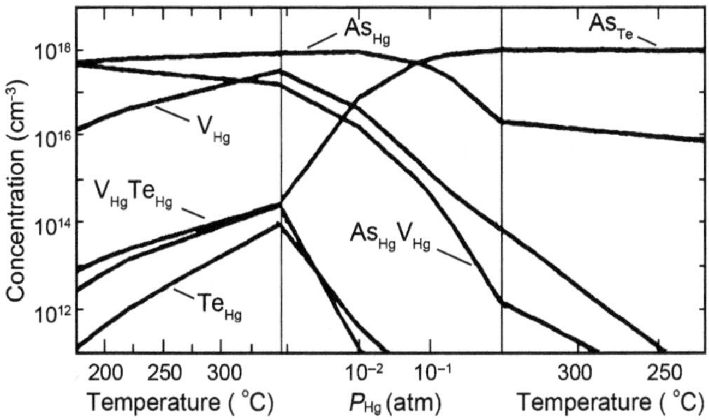

Fig. 2.16 Defect concentrations as a function of temperature and Hg partial pressure in a higher temperature annealing process for $Hg_{0.7}Cd_{0.3}Te$

that the As_{Te} population sharply increases and the concentrations of the As_{Hg}, and the As_{Hg}–V_{Hg} complex sharply decrease. In Fig. 2.16, starting from MBE growth conditions at 175°C, the material is heated to 350°C under a Te-saturated condition, and is then subjected to increasing Hg partial pressures from 10^{-3} to 1 atm while the temperature is held at 350°C; then the temperature is reduced to 220°C under a Hg-saturated condition. Then the As_{Te} population rises to $10^{18}\,cm^{-3}$, the As_{Hg} population goes to $10^{16}\,cm^{-3}$, and the concentration of the V_{Hg} and the As_{Hg}–V_{Hg} complex decrease to below $10^{11}\,cm^{-3}$. For both of the paths in Figs. 2.15 and 2.16, the As_{Hg} population sharply decreases and the As_{Te} population sharply increases. In the second process there is more thermal energy to surmount the activation barrier to the transfer of the As atom from the Hg to the Te sublattice. At the same time, there are sufficient vacancies present to promote the reactions in the annealing process. In addition, in both processes one needs to traverse the phase field from Te- to Hg-saturated conditions slowly enough to allow enough time for the As transfer to occur before the Hg vacancies are depleted.

The above model has also been used to discuss HgCdTe with Li, Na, and Cu doping by LPE growth. Berding et al. (1998b) have discussed the $Hg_{1-x}Cd_xTe$, $x = 0.22$, by LPE growth, and calculated the Li, Na, and Cu defect density as a function of the Hg partial pressure. They also treated these element's behavior in CdTe substrates. It was found that all three impurities are preferentially incorporate on the cation sublattice, where they are acceptors. Then keeping the Hg partial pressure fixed, and reducing the temperature, the concentration of Li, Na, and Cu interstitial atoms increases, while the concentration of substitutional impurities decreases. Under Hg-saturated low temperature annealing conditions, Li and Na diffuse out of the CdTe substrate toward a HgCdTe layer because of a free energy difference. Thereby, CdTe and CdZnTe substrates can be preannealed to reduce the group I impurities greatly. Then the sacrificial HgCdTe layer is etched off leaving a purified substrate behind. This substrate is then used to grow a group I element free HgCdTe epitaxial layer by LPE. Details of the calculated results can be found in Berding et al. (1998b).

2.1.5 Experimental Methods

2.1.5.1 High-Frequency and Low-Frequency Capacitance Measurement Principles

For a planar device semiconductor, generally speaking, a passivation layer or a dielectric layer, as a protective film on its surface, is needed to promote the operation of this class of devices. However, interface defect states are inevitably introduced. These defect states will become carrier scattering centers, and will reduce the mobility and lifetime of carriers and the signal-to-noise ratio (SNR) of devices. Therefore, it is necessary to detect and control these defect states, and the usual method is a combination of high- and low-frequency capacitance measurements. The procedure starts by photoengraving a metallic gate electrode on the insulating layer placed on the surface of a semiconductor, to make a MIS structure. Then measure the high- and low-frequency capacitance of this structure. From these measurements the distribution of interface defect states in the band gap can be deduced. Usually the interface state energies are distributed continuously throughout band gap in what is referred to a U-shaped distribution with its minimum near mid-gap, and the filled electron interface state populations determined by the Fermi level.

As the surface potential of a semiconductor changes, the Fermi level at the interface will change accordingly. Then the number of electrons filling the interface states will change, resulting in an electron exchange between interface states and the bulk semiconductor. When a direct current bias and alternating-current small-signal are applied, the interface will charge and discharge as the surface potential changes. The interface states produce a response to the changing surface potential equivalent to a capacitance, C_{ss}, called the interface-state capacitance. When the surface potential change is $d\varphi_s$, the interface charge concentration will change by an amount dQ_{ss}. So we have:

$$C_{ss} = dQ_{ss}/d\varphi_s. \tag{2.39}$$

For an MIS structure, the measured capacitance C_m is the series combination capacitance of its insulating layer capacitance C_i and the total interface capacitance, C_{it};

$$(C_m)^{-1} = (C_i)^{-1} + (C_{it})^{-1}. \tag{2.40}$$

Since the time constant for the exchange of electrons between interface states and the bulk is relatively long, the change of the interface state charge can keep up with the variation of an alternating-current small signal only when the frequency of alternating-current small signal is very low so it can contribute to the capacitance. Then the measured capacitance C_m is just C_{LF}, and C_{it} consists of surface depletion/accumulation layer capacitance, C_s and an interface state capacitance, C_{ss}. When the alternating-current small signal frequency is very high, the interface state charge can not keep up with the variation rate to exchange electrons with the bulk, so does not contribute to the capacitance. Consequently, the measured capacitance

C_m is just C_{HF}, and C_{it} is only the semiconductor surface capacitance C_s, so we can get:

$$(C_{LF})^{-1} = (C_i)^{-1} + (C_s + C_{ss})^{-1}, \quad (2.41)$$
$$(C_{HF})^{-1} = (C_i)^{-1} + (C_s)^{-1}. \quad (2.42)$$

Eliminating, C_s from (2.41) to (2.42), yields an expression for the interface state capacitance:

$$C_{ss} = C_i \times [(C_i/C_{LF} - 1)^{-1} - (C_i/C_{HF} - 1)^{-1}]. \quad (2.43)$$

Again from (2.39), the interface state density as a function of the surface potential can be found:

$$N_{ss} = C_{ss}/e \quad (2.44)$$

Therefore, the interface state density can be obtained from the high- and low-frequency capacitance spectra and (2.43) and (2.44). The energy level locations of the interface states in the band gap can be obtained from an energy band potential model of a two-dimensional electron gas and the low-frequency capacitance spectra (Nicollian and Brews 1982).

As examples, now we will introduce the measured results of the density of states of a ZnS/MCT, and an SiO/SiO_2/InSb interface. MCT and InSb are made into MIS structures, and their high- and low-frequency capacitance spectra are measured (shown in Figs. 2.17 and 2.18). The distribution of interface states was calculated using the method mentioned above, and shown in Figs. 2.19 and 2.20. The parameters of the samples, and the test condition are as follows: (1) an n-type $Hg_{1-x}Cd_xTe$, $x = 0.30$, donor concentration $N_D \sim 1.7 \times 10^{14}\,cm^{-3}$, an insulator layer thickness of the ZnS is 200 nm, the area of metallic gate electrode is $0.002\,cm^2$, the testing temperature is 80 K, and the testing frequencies are 20 Hz and 10 MHz respectively;

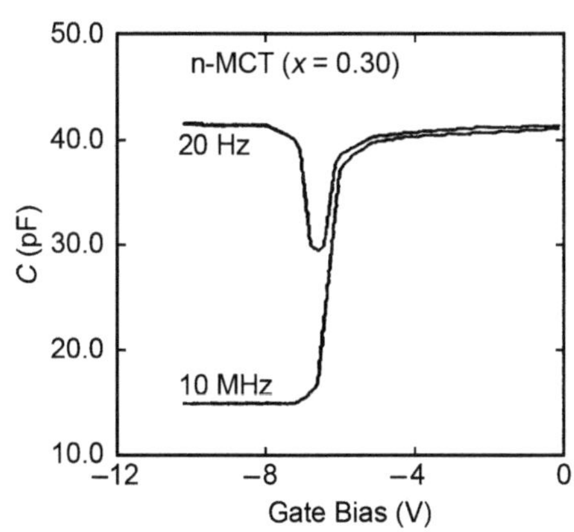

Fig. 2.17 The high- and low-frequency capacitance–voltage (C–V) spectra of a mercury cadmium tellurium (MCT) ($x = 0.30$) metal–insulator–semiconductor (MIS) device

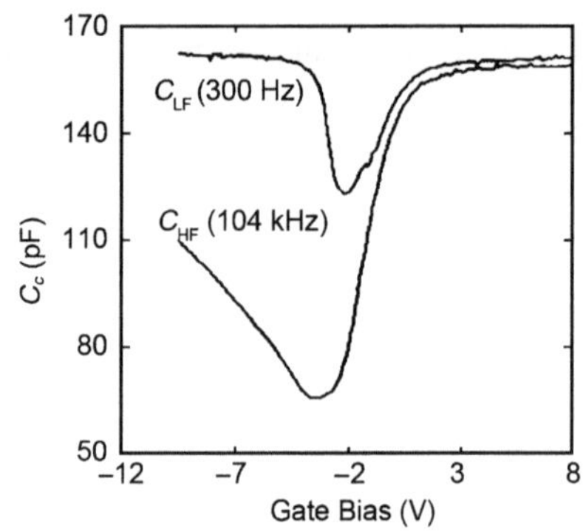

Fig. 2.18 The high- and low-frequency capacitance–voltage (C–V) spectra of an InSb MIS device

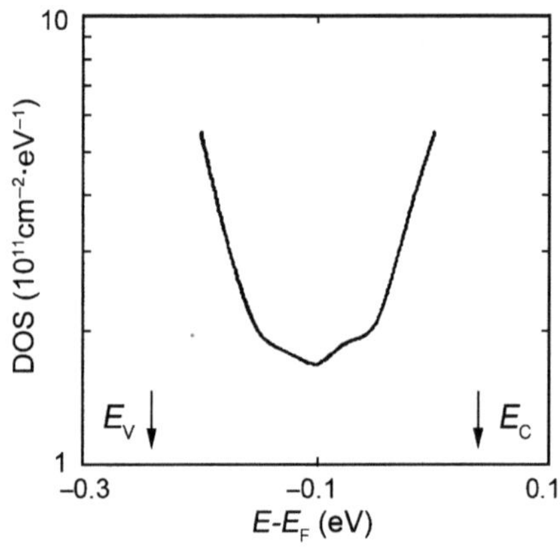

Fig. 2.19 The distribution of interface states in an MCT MIS device

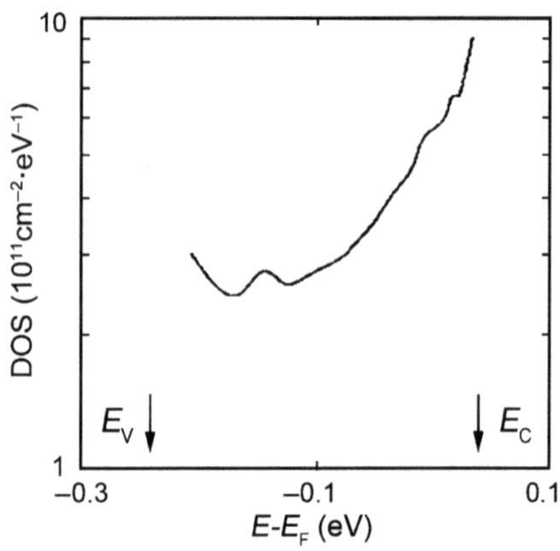

Fig. 2.20 The distribution of interface states in an InSb MIS device

(2) an n-type InSb, donor concentration $N_D \sim 2.0 \times 10^{14}\,cm^{-3}$, insulator layer thickness of SiO/SiO_2 is about 180 nm, the area of the metallic gate electrode is 0.00785 cm^2, the testing temperature is 80 K, and the testing frequencies are 300 Hz and 104 kHz, respectively.

From Figs. 2.19 to 2.20, we can see that the distributions of interface state densities for the two kinds interfaces exhibit the usual "U" shape, the minimum of the interface state density (located at about the middle of band gap) is of the order of $10^{11}\,cm^{-2}$. Further experiments and calculations indicate that there is a fixed positive charge with a density $8.2 \times 10^{11}\,cm^2$ and a slow hole trap density of $\sim 4.6 \times 10^{10}\,cm^2$ in the ZnS dielectric layer. There is also a fixed positive charge density of $1.0 \times 10^{12}\,cm^2$, and a slow hole trap density of $2.4 \times 10^{12}\,cm^2$ in the SiO/SiO_2 dielectric layer.

2.1.5.2 Deep Level Transient Spectroscopy

Deep level transient spectroscopy (DLTS) is an important method to determine deep level energies and their response properties in semiconductors. This method measures the temperature dependence of the admittance spectrum, or the temperature dependence of the capacitance spectrum. At a given temperature, if a deep level releases (or capture) carriers, the admittance will change. The principle is the same as a measurement of the capacitance. A scanning DLTS spectrum at different temperatures can separately detect all kinds of deep levels depending on their energy distributions in the band gap (each has a different DLTS peak). From the DLTS peak locations and peak heights, one can obtain important parameters, such as the energies of the deep levels, carrier densities, and the cross sections of the trapped carriers. Detailed experimental methods and results are presented in Sect. 2.3.

2.1.5.3 Photoluminescence Spectroscopy

Photoluminescence (PL) spectroscopy is a measurement tool, that can be used not only to deduce the conduction and valence band shapes and the exciton electronic states in semiconductors, but also the energy shapes of the impurity defect states. In a sense, the PL process is the inverse of the light absorption process. It can be divided into three subprocess:

1. An external exciting light (or an external injection current, or irradiation beam) excites carriers in the material, forming a nonequilibrium electron–hole pair distribution.
2. The nonequilibrium electron–hole pair reaches a relatively low energy state in a radiative or a nonradiative recombination process.
3. The light generated by a radiative recombination process propagates within the semiconductors.

Since any light emitted in the radiative recombination process will be reabsorbed again during propagation (radiation trapping), photoluminescence can occur only from the range of several electron diffusion lengths of the irradiated surface of a semiconductor.

Therefore, a PL experiment has certain limitations on the experimental technique and material structures to which it can be applied. For relatively thick samples, the PL detector should be placed above the irradiated surface of the semiconductor; and for relatively thin samples, the PL detector can be placed at the back of the irradiated surface. Furthermore, the surface perfection of the material is important because all kinds of defect states, such as surface states and dislocations which may induce nonradiative recombination, should be minimized. Otherwise, a PL experiment is insensitive. The dynamics of the PL process can be judged by the measured peak position of the spectrum and the curves' shapes compared to those of a theoretical analysis. Thereby, the impurity state density in a material can be deduced.

Using a PL spectrum to investigate impurities in HgCdTe materials encounters a new difficulty, which is mainly due to the HgCdTe material properties. The band gap width of HgCdTe is relatively small, and the ionization energy of donors is only $\sim$0.04 eV, hardly binding electrons and leading to the materials being in a completely ionization state even at relatively low-temperatures. The PL spectrum basically is in the wave length range of the mid- to far-infrared. In this range, the responsivity of detectors is relatively low, and the background radiation is rather strong. Furthermore, nonradiative recombination in HgCdTe is relatively fast, so a PL signal is also small. All these factors degrade the detection efficiency of the PL spectra, and the difficulty is bad especially for low alloy compositions. So, there are few research reports about the PL of MCT, and they mainly concentrate on higher alloy composition HgCdTe materials. To investigate small alloy compositions with PL, employment of far infrared bands is necessary. This topic will be discussed again in another section.

2.1.5.4 Photothermal Ionization Spectrum Principles

Since Lifshits et al. discovered the photothermal ionization phenomenon with a semiconductor infrared-photoconductivity experiment in the 1980s, photothermal ionization spectroscopy (PTIS) has been used widely to investigate the shallow impurity behavior in semiconductors, especially for impurities in high-purity semiconductors. At low-temperature, impurity ionization can be divided into two steps:

1. The ground state of the impurity is excited by a photon to an excited state.
2. Then it ionizes into the conduction band to provide conductance (shown in Fig. 2.21).

Thus the photothermal ionization process combines the high-resolution of optics with the high sensitivity of electricity, to provide a useful means to observe micro-impurities in high-purity semiconductors. Obviously, PTIS has rigorous requirements for experimental conditions. First, the temperature cannot be too high to enable the impurity to stay in its ground state; second, the temperature also cannot be too low so there are enough phonons to help the impurities ionize from the excited state into the conduction band. Furthermore, good ohmic contact electrodes are required. It is difficult to investigate the shallow impurities in narrow-band

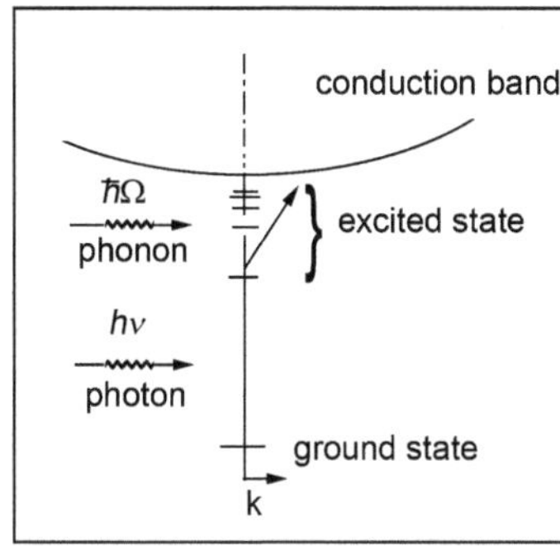

Fig. 2.21 A sketch of photothermal ionization process

gap semiconductors InSb and MCT using this method, because it is difficult to fabricate sufficiently high-purity material. Furthermore, there are always distinct band tails in these materials, so shallow impurity energy levels are slightly removed from the band-edge. Under these conditions, localization can occur only at very low-temperatures or by the application of very strong magnetic fields. Consequently, it is hard to satisfy these rigorous experimental conditions. So this method is usually only used to observe shallow impurities in high-purity GaAs, Si, and Ge.

2.1.5.5 Quantum Capacitance Spectrum Technology

The several of the experimental methods mentioned earlier can be used to investigate impurity defects with energy levels located in the band gap, but cannot test for resonant defect states. In narrow gap semiconductor materials, many "deep" levels caused by short-range interactions are located in the conduction band. These "deep" levels interact with the continuum states in the conduction band, forming resonant defect states. Generally speaking, it is comparatively troublesome to investigate resonant defect states, since it is hard to separate the resonant states from continuum conduction band states. There have been many theoretical and experimental efforts devoted to these aspects of the problem, such as the chemical trends theory of sp^3 binding of deep trap levels, the measurement of transport under pressure, far infrared spectroscopy, etc. Here a new model for investigating resonant impurity defect states will be established based on the capacitance characteristics of narrow gap semiconductors.

Band bending on the surface of semiconductors can be investigated with a capacitance–voltage (CV) measurement on MIS structures. For narrow band semiconductor MIS devices, quantization phenomenon can be found within the surface inversion layer. A ground state sub-band quantization level is located at a position E_0, which lies above conduction band minimum (shown in Fig. 2.22). When the magnitude of band bending is equal to E_g, that is when the conduction band minimum bends below the Fermi level, the inversion layer channel still cannot be filled by electrons. To the contrary, only when the energy band bending reaches $E_g + E_0$, i.e., when the ground state of the quantized sub-band lies below the Fermi level, can

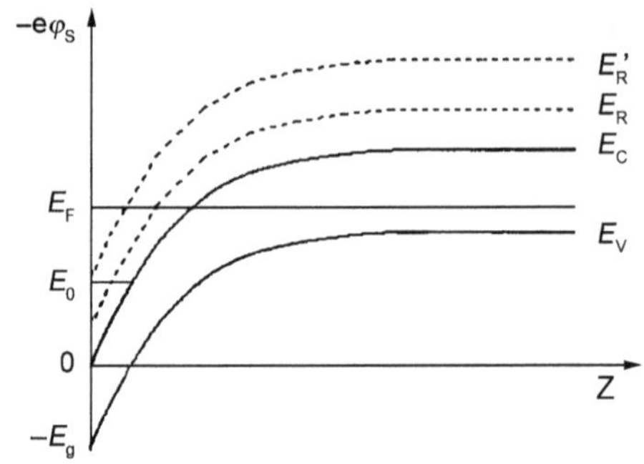

Fig. 2.22 An energy band bending sketch for a narrow band gap semiconductor MIS structure with resonant defect states

the inversion layer channel hold electrons. Therefore, the threshold voltage for the inversion layer formation is retarded. If there is a resonant defect state between the conduction band minimum and E_0, at E_R as shown if Fig. 2.22, it will be reflected in the capacitance CV spectrum. If starting from the case of a flat band, with an impurity state lying far above the Fermi level so that it cannot bind electrons, then when the impurity state falls below the Fermi level because of energy band bending, electrons will fill the resonant defect energy levels. If the states' density is high enough, this kind of electron filling will influence the charging and discharging process of the inversion layer and hence contribute to the capacitance. So, an extra peak will occur in the capacitance spectra before the MIS structure inverts. Some relevant information about the resonant defect state, such as its concentration, and the location of its energy level, can be obtained from its peak position and peak amplitude.

When a resonant defect state is located at above E_0 and is an acceptor-type (see E_R' in Fig. 2.22), then E_R' also can be observed from a capacitance spectrum. Since the trapping and release times of electrons from impurity defect states is usually longer than that for continuum states, when an MIS device is inverted, the acceptor-type resonant defect states will repel other electrons occupying continuum states after it captures electrons, leading to a decrease of the number of electrons that contribute to the capacitance. So, the rate of the capacitance increase slows, or even decreases. With an increase of test frequency, the probability of resonant defect states capturing electrons becomes smaller, and their influence on the capacitance also gradually weakens. Thus information about resonant defect states, their concentrations, the location of their energy levels, and the lifetime of trapped electrons, can be obtained from this frequency-conversion capacitance spectroscopy (FCS) measurement.

2.1.5.6 Positron Annihilation Spectra for MCT

Positron annihilation spectra technology (PAT) was established based on the microscopic distribution of positrons in solids being related to the interaction between a positron and an ion core, as well as an electron. Since the thermal momentum of a positron is almost zero, and a positron is a particle with a light mass and a positive charge, it will suffer a strong repulsive interaction with an ion core. Therefore, when positrons are in a perfect crystal, they are located mainly in interstitial sites and can diffuse freely. Under this condition, positrons are free particles. When a positron is in an imperfect crystal with a lattice vacancy, or in the vicinity of a dislocation core, these defects will strongly attract the positrons since these vacancy-type defects lack ion cores and electron redistributions will induce a negative potential at such sites. So, positrons will tend to be in bound states around these sites. All the positrons, whether they are in free or bound states, will be annihilated with electrons. The densities of valence electrons in such defect sites is lower than that of lattice atoms in a perfect crystal, so the probability of positron annihilation at defect sites will be reduced, thereby leading prolonged lifetimes. The lifetime of positrons in a perfect crystal lattice, is shorter. Thus the defects in crystal can be detected quite sensitively with positron lifetime measurements. The measurement proceeds as follows. A positron source ^{22}Na releases a positron and at the same time, radiates a γ photon with energy of 1.28 MeV. This photon sets the zero of time. The positron later annihilates with an electron in the sample after a time Δt releasing two γ photons each with energy of about 0.511 MeV. Measuring the interval Δt between the 1.28 and a 0.511 MeV γ photon, determines the lifetime of the positron in the sample. This kind of technology has been widely used for the past 20 years to study thermodynamics problems of hole-type defects in metals. In recent years, the technology has been used to investigate defects in semiconductors, but was seldom used to investigate defects in MCT. As an example, we will introduce measurement results of a positron annihilation spectrum for a highly doped p-type MCT sample.

Choose a p-type MCT ($x = 0.5$) bulk material with a concentration of $N_{\mathrm{AD}} = 1.5 \times 10^{18}\,\mathrm{cm}^{-3}$, then do an anodic sulfurisation on its surface, cut the sample into two pieces, fix them in a sample holder of the measurement system, and positron an ^{22}NaCl source on a site between two samples. The measurement system is a fast capture positron annihilation lifetime apparatus made by the ORFEC company with a resolution of 24 ps. The PAT spectrum of the sample is measured at room temperature. Every measurement lasts ~8 h, and records about 2×10^6 positron annihilation events to build a spectrum. The system is scaled to a pair of single crystal Si wafers, and the measurement results are fitted by universal Positronfit software.

Figure 2.23a shows the PAT spectrum of the sample mentioned above at room-temperature. Figure 2.23b shows the measured PAT spectrum of the same sample at room-temperature after being stored at atmospheric pressure for 3-months. The fits to the two spectral lines are shown in Table 2.6.

From Table 2.6, we find that there are three positron annihilation lifetimes observed in the sample. They are $\tau_1 = 237 \sim 288 \pm 14$ ps, $\tau_2 = 366 \sim 400 \pm 22$ ps, and $\tau_3 = 1{,}996 \sim 2{,}099 \pm 43$ ps. Generally speaking, the number of measured lifetimes reflects the number of possible defect species in the sample. Thus, we conclude there are three kinds of trapping mechanisms in the measured sample. The results of the second measurement following the 3-month interval are close to the original values lying within the statistical error. This result may be caused by the anodic sulfide stabilization, since an anodic sulfide can serve as a good CdS protective film on the surface of MCT. It will especially enhance the stability of p-type MCT

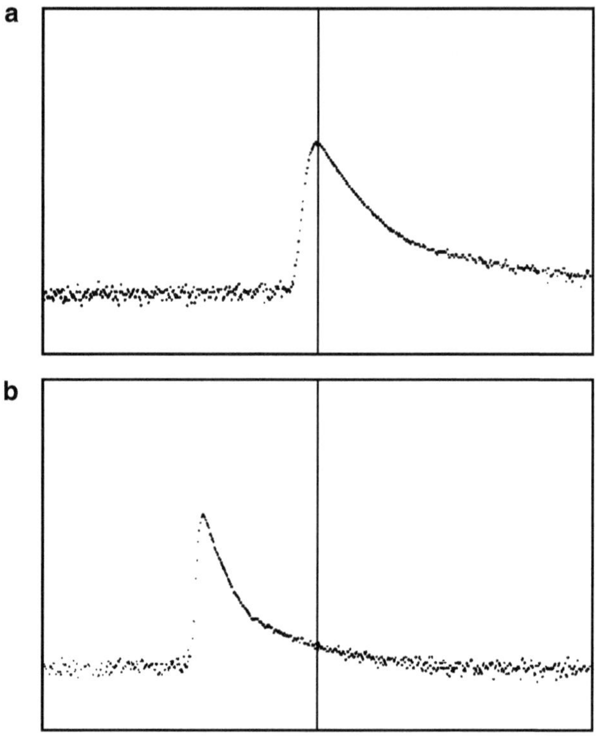

Fig. 2.23 The positron annihilation spectra technology (PAT) spectrum of a p-type HgCdTe sample (**a**) and that for the same sample three months later (**b**)

Table 2.6 The results obtained by fitting to the positron annihilation spectra technology (PAT) spectrum at room-temperature

	Lifetime of positron in ps and the relative intensity of the apparent spectrum	
	The first measurement	Three months later
τ_1	237 ± 14	253 ± 10
I_1	42.75 ± 10.39	59.02 ± 8.92
τ_2	366 ± 16	400 ± 22
I_2	52.82 ± 10.31	37.9 ± 8.83
τ_3	$1{,}996 \pm 37$	$2{,}099 \pm 43$
I_3	4.43 ± 0.12	4.07 ± 0.12

materials. Gely et al. (1990) have measured the PAT spectra of p-type MCT with different doping concentrations, and found that the positron lifetime τ remains almost unchanged and is about 309 ± 1 ps almost independent of temperature, when $N_{AD} > 1.709 \times 10^{16}\,\text{cm}^{-3}$. Gely concluded that the lowest value of a positron lifetime is about 270 ± 10 ps in a perfect MCT crystal, which is very different from the example mentioned above. With high p-type doping the measured positron lifetime

τ_1 is shorter than 270 ps. Evidently, τ_1 results from a near perfect CdS protective film and on an MCT crystal. But because the thin CdS protective film thickness is $\sim$1 μm, and the positron penetration depth is $\sim$0.3 mm, τ_1 results from a substantial thickness of perfect crystal. Thus the new positron lifetime for a perfect crystal is shorter than that of prior measurements, and we conclude that the earlier measurements were made on slightly imperfect crystals. Second, the measured τ_2 is about 30% larger than Gely's value of 309 ps, a lifetime ascribed to Hg^{2+} vacancies. The measured τ_3 is very large, and results from positron annihilation on the surface of the sample. This lifetime value and its intensity suggest this conclusion. There are theoretical calculations indicating that positron annihilation on a surface produces the longest lifetimes. Furthermore, the thickness of the surface layer is very small so positron annihilation on the surface, would lead to the observed very weak intensity.

2.1.5.7 Optical Hall Effect Measurements

For a p-HgCdTe sample with a concentration in the low range, $N_A \leqslant 5\times10^{15}\,\text{cm}^{-3}$, due to mixed conduction effects, one cannot deduce the acceptor concentration and energy level from the usual temperature dependent Hall coefficient. At the same time, in this range of acceptor concentrations the compensation from the residual donor becomes important. Bartoli et al. (1986) introduced an optical Hall measurement technology. They obtained acceptor concentrations, acceptor ionization energies, and the compensation degree using results from fits to the measured optical excitations at low-temperature. The light source used in these experiments is a variable temperature black body source when the implanted carrier concentration is low, otherwise it is a 50 W continuous CO_2 laser light emitter when the implanted carrier concentration is high. A chopped light device is used to yield a flat 25 ms pulse. The optically excited carrier concentration found by a Hall coefficient measurement is $n_m = 1/eR(B = 500\,\text{G})$. The $\mu_m - n_m$ curve can be obtained from the mobility $\mu_m = \sigma_m/n_m e$. σ_m is the measured conductivity. The mobility of electrons is limited by the concentration of charged scattering centers:

$$N_{cc} \approx N_{A^-} + N_{D^+} + p. \tag{2.45}$$

At high-temperatures, the acceptors are all ionized and the hole concentration in the valence band is $p = N_A - N_{D^+}$, then we have $N_{cc} \approx 2N_A$. While at low-temperatures, the holes are frozen out, and $p = 0$, $N_{A^-} = N_{D^+}$, so we have $N_{cc} \approx 2N_{D^+}$ (see Fig. 2.24). Therefore, the mobility is very sensitive to impurity compensation at low-temperature.

Setting $x = 0.225$ and for a sample thickness of 50 μm, Fig. 2.25 shows a theoretically calculated curve of the electron mobility as a function of the optically excited electron concentration for various donor concentrations N_D. It can be seen from Fig. 2.25 that the electron mobility as a function of the optically excited electron concentration is very sensitive to the donor concentration. Therefore, the level of compensation N_D can be obtained using N_D as a parameter to fit the $\mu - n$ curve.

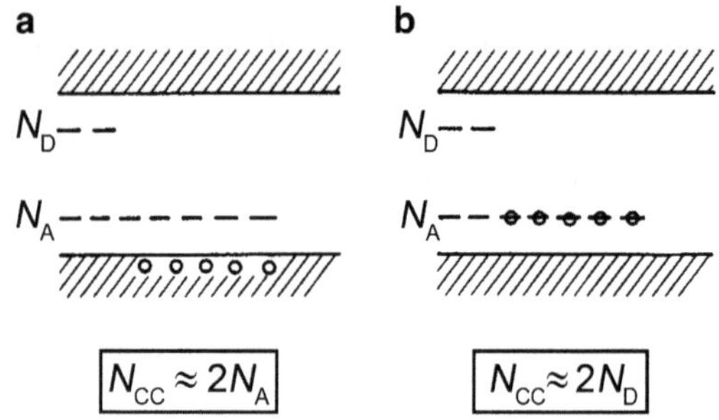

Fig. 2.24 A sketch of the charged scattering center concentrations. (**a**) depletion at high-temperature ($k_BT >> E_A$, $N_{cc} = 2N_A$); (**b**) hole frozen out at low-temperature ($k_BT << E_A$, $N_{cc} = 2N_D$)

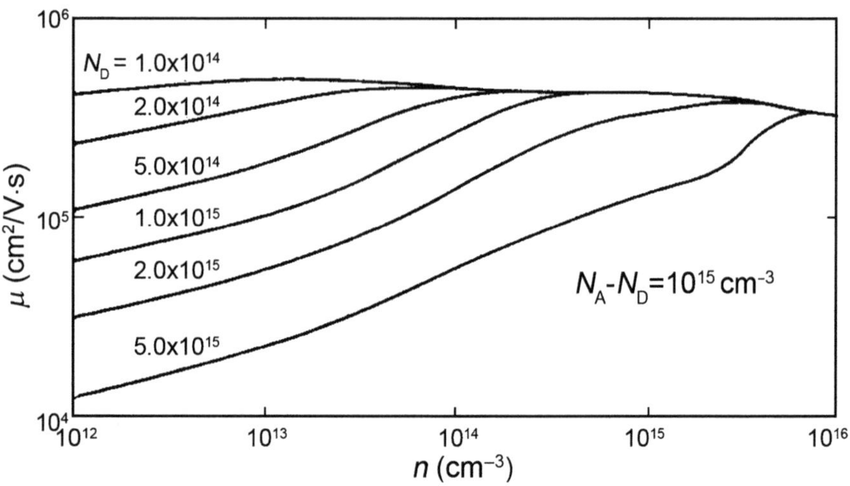

Fig. 2.25 The electron mobility as a function of the optically excited electron concentration and the donor concentration

Figure 2.26 shows experimental dots for $\mu_m \sim n_m$, and a theoretically fitted curve at a temperature of 11 K. N_D is the only adjustable parameter when the fitting is executed, and the result obtained is $N_D = 3.9 \times 10^{15}\,\text{cm}^{-3}$. After N_D is determined, the hole concentration and the donor ionization energy can be calculated from the magnetic field dependence of the Hall data from the following formula:

$$\frac{p(p + N_D)}{N_A - p - N_D} = \frac{1}{2} N_V \exp(-E_A/k_BT), \tag{2.46}$$

where N_D is already obtained, and N_A and E_A are fitting parameters. Using a fit to the calculated $p \sim T^{-1}$ curve, the measured the values of N_A and E_A can be extracted. Both the measured and the fitted curves are shown in Fig. 2.27. The values of fitting parameters are $N_A = 6 \times 10^{15}\,\text{cm}^{-3}$ and $E_A = 11\,\text{meV}$. The fitted result indicates that a single acceptor, most likely Cu, is the impurity in the sample. To fit the experimental $p \sim T^{-1}$ obtained from the magnetic field dependence of the Hall coefficient, if the optimal agreement cannot be obtained using a single acceptor model,

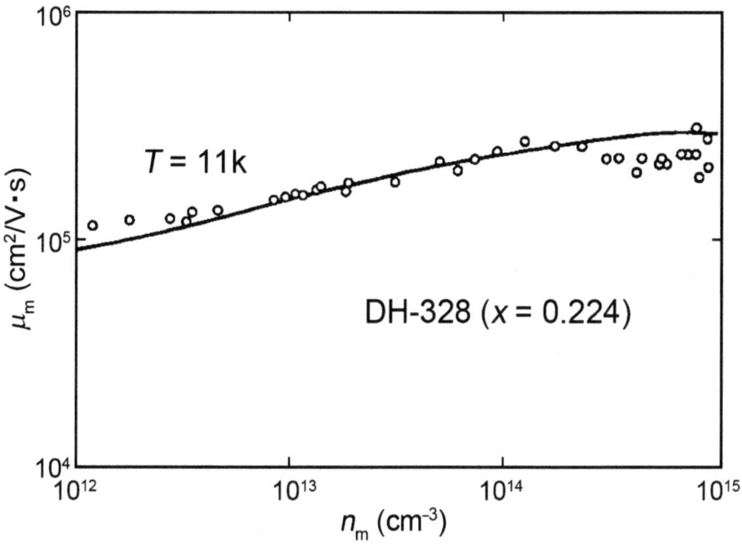

Fig. 2.26 The measured mobility (*experimental dots*) and a theoretically fitted curve at a temperature of 11 K

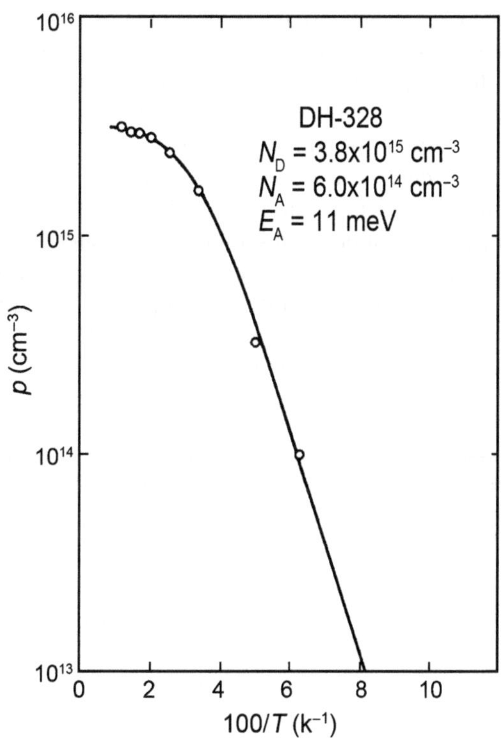

Fig. 2.27 The hole concentration as a function of the reciprocal temperature; the dots are the experimental data, and the curve is a fitted result

one can attempt to fit the experimental result using a double acceptor. So a distinction between single and double acceptor models can be deduced from the fitted results. Thus the optical Hall measurement is an effective method to observe the compensation degree, the acceptor concentration, and the acceptor ionization energy in a p-type HgCdTe.

2.2 Shallow Impurities

2.2.1 Introduction

Figure 2.28 illustrates the relative energy level positions of impurity defects in HgCdTe. The ionization energy is about 16 meV for the shallow acceptor level associated with the Hg vacancy, and its concentration depends on the density of the vacancies. The measured ionization energy of the shallow acceptor level associated with group V substitutional impurities on Te sites differs slightly from that of the Hg vacancy.

There are energy levels that resonate with conduction band above and near to the conduction band edge, which are the combination of Hg vacancies, V_{Hg}, with various impurity atoms or Te vacancies. In the band gap, deep levels are experimentally found in the energy positions $3/4E_g$, $2/4E_g$, and $1/4E_g$ above the top of valence band. The origins of these deep states will be discussed presently. Meanwhile, there are also other native point defects; Cd vacancies, and Hg, Cd, and Te interstitials, and antisite defects, as well as their compounds with other native point defects and impurities. To identify the impact of these levels, theoretical studies have been performed with first principals, tight-binding, and Green function methods. Experimental research has been carried out with various techniques, such as CV spectroscopy, high-pressure transport, DLTS, optical capacitance spectroscopy, far infrared spectroscopy, impurity cyclotron resonance (ICR), and far infrared photoconductivity as well as conventional transport methods.

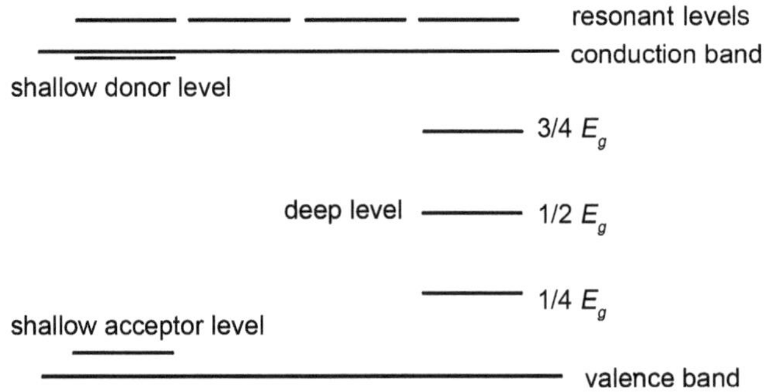

Fig. 2.28 Relative energy level positions of impurity defects in HgCdTe

As was shown in Sect. 2.5.4 of Vol. I (Berding et al. 1993, 1994), the combination of native point defects and impurities determine the electrical characteristics of HgCdTe alloys. In this study first principal's codes were used to determine the formation energies and densities of various native point defects and defect complexes in HgTe within the existence region as functions of the Hg partial pressure at three important temperatures, 500, 185, and 220°C. The 500°C temperature is typical of LPE growth, the 185°C is that of MBE growth, and 220°C is typical of a Hg saturated annealing temperature following MBE growth. The highest densities are for V_{Hg} vacancies, Te_{Hg} antisites, and a complex of the two $V_{Hg}Te_{Hg}$. Te vacancies, V_{Te}, while they have lower concentrations, also have important consequences on defect migration. The V_{Hg} is a shallow acceptor [possibly a double acceptor (Vydyanath et al. 1981)], and its density dominates over most of the Hg partial pressure range. However, at low Hg partial pressures where the V_{Te} density approaches that of the V_{Hg}, the $V_{Hg}Te_{Hg}$ complex density produces a state that resonates in the conduction band, and may exceed the V_{Hg} density to cause the material to become n-type as observed under these conditions. This conversion is most apparent at the lower temperatures corresponding to MBE growth and Hg saturated annealing conditions. For low impurity concentration material, at high Hg partial pressures the material is observed to convert once again from n- to p-type. The donors responsible for this conversion, called the "residual donors," are unknown. They are present in highly purified material, and never have been successfully identified with an impurity. It has been speculated that they are due to a moderate concentration of $V_{Hg}Te_{Hg}$ frozen in because this complex has a slow diffusion rate. This study was done on HgTe and its relevance to HgCdTe alloys is only likely to correctly reflect their behavior at low Cd concentrations. Also V_{Hg} vacancies may become occupied by impurity atoms and become deep levels or resonant states. In addition, defect damage caused by ion implantation is also an interesting theme. All the impurities and defects affect electrical parameters in grown HgCdTe crystals (Bartlett et al. 1980; Dornhaus and Nimtz 1983; Capper 1989, 1991).

An n-type HgCdTe alloy, with a low carrier density and a high mobility as well as a long minority carrier lifetime, is needed for a long-wave photoconductive detector working at 77 K (Willardson and Beer 1981; Keyes 1977). The impurities in HgCdTe have a critical influence on carrier lifetimes and mobilities. High impurity densities induce numerous recombination centers in the material, which drastically reduces the minority carrier lifetime. Ionization impurities and neutral impurities scatter carriers and depress the mobility. Therefore, it is very important to reduce the impurity density in HgCdTe crystals. Yet, to obtain photovoltaic devices with stable electrical properties, the HgCdTe material needs to be intentionally doped. A crucial step in fabricating HgCdTe focal-plane devices is to make p–n junction arrays in the active region of HgCdTe, by using ion implantation (Wang et al. 1991), diffusion methods, or by in-situ doping during MBE growth. Typically In ions are implanted, which manifest donor characteristics (Destefanis 1985; Schaake 1986; Vodop'yanov et al. 1982a, b), into p-type HgCdTe substrates to realize p–n junctions. While V_{Hg} are acceptors, they also are more effective scatters and life time killers than As-doped material, as long as the As can be placed on Te sites. Because As is amphoteric and is a donor when substituted on a Hg site, in LPE growth to locate it on a Te site it must be grown from the Hg side of the existence curve where the Hg vacancy density is a minimum (Berding et al. 1997). MBE growth always produces material at the Te side of the existence curve, and therefore as-grown, As resides mostly on the Hg sites. To get it to transfer to the Te sites requires a complex annealing procedure (Berding et al. 1998b).

HgCdTe is a semiconductor material whose electrical properties are dominated by alloy composition and native point defects. If in the intrinsic, or even extrinsic, region of HgCdTe material the minority carrier lifetime is mainly determined by Auger recombination, the material is considered to be of high purity. For high-purity HgCdTe, the electrical parameters depend on the composition and the density of point defects associated with the stoichiometry deviation. Once the composition is fixed, it is possible to control the density of point defects, and hence the electrical parameters of HgCdTe crystals, by optimizing the heat treatment condition (Tang 1974, 1976). For high-purity HgCdTe bulk material, good agreement has been established for electrical parameters between theory and experiment. In Figs. 2.29 and 2.30, composition-dependent electron density and mobility are illustrated for high-purity n-type HgCdTe at 77 K (Higgins et al. 1989). It is clear that the electron density is about 6–15 × 10^{13} cm^{-3} and the electron mobility is 3–5 × 10^{5} cm^2/Vs. Figure 2.31 depicts the lifetime of minority carriers as a function of temperature for a HgCdTe sample (Kinch and Borrello 1975). The lifetime of the minority carriers is approximately 4–6 μs at 77 K.

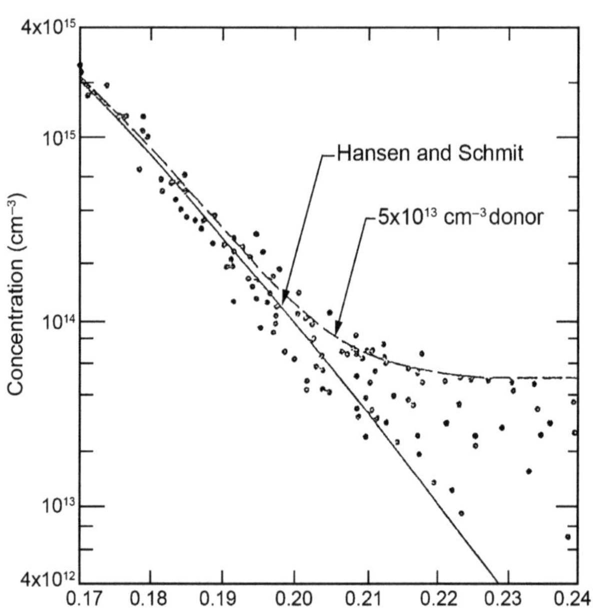

Fig. 2.29 Composition dependent carrier density of a high-purity n-type HgCdTe bulk crystal

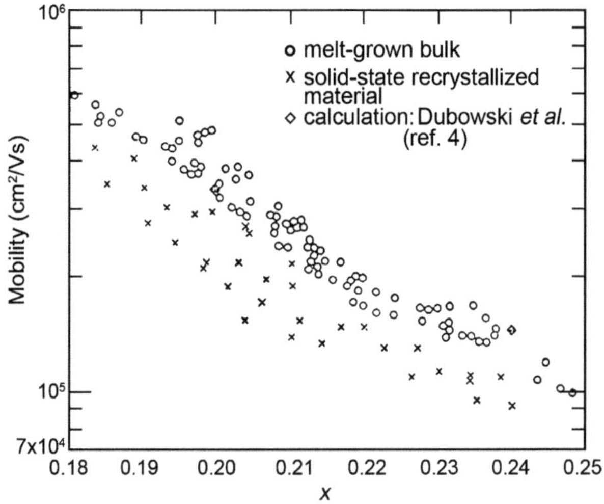

Fig. 2.30 Composition dependent carrier mobility of a high-purity n-type HgCdTe bulk crystal

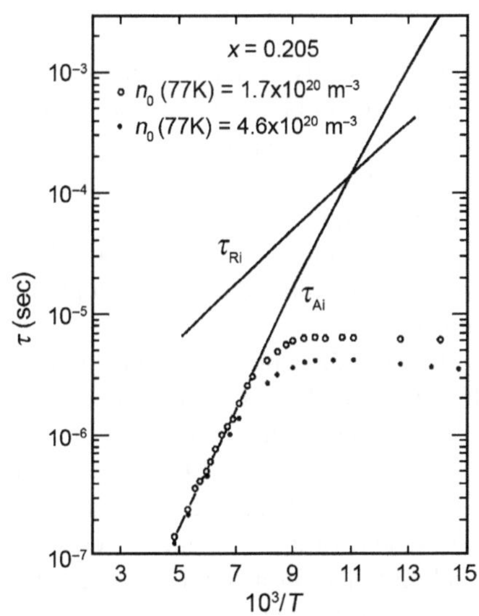

Fig. 2.31 Lifetimes of minority carriers as a function of temperature for a high-purity n-type HgCdTe bulk crystal

2.2.2 Shallow Donor Impurities

Because for low x values the band gap of $Hg_{1-x}Cd_xTe$ is narrow, the donor ionization energy is very small. For example, for a $Hg_{1-x}Cd_xTe$ sample with $x \approx 0.2$, its band gap energy is $E_g \approx 0.1$ eV, and correspondingly, for shallow donors its ionization energy E_d is only about 0.5 meV. This small energy is very difficult to deduce from a conventional Hall measurement. Shallow donor levels in

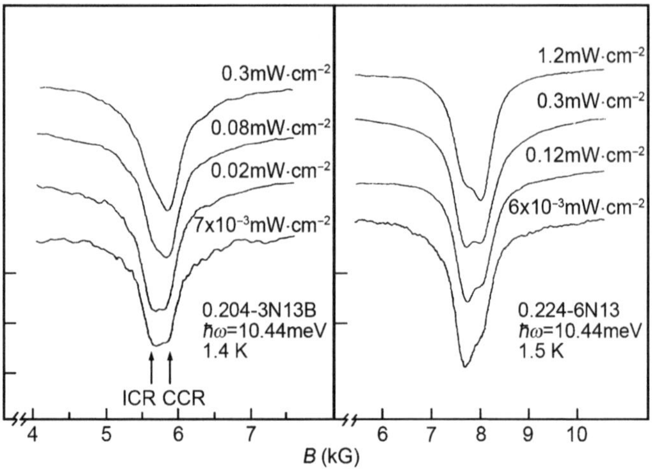

Fig. 2.32 Magneto-transmission spectra of two n-type $Hg_{1-x}Cd_xTe$ samples at different temperatures (The *arrows* indicate two absorption peaks.) respectively. They are n-type doped with carrier densities of $3 \times 10^{13}\,cm^{-3}$

HgCdTe usually nearly coincide with the bottom of conduction band, and hence it is also impossible to investigate the impurity optical transition from its ground to excited state using photothermal ionization photoconductivity spectroscopy, a technique that has had great success in studying high-purity Ge material. Indeed, it is an interesting question to determine if the shallow donor levels can be separated from the bottom of the conduction band. Therefore, it is meaningful to study the magnetic freeze-out effect of the shallow donors at low-temperature. ICR facilitates this kind of study.

Figure 2.32 depicts the conduction band electron cyclotron resonance spectra and ICR for two $Hg_{1-x}Cd_xTe$ samples in magnetic fields (Goldman et al. 1986). The compositions of the samples are $x = 0.204$ and 0.224, and the thicknesses are 290 and 260 μm, and $6 \times 10^{13}\,cm^3$, and carrier mobilities of $2.7 \times 10^5\,cm^2/Vs$ and $1.2 \times 10^5\,cm^2/Vs$, respectively, at a temperature of 77 K. The laser energy used is 10.44 meV. In these magnetic fields the conduction band electrons are quantized, leading to the formation of a series of Landau levels. As the magnetic field grows from 0, the energetic separation between the ground state Landau level, 0^+, and the first excited Landau level, 1^+, increases. When the energy separation becomes equal to the incident photon energy of the laser, a resonant absorption of the light takes place, and results in an absorption peak in a magneto-transmission spectrum, which is denoted as "CCR" in Fig. 2.32. Another absorption peak is also observed, that is caused by the impurity electron transition from the ground to the first excited state, and occurs at slightly lower magnetic field marked "ICR" in Fig. 2.32.

The energetic separation between the CCR and ICR is given by

$$\Delta E_B = (E_{110} - E_{000}) - (E_{1^+} - E_{0^+}). \tag{2.47}$$

This expression can be derived from:

$$\Delta E_{\mathrm{B}} = \left.\frac{\mathrm{d}(E_{1+} - E_{0+})}{\mathrm{d}B}\right|_{B=B_{\mathrm{CCR}}} \times (B_{\mathrm{CCR}} - B_{\mathrm{ICR}}) \quad (2.48)$$

where E_{1+} and E_{0+} represent the Landau levels of 1^+ and 0^+ respectively, and are functions of B. It can be either theoretically calculated or experimentally evaluated. Experimentally, the absorption peaks will appear at different magnetic fields if far infrared laser light with different energies, $\hbar\omega$, are used as the excitation source. Figure 2.33a plots the relation between the values of $\hbar\omega$ and B at which the CCR and ICR absorption peaks occur. The curves are obtained using a Bowers–Yafet model (Bowers and Yafet 1959) with the conduction band bottom effective mass used as a fitting parameter. It is also possible to connect the experimental points into two smooth curves for the two different composition samples. Then the slope of the curves at $B = B_{\mathrm{CCR}}$ is $\left.\frac{\mathrm{d}(E_{1+} - E_{0+})}{\mathrm{d}B}\right|_{B=B_{\mathrm{CCR}}}$. By multiplying $B_{\mathrm{CCR}} - B_{\mathrm{ICR}}$, the energy difference ΔE_{B} of the two transitions is obtained. The relation between B and ΔE_{B} is shown in Fig. 2.33b. The shallow impurity ionization energy can be obtained by extrapolating the curves to $B = 0$, and is approximately 0.3 meV. The theoretical values for the samples of $x = 0.204$ and 0.224, on the other hand, are 0.25 and 0.38 meV, respectively, according to an effective Rydberg energy calculation.

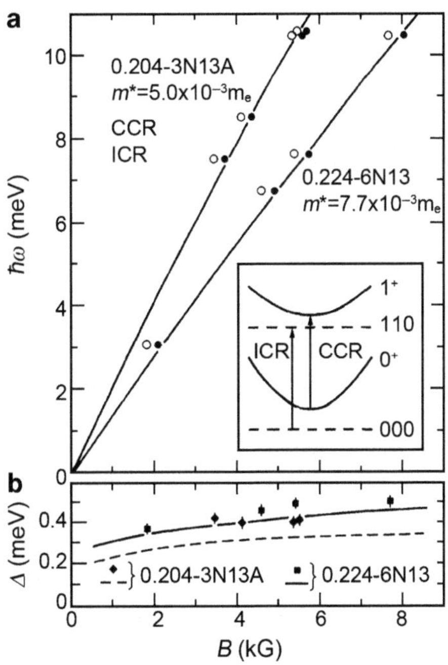

Fig. 2.33 (**a**) Resonance magnetic fields at several different photon energies ($\hbar\omega$). The *curves* are nonparabolic fits to the $0^+ \rightarrow 1^+$ transition, with the effective mass as a fitting parameter. The *inset* shows the relevant energy-level scheme. (**b**) The energy splitting $\Delta = \hbar\omega_{\mathrm{ICR}} - \hbar\omega_{\mathrm{CCR}}$ vs. magnetic field (B) for two $\mathrm{Hg}_{1-x}\mathrm{Cd}_x\mathrm{Te}$ samples with $x = 0.204$ and 0.224, respectively

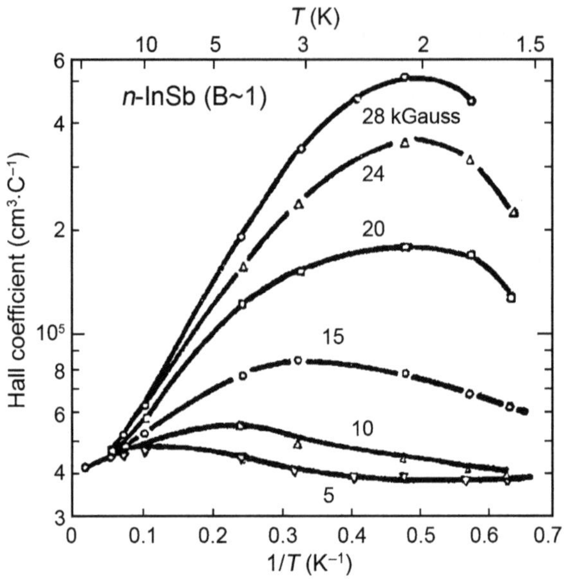

Fig. 2.34 Hall coefficients of an n-type InSb sample vs. temperature measured at different magnetic fields

A similar magnetic freeze-out phenomenon was also observed in InSb at low-temperature (Sladek 1958). In Fig. 2.34, the Hall coefficients of n-type InSb are plotted as a function of temperature from 10 to 1.6 K for different magnetic fields. It is clear that as the magnetic field gets higher, the Hall coefficient shows a sudden decrease above a critical temperature, which suggests a sudden reduction in the carrier density. The reason is that the electrons in the conduction band freeze-out as the donor levels deepen at the higher the magnetic fields. At a low-temperature, the frozen-out electrons may be re-excited into the conduction band by an electric field. Figure 2.35 illustrates such a case for n-type InSb in a magnetic field of 29 kilogauss at a temperature of 2.45 K. A drastic decrease of the Hall coefficient occurs as the electric field increases from zero to about $1\,\mathrm{V\,cm^{-1}}$, indicating a rapid increase of the carrier density in the conduction band as the electron distribution is heated by the electric field.

Chen et al. have observed the magnetic freeze-out effect of donor impurities by measuring the field-dependence of the electron temperature's impact on the electrical conductivity in HgCdTe (Chen 1990). The electrical conductivity shows a sharp jump at higher magnetic fields that is attributed to impact ionization induced by the electric field. Not surprisingly the higher the magnetic field, the larger the electric field required to initiate the jump of the electrical conductivity. This occurs because the magnetic-induced Landau level deepens as the magnetic field gets higher, resulting in an increase of the impurity ionization energy.

Zheng et al. have also studied the transport behavior of the electrons associated with localized shallow donor levels in wider gap $Hg_{0.58}Cd_{0.42}Te$ (Zheng et al. 1994). The result is illustrated in Fig. 2.36. At zero magnetic field the data indicates

Fig. 2.35 The Hall coefficient as a function of electric field for an n-type InSb sample measured at different magnetic fields

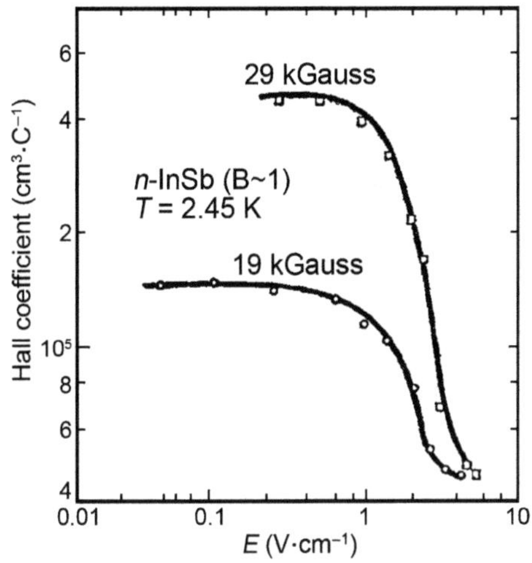

Fig. 2.36 Conductance vs. electric field for a $Hg_{0.58}Cd_{0.42}Te$ sample in different magnetic fields

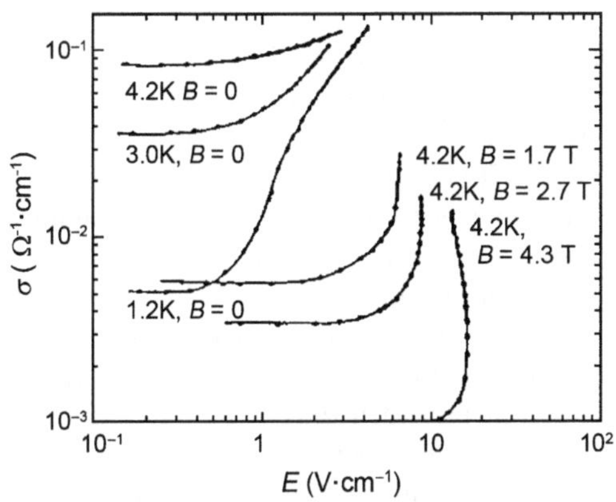

no evidence for a magnetic freeze-out effect of the donor levels. In general, impurity atoms are assumed to be completely ionized in HgCdTe at 77 K because the donor levels are quite shallow, nearly degenerate with the bottom of the conduction band. Therefore, the carrier density determined by measuring the Hall coefficient of a sample at 77 K can be taken as the effective donor concentration N_D*. The carrier density at any temperature can be easily calculated by using the relation $n_i^2 = n(n - ND*)$. The Fermi energy can also be deduced from the distribution of electronic states in the conduction band.

Figure 2.37 depicts the Fermi levels of $Hg_{1-x}Cd_xTe$ samples ($x = 0.194$) as a function of temperature with different donor concentrations. It is clear that the location of actual Fermi energy relative to the bottom of conduction band varies with an increase of temperature from 77 to 300 K, and gradually approaches the intrinsic Fermi energy as the temperature gets close to room temperature. The Fermi energy is higher for higher effective donor concentration, and this effect is most significant at low-temperature.

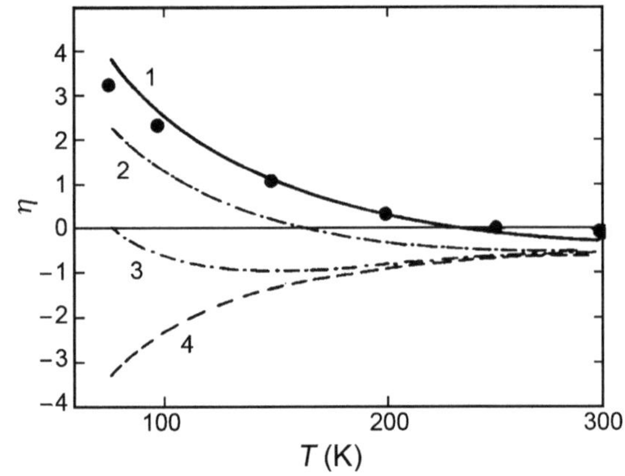

Fig. 2.37 Fermi energy levels of a series of HgCdTe samples ($x = 0.194$) with different donor concentrations. *curve 1*: $N_D = 2.2 \times 10^{16}\,\text{cm}^{-3}$, *curve 2*: $N_D = 1 \times 10^{16}\,\text{cm}^{-3}$, *curve 3*: $N_D = 2.26 \times 10^{15}\,\text{cm}^{-3}$, and *curve 4*: intrinsic condition

2.2.3 Shallow Acceptor Impurities

Shallow acceptor levels in HgCdTe pose an interesting problem. For unintentionally doped HgCdTe, the shallow acceptor levels observed are mainly due to Hg vacancies. The energy position of Hg vacancy is usually experimentally determined. Scott et al. (1976) have carried out Hall, far infrared transmission, and photoconductivity measurements on p-type $Hg_{1-x}Cd_xTe$ samples with $x = 0.4$ and doping concentration $p = 4 \times 10^{15}$, and $1 \times 10^{17}\,\text{cm}^{-3}$. In the photoconductivity measurement, a sharp peak occurs at 13.4 meV. In the far infrared transmission spectrum recorded at 8 K, a sharp absorption peak also appears at about 13.4 meV ($107\,\text{cm}^{-1}$), as illustrated in Fig. 2.38. This absorption peak was also evidenced in transmission measurements (Shen 1994).

According to experimental electrical results, a simple relation can be established between the shallow acceptor ionization energy and the hole density at 77 K (Scott et al. 1976),

$$E_A = E_0 - \alpha p_0^{1/3} \tag{2.49}$$

where $E_0 \approx 17\,\text{meV}$, $\alpha = 3 \times 10^{-8}\,\text{eV cm}$ and p_0 is the hole density at 77 K. Similar results of $E_0 \approx 17.5\,\text{meV}$, $\alpha = 2.4 \times 10^{-8}\,\text{eV cm}$ have also obtained by Hall measurements and corresponding fitting procedures (Yuan et al. 1990).

Li Biao et al. have published a direct experimental result for the Hg vacancy energy level (Li et al. 1998) from a far infrared spectrum. For a thick sample, it was found to be very difficult to separate the shallow acceptor levels from reststrahlen spectra and two-phonon absorption bands, because the shallow acceptor energies vary from 1 to 30 meV (or equivalently, 10–250 cm^{-1}), which are very close to the reststrahlen spectra energies and two-phonon absorption bands. For a sample grown by MBE or LPE, however, the thickness is very thin, making it possible to investigate the shallow acceptor levels by far infrared absorption measurements. The samples studied are 20-μm-thick undoped and Sb-doped materials prepared by LPE,

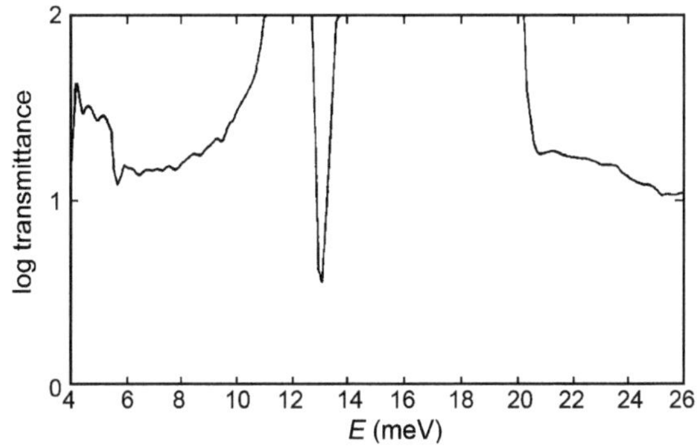

Fig. 2.38 The transmission spectrum of a $Hg_{0.6}Cd_{0.4}Te$ sample at a temperature of 8 K

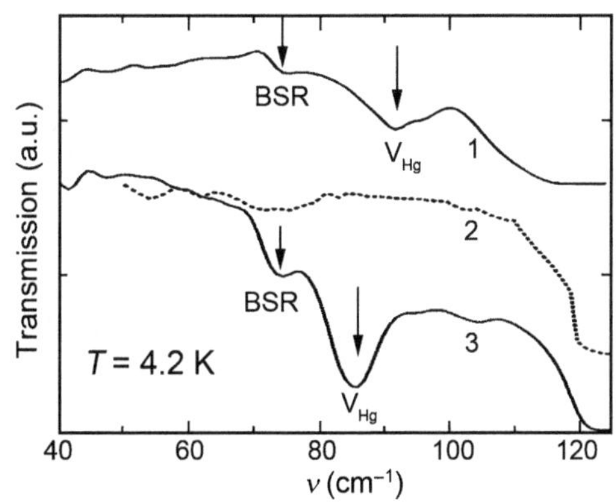

Fig. 2.39 Far infrared transmission spectra of the HgCdTe epitaxial layers at 4.2 K. *curve 1* is the p-type MBE thin film ($x = 0.285$), *curve 3* and *curve 2* are the transmission spectra of a before annealing p-type ($x = 0.37$) and an after annealing n-type LPE sample, respectively

and 10-μm-thick thin films prepared by MBE. The hole density falls in the range $1 \times 10^{15} - 7 \times 10^{16}\,cm^{-3}$. The measured spectroscopic range is 20–250 cm^{-1}.

Figure 2.39 shows typical far infrared transmission spectra measured at a temperature of 4.2 K. The absorption peak marked with BSR is an feature related to the beam splitter. Curve 1 is an absorption spectrum of a p-type HgCdTe ($x = 0.285$) sample grown by MBE. It has an absorption peak, marked with V_{Hg}, clearly seen at 92 cm^{-1}, or equivalently at 11.4 meV. Curve 3 is a far infrared absorption spectrum of a p-type HgCdTe ($x = 0.37$) sample grown by LPE. It has an obvious absorption peak, marked with V_{Hg}, at 86 cm^{-1}, or equivalently at 10.6 meV. Curve 2 is the far infrared transmission spectrum of the LPE sample after an n-type conversion annealing treatment. The annealing changes the sample from p-type to n-type. It is clear that the absorption peak previously appearing at 86 cm^{-1} in curve 3, is missing in curve 2. Further measurements indicate that the intensity of the absorption peak does not show a drastic decrease as the temperature increases. All these findings suggest that the absorption peak is not due to lattice absorption but are caused by Hg vacancies.

Figure 2.40 illustrates the results of Hall measurements and integrated absorption intensities (IAI) as well as absorption pair-peak heights (PPH) for a p-type sample prepared by LPE, and measured at different temperatures. From the Hall

Fig. 2.40 The hole density (Hall measurement), integrated absorption intensity (IAI) and pair-peak height (PPH) at different temperatures for a p-type $Hg_{0.63}Cd_{0.37}Te$ sample. The *inset* gives the far infrared transmission spectra at different temperatures

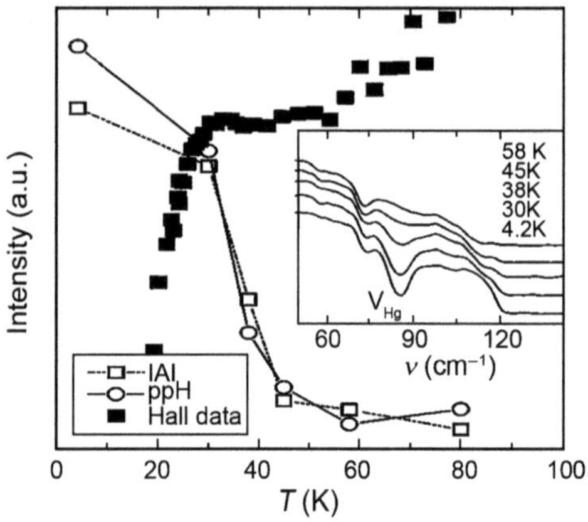

curve the ionization energy E_A and the concentration of the Hg vacancies N_A are found to be 9.7 meV and 7.6×10^{14} cm^{-3}, respectively. The ionization energy value is close to the result obtained by absorption spectroscopy. It is clear that as temperature decreases, the PPH and IAI increase and the hole concentration decreases, indicating the presence of a carrier freeze-out effect. On the assumption that the Hg vacancy concentration is proportional to the IAI, a relation can be deduced (Klauer et al. 1992):

$$C_{V_{Hg}} = \frac{\int \alpha \, dv}{a_{VHg} \ln 10} \tag{2.50}$$

where a_{VHg} is the absorption strength per Hg vacancy. From the Hall measurement, it is deduced that $C_{VHg} \approx N_A = 7.6\times10^{14}$ cm^{-3}, and from the absorption spectrum, $\int\alpha dv = \text{IAI} = 6\times10^3$ cm^{-2}. Therefore, the result derived is $a_{VHg} \approx 3.4\times10^{-12}$ cm. This value is very important for estimating the positively charged vacancy concentration. For example, for the undoped MBE sample involved in Fig. 2.39, the vacancy concentration is estimated to be 3.8×10^{14} cm^{-3} by multiplying absorption area with absorption strength per Hg vacancy, and is similar to the Hall data result, 5.5×10^{14} cm^{-3}.

Figure 2.41 depicts the Hg vacancy ionization energies E_A of LPE grown p-type and MBE grown n-type samples with different compositions. Obviously, most of the samples manifest ionization energies around 10–12 meV, independence of the composition. This value is close to the result obtained by Sasaki et al. (1992) on MBE samples, and that by Shin et al. (1980) on LPE samples. However, theoretical calculations suggest that the ionization energy of the Hg vacancy depends on the forbidden band gap E_g. The discrepancy may be introduced by the influence of a longitudinal compositional distribution, or an interface lattice mismatch, that affect the acceptor levels of HgCdTe epilayers.

No absorption peak is observed from As impurities in As-ion-implanted HgCdTe samples. The reason is that the As impurity is amphoteric leading to highly compensated material with the As mostly residing on the cation sublattice where it

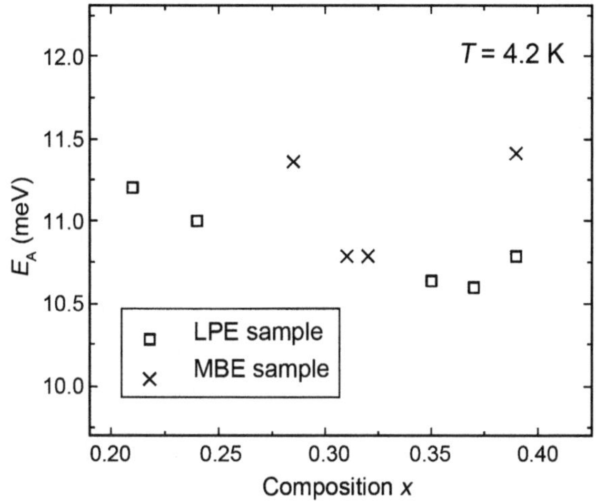

Fig. 2.41 The Hg vacancy ionization energies E_A of HgCdTe LPE grown and MBE grown samples with different compositions, obtained from far infrared absorption spectra at 4.2 K

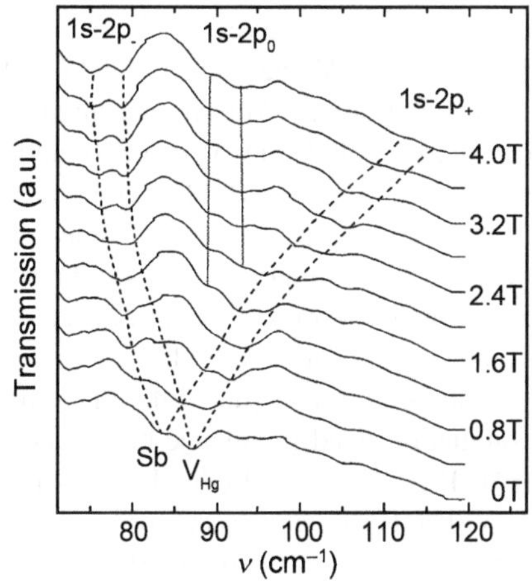

Fig. 2.42 The magneto-optical far infrared spectra of an $Hg_{0.63}Cd_{0.37}Te$ sample prepared by LPE and measured at a temperature of 4.2 K

is a shallow donor (Berding et al. 1997; Berding and Sher 1998a). For Sb-doped $Hg_{1-x}Cd_xTe$ ($x = 0.39$) samples prepared by LPE, besides the Hg vacancy absorption that appears at 87 cm^{-1}, an additional absorption peak is at 83 cm^{-1}, or equivalently, 10.5 meV, which is Sb impurity related. This is consistent with the reported value of 11 meV for the acceptor level of Sb-doped $Hg_{1-x}Cd_xTe$ ($x = 0.22$) samples (Chen and Dodge 1986). Figure 2.42 plots the magneto-optical far infrared spectra of acceptor states in HgCdTe at a temperature of 4.2 K. Due to the Zeeman effect, the excited states of the Sb impurity and the Hg vacancy split in magnetic fields. In Fig. 2.43, the Zeeman splitting is plotted as a function of the magnetic field.

The energy positions of acceptor levels can also be investigated by PL spectroscopy. Richard and Guldner et al. have performed PL measurements on

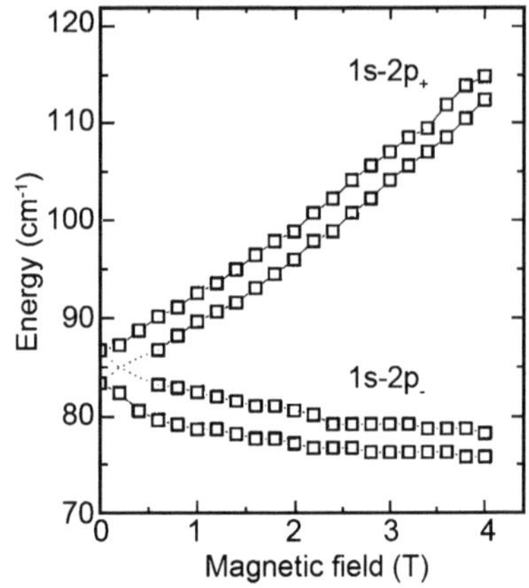

Fig. 2.43 The Zeeman splitting as a function of magnetic field for the Sb acceptor state in a $Hg_{0.63}Cd_{0.37}Te$ sample

$Hg_{1-x}Cd_xTe$ samples ($x = 0.285$) at a temperature of 18.6 K by using a CO laser as the excitation source. The excitation energy is 244.6 meV, and the power is 140 mW. Recombination emission from the conduction band to acceptor levels have been observed, with $\Delta = 3.2$ meV, $E_g - E_A = 202.7$ meV. Because $E_A = 14.3$ meV, we find $E_g = 217$ meV.

Kurtz et al. (1993) have measured infrared PL spectra of $Hg_{1-x}Cd_xTe$ samples prepared by MOCVD ($x = 0.216$) and by LPE ($x = 0.234$), respectively, by using a double-modulation (DM) technique. In Fig. 2.44, the PL spectra are depicted for the samples before and after an annealing process. Before annealing, the PL peak reflects the recombination processes between the conduction band and acceptor levels, and reveals several Hg vacancy acceptor levels. The band-edge PL peak is relatively broad and appears at a higher energy. After annealing in a Hg atmosphere, the Hg vacancies are removed and the PL peak reflects the luminescence of radiation from conduction band to valence band recombination. The full-width at half-maximum (FWHM) of this band-edge PL peak is relatively small, and appears at a higher energy. The Hg vacancy acceptor level is found to be 12 and 19 meV for $Hg_{1-x}Cd_xTe$ samples with $x = 0.216$ and 0.234, respectively.

From the measurements, the Hg vacancy acceptor level is found to lie in the range of 10–15 meV above the valence band edge and it moves down to the valence band edge as the acceptor concentration increases. According to Scott the relationship is $E_A = 0$ for $N_A - 3.2 \times 10^{17}\,cm^{-3}$. Therefore, the PL peaks of the conduction-band to acceptor levels are expected to show a blue shift with an increase of the acceptor concentration.

Hunter and McGill (1981) have measured the PL spectra of two $Hg_{1-x}Cd_xTe$ samples with $x = 0.32$ and 0.48, and observed recombination emissions of the conduction band to valence band, between acceptor levels, the donor to acceptor levels, and among bound excitons and other levels. Both the line shape and the intensity of the conduction band-valence band transition depend on the excitation power. Figure 2.45 illustrates a PL spectra of a $Hg_{1-x}Cd_xTe(x = 0.48)$ sample at

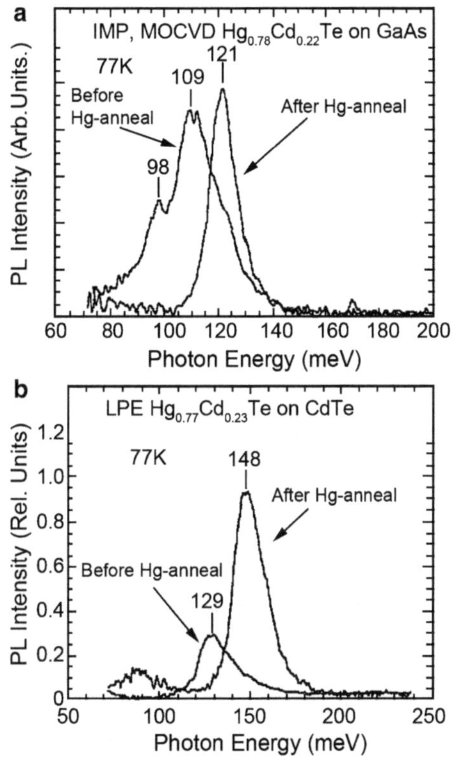

Fig. 2.44 Photoluminescence (PL) spectra of two HgCdTe samples prepared by MOCVD (**a**) and LPE (**b**), measured before and after an annealing process

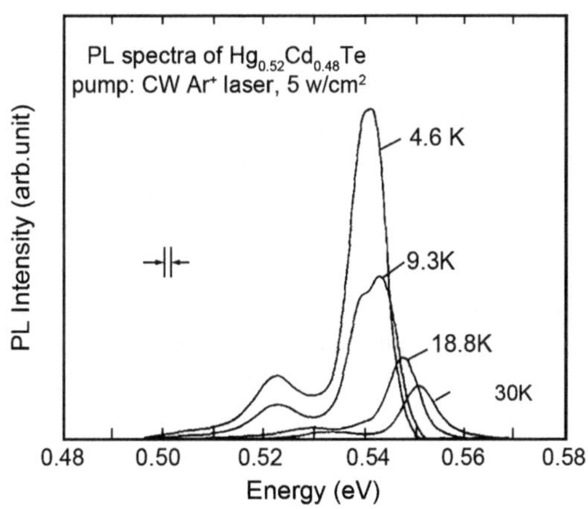

Fig. 2.45 The PL spectra of a $Hg_{0.52}Cd_{0.48}Te$ sample measured at 4.6, 9.3, 18.8 and 30 K, respectively

different temperatures. The peak at the lower energy is due to the recombination between the conduction band and acceptor levels for $T > 10\,K$, and between the donor and acceptor levels at lower temperature. The peak at high energy corresponds to the PL transition between the conduction and valence band, and the middle

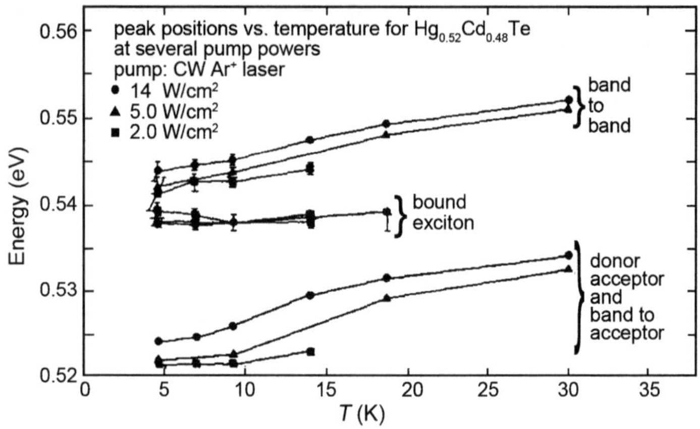

Fig. 2.46 The PL peak energy as a function of temperature for an $Hg_{0.52}Cd_{0.48}Te$ sample at different temperatures

peak position is due to the transition with bound excitons (especially obvious at $T = 9.3$ K). In Fig. 2.46, the PL peak energy is plotted as a function of temperature for a $Hg_{1-x}Cd_xTe$ sample with $x = 0.48$. It can be determined from the PL peak energy that the acceptor ionization energy, E_A, is 14.0 ± 1.5 meV for $x = 0.32$ and 15.5 ± 2.0 meV for $x = 0.48$. The donor ionization energy, E_D, is also obtained to be 1.0 ± 1.0 meV for $x = 0.32$ and 4.5 ± 2.0 meV for $x = 0.48$. The acceptor may be an Au-substitutional impurity on a cation site or a cation vacancy.

The experimental acceptor ionization energy of an Sb-doped HgCdTe sample is slightly smaller than that of the Hg vacancy. Chen et al. (1990) and Li et al. (1998) reported a value of about 10 meV. Smaller values were also reported. For example, Wang (1989) has studied a series of Sb-doped HgCdTe bulk crystal samples. The samples became p-type after a conventional Hg-saturated, n-type anneal. The Hg vacancy concentration is Sb independent, and can be deduced from the relation between the Hg vacancy concentration and the heat treatment, to be 10^{14} cm^{-3}. Meanwhile, the p-type doping due to the Sb concentration is larger than 10^{15} cm^{-3}. The reason for the p-type characteristic of an Sb-doped HgCdTe sample after low-temperature processing is that Sb atoms replace the Te in the lattice (Berding et al. 1997; Berding and Sher 1998a), or are in interstitial sites, where they are acceptors.

The Hall coefficient, electric conductivity, and mobility of the samples have been measured. In the low-temperature ionization region the resistivity can be expressed as:

$$\rho \propto T^{-3/4} e^{(E_A/2k_B T)}, \tag{2.51}$$

therefore, the slope of a $\ln(\rho T^{3/4}) - 1/T$ curve gives the acceptor ionization energy E_A. The acceptor ionization energies determined with this method are listed in Table 2.7 for a series of Sb-doped $Hg_{1-x}Cd_xTe$ samples.

Table 2.7 The acceptor ionization energies of p-HgCdTe

Acceptor type	x	E_A(meV)	P(cm^{-3})	Reference
Sb	0.36	6	3.2×10^{16}	Wang (1989)
Sb	0.21	7	6×10^{16}	Capper et al. (1985)
Sb	0.22	2	7.2×10^{16}	Gold and Nelson (1986), Chen and Tregilgas (1987)
Sb	0.22	11		Chen and Dodge (1986)
Sb	0.39	10.5	7.6×10^{14}	Li et al. (1998)
V_{Hg}	0.26–0.33	15–18	5×10^{16}	Capper et al. (1985), Chen and Tregilgas (1987)
V_{Hg}	0.32–0.48	15	-10^{16}	Hunter and McGill (1981)
V_{Hg}	0.20–0.39	11.5	7.6×10^{14}	Li et al. (1998)

2.3 Deep Levels

2.3.1 *Deep Level Transient Spectroscopy of HgCdTe*

A deep level is an energy state located in the forbidden band near the center of the gap that is produced by the short-range primitive cell potential of a defect center. The defect can be due to an impurity, a native point defect, a complex of native point defects, or a complex of a native point defect with an impurity. The deep level can capture holes or electrons and hence acts as an electron/hole trap. It reduces carrier lifetimes to affect device performance.

To study the deep energy levels of defects in bulk material, a junction space charge technique is often adopted. The optical methods (Hall, PL, absorption, and PTIS) have an obvious advantage of high spectral resolution, which is very helpful for exact determination of the energy positions of defect levels. However, they cannot determine accurate information about other electrical parameters impacting the carrier density, and the capture and the emission rates. The junction space charge method has a lower spectral resolution, about 10 meV, but it can be used to get the absolute value of the electrical parameters. One major junction space charge method effective in studying the deep levels in HgCdTe, is DLTS. Other experimental methods besides the DLTS, that can be used in the study of deep levels include, admittance spectroscopy (AS), low-frequency conductance spectroscopy (LFCS), and thermally stimulated current (TSC).

DLTS was first established in 1974 (Lang 1974). It is an effective method for studying deep levels in semiconductors. With this method, different parameters can be measured, which include the concentration of deep level traps (N_T), the energy levels of the deep level traps (E_t), the electron or hole capture cross section (σ),

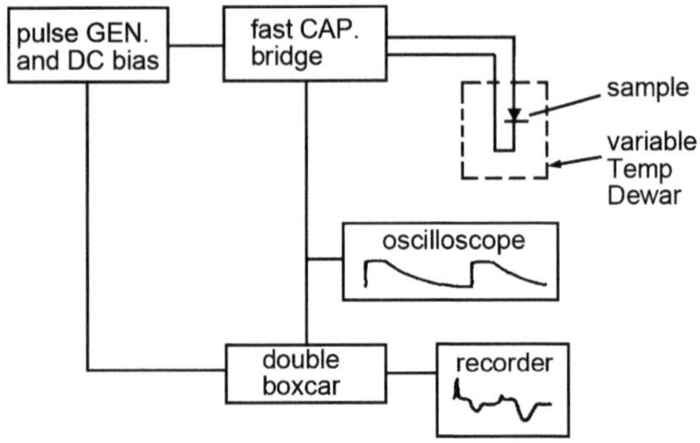

Fig. 2.47 A schematic of the deep level transient spectroscopy (DLTS) setup

the optical cross section (σ^o), the distribution of random traps in the diode junction area ($N_T(z)$), and the carrier emission rate of traps. One can also distinguish if the trap captures majority or minority carriers. The schematic configuration for DLTS is plotted in Fig. 2.47.

Samples that can be measured by DLTS include a Schottky potential barrier diode, an MOS device, or more generally any structure for which a depletion layer can be established. For an n^+-p photoelectric diode in reverse bias, the depletion layer is mainly on the p side of the diode. The sample is placed in a variable-temperature cryostat and its capacitance is measured with a fast capacitance bridge. The sample is first set in reverse bias for a long enough time to ensure the equilibration of a depletion layer in the p region. In the depletion layer any carriers captured in the traps are removed and thus the traps are initially empty.

Now two operations are performed. One operation is to apply a pulse signal with its height being smaller than the reverse bias voltage, so that even at the pulse maximum the device is still in reverse bias, as illustrated in Fig. 2.48. Under the influence of the pulse, some holes return to the depletion layer in the p region, while others return after the pulse is over. A few holes will be captured in this process by the traps in the depletion layer. The other operation is to apply a pulse signal with its height higher than the reverse bias voltage. As the pulse reaches its maximum, the device will be driven into forward bias, so that not only holes, but also electrons are swept into the depletion layer of junction region. After the pulse, the holes/electrons will be captured if there are corresponding traps in the region. Obviously, hole traps can be detected in the first operation, while in the second operation the hole and electron traps can be detected together. By combining the two operations, the behavior of the electron and hole traps can be distinguished.

After the pulse, the electrons or holes trapped in the deep level traps can via thermal excitation return to their equilibrium from the initial nonequilibrium condition. This gives rise to a capacitance transient of the device whose response is determined by the thermal emission rates of electrons or holes in the traps, the temperature, and the location of the deep levels. Measuring the capacitance transients at different temperatures leads to a determination of the behavior of the corresponding

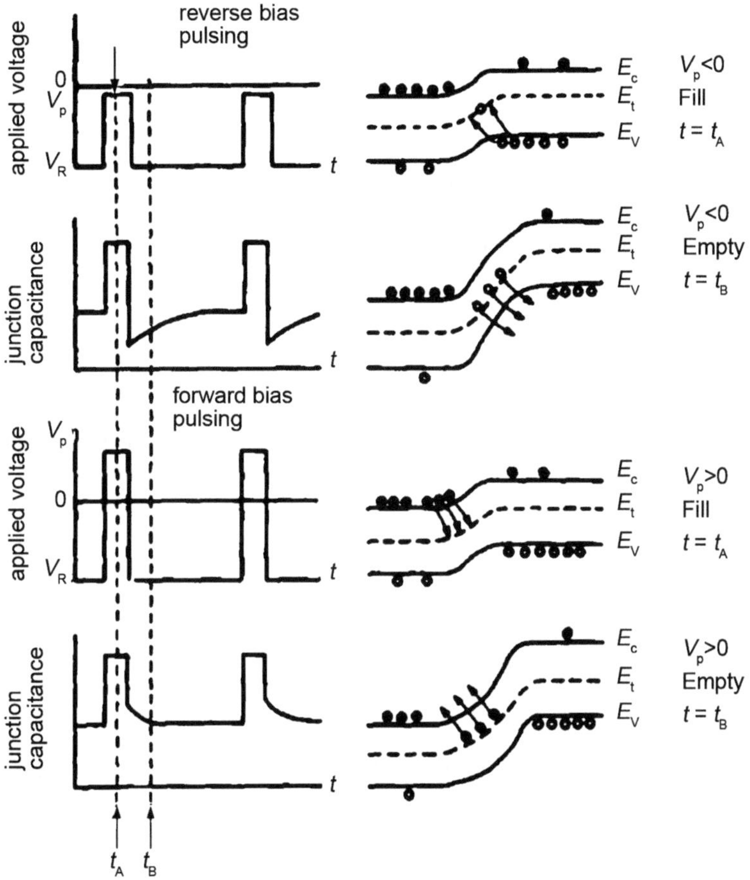

Fig. 2.48 A schematic of the capacitance transient change caused by the emission of the trapped carriers

deep level traps. The capacitance transient of deep levels as a function of temperature is determined experimentally mainly by measuring the transients due to thermal emission rates of the trapped electrons and/or holes, and is related to a space-charge dynamical process of capture and thermal emission.

If the depletion layer of an n^+-p junction lies on the p-region side, and is step-like with the thickness W, then the junction capacitance can be expressed as:

$$C(t) = \frac{A\varepsilon}{W} = \left(\frac{A^2 q\varepsilon}{2(V_{bi} + V_R)} N_I\right)^{1/2}, \tag{2.52}$$

where A is the junction area, V_{bi} is the build-in electric potential of the junction, V_R is the applied reverse bias voltage, and $N_I = N_A - N_D$ is the net density of charge centers in the p-type depletion layer. The maximum electric field $|F_m|$ across the junction is:

$$|F_{\mathrm{m}}| = \frac{V}{W} = \frac{Q/C(t)}{W} = \frac{qN_I W}{\varepsilon}. \tag{2.53}$$

Since the electric field $|F_{\mathrm{m}}|$ is normally very strong, the thermally excited carriers will be swept out of the space charge layer very rapidly, and because the trap capture rates are comparatively slow it is safe to assume that the swept-out carriers will not be re-captured by the traps in this process.

The capture rates of electrons and holes can be expressed, respectively, as:

$$C_{\mathrm{n}} = \sigma_{\mathrm{n}} \langle v_{\mathrm{n}} \rangle n, \tag{2.54}$$

$$C_{\mathrm{p}} = \sigma_{\mathrm{p}} \langle v_{\mathrm{p}} \rangle p, \tag{2.55}$$

where σ_{n} and σ_{p} are capture cross sections, n and p the densities, and $\langle v_{\mathrm{n}} \rangle$ and $\langle v_{\mathrm{p}} \rangle$ the average thermal velocities of electrons and holes. These average velocities are determined by:

$$\langle v_{\mathrm{p}} \rangle = \left(\frac{8k_{\mathrm{B}}T}{\pi m^*_{\mathrm{hh}}} \right)^{1/2}, \tag{2.56}$$

$$\langle v_{\mathrm{n}} \rangle = \left(\frac{8k_{\mathrm{B}}T}{\pi m^*} \right)^{1/2}, \tag{2.57}$$

where m^* and m^*_{hh} are the electron and hole effective masses. In thermal equilibrium, the thermal emission rate of each electron state is equal to the corresponding capture rate. According to Boltzmann statistics, the emission rates for electrons and holes are:

$$e_{\mathrm{n}} = (\sigma_{\mathrm{n}} \langle v_{\mathrm{n}} \rangle / g_{\mathrm{n}}) N_{\mathrm{c}} e^{(-\Delta E/k_{\mathrm{B}}T)}, \tag{2.58}$$

$$e_p = (\sigma_{\mathrm{p}} \langle v_{\mathrm{p}} \rangle / g_{\mathrm{p}}) N_{\mathrm{v}} e^{(-\Delta E/k_{\mathrm{B}}T)}, \tag{2.59}$$

where g_{n} and g_{p} are the ground state degeneracy factors of the electron and the hole traps, with $g_{\mathrm{p}} = 4$, and $g_{\mathrm{n}} = 2$, and:

$$\Delta E = E_{\mathrm{c}} - E_{\mathrm{t}} \quad \text{(for electrons)}, \tag{2.60}$$

$$\Delta E = E_{\mathrm{t}} - E_{\mathrm{v}} \quad \text{(for holes)}. \tag{2.61}$$

For a deep level trap that can capture and emit electrons and holes, the following expression holds:

$$\frac{\mathrm{d}N(t)}{\mathrm{d}t} = (C_n + e_p)(N_{\mathrm{T}} - N(t)) - (C_p + e_e)N(t), \tag{2.62}$$

where N_T is the total defect concentration, $N(t)$ the defect concentration occupied by electrons, and $N_T - N$ the defect concentration occupied by holes.

By assuming

$$\begin{cases} a = C_n + e_p = \sigma_n \langle v_n \rangle n + e_p, \\ b = C_p + e_n = \sigma_p \langle v_p \rangle p + e_n, \end{cases} \tag{2.63}$$

Equation 2.62 can be simplified further. For a trap filled with electrons at initial time $t = 0$, that is $N(0) = N_T$ integrating (2.62) leads to:

$$N(t) = \begin{cases} N_T & (t \leqslant 0) \\ \dfrac{a}{a+b} N_T \left\{1 + e^{[-(a+b)t]}\right\} & (t > 0). \end{cases} \tag{2.64}$$

Correspondingly, for a trap filled with holes at the initial time $t = 0$, the integration becomes:

$$N(t) = \begin{cases} 0 & (t \leqslant 0) \\ \dfrac{a}{a+b} N_T \left\{1 - e^{[-(a+b)t]}\right\} & (t > 0) \end{cases}. \tag{2.65}$$

For either of the two cases, one finds from (2.64) to (2.65) for $t \to \infty$:

$$N(t \to \infty) = \frac{a}{a+b} N_T, \tag{2.66}$$

which is the highest occupied steady-state concentration of a trap that can be reached for a given trap energy level.

Figure 2.49 presents the DLTS spectra of a p-type $Hg_{1-x}Cd_xTe$ ($x > 0.215$, $N_A = 9.1 \times 10^{15}\,cm^{-3}$) photodiode (Polla and Jones 1981). The lower spectrum was measured with the photodiode in reverse bias, $V_R = 0.60\,V$, and the applied

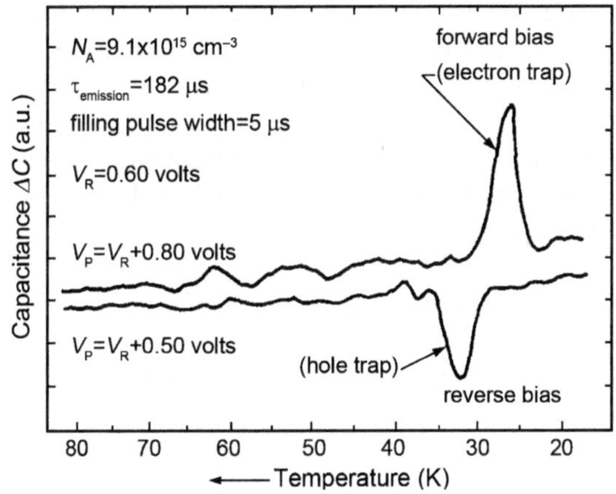

Fig. 2.49 DLTS spectra of a p-type $Hg_{0.785}Cd_{0.215}Te$ photodiode, where single hole- and electron-traps can be detected

pulse voltage, V_{pulse}, is $V_R + 0.50\,\text{V}$, with a pulse width of 5 µs. It is equivalent to the first operation mentioned above, and only the capture of holes is observed. A hole–trap related absorption peak obviously appears at a temperature of 32 K, with an emission time constant of 55 µs. The upper spectrum is a measurement for an applied pulse voltage of $V_{pulse} = V_R + 0.80\,\text{V}$. In this range of pulse times, the diode is just in forward bias. And neither electrons nor holes can enter the depletion region. Clearly an electron–trap related peak occurs at a temperature of 27 K, with an emission time constant of 182 µs. The peak will change its position on the temperature axis with the pulse width that is either higher or lower with a spectrum illustrated in Fig. 2.49. From (2.58) to (2.59), the reciprocal of the thermal emission rate is the time constant (τ), and both the $\langle v_n \rangle$ and $\langle v_p \rangle$ are proportional to $T^{1/2}$, while N_c and N_v are proportional to $T^{3/2}$, as shown in (2.67) and (2.68),

$$N_c = 2\left(\frac{m_c^* k_B T}{2\pi\hbar^2}\right)^{3/2}, \tag{2.67}$$

$$N_v = 2\left(\frac{m_{hh}^* k_B T}{2\pi\hbar^2}\right)^{3/2}. \tag{2.68}$$

Therefore, we have:

$$\frac{1}{\tau_e} = e_p = \sigma_p \frac{(k_B T)^2 m_{hh}^* e^{-\Delta E/k_B T}}{2\pi^2\hbar^3}, \tag{2.69}$$

and

$$(\tau_e T^2) \propto e^{(\Delta E/k_B T)}. \tag{2.70}$$

It is obvious that by drawing a straight line in the $\ln(\tau T^2) - 1/kT$ coordinates, the slope of the lines, $\Delta E = E_c - E_t$ or $\Delta E = E_t - E_v$ can be determined. In Fig. 2.50, the corresponding curves are given for the sample shown in Fig. 2.49, from which the hole and electron traps are determined to be at $E_t = E_v + 0.035\,\text{eV}$ and $E_t = E_c - 0.043\,\text{eV} = E_v + 0.043\,\text{eV}$, respectively.

The traps will not have enough time to capture carriers if the injection pulse time is very short. Meanwhile, most of the traps will not stop capturing carriers until saturation is reached if a wide injection pulse is applied. This leads to an exponential increase of the DLTS spectrum:

$$A(t) = A_{max}[1 - e^{(-t/\tau)}]. \tag{2.71}$$

Figure 2.51 depicts experimental results for an $Hg_{1-x}Cd_xTe$ sample with $x = 0.215$ reported by Polla and Jones (1981).

For $t = 0$, $A(0) = 0$; and for $t = \infty$, $A(\infty) = A_{max}$ is saturated. Therefore, one finds:

$$\ln\left(\frac{A_{max} - A(t)}{A_{max}}\right) = -\frac{t}{\tau}. \tag{2.72}$$

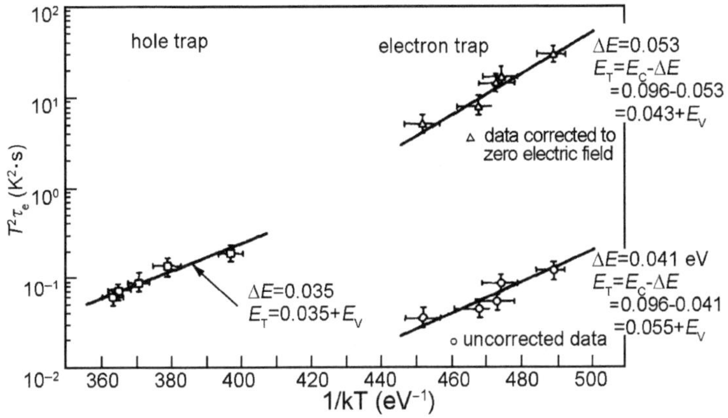

Fig. 2.50 Activation energies of the hole- and electron-traps calculated from the peak values illustrated in Fig. 2.49

Fig. 2.51 Capture cross sections of minority and majority carriers determined by the peak value of the DLTS spectra in Fig. 2.49

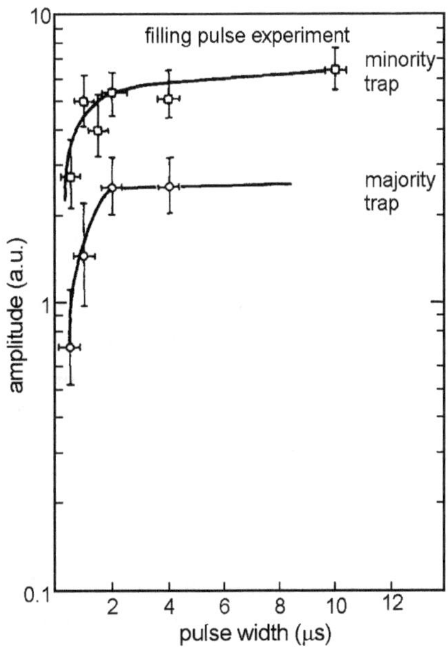

For the hole traps, the expression is:

$$\frac{1}{\tau_p} = \sigma_p \langle v_p \rangle p. \tag{2.73}$$

The $\ln([A_{max} - A(t)]/A_{max}) \sim t$ plot is a straight line, and from its slope, $\tau_p = (\sigma_p \langle v_p \rangle p)^{-1}$ is determined, which is the time needed to fill the traps fully. From

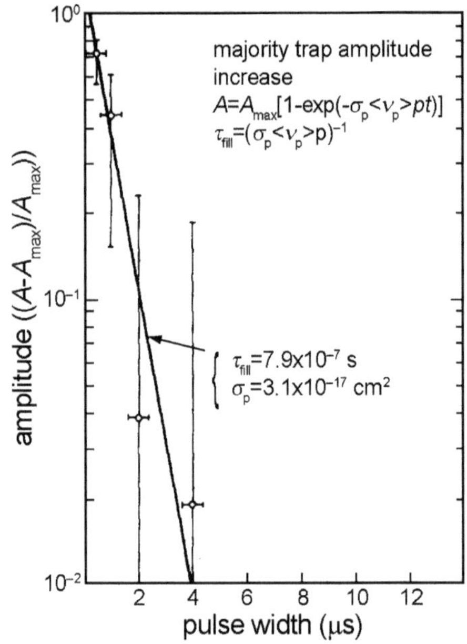

Fig. 2.52 Filling time constants of the majority carriers determined from the experimental results illustrated in Fig. 2.51

the thermal velocity $\langle v_p \rangle$ and the hole density p, the capture cross section σ_p can also be obtained. Figure 2.52 depicts the experimental relation between $\ln([A_{max} - A(t)]/A_{max})$ and t for the sample shown in Fig. 2.50. It can be easily derived from the slope that τ_p and σ_p for the hole traps at 30 K are 7.9×10^{-7} s and 3.1×10^{-17} cm^2, respectively; and σ_p for the electron traps is 2.1×10^{-15} cm^2. The capture cross sections of the traps are temperature independent.

Polla and Jones (1981) have employed this method in studying deep level spectra of undoped and Cu- and As-doped samples. Figure 2.53 plots the DLTS spectra of intrinsic $Hg_{1-x}Cd_xTe(x = 0.39)$ and a Cu-doped one, respectively (Jones et al. 1982). The hole density of the undoped sample is $p \approx 1 \times 10^{16}$ cm^{-3}. With an increase of the forward-bias voltage, the electron trap signal appearing at 150 K disappears gradually, and the electron trap signal at 45 K get stronger. The electron traps at these two temperatures suggest the assignments $E_C - 0.1$ eV and $E_C - 0.24$ eV, respectively. Under reverse bias, a trap at $E_V + 0.28$ eV shows up. For the Cu-doped sample illustrated in Fig. 2.53b, two Cu impurity-related traps are observed at 0.07 and 0.15 eV, respectively, above the top of the valence band. These deep level locations are concentration dependent, and the dependence is shown in Fig. 2.54. For the undoped $Hg_{1-x}Cd_xTe(x = 0.2\text{–}0.4)$ samples, the deep center levels are located at $E_V + 0.4E_g$ and $E_V + 0.75E_g$ in the forbidden band, respectively. The reported deep levels of CdTe are at $E_V + 0.15$ eV and $E_V + 0.36$ eV, respectively (Zanio 1978), which are also depicted in the figure.

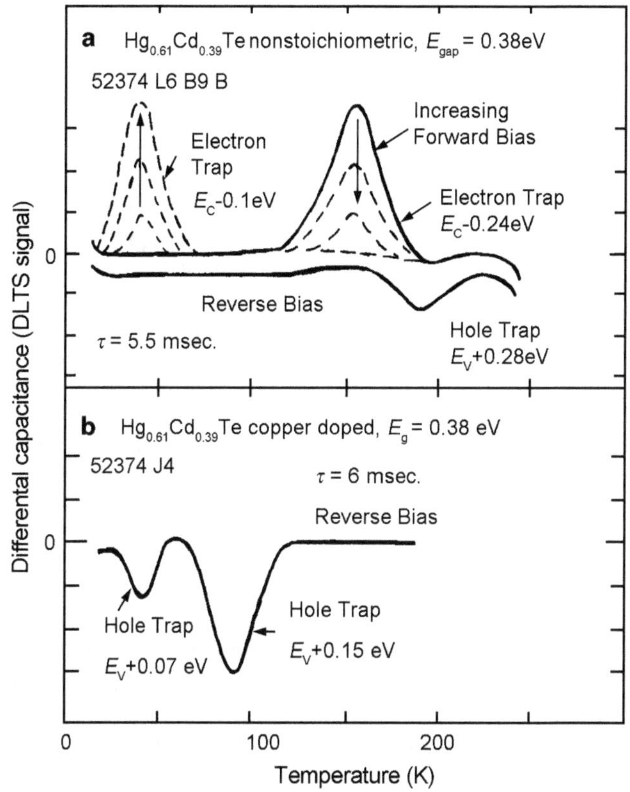

Fig. 2.53 DLTS spectra of undoped and Cu-doped HgCdTe samples. The electron traps decrease to zero at 150 K and increase with the forward-biased voltage at 40 K

The DLTS measurements also provide information on the electron and hole capture cross sections of the deep level centers. For the undoped $Hg_{1-x}Cd_xTe$ ($x = 0.2 \sim 0.4$) samples, the electron and hole capture cross sections are $\sigma_n \approx 10^{-15} \sim 10^{-16}\,cm^2$ and $\sigma_p \approx 10^{-17} \sim 10^{-18}\,cm^2$ for the energy level at $E_V + 0.4E_g$, and $\sigma_n \approx 10^{-16}\,cm^2$ and $\sigma_p \approx 10^{-17} \sim 10^{-20}\,cm^2$ for the deep level at $E_V + 0.75E_g$, respectively. For the Cu-doped $Hg_{1-x}Cd_xTe(x = 0.39)$ sample, the values are $\sigma_n \approx 10^{-16}\,cm^2$ and $\sigma_p \approx 10^{-18} \sim 10^{-19}\,cm^2$, respectively.

2.3.2 Deep Level Admittance Spectroscopy of HgCdTe

This technique is an important supplement to the thermal excitation current spectroscopy, thermal excitation capacitance spectroscopy, frequency swept conductance spectroscopy, and DLTS, and is suitable for narrow band semiconductors.

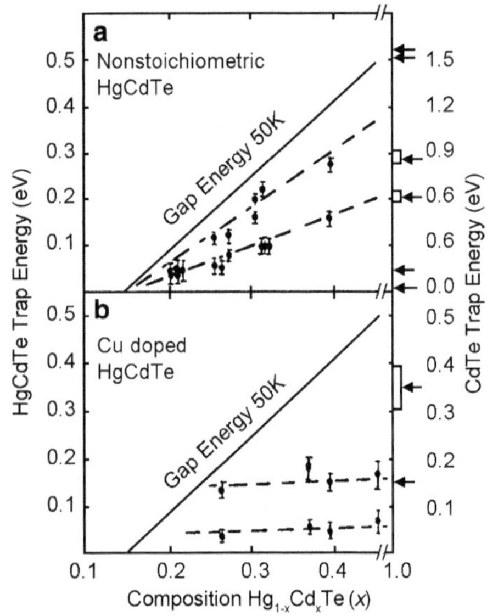

Fig. 2.54 Trap energy as a function of composition determined by DLTS measurements for an undoped (**a**) and Cu-doped (**b**) HgCdTe samples. The right-hand side is the value of CdTe according to Zanio (1978)

It is used mainly to measure the change of admittance with temperature (Losee 1972, 1975) but also details of the interface structure between the insulator and the active material (Tsau et al. 1986; Sher et al. 1983).

For an n^+-p junction containing deep levels, the conductance G_T is:

$$G_T = [e_p\omega^2/(e_p^2 + \omega^2)(N_T/p)]C_0, \tag{2.74}$$

where N_T is the concentration of the deep levels; ω the angular frequency of the driven electric current; p the carrier density, and e_p the hole thermal emission rate, which is given by:

$$e_p = q^{-1}\sigma_p \langle v_p \rangle N_v e^{[(E_v - E_t)/k_B T]}. \tag{2.75}$$

In this expression q ($q = 4$) is the degeneracy of the trap ground state, $\langle v_p \rangle$ is the average hole thermal velocity, σ_p is the hole capture cross section, N_ν is the effective density of states of the valence band, E_ν is the energy of the top of valence band; E_t is the trap energy, C_0 is the high-frequency junction capacitance determined by:

$$C_0 = \varepsilon A/W, \tag{2.76}$$

where ε is the dielectric constant of the semiconductor, A the area of the junction, and W is the width of the depletion layer. When the deep level response can follow the alternating signal, there is an additional capacitance C_T,

$$C_T = [e_p^2/(e_p^2 + \omega^2)](N_T/p)C_0. \tag{2.77}$$

The total capacitance of the junction C is therefore $C_0 + C_T$.

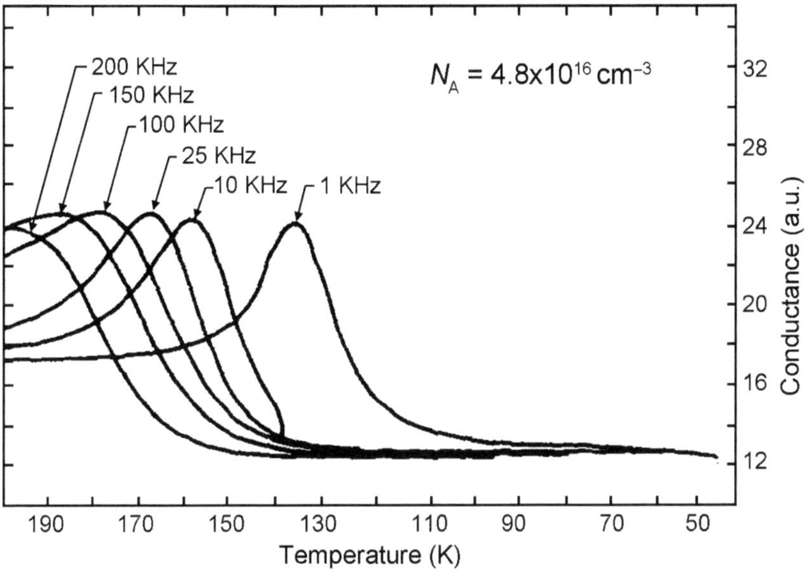

Fig. 2.55 The conductance as a function of temperature for a $Hg_{0.695}Cd_{0.305}Te$ photodiode measured at different frequencies

If both the p and C_0 change little with temperature, the temperature dependence of the admittance is determined mainly by $e_p(T)$. G_T will hence reach its maximum when $e_p = \omega$.

$$G_T|_{max} = \frac{1}{2}\omega(N_T/p)C_0. \tag{2.78}$$

If C_0 and p are obtained from C–V and Hall measurements, respectively, then N_T can be deduced from (2.78). Changing ω will produce a set of maximum values occurring at different temperatures.

Figure 2.55 illustrates the relation between the conductance and the temperature at different frequencies (Polla and Jones 1980a). The acceptor concentration N_A is $4.8 \times 10^{16}\,\text{cm}^{-3}$. Each curve has a maximum, which corresponds to a temperature determined by $\omega = e_p(T)$. The thermal emission rate $e_p(T)$ is given by (2.75), where $\langle v_p \rangle$ is proportional to $T^{1/2}$, and N_v is proportional to $T^{3/2}$. Thus from the relation:

$$\omega = e_p = q^{-1}\sigma_p \langle v_p \rangle N_v e^{[-(E_v-E_t)/k_B T]} \propto T^2 e^{[(E_v-E_t)/k_B T]}, \tag{2.79}$$

the slope of the $\ln(\omega^{-1}T^2) \sim 1/kT$ curve is deduced to be $\Delta E = E_t - E_v$, from which the location of the deep level can be determined. In addition, it is obvious

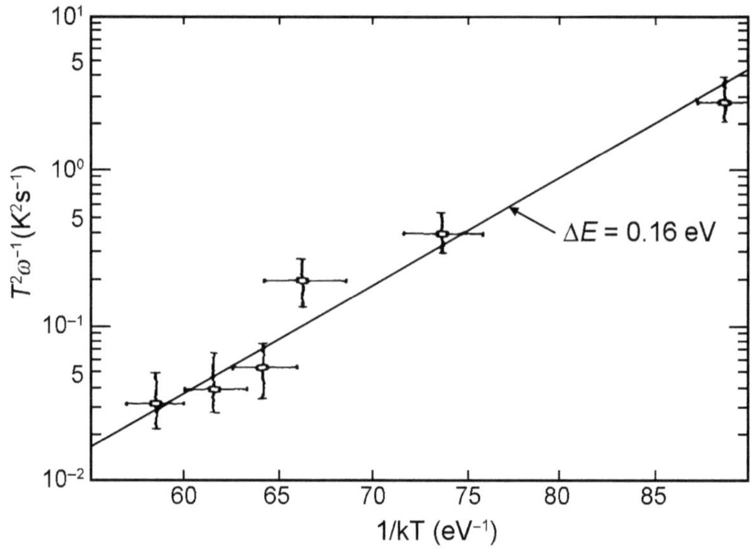

Fig. 2.56 The trap activation energy determined from the conductance peaks in Fig. 2.55. The trap activation energy corresponds to a hole trap located at 0.16 eV above the top of the valence band

from (2.74) to (2.78) that $C = C_0$ if $e_p(T_1) << \omega$, and $C = C_0 + (N_T/p)C_0$ if $e_p(T_2) >> \omega$. Therefore, we find:

$$\Delta C = (N_T/p)C_0, \tag{2.80}$$

where ΔC is the change of the junction capacitance between the temperatures T_1 and T_2, and C_0 is the high-frequency junction capacitance. The deep level concentration N_T can therefore be obtained from (2.80).

From the slope of the straight line $\ln(\omega^{-1}T^2) \sim 1/k_BT$ given in Fig. 2.56, the location of deep level is determined to be $E_t - E_\nu = 0.16$ eV. The capacitance difference ΔC between the two capacitance plateaus at temperatures T_1 and T_2 can be obtained from the capacitance-temperature curves measured at different frequencies, leading to a determination of $N_T = 4.4 \times 10^{14}\,cm^{-3}$. From (2.75) the hole capture cross section in the temperature range between $130 < T < 200$ K is $3 \times 10^{-16}\,cm^2$. For a sample with $x = 0.219$ and $N_A = 9.1 \times 10^{15}\,cm^{-3}$, it has been shown that $N_T = 2.3 \times 10^{15}\,cm^{-3}$ and $E_t = E_\nu + 0.046$ eV, and the hole capture cross section is $7 \times 10^{16}\,cm^2$ in the temperature range between $74 < T < 98$ K. By using this technique, the hole density, its energy level, and the capture cross section of any p-type HgCdTe alloy can also be determined.

Photon-modulation spectroscopy provides another pathway to investigate the deep levels in HgCdTe. The method adopts a dc probe light beam of tunable wavelength ($E_p < E_g$), and a modulated light beam at frequency $\omega(\hbar\omega > E_g)$ but of fixed wavelength, and employs a lock-in technique to probe the change of the emission rate ($\Delta I_p/I_p$). The modulation spectrum, $\Delta I_p/I_p \sim E_p$ is then obtained by scanning E_p. Figure 2.57 shows a schematic of the experimental setup.

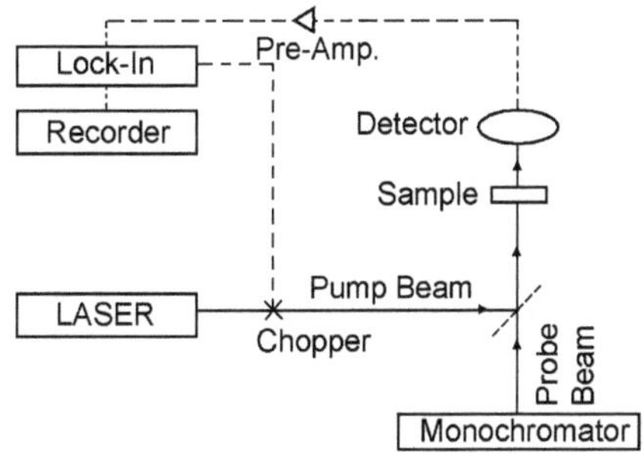

Fig. 2.57 A schematic of the photon-modulation spectroscopy set up

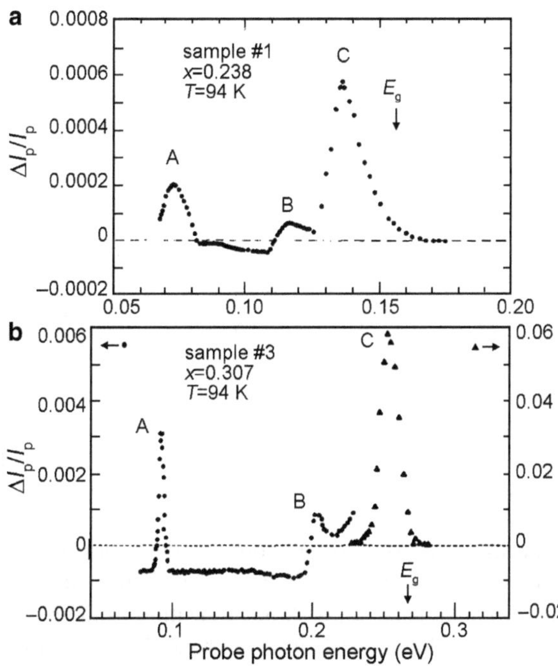

Fig. 2.58 $\Delta I_p/I_p$ as a function of the probe photon energy $\hbar\omega_p$ at $T = 49$ K for $Hg_{1-x}Cd_xTe$ ($x = 0.238$ and 0.307) samples

Polla measured the deep levels in $Hg_{1-x}Cd_xTe(x = 0.24$–$0.37)$ samples prepared by LPE using the photon-modulation spectroscopy method (Polla et al. 1982). The thicknesses of the samples were in the range between 15 and 40 μm. They were placed in a variable-temperature cryostat. The pump light was from an Ar^+ laser set at a wavelength of 514.5 nm and had a light intensity of $8.7\,W/cm^2$. The modulation frequency was 500 Hz. A Ge:Cu detector was used to sense the signal.

The modulation transmission spectra measured at a temperature of 94 K are depicted in Fig. 2.58 for $x = 0.238$ and 0.307 samples, respectively. Three peaks, labeled A, B, and C, are present for all compositions, and their energies are plotted

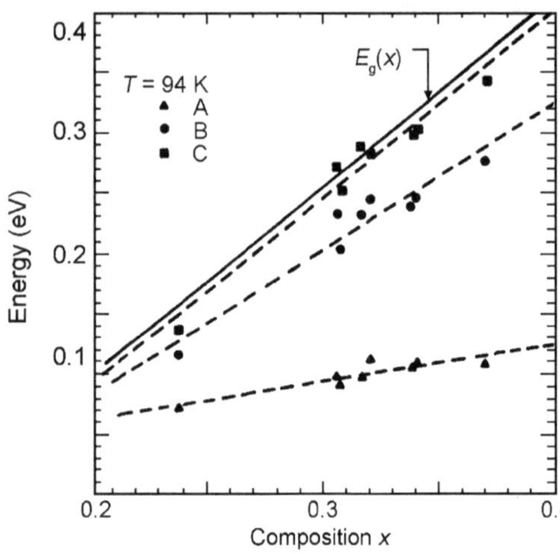

Fig. 2.59 Energy positions of peaks A, B, and C as functions of the composition for $Hg_{1-x}Cd_xTe$ samples. The *dotted lines* are the fits to the experimental data, and the *solid line* is calculated results by Schmit and Stelzer (1969)

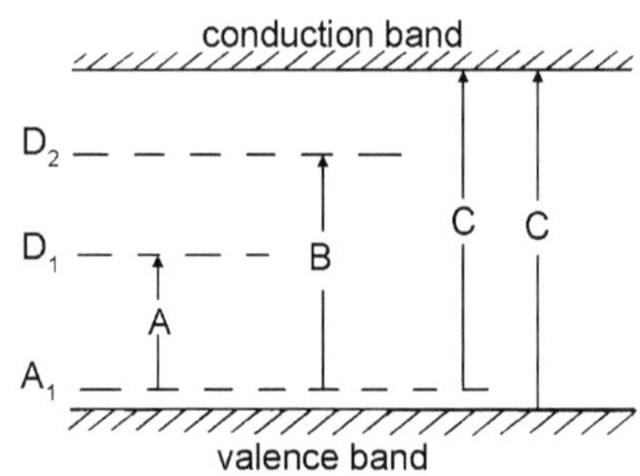

Fig. 2.60 A schematic diagram of the optical transitions corresponding to the peaks A, B, and C

in Fig. 2.59. The peaks correspond to three optical transitions, whose initial states are shallow acceptor states (Scott et al. 1976), and their final states are donor-like deep levels in the forbidden band. Peak A is a transition from an acceptor level to a deep level (D1) in the middle of the forbidden band. Peak B is from an acceptor level to the deep level at $3/4E_g$. And Peak C is from an acceptor level or the valence band edge to the conduction band. The alloy composition variations of these peaks are given in Fig. 2.59, and a schematic of their positions are in Fig. 2.60.

From these results, the deep levels corresponding to the transitions of A and B are located at $E_V + 0.4E_g$ and $E_V + 0.75E_g$, respectively, which are consistent with the Jones et al. results. It can therefore be concluded that these two deep levels are universal in HgCdTe. They may be the cation vacancy related defect states; the Kobayashi theoretical estimation which was mentioned in Sect. 2.1.

The DLTS method has also been used to measure the deep levels of Au-doped samples that had no As dopants added. Cotton and Wilson (1986) have measured the trap energy levels of these n-type $Hg_{1-x}Cd_xTe(x = 0.3)$ ion implanted samples with the DLTS method and found a defect-related trap energy level at $E_c - 0.19$ eV with a very low density, 5×10^{-21} cm^2. Chu using the DLTS method, has reported an electron trap at E_c−85 meV for a $Hg_{1-x}Cd_xTe(x = 0.313)$ sample prepared by LPE.

Huang et al. have systematically studied various deep levels of p- and n-type $Hg_{1-x}Cd_xTe$ samples with large Cd concentrations (Huang et al. 1988), and found that:

1. Two deep level traps are observed due to residual impurities in an n-type $Hg_{1-x}Cd_xTe(x = 0.403)$ sample. They are acceptor-like electron traps locate at $0.3E_g$ and $0.5E_g$, respectively, below the bottom of the conduction band. Their electron capture cross sections are $1.1 \times 10^{-17}\,cm^2$ at 130 K and $1.0 \times 10^{-16}\,cm^2$ at 172 K, and the concentrations are $0.61N_D$ and $0.072N_D$, respectively. The n-type $Hg_{1-x}Cd_xTe$ sample with $x = 0.804$ manifests similar results.
2. Two deep level traps are observed in an undoped p-type $Hg_{1-x}Cd_xTe$ $(x = 0.403)$ sample. They are donor-like hole traps locate at 55 meV and $0.4E_g$ above the top of the valence band, respectively. Their hole capture cross sections are $1.0 \times 10^{-18}\,cm^2$ at 90 K and $7.3 \times 10^{-18}\,cm^2$ at 250 K, and their concentrations are $0.29N_{AD}$ and $0.130N_{AD}$ respectively. No deep level traps are observed at 55 meV above the top of the valence band in this p-type $Hg_{1-x}Cd_xTe(x = 0.361)$ sample; only a deep level trap at $0.4E_g$ above the top of the valence band is observed.
3. The $Hg_{1-x}Cd_xTe$ samples grown by different methods have different types of deep levels, and hence different trap densities and locations, which significantly restrict the performance of the materials and detector devices.

2.3.3 Frequency Swept Conductance Spectroscopy

LFCS is another powerful way to examine the interface and trap properties of HgCdTe alloy to insulator contacts (Tsau et al. 1986). Unlike DLTS which is a method that initially takes the sample well away from equilibrium and studies the transient as it returns, the LFCS method applies a varying gate voltage and at each voltage setting allows the sample to equilibrate. Then the frequency could be swept from 10^{-4} Hz to 10 MHz (Sher et al. 1983). At each frequency a 4 cycle sequence is done and the signal detected with an automatically tuned lock-in amplifier. Each 4 cycle sequence is repeated at least 100 times to improve the signal to noise ratio. This process is accomplished with two devices, a Solartron model 1250 that operates from 10^{-4} Hz to 10 KHz and an H-P model 4,061 A semiconductor testing system, a frequency response amplifier that operates from 10 KHz up to 10 MHz. The lowest frequency used in this experiment was 10^{-2} Hz so there were eight decades of frequency employed. When this method was applied to an Si/SiO_2 interface (Sher et al. 1983) the usual U-shaped density of states observed in the band gap was decomposed into conduction and valence band tails that were observed to hybridize where they crossed near the center of the gap. In addition deep state traps were identified and characterized.

This method was applied to a $Hg_{0.7}Cd_{0.3}Te$ sample with a 1,660 A thick Photox SiO_2 layer deposited on its surface to form an MIS device structure (Tsau et al. 1986). The metal gate was made up of a thin 400 A thick layer of Ti, followed

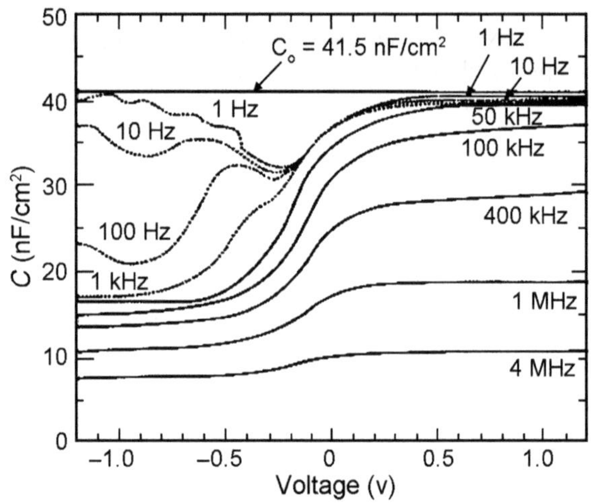

Fig. 2.61 The effective series capacitance vs. gate voltage for several frequencies

by 4,000 A of Au. The sample was n-type with a carrier concentration of $\sim 6 \times 10^{14}\,\mathrm{cm}^{-3}$, and was held at 77 K during the measurements. Figure 2.61 is a plot of the series capacitance C (nF/cm^2) vs. gate voltage V (volts) at different frequencies ranging from 1 Hz to 4 MHz. Notice that the dielectric constant of the Photox SiO_2 layer changes from 7.8 at 1 Hz to only 2.8 at 4 MHz. This brackets the 3.9 value reported for a good SiO_2 layer, being higher at low frequencies where there are ionic contributions, and lower at high-frequency because of voids in the layer.

Following the analysis method in Sher et al. (1983) an $\ln G_P/\omega$ vs. $\ln \omega$ plot is presented in Fig. 2.62. Once the insulator capacitance is subtracted out the equivalent circuit of the device response is well modeled as a parallel combination of series capacitance/resistance combinations. The impedance is then given by:

$$C_\mathrm{p} = \sum_i \frac{C_i}{1 + \omega^2 \tau_i^2}, \tag{2.81}$$

and

$$\frac{G_p}{\omega} = \sum_i \frac{C_i \omega \tau_i}{1 + \omega^2 \tau_i^2}. \tag{2.82}$$

The voltages at which the peaks occur identify the location of states in the band gap and the width the lifetime of those states. The lifetime is related to the effective series resistance R_i of the ith branch through $\tau_i = R_i C_i$. Because the model satisfies the Kramers–Kronig relations all the information resides in either the capacitance C_p or the G_p/ω relations. Actually both are used since there are times when it is easier to extract information from one rather than the other. Because of equipment difficulties the data taken between $\sim$1 and 10 KHz is less reliable than that in other frequency ranges, so that the results are shown as dotted curves.

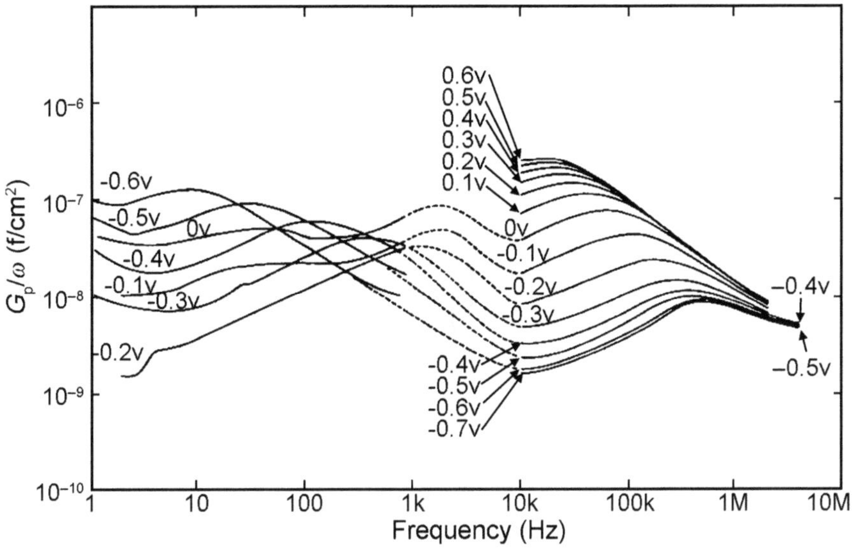

Fig. 2.62 The ln G_p/ω curves vs. ω for different gate voltages

The variation of the surface potential with gate voltage is extracted following the Berglund method (Berglund 1966; Sher et al. 1983). From the measured carrier concentration at 77 K, $n = 6.2 \times 10^{14}/\text{cm}^3$, the intrinsic carrier concentration of $n_i = 1.15 \times 10^9/\text{cm}^3$, a conduction band effective mass of $m_c^* = 0.0182$, and a valence band effective mass of $m_v^* = 0.55$, one gets a Fermi energy $E_F = 0.226\,\text{eV}$. Adding the surface potential $e\psi_s$ yields a surface Fermi energy $E_{Fs} = 0.181\,\text{eV}$. The flat band voltage for this sample is $-0.10\,\text{V}$. Subtracting out the work function difference between Ti and $x = 0.3$ HgCdTe leaves a surface charge density-induced potential across the insulator capacitance of $Q_o = 3 \times 10^9\,\text{C/cm}^2$ for a surface density of 2×10^{10} charges/cm^2. Because the average atom density of this surface is 3.2×10^{14} atoms/cm^2, this corresponds to one charge every $\sim 10^4$ atoms.

Figure 2.63 is a plot of the surface density N_{ss} vs. energy E relative to the valence band edge (dot-dashed curve). The total surface density is obtained from the relation $N_{ss} = C_p/e$ where C_p is given by (2.81). The other curves are:

1. A curve extracted from the data that has time constants τ_1 in the range between $\sim$1 Hz and 1 KHz. All these data (dashed curve) all form a sequence of points with continuously varying time constants as shown in Fig. 2.64. They are associated with the valence band tail.
2. A curve extracted from data that exhibits faster responses, between 10 and 100 KHz. These points (connected by a solid line) are associated with the conduction band tail. Their time constant variation is also in Fig. 2.64.
3. The remaining points connected by dotted curves, are the differences between the total curve and the solid and dot-dashed curves. They are localized trap states responses.

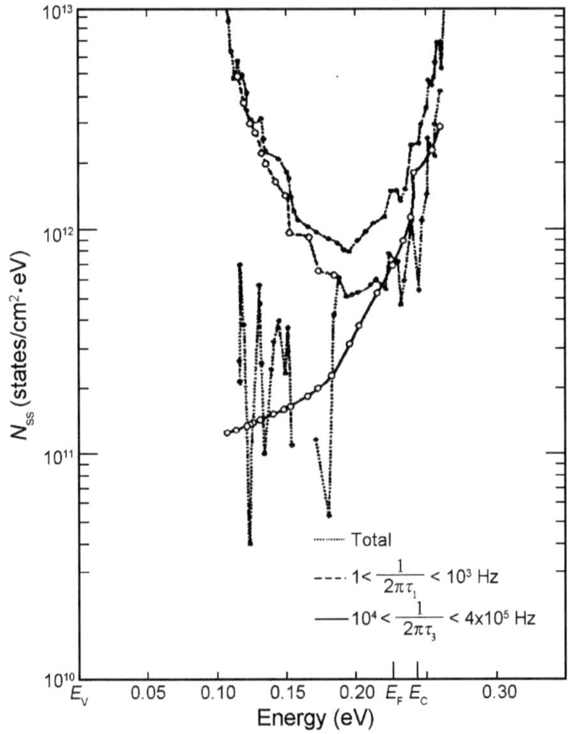

Fig. 2.63 Interface densities at 77 K as a function of energy relative to the top of the valence band edge. The curves are (**a**) the total interface state density curve obtained from C_p at low frequency (*dot-dashed curve*), (**b**) valence band tail curve obtained from the G_p/ω responses corresponding to the time constant set τ_1 (*dashed curve*), (**c**) the conduction band tail curve corresponding to the time constant set τ_3 (*solid curve*), and (**d**) the difference between the band tail curves and the total capacitance curve corresponding to the localized trap states

Notice that the valence band tail extends further into the gap than the conduction band tail so the usual U-shaped curve in this case has its minimum on the conduction band side of the gap. This is to be expected because the valence band effective mass is large compared to that of the conduction band. The integrated area under the valence band tail from 105 to 185 eV is $\sim 3 \times 10^{11}$ states/cm^2, which corresponds to only one state every 10^3 surface atom.

The band gap of an $x = 0.3$ alloy at 77 K is $E_C = 0.24$ eV. The apparent conduction band tail actually extends ~ 0.15 eV above the bulk material's conduction band edge. This is possible because of the low conduction band effective mass. While the density of states in the conduction band tail rises rapidly below E_C, it rises even more rapidly above E_C.

The set of response times for the valence band states fall on a curve characteristic of a Shockley–Reed center with a cross section of $\sim 1.2 \times 10^{-15}$ cm^2. However the conduction band tail responses do not follow such a curve and must be due to some other mechanism.

Fig. 2.64 The set, $\{\tau_1\}$, of time constant points for the valence band tail state responses, and the set, $\{\tau_3\}$, for the conduction band tail responses

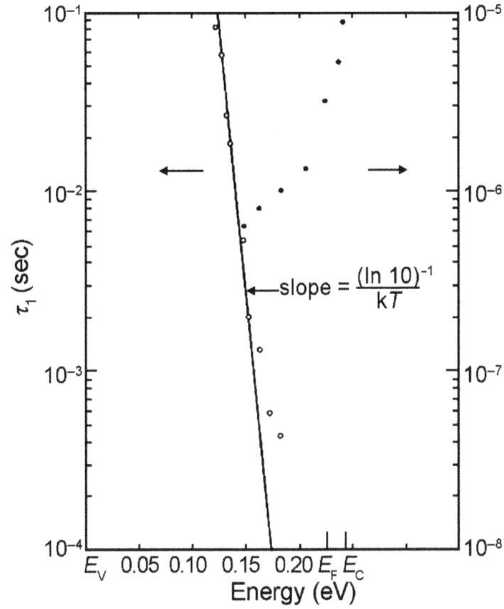

There are nine discrete states visible in Fig. 2.63. The origin of the four state clusters between 0.12 and 0.15 eV is likely due to dangling Te bonds with the other three bonds attached to either three Hg atoms, two Hg and one Cd, one Hg and two Cd, or three Cd. The lowest peak for example has a height of $\sim 7 \times 10^{11}$ states/eV $\cdot$ cm^2 and a half width at half maximum of $\sim$5 meV. Thus it has contributing to it $\sim 4 \times 10^9$ states/cm^2, or about one state per 10^5 surface atoms. The peaks are nearly equally spaced with an average spacing of $\sim$15 meV. A theory (Casselman et al. 1983) suggests that the spacing between these micro-clusters should be $\sim$25 meV in the bulk alloy. Small modifications are to be expected due to the fact that these clusters are at an interface.

Since this is a one sample study there is no explanation for the other five discrete states. It would require a multisample investigation in which systematic changes were made to modify the defect structure. The main lesson from this study is that it is an extremely powerful method that deserves to be used more.

2.4 Resonant Defect States

Though the usual DLTS can be employed in the study of semiconductor deep level impurities and defects (Polla and Jones 1980b; Jones et al. 1981), it is, however, powerful only to research the energy levels located in the band gap, but not to that of resonant states. In narrow-band semiconductor materials, many levels introduced by the primitive cell potentials of short-range defect centers are located in the valence or conduction band. They are called resonant states because of their interaction with the continuous energy band states. In the past several years, the resonant states in HgCdTe have attracted a lot of attention. For example, a theory of chemical trends was presented for sp^3-bonded substitutional deep-trap energy levels in $Hg_{1-x}Cd_xTe$ materials ($x = 0$–1) within the framework of the tight binding method (Kobayashi et al. 1982) or more accurately by an LMTO method (Berding et al. 1993). Consideration of the energies introduced by the different charge states of cation substitutional impurity defect states were conducted in $Hg_{1-x}Cd_xTe$ and detailed results of these energies were given for neutral and single ionization states (Myles 1987).

Experimentally, the resonant states in $Hg_{1-x}Cd_xTe$ can be investigated by transport and optical methods. For example, Ghenim et al. have carried out pressure-dependent transport measurements and found a resonant state energy level at 150 meV above the bottom of the conduction band with a concentration in the range of 1.4–$7.85 \times 10^{17}\,\mathrm{cm}^{-3}$ (Ghenim et al. 1985). Dornhaus et al. have suggested there is a resonant state energy level at 10 meV above the bottom of the conduction band by far infrared spectroscopy and transport techniques (Dornhaus et al. 1975). For some narrow-band semiconductor MIS samples, resonant-state energy levels in the conduction band have also been identified with C–V measurements. For example, a resonant state was found at 45 meV above the bottom of the conduction band in heavily p-type-doped $Hg_{1-x}Cd_xTe$ ($x = 0.21$) (Chu 1988; Chu et al. 1992).

2.4.1 Capacitance Spectroscopy of Resonant Defect States

The C–V spectroscopy of semiconductor MIS structures can detect the band bending at a semiconductor surface (Sze 1969). Recently, Mosser et al. have found electron quantization features in a surface inversion layer in a C–V study of narrow-band semiconductors (Mosser et al. 1988). Due to the potential well caused by surface band bending, the inversion layer electron energy is quantized, with a ground state energy level at E_0, comparable to the band gap E_g. Therefore, no quantized energy level will be filled in the inversion layer until the bending energy reaches E_g, so that the conduction band edge becomes lower than the Fermi energy. However, if the bending energy reaches $E_g + E_0$ so the ground state level is lower than the Fermi energy, the inversion layer states begin to fill. Therefore, the inversion threshold voltage in a C–V curve is delayed, providing the possibility of yielding evidence of the existence of resonant states between the bottom of the conduction band and the ground state level.

The Hg vacancy is usually regarded as a shallow acceptor in p-type HgCdTe (Zanio 1978). Its density is especially high in strongly p-type HgCdTe and the opportunity for the vacancies being occupied by impurities is also larger. Theoretical analysis predicts that many impurities located on cation sites form resonant defect states in the conduction band $Hg_{1-x}Cd_xTe$ samples with x in the range of 0.165–0.22 (Kobayashi et al. 1982; Myles 1987). That compensates the material and reduces N_a-N_d. Figure 2.65 illustrates the band bending of a sample with $x = 0.21$ and $N_a = 1.87 \times 10^{17}\,\mathrm{cm}^{-3}$, in which $E_g = 86\,\mathrm{meV}$ is the band gap, $E_0 = 143\,\mathrm{meV}$ is the zero-level energy of the electron subband, E_R is a resonant defect state energy level above the bottom of the conduction band, $Z_0 = 15\,\mathrm{nm}$ is the thickness of the inversion layer and $Z_d = 48\,\mathrm{nm}$ is the thickness of the depletion

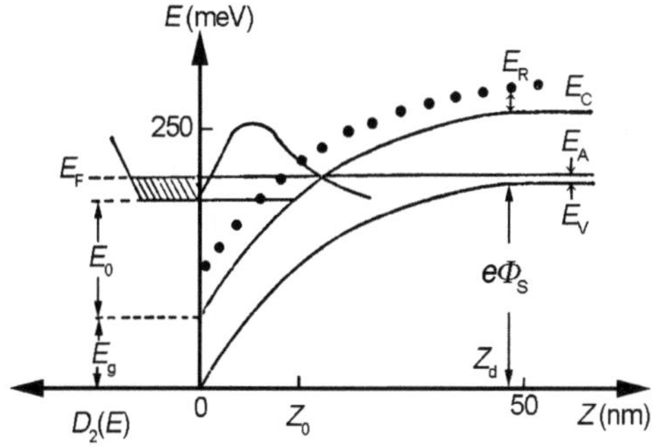

Fig. 2.65 Band bending of a $Hg_{0.79}Cd_{0.21}Te(N_a - N_d = 1.87 \times 10^{17}\,cm^{-3})$ sample with a surface electron density of $N_S = 2 \times 10^{11}\,cm^{-2}$ at 4.2 K

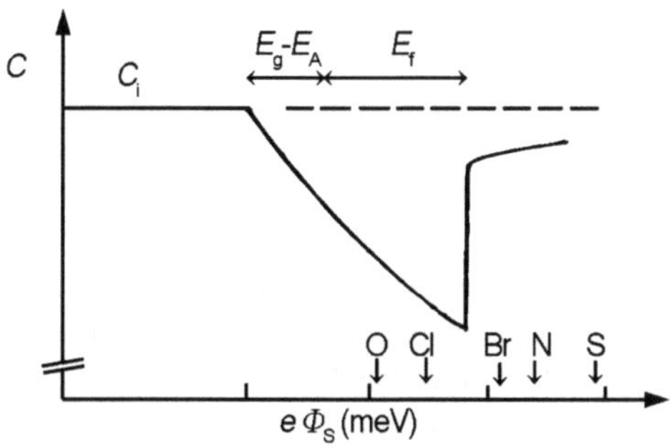

Fig. 2.66 Capacitance as a function of surface potential for HgCdTe MIS samples (*Vertical arrows* mark proposed energies of several cation substituted impurities)

layer. Figure 2.66 depicts energy positions of several substitutional impurities occupying Hg vacancies according to a rough theoretical estimate of Kobayashi et al. (1982) and another by Myles (1987) in an ideal capacitance spectrum. Obviously, if the band bending makes the resonant energy level lie below the Fermi level, electrons will fill this energy level. If the density of defect states is high enough, this energy level will affect the charging and discharging processes of the surface inversion layer and contribute to the surface capacitance C–V spectrum.

In the C–V spectrum of the HgCdTe bulk material fabricated into an MIS structure, an additional peak is observed in the region before the subband threshold voltage. Calculations indicate the existence of a resonant state at 45 meV above the conduction band edge. In the following, a preliminary discussion of the origin of this resonant state is presented.

The samples used were slices of a solid state recrystallized, p-type $Hg_{1-x}Cd_xTe$ with x values of ~0.22. They had been mechanically polished and etched in a 1/8% bromine in methanol solution. The MIS structures were produced by first growing ~70 nm of anodic oxide in a KOH electrolyte (0.1 N KOH, 90% ethylene glycol, 10% H_2O) with a constant current density of 0.3 mA/cm^2. An ~200 nm-thick ZnS layer was then deposited onto the anodic oxide, and finally an Al gate electrode was deposited. This kind of sample structure had a low interface trap state density ($<10^9\,cm^{-2}\,eV^{-1}$), and a small fixed positive charge ($<10^{12}\,cm^{-2}$) (Stahle et al. 1987).

Many C–V measurements have been reported on MIS structures of narrow-band semiconductors (Stahle et al. 1987; Beck et al. 1982; Rosback and Harper 1987).

Using a differential capacitance method, the accuracy can be as high as 0.01 pF/V. The experimental frequency is 233 Hz and the signal U_{AB} is proportional to the difference between the sample and a reference capacitor, C-C_{ref}, as illustrated in Fig. 2.67. At a temperature of 4.2 K, the capacitance of the sample (C) is the series combination of the sample insulation-layer capacitance (C_i), and the semiconductor surface-layer capacitance (C_s) (Sze 1969), which can be expressed as:

$$C = \frac{C_i C_S}{C_i + C_S}. \tag{2.83}$$

Investigated $Hg_{0.79}Ce_{0.21}Te$ bulk samples have an effective acceptor concentration of 3–4 $\times$ 10^{16} cm^{-3}. The capacitance spectra at 4.2 K show an additional peak before inversion occurs for all the samples with $N_{A^*} \sim 10^{17}$ cm^{-3}. Figure 2.68 depicts a typical C–V spectrum for a sample taking the anodic oxide and ZnS as dielectric layers, in which V_{fb} and V_T are the flat band voltage and subband threshold voltage, respectively. The regions $V_g < V_{fb}$, $V_{fb} < V_g < V_T$ and $V_T < V_g$ correspond to accumulation, depletion and inversion regions, respectively. The lower curve in Fig. 2.68 is a surface conductance-voltage spectrum, which exhibits a drastic increase of the surface conductance and surface capacitance at V_T, corresponding to the subband threshold voltage. The flat band voltage V_{fb} is determined by the boundary between the flat part of accumulation region and the steep part of depletion region in the

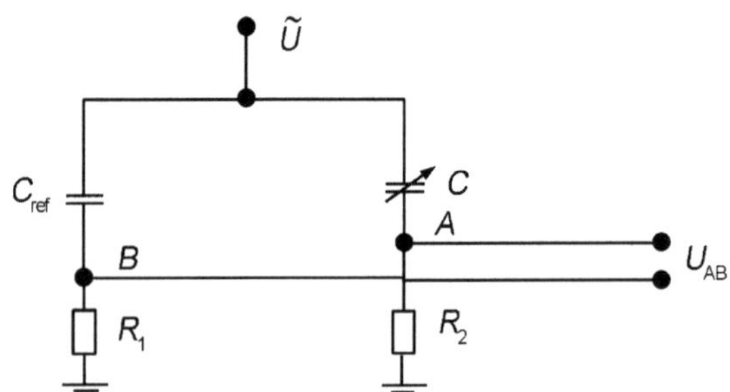

Fig. 2.67 A schematic of a differential capacitance electrical bridge

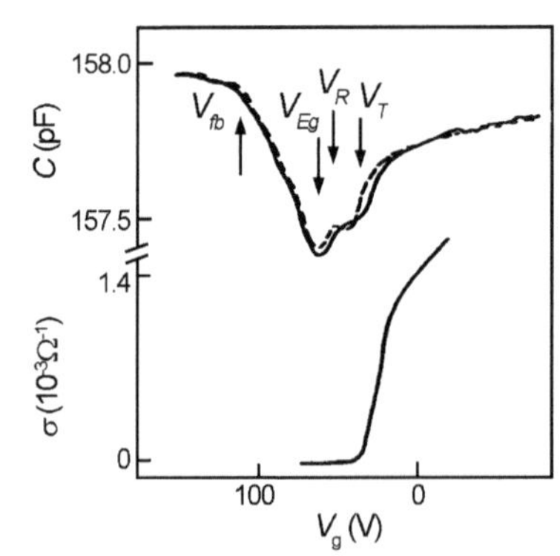

Fig. 2.68 A surface capacitance spectrum and a surface conductance spectrum vs. gate voltage for a $Hg_{0.79}Cd_{0.21}Te$ sample ($N_{A^*} \approx 1.87 \times 10^{17}$ cm^{-3}) at 4.2 K

C–V curve. To determine accurately V_{fb}, C–V, magneto-resistance oscillation, and cyclotron resonance measurements are required. The V_{Eg} in Fig. 2.68 is the energy due to the band bending at which the bottom of the conduction band coincides with the Fermi level. Because there is an additional peak located between the V_{Eg} and V_{T}, suggests that a resonant state lies above the conduction band edge. As the energy band bends, electrons will begin to fill in this energy level once the resonant state's threshold energy level falls below the Fermi level. Given this condition, an increase in the surface capacitance occurs, adding to the contribution from the depletion layer. The effect is equivalent to the addition of an extra surface depletion layer.

2.4.2 Theoretical Model

By fitting the capacitance spectrum, the energy level position and the density of resonant states can be deduced. The surface capacitance per unit area of semiconductors is defined as (Sze 1969):

$$C_S = \frac{\partial Q_S}{\partial \phi_S}, \tag{2.84}$$

where ϕ_s is the surface potential. Q_s is the surface charge density, which can be written as:

$$Q_S = -e(N_A^* z_d + N_R z_R + N_S), \tag{2.85}$$

where z_d is the depth of the depletion layer, z_R the distance from the resonant state energy level to the Fermi level, that is the resonant level lies energetically above the Fermi level if $z_R > 0$. $N_A^* = N_A - N_d - N_R$ is the effective acceptor concentration, N_R is the density of resonant states, and N_S (cm^{-2}) is the electron density in the inversion layer. The surface potential of semiconductor material can be derived from the Poisson equation (Sze 1969):

$$\frac{\partial^2 \phi}{\partial z^2} = -\frac{\rho}{\varepsilon \varepsilon_0}, \tag{2.86}$$

where ρ is the density of surface charges. Before the subband is filled, the distribution is given by:

$$\rho = \begin{cases} -e(N_A^* + N_R), & (0 < z < z_R), \\ -eN_A^*, & (z_R < z < z_d), \\ 0, & (z > z_d). \end{cases} \tag{2.87}$$

Assuming that the Fermi level at 4.2 K is pinned to the bulk material acceptor level, the total band bending is then:

$$e\phi_s = E_f + (E_g - E_A), \tag{2.88}$$

where E_A is the acceptor level, and E_g is the band-gap energy, whose values can be found from the literature (Scott et al. 1976; Chu et al. 1983; Yuan et al. 1990).

E_f is the Fermi energy relative to the conduction band edge. The effective acceptor concentration is determined from the slope |K| of that part of the C–V curve corresponding to the depletion layer (Rosback and Harper 1987), and is:

$$N_\mathrm{A}^* = \frac{C_\mathrm{i}^3}{\varepsilon\varepsilon_0 e A^2 \left|K\right|}, \tag{2.89}$$

where A is the area of capacitor. From this relation, the surface capacitance of the depletion layer in the region $V_\mathrm{fb} \leqslant V_\mathrm{g} \leqslant V_\mathrm{R}$ is given by:

$$C_\mathrm{S} = e\sqrt{\frac{\varepsilon\varepsilon_0 N_\mathrm{A}^*}{2}\frac{1}{E_\mathrm{g} - E_\mathrm{a} + E_\mathrm{f}}}. \tag{2.90}$$

And in the region $V_\mathrm{R} \leqslant V_\mathrm{g} \leqslant V_\mathrm{T}$, the resonant state contributes, and the surface capacitance is:

$$C_\mathrm{S}' = \frac{e\sqrt{\frac{\varepsilon\varepsilon_0}{2}(N_\mathrm{A}^* + N_\mathrm{R})}}{\sqrt{(N_\mathrm{A}^* + N_\mathrm{R})(E_\mathrm{g} - E_\mathrm{a} + E_\mathrm{f}) - N_\mathrm{R}(E_\mathrm{g} - E_\mathrm{a} + E_\mathrm{R})}}, \tag{2.91}$$

where E_R is the resonant state energy counted from the conduction band edge, and can be deduced from the threshold voltage V_R of the resonant peak:

$$E_\mathrm{R} = \frac{C_\mathrm{i}^2(V_\mathrm{R} - V_{fb})^2}{2\varepsilon\varepsilon_0 \times N_\mathrm{A}^*} - (E_\mathrm{g} - E_\mathrm{a}). \tag{2.92}$$

It is obvious from (2.90) to (2.91) that for $V_\mathrm{G} = V_\mathrm{R}$ the band bending makes the resonant state energy E_R meet E_F, leading to a jump to the surface capacitance, $\Delta C_\mathrm{S} = C_\mathrm{S}' - C_\mathrm{S}$, which is given by:

$$\Delta C_\mathrm{S} = e\sqrt{\frac{\varepsilon\varepsilon_0}{2}\frac{1}{E_\mathrm{g} - E_\mathrm{a} + E_\mathrm{R}}} \times \frac{N_\mathrm{R}}{\sqrt{N_\mathrm{A}^*}}. \tag{2.93}$$

The measured sample capacitance manifests a corresponding change,

$$\Delta C = \frac{C_\mathrm{i}^2 \Delta C_\mathrm{S}}{(C_\mathrm{i} + C_\mathrm{S}')\left(C_\mathrm{i}' + C_\mathrm{S}\right)}, \tag{2.94}$$

thereby adding a peak to the capacitance spectrum. The density of resonant defect states N_R can be derived from ΔC together with (2.93) and (2.94),

$$N_\mathrm{R} = \Delta C_\mathrm{s} \times [N_\mathrm{AD} \times (E_\mathrm{g} - E_\mathrm{A} + E_\mathrm{R})]^{1/2}/[e(\varepsilon_s\varepsilon_0/2)^{1/2}]. \tag{2.95}$$

Table 2.8 Energy locations and densities of resonant defect states

Sample	Insulating layer	x	$N_A{}^*(10^{17}\,cm^{-3})$	$N_R(10^{17}\,cm^{-3})$	E_R (eV)
3a	SiO_2	0.21	1.65	1.1	0.041
4a	A.O. + ZnS	0.21	1.87	0.9	0.045
5b	A.O. + ZnS	0.21	3.17	1.8	0.047
5b3	A.O. + ZnS	0.21	3.96	2.5	0.045

The density as well as the energy of resonant defect states can also be calculated by fitting the C–V curve. In Fig. 2.68, the dotted curve is a theoretical fit. From the best-fit one finds that $E_R = 0.045\,eV$, and $N_R = 9 \times 10^{16}\,cm^{-3}$. Fitted results are listed in Table 2.8 for several $Hg_{1-x}Cd_xTe$ samples with $x = 0.21$. The experimental results suggest that for the samples with $N_A{}^* \sim 10^{17}\,cm^{-3}$ an additional peak is always observable, while for the samples with lower $N_A{}^*$, e.g., about $10^{16}\,cm^{-3}$, no additional peak can be detected.

2.4.3 Resonant States of Cation Substitutional Impurities

The origin of this type of resonant state possibly, is the defect state introduced by oxygen occupying a cation site. Theoretical analyses indicates that the energy levels of the neutral and single ionization states of an oxygen atom are located at 50 and 20 meV above the conduction band edge, respectively, for a $Hg_{0.79}Cd_{0.21}Te$ alloy (Kobayashi et al. 1982). This result is similar to the $C - V$ experimental value. The possibility of oxygen atoms occupying cation sites can be estimated qualitatively. If the sample is exposed to air after annealing or etching, or in the process of anodic oxidation, oxygen atoms may occupy cation sites. This can lead to pollution of the bulk material. In the standard Br–alcohol chemical etching process, the surface damage is normally as deep as several tens of nanometers, and an activated Te-rich layer at the surface is easily oxidized. Then oxygen atoms easily diffuse into the bulk via a Hg vacancy assisted mechanism (Herman and Pessa 1985). In the anodic oxidation process, the surface oxygen concentration is about 50% (Stahle et al. 1987), thus a large number of oxygen atoms are present and combine with Hg, Te, and Cd to produce the oxidized films. In this process, it is inevitable that some of the Hg vacancies will be occupied by oxygen atoms.

It has been determined experimentally that the process of the replacement of Hg vacancies with impurity atoms will not go on without limit, and saturates at a probability, η. Thus impurity atoms can only replace a part of the Hg vacancies. If the total Hg vacancy density is N_A, the number of Hg vacancies being replaced by impurities, N_R, is then ηN_A when the system reaches an equilibrium state. It was deduced from the fitted results of several samples that η for oxygen is about 0.3–0.4. According to this analysis, it is understandable why only for samples with high initial Hg vacancy densities, is the additional peak observed. The absolute value of the density of oxygen atoms replacing Hg vacancies in low initial density samples is low. While even for initially low Hg vacancy densities a jump of the surface capacitance can be introduced, it is clear from (2.93) that:

$$\Delta C \propto \frac{\eta\sqrt{N_A}}{\sqrt{1-\eta}}. \tag{2.96}$$

This means for samples with low N_A, the jump in the capacitance is small and is likely to be masked by interfacial and inhomogeneous effects. An additional reason is that the threshold energy E_0 (near E_R) of the ground subband state is smaller for a sample with identical composition but lower N_A. This indicates that the inversion threshold voltage V_T is close to V_R and the capacitance jump in inversion due to a resonant state makes only a minor contribution that may not be detectable.

To investigate the mechanism further, the change of the defect state densities induced by annealing is studied along with the impact of an Au doping treatment. If the defect states are due to oxygen atoms occupying Hg vacancies, in a low-temperature, Hg saturated annealing process the vacancies will be filled with Hg atoms and the oxygen atoms will be driven out. Hence the density of resonant states will decrease. On the other hand, if the p-type acceptor initially in the sample is not a Hg vacancy but Au, the resonant state will not be detectable even if the initial acceptor concentration is very high. If a sample, in which initially the resonant state peak is obviously observed, is then exposed to a Hg saturated low-temperature anneal to reduce the impurity density and its C–V curve is measure again. According to this model the resonant peak should be reduced. The sample is then doped with

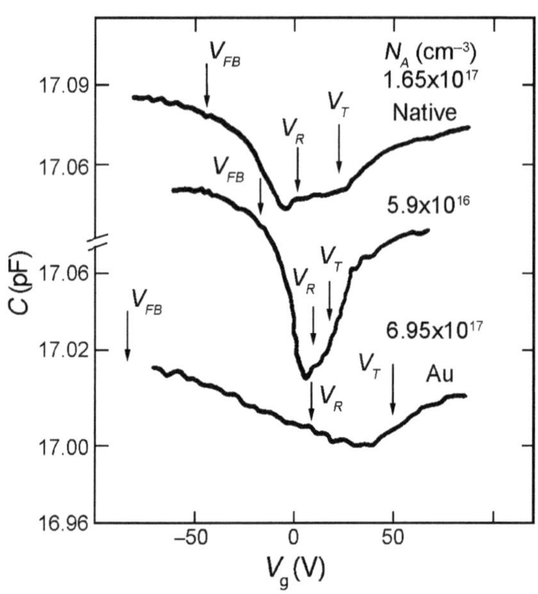

Fig. 2.69 Capacitance spectra of a $Hg_{0.79}Cd_{0.21}Te$ sample ($N_A^* \approx 1.65 \times 10^{17}\,cm^{-3}$) before and after annealing, and after an Au doping process

Au to make it an initially high p-type density sample again and the C–V measurement is repeated. The results are depicted in Fig. 2.69. The top spectrum is that of a $HgCdTe/SiO_2$ dielectric film sample ($N_A{}^* = 1.65 \times 10^{17}\,cm^{-2}$), which clearly shows the resonant state peak. A theoretical fit indicates the density of resonant states is $\sim 1.1 \times 10^{17}\,cm^{-3}$ and E_R is $\sim 0.041\,eV$. The middle spectrum is that of the same sample after a 10 h anneal at a temperature of 250°C. The acceptor concentration is $\sim 5.9 \times 10^{16}\,cm^{-3}$ and the resonant state peak gets weaker. A theoretical fit suggests that the density of the $E_R = 0.041\,eV$ resonant state reduces to $4 \times 10^{16}\,cm^{-3}$. The bottom spectrum represents the result after Au doping and 10 h anneal at a temperature of 250°C. Following this process, the low-density sample becomes a high-density one, $\sim 6.95 \times 10^{17}$. However, no obvious resonant state peak appears. A theoretical fit indicates that the resonant state density is still $4 \times 10^{16}\,cm^{-3}$. This illustrates that the resonant state does not depend on the acceptor concentration but on the Hg vacancy concentration. It therefore suggests that this resonant state originates from oxygen atoms replacing Hg vacancies and that annealing can reduce or eliminate its influence.

2.5 Photoluminescence Spectroscopy of Impurities and Defects

2.5.1 Introduction

The optical techniques that can be used to explore the nature of impurities and defects are generally comprised of PL spectroscopy, infrared transmission/reflection spectroscopy, magneto-optical spectroscopy, and PTIS, among which the PL spectroscopy is deemed to be an important and fundamental one. PL spectra from semiconductors offer information about not only intrinsic optical processes, but also impurities and defects. In fact, it has been advanced to be a standard technique for the study of semiconductor material properties (Göbel 1982). A PL measurement generally requires almost neither special sample preparation nor complicated instrumentation, and most importantly, it is nondestructive. It therefore attracted wide attention in the last few decades, and as a consequence, a huge related literature was published in journals and conferences in the field of semiconductors and low-dimensional systems. For wide bandgap semiconductors, PL spectroscopy is usually a valuable tool for monitoring the quality of material growth, evaluating the properties of heterojunction structures, and estimating indirectly the density of minority carriers and defects so as to predict device performance. However, PL study of narrow-gap semiconductors was seldom performed, mostly due to the following reasons:

1. A PL signal decreases dramatically with the energy gap, due to the enhancement of the nonradiative Auger recombination process.
2. At room-temperature environmental thermal emission in the mid-infrared spectral region is about 200 times stronger than the general PL signal.

3. The solid state detects available for PL spectroscopy studies require a much higher detectivity than photomultiplier tubes (PMT); and in the mid- and far-infrared spectral regions, no PMT is available.
4. Disturbances from the absorption of atmospheric and optical materials is serious in the mid- and far-infrared regions.

Thanks to the revolutionary progress made in computer science and technology, Fourier transform infrared (FTIR) spectrometer-based PL techniques have been developed. This significantly enhances the sensitivity and the spectral resolution and therefore dominates the study of the PL processes in semiconductors in the near- to mid-infrared. To extend the measurable spectral range to mid- and even far-infrared, suppression of the environmental thermal emission became crucial. This gave birth to the so-called DM-PL technique. Depending on the particular choice of the "modulation," the DM-PL technique may yield a measurement in the mid- and far-infrared regions with a reasonable SNR and spectral resolution. This makes DM a distinguishable technique relative to conventional PL measurements.

Since the first PL data for HgCdTe was reported by Elliott et al. (1972) in 1972, quite a lot HgCdTe samples with different Cd concentrations from 0.197 to 1 (CdTe) have been characterized. However, apparent inconsistencies exist among the data as well as its explanations, which most probably, stem from the following facts: (1) the samples used by different authors' groups were prepared by different methods and hence were of different quality and (2) the inaccuracy of energy gap determinations of HgCdTe is of the order of a few milielectron volts, which is the same order as the exciton binding energies, making the uncertainty of the PL signal identification even worse.

It was expected that a systematic PL spectroscopic study could be performed by three different approaches (Tang et al. 1995): (1) starting from the high Cd concentration end (2) in the mid concentration limit near $x = 0.4$, or (3) at the low concentration end near the range where IR applications are prevalent. At higher concentrations standard PL techniques are adequate, but as the concentration is lowered it is necessary to go to DM-PL methods.

Generally, the PL signal from a high quality $Hg_{1-x}Cd_xTe$ sample with Cd concentrations in the range of 1.0–0.7 are dominated by (1) nearly free but still localized exciton transitions, (2) a bound localized exciton, and (3) phonon replicas of localized or bound excitons. Meanwhile, the PL signal from $Hg_{1-x}Cd_xTe$ samples with a Cd concentration less than 0.26 appears usually as only a simple single peak. Due to the small binding energy of a nearly free exciton and the transition oscillator strength of low Cd concentration $Hg_{1-x}Cd_xTe$, it is usually difficult to distinguish whether a peak is from a band-to-band transition or due to a free exciton. Furthermore, the difference between the spectral line shapes of these two transition mechanisms is so small, that a lineshape fitting process for the PL spectrum may not work well (Tomm et al. 1990). Considering the exciton binding energy is very small in low Cd concentration HgCdTe samples, as well as the weak Hg–Te bond that enables the formation of defects and may introduce disorder fields to dissociate the excitons, the PL peak at 4.2 K is quite possibly caused by a band-to-band transition. The localization of excitons result from the alloy disorder-induced fluctuations of potential fields whose energy has been experimentally determined using thermally assisted dissociation experiments to follow a $x(1-x)$ behavior. When x is around 0.5, the energy takes on its maximum value which is as large as 10 meV.

2.5.2 *Theoretical Background for Photoluminescence*

In 1990, Tomm et al. (1990) reviewed systematically infrared PL in narrow-gap semiconductors. In this section, the theoretical background will be briefly introduced to further the understanding of the forthcoming materials.

The process of PL can be simply described as follows; when a light beam with intensity I_0 and photon energy $\hbar\omega > E_g$ impinges on the surface of a semiconductor with a band gap of E_g, it penetrates into the sample and will be absorbed with an absorption coefficient α far some distance below the surface. If the reflection and/or scattering at the interface are ignored, the intensity is then reduced to $I_x = I_0 \exp(-\alpha x)$.

The absorption of these photons leads to a nonequilibrium carrier generation process. In general, these photon-generated nonequilibrium carriers will relax rapidly to the extremes of the conduction and valence bands, respectively, so as to reach a minimum energy. This energy configuration can be described by a quasi-equilibrium Fermi level distribution. The effective temperatures of electrons and holes are normally higher than the measured sample temperature due to a heating effect. The process of carrier generation is faster than that of diffusion. Thus, the distribution of the photon-generated nonequilibrium carriers is nonuniform, which leads to the existence of diffusion to reduce the concentration gradients.

There are several recombination mechanisms inside the semiconductors and at surfaces. For a constant excitation power, the nonequilibrium carriers will form a quasistatic spatial distribution. In the case of a weak excitation, the lifetime of these nonequilibrium carriers is basically independent of their concentration. If the recombination of these carriers is realized by radiative recombination, the accompanying photon emission is the PL process. The PL internal process results from three different but interconnected processes: light absorption causing nonequilibrium carriers generation; diffusion of these carriers and radiative recombination of resulting electron–hole pairs; and the propagation of the radiative recombination induced photons in the sample and its subsequent emission from the sample (Shen 1992).

Because the PL process is due to a recombination process, it is necessary to examine the different recombination process in HgCdTe materials. It is known that there are both radiative and nonradiative recombination processes in narrow gap semiconductors. Surfaces influence the whole recombination process. However, the surface can be carefully treated before a PL measurement so its influence is reduced significantly to a point where it can be ignored. When the PL processes in HgCdTe are analyzed, the features of narrow gap semiconductors must be taken into account. The dielectric constant is large and the electron effective mass is small, resulting in a very weak exciton binding energy. For example, for $E_g = 100$ meV gap, the binding energy of a free exciton is only about 0.3 meV. (For the same reason a hydrogenic donor impurity is shallow in HgCdTe.) Another remarkable feature of narrow gap HgCdTe is the intensity of the Auger recombination process, which is an important nonradiative process (Petersen 1970; Kinch et al. 1973; Pratt et al. 1983), and gets stronger with a decrease of the band gap. The relation between the recombination rate R_{A} and E_{g} is given by:

$$R_{\mathrm{A}} \approx E_{\mathrm{g}}^{-\beta}, \tag{2.97}$$

where β takes on a value between 3 and 5.5 (Ziep et al. 1980).

For n-type semiconductors, the energy released in an Auger-1 process, one in which the recombination of an electron in the conduction band and a hole in the valance band, is transferred to another electron so that its energy is increased to conserve energy and momentum in the overall process. The lifetime of this Auger recombination in n-type HgCdTe can be written as (Lopes et al. 1993):

$$\begin{aligned} \tau_{A1} &= \frac{2n_i^2 \tau_{A1}^i}{(n_0 + p_0)n_0} \\ \tau_{A1}^i &= 3.8 \times 10^{-18} \varepsilon_\infty^2 (1+\mu)^{\frac{1}{2}} (1+2\mu) \exp\left(\frac{(1+2\mu)E_{\mathrm{g}}}{(1+\mu)k_{\mathrm{B}}T}\right) \left(\frac{m_0}{m_e^*} |F_1 F_2| \frac{k_{\mathrm{B}}T}{E_{\mathrm{g}}}\right)^{\frac{3}{2}}, \end{aligned} \tag{2.98}$$

where n_{i} is the intrinsic electron concentration, n_0 and p_0 the thermal equilibrium concentrations of electrons and holes, respectively, μ is the ratio of the effective masses of an electron and a hole, ε_∞ the high-frequency dielectric constant, E_{g} the band gap, k_{B} the Boltzmann constant, and T is the temperature. $|F_1 F_2|$ are overlap integrals of Bloch wave functions, F_1 between the conduction and valance bands, and F_2 between the conduction band edge and an excited conduction band state. The product when fit to experiment is taken to have a value is in the range of 0.1–0.3. However, in a recent paper where good band structures are calculated, it has been shown that the product is not well approximated by a constant but depends on the momentum $\boldsymbol{k}$, the alloy concentration, and the temperature (Krishnamurthy 2006).

For p-type HgCdTe samples, just considering an Auger-1 process is not sufficient, an Auger-7 process is also important (Lopes et al. 1993; Petersen 1983). An Auger-7 process is one in which the recombination energy and momentum of an electron in the conduction band and a heavy hole is taken up by the excitation of an electron from the light hole subband to the heavy hole subband. The corresponding lifetime is given by (Lopes et al. 1993):

$$\tau_{\mathrm{A7}} = \frac{2n_i^2 \tau_{\mathrm{A7}}^i}{(n_0 + p_0)p_0}. \tag{2.99}$$

Casselman (1981) calculated the ratio γ of τ_{A7}^i and τ_{A1}^i, which is a function of temperature and alloy concentration. For HgCdTe with a concentration x in the range 0.16–0.3, at temperatures in the range 50–300 K, the value of γ lies between 0.5 and 6. Specifically, for $x = 0.22$ and 0.3, γ is independent of temperature and has a value of about 1.5 and 0.5, respectively. Combining these two mechanisms together, the total Auger lifetime τ_A of p-type HgCdTe is given by (Lopes et al. 1993):

$$\tau_A = \frac{\tau_{A1}\tau_{A7}}{\tau_{A1} + \tau_{A7}}. \tag{2.100}$$

Similar to the recombination processes in wide gap semiconductors, the Schockly–Read process also exists in the nonequilibrium carrier recombination of narrow gap semiconductors. It too is a type of nonradiative recombination. In this kind of recombination process, nonequilibrium carriers recombine through a pathway provided by deep impurity states and/or dislocations providing energy steps to enhance the recombination probability. If the density of Shockley–Read centers is much lower than the carrier concentration, the Shockley–Read recombination lifetime τ_{SR} is determined by (Lopes et al. 1993):

$$\tau_{SR} = \frac{(n_0 + n_1)\tau_{p0} + (p_0 + p_1)\tau_{n0}}{(n_0 + p_0)} \tag{2.101}$$

with

$$\begin{aligned} n_1 &= N_C \exp\left(\frac{E_T - E_C}{k_B T}\right) \\ p_1 &= N_V \exp\left(\frac{E_V - E_T}{k_B T}\right), \\ \tau_{p0} &= (v_p \sigma_p N_t)^{-1} \\ \tau_{n0} &= (v_n \sigma_n N_t)^{-1} \end{aligned} \tag{2.102}$$

where σ_n and σ_p are the capture cross sections of electrons and holes, and υ_n and υ_p are their thermal velocities, respectively. E_T is the trap energy.

According to the calculation by Schacham and Finkman (1985), the inter-band radiative recombination e lifetime is given by:

$$\tau_r = \frac{1}{B(n + p)}, \tag{2.103}$$

where

$$\begin{aligned} B = 5.8 \times 10^{-13} \sqrt{\varepsilon_\infty} \left(\frac{m_0}{m_e^* + m_h^*}\right)^{\frac{3}{2}} \times \left(1 + \frac{m_0}{m_e^*} + \frac{m_0}{m_h^*}\right), \\ \times (300/T)^{\frac{3}{2}} \left[E_g^2 + 3k_B T E_g + 3.75(k_B T)^2\right] \end{aligned} \tag{2.104}$$

and where ε_∞ is the high-frequency dielectric constant, m_e^* and m_h^* electron and hole effective masses, m_0 the mass of free electron, and k_B is the Boltzmann constant.

For the nonequilibrium carriers in narrow gap HgCdTe, there are three main recombination mechanisms: radiative, Shockley–Read, and Auger recombination. These recombination mechanisms determine the lifetime of the nonequilibrium carriers. In the case of low injection, the relation between the corresponding lifetimes (τ_R, τ_A, and τ_{SR}) and the lifetime of a nonequilibrium carrier in a material is:

$$\frac{1}{\tau} = \frac{1}{\tau_A} + \frac{1}{\tau_R} + \frac{1}{\tau_{SR}}. \tag{2.105}$$

For a given density of nonequilibrium carriers in HgCdTe, the recombination can be realized by a radiative recombination process or an Auger and/or Shockley–Read nonradiative recombination processes. The radiation efficiency is the probability of a photon emission in a radiative recombination, which is given by

$$\eta \equiv \frac{1/\tau_R}{1/\tau} = \frac{1}{1 + \frac{\tau_R}{\tau_A} + \frac{\tau_R}{\tau_{SR}}}. \tag{2.106}$$

Therefore when the excitation power is constant, the PL signal strength is determined by internal recombination mechanisms, and the corresponding PL spectrum depends on a competition between internal radiative recombination and the other nonradiative recombination processes. To enhance the PL signal intensity and improve the SNR, it is necessary to suppress the probability of nonradiative recombination.

Now let us consider the major PL processes occurring near the band-edge of HgCdTe. As is well known, HgCdTe has a zincblende cubic structure. The extremes of the valence- and conduction-bands are in the center of the first Brillouin zone, and it is therefore a direct-gap semiconductor. The optical properties, especially the PL near the band edge, are mostly determined by the band edge structure. For a direct-gap semiconductor, nonequilibrium carriers at the band edge can recombine through

1. An inter-band transition (intrinsic transitions)
2. Free-exciton recombination
3. Bound exciton recombination
4. Radiative recombination of a free electron and a neutral acceptor (eA^0) or a free hole and a neutral donor (hD^0)
5. A donor–acceptor pair combination.

For narrow-gap HgCdTe, possibility number 3 is the main contributor through combinations of donor-bound exciton (D^0X) and acceptor-bound exciton (A^0X) processes. These recombination processes are illustrated in Fig. 2.70.

However, not all the radiative recombination processes in Fig. 2.70 will be observed for a given set of circumstances. Under a similar excitation condition, the PL spectra change with a sample's temperature. For example, at high-temperature, the direct band-to-band transition is always observable. But it is weak, even difficult to detect, at low-temperature. In the latter case the recombination through an

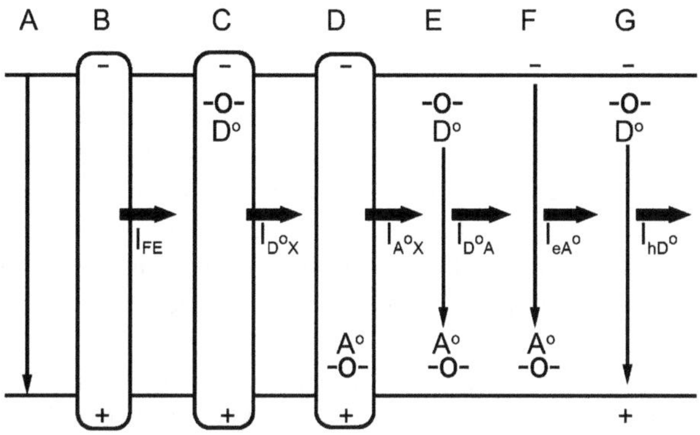

Fig. 2.70 Radiative recombination mechanisms of nonequilibrium carriers near the band edge of HgCdTe. (**a**) direct band-to-band transitions, (**b**) free-exciton recombination (*FE*), (**c**) donor-bound exciton recombination (D^0X), (**d**) acceptor-bound exciton recombination (A^0X), (**e**) donor–acceptor pair recombination (D^0A^0), (**f**) free electron–neutral acceptor recombination (eA^0), and (**g**) free hole–neutral donor recombination (hD^0)

exciton or impurity generally dominates because impurity and exciton states have large populations at low-temperature. The impurity assisted PL obviously changes drastically with the type and density of impurities. The PL spectrum also depends on the experimental conditions, such as excitation power. These conditions therefore provide a criterion for clarifying the mechanisms responsible for different PL peaks (Schmidt et al. 1992a, b).

As mentioned before, the light emission process is related to the light absorption process. For the PL process in the case of equilibrium and quasiequilibrium, the generation rate of photoelectron/hole pairs is equal to their recombination rate, which is known as von-Roosbröck–Shockley relation. It can be written as:

$$R_{sp}(\hbar\omega) = v_{en} \times G(\hbar\omega) \times \alpha(\hbar\omega) \times \frac{n_u n_l'}{n_l n_u' - n_u n_l'}, \qquad (2.107)$$

where R_{sp} is the spontaneous emission transition rate per unit volume in the energy space element $d(\hbar\omega)$ at $\hbar\omega$, υ_{en} the propagation velocity of energy, n_l the occupation probability of the lower state, n_l' the probability that the lower state is empty, n_u the occupation probability of the upper state, n_u' the probability that the upper state is empty, $\alpha(\hbar\omega)$ the absorption coefficient, $G(\hbar\omega)$ the density of the radiation field in the energy space element $d(\hbar\omega)$ at $\hbar\omega$. This relation can be simplified for a direct-gap transition using parabolic energy bands:

$$R_{sp}(\hbar\omega) = \frac{(\eta\hbar\omega)^2}{(\pi c)^2\hbar^3} \times \alpha(\hbar\omega) \times \left[\exp\left(\frac{\hbar\omega - E_g - E_F^n - E_F^p}{k_B T}\right)\right]^{-1}, \qquad (2.108)$$

where η is the refractive index, E_F^N and E_F^P are the quasi-Fermi levels of electrons and holes, respectively, E_g is the band gap.

Because PL is a process involving generation and recombination of nonequilibrium carriers, the quasi-equilibrium distribution of carriers described by quasi-Fermi levels is appropriate. In principle, the radiative recombination rate and line shape of the emission spectrum can be derived from the corresponding absorption spectrum taken for the same sample with a given refractive index η. This is important when an experimental PL spectrum is to be understood, such as the emission spectral band width in bulk GaAs (Griffiths and Haseth 1986). For the band-to-band PL transition in HgCdTe at high-temperature, a similar method can be employed to characterize the PL peaks.

For a PL study of narrow-gap HgCdTe semiconductors in the infrared region, a FTIR spectrometer is an obvious choice. The FTIR spectroscopic technique has been developed to such a level that it not only provides the two advantages of through-put and multi-channels, but also may enable the possibility of incorporating a modulation process into the PL measurement. The FTIR spectrometer makes use of the relation between an interferogram of a Michelson interferometer and an optical spectrum. The spectrum is deduced by first recording the interferogram, and then Fourier transforming it to obtain the final spectrum. In the following, the basic principle employed is described (Griffiths and Haseth 1986; Fuchs et al. 1989).

Considering a case in which a monochromatic light beam, with amplitude a and wavenumber υ, passes through a beam splitter, the reflection coefficient is denoted by r and the transmission coefficient by t. The beam splitter splits the light into two parts: one is a reflected beam with amplitude ra and transmitted beam with amplitude ta. The transmitted beam goes to the fixed mirror M1 of the interferometer, and the reflected beam goes to the moving mirror M2. Then the beams are reflected back to the beam splitter and recombined. The motion of the moving mirror changes the total path length of the reflected beam, and the optical path difference between the two beams results in temporal constructive and destructive interference. The jointed beam is collected by a detector. The signal amplitude collected by the detector is:

$$A_\mathrm{D} = r \times a \times t(1 + e^{-\mathrm{i}\varphi}). \tag{2.109}$$

The intensity of this signal is:

$$I_\mathrm{D}(x, \nu) = A_\mathrm{D} * A_\mathrm{D}^* = 2RTB_0(\nu)(1 + \cos\varphi), \tag{2.110}$$

where R, T are the reflectance ratio and transmittance ratio of the beam splitter, $B_0(\upsilon) = a \times a^*$ is the intensity of the incident beam, x is the optical path difference, and φ is the phase difference between the two beams from the fixed and moving mirrors.

$$\varphi = 2\pi\frac{x}{\lambda} = 2\pi\nu x, \tag{2.111}$$

When the incident beam has a random spectral distribution, $I_\mathrm{D}(x, \nu) = \mathrm{d}I_\mathrm{D}(x)/\mathrm{d}\nu$ is an integral with the spectral element of an infinitely narrow line width $\mathrm{d}\nu$:

$$I_\mathrm{D}(x) = \int_0^\infty 2RTB_0(\nu)\,[1 + \cos(2\pi\nu x)]\,\mathrm{d}\nu. \tag{2.112}$$

The final spectrum, B_ν, can be obtained by doing a Fourier transformation of the interferogram:

$$B_\nu = \int_{-\infty}^{\infty} I_D(x) e^{-i2\pi\nu x} dx. \tag{2.113}$$

Therefore, for any wave number, the spectral density B_ν can be obtained from the Fourier transform after the interferogram is known.

However, the above method is difficult to exercise in practice. In fact, the sampling points are limited and the arm length of the Michelson interferometer in the FTIR spectrometer is finite. The limitation this imposes on sampling points and the sampling range results in side lobe oscillations. These can be suppressed, nevertheless, by an apodictic treatment. That is, the Fourier transformation is performed on the resultant interferogram after having been multiplied by an apodictic function.

The Happ–Genzel function and the Boxcar function are both common apodictic functions in the Fourier transformation of a spectrum. The suppression by the apodictic treatment may, in addition, result in a line broadening and degradation of the spectral resolution. Thus a compromise has to be made between the spectrum distortion and the reduction of the resolution. The Happ–Genzel function can effectively suppress the side lobes and at the same time not decrease the resolution so remarkably. Thus it is frequently used, and is especially useful for high-resolution spectrum measurements.

In an ideal experiment, the system noise arises from the infrared detector. To reduce the noise, the noise component outside the measured spectral range is always eliminated by a suitable electrical filter. However, the electrical filter introduces an additive phase shift to the signal. The optical path, data acquisition and other electronic circuits may also introduce additive phase shifts (Griffiths and Haseth 1986). Thus, it is necessary to use a phase correction to eliminate these additive phase shifts. The spectrum after phase correction can be written as:

$$B(\nu) = \int_{0}^{\infty} A(x, \varepsilon) I_D(x + \varepsilon) \cos 2\pi\nu(x + \varepsilon), \tag{2.114}$$

where $A(x, \varepsilon)$ is an apodictic function.

Thanks to the development and application of the algorithm for fast Fourier transformation (FFT), the rapid-scan mode of a modern FTIR spectrometer ensures a spectrum will be obtained in just several seconds or minutes with high resolution and SNR.

Briefly, the major advantages of the FTIR spectrometer are as follows (Griffiths and Haseth 1986):

1. Multi-channel advantage.

By using this interference method, the spectral signals from all the frequencies can be obtained in a single scan of the moving mirror. Thus the measuring time can be significantly reduced with a similar SNR and resolution. In contrast to the conventional dispersive technique, a spectrum with significantly enhanced SNR can be recorded for a given measuring time.

For the dispersive technique, on the other hand, the intensity of a received signal with wavenumber ν_1, is given by:

$$\int_t^{t+\Delta T} B_0(\nu_1)\mathrm{d}t = B_0(\nu_1)\Delta T, \tag{2.115}$$

where v is the velocity of the moving mirror.

For a spectral element with wavenumber υ_1, the FTIR technique uses $\cos(2\pi\nu_1 x) = \cos[2\pi\nu_1(\nu t)]$ to encode it. Then N spectral elements are coded and their signals are observed simultaneously in time T (N elements in time ΔT). Thus the intensity of the received signal for a given spectral FTIR element is given by:

$$\int_0^{t=N\Delta T} B_0(\nu_i)\mathrm{d}t = N\left[B_0(\nu_1)\Delta T\right]. \tag{2.116}$$

It is obvious that the signal intensity increases by a factor of N using this FTIR technique and thus the SNR increases by a factor of $\sqrt{N}$!.

2. High throughput advantage.

In contrast to the slit of dispersive spectrometers, all the radiative power from the light source passing through the aperture is received by the detector of the FTIR spectrometer. Thus its throughput is improved many times relative to that of dispersive spectrometers. Another obvious advantage of the FTIR technique is its spectral resolution, which is determined by the arm length of the Michelson interferometer from the mirrors to the beam splitter. Therefore, the energy resolution is identical over the whole spectral range being measured and is easy to upgrade. The disadvantage of the FTIR technique, if there is any, is its relatively high price and because it obtains a spectrum indirectly. The spectrum may be distorted if any factor in the data acquisition and/or processing is not implemented carefully.

The preferred excitation source for a PL measurement is the 514.5 or 488.0 nm spectral line of an Ar^+ laser, even in the measurement of narrow-gap semiconductors. A HgCdTe sample to be studied is mounted on the cold finger of an optical Dewar to reach low enough temperatures for the measurements, where weak radiative recombination processes exist in these narrow-gap systems. The sample is usually immersed in liquid Helium or a Helium gas during the measurement. To control and determine the sample's temperature accurately, a sensor and a heater are installed just above the sample. The accuracy of the temperature can be controlled within ± 0.1 K, so that a study of temperature effects is reliable in a range of temperatures from $\sim$4 to 77 K. The PL signals are collected by a parabolic mirror and sent to an FTIR spectrometer to record the spectrum. The choice of detectors depends on the spectral region of the PL measurement. Liquid-nitrogen cooled InSb and HgCdTe detectors are commonly used. At times $Hg_{1-x}Cd_xTe$ with a large x, InAs, or Ge detectors can also be a proper choice. The matching of the spectral region to the detector of the FTIR is important to ensure a good measurement.

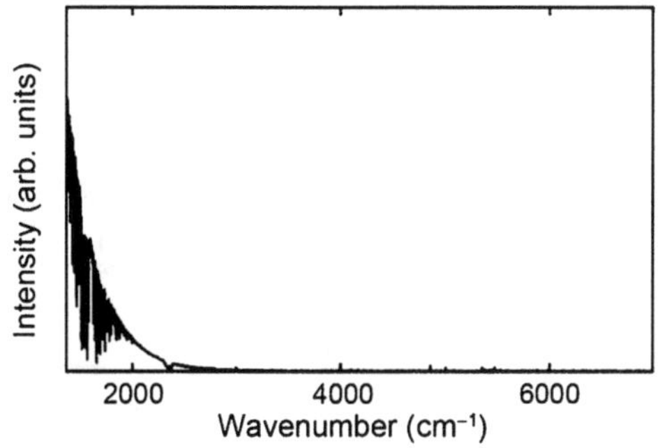

Fig. 2.71 Environmental thermal emission in a conventional PL measurement (with a liquid-nitrogen cooled HgCdTe detector and a KBr bean splitter)

To realize PL measurements in the mid- and far-infrared spectral regions, it is crucial to advance the conventional FTIR PL technique to the DM-PL method. In fact, even in the 4–5 μm spectral region, PL spectra measurements are already very difficult due to strong environmental background thermal emission, which is illustrated in Fig. 2.71. This noise is generally at least 200 times stronger than the PL signal. The double-modulated PL (DM-PL) technique eliminates the influence of the environmental thermal background, and is thus important for the nondestructive detection and evaluation of materials working in these regions.

To eliminate the influence of the background thermal emission, the main possible improvements to the conventional FTIR PL technique are as follows:

Frame subtraction method After a normal PL measurement, the background spectrum is measured. Then the background is subtracted from the PL spectrum. It is a convenient method requiring no hardware modification, and is effective for samples with strong PL signals that are larger than, or at least of the same order of magnitude as the background signals. However, for most samples the PL signal is not strong enough to ensure such a treatment will be effective. A more efficient technique is thus called for.

Phase sensitive excitation (PSE) technique The realization and application of this method to HgCdTe PL measurement was first developed by Fuchs et al. (1989), Fuchs and Koidl (1991). The technique is illustrated in Fig. 2.72.

According to the principles behind the FTIR technique, the inter-ferogram of the detected light passed by a Michelson interferometer, can be written:

$$I(x) = \frac{1}{2}\sum_i I_i + \frac{1}{2}m\sum_i I_i \cos(2\pi \nu_i x), \tag{2.117}$$

where ν_i is the wavenumber, I_i is the density of the monochromatic light at this wavenumber, $x(t) = \nu t$ is the optical path difference of the interferometer, and m is the modulation efficiency. For an ideal interferometer, m is taken as 1.

$$I_{\mathrm{PL}} = I_{\mathrm{PL0}}\left[\frac{1}{2} + \frac{1}{2}\cos(f_0 t + \phi_1)\right], \tag{2.118}$$

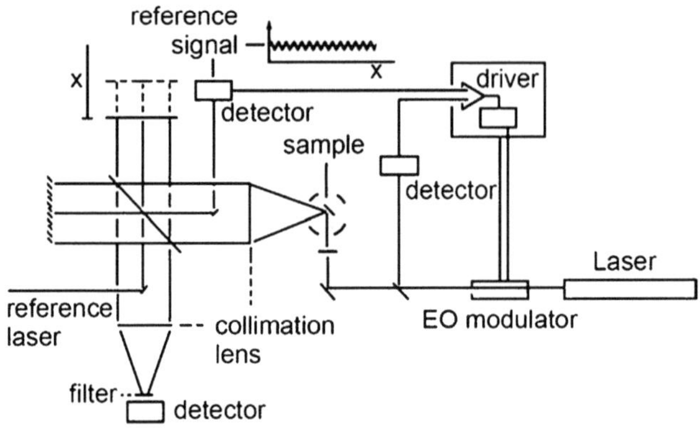

Fig. 2.72 Schematic of a phase sensitive excitation (PSE) Fourier transform infrared (FTIR) spectrometer-based PL technique

where I_{PL0} is the output power of a CW laser, ϕ_1 is the generated phase difference in this process, f_0 is given by the alignment the He–Ne laser incident on the FTIR, and is given by:

$$f_0 = 2\pi\nu_0\nu. \tag{2.119}$$

According to Fuchs et al., the PL signal is proportional to the power density of the excitation laser, $I_i(t) = CI_{\mathrm{PL}}(t)$, then

$$I_i(t) = I_i\left[\frac{1}{2} + \frac{1}{2}\cos(f_0 t + \phi)\right]. \tag{2.120}$$

Thus we find:

$$\begin{aligned} I_i(x) &= \frac{1}{2}I_i(2-m)\left[\frac{1}{2} + \frac{1}{2}\cos(f_0 t + \phi)\right] \\ &\quad + \frac{1}{2}I_i m\left[\frac{1}{2} + \frac{1}{2}\cos(f_0 t + \phi)\right]\cos(2\pi\nu_i x_1). \end{aligned} \tag{2.121}$$

By making a transformation, the appropriate difference equation becomes:

$$\begin{aligned} I_i(x) &= \frac{1}{2}I_i(2-m)\left[\frac{1}{2} + \frac{1}{2}\cos(f_0 t + \phi)\right] + \frac{1}{2}I_i m\cdot \\ &\times\left\{\frac{1}{2}\cos(2\pi\nu_0 t+\phi)+\frac{1}{4}\cos[2\pi(\nu_0+\nu_i)t+\phi]+\frac{1}{4}\cos[2\pi(\nu_0-\nu_i)t+\phi]\right\}. \end{aligned} \tag{2.122}$$

The modulated PL signals transfer 50% of the intensity into each of the two side band of the modulation frequency f_0. For a commercial FTIR spectrometer, an He–Ne laser is used as the alignment laser, and hence f_0 is approximately 15,798 cm^{-1}. The resulting modulated PL signals are therefore kept away from the region of the background thermal emission and a PL spectrum of relatively high quality is attained. The problem is, however, that the assumption, $I_i(t) = C I_{\mathrm{PL}}(t)$, is not always correct. For the spectral region of narrow gap semiconductors, the possible recombination mechanisms for nonequilibrium photogenerated carriers consist of band-to-band transition; bound-to-free recombination; donor-bound exciton recombination (D^0X) and acceptor-bound exciton recombination (A^0X); free-exciton recombination (*FE*); and donor–acceptor pair recombination (D^0A^0). Among these mechanisms, only the PL between *FE* and bound excitons strictly fulfills this assumption (Schmidt et al. 1992a, b). As a result, this technique may lead to a distortion of the spectrum. Simulation results also support this contention.

Phase sensitive detection (PSD), or namely, DM-PL technique DM-PL spectroscopy was first proposed by Griffiths and Haseth (1986), followed by Reisinger et al. (1989). The thought behind this method is as follows. Besides the modulation introduced by the Michelson interferometer in an FTIR spectrometer, an additional modulation with higher frequency is applied to the CW laser by, for example, a mechanical chopper. This chopped light is sent into the sample and then into the FTIR and the signal is detected with a lock-in-amplifier (LIA) referenced to the chopper frequency. The demodulated light is then passed through a preamplifier, the inverse Fourier transform unit and a filter. The final spectrum is then obtained (Chang 1995). For a high quality spectrum the modulation frequency of the excitation laser light should be much higher than the highest frequency component of the interferogram and its period must be at the same time much smaller than the integration time constant of the LIA. This integration time is limited by the sampling frequency of the FTIR. The integration time should be slightly less than or equal to the sampling cycle time in order to obtain a SNR as large as possible. Formerly the DM-PL technique was limited by the chopper frequency. To mitigate this problem the scanning speed could be decreased. On the other hand, a low scanning frequency would decrease the SNR. A better solution is to increase the modulated frequency. Chang (1995) increased the modulation frequency by using an acousto-optic modulator (AOM) and in so doing the quality of the PL spectrum was greatly improved.

The experimental setup for the DM-PL spectroscopy is illustrated in Fig. 2.73. The light from the CW laser irradiates the AOM at the Bragg angle. The first-order

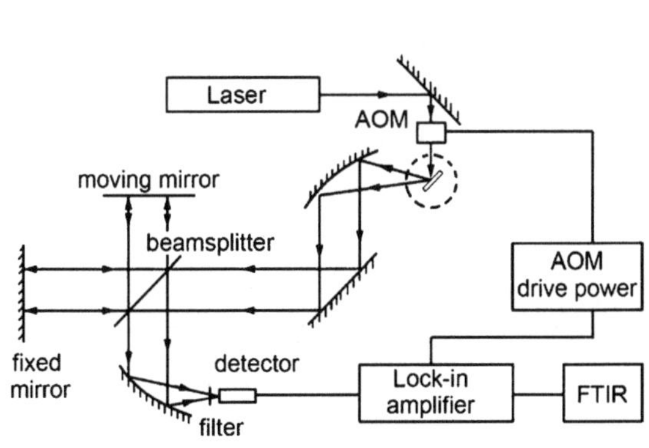

Fig. 2.73 An experimental setup schematic for a DM-PL measurement

diffraction beam is used as an output, and now due to the modulation of the AOM it is a "square wave" with a frequency of 100 KHz. It then impinges onto the sample in a Dewar and the PL signal from the sample with the AOM modulation frequency is collected by a parabolic mirror and fed into the entrance of the FTIR spectrometer. After passing through the Michelson interferometer of the FTIR spectrometer, the PL signal is focused onto an infrared detector, the output of which is fed to an LIA. Meanwhile, the modulation frequency of the AOM is applied to the LIA as its reference. The output signal of the LIA is fed back into an amplifier, a filter, and an inverse Fourier transformation unit. In this way, only the PL signals at the modulation frequency and at a fixed phase difference relative to the modulated excitation laser light is detected and amplified, and the background thermal emission is totally suppressed. A PL spectrum is thereby obtained, free of the background thermal emission disturbance.

To ensure a successful PL measurement with the DM-PL technique, particular problems should be treated with great care:

a. *The choice of the AOM frequency*. Theoretically, the increase of the modulation frequency is helpful for improving the SNR of the spectrum in the case of the LIA working at a particular integration time constant. However, the increase of the frequency will result in a decrease of the directivity D^* due to a limited response time of the detector. Figure 2.74 shows the D^* as a function of frequency for a photovoltaic HgCdTe detector and a TGS (TGS: triglycine sulfate detector operated at room temperature) detector, respectively. It is clear that the TGS detector cannot be used for this purpose due to its slow response. The response time of a photovoltaic HgCdTe detector is fast and hence is the best choice in this case. Nevertheless, for an FTIR spectrometer equipped with a photoconductive HgCdTe detector whose response time is a bit longer, it is experimentally shown that a quite nice SNR may be ensured if the modulation frequency is taken to be no higher than 100 KHz.

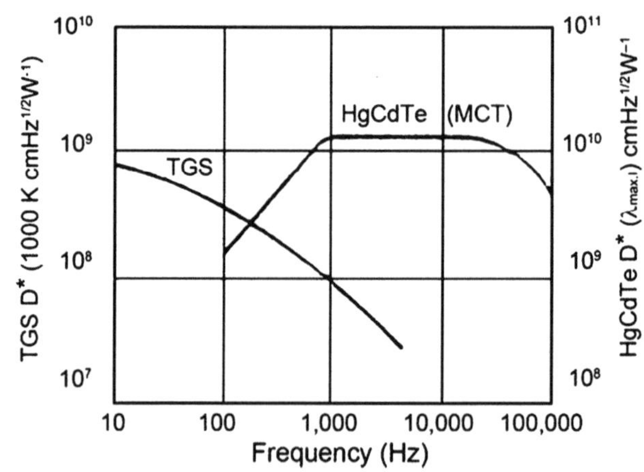

Fig. 2.74 Detectivities D^* as a function of frequency for HgCdTe and TGS detectors, respectively

b. *The choice of the cutoff frequency of the band-pass filter.* A monochromatic beam with wavenumber σ corresponds to a sinusoidal interferogram with a Fourier frequency $2\sigma v$ after being modulated by the Michelson interferometer. Given the velocity of the moving mirror, the sinusoidal wave frequencies can be calculated for monochromatic beams with different wave numbers in this way. In a real PL measurement, the spectral signal only shows up in a certain region. Thus the interferogram can be filtered with an electrical band-pass filter so as to let only the useful signal within the spectral range pass through. The SNR in this manner is improved.
c. *The choice of a suitable phase correction.* To eliminate the additive phase shifts, a phase correction is applied in the optical measurement to the FTIR spectrometer (Griffiths and Haseth 1986). In general, an autocorrection method is used. Data points around the zero path-difference point (a typical value is 256) are chosen for the transformation. For a conventional optical spectrum measurement, these sampling points are strong enough to produce a good SNR. The array θ can be written:

$$\theta = \arctan(\mathrm{Im}_i/\mathrm{Re}_i), \tag{2.123}$$

where i labels the points, Im and Re refer to the imaginary and the real part of transforms, respectively. In general, θ is basically thought of as removing the noise in the real phase array of the spectrum. However, it is not appropriate for a PL measurement using DM-PL spectroscopy due to the weak signal and large noise. On one hand, they lead to errors in the choice of the zero path point. On other hand, even for points around the zero path point, their SNR is not high and results in large error in the phase array calculation. The wrong choice of the zero path point directly causes a large distortion of signals. Thus, this type of autocorrection method is not suitable in a PL measurement of HgCdTe. One solution is to choose the zero path point manually and perform a phase correction. Another solution is to make use of background in high- and low-pass filtering conditions and perform a phase correction from a calculation.

According to a brief comparison (Shao et al. 2006), the double modulation was superior to a frame-to-frame subtraction, though the former experienced a rather tough limitation on the choice of the sampling time constant of the LIA for short-wavelength PL measurements. The PSE modulation PL method, on the other hand, relaxes the restriction on the sampling time constant of the LIA. While it separated the PL from the background thermal emission, the dynamic range of the detector, and hence the SNR of the PL signal as well, were limited. Also it can suffer from interference with the intense line at the carrier frequency and could not be run safely with high spectral resolution.

Furthermore, both the PSD and PSE modulation techniques were based on a continuous-scan (also known as a rapid-scan) FTIR spectrometer, and can not be used to eliminate the disturbance of the internal He–Ne laser line for the PL spectra around 630 nm, because of either a difficulty of realization or the presence of the He–Ne laser line as an intrinsic characteristic. For this reason, and especially to significantly improve the SNR and reduce time cost of a single measurement, Shao et al. designed and demonstrated a new type of DM-PL based on a modern FTIR spectrometer with a step-scan operational mode (Shao et al. 2006).

Step-scan FTIR spectrometer-based modulated PL technique The experimental setup for a step-scan FTIR spectrometer-based modulated PL technique is similar to the previous DM-PL one, except for the fact that (1) the FTIR spectrometer runs in step-scan mode, in which during each step the laser is amplitude-modulated a process that fully relaxes the limitation on the choice of the modulation frequency without any trade off, (2) a combination of a mechanical chopper and an LIA is used to replace the combination of an AOM and an LIA, and (3) the pumping laser is amplitude-modulated by a mechanical chopper whose frequency serves an a reference for the LIA.

A typical result established for a narrow-gap HgCdTe sample is depicted in Fig. 2.75 together with a representative PL spectrum obtained by Chang (1995) using a continuous-scan DM-PL technique. The significantly improved SNR of the

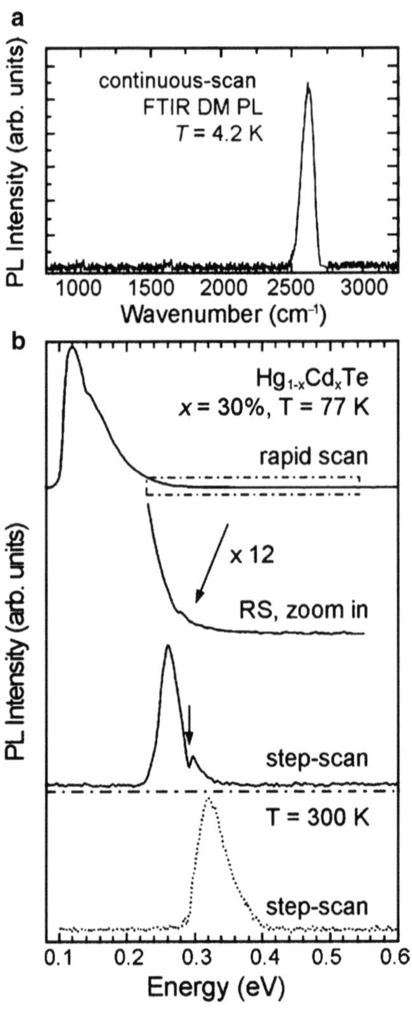

Fig. 2.75 Comparison of PL spectra taken by: (**a**) a continuous-scan FTIR DM-PL, and (**b**) a newly developed step-scan FTIR modulated PL technique. The upper PL spectrum (**a**) was measured at a temperature of 4.2 K, while the others (**b**) were taken at either 77 K or room temperature

PL spectrum recorded by the new technique is amazing and will surely warrant continuing the study of PL process in the narrow gap semiconductors in the coming days.

As pointed out (Shao et al. 2006), the most distinct feature distinguishing the modulated step-scan FTIR PL technique from both previous continuous-scan FTIR-based PSD and PSE modulation methods is that, in the step-scan mode there is no Fourier transform required following the data acquisition. It therefore totally resolves the difficulty arising from compromises among the modulation frequency, the scan velocity, and the LIA sampling time constant. The direct results are the obvious advances in either the applicable spectral region, or the experimental time required for a single PL spectrum with a reasonable spectral SNR. To facilitate a simple comparison (Shao et al. 2006), the previous work by Reisinger et al. (1989) and Fuchs et al. (1989), respectively, were taken as good examples of the continuous-scan FTIR-based PSD and PSE DM-PL methods.

It was noted by Shao et al. (2006) that with a configuration consisting of the 4-kHz chopper frequency and a 0.0355 cm/s scanner velocity employed by Reisinger et al. (1989), the highest applicable spectral range for the continuous-scan FTIR-based PSD technique was up to about 5,700 cm^{-1}, which is far from the demands for application to the NIR region. Alternatively, if the entire NIR spectral region were insisted on, the scan velocity must be slowed to $\sim$0.0127 cm/s, which is too slow for the scanner to be well controlled. In both cases, the maximal sampling time constant of the LIA is only of the order of magnitude of 1 ms, which for a weak signal is simply too short to ensure phase locking of the LIA! For the continuous-scan FTIR-based PSE method, the internal He–Ne laser line Fourier frequency was normally chosen as the carrier frequency for the sake of securing precise phase stability. This means such an arrangement can not be directly applied to a measurement in the spectral region around 630 nm.

The experimental time consumed by Reisinger et al. (1989) recording a PL spectrum involves 500 repeated scans for a single spectrum, which requires about 25 min for a scanner velocity of 0.0355 cm/s and a spectral resolution of 10 cm^{-1}. A similar situation was also found for the continuous-scan FTIR-based PSE modulation technique. It produced a PL spectrum with a similar SNR to that of the PSD measurement in a similar duration. 8,000 scans taking a total duration of 140 min were consumed to establish a spectrum with a resolution of 12 cm^{-1} and a reasonable SNR.

2.5.3 Infrared PL from an Sb-Doped HgCdTe

Photovoltaic opto-electronic applications of HgCdTe require the formation of a p–n junction. The group V element Sb is expected to act as acceptor in HgCdTe to convert an n-type layer to p-type to achieve a p–n junction. The successful realization of such a conversion processes relies on a knowledge of the doping impurity effectiveness; consequently a study of the Sb impurity doping behavior in HgCdTe is necessary (Chang 1995).

A series of HgCdTe ($x = 0.39$) samples with different Sb doping concentrations were used in PL measurements conducted by Chang, et al. Hall measurements were performed with a van Der Pauw configuration under 0.1 T at temperatures of 77 and 300 K. The p-type sample DH04 had a higher doping level. The 77 K carrier concentration measured by the Hall effect was $3.9 \times 10^{14}\,\text{cm}^{-3}$. Sample DH01 has a lower doping level of only $1.3 \times 10^{13}\,\text{cm}^{-3}$ again determined by a 77 K Hall measurement. PL and transmission measurements were performed by a Nicolet 800 FTIR spectrometer in the temperature range 4.0–115.0 K. For the PL measurements, an Ar-ion ($\lambda = 514.5\,\text{nm}$) laser was used for excitation and a liquid-nitrogen-cooled HgCdTe detector was employed. For the transmission measurements, the signal was detected by either an InSb or a HgCdTe photodiode, with the light beam from a Globar source focused onto the samples' surface. The spectral resolution of the optical measurements was $2\,\text{cm}^{-1}$ in wave numbers (or equivalently about 0.25 meV).

Although an infrared transmission spectrum can offer information about the energy band gap, for an accurate determination of E_g the influence of the sample thickness has to be considered. For thin samples near band edge information can be obtained from infrared transmission. Chu et al. (1994) proposed a relation between the absorption coefficient $\alpha(E_g)$ at the energy gap E_g:

$$\alpha_{E_g} = -65.0 + 1.887 + (8694.0 - 10.31T)x. \tag{2.124}$$

The values of E_g and x of samples can be then derived from the transmission spectra and the aforementioned equation. The composition of the sample DH04 is estimated to be $x = 0.38$.

Figure 2.76 shows the temperature-dependent PL spectra of a p-type Sb-doped HgCdTe sample with a carrier concentration of $3.9 \times 10^{14}\,\text{cm}^{-3}$ at 77 K (DH04), under the same laser excitation density (I_0) of about $3\,\text{W/cm}^2$. At 4.2 K, the PL spectrum shown in Fig. 2.76 is dominated by two peaks (labeled A and B). A similar set of PL signals were measured for the sample with the lower doping level

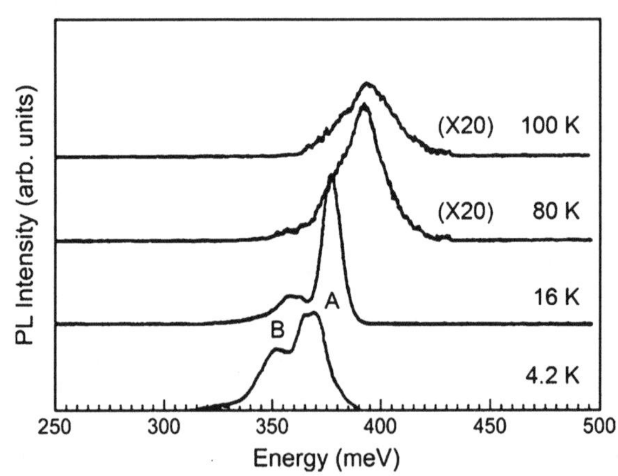

Fig. 2.76 Temperature-dependent PL spectra for an Sb-doped HgCdTe sample

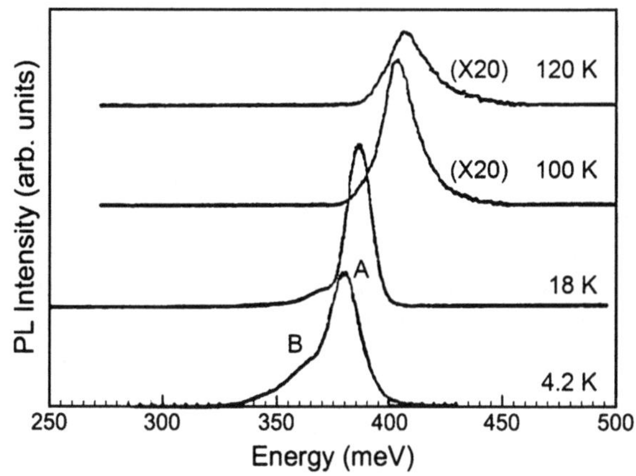

Fig. 2.77 Temperature-dependent PL spectra for an Sb-doped HgCdTe sample with a low doping concentration

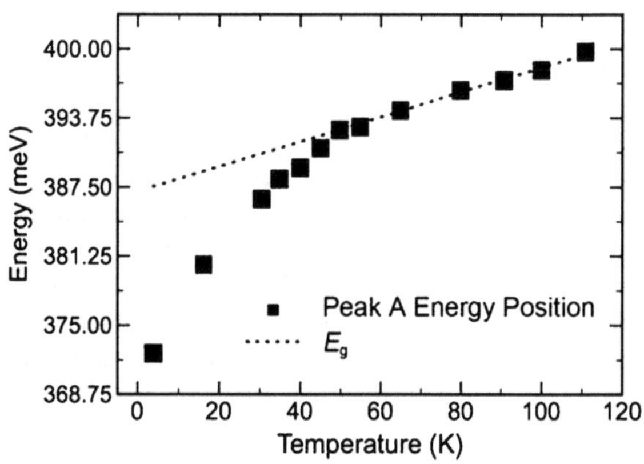

Fig. 2.78 The A peak's energy as a function of the sample's temperature. The *dotted line* is fitted to experimental data by using the formula for $E_g(x, T)$

(DH01) as shown in Fig. 2.77. The A structure has a sharper PL peak at the higher energy side, while peak B is broader with a FWHM of ~18 meV. When the temperature increases, peak B weakens and gradually disappears while peak A is enhanced. Increasing the excitation power intensity causes the integrated emission intensity of peak A to increase much faster than peak B. All our Sb-doped HgCdTe samples exhibit a similar PL structure and similar evolution with temperature and excitation power intensity, therefore their PL peaks can be assigned to the same origin.

The dependence of the A peak's energy as well as the band gap's energy on temperature is illustrated in Fig. 2.78. It clearly demonstrates that peak A is due to a band-to-band transition above about 50 K where its energy tracts well with the energy gap. But the obvious deviation from the fundamental energy gap below 50 K shows that there must be another transition mechanism responsible for the lower temperature PL structure. Since HgCdTe is a pseudobinary II–VI compound semiconductor, local composition fluctuations may exist, which result in local states with an energy E_{loc} depending on the concentration x as $E_{\text{Loc}} \sim x(1-x)$, and therefore have a maximum at $x = 0.5$. We assign the peak A at low-temperatures (below about 50 K) to a localized exciton transition, and obtain an extra localized energy of about 13 meV at a temperature of 4.2 K, which is in good agreement with the earlier results (Lusson et al. 1990). An anomalous behavior of peak A has been found: the integrated intensity of peak A increases from 4.2 to about 20 K, and then decreases with a further temperature increase, which can also be explained within the framework of alloy disorder and potential fluctuations. Therefore, peak A is due to a localized exciton PL below about 50 K, and when the temperature is increased the exciton is gradually delocalized and decomposed, and this peak gradually turns into a band-to-band transition.

The peak B is only present at lower temperatures below about 20 K, and its relative integrated PL intensity (I_{B}/I) under the same experimental conditions increases with the Sb doping concentration. For example, the PL intensity ratio is 0.14 and 0.19 for samples with effective carrier concentrations $1.3 \times 10^{13}\,\text{cm}^{-3}$ (DH01) and $3.9 \times 10^{14}\,\text{cm}^{-3}$ (DH04) (measured at 77 K) at 15 K and an excitation laser power intensity of $7\,\text{W/cm}^2$. The energy and integrated PL intensity of the peak B as a function of excitation power are depicted in Figs. 2.79 and 2.80.

The following characteristics can be observed only under stronger excitation intensities: (1) the peak shifts to higher energy and (2) the integrated intensity of the peak exhibits a tendency to saturate. To our knowledge, these two characteristics can only be fully explained within the theoretical framework of D°A° pair transitions. If the small influence of polarization from D° and A° is not taken into account, the peak energy for a transition involving randomly distributed donors and acceptors is given to first order by (Lannoo and Bourgoin 1981):

$$E = E_{\text{g}} - [E_{\text{D}} + E_{\text{A}} - e^2/(4\pi\varepsilon R)], \tag{2.125}$$

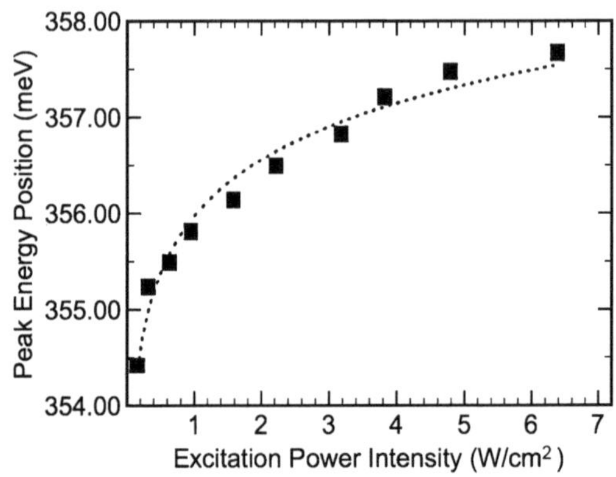

Fig. 2.79 The B peak's energy as a function of the excitation power intensity. The *dotted line* is fitted to the experimental data by using (2.125)

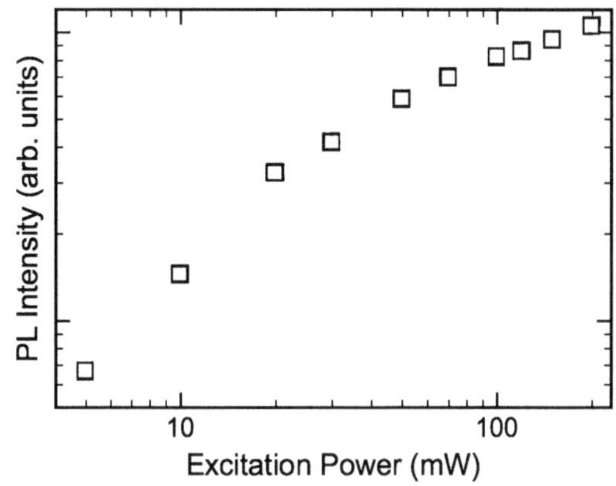

Fig. 2.80 Intensity of the B peak as a function of excitation power

where E_D and E_A are the donor and acceptor levels referenced to the conduction and valence bands, respectively, ε is the dielectric constant, and R is the separation of the donor–acceptor pair involved in the transition. The final term in parentheses accounts for the Coulomb interaction between the donor and acceptor. Equation 2.125 is valid for donor–acceptor separations greater than about three times of the effective Bohr radius of the bound electron or hole. Minor corrections are needed for smaller separations to account for wave function overlap, but they are not important in this case and will be neglected. If the separation were less than the Bohr radius of isolated carriers, the ionized D^+A^- pair could not bind a free carrier, so the transition probability would rapidly decrease. The characteristics of the B peak can be interpreted from (2.125) following some simple kinetics. The recombination occurs via tunneling; therefore the transition probability decreases exponentially with increasing separation. The photocarriers can recombine through other channels including nonradiative recombination with randomly located donors or acceptors. On the one hand, the PL intensity decreases as R increases, on the other hand the number of carriers, which are located in the same shell dR and emit the same energy, increases as R^2, so there exists a maximum PL intensity when R changes. At low-temperature and low excitation intensity, the term accounting for the Coulomb interaction can be neglected, and the peak energy is $E = E_g - (E_D + E_A)$. Increasing the excitation intensity, the concentration of neutral impurities D° and A° that are excited is increased, which results in a decreased average separation between A° and D°. The transition probability and energy of the emitted photon increases as the separation R is increased, so the PL peak shifts to higher energy. If the excitation intensity is high enough, the average separation between D° and A° is a minimum and the PL peak energy position will be a maximum and exhibit a tendency to saturate. The peak energy position dependence on the excitation intensity is (Yu 1977; Binsma et al. 1982):

$$E_P = E_{P_0} + \log(P_I/P_{I_0}), \tag{2.126}$$

where E_p and E_{p0} are the peak energy positions when the excitation intensity is P_I or P_{I0}, and β is a parameter that increases with an increasing degree of compensation. β is 1.98 for the sample shown in Fig. 2.79.

The saturation tendency of the PL intensity with increasing excitation intensity at a given temperature shows that this PL peak is related to centers that can saturate. Schmidt et al. (1992a, b) calculated the dependence of the PL intensity on excitation power intensity in direct gap semiconductors, that shows only D°A° PL manifests this obvious saturation tendency providing further evidence of the origin given above for the B peak. The B peak is clearly Sb-doping-related because at 77 K the relative intensity of the B peak increases with the Sb doping concentration, and hence the carrier concentration. Since Sb is a group Va element and the doped samples are p-type, considering the usual hydrogen-atom-like donor level is about 1.5 meV in $Hg_{1-x}Cd_xTe$ ($x = 0.39$), an Sb-doping-related acceptor level at about 30 meV, relative to the top of valence band, is deduced.

For the PL spectra recorded at higher temperature (above about 77 K), a weak feature on the low energy side of the A peak can be identified, which has been assigned as due to a bound-to-free transition. The impurity level involved is at about 30 meV. Because the hydrogen-atom-like shallow donor level is only about 2–3 meV in $Hg_{1-x}Cd_xTe$ ($x = 0.39$), it can no longer bind electrons when the temperature is increased. Then the D°A° transition is replaced by a bound-to-free transition. The transition from an Sb acceptor to the conduction band has been observed in PL spectra. This result is in agreement with that obtained from D°A° and also with that estimated from a hydrogen atom analogy.

In summary, Sb-doped HgCdTe samples have been investigated by infrared FTIR PL spectroscopy and a localized exciton, D°A°, and band-to-band and bound-to-free transition related PL structures are observed. An Sb-doping-related acceptor level of about 30 meV at 4.2 K is obtained from these PL spectra.

2.5.4 Infrared PL in As-doped HgCdTe Epilayers

MBE-grown $Hg_{1-x}Cd_xTe$ thin films are now a key material system for infrared photodetector applications due to their high performance. Photovoltaic infrared photodetector structures possess rapid responsivities and hence ensure high-speed direct detection and even heterodyne detection. Because this structure can work under zero bias, it reduces power consumption. In addition this structure is suitable for realizing focal plane arrays.

For a p-on-n HgCdTe structure, a low surface leakage current and n-type minority carriers with long lifetime are predicted. Excellent detection efficiency is found for these structures (Bubulac et al. 1987; Arias et al. 1989; Harris et al. 1991; Shin et al. 1993). By choosing proper growth conditions, such as decreasing the growth temperature and increasing the Hg partial pressure, n-type $Hg_{1-x}Cd_xTe$ thin films with superior properties can be obtained. To fabricate photovoltaic infrared detectors, As ion implantation is often used for doping. Under a high Hg partial pressure and with a suitable annealing temperature, As will occupy Te sites (Maxey et al. 1993) to yield the p-type surface layer of a p-on-n structure. Therefore, it is necessary to investigate the As-doped properties of $Hg_{1-x}Cd_xTe$.

The samples used by Chang (1995) were prepared by first growing a 5-μm-thick CdTe buffer layer on a GaAs substrate, then growing an n-type $Hg_{1-x}Cd_xTe$ film on the buffer layer. Then As-ions were implanted and the sample annealed under suitable time and temperature conditions to produce a 3-μm-thick p-type region at the surface. The samples were put into a dewar filled with liquid Helium. An Ar^+ laser ($\lambda = 514.5\,\text{nm}$) illuminated the sample. PL signals were collected by a parabolic mirror and sent into an FTIR spectrometer for analysis. Photocurrent and transmission experiments were also carried out using the same FTIR spectrometer. To make a comparison, the samples were measured again after a treatment by dilute Hydrobromil acid solutions for 1.5 min. Then a second comparative experiment was conducted. The PL signals from two $Hg_{1-x}Cd_xTe$ samples annealed at temperatures of 400 and 550°C but with a similar x and ion implantation conditions as before, were also recorded. The transmission spectra of another sample (450°C annealing temperature) before and after etching are displayed in Fig. 2.81.

From the energetic positions of the interference peaks, the thickness d of the $Hg_{1-x}Cd_xTe$ epitaxial layers can be estimated and are 10 and 7 μm, respectively. Furthermore, the concentration x of the samples near their surfaces can also be determined to be approximately 0.39 (Chu et al. 1983). By using the relation between the band gap E_g and temperature T (4.28 K) (Chu et al. 1994), the band gap at different temperatures can be calculated, which locates the position of the PL peak corresponding to a band-to-band transition. Figure 2.82 shows the photocurrent spectra of an infrared detector linear array made on this HgCdTe epilayer. These results are similar to those of the transmission spectra. Figure 2.83 shows the PL spectra of a sample taken at an identical excitation power but at different temperatures. Four PL peaks are clearly seen, which are labeled A, B, C, and D.

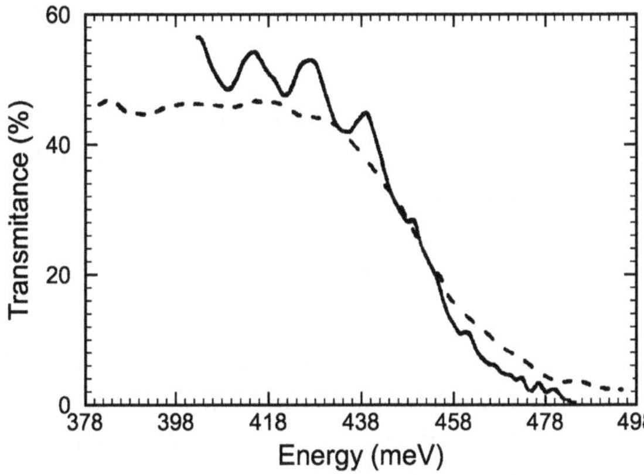

Fig. 2.81 Transmission spectra of a sample recorded before (*solid lines*) and after (*dotted line*) etching

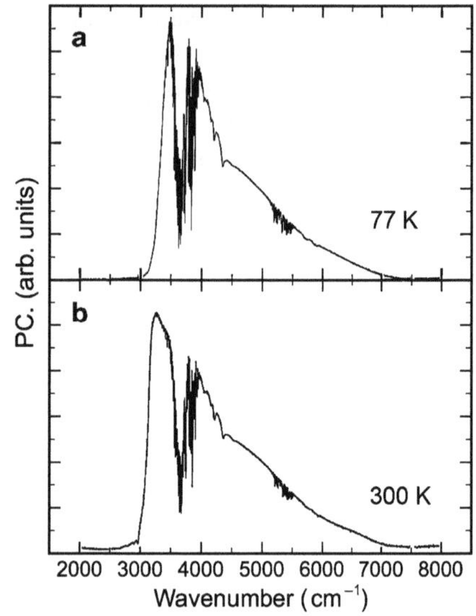

Fig. 2.82 Photocurrent spectra of a photosensitive unit of a p-on-n infrared detector linear array under zero bias. The unit was made on a HgCdTe epilayer after an As-ion implantation and subsequent anneal

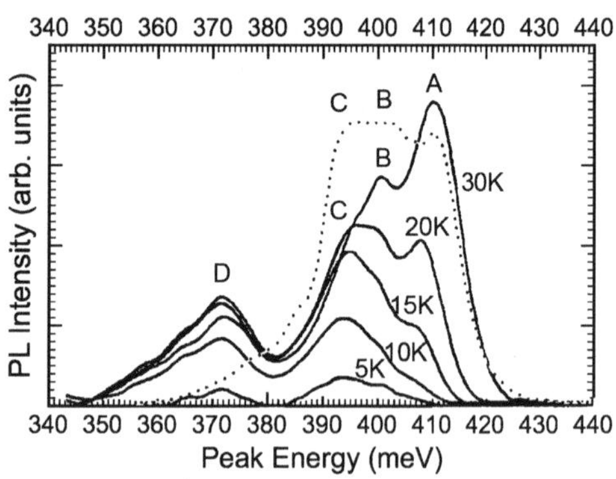

Fig. 2.83 Temperature-dependent PL spectra of a HgCdTe sample annealed at 450°C after As-ion implantation. The dotted line refers to the PL spectrum of a HgCdTe sample annealed at 400°C

In Fig. 2.84, the relation between the peak intensity $I_{\mathrm{P.PL}}$ and the excitation power P_{I} is given for the spectra depicted in Fig. 2.83. It is shown that the peak intensity as a function of the excitation power is:

$$I_{\mathrm{P \times PL}} = C P_{\mathrm{I}}^{\mathrm{v}}, \tag{2.127}$$

where C is a constant, and v is the power exponent. From the value of v, we can determine approximately the transition mechanisms responsible for the PL peaks.

The energy of peak A is close to E_{g}. Its intensity is weak at low-temperatures ($T < 20$ K) and gets stronger at higher temperatures ($T > 20$ K). There are no further PL peaks appearing on the high energy side of peak A. Therefore, peak A can be interpreted as the PL peak due to band-to-band transitions. The other peaks are identified based on the interpretation of peak A. The PL spectrum of sample 1 after

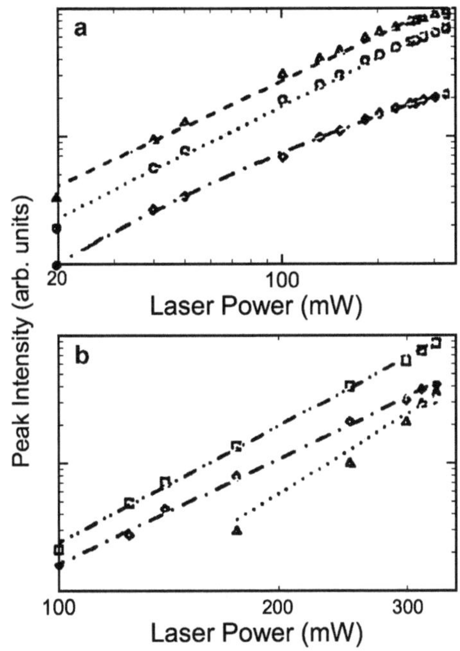

Fig. 2.84 Integrated intensity of the main PL peaks as a function of the excitation power. The results for the sample before and after etching are given in (**b**) and (**a**), respectively

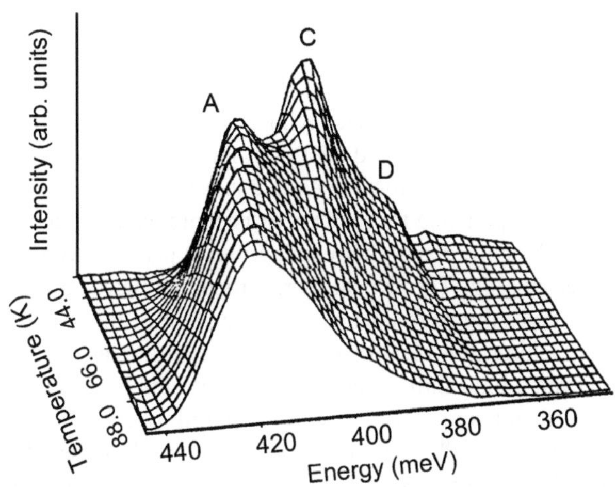

Fig. 2.85 Temperature-dependent PL spectra for samples after etching

etching is given in Fig. 2.85. The thickness of the sample was reduced to 3 μm. In general, for $Hg_{1-x}Cd_xTe$ using the same injection dose, but after annealing, according to secondary ion mass spectrometry (SIMS) data, the impurity density in a layer 3 μm under the surface decreases by 2–3 orders relative to the injection dose. Under this condition, the dose injected impurity layer is basically removed. It is obvious from Fig. 2.85 that the peaks B (unobservable) and D are reduced and peak C is enhanced relative to peak A. Thus the peaks B and D are related to the As impurity and the peak C is unrelated to the As impurity. Figure 2.84 gives the integrated intensity of main PL peaks before and after etching, as a function of the excitation power. The energy and the feature of the peak A are almost unchanged before and after etching. Therefore, it is still related to a band-to-band transition. The energy of peak C as a function of temperature is similar to that of peak A. Moreover, as the excitation power increases, the energy of the peak C does not change appreciably. The peak intensity as a function of excitation power is depicted in Fig. 2.84. The value of ν for peak C is slightly smaller than that of peak A. As a result, peak C may originate from a bound exciton. Its peak energy is 16–17 meV lower than that of the peak A. For $Hg_{1-x}Cd_xTe$ with $x = 0.39$, the binding energy of an exciton is approximately 1.5 meV, which does not account for the observations. Thus, there is an impurity level involved and its energy is in the range of about 14.5–15.5 meV. By comparing the above result to those reported in the other references (Hunter and McGill 1981, 1982; Tomm et al. 1990, 1994) the peak C can be ascribed to the PL of an acceptor level bound exciton due to a Hg vacancy.

From the temperature and excitation power dependences of their peak energy and intensity, the B and D peaks are found to be related to bound excitons. The B peak is strong for the sample before etching, but not distinguishable after etching. The C peak is not so strong before etching, but enhanced significantly after etching. It is believed that before removing the implanted layer by etching, As impurities occupy Hg vacancy sites and hence the density of Hg vacancies is reduced. Therefore, the intensity of the C peak is enhanced and the peak B is reduced by etching. It is concluded that the residual low intensity B PL peak detected after annealing is a donor level bound exciton of an As occupying a cation site. According to the energy of the B peak and the exciton bound energy in $Hg_{1-x}Cd_xTe(x = 0.39)$, it can be inferred that As occupying a cation site has a donor level of 8.5 meV.

The mechanism responsible for the D peak is complicated. According to a comparison of the results before and after etching, the D peak is strong before etching and reduced after etching. The D peak is regarded as related to the ion implantation. Figure 2.84 gives the peak intensity as a function of the excitation power for the sample after etching. It is clear that the D peak saturates as the excitation power increases. However, a blue shift of the D peak is detected for low excitation power. These are typical features of a D–A pair PL (Schmidt et al. 1992a, b). Thus the D peak can be ascribed to a PL peak due to a D–A pair. However, if the donor and acceptor are highly compensated in these samples, and the change with excitation power is not significant, the D–A pair PL would probably not be observable. After being etched, the concentration of As decreases by 2–3 orders, which results in a decrease of the compensation, and hence the D peak Fig. 2.84 may indeed represent a D–A pair feature.

A further comparison was conducted in an effort to validate the proposed origin of the D peak. The PL spectra of samples annealed at a temperature of 400°C are depicted in Fig. 2.83. From these spectra, only the peaks A, B, and C are observable. The D peak is almost undistinguishable. It may be concluded that the D peak is related not only to the ion implanted As, but also to the annealing temperature.

For the implanted As in $Hg_{1-x}Cd_xTe$, the As impurity will occupy a Te site as an acceptor with large probability when the samples are annealed at a high annealing temperature and Hg partial pressure. Therefore, the D peak is a D–A pair PL related to the As donor when occupying a cation site, and becomes an acceptor level when it occupies a Te site after the As ion implantation and an annealing process. The contributing donor level for the PL is supplied by some As still occupying cation sites.

From the energy of the D peak and the donor level due to an As occupying a cation site (8.5 meV), it is concluded that the As acceptor on a Te site in $Hg_{1-x}Cd_xTe(x = 0.39)$ results in an acceptor level at approximately 31.5 meV.

In summary, the impurity behavior of As-doped MBE-grown HgCdTe films were investigated by using PL spectroscopy. For As ion-doped $Hg_{1-x}Cd_xTe(x = 0.39)$ samples with different annealing temperatures, the PL peaks related to band-to-band, bound exciton, and D–A pair transitions were detected. Two energy levels related to As ion implantation were determined. The donor level due to an As occupying a cation site is located at approximately 8.5 meV below the bottom of the conduction band. The acceptor level due to an As occupying a Te site is located at approximately 31.5 meV above the top of the valance band. With the help of these PL spectra, the amphoteric behavior of high concentrations of implanted As ions were directly observed.

2.5.5 *Behavior of Fe as an Impurity in HgCdTe*

Recently, doping of transition and rare earth elements into $Hg_{1-x}Cd_xTe$, has been carried out. By using heavy doping to form four element alloys, we can strengthen the weak Hg bond to be more appropriate for device applications. Also, in device technology of $Hg_{1-x}Cd_xTe$, some Fe contamination is unavoidable. The behavior of Fe-doped $Hg_{1-x}Cd_xTe$ has been examined by infrared transmission spectroscopy, PL spectroscopy and Hall measurements. The results indicate that Fe impurities lead to a donor level which is about 80 meV below the bottom of the conduction band. In spite of exhibiting no electrical activity at low-temperature ($T < 180$ K), Fe impurity centers can become the main nonradiative recombination mechanism for nonequilibrium carriers. Thus, it can reduce the lifetime of nonequilibrium carriers to adversely influence $Hg_{1-x}Cd_xTe$ device performance.

The Fe-doped $Hg_{1-x}Cd_xTe(x = 0.31)$ has been cut from bulk material grown by the moving heater method (Chang et al. 1997). In these samples the doping concentration of Fe can be above 10^{18} cm^{-3}. The Van der Paul method was applied in a Hall measurement. The measurement was done from 77 to 300 K and with a magnetic density of 0.1 T. Infrared transmission spectroscopy was measured from 1.9 to 300 K by using an Oxford 104 liquid Helium Dewar and a globar light source. To reduce the thermal background interference, the double modulation technique was used to obtain highly sensitive PL spectra. PL spectra were measured from 3.9 to 300 K.

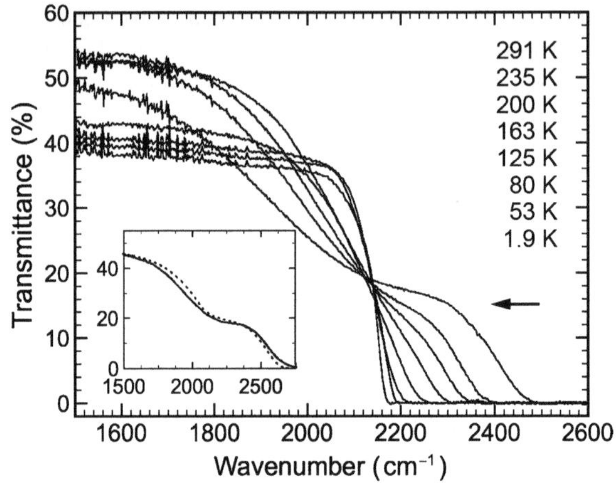

Fig. 2.86 Infrared transmission spectra of an Fe-doped $Hg_{1-x}Cd_xTe(x = 0.31)$ bulk sample at different temperatures. The *dotted line* in the *inset* shows the calculated transmittance taking into account the Urbach tail absorption and ionized impurity absorption

Figure 2.86 shows the infrared transmission spectra of an Fe-doped $Hg_{1-x}Cd_xTe$ at different temperatures. We can clearly observe impurity related absorption below the Urbach tail absorption at higher temperatures. This absorption structure is obvious at room temperature, becomes weaker with decreasing temperature, and disappears by a temperature of $\sim 3.9\,K$. From this temperature dependence, we can conclude it is due to an ionized impurity. Absorption is enhanced as the temperature increases and the impurities thermally ionize. Because the density of states of impurities is much smaller than the density of states of the conduction and the valence bands, its absorption coefficient is much smaller than that of band-to-band transitions. It only takes the form of an absorption shoulder appended onto the Urbach absorption tail. From a combination of the transitions from ionized donor impurity to the conduction band and band-to-band absorption, the infrared transmission spectra can be fitted by using the formula (4.2–300 K):

$$T_r = \frac{(1-R)^2 \exp(-\alpha d)}{1 - R^2 \exp(-2\alpha d)}, \tag{2.128}$$

where T_r is the transmittance, α the absorption coefficient, d the thickness of the sample, $d = 0.75$ mm, and R is reflectance. R for normal incidence is written as:

$$R = \frac{(n-1)^2 + k^2}{(n+1)^2 + k^2}, \tag{2.129}$$

where k is the extinction coefficient and n is the index of refraction. n is given by the empirical formula (Liu et al. 1994):

$$n(\lambda, T) = A + \frac{B}{1 - (C/\lambda)^2} + D\lambda^2. \tag{2.130}$$

A, B, C, and D are given in Eq. (4.30) of Vol. I, Sect. 1, Chap. 4. $\lambda(\mu\text{m})$ is the wavelength related to the photon energy $\hbar\omega$ (meV) via:

$$\lambda = 1.24 \times 10^3/(\hbar\omega), \tag{2.131}$$

and the extinction coefficient k is written as:

$$k = \frac{\lambda}{4\pi}\alpha. \tag{2.132}$$

In fact, the influence of k is very small.

In the impurity absorption distribution the main parameter in the fitting calculation is α and it can be regarded as a superposition of two absorption mechanisms:

$$\alpha = \alpha_{im} + \alpha_{in}, \tag{2.133}$$

where $\alpha_{\rm in}$ is the intrinsic inter-band absorption coefficient, and $\alpha_{\rm im}$ is the ionized impurity absorption coefficient. $\alpha_{\rm in}$ near the absorption edge is given by an empirical formula:

$$\alpha_{in} = \alpha_0 \exp\left[\frac{\delta}{k_B T}(E - E_0)\right]. \tag{2.134}$$

The parameters of (2.134) are in Eq. (4.120) of Vol. I, Sect. 4.3.

For the transition from an ionized donor impurity to the valence band, the absorption coefficient $\alpha_{\rm im}$ can be obtained from the proportionality expression (Seeger 1985):

$$\alpha_{im}\hbar\omega \propto \frac{N_D\sqrt{\hbar\omega - (E_g - \Delta\varepsilon_D)}}{g_D \exp\left[-(\varepsilon_D - E_F)/k_B T\right]}. \tag{2.135}$$

Furthermore, (2.135) can be written as a equality:

$$\alpha_{im} = \frac{C\sqrt{\hbar\omega - (Eg - \Delta\varepsilon_D)}}{\hbar\omega\left(g_D \exp\left[-(\varepsilon_D - E_F)/k_B T\right]\right)}, \tag{2.136}$$

where $N_{\rm D}$ is the concentration of donor impurities, $\hbar\omega$ the photon energy, $E_{\rm g}$ the band gap, $k_{\rm B}$ the Boltzmann constant, $g_{\rm D}$ the energy level degeneracy of an Fe donor impurity, $\varepsilon_{\rm D}$ the donor energy, $\Delta\varepsilon_{\rm D}$ is the donor ionization energy, $\Delta\varepsilon_{\rm D} = E_{\rm g} - \varepsilon_{\rm D}$, and C is a fitting parameter. The Fermi level $E_{\rm F}$ can be calculated from the impurity concentration obtained from a Hall measurement. The calculated results from transmission spectra at 300 K are given in the inset to Fig. 2.86. In the fitting calculation, only $\Delta\varepsilon_{\rm D}, \varepsilon_{\rm D}, x, C$, and $g_{\rm D}$ are adjustable parameters. The absorption edge of the transmission spectra is determined by the concentration x. The impurity absorption portion is mainly due to the impurity ionization energy. C is a proportionality parameter. The calculation is not sensitive to $g_{\rm D}$. In general, $g_{\rm D} = 2$. After the fitting is done, the composition x is 0.31 and the ionization energy $\Delta\varepsilon_D$ is 80 meV. As a result, the donor level deduced for the Fe impurity in $Hg_{1-x}Cd_xTe$ ($x = 0.31$) lies at about 80 meV below the conduction band edge. This energy is near $E_{\rm g}/4$ so it is a deep level.

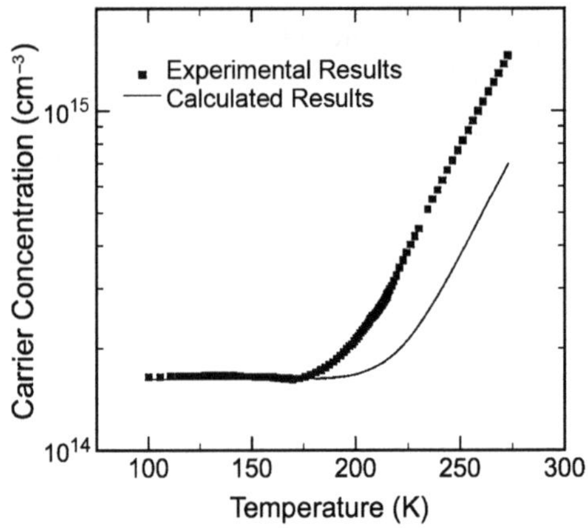

Fig. 2.87 Hall measurement results for an Fe-doped $Hg_{1-x}Cd_xTe(x = 0.31)$ at 77 and 300 K. The solid line is calculated using the measured intrinsic carrier concentration. The difference between the experimental and calculated results can be explained as due to the thermally induced carrier ionization of the Fe-related level

Because the Fe-related level is a deep level, it has little influence on the electrical conductivity of the material at low-temperature. The Hall measurement results support this position. Figure 2.87 gives the results of Hall measurements between 77 and 300 K. By using the Hall coefficients in the flat region at low-temperature, we can calculate the effective shallow donor concentration which is $1.06 \times 10^{14}\,\text{cm}^{-3}$. The carrier concentration is calculated using (2.137) for an Fe donor contribution of N_D^*:

$$n_i^2 = n(n - N_D^*), \tag{2.137}$$

where the intrinsic carrier concentration n_i is given by (2.138) (Chu et al. 1991). A detailed derivation of this formula is given in Sect. 3.1.

$$n_i = 9.56 \times 10^{14}\left(1 + \frac{3.25k_BT}{E_g}\right)E_g^{3/2}T^{3/2}\left[1 + 1.9E_g^{3/4}\exp\left(\frac{E_g}{2k_BT}\right)\right]^{-1} \tag{2.138}$$

From the result calculated in Fig. 2.87, we find that when the temperature exceeds 180 K, the carrier concentration obtained from experiments is higher than that calculated. This occurs because the Fe impurities ionize at the higher temperature. Then electrons are excited from the impurity levels to the conduction band where they contribute to the electrical transport. This allows some electrons to transit from the valence band to the ionized Fe levels. Thus, ionized impurity absorption will occur at these higher temperatures. In this case, we can observe an absorption shoulder in the infrared transmission spectra. At lower temperatures, this absorption is weak and not observed because the chance of impurity ionization is small.

While an Fe impurity has very little effect on electrical transport at low-temperature and is not clearly observed in the infrared transmission spectra, it could have an effect on the lifetime of nonequilibrium carriers and in the PL spectra taken at low-temperature. A DM-PL spectrum is given in Fig. 2.88. Only one PL

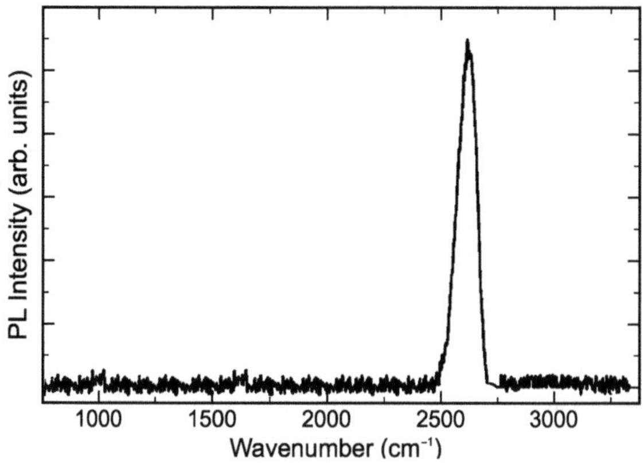

Fig. 2.88 DM-PL spectra of an Fe-doped HgCdTe sample at 4.2 K. The excitation power density is 8 W/cm^2

peak can be seen and its energy is independent of the excitation power. Furthermore, the integrated intensity of the PL peak shows no tendency to saturate with increased excitation power. Therefore, we think this single peak with its high energy is due to a band-to-band transition or a free exciton (Tomm et al. 1990, 1994). Because the bond energy of the free exciton in HgCdTe is small, less than 3 meV (Brice and Capper 1987) it lies within the band width of the observed peak. In any case, the PL peak is unrelated to the Fe doping because the Fe-related level is deep, equal to 1/4 of the energy gap it would produce a peak well below the one that is observed. No other PL peak is found in this energy position. However, from the following analysis it can be seen that Fe at low-temperature contributes a nonradiative recombination mechanism for nonequilibrium carriers, which has a significant influence on device applications. The integrated intensity of the PL peak decreases as the temperature is increased higher than 3.9 K. It indicates that there is a thermally activated nonradiative recombination mechanism. To investigate the thermally activated quenching mechanism of the PL peak, we fit the integrated intensity of the PL peak to a function of temperature when the temperature exceeds 60 K. The fitting formula used is:

$$I_{PL} = I_0\left[1 + C\exp\left(-\frac{E}{k_B T}\right)\right]^{-1}, \tag{2.139}$$

where $I_{\rm PL}$ is PL intensity, I_0 and C constants, and E is the thermal activation energy of the PL quenching process. The best-fit yields an activation energy E of 78.3 meV, which is consistent with the results calculated from infrared transmission spectra (80 meV). It indicates that an Fe-related level is the main cause of the PL quenching. Because the Fe-related level is a deep level, it can be the intermediate step of an indirect recombination mechanism for the nonequilibrium carriers. Then, it serves as an effective nonradiative center and has a significant effect on the rate of the recombination of photo-excited nonequilibrium carriers.

Ye (1995) measured the lifetime of the nonequilibrium carriers of an Fe-doped HgCdTe sample by using a microwave reflectance technique. They found the averaged lifetime is 0.39 μs, which is shorter than that of n-type materials of high quality. When they measured a surface distribution of the lifetimes, they found the maximum is 1.84 μs, and the minimum is 1.08 μs. The averaged relative deviation of all the test points is less than 10%, which represents a good uniformity (Ye 1995). Based on a topography analysis technique, Cai et al. (1994, 1995) found dislocations in bulk HgCdTe crystals resulting from stress introduced in the process of growth and post-growth treatments. These processes destroy the crystal symmetry and induce deep levels in the band gap to form recombination centers. The distribution of the dislocations is generally nonuniform, which leads to a discreteness of the lifetime distribution of the samples. Doping can improve the yield stress of semiconductor materials (Farges 1990). Fe-doping results in a strong interaction between the Fe atoms and the atoms in HgCdTe, which improves the yield stress of HgCdTe materials. It reduces the dislocation occurrence probability and improves the lifetime in the samples. These problems need further study.

By using optical and electrical methods, the behavior of an Fe-doped $Hg_{0.31}Cd_{0.69}Te$ crystal was investigated. The relation between the PL intensity and temperature was established as shown in Fig. 2.89. With this relation, a thermally activated nonradiative recombination centers properties were clarified in the Fe-doped samples, which has a donor level at about $1/4\ E_g$. This value is consistent with that of the deep donor level observed in the infrared transmission spectra. The results of Hall measurements are also explained. The experimental results indicate that Fe-doping has little effect on the low-temperature electrical properties of these materials. However, the Fe-related level ($E_g/4$) is the main recombination center at higher temperatures and leads to recombination of the nonequilibrium carriers in these materials.

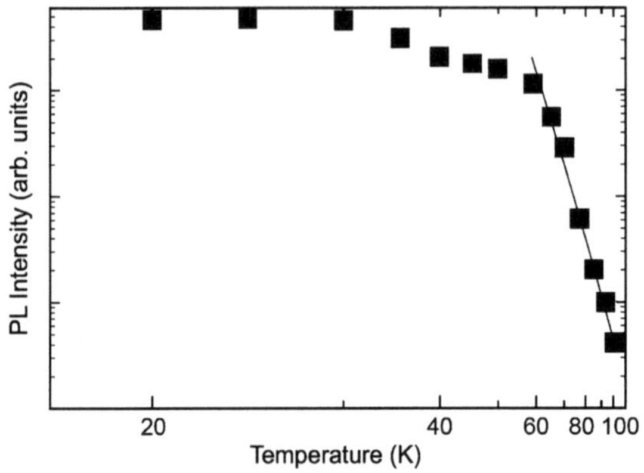

Fig. 2.89 The integrated PL intensity vs. temperature for an Fe-doped HgCdTe sample (square points). A fitted result for temperatures higher than 60 K is give by the solid line. The slope of the solid line is the thermal activation energy related to PL quenching. A best fit yields an energy of 78.3 meV

References

Arias JM, Shin SH, Pasko JG, DeWames RE, Gertner ER (1989) Long and middle wavelength infrared photodiodes fabricated with $Hg_{1-x}Cd_xTe$ grown by molecular-beam epitaxy. J Appl Phys 65:1747

Baars J, Seelewind H, Fritzsche Ch, Kaiser U, Ziegler J (1988) Arsenic ion implantation in $Hg_{1-x}Cd_xTe$. J Cryst Growth 86:762–767

Bartlett BE, Capper P, Harris JE, Quelch MJT (1980) Factors affecting the electrical characteristics of cadmium mercury telluride crystals. J Cryst Growth 49:600–606

Bartoli FJ, Hoffman CA, Meyer JR (1986) Characterization of impurities in p-type HgCdTe by photo-Hall techniques. J Vac Sci Technol A 4:2047–2050

Beck JD, Kinch MA, Esposito EJ, Chapman RA (1982) The MIS physics of the native oxide–$Hg_{1-x}Cd_xTe$ interface. J Vac Sci Technol 21:172–177

Berding MA, Sher A (1998a) Electronic quasichemical formalism: application to arsenic deactivation in silicon. Phys Rev B 58:3853–3864

Berding MA, Sher A (1999a) Arsenic incorporation during MBE growth of HgCdTe. J Electron Mater 28:799–803

Berding MA, Sher A (1999b) Amphoteric behavior of arsenic in HgCdTe. Appl Phys Lett 74: 685–687

Berding MA, van Schilfgaarde M, Sher A (1993) $Hg_{0.8}Cd_{0.2}Te$ native defects: densities and dopant properties. J Electron Mater 22:1005–1010

Berding MA, van Schilfgaarde M, Sher A (1994) First-principles calculation of native defect densities in $Hg_{0.8}Cd_{0.2}Te$. Phys Rev B 50:1519–1534

Berding MA, Sher A, van Schilfgaarde M (1995) Defect modeling studies in HgCdTe and CdTe. J Electron Mater 24:1127–1135

Berding MA, Sher A, van Schilfgaarde M (1997) Behavior of p-type dopants in HgCdTe. J Electron Mater 26:625–628

Berding MA, Sher A, van Schlifgaarde M (1998b) Lithium, sodium, and copper in $Hg_{0.78}Cd_{0.22}Te$ and CdTe-based substrates. J Electron Mater 27:573–578

Berding MA, Sher A, van Schilfgaarde M, Chen AB, Arias J (1998c) Modeling of arsenic activation in HgCdTe. J Electron Mater 27:605–609

Berglund CN (1966) Surface states at steam-grown silicon-silicon dioxide interfaces. IEEE Trans Electron Devices 13:701–705

Binsma JJM, Giling LJ, Bloem J (1982) Luminescence of $CuInS_2$ I: the broad band emission and its dependence on the defect chemistry. J Lumin 27:35–53

Bowers R, Yafet Y (1959) Magnetic susceptibility of InSb. Phys Rev 115:1165–1172

Brice JC, Capper P (1987) In: Brice JC, Capper P (eds) EMIS datareviews series. INSPEC Inst of Electrical Eng, London

Bubulac LO, Tennant WE, Edwall DD, Gertner ER, Robinson JC (1985) Cathodoluminescence of HgCdTe and CdTe on CdTe and sapphire. J Vac Sci Technol A 3:163–170

Bubulac LO, Lo DS, Tennant WE, Edwall DD, Chen JC, Ratusnik J, Robinson JC, Bostrup G (1987) p on n ion-implanted junctions in liquid phase epitaxy HgCdTe layers on CdTe substrates. Appl Phys Lett 50:1586–1588

Bye KL (1979) An X-ray topographic assessment of cadmium mercury telluride. J Mater Sci 14:619–625

Cai Y (1986) An X-ray topographic measurement of $Hg_{1-x}Cd_xTe$ defects. Infrared Technol 8: 47–48

Cai Y, Zheng GZ, Zhu XC, Jiang JH, Tang DY (1994) Synchrotron radiation topography analysis of multi-element HgCdTe detector linear arrays. J Infrared Millimeter Waves 13:385–390

Cai Y, Zheng GZ, Zhu XC, Jiang JH, Tang DY (1995) Synchrotron radiation topography analysis of defects in small CMT slices. Infrared Technol 17:9–13

Capper P (1982) The behaviour of selected impurities in $Cd_xHg_{1-x}Te$. J Cryst Growth 57:280–299

Capper P (1989) Bridgman growth of $Hg_{1-x}Cd_xTe$:A review. Prog. Cryst Growth Charact 19:259–293

Capper P (1991) A review of impurity behavior in bulk and epitaxial. J Vac Sci Technol B 9:1667

Capper P, Gosney JJG, Jones CL, Kenworthy I, Roberts JA (1985) Acceptor doping of bridgman-grown $Cd_xHg_{1-x}Te$. J Cryst Growth 71:57–65

Casselman TN (1981) Calculation of the auger lifetime in p-type $Hg_{1-x}Cd_xTe$. J Appl Phys 52:848–584

Casselman TN, Sher A, Silberman J, Spicer WE, Chen AB (1983) On the determination of the energy band offsets in $Hg_{1-x}Cd_xTe$ heterojunctions. J Vac Sci Technol 1:1692–1695

Chadi J (1977) Localized-orbital description of wave functions and energy bands in semiconductors. Phys Rev B 16:3572–3578

Chang Y (1995) Ph.D. thesis, Shanghai Institute of Technical Physics

Chang Y, Chu JH, Tang WG, Shen WZ, Tang DY (1997) Spectroscopic and electrical investigation on Fe-doped $Hg_{1-x}Cd_xTe$. Chin J Semicond 18:258–263

Chen AB (1977) Simple brillouin-zone scheme for the spectral properties of solids. Phys Rev B 16:3291–3302

Chen AB, Sher A (1980) Electronic structure of III–V semiconductors and alloys using simple orbitals. Phys Rev B 22:3886–3896

Chen AB, Sher A (1981a) Electronic structure of pseudobinary semi-conductor alloys $Al_xGa_{1-x}As$, GaP_xAs_{1-x}, and $Ga_xIn_{1-x}P$. Phys Rev B 23:5360–5374

Chen AB, Sher A (1981b) Calculation of the electronic properties of pseudobinary semiconductor alloys. Phys Rev B 23:5645–5648

Chen AB, Sher A (1985) Sensitivity of defect energy levels to host band structures and impurity potentials in CdTe. Phys Rev B 31:6490–6497

Chen MC, Dodge JA (1986) Electrical properties of antimony-doped p-type $Hg_{0.78}Cd_{0.22}Te$ liquid-phase-epitaxy films. Solid State Commun 59:449–452

Chen MC, Tregilgas JH (1987) The activation energy of copper shallow acceptors in mercury cadmium telluride. J Appl Phys 61:787–789

Chen W (1990) Master thesis, Shanghai Institute of Technical Physics

Chen YP, Zheng GZ, Gong YQ, Guo SL, Chen JX, Tang DY (1990) Composition dependence of hot electron effect in $Hg_{1-x}Cd_xTe$. Semicond Sci Technol 5:S304–306

Cheung DT (1985) An overview on defect studies in MCT. J Vac Sci Technol A 3:128–130

Chu JH (1988) Fachkolloquium "II-VI-Halbleiter". Jahrestagung der DGKK, Karlsruhe, Marz

Chu JH, Xu SC, Tang DY (1983) Energy gap versus alloy composition and temperature in $Hg_{1-x}Cd_xTe$. Appl Phys Lett 43:1064–1066

Chu JH, Mi ZY, Tang DY (1991) Intrinsic absorption spectroscopy and related physical quantities of narrow-gap semiconductors $Hg_{1-x}Cd_xTe$. Infrared Phys 32:195–211

Chu JH, Mi ZY, Sizmann R, Koch F, Wollrab R, Ziegler J, Maier H (1992) Influence of resonant defect states on subband structures in $Hg_{1-x}Cd_xTe$. J Vac Sci Tech B 10:1569–1573

Chu JH, Li B, Liu K, Tang DY (1994) Empirical rule of intrinsic absorption spectroscopy in $Hg_{1-x}Cd_xTe$. J Appl Phys 75:1234–1235

Cole A, Carey GP, Silberman JA, Spicer WE, Wilson JA (1985) Surface and bulk structural defects in $Hg_{1-x}Cd_xTe$. J Vac Sci Technol A 3:206–211

Collins RT, McGill TC (1983) Electronic properties of deep levels in p-type CdTe. J Vac Soc Techol A 1:1633

Cotton VA, Wilson JA (1986) Effects of ion implantation on deep electron traps in $Hg_{0.7}Cd_{0.3}Te$. J Vac Sci Technol A 4:2177–2180

Datsenko LI, Skorokhod M, Kislovski EN et al (1985) Inorg Mater 20:1575

Dean BE, Johnson CJ, McDevitt SC, Neugebauer GT, Sepich JL, Dobbyn RC, Kuriyama M, Ellsworth J, Vydyanath HR, Kennedy JJ (1991) Correlation of HgCdTe epilayer defects with underlying substrate defects by synchrotron X-ray topography. J Vac Sci Technol B 9: 1840–1846

Destefanis GL (1985) Indium ion implantation in $Hg_{0.78}Cd_{0.22}Te/CdTe$. J Vac Sci Technol A 3:171–175

Destefanis GL (1988) Electrical doping of HgCdTe by ion implantation and heat treatment. J Cryst Growth 86:700–722

Dornhaus R, Nimtz G (1983) Narrow-gap semiconductors. Springer tracts in modern phys, vol 98. Springer, Heidelberg, p130

Dornhaus R, Nimtz G, Schlabitz W, Burkhard H (1975) Resonant level in semiconducting $Hg_{1-x}Cd_xTe$. Solid State Commun 17:837–841

Easton BC, Maxey CD, Whiffin PAC, Roberts JA, Gale IG, Grainger F, Capper P (1991) Impurities and metalorganic chemical-vapor deposition growth of mercury cadmium telluride. J Vac Sci Technol B 9:1682

Elliott CT, Melngailis I, Harman TC, Foyt AG (1972) Carrier freeze-out and acceptor energies in nonstoichiometric p-type HgCdTe. J Phys Chem Solids 33:1527–1531

Farges JP (1990) In: Willardson RK, Beer AC (eds) Semicond and semimetals, vol 31. Academic, New York

Fuchs F, Koidl P (1991) Carrier localization in low-bandgap $Hg_{1-x}Cd_xTe$ crystals, studied by photoluminescence. Semicond Sci Technol 6:C71–C75

Fuchs F, Lusson A, Wagner J, Koidl P (1989) Double modulation techniques in FTIR photoluminescence. SPIE 1145:323

Gely C, Corbel C, Triboulet R (1990) A study of vacancy-type defects by positron-lifetime measurements in a II–VI semiconductor: $Cd_{0.2}Hg_{0.8}Te$. J Phys: Condens Matter 2:4763–4767

Ghenim L, Robert JL, Bousquet C, Raymond A, Destefanis GL (1985) Evidence, by transport measurements under hydrostatic pressure, of resonant level induced by implantation defects in expitaxial layers of $Hg_{1-x}Cd_xTe$. J Crystal Growth 72:448–452

Glodeanu A (1967) Helium-like impurities in semiconductors. Phys Status Solidi 19:K43–K46

Göbel EO (1982) In: Pearsall TP (ed) InGaAs alloy semiconductors. Wiley, New York, Chapter 13

Gold MC, Nelson DA (1986) Variable magnetic field hall effect measurements and analyses of high purity, Hg vacancy (p-type) HgCdTe. J Vac Sci Technol A 4:2040–2046

Goldman VJ, Drew HD, Shayegan M, Nelson DA (1986) Observation of impurity cyclotron resonance in $Hg_{1-x}Cd_xTe$. Phys Rev Lett 56:968–971

Gorshkov AV, Zaitov FA, Shangin SB, Shalyapina GM (1984) Sov Phys Solid State (Engl Transl) 26:1787

Gough JS, Houlton MR, Irvine SJC, Shaw N, Young ML, Astles MG (1991) The growth and properties of In-doped metalorganic vapor phase epitaxy interdiffused multilayer process (HgCd)Te. J Vac Sci Technol B 9:1687–1690

Griffiths PR, Haseth JA (1986) Fourier transform infrared spectrometry. Wiley, New York, Chapter 1

Harris KA, Myers TH, Yanka RW, Mohnkern LM, Otsuka N (1991) A high quantum efficiency in situ doped mid-wavelength infrared *p*-on-*n* homojunction superlattice detector grown by photoassisted molecular-beam epitaxy. J Vac Sci Technol B 9:1752–1758

Hass KC, Ehrenreich E, Velicky B (1983) Electronic structure of $Hg_{1-x}Cd_xTe$. Phys Rev B 27:1088–1100

Herman MA, Pessa M (1985) $Hg_{1-x}Cd_xTe - Hg_{1-y}Cd_yTe$ ($0 \leqslant x, y \leqslant 1$) heterostructures: properties, epitaxy, and applications. J Appl Phys 57:2671

Higgins WM, Pultz GN, Roy RG, Lancaster RA, Schmit JL (1989) Standard relationships in the properties of $Hg_{1-x}Cd_xTe$. J Vac Sci Technol A 7:271–275

Huang H, Tong FM, Tang DY (1988) Infrared Research 7A:89

Hughes WC, Swanson ML, Austin JC (1991) Indium-Hg vacancy interactions in $Hg_{1-x}Cd_xTe$ measured by perturbed angular correlation. Appl Phys Lett 59:938

Hughes WC, Austin JC, Swanson ML (1994) The thermodynamics of indium-vacancy pairs in $Hg_{0.79}Cd_{0.21}Te$. J Cryst Growth 138:1040–1045

Hunter AT, McGill TC (1981) Luminescence from HgCdTe alloys. J Appl Phys 52:5779

Hunter AT, McGill TC (1982) Luminescence studies of HgCdTe alloys. J Vac Sci Technol 21: 205–207

Jones CE, Nair V, Polla DL (1981) Generation-recombination centers in p-type $Hg_{1-x}Cd_xTe$. Appl Phys Lett 39:248

Jones CE, Nair V, Lindquist JL, Polla DL (1982) Effects of deep-level defects in $Hg_{1-x}Cd_xTe$ provided by DLTS. J Vac Sci Technol 21:187–190
Jones CL, Capper P, Quelch MJT (1983) The properties of gold in Bridgman grown $Cd_xHg_{1-x}Te$. J Cryst Growth 64:417–432
Jones H (1934) Proc Roy Soc A 144:225
Kane EO (1976) Gaussian representations of covalent wave functions; silicon. Phys Rev B 13:3478–3483
Keyes RJ (1977) Optical and infrared detectors. Springer, Berlin
Kinch MA, Borrello SR (1975) 0.1 eV HgCdTe photodetectors. Infrared Physics 15:111–124
Kinch MA, Brau MJ, Simmons A (1973) Recombination mechanisms in 8–14-μ HgCdTe. J Appl Phys 44:1649
Klauer S, Wöhlecke M, Kapphan S (1992) Influence of H-D isotopic substitution on the protonic conductivity of $LiNbO_3$. Phys Rev B 45:2786
Kobayashi A, Sankey OF, Dow JD (1982) Chemical trends for defect energy levels in $Hg_{1-x}Cd_xTe$. Phys Rev B 25:6367–6379
Krishnamurthy S, Berding MA, Yu ZG (2006) Minority carrier lifetimes in HgCdTe alloys. J Electron Mater 35:1369–1378
Kurilo IV, Kuchma VI (1982) Twinning in CdTe and $Cd_xHg_{1-x}Te$ crystals. Inorg Mater 18:479
Kurtz SR, Bajaj J, Edwall DD, Irvine SJC (1993) Infrared photoluminescence characterization of long-wavelength HgCdTe detector materials. Semicond Sci Technol 8:941–945
Lang DV (1974) Deep-level transient spectroscopy: a new method to characterize traps in semiconductors. J Appl Phys 45:3023
Lannoo M, Bourgoin J (1981) Point defects in semiconductors I: theoretical aspects. Springer, New York, Chap. 2
Li B, Gui YS, Chen ZH, Ye HJ, Chu JH, Wang SL, Ji RB, He L (1998) Study of impurity states in p-type $Hg_{1-x}Cd_xTe$ using far-infrared spectroscopy. Appl Phys Lett 73:1538
Liu K, Chu JH, Tang DY (1994) Composition and temperature dependence of the refractive index in $Hg_{1-x}Cd_xTe$. J Appl Phys 75:4176
Lopes VC, Syllaios AJ, Chen MC (1993) Minority carrier lifetime in mercury cadmium telluride. Semicond Sci Technol 8:824–841
Losee DL (1972) Admittance spectroscopy of deep impurity levels: ZnTe Schottky barriers. Appl Phys Lett 21:54
Losee DL (1975) Admittance spectroscopy of impurity levels in Schottky barriers. J Appl Phys 46:2204
Lucovsky G (1965) On the photoionization of deep impurity centers in semiconductors. Solid State Commun 3:299–302
Lusson A, Fuchs F, Marfaing Y (1990) Systematic photoluminescence study of $Cd_xHg_{1-x}Te$ alloys in a wide composition range. J Crystal Growth 101:673–677
Maxey CD, Whiffin PAC, Easton BC (1991) MOVPE growth and characterization of doped $Cd_xHg_{1-x}Te$ structures. Semicond Sci Technol 6:c26–c30
Maxey CD, Gale LG, Clegg JB, Whiffin PAC (1993) Doping studies in MOVPE-grown $Cd_xHg_{1-x}Te$. Semicond Sci Technol 8:s183–s196
Mirsky U, Shechtman D (1980) Transmission electron microscopy and X-ray topography study of mercury-cadmium-telluride. J Electron Mater 9:933–943
Mosser V, Sizmann R, Koch F, Ziegler J, Maier H (1988) Quantum physics of the surface layer capacitance for narrow-gap HgCdTe. Semicond Sci Technol 3:808–812
Mott N, Jones H (1936) The theory of the properties of metals and alloys. Clarendon, Oxford
Murakami S, Okamoto T, Maruyama K, Takiqawa H (1993) Iodine doping in mercury cadmium telluride ($Hg_{1-x}Cd_xTe$) grown by direct alloy growth using metalorganic chemical vapor deposition. Appl Phys Lett 63:899
Myles CW (1987) The 1987 US workshop on the Physics and Chemistry of HgCdTe. New Orleans, LA, PO/DI-31
Nicollian EH, Brews TR (1982) MOS physics and technology. Wiley, New York
Petersen PE (1970) Auger recombination in $Hg_{1-x}Cd_xTe$. J Appl Phys 41:3465

Petersen PE (1983) In: Willardson RK, Beer AC (eds) Semiconductors and semimetals, vol 18. Academic, New York, chap. 6

Petrov VI, Gareeva AR (1988) Bull Acad Sci USSR Phys Ser (USA) 52:110

Polla DL, Jones CE (1980a) Admittance spectroscopy of deep levels in $Hg_{1-x}Cd_xTe$. J Appl Phys 51:6233

Polla DL, Jones CE (1980b) Deep level transient spectroscopy in $Hg_{1-x}Cd_xTe$. Solid State Commun 36:809

Polla DL, Jones CE (1981) Deep level studies of $Hg_{1-x}Cd_xTe$. I: narrow-band-gap space-charge spectroscopy. J Appl Phys 52:5118

Polla DL, Aggarwal RL, Mroczkowski JA, Shanley JF, Reine MB (1982) Observation of deep levels in $Hg_{1-x}Cd_xTe$ with optical modulation spectroscopy. Appl Phys Lett 40:338

Pratt RG, Hewett J, Capper P, Jones CL, Quelch MJ (1983) Minority carrier lifetime in n-type Bridgman grown $Hg_{1-x}Cd_xTe$. J Appl Phys 54:5152

Pratt RG, Hewett J, Capper P (1986) Minority-carrier lifetime in doped and undoped n-type $Cd_xHg_{1-x}Te$. J Appl Phys 60:2377

Reisinger AR, Roberts RN, Chinn SR, Myersll TH (1989) Photoluminescence of infrared-sensing materials using an FTIR spectrometer. Rev Sci Instrum 60:82

Rosback JP, Harper ME (1987) Doping and composition profiling in $Hg_{1-x}Cd_xTe$ by the graded capacitance–voltage method. J Appl Phys 62:1717

Rosemeier RG (1983) Real-time X-ray topography: applications to bulk HgCdTe materials. J Vac Sci Technol A 1:1656–1660

Ryssel H, Lang G, Biersack JP, Muller K, Kruger W (1980) Ion implantation doping of $Cd_{0.2}Hg_{0.8}Te$ for infrared detectors. IEEE Trans Electron Devices 27:58–62

Sasaki T, Oda N, Kawano M, Sone S, Kanno K, Saga M (1992) Mercury annealing effect on the electrical properties of HgCdTe grown by molecular beam epitaxy. J Cryst Growth 117: 222–226

Schaake HF (1986) Ion implantation damage in $Hg_{0.8}Cd_{0.2}Te$. J Vac Sci Technol A 4:2174–2176

Schaake HF (1988) Proc SPIT-Int Soc Opt Eng (USA) 946:186

Schacham SE, Finkman E (1985) Recombination mechanisms in p-type HgCdTe: freezeout and background flux effects. J Appl Phys 57:2001

Schmidt T, Lischka K, Zulehner W (1992a) Excitation-power dependence of the near-band-edge photoluminescence of semiconductors. Phys Rev B 45:8989–8994

Schmidt T, Daniel G, Lischka K (1992b) The excitation power dependence of the near band edge photoluminescence of II–VI semiconductors. J Cryst Growth 117:748–752

Schmit JL, Stelzer EL (1969) Temperature and alloy compositional dependences of the energy gap of $Hg_{1-x}Cd_xTe$. J Appl Phys 40:4865

Scott W, Stelzer EL, Hager RJ (1976) Electrical and far-infrared optical properties of p-type $Hg_{1-x}Cd_xTe$. J Appl Phys 47:1408

Seeger K (1985) Semiconductor physics: an introduction. Springer, Berlin

Shao J, Lu W, Lü X, Yue FY, Li ZF, Guo SL, Chu JH (2006) Modulated photoluminescence spectroscopy with a step-scan Fourier transform infrared spectrometer. Rev Sci Instrum 77:063104

Shen J, Chen JZ, Chen YQ (1980) Infrared Phys Technol 1:1

Shen XC (1992) Optical properties of semiconductor. Science, Beijing

Shen XC (1994) Prog Phys (Chinese) 4:452

Sher A, Hoffman HJ, Su P, Tsuo YH (1983) Techniques for improving the Si–SiO_2 interface characterization. J Appl Phys 54:5183

Shin SH, Chu M, Vanderwyck AHB, Lanir M, Wang CC (1980) Electrical properties of as-grown $Hg_{1-x}Cd_xTe$ epitaxial layers. J Appl Phys 51:3772

Shin SH, Arias JM, Edwell DD et al (1991) Proc 1991 US workshop on chemistry of MCT and II–VI compounds 69

Shin SH, Arias JM, Zandian M, Pasko JG, Bubulac LO, DeWames RE (1993) Annealing effect on the P-type carrier concentration in low-temperature processed arsenic-doped HgCdTe. J Electron Mater 22:1039–1047

Sladek RJ (1958) Magnetoresistance oscillations in single-crystal and polycrystalline indium arsenide. Phys Rev 110:817–826

Stahle CM, Helms CR, Simmons A (1987) Thermal stability of the anodic oxide/$Hg_{1-x}Cd_xTe$ interface. J Vac Sci Technol B 5:1092–1096

Swarts CA, Daw MS, McGill TC (1982) Bulk vacancies in $Cd_xHg_{1-x}Te$. J Vac Sci Technol 21:198

Swink LW, Brau MJ (1970) Rapid, nondestructive evaluation of macroscopic defects in crystalline materials: the laue topography of (Hg, Cd) Te. Metall Mater Trans B 1: 629–634

Sze SM (1969) Physics of semiconducting devices. Wiley, New York, p 429

Tang DY (1974) Infrared Phys Technol 16:345

Tang DY (1976) Infrared Phys Technol 25–26:53

Tang WG, Chang Y, Shen WZ (1995) Basic applications studies of HgCdTe materials and devices, 1995 collections of the theses. Shanghai Institute of Technical Physics, Shanghai, p 108

Tomm JW, Herrmann KH, Yunovich AE (1990) Infrared photoluminescence in narrow-gap semiconductors. Phys Stat Sol A 122:11–42

Tomm JW, Herrmann KH, Hoerstel W, Lindstaedt M, Kissel H, Fuchs F (1994) On the nature of the excitonic luminescence in narrow-gap $Hg_{1-x}Cd_xTe$ ($x \cong 0.3$). J Cryst Growth 138:175–181

Tsau GH, Sher A, Madou M, Wilson JA, Cotton VA, Jones CE (1986) Low-frequency admittance measurements on the HgCdTe/Photox SiO_2 interface. J Appl Phys 59:1238

Vodop'yanov LK, Kozyrev SP, Spitsyn AV (1982a) Ion implantation of impurities in n-type $Hg_{0.8}Cd_{0.2}Te$. I Group II ions Mg^+, Zn^+, Cd^+. Sov Phys Semicond (USA) 16:502–506

Vodop'yanov LK, Kozyrev SP, Spitsyn AV (1982b) Ion implantation of impurities in n-type $Hg_{0.8}Cd_{0.2}Te$. II. group III ions Al^+ and Ga^+. Sov Phys Semicond (USA) 16:626–629

Vydyanath HR (1990) Amphoteric behaviour of group V dopants in (Hg, Cd)Te. Semicond Sci Technol 5:S213–S216

Vydyanath HR (1991) Mechanisms of incorporation of donor and acceptor dopants in (Hg, Cd)Te alloys. J Vac Sci Technol B 9:1716

Vydyanath HR, Kröger FA (1982) Doping behavior of iodine in $Hg_{0.8}Cd_{0.2}Te^+$. J Electron Mater 11:111–131

Vydyanath HR, Donovan JC, Nelson DA (1981) Lattice defects in semiconducting $Hg_{1-x}Cd_xTe$ Alloys III–defect structure of undoped $Hg_{0.6}Cd_{0.4}Te$. J Electrochem Soc 128:2625–2629

Wang J (1989) PhD thesis. Shanghai Institute of Technical Physics

Wang SW, Tang DY, Tong FM (1991) Research and progress in semiconductor devices II. Science, Beijing

Wang Y, Cai Y, He YC et al (1992) Infrared Technol 14:1

Wang ZX, Wei XJ, Lai DS et al (1984) Laser & Infrared 1:53

Willardson RK, Beer AC (1981) mercury cadmium telluride. Semiconductors and semimetals, vol 18. Academic, New York

Yang JR (1988) PhD thesis. Shanghai Institute of Technical Physics

Ye RQ (1995) Basic applications studies of HgCdTe materials and devices, 1995 collections of the theses. Shanghai Institute of Technical Physics, Shanghai, p 165

Yoshikawa M, Ueda S, Maruyama K, Takigawa H (1985) The behavior of oxygen in HgCdTe. J Vac Sci Technol A 3:153–155

Yu F, Xu S, Zhang S (1990) Infrared Phys 30:61

Yu FJ (1976) Infrared Phys Technol 25–26:30

Yu PW (1977) Excitation-dependent emission in Mg-, Be-, Cd-, and Zn-implanted GaAs. J Appl Phys 48:5043

Yuan HX, Tong FM, Tang DY (1990) Analysis of reverse leakage current mechanisms in $Hg_{1-x}Cd_xTe$ photodiodes. Chin J Infrared Res 9:415–424

Zanio K (1978) In: Willardson RK, Beer AC (eds) Semicond and semimetals, vol 13. Academic, New York

Zheng GZ, Wei YY, Shen JX et al (1994) J Infrared Millimeter Waves 13:347

Ziep O, Genzow D, Mocker M, Herrmann KH (1980) Phys Stat Sol B 99:129

Chapter 3
Recombination

3.1 Recombination Mechanisms and Life Times

3.1.1 Recombination Mechanisms

At room temperature, semiconductors can be in a stable equilibrium, but most doped semiconductors or alloys are in metastable states with compositions that are frozen at a temperature where diffusion stops. Under these conditions, the carrier densities of semiconductors are in quasi-equilibrium and remain constant. However, external effects can destroy the balance between thermal excitation and recombination to excite extra carriers. Often the extra carriers that play an important roll in devices are the minority carriers, holes in n-type and electrons in p-type material. If the source of the external effects disappears, the nonequilibrium carrier population will also decrease because the probability of carrier recombination is larger than that of carrier generation. Normally, we call the time duration of the return to equilibrium, the minority carrier life time. As is well known, the carrier life time is very different for different semiconductor materials. For example, it can exceed 10^{-2}–10^{-3} μs for single crystal Ge, and the values can be in the range of 10 μs for high-purity Si. However, the carrier life time is quite long, about 10^{-2}–10^{-3} μs or longer for GaAs semiconductor materials. The value is about 1 μs for HgCdTe materials. Note that the life time can change over a wide range for different alloy concentrations of the same semiconductor material.

HgCdTe infrared detectors are intrinsic photoconductivity or photovoltaic devices. However, the life time of the carriers excited by photons, play an important role in both classes of detectors.

The extra carriers can be exited by many methods, such as the photon, X-ray, γ-ray, and other particles. In addition, electrical methods also can generate hot extra carriers in the semiconductors. Most often, the extra carriers are generated in the form of electron–hole pairs. But it is also possible to photoexcite donors or acceptors to produce the extra carriers. Once the excitation source is removed, equilibrium will be reestablished.

J. Chu and A. Sher, *Device Physics of Narrow Gap Semiconductors*, Microdevices,
DOI 10.1007/978-1-4419-1040-0_3, © Springer Science+Business Media, LLC 2010

There are three principal recombination mechanisms in semi-conductors: Shockley–Read, Radiative, and Auger recombination. Figure 3.1 shows three recombination mechanisms.

Shockley–Read recombination is assisted by a deep energy center in the bandgap. The electrons in the conduction band recombine with the holes in the valance band with the help of the recombination center and the emission of phonons. This recombination mechanism is not intrinsic and can be suppressed by improving the material growth techniques to reduce the number of recombination centers.

Radiative recombination is one in which an electron in the conduction band transitions to the valance band to recombine with a hole and emits a photon. It generates a photon with the energy $h\nu$, and can be observed in a photoluminescence (PL) measurement. This recombination is intrinsic and is the opposite of photon absorption. Roosbroeck and Shockley (1954) reported the principle mechanism behind radiative recombination in 1954.

Auger recombination is shown in Fig. 3.1c. An electron in the conduction band transitions to the valance band to recombine with a hole, and the energy released causes another electron to transition to a higher state in the conduction band or another hole in the valance band to transition to a lower state in the valance band. Finally, the excited electron or hole can relax to their respective band edges and generate a phonon. This is also an intrinsic process and the opposite of a "*hot electron* (*or hole*)" process in which an excited electron (or hole) drops into a lower conduction band state (or higher valance band state) and the resulting energy excites an electron–hole pair. The mechanism for a hot electron is shown in Fig. 3.2.

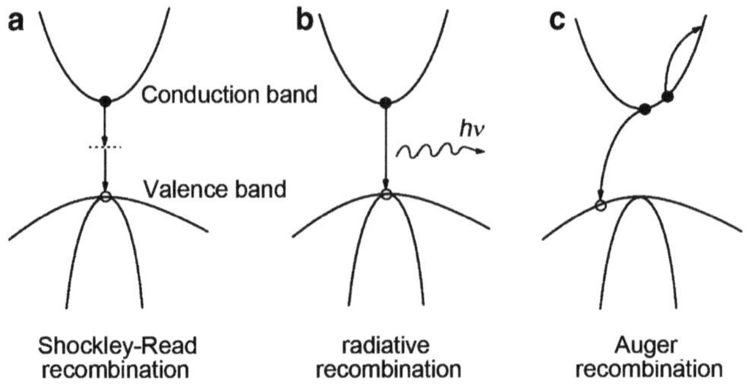

Fig. 3.1 Three recombination mechanisms

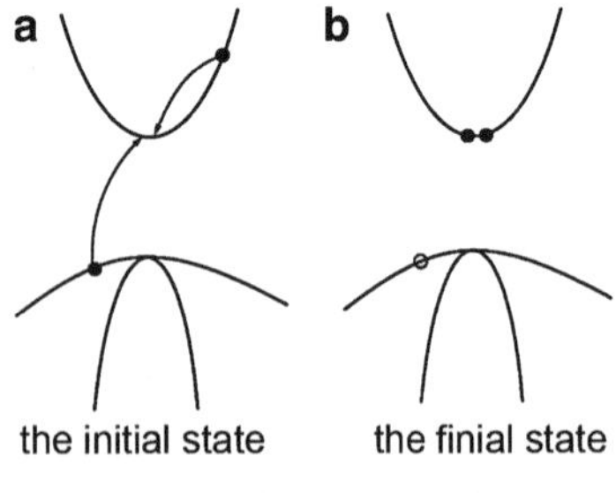

Fig. 3.2 The hot electron process

In the beginning, the initial state is a hot electron and the finial state contains two electrons in the conduction band and a hole in the valance band. It is the opposite of Auger recombination. However, for Auger recombination the initial state contains three particles and the finial state contains an electron with a higher energy.

3.1.2 *The Continuity Equation and Lifetimes*

The generation and recombination of carriers can be attributed to both internal and external factors. Assuming: $g_{\mathrm{i}}^{(\mathrm{n})}, g_{\mathrm{i}}^{(\mathrm{p})}, r_{\mathrm{i}}^{(\mathrm{n})}, r_{\mathrm{i}}^{(\mathrm{p})}$ are the generation and recombination rates of electrons or holes by internal factors, and $g_{\mathrm{e}}^{(\mathrm{n})}, g_{\mathrm{e}}^{(\mathrm{p})}, r_{\mathrm{e}}^{(\mathrm{n})}, r_{\mathrm{e}}^{(\mathrm{p})}$ are the generation and recombination rates of electrons or holes by external factors, the continuity equation becomes:

$$\frac{\partial n}{\partial t} = -\frac{1}{e}\nabla \cdot J_{\mathrm{n}} + g_{\mathrm{i}}^{(\mathrm{n})} + g_{\mathrm{e}}^{(\mathrm{n})} - r_{\mathrm{i}}^{(\mathrm{n})} - r_{\mathrm{e}}^{(\mathrm{n})}, \tag{3.1}$$

$$\frac{\partial p}{\partial t} = -\frac{1}{e}\nabla \cdot J_{\mathrm{p}} + g_{\mathrm{i}}^{(\mathrm{p})} + g_{\mathrm{e}}^{(\mathrm{p})} - r_{\mathrm{i}}^{(\mathrm{p})} - r_{\mathrm{e}}^{(\mathrm{p})}. \tag{3.2}$$

For an approximation where, $g_{\mathrm{e}} = r_{\mathrm{e}} = 0$, and if there is no current, $J = 0$, the above equation simplifies to:

$$\frac{\partial n}{\partial t} = g_{\mathrm{i}}^{(\mathrm{n})} - r_{\mathrm{i}}^{(\mathrm{n})}. \tag{3.3}$$

Under a steady-state condition, $\frac{\partial n}{\partial t} = 0$, we have

$$g_{\mathrm{i}}^{(\mathrm{n})} = r_{\mathrm{i}}^{(\mathrm{n})}. \tag{3.4}$$

It states that in equilibrium the carrier generation rate is equal to the recombination rate. Let g_{i} and r_{i} in equilibrium be expressed as g_0 and r_0. The equilibrium carrier densities for electrons, n_0, and holes, p_0, balance the rates in (3.4). They are responsible for the dark current. In the nonparabolic band approximation, we find:

$$\begin{cases} n_0 = N_{\mathrm{c}} e^{-(E_c - E_f)/kT} \\ p_0 = N_{\mathrm{v}} e^{-(E_f - E_v)/kT} \\ n_0 p_0 = N_{\mathrm{c}} N_{\mathrm{v}} e^{-(E_c - E_v)/kT} = n_{\mathrm{i}}^2 \end{cases}. \tag{3.5}$$

If $g_{\mathrm{i}} > r_{\mathrm{i}}$, then $\frac{\partial n}{\partial t} > 0$, and the carrier density increases with time. If the reverse is true $g_{\mathrm{i}} < r_{\mathrm{i}}$, then $\frac{\partial n}{\partial t} < 0$, and the carrier density decreases with the time. Under an adiabatic approximation, there is no current in the semiconductor and no exchange of energy. The continuity equation, $\frac{\partial n}{\partial t} = g_{\mathrm{i}} - r_{\mathrm{i}} \neq 0$, governs the relaxation of the excess carriers. If the external factors result in the carrier density increase, $n–n_0 = \delta n > 0$, then the generation rate dominates, $r_i < g_i$. If $n–n_0 = \delta n < 0$, then recombination dominates the relaxation, $g_i > r_i$. Define:

$$R = r_i - g_i \tag{3.6}$$

in equilibrium, $r_i = g_i = 0$, thus $R = 0$. If there are excess carriers, then $R \neq 0$. Therefore, R reflects the character of the generation/recombination process of the excess carriers in a unit time and volume. The continuity equation with no net current becomes:

$$\frac{\partial n}{\partial t} = -R(r,t). \tag{3.7}$$

Assuming the recombination probability of the free carriers/unit time and in a unit volume is approximated by $1/\tau_n$, then $R = (n - n_0)/\tau_n$, (3.7) becomes:

$$\frac{\partial n}{\partial t} = g_i - \frac{n - n_0}{\tau_n}, \tag{3.8}$$

so $\delta n(t) = n(t) - n_0 \cong g_i\left[1 - e^{-t/\tau_n}\right]$. When g_i is 0, the deviation from equilibrium vanishes. For finite g_i, the deviation relaxes back to equilibrium with a time constant τ_n. However, if initially the system starts in a nonequilibrium state with population $n(0)$ and relaxes back to equilibrium, then $\delta n = (n(0) - n_0)e^{-t/\tau_n}$.

Similarly, (3.8) also applies to holes:

$$\frac{\partial p}{\partial t} = g_p - \frac{p - p_0}{\tau_p}. \tag{3.9}$$

τ_p is the lifetime of the excess holes in the relaxation time approximation. Note the electron and hole recombination rates need not be the same.

3.1.3 The Principle Recombination Mechanisms and the Resulting Lifetimes of HgCdTe

The bulk lifetime of semiconductor materials is mainly determined by recombination. Auger recombination is the dominate mechanism in pure narrow gap HgCdTe materials compared to some wider band gap materials such as GaAs where radiative recombination is also important. Even in wider gap HgCdTe, Auger recombination still plays an important roll. In indirect gap semiconductors like Ge and Si, the Auger process and radiative recombination are complicated by the need for phonon assistance.

For the excess carriers in HgCdTe, there are three recombination mechanisms: inter-band direct Auger recombination, inter-band direct radiative recombination, and Shockley–Read recombination assisted by a defect energy level center.

1. Auger recombination (Kinch 1981)

Auger recombination mechanisms of types 1 and 7 are the main mechanisms in narrow band materials with a low carrier density (Casselman and Petersen 1980; Casselman 1981). An analytical expression for the recombination time of intrinsic Auger 1 recombination is:

$$\tau_{A1}^{\mathrm{i}} = 3.8 \times 10^{-18} \varepsilon_{\infty}^{2} \frac{m_0}{m_{\mathrm{e}}^{*}} \left(1 + \frac{m_e^{*}}{m_{\mathrm{h}}^{*}}\right)^{1/2} \left(1 + \frac{2m_{\mathrm{e}}^{*}}{m_{\mathrm{h}}^{*}}\right) \left(\frac{E_{\mathrm{g}}}{k_{\mathrm{B}}T}\right)^{3/2}$$
$$\times \exp\left[\left(1 + \frac{2m_{\mathrm{e}}^{*}}{m_{\mathrm{h}}^{*}}\right)\left(1 + \frac{m_{\mathrm{e}}^{*}}{m_{\mathrm{h}}^{*}}\right)^{-1} \frac{E_{\mathrm{g}}}{k_{\mathrm{B}}T}\right] |F_1 F_2|^{-2}. \quad (3.10)$$

Here, $|F_1 F_2|$ is a product of the Bloch function overlap between the conduction and valance band, and the conduction band edge and an excited conduction band state. Typically, we take $|F_1 F_2| = 0.20$ in order to fit the experimental HgCdTe data for $x = 0.2$ over a limited temperature range. In fact, when this quantity is calculated using an accurate band structure it turns out that the constant chosen is a poor approximation to its actual $\boldsymbol{k}$ and T dependent behavior for a range of x values (Krishnamurthy et al. 2006). This occurs because band structures, both the gap and wave functions, are temperature dependent.

The recombination times of Auger 1 and 7 for HgCdTe have the following form:

$$\tau_{\mathrm{A1}} = 2\tau_{\mathrm{A1}}^{\mathrm{i}} z^2/(1 + z^2), \quad (3.11)$$
$$\tau_{\mathrm{A7}} = 2\tau_{\mathrm{A7}}^{\mathrm{i}} z^2/(z + z^2). \quad (3.12)$$

Here, z is defined as $z \equiv p/n_{\mathrm{i}}$. The ratio of the recombination time for Auger 7and 1 is:

$$\Gamma = \frac{\tau_{\mathrm{A_i}}^{(7)}}{\tau_{\mathrm{A_i}}^{(1)}} = 2\frac{m_{\mathrm{c}}^{*}(E_{\mathrm{th}})}{m_0^{*}} \frac{(1 - 5E_{\mathrm{g}}/4k_{\mathrm{B}}T)}{(1 - 3E_{\mathrm{g}}/2k_{\mathrm{B}}T)}, \quad (3.13)$$

where m_0^{*} and $m_{\mathrm{c}}^{*}(E_{\mathrm{th}})$ are the effective mass of electrons near the bottom of the conduction band and those at energy E_{th}, and $m_{\mathrm{hh}}^{*} = 0.5$ has been taken.

The net Auger lifetime of a nonequilibrium carrier population is $\tau_{\mathrm{A}} = \tau_{\mathrm{A1}}\tau_{\mathrm{A7}}/(\tau_{\mathrm{A1}} + \tau_{\mathrm{A7}})$. Figure 3.3 shows the relation between the Auger lifetime and temperature T. Also, Auger 3 and 7 type recombination play an important role in p-type materials. A detailed discussion will follow later in the text.

2. S–R recombination (Pines and Stafsudd 1980)

All semiconductor materials have some impurities although they are nearly pure. Some of these impurities have energy levels in the band-gap. These energy levels can assist electron and hole recombination. Therefore, they will increase the recombination probability of a nonequilibrium carrier concentration.

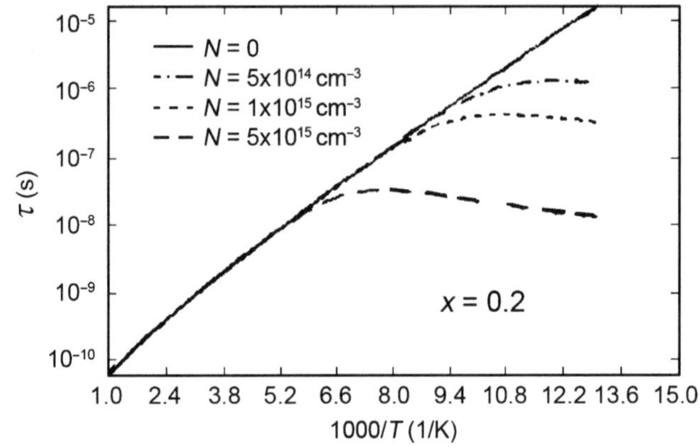

Fig. 3.3 A relation between the Auger lifetime and temperature T for different carrier concentrations

The S–R recombination lifetime can be calculated using the Shockley–Read equation:

$$\tau_{\text{S-R}} = \frac{\sigma_n(n_0 + n_1 + \Delta n) + \sigma_p(p_0 + p_1 + \Delta p)}{N_t\sigma_p\sigma_n v_T(n_0 + p_0 + \Delta p)}, \quad (3.14)$$

which for a small carrier density becomes:

$$\tau_{\text{S-R}} = \frac{\sigma_n\left(n_0 + n_i\, e^{\left(\frac{E_t - E_F}{k_B T}\right)}\right) + \sigma_p\left(p_0 + n_i\, e^{\left(\frac{E_F - E_t}{k_B T}\right)}\right)}{\sigma_n\sigma_p v_T N_t(n_0 + p_0)}. \quad (3.15)$$

Here, σ_p and σ_n are the recombination cross sections of the hole and electron, respectively. The recombination cross sections are in the range of 10^{-13}–$10^{-17}\,\text{cm}^2$, and v_T is the thermal speed of electrons in HgCdTe, $v_T = \sqrt{3k_BT/m_e^*}$. If $N_t = 10^{14}\,\text{cm}^{-3}$ for HgCdTe with $x = 0.2$, and suppose $E_t = E_c - E_t = 21$ meV and $\sigma_p = \sigma_n = 10^{-16}\,\text{cm}^2$ then Fig. 3.4 presents the relation between the S–R recombination lifetime and temperature T. $\tau_{\text{S-R}}$ is interdependent of T for low temperatures. In the mid temperature range where $\exp[-(E_g - E_t)/k_BT]$ does not vanish, this recombination mechanism plays an important role.

The S–R recombination rate is maximum when E_t is close to E_i, and hardly present because of the exponential relation shown in Fig. 3.5. Therefore, only deep energy level impurities affect the lifetime. Equation (3.15) was derived using Boltzmann statistics which is proper for deep levels but not in general. When this type of equation is derived using full Fermi–Dirac statistics, this constraint is removed. Then, the S–R recombination rate is maximized, the life time is minimized, at a temperature when the Fermi level is equal to the trap level even if it is shallow (Krishnamurthy et al. 2006).

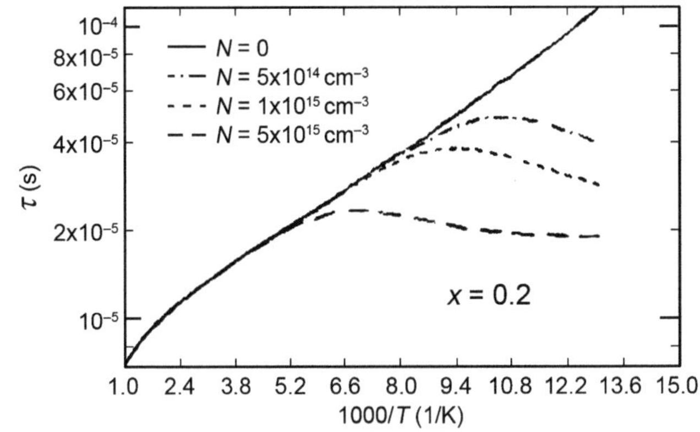

Fig. 3.4 The relation between the R–S recombination lifetime and temperature T

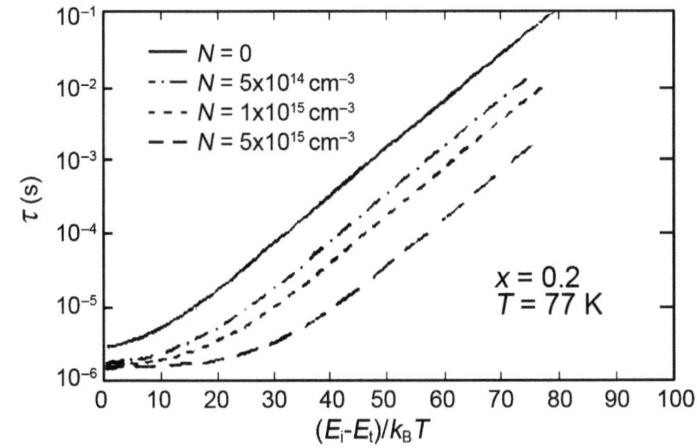

Fig. 3.5 The relation between the R–S recombination lifetime and $(E_i - E_t)/k_B T$

3. Radiative recombination (Schacham and Finkman 1985)

The lifetime of the inter-band direct recombination can be determined by the Roosbrock–Shockley method:

$$\tau_R = \frac{1}{B(n_0 + p_0 + \Delta p)}, \tag{3.16}$$

$$B = 5.8 \times 10^{-13} \varepsilon_\infty^{1/2} \left(\frac{m_0}{m_e^* + m_h^*} \right)^{3/2} \left(1 + \frac{m_0}{m_e^*} + \frac{m_0}{m_h^*} \right) \left(\frac{300}{T} \right)^{3/2} \left(E_g^2 + 3k_B T E_g + 3.75 (k_B T)^2 \right). \tag{3.17}$$

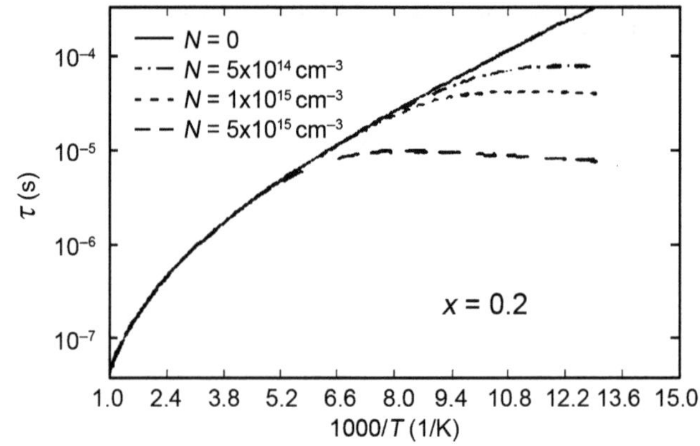

Fig. 3.6 The relation between the radiative recombination lifetime and temperature T

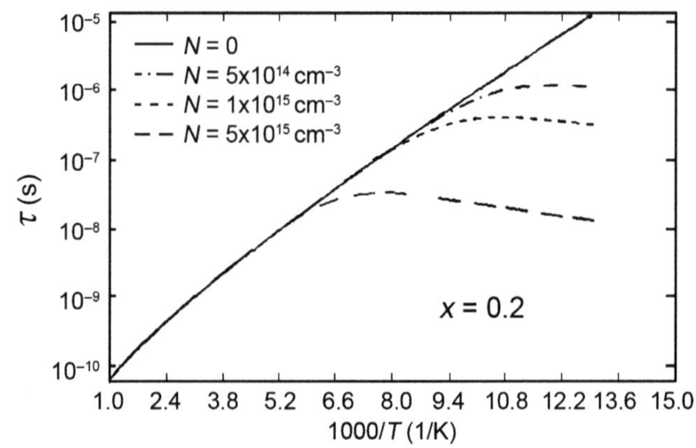

Fig. 3.7 The relation between the net lifetime and temperature T

Figure 3.6 shows the relation between the radiative recombination lifetime and temperature T. At low temperature and small carrier density, the lifetime is interdependent of the temperature and is proportional to $\exp\sqrt{T^*/T}$ with increasing temperature.

4. The bulk radiative recombination lifetime

The three recombination mechanisms each have an effect on HgCdTe materials, thus the net lifetime is:

$$\tau = \left(\frac{1}{\tau_A} + \frac{1}{\tau_{S\text{–}R}} + \frac{1}{\tau_R}\right)^{-1}. \tag{3.18}$$

If we compare Fig. 3.7 with Fig. 3.3, it is found that the net lifetime is mainly determined by Auger recombination in the low and mid temperature ranges.

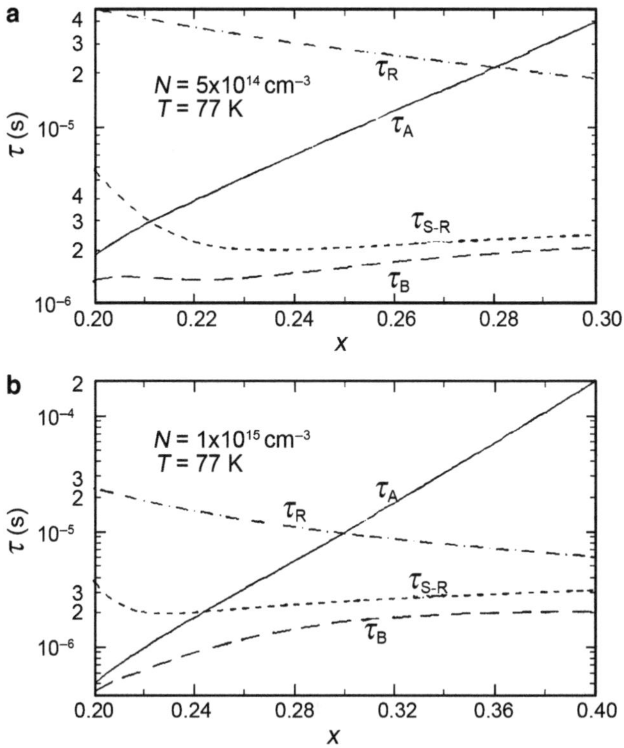

Fig. 3.8 The distribution of lifetimes with composition x at 77K for (**a**) $N_d = 5 \times 10^{14}\,\text{cm}^{-3}$; (**b**) $N_d = 1 \times 10^{15}\,\text{cm}^{-3}$

Figure 3.8 presents the calculated net lifetime. At low temperature, the thermal energy of electrons is very low, and Auger recombination dominates for HgCdTe with alloy compositions $x < 0.2$ and the doping concentration $N_d = 5\times10^{14}\,\text{cm}^{-3}$. The reason is that the band-gap of this material is small and the radiative recombination rate is inversely proportional to E_g^2. However, the Auger intrinsic recombination process rate decreases with an increasing alloy fraction and the subsequent band-gap increases. The radiative recombination rate increases, but in narrow gap material is usually overcome by S–R recombination. Auger recombination plays an important role in HgCdTe with alloy compositions $x < 0.23$ and the doping concentrations $N_d \leqslant 1 \times 10^{15}\,\text{cm}^{-3}$, and S–R recombination, except for very pure material, plays an important role for alloy compositions $x > 0.26$.

At room temperature, the probability of Auger recombination increases because the electron thermal energy increases as seen in Fig. 3.9a, b. The bulk lifetime is mainly determined by Auger recombination.

To conclude, Auger recombination is not only special but also the principal recombination mechanism for HgCdTe with low alloy compositions.

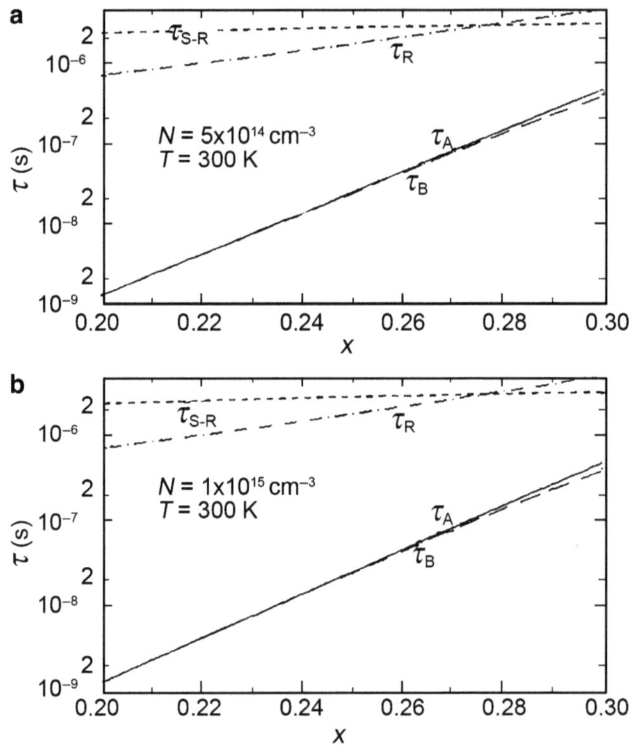

Fig. 3.9 The distribution of lifetimes with the composition x at 300 K for (**a**) $N_d = 5\times10^{14}\,\text{cm}^{-3}$; (**b**) $N_d = 1\times10^{15}\,\text{cm}^{-3}$

3.2 Auger Recombination

3.2.1 The Types of Auger Recombination

Auger recombination can be caused by several different mechanisms consistent with the energy band structure of HgCdTe. Beattie (1962) reported ten possible Auger recombination mechanisms for the conduction band, the heavy-hole band, and the light-hole band. The each of the ten possibilities is consistent with energy and momentum conservation. In each mechanism, the electron transition from the conductive band to recombine with hole in the valance band. The extra energy and momentum is taken up by a second electron or hole transferring to a higher energy state. There are ten possible processes because there is one conduction band and two subvalence bands. Figure 3.10 shows the ten possible processes.

Auger recombination mechanism 1 (AM-1) is the main process in the n-type HgCdTe materials because there are many electrons in conduction band and the holes are mainly located in the heavy-hole band. However, AM-3 and AM-7 are the main processes for the p-type HgCdTe materials because there are many heavy

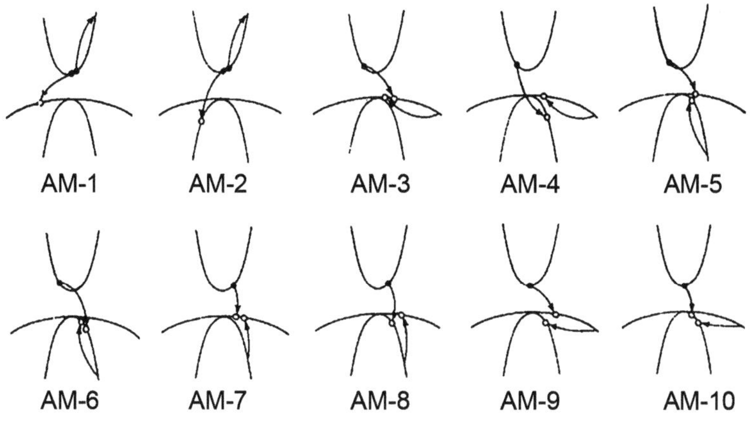

Fig. 3.10 Ten possible Auger recombination processes in HgCdTe (Arrows indicate the electron transitions (Beattie 1962))

holes in the valence band and the not many electrons. The other mechanisms are related to the light-hole band and have smaller probabilities compared to the above three processes.

3.2.2 Auger Lifetime

The lifetime of the minority carriers is determined by the rate at which carriers generated by an external stimulus relax back to equilibrium. From the electron and hole continuity equations, one can approximate the lifetime:

$$\frac{\partial n}{\partial t} = [g_{\mathrm{n}} - r_{\mathrm{n}}] + U_{\mathrm{ext}}^{\mathrm{n}} + \frac{1}{e}\nabla \cdot J_{\mathrm{n}}, \tag{3.19a}$$

$$\frac{\partial p}{\partial t} = [g_{\mathrm{p}} - r_{\mathrm{p}}] + U_{\mathrm{ext}}^{\mathrm{p}} + \frac{1}{e}\nabla \cdot J_{\mathrm{p}}. \tag{3.19b}$$

Here, n and p are the concentrations of the electrons and the holes, respectively. g_{n} and g_{p} are the generation rates when there is no external influence, r_{n} and r_{p} are the recombination rates, $U_{\mathrm{ext}}^{\mathrm{n}}$ and $U_{\mathrm{ext}}^{\mathrm{p}}$ are the generation rates due to external effects, and J_{n} and $\mathrm{J_p}$ are the drift and diffusion current densities caused by electrons and holes. Equations (3.19a) and (3.19b) are the basic formalisms for a current calculation in a p–n junction (Blakemore 1962; Many et al. 1965).

Let

$$n = n_0 + \nabla n \tag{3.20}$$

and

$$p = p_0 + \nabla p, \tag{3.21}$$

where n_0 and p_0 are the equilibrium electron and hole densities and ∇n and ∇p are the deviation from these densities due to extrinsic effects. In the relaxation time approximation, we have for electrons:

$$r_{\mathrm{i}}^{\mathrm{n}} = -\frac{n_0 + \nabla n}{\tau_{\mathrm{n}}}. \tag{3.22}$$

Now, let the current densities and extrinsic generation and recombination terms vanish. Then, in equilibrium where $\frac{\partial n}{\partial t} = \frac{\partial p}{\partial t} = 0$, and $\nabla n = \nabla p = 0$, we find for electrons:

$$n_0 = g_{\mathrm{n}} \tau_{\mathrm{n}}. \tag{3.23}$$

The continuity equation for electrons now simplifies to:

$$\frac{\partial \nabla n}{\partial t} = \frac{1}{e} \nabla \cdot \mathbf{J}_{\mathrm{n}} + U_{\mathrm{ext}}^{\mathrm{n}} - \frac{\nabla n}{\tau_{\mathrm{n}}}. \tag{3.24}$$

A similar treatment applies to holes, so the hole continuity equation is:

$$\frac{\partial \nabla p}{\partial t} = \frac{1}{e} \nabla \cdot \mathbf{J}_{\mathrm{p}} + U_{\mathrm{ext}}^{\mathrm{p}} - \frac{\nabla p}{\tau_{\mathrm{p}}}. \tag{3.25}$$

If the divergence of the current density $\nabla \cdot J_n$ is zero, then the continuity equation simplifies to:

$$\frac{\partial \Delta n}{\partial t} = U_{\mathrm{ext}}^{\mathrm{n}} - \frac{\Delta n}{\tau_{\mathrm{n}}}. \tag{3.26}$$

If the relaxation lifetime τ_{n} is independent of time then the solution to (3.26) for Δn is an exponentially decreasing function. Generally speaking, τ_{n} and τ_{p} depend on the carrier concentrations. However, (3.26) is still effective.

For an Auger process:

$$\tau_{\mathrm{n,A}} = \frac{n - n_0}{(r_{\mathrm{n,A}} - g_{\mathrm{n}})}, \tag{3.27}$$

here n_0 is the equilibrium carrier concentration. Subscript A denotes an Auger processes. Equation (3.26) can be derived from the transition and recombination probability of an electron and a hole per unit volume and including Fermi–Dirac statistics.

In the parabolic approximation for the bands:

$$\begin{cases} g^{(1)} = \frac{n}{n_0} g_0^{(1)} \\ r^{(1)} = \frac{n^2 p}{n_0^2 p_0} r_0^{(1)} \end{cases}. \tag{3.28}$$

Here, superscript 1 indicates an AM-1 process. The generation rate must be equal to the recombination rate in the steady state. If we have $\Delta n = \Delta p$, then:

$$\tau_{\mathrm{A}}^{(1)} = \frac{n_{\mathrm{i}}^4}{n p_0 (n + p_0)} \frac{1}{g_0^{(1)}}. \tag{3.29}$$

The above equation indicates that the Auger lifetime $\tau_A^{(1)}$ can be determined when $g_0^{(1)}$ is known. Considering an integral over the transition probabilities per unit time for all electrons, the generation rate in the absence of an external excitation mechanism, can be calculated. The transition probability from an initial to a final state during the time period t is reported by Schiff (1955):

$$T_{\rm if} = \frac{2t^2}{\hbar^2} |U_{\rm if}|^2 \frac{1-\cos x}{x^2}, \tag{3.30}$$

Here, $x = (t/\hbar)|E_{\rm f} - E_{\rm i}|$, $E_{\rm i}$ and $E_{\rm f}$ are, respectively, the initial and final states energies. $U_{\rm if}$ is the interaction matrix element. The appropriate wave functions are the products of space functions and spin–orbit functions $\phi_{\rm S,T}$:

$$|U_{\rm if}| = \left\langle \varphi_{\rm S,T}^* \middle| H^{(1)} \middle| \phi_{\rm S,T} \right\rangle, \tag{3.31}$$

$H^{(1)}$ is a two electron state interaction perturbation:

$$H^{(1)}(r_1, r_2) = \frac{e^2 \exp(-\lambda |r_1 - r_2|)}{\varepsilon |r_1 - r_2|}. \tag{3.32}$$

λ is the screening length and ε is the dielectric constant. Thus, we find:

$$g_0^{(1)} = \left[\frac{1}{tv}\right]\left[\frac{V}{8\pi^3}\right]^4 \iiiint T_{\rm if}\,[f(k_1)]\,[f(k_2)]\,[1 - f(k_1')]\,[1 - f(k_2')]\,{\rm d}k_1\,{\rm d}k_2\,{\rm d}k_1'\,{\rm d}k_2'. \tag{3.33}$$

The detailed calculation can be found in a paper by Beattie and Landsberg (1959) and another review by Petersen (1970). If the screening effect is neglected and under the parabolic nongenerate band approximation we find:

$$g_0^{(1)} = \frac{8(2\pi)^{5/2} e^4 m_0}{h^3} \frac{(m_0^*/m_0)\,|F_1 F_2|^2}{\varepsilon^2 (1+\mu)^{1/2} [1+2\mu]} n_0 \left(\frac{k_{\rm B} T}{E_{\rm g}}\right)^{3/2} e^{\left[-\left(\frac{1+2\mu}{1+\mu}\right)\frac{E_{\rm g}}{k_{\rm B}T}\right]}. \tag{3.34}$$

$|F_1 F_2|$ is the product of the overlap integrals between the bottom of the conduction band and the valence band and the bottom of the conduction band and an excited conduction band state with energy $E_e \cong E_{\rm g}$. This product is approximated by a constant lying in the range $0.20 \sim 0.25$. μ is the ratio of the electron effective mass in the conductive band to the effective heavy hole effective mass in the valence band. Equation (3.34) can be used to calculate the Auger 1 (AM-1) recombination rate. Then, one can calculate the carrier lifetime for n-type HgCdTe from (3.29). One simplified equation form for $\tau_A^{(1)}$ can be written as (3.28).

If the contribution from the light hole is neglected, the Auger recombination 3 (AM-3) in p-type HgCdTe can be expressed as:

$$g_0^{(3)} = \frac{8(2\pi)^{5/2}e^4 m_0}{h^3} \frac{(m_{\mathrm{hh}}^*/m_0)|F_1F_2|^2}{\varepsilon^2(1+1/\mu)^{1/2}[1+2/\mu]} p_0 \left(\frac{kT}{E_g}\right)^{3/2} e^{\left[-\left(\frac{2+\mu}{1+\mu}\right)\frac{E_g}{kT}\right]}. \tag{3.35}$$

Then, the generation and recombination determined by AM-3 is:

$$g^{(3)} = \frac{p}{p_0} g_0^{(3)}, r^{(3)} = \frac{p^2 n}{p_0^2 n_0} r_0^{(3)}. \tag{3.36}$$

Thus, one can calculate the total lifetime due to the combination of AM-1 and AM-3 determined from (3.36) to be:

$$\begin{aligned} \tau_A &= \frac{n_i^4}{(n_0+p_0+\Delta n)\left(g_0^{(1)} p_0 n + g_0^{(3)} p_0 n\right)} \\ &= \frac{2n_i^2\left[n_0/2g_0^{(1)}\right]}{(n_0+p_0+\Delta n)[(n_0+\Delta n)+\beta(p_0+\Delta n)]}. \end{aligned} \tag{3.37}$$

Here, $\Delta n = \Delta p$ has been assumed. The parameter β is:

$$\beta = \frac{n_0 g_0^{(3)}}{p_0 g_0^{(1)}} = \frac{\mu^{1/2}(1+2\mu)}{2+\mu} \exp\left[-\left(\frac{1-\mu}{1+\mu}\right)\frac{E_g}{k_B T}\right]. \tag{3.38}$$

Equation (3.37) enables a study of the relations among the Auger lifetime temperature and carrier concentration.

For the lifetime of intrinsic HgCdTe materials, $\mu \ll 1$ and the lifetime from (3.37) or (3.29) yields a relation:

$$\tau_{Ai} = \frac{n_i}{2g_0^{(1)}} = \frac{n_0}{2g_0^{(1)}}. \tag{3.39}$$

Then, the lifetime τ_A can be expressed in terms of the function τ_{Ai}. For n-type materials with $\Delta n \ll n_0$, and $n_0 \gg \beta p_0$:

$$\tau_A \cong 2\frac{n_i^2}{n_0^2}\tau_{Ai} \propto \frac{1}{n_0^2}\exp\left[\frac{\mu}{1+\mu}\frac{E_g}{k_B T}\right]. \tag{3.40}$$

Thus, we find:

$$2n_i^2\tau_{Ai} \propto \exp\left[\frac{\mu}{1+\mu}\frac{E_g}{k_B T}\right]. \tag{3.41}$$

From the above form, one can conclude that the Auger lifetime depends on the band-gap energy with which it has an exponential relationship. When the effective mass fraction, μ, and the band-gap decreases the carrier lifetime also decreases.

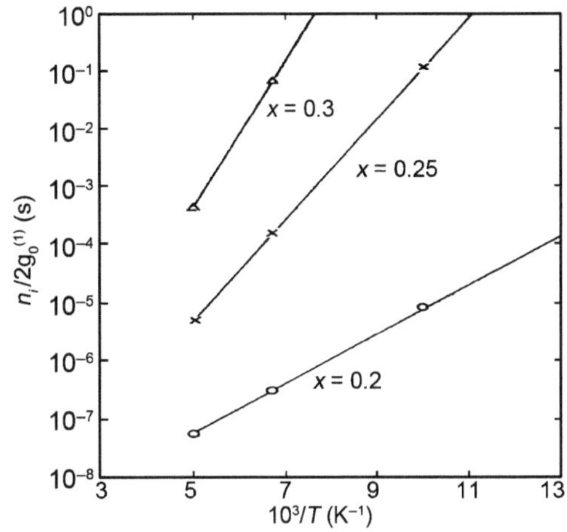

Fig. 3.11 The intrinsic Auger lifetime depends on the temperature for different alloy fractions of HgCdTe material with $|F_1F_2| = 0.25$

For p-type materials, $\beta p_0 \gg n_0$ we have:

$$\tau_{\mathrm{A}} = \frac{2}{\beta}\frac{n_{\mathrm{i}}^2}{p_0^2}\tau_{\mathrm{Ai}} \propto \frac{1}{\beta p_0^2}\exp\left[\frac{\mu}{1+\mu}\frac{E_{\mathrm{g}}}{k_{\mathrm{B}}T}\right]. \tag{3.42}$$

Comparing (3.40) with (3.42), the carrier lifetime of a p-type material is longer than that of an n-type material due to $\beta \ll 1$ for the same carrier concentration. In addition, one can calculate the Auger lifetime of p-type or n-type materials if the parameters, β, τ_{Ai} and the carrier concentrations are known. For an intrinsic Auger lifetime $\tau_{\mathrm{Ai}} = n_{\mathrm{i}}/2g_0^{(1)}$, $g_0^{(1)}$ can be calculated from (3.35). Figure 3.11 shows that the intrinsic Auger lifetime depends on the temperature for different compositional fractions, x, in $\mathrm{Hg}_{1-x}\mathrm{Cd}_x\mathrm{Te}$ because the band-gap and wave functions vary with x. Because of this fact taking a constant, $|F_1F_2| = 0.25$, is only an approximation (Krishnamurthy et al. 2006). The relationship between parameter β and the HgCdTe alloy composition fraction x at different temperatures is shown in Fig. 3.12. Evidently $\beta \ll 1$ is found for most situations. Note that the above calculation contains the following approximations: a two-band model, neglecting the light hole and spin–orbit bands, taking parabolic bands, letting $|F_1F_2|$ be a constant, and assuming nongenerate distributions. For n-type HgCdTe materials with low concentrations, the above approximations are reasonable and the theoretical calculations are in agreement with the experimental results. However, for p-type HgCdTe materials, the agreement is not good and the lifetime decreases when account is taken of the light-hole band.

If we consider the effect of a nonparabolic band and the $|F_1F_2|$ dependence on $\boldsymbol{k}$, and $\beta \ll 1$ then the Auger lifetime for n-type HgCdTe materials can be calculated. Equation (3.37) can be simplified to:

$$\tau_{\mathrm{A}} = \frac{2n_{\mathrm{i}}^2 n_0/2g_0^{(1)}}{(n_0+p_0+\Delta n)(n_0+\Delta n)}. \tag{3.43}$$

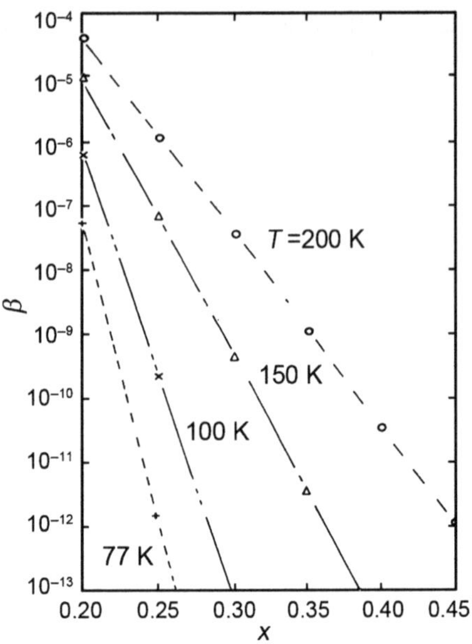

Fig. 3.12 The relationship between the parameter β and HgCdTe alloy fraction at different temperatures

$g_0^{(1)}$ has been obtained under the above generalized conditions by Petersen (1970) and the results indicate that τ'_{A} is the same as τ_{A} for temperatures between 100 and 300 K derived with none above effects included. However, τ'_{A} is shorter than τ_{A} between 77 and 100 K ($x = 0.2$, and $N_{\mathrm{D}} = 10^{14}\,\mathrm{cm}^{-3}$). Evidently, (3.43) can be used to approximate the Auger lifetimes of n-type HgCdTe materials.

In (3.34) and (3.35), the exponential terms in $g_0^{(1)}$ and $g_0^{(3)}$ take the following forms:

$$g_0^{(1)} \propto \exp\left[-\left(\frac{1+2\mu}{1+\mu}\right)\frac{E_{\mathrm{g}}}{k_{\mathrm{B}}T}\right], \tag{3.44a}$$

and

$$g_0^{(3)} \propto \exp\left[-\left(\frac{2+\mu}{1+\mu}\right)\frac{E_{\mathrm{g}}}{k_{\mathrm{B}}T}\right]. \tag{3.44b}$$

For an AM-1 process, define:

$$E_{\mathrm{th}}^{1} \equiv -\left(\frac{1+2\mu}{1+\mu}\right)E_{\mathrm{g}}. \tag{3.45a}$$

For an AM-3 process, define:

$$E_{\mathrm{th}}^{3} \equiv -\left(\frac{2+\mu}{1+\mu}\right)E_{\mathrm{g}}. \tag{3.45b}$$

E_{th} are the threshold energies for these processes. These relationships can be derived from energy and momentum conversation in their respective Auger recombination processes. The physical meaning of $E_{\rm th}$ is that it is the minimum energy required for an electron/hole recombination to occur via an Auger AM-1 or AM-3 process. Because $\mu = m_{\rm e}^*/m_{\rm hh}^* \ll 1, E_{\rm th}^1 \cong E_{\rm g}$, so the AM-1 recombination rate can be simplified to:

$$g_0^{(1)} \propto \exp\left[-\frac{E_{\rm g}}{k_{\rm B}T}\right]. \tag{3.46}$$

Similarly, we find $E_{\rm th}^{(3)} \cong 2E_{\rm g}$ for an AM-3 process; $E_{\rm th}^{(2)} \cong \frac{1+2m_{\rm e}^*/m_{\rm lh}^*}{1+m_{\rm e}^*/m_{\rm lh}^*}E_{\rm g} \cong \frac{3}{2}E_{\rm g}$ for an AM-2 process; and $E_{\rm th}^{(7)} \cong \frac{2+m_{\rm e}^*/m_{\rm hh}^*}{2+m_{\rm e}^*/m_{\rm hh}^*-m_{\rm lh}^*/m_{\rm hh}^*}E_{\rm g} \cong E_{\rm g}$ for an AM-7 process.

From (3.46), the steady-state population is proportional to an exponential of the threshold energy. Therefore, in n-type material, if $E_{\rm th} > E_{\rm g}$ in an Auger process AM-3 is playing a role, while if $E_{\rm th}{\sim}E_{\rm g}$ Auger MA-1 dominates. Moreover, in p-type, Auger 1 and 7 will dominate because the heavy-hole band plays an important role because the state density of the light-hole bands is less than that of the heavy-hole bands. The Auger 3 process is also very important in heavily doped p-type materials.

For an AM-7 process, the recombination between an electron in the conductive band and a hole in the heavy-hole band is assisted by exciting an electron in light-hole band into the heavy-hole band. By considering nonparabolic bands, the $|F_1F_2|$ dependence on $\boldsymbol{k}$, and nongenerate statistics, Beattie and Smith (1967) deduced the relation:

$$\tau_{\rm A}^{(7)} = D(E_{\rm th})\left[1+\frac{2n_0\Delta p_{\rm hh}}{p_{0,\rm hh}\Delta n}-\frac{\Delta p_{\rm lh}n_0}{\Delta n p_{0,\rm lh}}\right]^{-1}\frac{\eta_{\rm th}^{-3/2}}{I(\eta_{\rm th})}, \tag{3.47}$$

where $\eta_{\rm th} = E_{\rm th}/kT$, and $D(E_{\rm th})$ is a function of the effective mass, $|F_1F_2|$, and threshold $E_{\rm th}$. $I(\eta_{\rm th})$ is an integral over the conduction band. Assuming the same Fermi level holds for the light- and heavy-hole bands, and the effective mass are constants, (3.47) can be simplified as:

$$\tau_{\rm A}^{(7)} = \frac{n_0/p_0}{1+n_0/p_0}\cdot 2\tau_{\rm Ai}^{(7)} = \frac{2n_{\rm i}^2\tau_{\rm Ai}^{(7)}}{(n_0+p_0)p_0}, \tag{3.48}$$

where $\tau_{\rm Ai}^{(7)}$ is the intrinsic lifetime for AM-7. For intrinsic materials, the bracketed item in (3.47) is 1/2, thus:

$$\tau_{\rm Ai}^{(7)} = \frac{D(E_{\rm th})}{2}\frac{\eta_{\rm th}^{-3/2}}{I(\eta_{\rm th})}. \tag{3.49}$$

For an AM-1 process, $\tau_{\rm A}^{(1)}$ is given by (3.37) and (3.40), and the intrinsic limit by (3.39). When Δn is very small:

$$\tau_{\rm A}^{(1)} = \frac{1}{1+n_0/p_0}\cdot 2\tau_{\rm Ai}^{(1)} = \frac{2n_{\rm i}^2\tau_{\rm Ai}^{(1)}}{(n_0+p_0)n_0}. \tag{3.50}$$

Thus, one can compare $\tau_{\text{Ai}}^{(7)}$ with $\tau_{\text{Ai}}^{(1)}$, from (3.48) and (3.50), we find:

$$\frac{\tau_{\text{A}}^{(7)}}{\tau_{\text{A}}^{(1)}} = \frac{n_0}{p_0}\frac{\tau_{\text{Ai}}^{(7)}}{\tau_{\text{Ai}}^{(1)}} = \frac{n_{\text{i}}^2}{p_0^2}\frac{\tau_{\text{Ai}}^{(7)}}{\tau_{\text{Ai}}^{(1)}}. \tag{3.51}$$

Casselman and Petersen (1980) calculated the results:

$$\gamma = \frac{\tau_{\text{Ai}}^{(7)}}{\tau_{\text{Ai}}^{(1)}} = 2\frac{m_{\text{c}}^*(E_{\text{th}})}{m_0^*}\frac{\left[1 - \frac{5}{4}\eta_{\text{th}}\right]}{\left[1 - \frac{3}{2}\eta_{\text{th}}\right]}, \tag{3.52}$$

where $m_{\text{c}}^*(E_{\text{th}})$ is the effective mass of the conduction band at E_{th}, and m_0^* is the effective mass at the conduction band edge. $E_{\text{th}}^{(1)} \cong E_{\text{th}}^{(7)} = E_{\text{g}}$, for the alloy concentrations, $0.16 < x < 0.3$, temperatures $50\,K < T < 100\,\text{K}$, and $0.1 < \gamma < 6$ (Casselman and Petersen 1980). So, one can conclude that AM-1 and AM-7 processes play important roles in both n-type and p-type materials. Similarly, we can compare AM-7 with AM-3, which are important in p-type semiconductors:

$$\frac{\tau_{\text{A}}^{(7)}}{\tau_{\text{A}}^{(3)}} = \gamma\frac{1}{2}\left(\frac{m_0^*}{m_{\text{hh}}^*}\right)\exp\left(-\frac{E_{\text{g}}}{k_{\text{B}}T}\right). \tag{3.53}$$

The γ value and the other band parameters must be assigned to complete the above expression.

Note that the above discussion is only valid for a nondegenerate population distribution. Degeneracy can always exist in a small fraction of a HgCdTe material or at low temperatures even for $x = 0.2$ samples. In a case where the band-gap is small and the electron effective mass in the conduction band is also very small, then the Fermi level lies above the conduction band edge. In such cases, the Auger process is modified. Other abnormal cases occur, at low temperature when a local concentration fluctuation causes the band-gap and the effective mass in the conduction band to be very small, or if there is a local carrier concentration fluctuation to a higher value. Then, space charge layers are formed causing the bands to bend so that the Fermi level lies above the conduction band edge. In such local regions, the Auger process is modified. Therefore for an AM-1 process, the electrons from the Fermi level transition from above the bottom of conduction band to the valence band to recombine with holes and the extra energy excites electrons from the Fermi level to the higher energy states. The detailed process is shown in Fig. 3.13.

For a degenerate case, Gerhardts et al. (1978) investigated the Auger lifetime of n-type HgCdTe and found:

$$\tau^{-1} = 3\frac{e^4 m_0}{\hbar^3}(2^4\pi^5\varepsilon^2 N_\nu)^{-1}(1 + p/k_{\text{B}}Tn')E_{\text{f}}\int \text{d}^3k_1 \int \text{d}^3k_2$$
$$f(E_1)f(E_2)[1 - f(E_1')]\exp(E_2'/k_BT)Mk_1'(E_1' - E_{\text{g}})/\eta E_{\text{p}}. \tag{3.54}$$

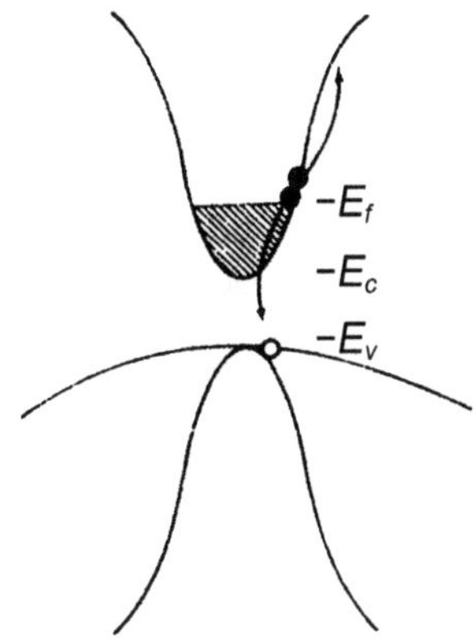

Fig. 3.13 An AM-1 transition processing in a degenerate case

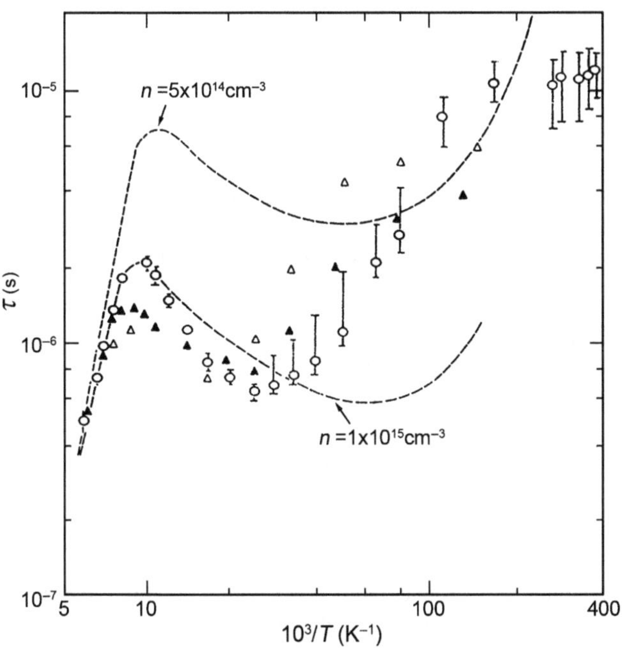

Fig. 3.14 The relation between the carrier lifetime and temperature for an n-type $Hg_{0.2}Cd_{0.8}Te$ sample with different carrier concentrations

Here, N_v is the effective valence band state density. $n' = dn/dE_1, \eta = (E_E - E_c)k_BT$, and M is a transition matrix element. The calculated results are given by the dashed curves in Fig. 3.14. The dots are the $\tau - 1/T$ experimental data for the different carrier concentrations.

3.3 Shockley–Read Recombination

3.3.1 *Single-Level Recombination Center*

Shockley–Read is a basic process for electron–hole recombination in a semiconductor, there are several classic text books dealing with the mechanism (Shockley and Read 1952; Hall 1952; Sah and Tasch 1967; Milnes 1973). Nonequilibrium electrons (holes) created by an external excitation in intrinsic semiconductors rapidly relax to the bottom (top) of conduction (valence) band and then recombine with holes (electrons) within a characteristic lifetime. If there are impurity-related energy levels in the band-gap, the recombination rate will be enhanced, and thus the carrier lifetime is shortened. As shown in Sect. 3.1, starting from a nonequilibrium distribution, no external excitation source, and in the relaxation time approximation we find:

$$n = \Delta n_0 \exp(-t/\tau_\mathrm{n}) + n_0 \tag{3.55}$$

and

$$p = \Delta p_0 \exp\left(-t/\tau_\mathrm{p}\right) + p_0. \tag{3.56}$$

Now, we consider the trap capture cross-section and occupation probability. According to the model proposed by Shockley–Read (Shockley and Read 1952), who studied carrier dynamics at a single energy level within the band-gap, there are four possible processes involving the deep energy center as shown in Fig. 3.15.

(a) An unoccupied neutral deep center, with a concentration of N_T^0, captures one electron from the conduction band, with a capture cross-section or recombination cross-section σ_n, and a capture rate (i.e., capture coefficient or recombination rate) $c_\mathrm{n} = \sigma_\mathrm{n} v_\mathrm{n}$, where v_n is the thermal speed of the electrons.

$$N_\mathrm{T}^0 + e \rightarrow N_\mathrm{T}^-. \tag{3.57}$$

(b) An occupied deep center emits electrons onto the conduction band with an emission rate e_n.
(c) A hole is captured by an occupied center with a capture cross section σ_p and a capture rate $c_\mathrm{p} = \sigma_\mathrm{p} v_\mathrm{p}$, where v_p is the hole thermal speed.

$$N_\mathrm{T}^- + h \rightarrow N_\mathrm{T}^0. \tag{3.58}$$

(d) A hole is emitted from an unoccupied center with an emission rate e_p.

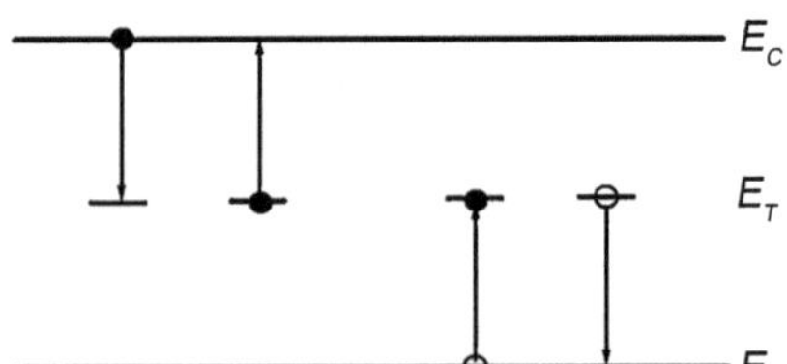

Fig. 3.15 Four possible capture and emission processes for electrons and holes

The capture rate of an electron is proportional to the electron's concentration n at bottom of conduction band and the unoccupied deep center concentration $N_T(1-f_T)$, where f_T is the Fermi–Dirac distribution function. If δn is small, then f_T is the equilibrium value of its distribution function, i.e.:

$$f_\mathrm{T} = \left[1 + \exp\left(\frac{E_\mathrm{T} - E_\mathrm{F}}{kT}\right)\right]^{-1}. \tag{3.59}$$

If $E_\mathrm{T} - E_\mathrm{F} \gg kT$, (3.59) simplifies to its Maxwell–Boltzmann form:

$$f_{\mathrm{T_{MB}}} = \left[\exp\left(\frac{E_\mathrm{T} - E_\mathrm{F}}{kT}\right)\right]^{-1}. \tag{3.60}$$

Obviously in the absence of an external generation mechanism, the electron emission process (b) depends on the rate e_n and the fraction of the occupied center concentration $N_\mathrm{T} f_\mathrm{T}$. In the general case, if δn is small, E_F is not changing, and the Fermi–Dirac distribution function f_T is only a function of temperature T, it does not depend on time t. So, the electron concentration changes as:

$$\frac{\mathrm{d}n(t)}{\mathrm{d}t} = -c_\mathrm{n} n\,(t)\, N_\mathrm{T}(1 - f_\mathrm{T}) + e_\mathrm{n} N_\mathrm{T} f_\mathrm{T}. \tag{3.61a}$$

Similarly, for holes the change is

$$\frac{\mathrm{d}p\,(t)}{\mathrm{d}t} = -c_\mathrm{p} p\,(t)\, N_\mathrm{T} f_\mathrm{T} + e_\mathrm{p} N_\mathrm{T}(1 - f_\mathrm{T}). \tag{3.61b}$$

However, if in the critical theoretical consideration, any change of electron density δn in the conduction band, it must be related to the change of Fermi level E_F, then the Fermi–Dirac distribution function f_T is not only a function of temperature T but also a function of time t. In this case, the situation becomes complicated, (3.61a) and (3.61b) become:

$$\frac{\mathrm{d}n(t)}{\mathrm{d}t} = -c_\mathrm{n} n\,(t)\, N_\mathrm{T}(1 - f\,(t)) + e_\mathrm{n} N_\mathrm{T} f\,(t), \tag{3.62a}$$

and

$$\frac{\mathrm{d}p\,(t)}{\mathrm{d}t} = -c_\mathrm{p} p\,(t)\, N_\mathrm{T} f\,(t) + e_\mathrm{p} N_\mathrm{T}(1 - f\,(t)). \tag{3.62b}$$

Equations (3.62a) and (3.62b) constitute two equations in three unknowns, $n\,(t)$, $p\,(t)$, and $f\,(t)$. The third equation is charge neutrality:

$$n(t) + N_\mathrm{T} f\,(t) = p\,(t)\,. \tag{3.62c}$$

These equations constitute to a nonlinear set of first-order differential equations.

Under steady-conditions equilibrium ($\mathrm{d}n/\mathrm{d}t = 0$, $\mathrm{d}p/\mathrm{d}t = 0$, $n = n_0$, and $p = p_0$), we obtain from (3.61a):

$$\frac{e_{\mathrm{ne}}}{c_{\mathrm{ne}}} = n_0 \exp\left(\frac{E_{\mathrm{T}} - E_{\mathrm{F}}}{kT}\right) \equiv n_1, \tag{3.63}$$

where e_{ne} and c_{ne} are the emission rate and the capture rate at equilibrium. In fact, n_1 is equal to $N_{\mathrm{c}} \exp\left(\frac{E_{\mathrm{c}}-E_{\mathrm{T}}}{kT}\right)$, which is the electron concentration when the Fermi level overlaps with the impurity level E_{T}. Similarly, from (3.61b), we obtain:

$$\frac{e_{\mathrm{pe}}}{c_{\mathrm{pe}}} = p_0 \exp\left(\frac{E_{\mathrm{F}} - E_{\mathrm{T}}}{kT}\right) \equiv p_1. \tag{3.64}$$

We can also rewrite (3.61a) and (3.61b) into:

$$\frac{\mathrm{d}n}{\mathrm{d}t} = -c_{\mathrm{n}}(nN_{\mathrm{T}}^0 - n_1 N_{\mathrm{T}}^-) \tag{3.65}$$

and

$$\frac{\mathrm{d}p}{\mathrm{d}t} = -c_{\mathrm{p}}(pN_{\mathrm{T}}^- - p_1 N_{\mathrm{T}}^0), \tag{3.66}$$

where the capture trapping center for an n-type material is in one of the two time-dependent acceptor states, N_{T}^- or N_{T}^0. For a p-type material, there is a larger chance if it can do so, the trap will first capture a hole to become N_{T}^+ and then capture an electron to assist the recombination. But here we assume the trap is such that it has only one empty state that can accommodate an electron but not a hole. Then, $N_{\mathrm{T}}^0 = N_{\mathrm{T}}(1-f)$, $N_{\mathrm{T}}^- = N_{\mathrm{T}} f$, and $N_{\mathrm{T}}^0 + N_{\mathrm{T}}^- = N_{\mathrm{T}}$, where N_{T} is the total density of traps. Obviously, the rate at which N_{T}^- traps are formed is:

$$\frac{\mathrm{d}N_{\mathrm{T}}^-}{\mathrm{d}t} = -\frac{\mathrm{d}n}{\mathrm{d}t} + \frac{\mathrm{d}p}{\mathrm{d}t} = c_{\mathrm{n}}(nN_{\mathrm{T}}^0 - n_1 N_{\mathrm{T}}^-) - c_{\mathrm{p}}(pN_{\mathrm{T}}^- - p_1 N_{\mathrm{T}}^0). \tag{3.67}$$

When all the trap centers are neutral, the maximum rate at which neutral traps are formed is the minimum recombination lifetime and it is:

$$\tau_{\mathrm{n}0} = \frac{-n}{\mathrm{d}n/\mathrm{d}t|_{N_{\mathrm{T}}^0 \to N_{\mathrm{T}}}} = \frac{1}{c_{\mathrm{n}} N_{\mathrm{T}}} = (\sigma_{\mathrm{n}} v_{\mathrm{n}} N_{\mathrm{T}})^{-1}. \tag{3.68}$$

Similarly, when $N_{\mathrm{T}}^- \to N_{\mathrm{T}}$, the minimum possible lifetime of holes is:

$$\tau_{\mathrm{p}0} = \frac{1}{c_{\mathrm{p}} N_{\mathrm{T}}} = \left(\sigma_{\mathrm{p}} v_{\mathrm{p}} N_{\mathrm{T}}\right)^{-1}. \tag{3.69}$$

The results of (3.68) and (3.69) can be also obtained from an analysis in which the trap population is assumed to remain nearly constant in a Fermi distribution as the system relaxes from an initial nonequilibrium to its equilibrium distribution. See Appendix 3.A.

In the case of equilibrium, it means there are no electrons captured or emitted by capture centers, i.e. $\frac{dN_T^-}{dt} = 0$, and we obtain N_T^- from (3.67):

$$\frac{N_T^-}{N_T} = \frac{c_n n_0 + c_p p_1}{c_n(n_0 + n_1) + c_p(p_0 + p_1)}. \tag{3.70}$$

So, N_T^o is:

$$\frac{N_T^0}{N_T} = \frac{c_n n_1 + c_p p_0}{c_n(n_0 + n_1) + c_p(p_0 + p_1)}. \tag{3.71}$$

At this point, the Shockley–Read treatment makes an unjustified approximation. They say there is a steady-state established darning the entire time the system is returning to equilibrium, so in (3.70) and (3.71), n_0 and p_0 can be replaced by $n(t)$ and $p(t)$. Given this approximation and substituting N_T^- from (3.70) and N_T^0 from (3.71) into (3.65) and (3.66), we obtain:

$$\frac{dn}{dt} = \frac{dp}{dt} = -\frac{np - n_i^2}{\tau_{p0}(n + n_1) + \tau_{n0}(p + p_1)}. \tag{3.72}$$

In this equation, $\frac{dn}{dt} = \frac{dp}{dt}$, is treated as a consequence derived from the "steady-state" approximation but it could just as well be taken as the "steady-state" approximation. In that case, $f(t)$ is solved in terms of $n(t)$ and $p(t)$ from the condition $\frac{dn}{dt} = \frac{dp}{dt}$ and substituted into either (3.61) or (3.62) to obtain (3.72). Once again there is no justification for this approximation other than it makes the expressions more tractable. The proper way to proceed is to use the two continuity equations and the charge neutrality condition as the third equation to solve the three unknowns $n(t)$, $p(t)$, and $f(t)$.

Since many deep centers act as acceptors, we now discuss the effects on an n-type semiconductor with acceptor centers. Under external perturbations such as photon excitation or charge injection, the density of electrons (holes) increases from its equilibrium value $n_0(p_0)$ to $n_0 + \Delta n_0(p_0 + \Delta p_0)$. As soon as the external perturbation stops, the density of electrons and holes begin to decrease toward their equilibrium values again by following the equation:

$$\frac{dn}{dt} = \frac{dp}{dt} = -\frac{(n_0 + \Delta n_0)(p_0 + \Delta p_0) - n_i^2}{\tau_{p0}(n_0 + \Delta n_o + n_1) + \tau_{n0}(p_0 + \Delta p_o + p_1)}. \tag{3.73}$$

According to:

$$\frac{dn}{dt} = -\frac{\Delta n}{\tau_n}, \tag{3.74}$$

we have:

$$\tau_n = -\frac{\tau_{p0}(n_0 + \Delta n_0 + n_1) + \tau_{n0}(p_0 + \Delta p_0 + p_1)}{(n_0 + \Delta n_0)(p_0 + \Delta p_0) - n_i^2} \cdot \Delta n_0. \tag{3.75}$$

Substituting $\tau_{p0} = \left(\sigma_p v_p N_T\right)^{-1}$ and $\tau_{n0} = (\sigma_n v_n N_T)^{-1}$ into (3.75), we obtain:

$$\tau_n = -\frac{\sigma_n v_n(n_0 + \Delta n_0 + n_1) + \sigma_p v_p(p_0 + \Delta p_0 + p_1)}{\sigma_n \sigma_p v_n v_p N_T^2 \left[(n_0 + \Delta n_0)(p_0 + \Delta p_0) - n_i^2\right]} \cdot \Delta n_0. \tag{3.76}$$

An identical expression for τ_p can be obtained from $\frac{dp}{dt} = -\frac{\Delta p}{\tau_p}$ and the fact that in this case $\nabla n_0 = \nabla p_0$. Thus, the S–R lifetime can be expressed in a single formula by letting $\tau_n = \tau_{S\text{–}R}$. Neglecting Δn_0 and Δp_0 when they are small, (3.76) becomes:

$$\tau_{S\text{–}R} = \frac{\tau_{p0}(n_0 + n_1) + \tau_{n0}(p_0 + p_1)}{(n_0 + p_0)} = \tau_{p0}\frac{n_0 + n_1}{n_0 + p_0} + \tau_{n0}\frac{p_0 + p_1}{n_0 + p_0}. \tag{3.77}$$

As an example, examine a case of a Au doped n-type Si, in which the concentration of electrons is $10^{16}\,\text{cm}^{-3}$ and the density of Au atoms is $5 \times 10^{15}\,\text{cm}^{-3}$. The resistance of this Si is $1\,\Omega \cdot \text{cm}$ at 300 K. Supposing all Au atoms in Si are in the form of Au^- and form a deep level at $E_c - 0.54\,\text{eV}$; the observed capture cross-sections $\sigma_p \cong 10^{-15}\,\text{cm}^2$ and $\sigma_n \cong 5 \times 10^{-16}\,\text{cm}^2$ for the processes are $\text{Au}^- + h \to \text{Au}$ and $\text{Au} + e \to \text{Au}^-$, respectively; and the thermal velocity of electrons v_n is of order of magnitude of 10^7 cm/s and that of the holes $v_p \cong 10^6$ cm/s, then we obtain:

$$\tau_{p0} = \left(\sigma_p v_n N_T\right)^{-1} = (10^{-15} \times 10^7 \times 5 \times 10^{15})^{-1} = 2 \times 10^{-8}\,\text{s}, \tag{3.78}$$

$$\tau_{n0} = \left(\sigma_n v_p N_T\right)^{-1} = (5 \times 10^{-16} \times 10^6 \times 5 \times 10^{15})^{-1} = 4 \times 10^{-7}\,\text{s}, \tag{3.79}$$

from (3.68) and (3.69). Note that $n_1 \approx p_1/50 \approx 2 \times 10^{10}\,\text{cm}^{-3}$ but even though $p_1 \approx 50 p_0$, it can still be neglected with respect to $n_0(10^{16}\,\text{cm}^{-3})$. This is a case where the Fermi level lies well above the trap level, so τ_{n0} approximates $\tau_{S\text{–}R}$. So, the Au in Si is effective in decreasing the minority carrier lifetime for a high sensitivity device. In p-type Si, the donor deep level at $E_v + 0.35\,\text{eV}$ formed by Au plays a same role as Au in n-type Si.

3.3.2 General Lifetime Analysis

The prerequisites for the discussion in last section are, $\Delta n_0 = \Delta p_0$; zero external generation, a low density of recombination centers; and single-trap energy level. In a more general case, i.e. without these simplifications, the expressions in last section will be more complicated.

In the case of $\Delta n_0 \neq \Delta p_0$, which occurs when there is a low injection density, the lifetime derived by Shockley (1958) is:

$$\tau_p = \left[\frac{\tau_{n0}(p_0 + p_1) + \tau_{p0}[n_0 + n_1 + N_T(1 + n_0/n_1)^{-1}]}{n_0 + p_0 + N_T(1 + n_0/n_1)^{-1}(1 + n_1/n_0)^{-1}}\right]_{\Delta n \to 0} \tag{3.80}$$

and

$$\tau_n = \left[\frac{\tau_{p0}(n_0 + n_1) + \tau_{n0}[p_0 + p_1 + N_T(1 + p_0/p_1)^{-1}]}{n_0 + p_0 + N_T(1 + p_0/p_1)^{-1}(1 + p_1/p_0)^{-1}} \right]_{\Delta p \to 0}. \tag{3.81}$$

If N_T is small, (3.80) and (3.81) reduce to a single expression:

$$\tau_0 = \frac{\tau_{p0}(n_0 + n_1) + \tau_{n0}(p_0 + p_1)}{n_0 + p_0}. \tag{3.82}$$

In the case of a high carrier concentration, Shockley and Read (1952) derived the following expression:

$$\tau = \tau_0 \left[\frac{1 + \Delta n_0(\tau_{n0} + \tau_{p0})/[\tau_{p0}(n_0 + n_1) + \tau_{n0}(p_0 + p_1)]}{1 + \Delta n_0/(n_0 + p_0)} \right], \tag{3.83}$$

where τ_0 is given in (3.82), indicating the lifetime varies starting from τ_0 as Δn_0 increases. For heavily doped n-type samples (3.83) can be expressed in terms of the change of conductivity $\Delta\sigma$ and the ratio b of the electron mobility to the hole mobility:

$$\tau \cong \tau_0 \left[\frac{1 + \frac{\Delta\sigma b}{\sigma_0(1+b)}[(\tau_{p0} + \tau_{n0})/\tau_{p0}]}{1 + \frac{\Delta\sigma b}{\sigma_0(1+b)}} \right] \text{ or} \tag{3.84}$$

$$\tau \cong \tau_0 \left[\frac{1 + \delta[(\tau_{p0} + \tau_{n0})/\tau_{p0}]}{1 + \delta} \right], \tag{3.85}$$

where $\delta = \frac{\Delta\sigma b}{\sigma_0(1+b)}$.

Sandiford (1957) analyzed the role of Shockley–Read–Hall-type centers in the transient photoconductivity decay assuming that the trap populations are time independent. He combined (3.65) and (3.66), and found an expression:

$$\Delta p = Ae^{-t/\tau_+} + Be^{-t/\tau_-}, \tag{3.86}$$

where A and B are determined from the initial conditions. He treats a case (see Appendix 3.B) where the deviation from equilibrium are small, so $f(t)$ can be approximated by f_T (N_T^- and N_T^0 are constants), and N_T is small enough so the following approximation holds:

$$[c_p(N_T^- + p_0 + p_1) + c_n(N_T^0 + n_0 + n_1)]^2 >> 4c_pc_n[N_T^- N_T^0 + N_T^-(n_0 + n_1) + N_T^0(p_0 + p_1)]. \tag{3.87}$$

Following this logic, he finds:

$$\tau_+ = \left(c_p \left[p_0 + p_1 + N_T(1 + p_0/p_1)^{-1} \right] + c_n \left[n_0 + n_1 + N_T(1 + p_0/p_1)^{-1} \right] \right)^{-1} \tag{3.88}$$

and

$$\tau_- = \frac{\tau_{n0}\left(p_0 + p_1 + N_T\left(1 + p_0/p_1\right)^{-1}\right) + \tau_{p0}\left(n_0 + n_1 + N_T\left(1 + n_0/n_1\right)^{-1}\right)}{n_0 + p_0 + N_T\left(1 + n_0/n_1\right)^{-1}\left(1 + n_1/n_0\right)^{-1}} \tag{3.89}$$

Hence, the decay of Δp or Δn may not be expressed by a single exponential. In some circumstance, there is a big difference between τ_+ and τ_-. For instance, when $c_p = 10^{-9}\,\mathrm{cm^3 \cdot s^{-1}}$, $c_n = 10^{-7}\,\mathrm{cm^3\,s^{-1}}$, $N_T = 10^{13}\,\mathrm{cm^{-3}}$, $n_0 = 10^{14}\,\mathrm{cm^{-3}}$, and $p_1 = 3 \times 10^{16}\,\mathrm{cm^{-3}}$, he obtains $\tau_+ = 400\mu$s and $\tau_- = 0.025\mu$s. However, as shown in Appendix 3.B, in the case of an n-type semiconductor, $A \cong 0$ and $B \cong \Delta p(0)$, one cannot conclude that the fast decay dominates. There are other cases where $A > B$, so there is a fast decay with amplitude A followed by a slower decay with amplitude B. In still other cases, $A > 0$, is positive while, $B < 0$, is negative so there is an overshoot of the decay.

Wertheim (1958) studied transient recombination. He got the same expression as those in (3.88) and (3.89), but expressed them in a different form:

$$\tau_i = \left(c_p\left[p_0 + p_1 + N_T\right] + c_n\left[n_0 + n_1 + N_T\right]\right)^{-1}, \tag{3.90}$$

$$\tau_t = \frac{\tau_{n0}(p_0 + p_1 + N_T^-) + \tau_{p0}(n_0 + n_1 + N_T)}{n_0 + p_0 + (N_T^- N_T^0/N_T)} \tag{3.91}$$

Figure 3.16 shows the relation of lifetime and temperature for a surface bombarded n-Si. An energy level is produced at 0.27 eV above the valence band, which acts as a recombination center. The density of these centers can be created controllably. In this case, an expression (3.92) was obtained by Wertheim when condition (3.87) is satisfied.

$$\tau_t = \frac{1}{c_p N_T} + \frac{1 + p_1/N_T}{c_n n_0}. \tag{3.92}$$

Equation (3.92) indicates as the density of bombarding electrons increases to produce N_T^*, the lifetime reaches a minimum. But actually in Fig. 3.16a, a minimum is observed at $10^{17}\,\mathrm{cm^{-2}}$ which is hard to interpret since n_0 is not a constant and decreases during bombardment. Equation (3.92) describes the lifetime variations under various bombardment conditions.

The other issues Wertheim dealt with include (a) direct recombination and recombination via trap centers; (b) recombination in crystals with two recombination centers. For the late case, if $N_1 + N_2 < n_0 + p_0$, he obtained:

$$\frac{1}{\tau} = \frac{[(1 + \mu_1)/\tau_1] + [(1 + \mu_2)/\tau_2]}{1 + \mu_1(1 + \nu_1) + \mu_2(1 + \nu_2)}. \tag{3.93}$$

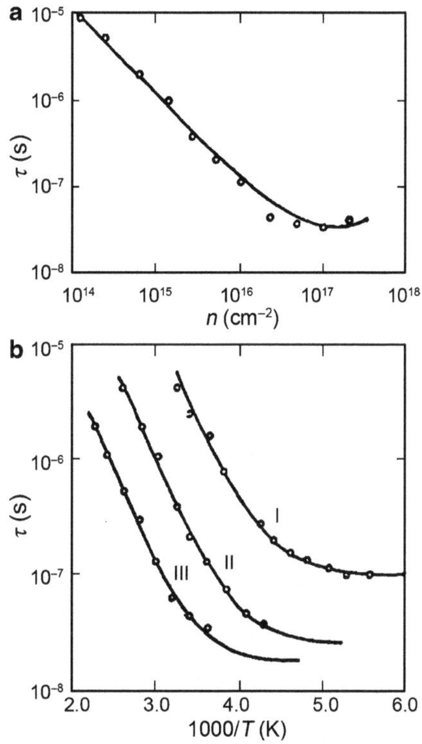

Fig. 3.16 The electron lifetime of an electron bombarded n-Si (7Ω·cm) in which an acceptor level is produced at $E_v + 0.27\,\text{eV}$. (**a**) The curve relating the lifetime to n, the density of electrons bombarding the sample. (**b**) The curves relating the electron lifetime and to the temperature under three bombardment conditions

In an n-type material, the quantities in this expression are defined as follows:

$$\begin{aligned}
\mu_1 &= N_1^-/[p_0 + p_{11} + (n_0 + n_{11})c_{n1}/c_{p1}],\\
\mu_2 &= N_2^-/[p_0 + p_{12} + (n_0 + n_{12})c_{n2}/c_{p2}],\\
v_1 &= c_{n1}(n_0 + n_{11})/[c_{n2}(n_0 + n_{12}) + c_{p2}(p_0 + p_{12})],\\
v_2 &= c_{n2}(n_0 + n_{12})/[c_{n1}(n_0 + n_{11}) + c_{p1}(p_0 + p_{11})],
\end{aligned} \tag{3.94}$$

where n_{11}, p_{11} are the concentrations of electrons and holes, respectively, when the Fermi level is located at the first impurity level. The n_{12} and p_{12} are the corresponding quantities related to the second impurity.

From (3.89) and (3.90), in general the reciprocal of the time constants are matched when $\mu_{\rm I} \ll 1$ and $\mu_i v_i \ll 1$, and only when the density of recombination centers is low enough.

When the recombination centers' levels are close to mid-gap, $\frac{c_{pi} N_i^-}{c_{ni} n_0} < 1, i = 1,\ 2$ is required, indicating a deviation will happen if one of the two centers is negatively charged since $c_p > c_n$ is possible.

Choo (1970) studied the carrier lifetimes in semiconductors containing two recombination energy levels with and without interactions between them. Baicker and Fang (1972) theoretically studied two single-valence trap centers, or one set of valence two trap centers in silicon. Similar issues were considered by Srour and

Curtis (1972) in studying impurities introduced by radiation. These analyses offer a reference in a study of the carrier lifetimes in narrow semiconductors in which the impurities are produced by high-energy radiation.

3.4 Radiative Recombination

3.4.1 Radiative Recombination Processes in Semiconductors

Investigations into radiative recombination are very important to semi-conductor physics and to luminescent devices such as LEDs and lasers. The excess carriers created by electron excitation, current injection, or optical pumping relax back to their equilibrium state driven by different recombination mechanisms. In an intrinsic semiconductor with a direct band-gap, an electron and a hole recombine radiatively to emit a photon, as shown in Fig. 3.17.

The photon emission spectrum depends on carrier distributions in the conduction and valence bands and the momentum conservation law. The momentum conservation law requires a near identical electron and hole momentum because the momentum of photon is much smaller than that of the electrons and holes. Supposing the transition matrix elements are constants with respect to energy, the emission intensity (Mooradian and Fan 1966) per unit photon energy, under a parabolic approximation, is:

$$I(h\nu) \propto \nu^2 (h\nu - E_g) f_e f_h, \tag{3.95}$$

where $h\nu$ is the photon energy, and f_e and f_h are Fermi distribution functions of electrons and holes, respectively. If electrons and holes obey a Maxwell–Boltzmann distribution, then:

$$I(h\nu) \propto \nu^2 (h\nu - E_g) \exp[-(h\nu - E_g)/k_B T], \tag{3.96}$$

where the Coulomb interaction among the free electrons and between the free electrons and holes are not included. At low temperature and low excitation density

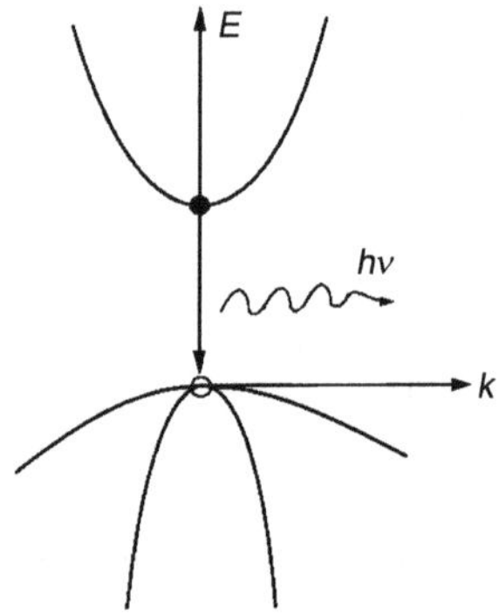

Fig. 3.17 Diagram of radiative recombination

the recombination happens starting from exciton states or bounded impurity states rather than band-to-band transitions since the free carriers may quickly relax to free exciton or be bound at impurity sites. At high temperature (i.e. $k_BT > E'$, with E' related to a bound state energy), then band-to-band transitions will dominate the recombination process because the bound carriers thermally disassociated to free states near the band edges. The PL of the narrow direct band-gap semiconductor HgCdTe is discussed in Sect. 3.6. For a PL study of InSb, the reader is referred to the early work done by Mooradian and Fan (1966).

The recombination rate of free carriers is proportional to square of the carrier concentration n and a factor B, $\sim n^2 B$, where B depends on the transition matrix elements. The factor B can be obtained from the above expressions under an equilibrium condition, where the recombination rate is equal to absorption rate.

3.4.2 *Lifetime of Radiative Recombination*

The radiative recombination lifetime is expressed as $1/\tau = nB$. The lifetime is inversely proportional to the carrier concentration. For GaAs, $B = 10^{-9}\,\mathrm{cm^3\,s^{-1}}$, and when n is less than $10^{16}\,\mathrm{cm^{-3}}$, the band-to-band radiative recombination lifetime is about 10^{-7} s. For such long radiative lifetimes, the shorter nonradiative lifetime will dominate the recombination processes.

In the case of indirect band-gap semiconductors, phonons in a second-order process must be involved in the recombination process in order to conserve momentum, as shown in Fig. 3.18. In this case, the recombination rate is quite slow. Various possible radiative recombination processes are shown in Fig. 3.19.

In thermal equilibrium, the recombination rate of electrons and holes is equal to the photon-induced generation rate. The recombination rate is proportional to the square of intrinsic carrier concentration:

$$G_R = Bn_i^2, \tag{3.97}$$

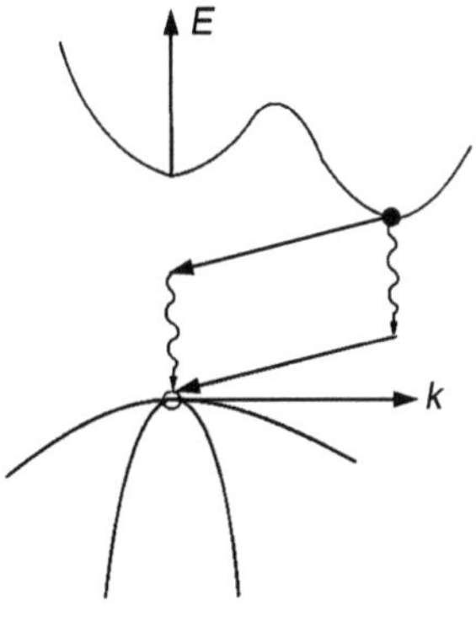

Fig. 3.18 Diagram of an indirect recombination process

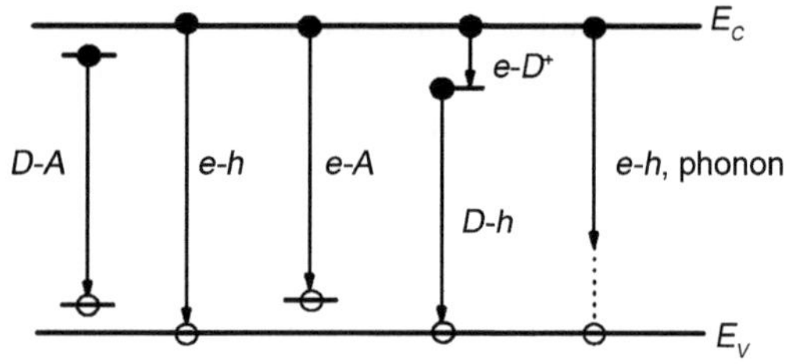

Fig. 3.19 Various possible radiative recombination processes

while the generation rate equals the photon absorption rate at temperature T, (Roosbroeck and Shockley 1954). Thus, we have:

$$G_{\mathrm{R}} = \frac{8\pi}{h^3c^2}\int_0^{\infty}\frac{\varepsilon(E)\alpha(E)E^2\mathrm{d}E}{\exp(E/k_{\mathrm{B}}T)-1}. \tag{3.98}$$

where G_{R} is radiative recombination rate at thermal equilibrium, $\varepsilon(E)$ is the relative dielectric constant, and $\alpha(E)$ is the absorption coefficient. The coefficient B is:

$$B = \frac{1}{n_i^2}\frac{8\pi}{h^3c^2}\int_0^{\infty}\frac{\varepsilon(E)\alpha(E)E^2\mathrm{d}E}{\exp(E/k_{\mathrm{B}}T)-1}. \tag{3.99}$$

B has dimensions $\mathrm{cm}^3.\mathrm{s}^{-1}$, G_{R} has dimensions $\mathrm{cm}^{-3}.\mathrm{s}^{-1}$, and the right side of (3.98) also has dimensions $\mathrm{cm}^{-3}.\mathrm{s}^{-1}$. At thermal equilibrium, a balance between creation and recombination of an electron–hole pair is required, i.e. $n_0 = p_0 = n_{\mathrm{i}}$ for an intrinsic semiconductor, where n_0 and p_0 are thermal equilibrium concentrations of electrons and holes, respectively. When semiconductors are excited externally, excess carriers are created, and thus recombination is a dominant process leading to an equilibrium state. In this case, the recombination rate of the excess carriers becomes (Tang 1974):

$$R = B(np - n_{\mathrm{i}}^2). \tag{3.100}$$

In the case of weak excitation, $n = n_0 + \Delta n$, $p = p_0 + \Delta n$, $\Delta n << n_0, p_0$, then the radiative lifetime is:

$$\tau_{\mathrm{n}} = \tau_{\mathrm{p}} = \tau_{\mathrm{R}} = \frac{\Delta n}{R} = \frac{\Delta n}{B(np - n_{\mathrm{i}}^2)}. \tag{3.101}$$

We have

$$\tau_{\mathrm{R}} = \frac{1}{B(n_0 + p_0 + \Delta n)} \approx \frac{1}{B(n_0 + p_0)}. \tag{3.102}$$

Substituting (3.97) into (3.102), we have:

$$\tau_{\mathrm{R}} = \frac{n_{\mathrm{i}}^2}{G_{\mathrm{R}}(n_0 + p_0 + \Delta n)} \approx \frac{n_{\mathrm{i}}^2}{G_{\mathrm{R}}(n_0 + p_0)} \tag{3.103}$$

G_R is determined by (8.105) and for intrinsic materials $n_0 = p_0 = n_i$, we find:

$$\tau_{Ri} = \frac{1}{2Bn_i}. \tag{3.104}$$

Substituting B into (3.102), we obtain:

$$\tau_R = \frac{2\tau_{Ri} n_i}{n_0 + p_0}. \tag{3.105}$$

For n-type semiconductors, $n_0 \gg p_0$, then:

$$\tau_R = \frac{1}{Bn_0} = 2\tau_{Ri}\left(\frac{n_i}{n_0}\right). \tag{3.106}$$

For p-type semiconductors, $p_0 \gg n_0$, then:

$$\tau_R = \frac{1}{Bp_0} = 2\tau_{Ri}\left(\frac{n_i}{p_0}\right). \tag{3.107}$$

From (3.101) and (3.102), the radiative recombination lifetime is inversely proportional to the majority carrier concentration. In (3.104), (3.106), and (3.107), the lifetimes are related to the coefficient B, which can be calculated from (3.99) after substituting the absorption spectrum and the dielectric response into it. The absorption spectrum and dielectric response has been discussed in detail in Chap. 4 of "Physics and Properties of Narrow Gap Semiconductors" (Chu and Sher 2007). Using an expression of Bardeen et al. (1956), Schacham and Finkman (1985) derived an expression for absorption coefficient:

$$\alpha = \frac{2^{2/3}}{3\varepsilon^{1/2}}\frac{m_0 q^2}{\hbar^2}\left(\frac{m_e^* m_h^*}{m_0(m_e^* + m_h^*)}\right)^{3/2}\left(1 + \frac{m_0}{m_e^*} + \frac{m_0}{m_h^*}\right)\left(\frac{E - E_g}{m_0 c^2}\right)^{1/2}. \tag{3.108}$$

If $E_g > kT$, then

$$B = 5.8 \times 10^{-13}\varepsilon^{1/2}\left(\frac{m_0}{m_e^* + m_h^*}\right)^{3/2}\left(1 + \frac{m_0}{m_e^*} + \frac{m_0}{m_h^*}\right)\left(\frac{300}{T}\right)^{3/2} \left(E_g^2 + 3k_B T E_g + 3.75k^2T^2\right) \tag{3.109}$$

or

$$B = \frac{1}{n_i^2}8.685 \times 10^{28}\varepsilon^{1/2}\left(\frac{m_e^* m_h^*}{(m_e^* + m_h^*)\, m_0}\right)^{3/2}\left(1 + \frac{m_0}{m_e^*} + \frac{m_0}{m_h^*}\right)(kT)^{3/2}. \cdot \exp\left(-\frac{E_g}{kT}\right)\left(E_g^2 + 3k_B T E_g + 3.75k^2T^2\right) \tag{3.110}$$

Considering the electrons that are involved in the recombination process are distributed at the bottom of the conduction band, substituting $\alpha(E)$ in (3.99), Schacham et al. obtained a simplified expression:

$$B = \frac{1}{n_i^2} 2.8 \times 10^{17} \varepsilon \beta T^{3/2} \exp\left(-\frac{E_g}{kT}\right)\left(E_g^2 + 3k_B T E_g + 3.75 k_B^2 T^2\right), \quad (3.111)$$

where

$$\beta = 2.109 \times 10^5 \left(\frac{1+x}{81.9+T}\right)^{1/2} (\mathrm{cm^{-1}eV^{-1/2}}). \quad (3.112)$$

The parabolic approximation used to derive (3.108) restricts its application to HgCdTe. A more rigorous treatment must use a nonparabolic expression. Pratt et al. (1983) found that for $Hg_{0.68}Cd_{0.32}Te$ $\alpha(E)$ can be expressed as:

$$a(E) = 2 \times 10^5 (E - E_g)^{3/2} \mathrm{cm^{-1}}. \quad (3.113)$$

Substituting (3.113) into (3.95), and considering that $\exp(E_g/kT) \gg 1$, B simplifies to:

$$B = \frac{1}{n_i^2} 1.29 \times 10^{30} (k_B T)^{9/2} \exp\left(-\frac{E_g}{k_B T}\right)\left(\left(\frac{E_g}{k_B T}\right)^2 + 5\frac{E_g}{k_B T} + 8.75\right) \mathrm{s^{-1}\,cm^{-3}}. \quad (3.114)$$

In addition, a simplified expression for B can also be obtained using the exponential fit to the absorption coefficient $\alpha(E) = \alpha_g \exp[\beta'(E - E_g)]^{1/2}$ (refer to (4.136) of "Physics and Properties of Narrow Gap Semiconductors" (Chu and Sher 2007)).

3.4.3 Radiative Recombination in p-Type HgCdTe Materials

From (3.105), we see that $B = G_R/n_i^2$ is a slow function of temperature, and therefore the concentration of majority carriers in n-type semiconductors can be taken to be a constant over quite a large range of temperatures in a nonintrinsic state. Consequently, the radiative recombination lifetime at moderate temperatures is also a slow function of temperature for n-type materials. However, this is not the case for p-type materials at lower temperatures, where the lifetime increases exponentially with temperature due to the carrier freeze-out effect in the low-temperature range. The relation between carrier concentration and temperature in this case can be obtained from the neutrality condition given by (3.115):

$$n_0 + N_a^- = p_0 + N_d^+, \quad (3.115)$$

where N_d^+ and N_a^- are concentrations of donor and acceptor ions, and n_0 and p_0 are the electron and hole concentrations at thermal equilibrium. Another factor is the compensation. At low temperatures, we have for a compensated semiconductor:

$$p_0 \cong \frac{N_a - N_d}{N_d}\frac{N_v}{g}\exp\left(-E_a/k_BT\right). \tag{3.116}$$

While for an uncompensated semiconductor, i.e., $p_0 \gg N_d$, one finds:

$$p_0 \cong \left(\frac{N_vN_a}{g}\right)^{1/2}\exp\left(-E_a/k_BT\right) \tag{3.117}$$

where N_v is the effective density of states at the top of valence band, and g is the degeneracy of the acceptors. Schacham and Finkman (1985) calculated the radiative lifetime of a p-type $Hg_{0.785}Cd_{0.215}Te$ sample with $E_a = 15\,\text{meV}$ and $N_a - N_d = 10^{15}, 10^{16}, 10^{17}\,\text{cm}^{-3}$ (Fig. 3.20).

By comparing curves (2) and (3), we see that the recombination rate is decreased and the carrier lifetime is increased as a result of the decrease of the hole concentration induced by carrier freeze-out. From curves (1) and (2), the lifetime of a compensated semiconductor relative to that of an uncompensated one is increased since the lifetime τ_R is proportional to $N_d/(N_a - N_d)$ in the case of compensation.

Besides the freeze-out effect, background radiation also has an impact on the lifetime at low temperatures for $Hg_{1-x}Cd_xTe$ with small x. If the background temperature is a constant, the lifetime is a function of view angle, composition, and temperature of the device. The change of composition and temperature produces a change of the band-gap, therefore the cut-off wavelength. This results in a change of integrated background radiation. Figure 3.21 shows the temperature dependence of the background radiation flux at a view angle of 180° and a background temperature

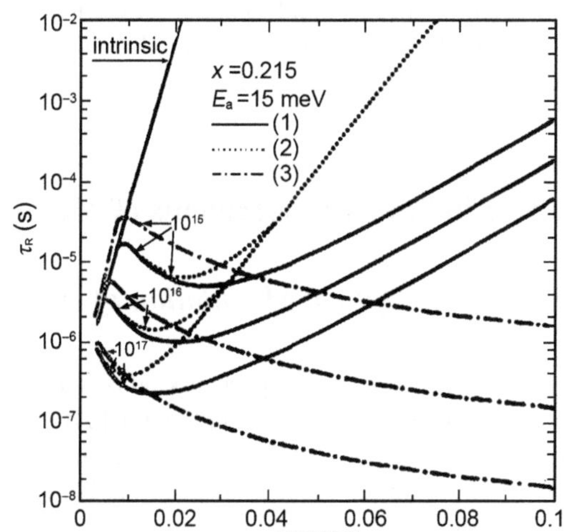

Fig. 3.20 Radiative lifetime of a p-type $Hg_{0.785}Cd_{0.215}Te$ sample with $E_a = 15\,\text{meV}$ and various doping concentrations: (1) $N_d = 0$, uncompensated semiconductor, (2) $N_d = 0.5N_a$, compensated semiconductor, and (3) without considering the freeze-out effect

Fig. 3.21 Variation of background radiation flux as a function of temperature of HgCdTe

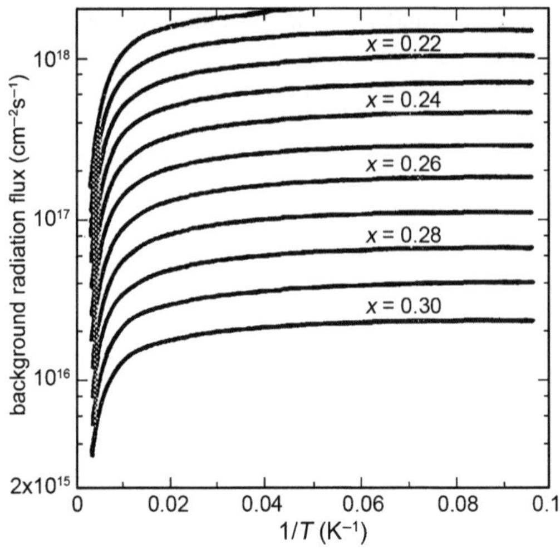

of 300 K for HgCdTe samples with different compositions (Schacham and Finkman 1985). We can see that for a $Hg_{1-x}Cd_xTe$ device with $x = 0.22$ the photon flux of a 300 K background radiation reaches a value of $10^{18}\,cm^{-2}\,s^{-1}$ when the device is held at low temperature. At low temperature p_0 decreases exponentially, and thus the generation of excess carriers and radiative recombination processes are dominated by this background radiation. Due to the generation of excess carriers from the background radiation, the recombination rate increases which results in a negative impact on increasing the lifetime. So, the recombination lifetime does not increase exponentially at low temperatures and it faces a repression, see Fig. 3.22. This phenomenon is clearer for small x, even at higher temperatures.

Figure 3.22 is a calculated result by Schacham. Figure 3.22a is the lifetime of HgCdTe samples with $x = 0.25$ and $x = 0.29$ at different doping concentrations, in which the background radiation effect has been considered. Figure 3.22b is the radiative lifetime of a HgCdTe sample with $x = 0.215$ at different doping concentrations and different view angles.

3.5 Lifetime Measurements of Minority Carriers

3.5.1 *The Optical Modulation of Infrared Absorption Method*

$Hg_{1-x}Cd_xTe$ is a very important kind of semiconductor material in the fabrication of infrared detectors (Long and Schmit 1970; Tang 1974). Lifetimes of nonequilibrium carriers in $Hg_{1-x}Cd_xTe$ are deterministic performance parameters (Petersen 1970; Kinch et al. 1973; Bartoli et al. 1974; Gerhardts et al. 1978; Baker et al. 1978). Most research has focused on n-type $Hg_{1-x}Cd_xTe$ with a composition near

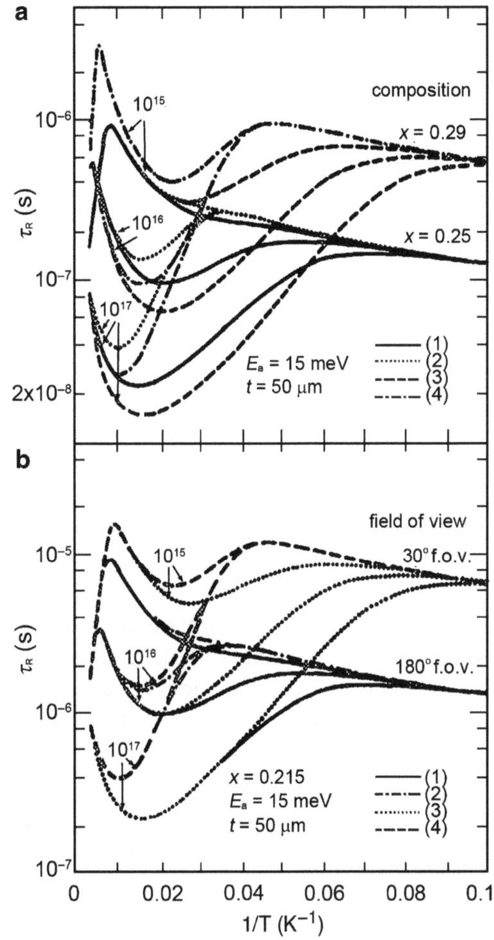

Fig. 3.22 (**a**) Radiative lifetime of $Hg_{1-x}Cd_xTe$ samples with $x = 0.25$ and $x = 0.29$ at different doping concentrations. The effects of background radiation have been taken into account. (**b**) Radiative lifetime of $Hg_{1-x}Cd_xTe$ with $x = 0.215$ at different doping concentrations and different view angles

$x = 0.2(E_g \sim 0.1\,\text{eV}, 77\,\text{K})$. The methods for measuring lifetimes of nonequilibrium carriers include the classical photoconductivity decay rate, and the steady-state photoconductivity. These methods employ the decay of the photoconductivity to determine the minority carrier lifetime. In addition, techniques such as pulse recovery and electron beam induced current (Polla et al. 1981a) are also adopted in the research on p-type $Hg_{1-x}Cd_xTe$. The carrier lifetime and the uniformity of its distribution are important in the fabrication of IR detectors, for example, in the manufacturing of SPRITE detectors (Elliott 1977 1981), where material having long carrier lifetimes exhibit the best performance. Therefore, in the selection of the best detectors, it is of great importance to use nondestructive techniques to measure the carrier lifetime.

Optical Modulation of Infrared Absorption (OMIA) is a noncontact, nondestructive measurement technique. By using this technique, Afromowitz and Didomenico (1971) first experimentally determined the carrier lifetime of GaP.

Then, Mroczkowski et al. (Mroczkowski et al. 1981; Polla et al. 1981b) reported the lifetime of carriers in near intrinsic and p-type $Hg_{0.7}Cd_{0.3}Te$ at low temperature.

The principle behind the operation of the OMIA technique for measuring the carrier lifetime of a semiconductor is that the optically generated carriers will affect the infrared absorption of free carriers, and thus the lifetime of excess carriers can be quantitatively determined.

There are two independent light beams used in an OMIA experiment: a modulated pump beam with photon energy $h\nu_{\mathrm{Pump}} > E_{\mathrm{g}}$ and a probe beam with photon energy $h\nu_{\mathrm{Probe}} < E_{\mathrm{g}}$. In the experiment, the probe-beam intensity passing through the sample without the modulated pump beam, I, is first measured, then the probe-beam intensity passing through the sample with the pump beam present, $I + \Delta I$, where ΔI is the peak amplitude of the modulated signal, is determined. The lifetime of photogenerated carriers can be obtained from the ratio $\Delta I/I$.

Additional electron–hole pairs are generated in the semiconductor sample when the pump beam is applied, and as a consequence the probe-beam absorption of the sample increases from $\alpha(E)$ to $\alpha(E) + \Delta\alpha(E)$, and if $\Delta n = \Delta p$, then:

$$\Delta\alpha(E) = \left[\sigma_{v2v1}(E) + \sigma_{\mathrm{n}}(E) + \sigma_{\mathrm{p}}(E)\right]\Delta p = \sigma(E)\Delta p. \tag{3.118}$$

Here, $\sigma_{v2v1}(E)$ is the absorption cross-section of inter-band transitions from the light-hole band to the heavy-hole band, and $\sigma_{\mathrm{n}}(E)$ and $\sigma_{\mathrm{p}}(E)$ are absorption cross-sections of intra-band transitions of electrons and holes, respectively.

When the pump beam is applied, the transmission intensity of the probe beam passing through a sample of thickness d is:

$$\begin{aligned} I + \Delta I &= I\left[1 - \sigma(E)\int_0^{\mathrm{d}} \Delta p(x)\mathrm{d}x\right], \\ (\Delta I)/I &= -\sigma(E)\int_0^{\mathrm{d}} \Delta p(x)\mathrm{d}x. \end{aligned} \tag{3.119}$$

If surface recombination can be ignored, and the pump beam is totally absorbed within a depth much smaller than the diffusion length, then the stationary distribution of excess carriers is:

$$\Delta p(x) = \left[(1 - R_\lambda)r_\lambda Q_{\mathrm{p}}\tau/L\right]\exp(-x/L), \tag{3.120}$$

where R_λ is the reflectivity of pump radiation at the sample surface, r_λ the quantum yield (the number of electrons excited per photon) for pump radiation of wavelength λ, Q_{p} is the photon flux density, τ the lifetime, and L is the diffusion length. For pump photon energies greater than the band-gap, the quantum yield can be greater than one. From (3.118) and (3.119), we get:

$$\Delta I/I = (1 - R_\lambda)r_\lambda\sigma(E)Q_{\mathrm{p}}\tau\left[1 - \exp(-d/L)\right]. \tag{3.121}$$

In a situation where the sample thickness d is much larger than the diffusion length L, then:

$$(\Delta I)/I = (1 - R_\lambda) r_\lambda \sigma(E) Q_p \tau. \tag{3.122}$$

Equation (3.122) indicates that if R_λ, r_λ and $\sigma(E)$ are known, then for a given Q_p, the carrier lifetime τ can be obtained by measuring $\Delta I/I$.

R_λ is obtained from the refractive indices of HgCdTe (Baars and Sorger 1972); for pump radiation at $\lambda = 0.6328\,\mu$m, then $r_\lambda = 3$; and σ_{v2v1} is given in the literature (Mroczkowski et al. 1981):

$$\sigma_{v2v1}(E) = \frac{e^2 p^2 k}{nE\hbar c}\left(\frac{\eta - E_g}{4p^2 - 3\eta\beta}\right)\frac{f\exp(E_{v1}/k_B T)}{N_v}, \tag{3.123}$$

where $\eta = (E_g^2 + 8P^2k^2/3)^{1/2}$, $\beta = \hbar^2(m_0^{-1} + m_{hh}^{-1})$, $E_{v_1} = -\hbar^2k^2/2m_{hh}$, P is the Kane momentum matrix element, n the refractive index, N_v the effective density of states in the heavy-hole band, f the fraction of holes in the heavy-hole band and is $\sim$0.75, m_{hh} the effective mass of the heavy-hole band, and $\hbar k$ is the lattice momentum of the energy difference E (the photo energy of the probe) between the heavy-hole band and the light-hole band:

$$k^2 = \left(\frac{4P^2}{3\beta^2} - \frac{E_g}{\beta}\right) - \frac{2E}{\beta} - \left[\left(\frac{4P^2}{3\beta^2} - \frac{E_g}{\beta}\right)^2 - \frac{16P^2E}{3\beta^2}\right]^{1/2}. \tag{3.124}$$

Classical free carrier absorption formulas are adopted for calculating σ_n and σ_p (Moss et al. 1973):

$$\sigma_n + \sigma_p = \frac{\lambda^2 e\hbar}{137\pi c^2 n}\left(\frac{1}{m_e^2\mu_e} + \frac{1}{m_h^2\mu_h}\right), \tag{3.125}$$

here μ_e and μ_h are electron and hole mobilities, respectively.

The calculated results for probe beams with wavelengths $\lambda = 10.6$ and $14.5\,\mu$m, are given in Table 3.1.

The configuration of an OMIA experiment is shown in Fig. 3.23. In the experiment, the probe beam is a CO_2 laser ($\lambda = 10.6\,\mu$m) and the pump beam is a He–Ne laser (0.6328 μm). The pathways of the probe beam and the pump beam are carefully adjusted so that the two beams aim at the same point on sample. The sample is placed in a container with windows, in which it is cooled by liquid nitrogen and the temperature can be controlled. A chopper is used to modulate the pump beam (or the probe beam). The transmitted intensity of the probe beam passing through the

Table 3.1 Absorption cross-section

Wavelength of the probe beam (μm)	$\sigma_{v_2v_1}(cm^2)$	$\sigma_n(cm^2)$	$\sigma_p(cm^2)$
10.6	0.234×10^{-15}	$5.05 \times 10^{-13}/\mu_e$	$2.02 \times 10^{-12}/\mu_h$
14.5	0.218×10^{-15}	$9.45 \times 10^{-13}/\mu_e$	$3.78 \times 10^{-12}/\mu_h$

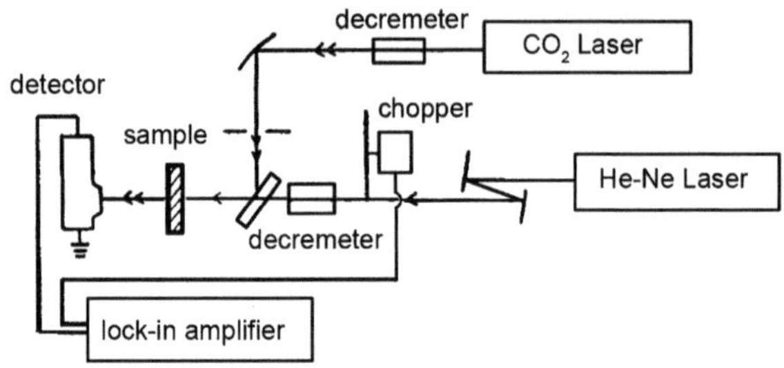

Fig. 3.23 The experimental setup for an OMIA measurement of the carrier lifetime

Table 3.2 The material parameters and measured lifetimes of four HgCdTe samples

			77 K	77 K	77 K	90 K	90 K
Sample No.	Composition (x)	$\left(\frac{\Delta x}{x}\right)$ %	E_g (eV)	$N_D - N_A$ (cm^{-3})	μ_e ($cm^2/v.s$)	τ (s)	Range of τ(s)
1	0.245	1.9	0.164	5.7×10^{15}	2.6×10^4	9.4×10^{-7}	2.5×10^{-7}–1.8×10^{-6}
2	0.325	1.3	0.292	1.1×10^{15}	6.9×10^3	6.3×10^{-8}	1.9×10^{-8}–1.6×10^{-7}
3	0.34	2.4	0.315	4.7×10^{14}	2.3×10^4	1.4×10^{-7}	8.6×10^{-8}–2.6×10^{-7}
4	0.265	3.5	0.195	5.7×10^{15}	3.7×10^4	1.6×10^{-7}	5.7×10^{-8}–2.8×10^{-7}

sample without a pump beam present is, I, and is measured first; then the modulated pump beam is applied and the transmission intensity of the resulting output of the now modulated probe beam passing through sample, $I + \nabla I$, is measured again. The modulation frequency is 800 Hz.

The transmitted intensity is recorded by a $Hg_{0.8}Cd_{0.2}Te$ photoelectric detector (cut-off wavelength: 12 μm) operated at 77 K. The output signal of the detector is fed into a lock-in amplifier. In an OMIA experiment, a probe light with a continuous spectra (with long enough wavelengths) can also be used, but then the computations become complex.

Ling and Lu (1984) using OMIA measured the carrier lifetime of n-type $Hg_{1-x}Cd_xTe$ with alloy concentrations between $0.24 < x < 0.35$. The samples were prepared by solid-state re-crystallization, and processed by mechanical polishing, followed by chemical etching in a mixed solution of bromine–methanol, to a final thickness of 0.3–0.5 mm. The concentrations are determined by density and electron-probe measurements. The uniformity of the alloy concentrations and electric parameters are listed in Table 3.2. The power of probe beam I_{probe} is limited to the order of a milliwatt, while the power of pump beam I_{pump} is 15 mW. Q_p is calculated from the measured beam power by assuming a Gaussian beam distribution.

The relation between ΔI and I under the condition of a fixed pump beam power $Q_p = 5.4 \times 10^{18}/s \cdot cm^2$ is shown in Fig. 3.24, while the relation between ΔI and I under the condition of fixed probe beam power I_{probe} is shown in Fig. 3.25. The linear relations in Figs. 3.24 and 3.25 are consistent with (3.122) which means that for the above experimental conditions, the lifetime of carriers is independent of both the intensity of probe beam and the intensity of pump beam.

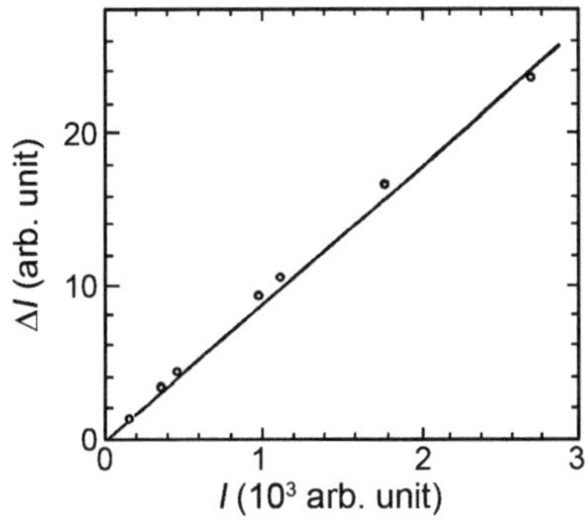

Fig. 3.24 The linear curve of ΔI vs. I for sample 3($x = 0.34$) under the condition of a fixed pump beam power I_{pumb} and temperature $T = 300\,\text{K}(Q_{\text{p}} = 5.4 \times 10^{18}/\text{s}\,\text{cm}^2, \sigma(E) = 1.5 \times 10 -^{15}$ cm^2, with slope $\Delta I/I = 8.8 \times 10^{-3}$, the probe beam is a CO_2 laser ($\lambda = 10.6\,\mu$m) and the pump beam is a He–Ne laser (0.6328 μm))

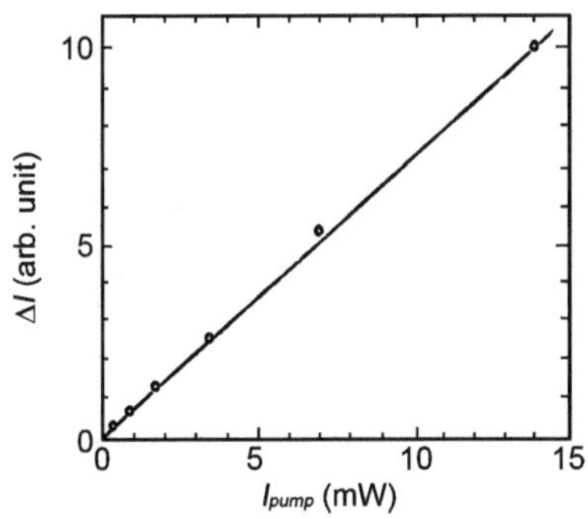

Fig. 3.25 The linear curve of ΔI vs. I_{pumb} for sample No. 3 ($x = 0.34$) under the condition of fixed-probe beam power I_{probe} and temperature $T = 300\,\text{K}\,(\lambda_{\text{probe}} = 10.6\,\mu\text{m}, \lambda_{\text{pump}} = 0.6328\,\mu\text{m})$ and in this case the result is the same as that obtained in Fig. 3.24 ($\tau = 3.6 \times 10^{-7}$ s)

In the condition $Q_{\text{p}} = 3.4 \times 10^{18}/\text{s} \cdot \text{cm}^2$, the ratio of $\Delta I/I$ is measured for samples labeled 1–4 within temperatures ranging from 90 to 330 K. One needs to know the temperature dependence of $\sigma(E)$ to obtain $\tau(T)$ from $\Delta I/I$.

According to (3.123) and (3.125), the relation between the absorption cross-section $\sigma(E)$ and the temperature is calculated in terms of the sample composition (determining E_{g}) and the CO_2 laser beam wavelength ($\lambda = 10.6\,\mu$m or $E = 0.117$ eV). In the calculation, $P = 8.5 \times 10^{-8}$ eV·cm (Mroczkowski et al. 1981) and $m_{\text{hh}} = 0.55 \cdot m_0$. Both m_{e} and E_{g} are cited from literature (Schmit 1970; Chu et al. 1982), and they are both alloy concentration and temperature dependent. For an n-type material, σ_{p} can be neglected, because $m_{\text{h}}^2 \mu_{\text{h}} >> m_{\text{e}}^2 \mu_{\text{e}}$. The electron mobility μ_{e} is obtained from the Hall effect and resistivity measurements. The calculated cross-section of sample No. 4 ($x = 0.265$) probed by radiation with a photon energy $E = 0.117\,\text{eV}(\lambda = 10.6\,\mu\text{m})$ is shown in Fig. 3.26. It is apparent that at low temperature the cross-section is mainly determined by $\sigma_{v_1 v_2}$, while within the intrinsic temperature range the cross-section is mainly determined by σ_{n}.

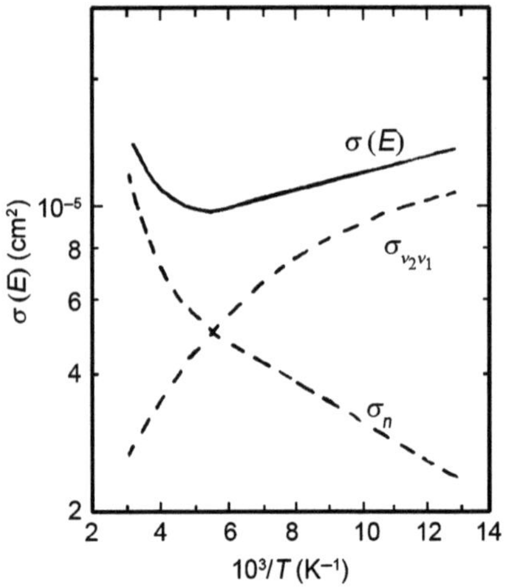

Fig. 3.26 The cross-section of sample No. 4 ($x = 0.265$) probed by a beam with $E = 0.117\,\mathrm{eV}(\lambda = 10.6\,\mu\mathrm{m})$

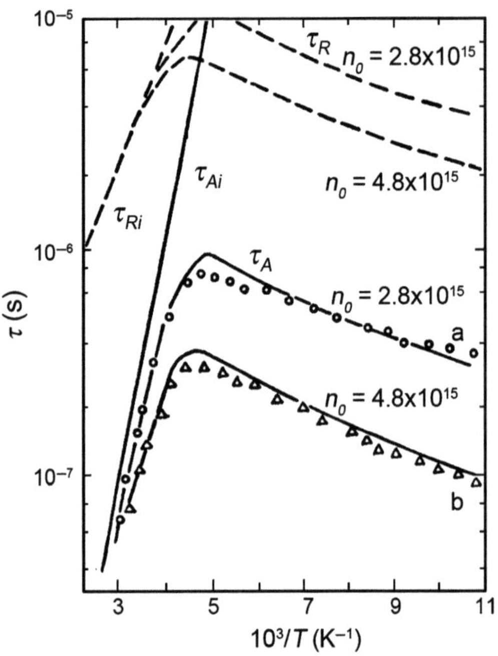

Fig. 3.27 The relation between lifetime and temperature of sample No. 4 ($x = 0.265$), in which lifetime is dominated by Auger recombination. The data designated by "a" and "b" were taken by different groups

The relation between lifetime τ and temperature T can be obtained from (3.122) with experimental $\Delta I/I$ data and calculated values of $\sigma(E)$. These relations for samples No. 4 ($x = 0.265$) and No. 2 ($x = 0.325$) are shown in Figs. 3.27 and 3.28, respectively. These two groups of experimental data correspond to two different test positions "a" and "b" on the samples.

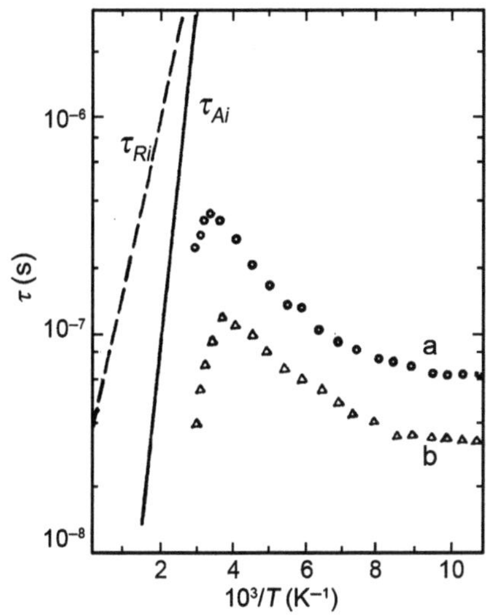

Fig. 3.28 The relation between lifetime and temperature of sample No. 2 ($x = 0.325$), in which the lifetime is dominated by Shockley–Read recombination. The data designated by "*a*" and "*b*" were taken by different groups

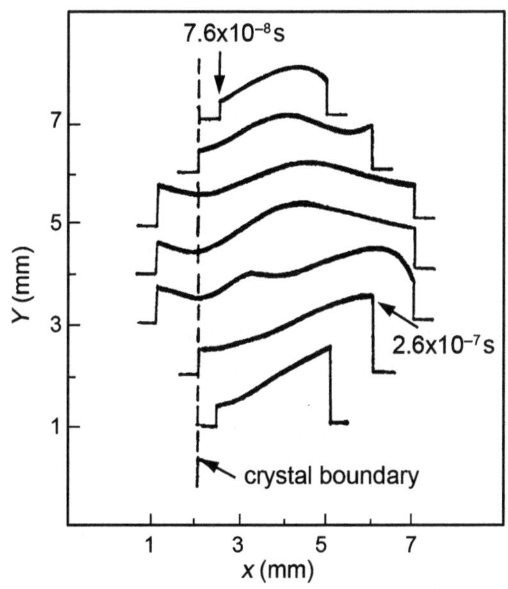

Fig. 3.29 The distribution of lifetimes over a plane with the probe spot moved in 1 mm steps (Sample No. 3, $x = 0.34$, $T = 90\,\text{K}$)

The distribution of carrier lifetimes over a plane has also been measured by scanning the probe spot with a step length of 1 mm at 300 and 90 K. The resulting distributions for sample No. 3 ($x = 0.34$) at 90 K is shown in Fig. 3.29.

Let's analyze the experimental results for various carrier recombination mechanisms. If there are no recombination centers, then the carrier recombination mechanisms include only radiative recombination and Auger recombination.

Under a weak excitation condition, the lifetime for radiative recombination is:

$$\tau_R = \frac{1}{B(n_0 + p_0)}. \tag{3.126}$$

For an intrinsic semiconductor, $n_0 = p_0 = n_i$, the radiative lifetime dependence reduces to:

$$\tau_{Ri} = \frac{1}{2Bn_i}. \tag{3.127}$$

Here, B is the capture probability for minority carriers. By assuming (a) the energy band structure is parabolic near $k = 0$, (b) the carriers obey Boltzmann statistics, and (c) the theoretical formula for the absorption coefficient is adopted, B can be written as (Hall 1959):

$$B = 5.8 \times 10^{-13} \varepsilon_\infty^{1/2} \left(\frac{m_0}{m_e^* + m_h^*}\right)^{3/2} \left(1 + \frac{m_0}{m_e^*} + \frac{m_0}{m_h^*}\right) \left(\frac{300}{T}\right)^{3/2}. \times \left(E_g^2 + 3k_B T E_g + 3.75 k_B^2 T^2\right) \tag{3.128}$$

Under a weak excitation condition, for a semiconductor material with $\mu = \frac{m_e^*}{m_h^*} \ll 1$, the Auger recombination lifetime is (Blakemore 1962):

$$\tau_{Ai} = 3.8 \times 10^{-18} \varepsilon_\infty^{1/2} (1+\mu)^{1/2} (1+2\mu) \times \left[\frac{m_e^*}{m_0} |F_1 F_2|^2 \left(\frac{k_B T}{E_g}\right)^{3/2}\right]^{-1} \exp\left[\frac{1+2\mu}{1+\mu} \frac{E_g}{k_B T}\right], \tag{3.129}$$

in which $F_1 F_2$ is the product of Bloch function overlap integrals between the conduction and valance bands, and the conduction band edge and an excited level in the conduction band (Auger 1).

For an n-type material, the Auger lifetime reduced to the expression:

$$\tau_A = \frac{2\tau_{Ai} n_i^2}{n_0(n_0 + p_0)}. \tag{3.130}$$

Both the radiation recombination lifetime τ_R and the inter-band collision recombination lifetime τ_A depend on the composition (x) of $Hg_{1-x}Cd_xTe$, the temperature (T) and the net donor concentration ($N_D - N_A$). According to the report of Baker et al. (1978), if x is small then $\tau_A \leqslant \tau_R$, i.e., the lifetime is dominated by collision induced recombination, but with an increase of x, τ_A increases rapidly while τ_R decreases slowly, which means there is a critical composition (x_c), for which if $x > x_c$ then $\tau_A \geq \tau_R$, and the lifetime is dominated by radiative recombination. The critical composition is related to temperature and the net donor concentration, for the case of $N_D - N_A = 1 \times 10^{15}\,\text{cm}^{-3}$ and $T = 90\,\text{K}$, then $x_c \approx 0.3$.

If there is a recombination center with energy level E_T and density N_T, existing within the energy gap, electrons and holes can recombine via this center. For a condition in which $N_T \ll n_0$ and weak excitation, the lifetime in the "steady-state approximation" can be expressed by the Shockley–Read formula (Shockley and Read 1952):

$$\tau = \tau_{p0}\frac{n_0 + n_1}{n_0 + p_0} + \tau_{n0}\frac{p_0 + p_1}{n_0 + p_0}, \tag{3.131}$$

where n_1 and p_1 are the concentrations of electrons and holes if the Fermi level coincides with the energy of the recombination center, while $\tau_{p0} = (N_T c_p)^{-1}$, $\tau_{n0} = (N_T c_n)^{-1}$, here c_n and c_p are, respectively, the recombination center's capture coefficients for electrons and holes.

For an n-type sample, within a nonintrinsic temperature range, because $n_0 \gg p_0$, $n_0 \gg n_1$, $p_1 \gg p_0$, then (3.131) can be simplified:

$$\tau_{S-R} = \tau_{p0} + \tau_{n0}\frac{p_1}{n_0} = \tau_{p0} + \tau_{n0}\frac{N_v \exp\left[(E_v - E_T)/k_B T\right]}{n_0}. \tag{3.132}$$

Within this range of temperatures, $n_0 = N_D - N_A$, and it is a temperature independent constant. If the lifetime is dominated by the recombination center mechanism, i.e., is smaller than τ_R or τ_A, then the lifetime varies exponentially with temperature following the relation in (3.132). We can analyze the experimental results by using above recombination theory.

The relation between the lifetime vs. temperature of sample No. 4 ($x = 0.265$) is shown in Fig. 3.25. The solid and dashed curves are theoretical lifetimes due to collision induced recombination and radiative recombination computed from (3.127) to (3.130). The results show that, for the same net donor concentration, τ_R is greater than τ_A by an order of magnitude. The two groups experimental data given in the figure can be fitted by theoretical values with $n_0 = 2.8 \times 10^{15}$ and $n_0 = 4.8 \times 10^{15}\,\text{cm}^{-3}$, respectively. The net donor concentrations n_0 used to fit the experimental data are in accord with the results of Hall-effect measurements. (See Table 3.2.) For the intrinsic temperature $T = 250\,\text{K}$, the experimental data lies close to the theoretical lifetime τ_{Ai} due to intrinsic collision recombination. Therefore, no matter whether the temperature is at the edge or above the intrinsic range, the lifetime of sample No. 4 is dominated by inter-band collision recombination. The difference between the two groups experimental data reflect that the experiments were conducted on different locations on the samples, which means the net donor concentration was nonuniform. Similar results were obtained on sample No. 1 ($x = 0.245$).

The lifetime vs. temperature of sample No. 2 ($x = 0.325$) is shown in Fig. 3.13. We find that at low temperature the experimental lifetime is lower than that of the theoretical radiative recombination prediction (90 K, $n_0 = 1.1 \times 10^{15}\,\text{cm}^{-3}$, $\tau_R = 8.2 \times 10^{-6}$ s) by two orders of magnitude. Within the nonintrinsic temperature range, the relation between the lifetime and temperature follows the characteristics of Shockley–Read recombination, and can be described by (3.130). For "group a" experimental data, given $\tau_{p0} = 66\,\text{ns}$, the energy level of the recombination center E_R can be obtained from the slope of the expression:

$$\ln(\tau - \tau_{p0})/T^{3/2} \sim \frac{1}{T}. \tag{3.133}$$

Extracted from (3.133), from which we get $E_T \approx E_V + 30\,\text{meV}$. Within the intrinsic temperature range, where the lifetime is dominated by recombination via a recombination center, the lifetime τ severally deviates from that of an intrinsic collision recombination process. The differences between the two groups' data in the figure reflect the fact that the density of recombination centers N_R differs at different positions. Similar results are obtained for sample No. 3 ($x = 0.34$), in which $\Delta E = |E_T - E_v| \approx 40\,\text{meV}$.

The distribution of a typical lifetime obtained by sweeping the probe beam spot over a plane is shown in Fig. 3.27, it is the data measured on sample No. 3 ($x = 0.34$) at $T = 90\,\text{K}$. It is obvious that the lifetimes differ from position to position, and decrease at grain boundaries. The planar distribution of the lifetimes at $T = 300\,\text{K}$ is nearly the same as that at $T = 90\,\text{K}$. This can also be seen from the relation between lifetimes and temperature for different sample positions shown in Figs. 3.25 and 3.26. The range of lifetimes for all samples at $T = 90\,\text{K}$ are listed in Table 3.2.

According to these lifetime vs. temperature measurement results, for different samples and probe spot positions, we conclude that the variation of lifetimes originates from fluctuations of the net donor concentration or a nonuniform distribution of recombination centers. According to (3.126) and (3.130), for radiative recombination τ_R is proportional to $1/n_0$, and for inter-band collision recombination τ_A is proportional to $1/n_0^2$. For recombination via a recombination center (Shockley–Read recombination), at low temperature, $\tau_{S\text{–}R}/\tau_{po} = (N_R c_p)^{-1}$, i.e., the lifetime $\tau_{S\text{–}R}$ is proportional to $1/N_T$.

There are various defects in a HgCdTe semiconductor, in which a donor (a Hg interstitial atom), an acceptor (a mercury vacancy), and the density of defect energy levels are related to thermal processing conditions (Tang 1976), so it is easy to understand why there are fluctuations of the net donor concentration, or a nonuniform distribution of recombination centers. In the measurement of carrier lifetimes of n-type $Hg_{1-x}Cd_xTe$ materials, compared to classical stationary state photoconductivity or photoconductivity decay, the OMIA technique has the following virtues: it offers a noncontact, nondestructive, point by point measurement technique through probe scanning.

If appropriate probe beams ($h\nu_{\text{Probe}} < E_g$) and pump beams ($h\nu_{\text{Pump}} > E_g$) are selected, the OMIA technique can also be used for other semiconductor materials. For example, the carrier lifetime in an n-type InSb sample with a net donor concentration $n = 4.5 \times 10^{-8}$ s measured by an OMIA technique at room temperature, resulted in $\tau = 4.5 \times 10^{-8}$ s (the absorption cross-section used in the calculation is cited in the literature (Kurnick and Powell 1959)). This value coincides with the lifetime of InSb determined to be due to intrinsic collision recombination at 300 K reported by Huang and Tang (1965).

3.5.2 The Investigation of Minority Carriers Lifetimes in Semiconductors by Microwave Reflection

Although photoconductivity decay is a simple technique for measuring the lifetime of minority carries, it has the disadvantage of requiring that two contact electrodes be added at both sides of wafer which can induce damage in the material. Microwave reflection is a noncontact technique for measuring the minority carrier lifetime. Using this technique, the contact electrodes are eliminated and any damage to the sample can be avoided.

When a light beam pulse with photon energy greater than energy gap irradiates a semiconductor, nonequilibrium carriers are generated. A sketch of the measurement system of the nonequilibrium carrier lifetime by direct current photoconductivity is shown in Fig. 3.30. The decay of the photoconductivity with time can be observed directly on the screen of oscilloscope (an exponential decay), thus the lifetime τ can be obtained from the photoconductivity vs. time curve.

A sketch of the system for measuring the minority carrier lifetime by microwave reflection is presented in Fig. 3.31. If a light-beam pulse and a high-frequency electromagnetic wave (microwave) simultaneously irradiate a semiconductor sample, the variation of microwave reflectivity induced by the change of the conductivity generated by a light beam pulse is:

$$\Delta R = R(\sigma + \Delta\sigma) - R(\sigma), \tag{3.134}$$

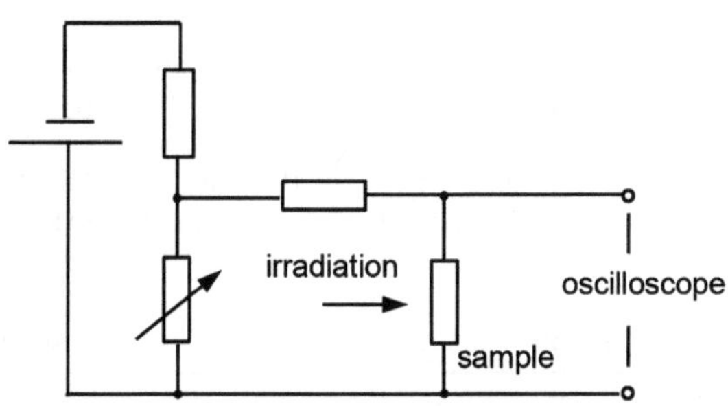

Fig. 3.30 A sketch of the system for the minority carrier photoconductivity decay lifetime measurement

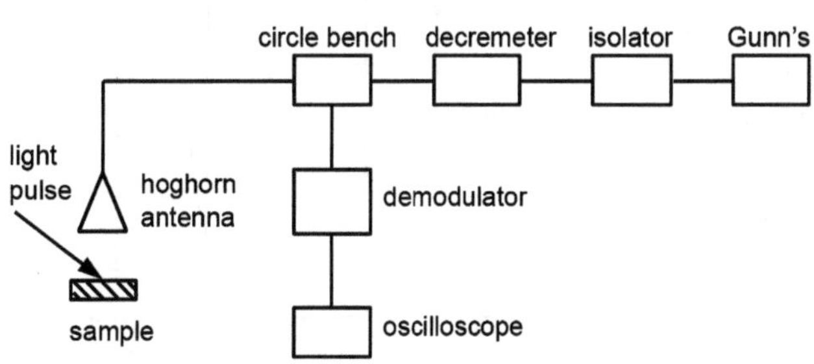

Fig. 3.31 The configuration of a microwave reflection technique for measuring the minority carrier lifetime

where σ is conductivity and $\Delta\sigma$ is photo-induced change in the conductivity. For a small light injection condition, (3.134) can be written as:

$$\Delta R = \frac{\partial R(\sigma)}{\partial \sigma} \cdot \Delta\sigma. \tag{3.135}$$

Within the microwave band, the reflectance is highly sensitive to the conductivity, therefore the variation of the reflectance can accurately measure its decay and the nonequilibrium carrier lifetime.

The configuration of a microwave reflection apparatus for measuring the minority carrier lifetime suggested by Mao and Chu (1993) is shown in Fig. 3.31. A pulsed light beam (wavelength 0.83 μm, power ~10 mW) emitted from a semiconductor laser irradiates a sample; while microwave radiation (wavelength 8 mm, power ~10 mW) generated by a Gunn's oscillator, after passing through isolator, attenuator and circulator, and transmitted out via a horn antenna also irradiates the sample. The microwave power irradiates the same position as the pulsed light beam. The reflected microwave power is received by the horn antenna, and via a circulator and a detector, is fed into an oscilloscope (TDS520). Information about the photoconductivity decay and the nonequilibrium carrier lifetime is obtained from the transient variation of the reflected power with time.

In order to compare the lifetimes measured by both photoconductivity decay (PCD) and microwave radiation (MR) under equivalent experimental conditions, contact electrodes are placed on both sides of samples. Then, both the PCD and MR methods can be applied at the same position.

These two methods were used to measure the lifetimes at different positions on two samples, the experimental results are shown in Figs. 3.32, 3.33, and 3.34. The parameters of the HgCdTe samples are listed below:

No. 1: $x = 0.4$, $n = 2.3 \times 10^{15}\,\mathrm{cm}^{-3}$, 300 K; $n = 1.38 \times 10^{14}\,\mathrm{cm}^{-3}$, 77 K.
No. 2: $x = 0.4$, $n = 1.78 \times 10^{15}\,\mathrm{cm}^{-3}$, 300 K; $n = 6.4 \times 10^{13}\,\mathrm{cm}^{-3}$, 77 K.

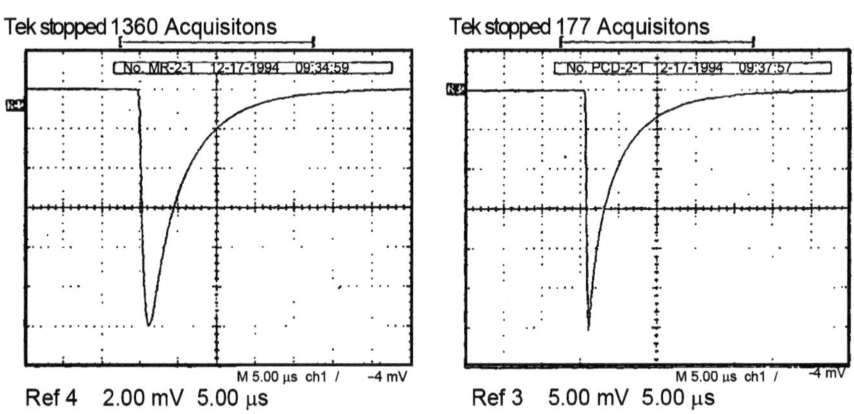

Fig. 3.32 The minority carrier lifetime in HgCdTe obtained by MR and PCD

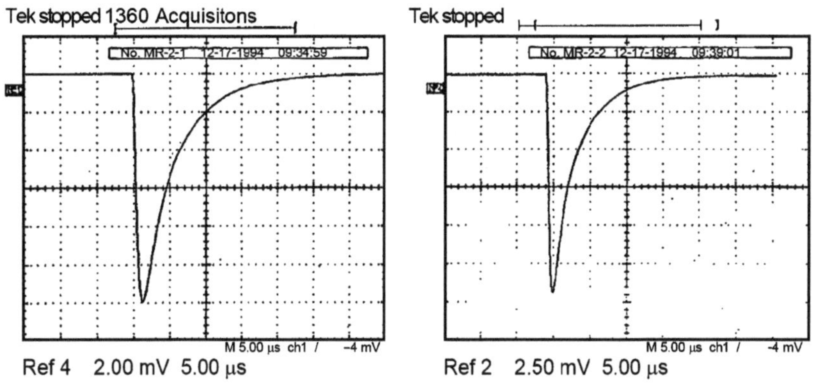

Fig. 3.33 The minority carrier lifetime for different positions on a HgCdTe sample measured by MR

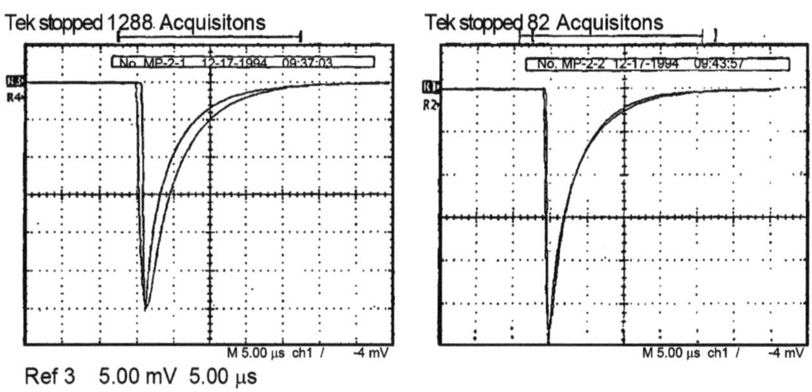

Fig. 3.34 A comparison of the minority carrier lifetimes measured by MR and PCD

It can be seen from the experimental results (see Fig. 3.32) that the minority carrier lifetime measured by MR basically coincides with that of PCD. The minority carrier lifetime for different positions on a sample was also measured by MR, and the results are shown in Fig. 3.33, where it is obvious that the lifetime varies with position. While there is a little difference between the lifetimes obtained by MR and PCD, the lifetimes obtained by MR are generally longer than those obtained by PCD. There are differences of the lifetimes measured at different positions on the same sample using MR and PCD, as shown in Fig. 3.34.

In a PCD measurement, two electrodes are necessary one on each side of the sample, and a bias voltage must be applied to the electrodes. Because the measured conductance responds not only to the region irradiated by the light beam pulse but also to adjacent regions, this can cause a relatively large error in the measured conductance. While in the MR measurement, the reflected microwave power transient is mainly determined by the photoconductive decay of the irradiated region. Therefore, the lifetimes measured by MR are more accurate than those of PCD.

3.5.3 The Application of Scanning Photoluminescence for Lifetime Uniformity Measurements

In the fabrication of $Hg_{1-x}Cd_xTe$ focal plane devices, rigorous material measurement techniques are required. Not only the net nature of material but also the property distribution of tiny regions must be tested. Therefore, many relevant test means have been proposed (Gong 1993; Bajaj et al. 1987; Kopanski et al. 1992; Mao and Chu 1993). These test means, however, often require the preparation of electrodes on samples, which restricts wide applications of such techniques. Because of its nondestructive and noncontact character, the spot scanning PL technique over a plane has been widely used to test semiconductor materials (Hovel 1992). But for $Hg_{1-x}Cd_xTe$, radiative recombination of the photo-generated nonequilibrium carriers causes only a small portion recombination process, while other factors (including thermal background radiation) have a comparatively large effect, which introduces some difficulties into the study. Here, we introduce a Fourier transformation infrared PL technique coupled to a scanning PL measurement. Adopting this technique enables both the surface composition and the nonequilibrium carrier lifetime distributions to be obtained.

Samples selected for scanning PL experiments have included Sb-doped and undoped $Hg_{1-x}Cd_xTe$ single crystal wafers. The selected samples are first etched in a methyl–alcohol solution in order to reduce negative effects due to surface recombination. Then, the sample is placed in a dewar with windows, in which the sample is liquid nitrogen cooled to a temperature of 77 K. The dewar is fixed on a scanning mechanical framework, which moves it in a plane parallel to the sample's surface. The precision of this movement is controllable within 10 μm. A continuous-wave (cw) Ar^+514.5 nm laser is used to excite the front surface of sample. The resulting luminescence is then collected by a parabola mirror and fed into Nicolet 800 Fourier transform spectrometer, after which the signal is finally received by an InSb or a HgCdTe detector operated at 77 K, that is included in Nicolet 800 spectrometer. The Laser can be focused according to different needs of devices, and the mechanical scan system can be set to different step lengths to get different spatial resolutions. The best spatial resolution of this system is 20 μm. It should be pointed out that the scanning is carried out by moving the sample, while the path of laser beam is unchanged during the whole experimental process. In this way, a comparison of the PL intensity from different areas of a sample's surface can be generated.

First, we examine the compositional uniformity of $Hg_{1-x}Cd_xTe$ samples. The PL spectra for different positions on a sample's surface are shown in Fig. 3.35. The sample is a part of an Sb-doped $Hg_{1-x}Cd_xTe$ wafer. Above 50 K, the high-energy side of the PL peak has the highest intensity, the relation between the peak's location and temperature is the same as that between the energy gap and temperature, which means this peak corresponds to a transition from the conduction band to the valence band. A two-dimensional distribution is shown in Fig. 3.36 of peaks'

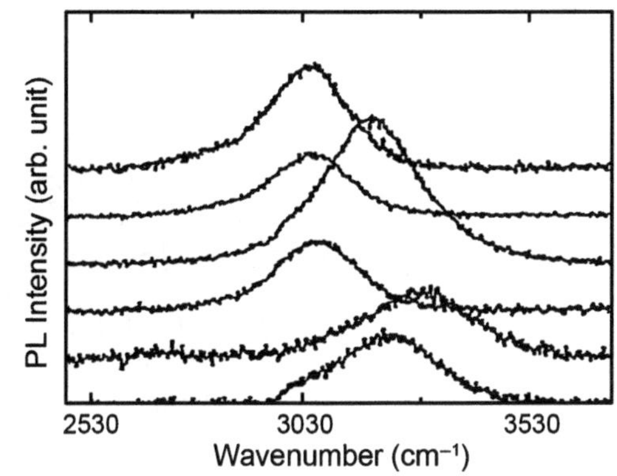

Fig. 3.35 The photoluminescence spectra of Sb-doped HgCdTe samples at 77 K

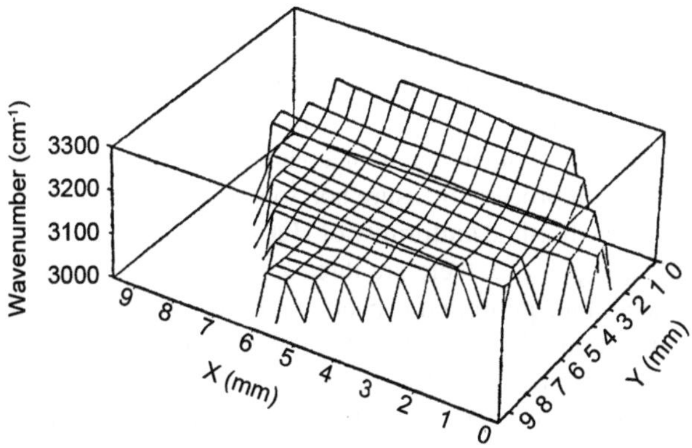

Fig. 3.36 A planar distribution of photoluminescence peak locations

locations after irradiation excited transitions among the bands. Ignoring a 1–2 meV deviation caused by self-absorption in the sample, the peak energies are equal to the band-gap (E_g).

Thus, from the empirical equation (Chu et al. 1983, 1994) that provides a relation among E_g and composition (x) and temperature (T), the composition of a sample and its distribution over a plane can be deduced. Moreover, the Urbach band tail has no effect on the planar location distribution of a PL peak, therefore, a more accurate compositional (x) distribution results as Fig. 3.37 indicates. From Fig. 3.37, we learn that the composition distribution is not homogeneous. There is a larger value on the rim of the wafer, while there is a smaller value near the center. The reasons for these observations are concluded to be: First, during the crystal growth process, if the solid–liquid interface is concave to the melt, the Cadmium composition x will condense first in the peripheral region and will occupy relatively more sites; Second, because the density of a HgTe solution is higher than that of HgCdTe, gravity in a vertical crucible will cause a surplus of HgTe to separate out during crystallization of melt, so it will aggregate in the middle of a concave interface. If we want a larger

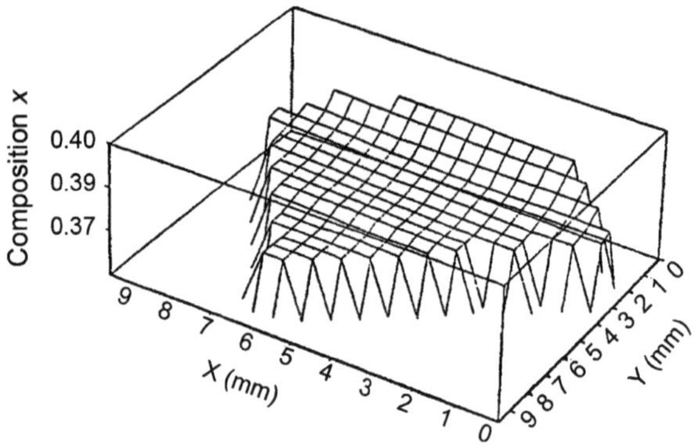

Fig. 3.37 A planar distribution of the composition

x value in a grown crystal, there will be a larger proportion of CdTe in the melt; consequently, the horizontal homogeneity will become worse (Jones et al. 1982). Therefore, it is quite important to maintain a planar solid–liquid interface during the HgCdTe crystal growth process to ensure horizontal homogeneity of the crystal.

We can analyze the lifetime uniformity of HgCdTe samples through scanning PL spectra. In bulk $Hg_{1-x}Cd_xTe$, there are three kinds of recombination mechanisms for photo-generated nonequilibrium carriers, radiative τ_r, Shockley–Read τ_{SR}, and Auger τ_A recombination. PL is directly produced by radiative recombination. At low temperature, the concentration of light-generated minority carriers is much larger than that of equilibrium carriers, for p-type $Hg_{1-x}Cd_xTe$ material ($n \approx \Delta n$, n and Δn are the concentrations of electrons and photo-generated nonequilibrium carriers, respectively). The recombination rate equation including all mechanisms mentioned above can be written as:

$$\frac{dn}{dt} = \frac{\eta \alpha I_0}{\hbar \omega_{ex}} - \frac{n}{\tau}, \tag{3.136}$$

in which

$$\frac{1}{\tau} = \frac{1}{\tau_r} + \frac{1}{\tau_{SR}} + \frac{1}{\tau_A}, \tag{3.137}$$

where α is the absorption coefficient of radiation with energy $\hbar\omega_{ex}$, I the power density of this radiation, η the quantum efficiency, and τ is the net lifetime of the photo-generated nonequilibrium carriers. The intensity of PL $I_{PL} \propto n/\tau_r$. Keeping the intensity of the laser constant, the concentration of carriers will not vary with time, thus $\frac{dn}{dt} = 0$, then:

$$n = \frac{\eta \alpha I_0 \tau}{\hbar \omega_{ex}}. \tag{3.138}$$

The intensity of PL is:

$$I_{PL} \propto n/\tau_r = \frac{\eta \alpha I_0 \tau}{\hbar \omega_{ex} \tau_r} = \frac{\eta \alpha I_0}{\hbar \omega_{ex}} \left(\frac{\tau}{\tau_r} \right). \tag{3.139}$$

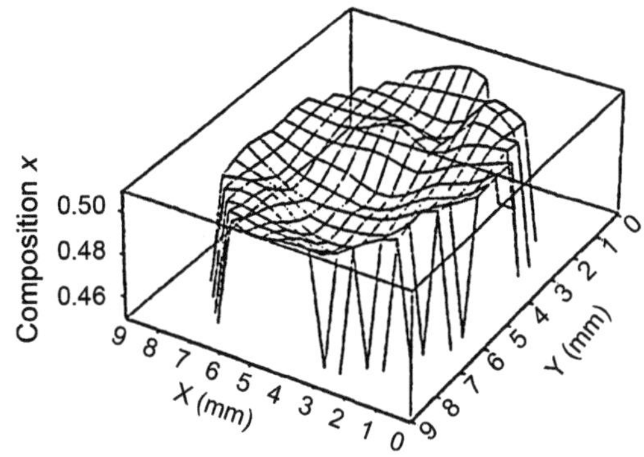

Fig. 3.38 This sample's measured two-dimensional photoluminescence intensity distribution shows a rather great discreteness

Equation (3.139) indicates that the fraction of radiation recombination to the overall recombination lifetime of nonequilibrium carriers can be calculated from the measured PL intensity. According to the calculated results of Schacham and Finkman (1985), the lifetime of inter-band direct radiation recombination is:

$$\tau_r = \frac{1}{B(n+p)}, \tag{3.140}$$

where B is the same as that in (3.99).

Some samples exhibit areas where strong intensity variations overwhelm the tiny differences caused by the compositional heterogeneity. For a sample of this sort, the two-dimensional PL intensity distribution is shown in Fig. 3.38 (the sample is the same as the one mentioned in Fig. 3.35). Here, there is a rather great discreteness in PL intensity, which can be explained by a highly dispersed distribution of nonequilibrium carrier lifetimes. Possible mechanisms within bulk $Hg_{1-x}Cd_{\underline{x}}Te$ include crystal defects such as dislocations, sublattice interfaces, lattice deformations, and glide planes. As mentioned previously, these defects could cause deep energy levels in the band-gap and decrease local nonradiation recombination lifetimes.

We see from Fig. 3.39 that the PL intensity in the middle of this wafer is higher than that at the edge. This means that the lattice exhibits better integrity in the middle of the wafer and correspondingly the lifetime of light-generated nonequilibrium carriers is longer. The extra nonradiative lifetime reduction is likely due to a surface-assisted recombination mechanism. The negative effect of lattice imperfections can be identified directly from the following experimental results. A nick is intentionally made on the sample examined in Fig. 3.39, and the PL intensity distribution near the nick is shown in Fig. 3.40. We can see in Fig. 3.40 an obvious PL intensity decrease of about 1–2 orders of magnitude relative to that shown in Fig. 3.39 at the nick. This indicates that the ratio of radiative recombination to the total recombination decreases greatly at the nick. If the resolution of the mechanical system is further improved, lattice defects such as dislocations, and slip planes can be distinguish. In fact, by using this method, the defects mentioned earlier have been observed in GaAs (Hovel 1992).

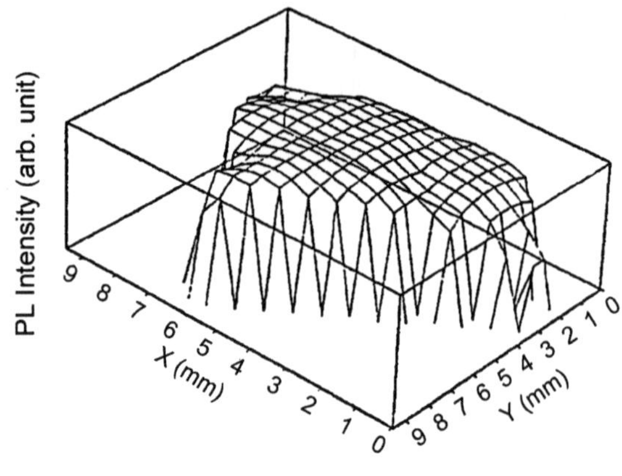

Fig. 3.39 Peak intensity distribution of band to band photoluminescence (the same sample as used in Fig. 3.38)

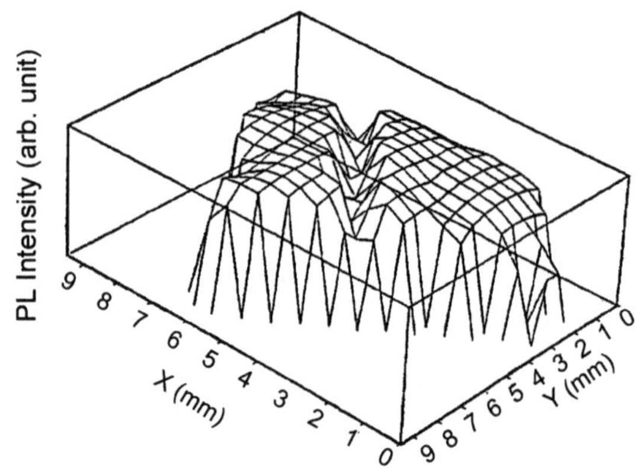

Fig. 3.40 The change of photoluminescence intensity after a nick is made on the surface of the same as used in Fig. 3.39

3.5.4 Experimental Investigation of Minority Carrier Lifetimes in Undoped and p-Type HgCdTe

There has been much experimental research reported on the minority carrier lifetimes of HgCdTe. Chen measured the minority carrier lifetimes of p-type HgCdTe with hole concentrations ranging from 1×10^{15} to $5 \times 10^{15}\,\text{cm}^{-3}$, as well as undoped LPE films (Chen et al. 1987). Tung measured minority carrier lifetimes of As-doped LPE films with hole concentrations ranging from $5{\times}10^{15}$ to $2{\times}10^{17}\,\text{cm}^{-3}$ (Tung et al. 1987). Lacklison and Capper (1987) measured sample lifetimes with hole concentrations ranging from 3×10^{15} to $2 \times 10^{17}\,\text{cm}^{-3}$. By simulating a temperature-dependence of lifetimes, Lacklison identified an SR recombination center at $E_v + 0.015\,\text{eV}$. Souza et al. (1990) measured the minority carrier lifetimes of un-doped MBE films with x ranging between $x = 0.2$ and 0.3, and found that they mainly depend on an SR recombination center near the middle of band-gap. Adomaitis et al. (1990) measured the minority carrier lifetimes of undoped LPE films, according to his result it is as short as 53 ps. Adomaitis believed that the minority carrier lifetime in the samples with hole concentrations higher than $4 \times 10^{16}\,\text{cm}^{-3}$ is dominated by the Auger 7 process. For further study on the relation between minority carrier lifetimes of HgCdTe films vs. temperature and hole concentration, as well as comparisons of the recombination mechanisms in doped and undoped HgCdTe samples, see the literature (Chen et al. 1995).

The minority carrier lifetime of p-type HgCdTe depends on Auger 1, Auger 7, radiative and S–R recombination process:

$$\frac{1}{\tau} = \frac{1}{\tau_{A_1}} + \frac{1}{\tau_{A_7}} + \frac{1}{\tau_R} + \frac{1}{\tau_{SR}}. \tag{3.141}$$

Under a situation where there is degeneracy:

$$\tau_{A_1} = \frac{2n_i^2 \tau_{A_1}^i}{(n_0 + p_0)n_0}, \tag{3.142}$$

$$\tau_{A_7} = \frac{2n_i^2 \tau_{A_7}^i}{(n_0 + p_0)p_0}, \tag{3.143}$$

$$r = \frac{\tau_{A_7}^i}{\tau_{A_1}^i} \cong 2\frac{m_e^*(E_T)}{m_0^*}\left(\frac{1-\eta_T^{5/4}}{1-\eta_T^{3/2}}\right). \tag{3.144}$$

Here, m_0^* is the electron effective mass at the bottom of the conduction band, $m_e^*(E_T)$ is the effective mass of an electron with energy E_T higher than the bottom of the conduction band, and $\eta_T = E_T/k_BT = E_g/k_BT$. Equations (3.142) and (3.143) fit data for carrier concentrations up to $1 \times 10^{18}\,\text{cm}^{-3}$ at 77 K. In general, for p-type HgCdTe, an Auger 1 process is important only at higher temperatures (generally higher than 150 K). Especially within the lower part of this range of intrinsic temperatures, the Auger 7 process plays a key role. The factor $|F_1F_2|$ for the Bloch wave-function overlap integrals, which is necessary for calculating the lifetime determined by an Auger recombination process, is obtained by fitting to experiments. Its value is about 0.15–0.2 (Chen and Colombo 1992). The value of $r = \tau_{A_7}^i/\tau_{A_1}^i$ is ~2 for $Hg_{1-x}Cd_xTe$ with $x = 0.2$ (Casselman 1981). It is also obtained by fitting experiment results to (3.134) where the result obtained is thought to be relative more exact. More precise calculations of $|F_1F_2|$ based on proper band structures show that in fact this factor depends on $\boldsymbol{k}$ and temperature T, and these constant values are only rough approximations (Krishnamurthy et al. 2006). Chen et al. (1995) measured at 77 K the relation between the lifetime vs. the carrier concentration in LPE grown $Hg_{1-x}Cd_xTe$ ($x = 0.225$) films. The experimental and fitted computational results are shown in Fig. 3.41. In this figure, the solid curve is a fitted lifetime considering only Auger 7 and radiative recombination processes. The circles, diamonds, and triangles denote experimental lifetime data for Au-doped, Cu-doped, and undoped samples, respectively. The undoped sample is p-type in which the acceptors are Hg vacancies. The dashed curve is fit to the variation of the lifetime of an undoped sample. It is obvious that when the concentration of carriers lies below

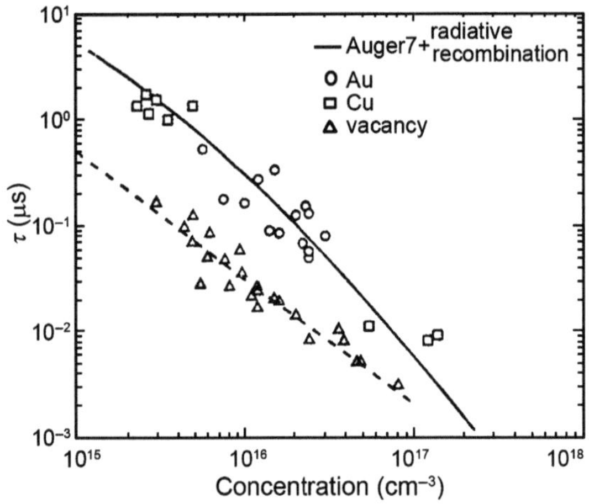

Fig. 3.41 Relations between the lifetimes and the carrier concentrations at 77 K for LPE grown $Hg_{1-x}Cd_xTe(x = 0.225)$ films. The solid curve is a calculated lifetime considering only Auger 7 and radiative recombination

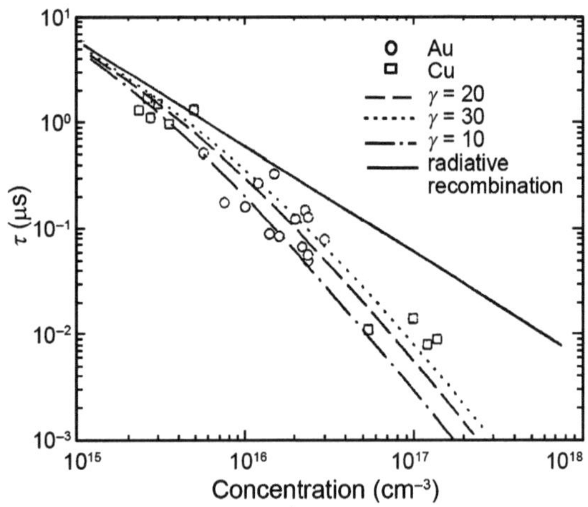

Fig. 3.42 The relation at 77 K between the lifetimes and carrier concentrations for LPE grown $Hg_{1-x}Cd_xTe(x = 0.225)$ films. The solid curve is a fitted lifetime only considering radiative recombination

$1 \times 10^{17}\,\text{cm}^{-3}$, the lifetime of doped samples (2–8 ns) is longer than that of an undoped sample (150–8 ns). A possible reason is due to S–R recombination centers associated with Hg vacancies which reduces the lifetime.

In order to determine the contribution from radiative, Auger 7 and Auger 1 recombination to the lifetime of doped samples ($x = 0.225$, the same as that used in Fig. 3.41) at 77 K, the value of $\gamma = \tau^{i}_{A_7}/\tau^{i}_{A_1}$ is set to 10, 20, 30, and the corresponding calculated results are shown in Fig. 3.42 (dashed lines). We find that $\gamma = 20$ provides the optimum fit to the experimental results (circles and diamonds denote experimental lifetime data of Au-doped and Cu-doped samples). In contrast, the calculated result considering only radiative recombination is also presented in the figure (the solid line). According to this result, we conclude that radiative recombination is the major recombination mechanism at 77 K if the concentration of carriers is lower than $5 \times 10^{15}\,\text{cm}^{-3}$.

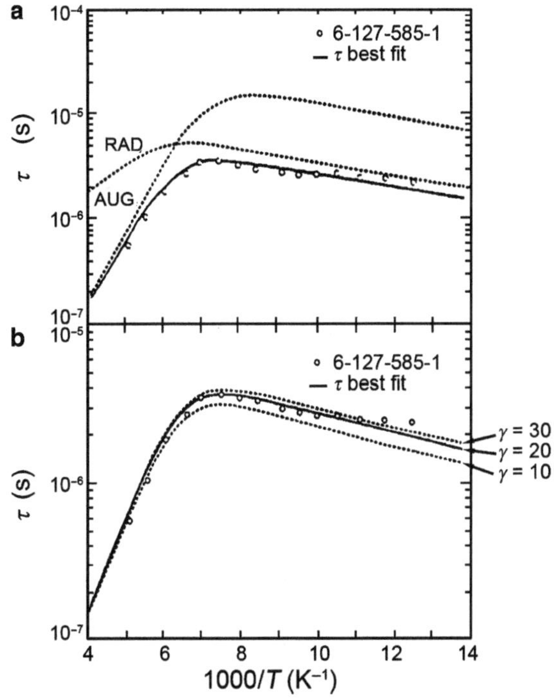

Fig. 3.43 (**a**) The relation between lifetime and reciprocal temperature for a Cu-doped LPE HgCdTe. The carrier concentration is $2.6 \times 10^{15}\,\mathrm{cm}^{-3}$. (**b**) The results including Auger 7 and Auger 1 processes and $\gamma = \tau^{i}_{A_7}/\tau^{i}_{A_1} = 20$

There is no S–R recombination contribution for the Au-doped and Cu-doped LPE samples presented in Fig. 3.41. This is also true in temperature-dependent lifetime measurements. The relation between the minority carrier lifetime and temperature for an $Hg_{1-x}Cd_xTe$ sample ($x = 0.222$) is shown in Fig. 3.43a. The dashed curve labeled AUG is the lifetime due to Auger 7 and Auger 1 processes, while the dashed curve labeled RAD is the lifetime due to radiative recombination. This sample is Cu-doped, and the carrier concentration is $2.6 \times 10^{15}\,\mathrm{cm}^{-3}$. In samples with lower carrier concentrations, radiative recombination dominates at lower temperatures (lower than 150 K), while Auger recombination dominates at higher temperatures. The solid curve is the net lifetime including Auger and radiative recombination. In the calculation, Auger recombination includes both Auger 7 and Auger 1 processes, and setting $\gamma = \tau^{i}_{A_7}/\tau^{i}_{A_1} = 20$; if $\gamma = 30$ or 10, there is a little difference at higher temperatures, as shown in Fig. 3.43b.

In undoped p-type LPE samples, with the p-type conduction mainly due to Hg vacancies, both the Hg vacancies and the impurities play roles as recombination centers. Therefore, at low temperature, the main recombination process changes from a radiative recombination mechanism to a S–R recombination mechanism. The calculated relation between the lifetime and temperature is shown in Fig. 3.44, where the three dashed lines labeled SRH, RAD, and AUG represent their respective contributions to the recombination lifetime. The solid line is the optimum simulated lifetime obtained by considering all contributions from the above three recombination mechanisms. The data points are experimental values. The samples are p-type

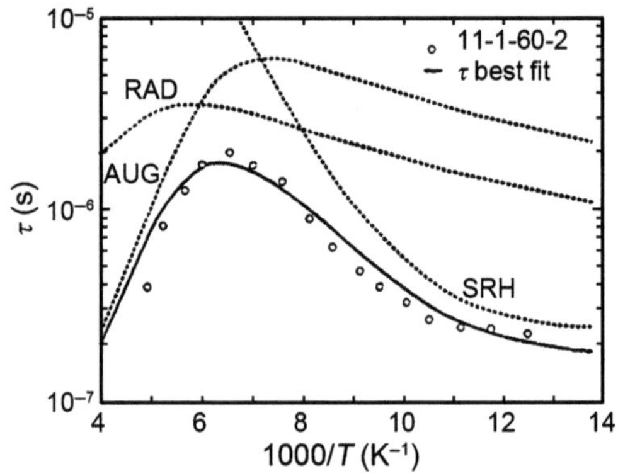

Fig. 3.44 The calculated curves for the lifetime vs. the reciprocal temperature. The three dashed curves labeled SRH, RAD and AUG are the contributions of SR, radiative and Auger recombination to the lifetime, respectively. The solid line is the optimum simulated lifetime obtained by considering all contributions from the above three recombination mechanisms

$Hg_{1-x}Cd_xTe$ films grown by LPE with $x = 0.225$, the carrier concentration is $4.8\times10^{15}\,cm^{-3}$. The parameters for obtaining an optimum simulation are: the cross-section of S–R recombination center for electrons and holes are $\sigma_n = 1\times10^{-16}\,cm^2$ and $\sigma_p = 5\times10^{-19}\,cm^2$, respectively, the recombination center concentration is $N_r = 8\times10^{14}\,cm^{-3}$, and the recombination center energy level is located in the middle of the energy gap.

We next offer comprehensive experimental results for the minority carrier lifetimes of n-type HgCdTe. Wijewarnasuriya et al. (1995) measured the minority carrier lifetimes of in In-doped HgCdTe films. Samples were grown by MBE on CdZnTe (211B) substrates. The indium is doped in situ, and the as-grown films are then annealed under an Hg saturated atmosphere at 250°C in order to decrease Hg vacancies. The minority carrier lifetime is measured using the photoconductivity decay method. In n-type samples, the major Auger recombination process is Auger 1:

$$\tau_{A_1} = \frac{2n_i^2\tau_{A_1}^i}{(n+p)n}, \tag{3.145}$$

the radiative recombination lifetime is:

$$\tau_R = \frac{2\tau_{R_i}n_i}{(n_0+p_0)}, \tag{3.146}$$

and the S–R recombination lifetime is:

$$\tau_{SR} = \frac{\tau_{p0}(n_0+n_1)}{n_0+p_0} + \frac{\tau_{n0}(p_0+p_1)}{n_0+p_0}, \tag{3.147}$$

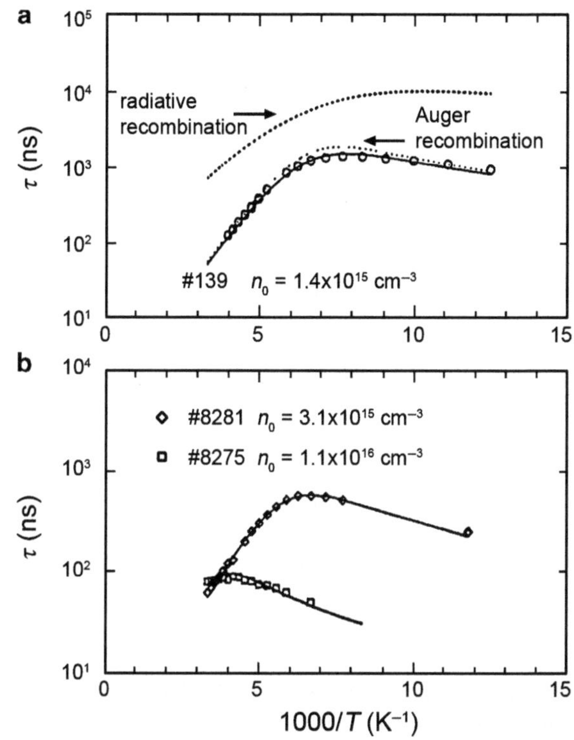

Fig. 3.45 The lifetime vs. the reciprocal temperature in Indium lightly doped HgCdTe. Circles and diamonds are experimental data, and curves are results calculated to fit the experimental data

where τ_{n0} and τ_{p0} are the shortest electron trap and hole trap time constants, $n_1 = n_0 \exp(E_T - E_F)/k_B T$, $p_1 = p_0 \exp(E_T - E_F)/k_B T$, and E_T is the energy level of a S–R recombination center.

The experimental and calculated results for the lifetimes vs. the reciprocal temperature are shown in Fig. 3.45. It is obvious that the lifetime increases up to a maximum with a decrease of the temperature, and then the lifetime decreases with a continuing decrease of the temperature. The parameters for the sample's data in Fig. 3.45a are as follows: $x = 0.217$, thickness $t = 8.7\,\mu$m, $N_D = 1.4\times10^{15}\,\mathrm{cm}^{-3}$, $\mu = 1.6\times10^5\,\mathrm{cm}^2/\mathrm{Vs}$, and $n_0 = 1.3\times10^{15}\,\mathrm{cm}^{-3}$. At 80 K, $\tau_{80\,K} = 940$ ns is found. The two dashed curves in the figure are the calculated net lifetimes due to radiative and Auger recombination. We see that at 130 K the lifetime has a maximum (1.4 μs). At lower temperatures, the lifetime decreases exponentially. The results for samples #8275 and #8281 are shown in Fig. 3.45b, in which the data points are experimental results and the solid curve is the calculated lifetime including both Auger and radiative recombination. The parameters for sample #8275 are as follows: $x = 0.244$, $t = 9.5\,\mu$m, $N_D = 2.1\times10^{16}\,\mathrm{cm}^{-3}$, $\mu = 1.0\times10^5\,\mathrm{cm}^2/\mathrm{Vs}$, $n_0 = 1.1\times10^{16}\,\mathrm{cm}^{-3}$, and $\tau = 20$ ns (80 K). The maximum lifetime is $\tau_{max} = 100$ ns. The parameters for

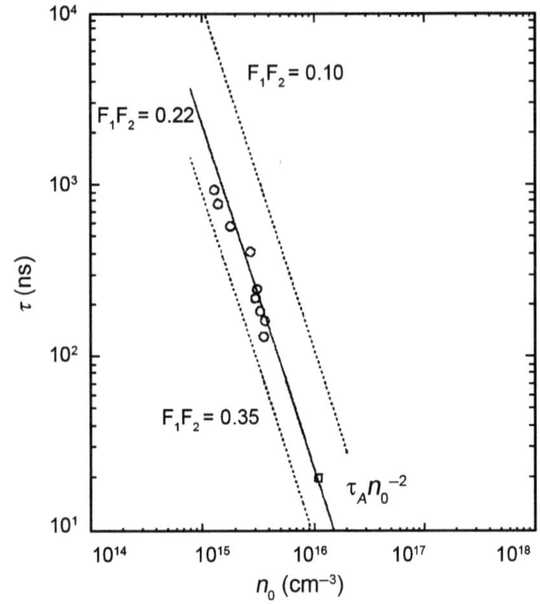

Fig. 3.46 The lifetime vs. the electron concentration at 80 K for an In-doped HgCdTe sample grown by MBE

sample #8281 are as follows: $x = 0.237$, $t = 11.5\,\mu$m, $N_D = 4.0 \times 10^{15}\,$cm^{-3}, $\mu = 1.0 \times 10^5\,$cm^2/Vs, $n_0 = 3.1 \times 10^{15}\,$cm^{-3}, and $\tau = 184\,$ns (80 K).

Generally, we calculated the net lifetime including Auger and radiative recombination and compare it to the measured $\tau \sim 1/T$ experimental data. At low temperatures, if the calculated values are greater than that of the experiment, the interpretation is that there is a contribution from S–R recombination.

In the calculation of the Auger recombination lifetime, the factor $|F_1 F_2|$ for the Bloch wave-function overlapped integrals is selected to fit the experimental lifetime results given the carrier concentration at low temperature (77 K).

The lifetime vs. the electron concentration of an In-doped HgCdTe grown by MBE is shown in Fig. 3.46 (80 K). When the electron concentration increases from 1.4×10^{15} to $1.0 \times 10^{16}\,$cm^{-3}, the lifetime decreases from 950 to 20 ns. Supposing there are no SR recombination centers in n-type samples, then the Auger recombination dominates at low temperature. According to (3.145), $\tau_A \propto n_0^{-2}$, by changing the value of $|F_1 F_2|$, the optimum simulation is obtained when $|F_1 F_2| = 0.22$.

For a HgCdTe device, there is an anodic oxide layer or passivation layer on its surface. This layer is degraded by being irradiated by an ultraviolet light when it is applied, because electrons in the passivation layer are then excited from the valence band to the conduction band and subsequently captured on otherwise neutral trapping centers, as showed in Fig. 3.47. This permanently changes them. Therefore, electrons in the accumulation layer of a HgCdTe surface will easily recombine with holes in valence band in the passivation layer, causing carrier lifetimes to decrease. The experimental results for the lifetime including the results before and after ultraviolet irradiation are showed in the following figure. Data points are experimental values, and real curves are fitted calculated results (Staszewski and Wollrab 1989).

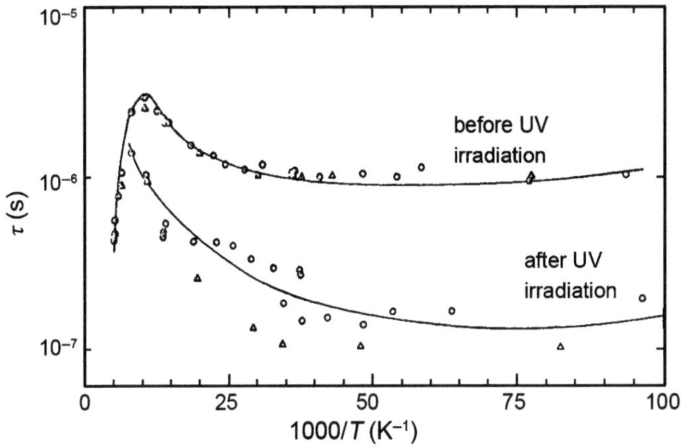

Fig. 3.47 The relation between carrier lifetimes and reciprocal temperatures before and after ultraviolet irradiation

3.6 Surface Recombination

3.6.1 The Effect of Surface Recombination

Due to crystal imperfections, the potential field periodicity is destroyed and additional energy levels are generated in the forbidden band. Tamm (1932) first pointed out in 1932 that: The existence of a crystal surface makes the periodic potential field cut off at the surface to cause additional energy levels. The situation at real surfaces is more complicated because of the existence of oxidization and impurities. Here, we only discuss a one-dimensional crystal surface under an ideal situation. Figure 3.48 shows the potential energy variation at an ideal one-dimensional crystal surface located at $x = 0$; $x > 0$ corresponds to inside the crystal where the potential energy changes periodically with x. A cycle length is a, so the potential energy function is $V(x + a) = V(x)$. $x < 0$ corresponds to the area beyond the crystal where the potential energy is a constant V_0.

For such a semi-infinite periodic potential field, the wave function of an electron obeys the Schrödinger equation:

$$-\frac{\hbar^2}{2m_0}\frac{\partial^2\psi}{\partial x^2} + V_0\psi = E\psi \quad (x \leq 0) \tag{3.148}$$

and

$$-\frac{\hbar^2}{2m_0}\frac{\partial^2\psi}{\partial x^2} + V(x)\psi = E\psi \quad (x \geq 0). \tag{3.149}$$

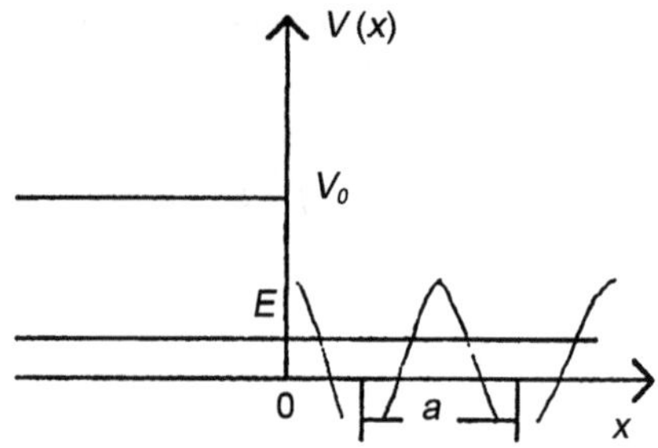

Fig. 3.48 The potential energy function for an ideal one-dimensional crystal

Consider the situation of an electron with energy $E < V_0$, and using the condition that the wave functions has a limited displacement at $\pm\infty$, then solving the above-mentioned two equations, we find:

$$x \leq 0 \quad \psi_1 = A \exp\left(\frac{\sqrt{2m_0(V_0 - E)}}{\hbar} x\right) \tag{3.150}$$

and

$$x \geq 0 \quad \psi_2 = A_1 u_k(x) \exp(\mathrm{i}2\pi kx) + A_2 u_{-k}(x) \exp(-\mathrm{i}2\pi kx). \tag{3.151}$$

Because of the continuity of wave functions and their first derivatives, we know that:

$$A_1 u_k(0) + A_2 u_{-k}(0) = A, \tag{3.152}$$

and

$$A_1[u_k'(0) + \mathrm{i}2\pi k u_k(0)] + A_2[u_{-k}'(0) - \mathrm{i}2\pi k u_{-k}(0)]. \quad = A\frac{\sqrt{2m_0(V_0 - E)}}{\hbar} \tag{3.153}$$

The three coefficients A_1, A_2, and A are determined by the above two equations. If k is a real number, the wave functions are limited at $x \to \pm\infty$, so both coefficients A_1 and A_2 can be nonzero. Equations (3.152) and (3.153) constitute two equations in three unknowns. The third equation is the normalization condition. These solutions are the permissible states in this potential field, while the corresponding levels are the allowed energies.

When k is a complex number, $k = k' + \mathrm{i}k''$, where both k' and k'' are real numbers, then (3.151) becomes:

$$\begin{aligned}\psi_2 &= A_1 u_k(x) \exp(\mathrm{i}2\pi k'x) \exp(-2\pi k''x)\\ &\quad + A_2 u_{-k}(x) \exp(-\mathrm{i}2\pi k'x) \exp(-2\pi k''x).\end{aligned} \tag{3.154}$$

We know that when $x \to \pm\infty$, ψ_2 must be finite, this means in a one-dimensional infinite periodic potential field, k can't be a complex. While in a one-dimensional semi-infinite periodic potential field, if A_1 or A_2 equals to zero, k can be complex. For example: if $A_2 = 0$, then:

$$\psi_2 = A_1 u_k(x) \exp(\mathrm{i}2\pi k' x) \exp(-2\pi k'' x). \tag{3.155}$$

We find that if $k > 0$ then ψ_2 is limited, so there are solutions. Solving (3.150) and (3.151), knowing that both A and A_1 are nonzero yields:

$$E = V - \frac{\hbar^2}{2m_0}\left(\frac{u_k'(0)}{u_k(0)} + \mathrm{i}2\pi k\right)^2. \tag{3.156}$$

In this instance, we find that on both sides of $x = 0$, the wave functions decay exponentially. This indicates that the electron distribution is concentrated around $x = 0$, indicating that the electron is bound near the surface. At the surface, k is complex and so is the surface energy. Tamm (1932) calculated a semi-infinite Kronig–Penney model, and proved that given a certain condition, each surface atom has a corresponding surface energy level, with a magnitude of about $10^{-15}\,\mathrm{cm}^{-2}$. For a "pure silicon surface" under an ultra-high vacuum the surface-state density was measured, and it indicated that the density of surface states coincides with this theory, and its magnitude is about proportional to the dangling-bond density of surface atoms. Therefore, this finding is explained because a crystal lattice terminates suddenly at a surface, producing the appearance of an unmated atom and a corresponding electron surface state.

The surface of HgCdTe is generally processed by a passivation technology (commonly, an anodic oxidation or a CdTe layer), so there is often an oxidized layer on its surface (including natural oxidation in ambient air). Then, the dangling bonds are saturated by oxygen atoms and the surface-state density is greatly reduced.

Surface recombination is a nonequilibrium carrier recombination process. We have shown previously that there are "deep-levels" on the surface, similar to those in the bulk through which nonequilibrium carriers can recombine. The recombination rate through these surface energy levels is according to an extension of the S–R theory:

$$U_\mathrm{z} = \frac{(n_0 + p_0)\Delta p}{\dfrac{n_{\mathrm{s}0} + n_{\mathrm{s}1}}{N_{\mathrm{ST}} c_\mathrm{p}} + \dfrac{p_{\mathrm{s}0} + p_{\mathrm{s}1}}{N_{\mathrm{ST}} c_\mathrm{n}}}. \tag{3.157}$$

Surface recombination rate is:

$$S = \frac{U_\mathrm{z}}{\Delta p} = \frac{(n_0 + p_0)}{\dfrac{n_{\mathrm{s}0} + n_{\mathrm{s}1}}{S_\mathrm{p}} + \dfrac{p_{\mathrm{s}0} + p_{\mathrm{s}1}}{S_\mathrm{n}}}. \tag{3.158}$$

The surface recombination rate of a hole is:

$$S_\mathrm{p} = \sigma_\mathrm{p} v_\mathrm{p} N_{\mathrm{sT}}. \tag{3.159}$$

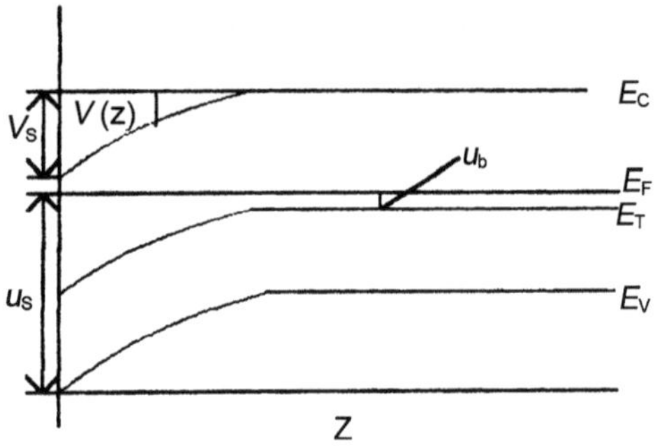

Fig. 3.49 The space charge area distribution of an n-type semiconductor

The surface recombination rate of an electron is:

$$S_n = \sigma_n v_n N_{sT}. \tag{3.160}$$

Hence, we have:

$$S = \frac{\sigma_p \sigma_n v_p v_n N_{sT}(n_0 + p_0)}{\sigma_n v_n (n_{s0} + n_{s1}) + \sigma_p v_p (p_{s0} + p_{s1})}. \tag{3.161}$$

Figure 3.49 shows a trap energy level, E_T, at the surface of an n-type semiconductor, in which E_C is the conduction band minimum, E_V the valance band maximum, E_F the Fermi energy, V_S the maximum of the surface-potential band bending at the conductive band edge. Under the thermal equilibrium condition, and if the electric current due to the recombination of surface electrons and holes is small, then the bulk and surface can be considered to be in an equilibrium state, so we get:

$$n(z) = n_i \exp\left(\frac{E_F - E_T}{k_B T} + \frac{eV(z)}{k_B T}\right) \tag{3.162}$$

and

$$p(z) = n_i \exp\left(-\frac{E_F - E_T}{k_B T} - \frac{eV(z)}{k_B T}\right). \tag{3.163}$$

Now, supposing that:

$$u_s = u_b + v_s = \frac{E_F - E_T}{k_B T} + \frac{eV(z)}{k_B T}, \tag{3.164}$$

we find:

$$n_0 = n_i \exp(u_b), \tag{3.165}$$

$$p_0 = n_i \exp(-u_b), \tag{3.166}$$

$$n_{s0} = n_i \exp(u_s) = n_0 \exp(eV_s/k_B T), \tag{3.167}$$

$$n_{s0} = n_i \exp(u_s) = n_0 \exp(eV_s/k_B T). \tag{3.168}$$

Let, $u_0 \equiv \ln(\sigma_p v_p/\sigma_n v_n)/2$, then:

$$S = \frac{\sigma_p\sigma_n v_p v_n N_{sT}(n_0+p_0)}{\left\{\begin{array}{l}\sigma_n v_n n_i\left[\exp(u_b+v_s)+\exp\left(\frac{E_T-E_F}{k_BT}\right)\right]\\ +\sigma_p v_p n_i\left[\exp(-u_b-v_s)+\exp\left(\frac{E_F-E_T}{k_BT}\right)\right]\end{array}\right\}}$$

$$= \frac{\sqrt{\sigma_p\sigma_n v_p v_n}N_{sT}(n_0+p_0)}{2n_i\left[\cosh(u_s-u_0)+\cosh\left(\frac{E_F-E_T}{k_BT}-u_0\right)\right]} \tag{3.169}$$

and

$$S_{max} = \frac{\sqrt{\sigma_p\sigma_n v_p v_n}N_{sT}(n_0+p_0)}{2n_i\left[1+\cosh\left(\frac{E_F-E_T}{k_BT}-u_0\right)\right]}. \tag{3.170}$$

To deduce a value for S, we first must know N_{sT}, and second the energy E_T. Figures 3.50 and 3.51 are sketches of S_{max}. From this figure, we find that the value of S_{max} lies in the range 10^2–10^4 cm/s. Therefore, we have:

$$\frac{S}{S_{max}} = \frac{1+\cosh\left(\frac{E_F-E_T}{k_BT}-u_0\right)}{\cosh(u_s-u_0)+\cosh\left(\frac{E_F-E_T}{k_BT}-u_0\right)}. \tag{3.171}$$

In an oxidized surface layer on HgCdTe, various kinds of charge states will appear. Due to the high density of fixed charge in HgCdTe, a detailed analysis of surface charge and surface potential is needed. Related computations (Gong 1993) indicated that when the thickness of an oxidized layer is about 700 Å, the rate of surface recombination is a minimum. Because the high fixed charge density (Q_{fc} is in the range of $5\times10^{11} - 1\times10^{12}$ cm^{-2} (Nemirovsky 1990)) in the surface-oxidized layer, the surface energy band of HgCdTe is seriously bent downward, this

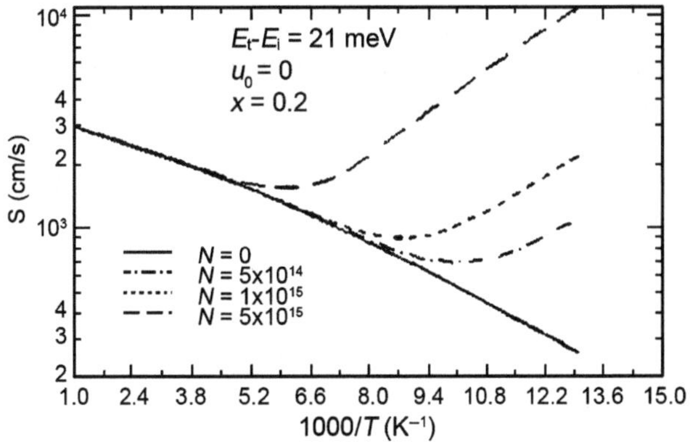

Fig. 3.50 Relation between S_{max} and temperature T

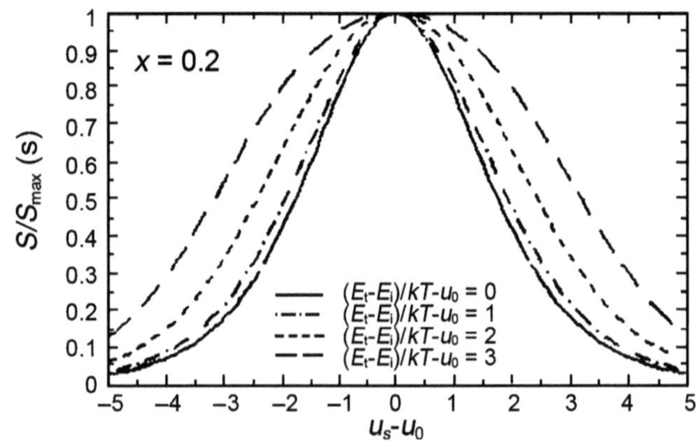

Fig. 3.51 The relation between $S/S_{\max}$ and $u_s - u_0$

means V_s is very large. Meanwhile, the forbidden band of HgCdTe is very narrow, so degenerate states readily appear at the surface, thus Boltzmann distributions must be replaced by Fermi distributions. In addition, similar to the bulk situation, both Auger recombination and radiative recombination must be included. Experimental lifetimes should actually be a comprehensive result of surface recombination and bulk recombination. If the two kinds of recombination processes do not influence each other, the effective lifetime of a nonequilibrium carrier will be: $\frac{1}{\tau} = \frac{1}{\tau_B} + \frac{1}{\tau_s}$.

3.6.2 Surface Recombination Rate

In the measurement of lifetimes, especially the measurement of the intensity variation of light-generated carriers, we must distinguish between the relative contributions of surface recombination and bulk recombination. During the measurement process, the light should be carefully regulated to be quasi-monochromatic and have a low intensity, to prevent the creation of a high density of electron–hole pairs and a heavily nonuniform carrier distribution in the surface area. In this case, the surface recombination current density can be expressed as $eS\delta p$, here δp is the density of surplus minority carriers near the surface, and S is a surface recombination rate constant with the unit cm/s.

Given these conditions, the observed lifetime can be expressed as:

$$\frac{1}{\tau_{\text{obs}}} = \frac{1}{\tau_B} + \frac{1}{\tau_s}, \tag{3.172}$$

in which τ_s is the surface recombination lifetime and τ_B is the bulk recombination lifetime. These two times constants can be distinguished, in principle, through experiments (including a change of a sample's cross sectional size). If the object studied has a rectangular-cross-section ($2A \times 2B$), and the carriers are uniformly generated, Shockley (1950) pointed out that the relation between τ_s and S, the recombination rate per unit volume, can be expressed as:

$$\frac{1}{\tau_S} = D\left(\frac{\eta^2}{A^2} + \frac{\xi^2}{B^2}\right), \tag{3.173}$$

where D is the minority carrier diffusion constant, $SA/D \equiv \eta \tan \eta$, and $SB/D \equiv \xi \tan \xi$. The pair of smallest solutions of these equations, η_0 and ξ_0, correspond to the longest lifetime, and set an important limit.

After carefully washing and etching, the surface recombination rates for germanium and silicon can be reduced to values lower than 10^3 cm/s, while that of compound semiconductors can be reduced to values lower than 10^5 cm/s.

Now, let's investigate an n-type semiconductor material. Suppose the surface area contains a uniform recombination center density N_r per unit area, with all the centers located at the E_i-$e\phi_s$ energy level (E_i is the energy of the gap's center). These energy states will often accept electrons, and produce a depletion layer near the surface, as shown in Fig. 3.52.

The recombination rate due to this kind of recombination center is determined by its capture cross-section and the acquisition rates of surface holes and electrons. When the hole and electron distribution is nondegenerate, we can get a recombination rate per unit area from a statistical analysis:

$$r_s = \frac{N_r c_p c_n (p_b + n_b)\delta p}{c_p(p_s + p_{s1}) + c_n(n_s + n_{s1})}. \tag{3.174}$$

In this equation c_p and c_n are, respectively, the probability per unit time and volume of a hole or an electron being captured by the recombination center, given by the

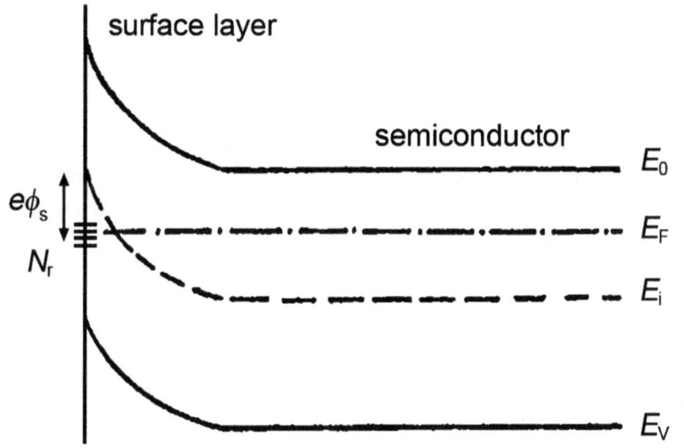

Fig. 3.52 Acceptor-like surface states below the mid-gap for an n-type semiconductor

product of corresponding capture cross-section and the velocities of their thermal motion; p_s and n_s are densities of free carriers on the surface; p_{s1} and n_{s1} are surface densities when the Fermi energy is at the energy E_i. p_b and n_b are densities of bulk carriers, δp is density of surplus minority carriers inside surface space charge layer. Supposing the densities p_s and n_s equal those of the bulk, then, according to (3.174), the surface recombination rate is given by:

$$s = \frac{r_s}{\delta p} = \frac{N_r c_p c_n (p_b + n_b)}{n_i c_n \exp\left[\dfrac{E_F - E_i + e\varphi_s}{k_B T}\right] + n_i c_p \exp\left[-\dfrac{E_F - E_i + e\varphi_s}{k_B T}\right]}, \quad (3.175)$$

$e\phi_s$ is negative if the level $e\phi_s$ lies bellow the center of the gap.

According to the Shockley–Read–Hall (Shockley and Read 1952; Hall 1952) discussion, the above formula applies to a recombination center having a single-energy-level. Gräfe (1971) has studied the validity of the above formula. Because the formula is a simple model of an extremely complicated situation, it can be only accepted to a certain extent.

3.6.3 *The Effect of Fixed Surface Charge on the Performance of HgCdTe Photoconductive Detectors*

Passivation technology in the fabricating of $Hg_{1-x}Cd_xTe$ photoconductive detectors are designed to form heavy accumulated layers at the surfaces of $Hg_{1-x}Cd_xTe$. Accumulated layers will reduce the surface recombination rate and improve the device performance; on the other hand, due to its area of high conductance, the resistance, and further, the responsivity and detectivity of the device are degraded. Therefore, a theoretically optimized passivation treatment of devices can guide the direction for improving device performance.

The importance of the passivation layer was noted early (Kinch 1981), and further investigations were reported (Nemirovsky and Bahir 1989; Nemirovsky 1990). An accumulated layer caused by passivation technology obviously has a two-dimension character. We can adopt Fang–Howard's variational method (Fang and Howard 1966) to calculate the distribution of the surface potential in a one-dimension model to calculate the distribution of light-induced carriers and their voltage response in a $Hg_{1-x}Cd_xTe$ photoconductive detector.

A large amount of fixed positive charge exists in the passivation layer on a $Hg_{1-x}Cd_xTe$ photoconductive detector, which can induce an equal density of electrons on the surface of a bulk material. These surface electrons distribute within a very thin thickness to form a quasi-two-dimensional electron gas. Many theoretical methods have been developed to calculate the properties of a two-dimensional electron gas, and models have been developed already for a single band and for multi-bands. Poisson and Schrödinger equations are the start of all these models. A variational self-consistent method developed by Fang and Howard is a relatively simple and practical method. In Fang's model, the density of surface electrons is assumed to be much higher than that of the bulk-electron density; therefore, the surface potential and the density of fixed positive charges in the surface have the following relation:

$$V_{\mathrm{H}} = \frac{3N_{\mathrm{I}}e^2}{2\varepsilon_{\mathrm{s}}\varepsilon_0 b}, \tag{3.176}$$

where

$$b = \left(\frac{33m_{\mathrm{n}}e^2N_{\mathrm{I}}}{8\varepsilon_{\mathrm{s}}\varepsilon_0\hbar^2}\right). \tag{3.177}$$

Here, m_{n} is the effective mass of the electrons, e the charge of an electron, ε_0 and ε_{s} are the vacuum dielectric constant and the low-frequency dielectric constant. Low-frequency dielectric constants (Yadava et al. 1994) have been related to the material composition:

$$\varepsilon_{\mathrm{s}} = 20.8 - 16.8x + 10.6x^2 - 9.4x^3 + 5.3x^4. \tag{3.178}$$

This surface potential arising from the accumulated layer in HgCdTe will prevent holes from moving toward the surface. In this case, the rate of the effective surface recombination (White 1981) is:

$$S_{\mathrm{eff}} = S_0 \exp(-V_{\mathrm{H}}/k_{\mathrm{B}}T), \tag{3.179}$$

where S_0 is the surface recombination rate in a flat band situation, k_{B} the Boltzmann constant, and T is the temperature.

Accumulated layers exist in both the top and bottom surfaces in a HgCdTe photoconductive detector, assuming the top and bottom surfaces are identical, then the resistance of the detector is:

$$R = \frac{1}{(N_{\mathrm{b}}\mu_{\mathrm{b}} + 2N_{\mathrm{I}}\mu_{\mathrm{s}}/d)e}\frac{1}{wd}. \tag{3.180}$$

Here, l, w, and d are the length, width, and height of detector, respectively, N_{b} the bulk electron density, μ_{s} mobility of the surface electrons (about $10^4\,\mathrm{cm^2/Vs}$ at 77 K), and μ_{b} is the mobility of the bulk electrons which can be expressed by Rosbeck's formula (Rosbeck et al. 1982). For a $\mathrm{Hg_{1-x}Cd_xTe}$ material with a composition of $x = 0.2$, μ_{b} is about $1\text{–}3 \times 10^5\,\mathrm{cm^2/Vs}$ at 77 K.

This surface recombination has important influence on the lifetime of minority carriers. Because of the existence of surface recombination, the real lifetime of material with bulk lifetime τ_{b} becomes:

$$1/\tau_{\mathrm{net}} = 1/\tau_{\mathrm{b}} + 1/\tau_{\mathrm{s}} = 1/\tau_{\mathrm{b}} + 2S_{\mathrm{eff}}/d. \tag{3.181}$$

Equation (3.183) captures the influence of the surface potential on the surface recombination rate. To calculate the real lifetime, we have adopted the effective surface recombination rate given in (3.181).

For photoconductive detector operating under constant current model, the voltage response to monochromatic radiation (wavelength = λ) is:

$$R_\mathrm{V} = (\lambda/hc)\eta q R(\mu_\mathrm{b} E \tau_\mathrm{net}/l). \tag{3.182}$$

In which, E is the intensity of the bias electric field, c the velocity of light, h the Plank constant, η the quantum efficiency, λ the wavelength, the R on the left is the voltage responsivity R_V, and the R on the right is the resistance.

In order to study the influence of surface fixed charge on the performance of a device (only the situation in which $T = 77\,\mathrm{K}$ is considered), we suppose the material for fabricating the photoconductive detector is n-type, the composition is $x = 0.214$, the bulk lifetime of minority carriers is 10 ns to 10 μs, the concentration of impurities is 5×10^{14} to $5 \times 10^{15}\,\mathrm{cm}^{-3}$, and the mobility of surface electrons is $2 \times 10^4\,\mathrm{cm}^2/\mathrm{Vs}$. The photosensitive area of the detector is $50 \times 50\,\mu$m, and the thickness is 88 μm. The wavelength of incident light is 10.6 μm, quantum efficiency is 0.6, and the bias electric field is 20 V/cm.

The relation of the detector resistance and surface fixed charge is shown in Fig. 3.53. The electron accumulated layer at surface caused by passivation is a high-conductance leakage path, and it will reduce the resistance of the device and deteriorate its performance. It can be seen from Fig. 3.53 that when the density of surface fixed positive charges changes between 10^{11} and $10^{12}\,\mathrm{cm}^{-2}$, the functional relation between the surface fixed charge and the resistance of a detector changes dramatically. If $N_\mathrm{I} < 10^{11}\,\mathrm{cm}^{-2}$, the resistance of the detector is mainly decided by that of the bulk material and is nearly independent of the surface passivation; while for the case of $N_\mathrm{I} > 10^{12}\,\mathrm{cm}^{-2}$, the resistance of the detector is mainly due to the density of surface electrons caused by the surface passivation.

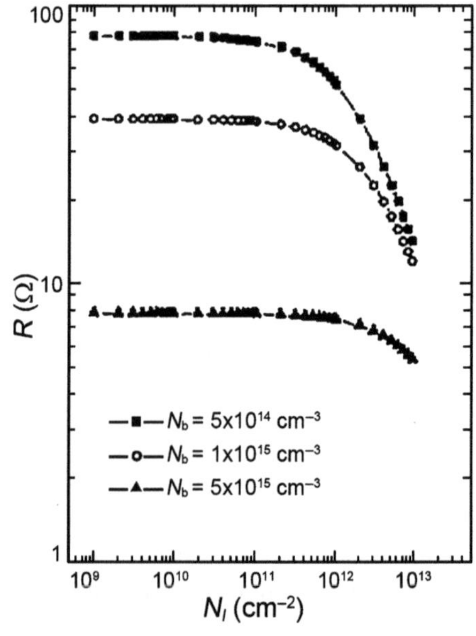

Fig. 3.53 Relation between the resistance R of devices which have different bulk carrier densities and the density N_I of surface fixed electric charges

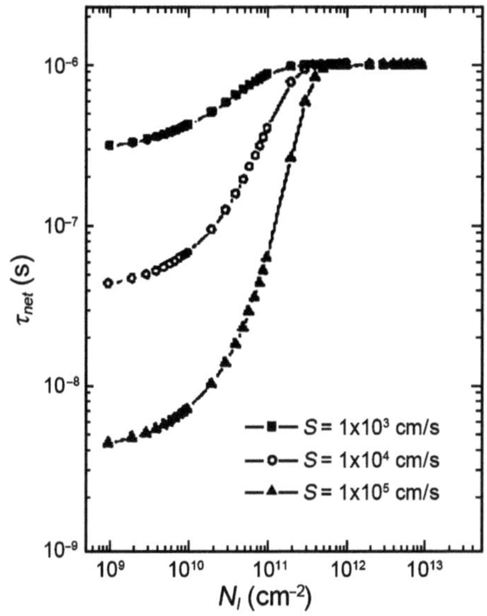

Fig. 3.54 Relations between the net lifetime τ_{net} of devices, that have difference surface recombination rates S_0, and the density of surface fixed electric charge N_{I}

The effect of surface fixed charges on the net lifetime τ_{net} of a detector is shown in Fig. 3.54. When N_{I} is very small, τ_{net} is also short and it increases rapidly with an increase of N_{I} until it reaches τ_{b}. An electron accumulated layer is formed in the surface region of an n-type detector due to the surface fixed charge, causing an energy band bending at the surface. This forms a potential well for electrons, which serves as a barrier for bulk minority carriers (holes), thus preventing the light-generated holes from diffusing to surface and recombining. Therefore, the existence of surface accumulated layer reduces the surface recombination rate. The repulsive surface potential for holes increases with an increase of the density of fixed charge, as described by (3.178). When the density of fixed charge exceeds $3 \times 10^{11}\,\text{cm}^{-2}$, the surface potential barrier to holes becomes high enough to prevent light-generated holes from passing through it to reach the detector surface, even for surface recombination rates as high as $1 \times 10^5\,\text{cm/s}$. This phenomenon is very favorable to the improvement of detector performance.

The relations between the voltage responsivity and the density of fixed charge for different doping concentrations of bulk material are shown in Fig. 3.55. The lifetime of minority carriers in these detectors is $\sim$1 μs, and the surface recombination rate is $\sim$1,000 cm/s, which means these are high-efficiency devices. When the density of fixed charges in the passivation layers exceeds $10^{11}\,\text{cm}^{-2}$, the existence of surface barriers reduces the recombination loss of light-generated holes, and the performance of devices is improved. Once the densities of fixed charges exceed $\sim 10^{12}\,\text{cm}^{-2}$, the increase of the conductivity due to the surface accumulate layer leads to a decrease of the intrinsic resistance and the performance of devices. With a further increase of the density of fixed charge, the performance drops quickly, especially for lightly doped devices.

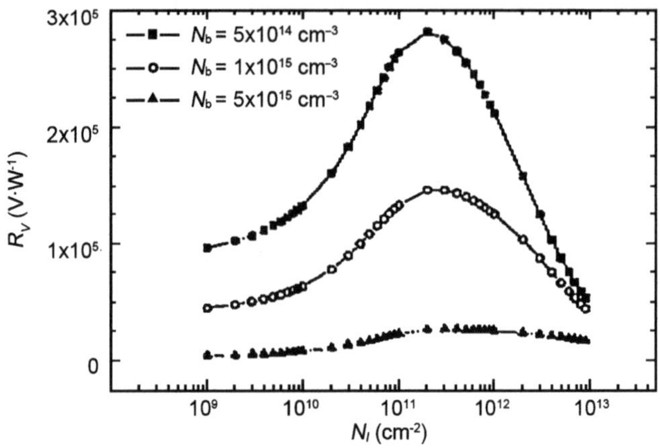

Fig. 3.55 Relations between voltage responsivity R_v of devices that have different doping densities N_b, and the density of surface fixed electric charge N_I

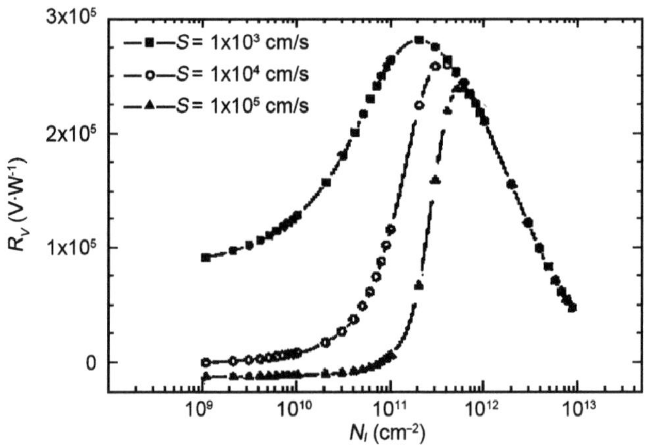

Fig. 3.56 Relations between voltage responsivity R_v of devices that have different surface recombination rates S_0, and the density of fixed electric surface charge N_I

The relation between the voltage responsivity and the density of fixed charge for different surface recombination rates are shown Fig. 3.56. When the lifetime of minority carriers in a device is $\sim$1 μs, and the doping concentration of the bulk material is $\sim 5 \times 10^{14}$ cm^{-3}, the voltage responsivity has a peak value when the density of fixed charge lies in the range of 10^{11}–10^{12} cm^{-2}. As the surface recombination rate increases, the peak shifts toward a higher density of fixed charge. The height of the peak is determined by the surface recombination rate. The lower the surface recombination rate, the higher is the voltage responsivity. When the density of fixed charge is higher than 6×10^{11} cm^{-2}, the performance of detectors is insensitive to the surface recombination rate, even though the surface recombination rate changes

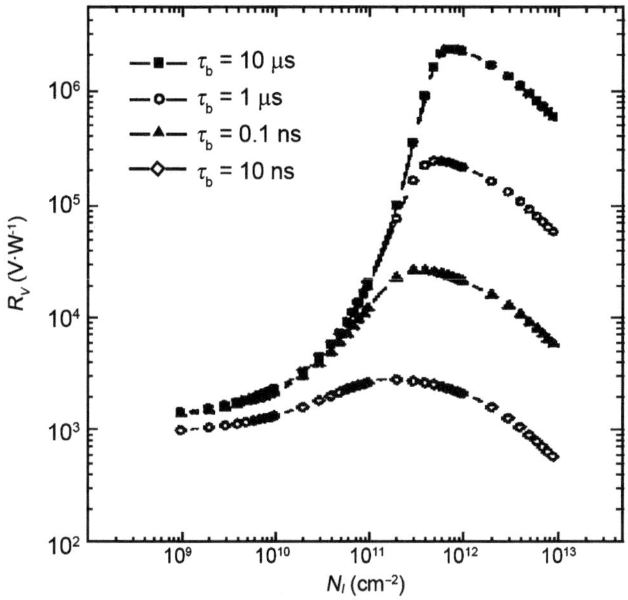

Fig. 3.57 Relation between voltage respond R_v of device which have different lifetime τ_b of few carriers and the density of surface fixed electric charge N_I

by two orders of magnitude. This indicates that by choosing an appropriate passivation technology, we can control the effective surface recombination rate and improve device performance.

The relation between the voltage responsivity and the density of fixed charge for different minority carrier lifetimes is shown in Fig. 3.57. In these calculations, the bulk material doping concentration is taken to be $5 \times 10^{14}\,\text{cm}^{-3}$, and the surface recombination rate is 10^5 cm/s. In this situation, if the density of the surface-fixed charge is lower than $\sim 10^{11}\,\text{cm}^{-2}$, due to the fact that the effective lifetime τ_{net} of a device is mainly determined by the surface recombination rate, the performance of devices is basically independent of the minority carrier lifetime. The performance of detectors changes greatly when the density of surface-fixed charges lies in the range of 10^{11}–$10^{12}\,\text{cm}^{-2}$. This is especially true for long-lifetime devices ($\tau_b = 10\,\mu\text{s}$), where the detector voltage responsivity changes by over two orders of magnitude for a small change in N_i.

According to the above discussion, we know that the performance of HgCdTe photoconductive detector is highly related to the density of surface-fixed charge caused by passivation technology. The existence of surface-fixed charge can reduce the loss of light-generated holes by surface recombination and improve the performance of the device, on the other hand, it will reduce the performance of the device by reducing the resistance of the device, these two aspects are contradictory to each other. In other words, we can optimize the performance of the device by choosing appropriate passivation technology to control the density of surface-fixed charge.

Appendix 3.A

The results of (3.69) and (3.70) can be also obtained from the Fermi level distribution analysis.

Equations (3.64) and (3.65) can be written as:

$$n_1 = n_0 \frac{1 - f_T^0}{f_T^0} \tag{3.a.1}$$

and

$$p_1 = p_0 \frac{f_T^0}{1 - f_T^0}, \tag{3.a.2}$$

where f_T is given by (3.60). Then, we can get:

$$\begin{aligned} N_T^- &= N_T f_T \cong N_T f_T^0 \\ N_T^0 &= N_T(1 - f_T) \cong N_T(1 - f_T^0). \end{aligned} \tag{3.a.3}$$

Equation (3.a.3) is well approximated in (3.66) and (3.67) because their equilibrium values are never far from equilibrium.

Now, we treat three simple cases:

$$\text{(a)} \frac{E_F - E_T}{kT} \gg 1.$$

In this case, generally the material is n-type, we have $f_T^0 \cong 1$, $1 - f_T^0 \cong 0$, $N_T^0 \cong 0$, and $n_1 N_T^- \cong 0$. Then, (3.66) reduced to $\frac{dn}{dt} \cong 0$. Due to $p_1 N_T^0 = p_0 \frac{f_T^0}{1-f_T^0} N_T \left(1 - f_T^0\right) \cong p_0 N_T$, (3.67) becomes $\frac{dp}{dt} \cong -c_p N_T (p - p_0)$. Its solution for the S–R relaxation time is given by:

$$\tau_{S\text{–}R}^{a} \cong \left(\sigma_p v_p N_T\right)^{-1}, \tag{3.a.4}$$

where σ_p is the hole capture cross section, and v_p is the speed of a hole. $\tau_{S\text{–}R}^{a}$ is limited by the relaxation of the hole concentration. For an n-type material there is a larger chance the traps will first capture more electrons and then capture holes to assist the recombination. So, the lifetime is determined by the hole-capture cross-section and velocity.

$$\text{(b) } E_T = E_F$$

In this case, we have $f_T^0 = \left(1 - f_T^0\right) = \frac{1}{2}$, $N_T^0 = N_T^- = \frac{N_T}{2}$, $n_1 = n_0$, and $p_1 = p_0$. Then, (3.66) and (3.67) become:

$$\frac{dn}{dt} \cong -c_n \frac{N_T}{2} (n - n_0), \tag{3.a.5}$$

$$\frac{\mathrm{d}p}{\mathrm{d}t} \cong -c_\mathrm{p}\frac{N_\mathrm{T}}{2}\left(p - p_0\right). \tag{3.a.6}$$

There are two S–R relaxation times ($\tau_{\text{S–R}}^{\text{bn}}$ and $\tau_{\text{S–R}}^{\text{bp}}$) associated with the solutions to (3.a.5) and (3.a.6). The solutions are given by

$$\tau_{\text{S–R}}^{\text{bn}} = 2\left(\sigma_\mathrm{n} v_\mathrm{n} N_\mathrm{T}\right)^{-1}, \tag{3.a.7}$$

$$\tau_{\text{S–R}}^{\text{bp}} \cong 2\left(\sigma_\mathrm{p} v_\mathrm{p} N_\mathrm{T}\right)^{-1}, \tag{3.a.8}$$

where σ_n is the electron capture cross-section, and v_n is the speed of electron. In any case, the time constant, due to the factor of 2, $\tau_{\text{S–R}}^{\text{bn}}$ or $\tau_{\text{S–R}}^{\text{bp}}$ is likely to be longer than the minimum S–R relaxation time and it occurs when $E_\mathrm{T} = E_\mathrm{F}$.

$$\text{(c)}\frac{E_\mathrm{T} - E_\mathrm{F}}{kT} \gg 1.$$

In this case, generally the material is p-type, we have $f_\mathrm{T}^0 \cong 0$, $1 - f_\mathrm{T}^0 \cong 1$, $n_1 N_\mathrm{T}^- \cong n_0 N_\mathrm{T}$, $p_1 N_\mathrm{T}^0 \cong 0$. Then, (3.66) and (3.67) become:

$$\frac{\mathrm{d}n}{\mathrm{d}t} \cong -c_\mathrm{n} N_\mathrm{T}\left(n - n_0\right), \tag{3.a.9}$$

$$\frac{\mathrm{d}p}{\mathrm{d}t} \cong 0. \tag{3.a.10}$$

The solution is given by

$$\tau_{\text{S–R}}^{\text{c}} \cong \left(\sigma_\mathrm{n} v_\mathrm{n} N_\mathrm{T}\right)^{-1}. \tag{3.a.11}$$

$\tau_{\text{S–R}}^{\text{c}}$ is limited by the relaxation of the electron concentration. For a p-type material there is a smaller chance an electron will be captured but once it is the hole will be captured quickly to assist the recombination. So, the lifetime is determined by the electron capture cross-section and velocity.

Appendix 3.B Sandiford Paper

He writes the continuity equations in the form (notation adjusted to agree with that used in this book):

$$-\frac{\mathrm{d}\Delta p}{\mathrm{d}t} = c_\mathrm{p}\left(N_\mathrm{T}^- + p_0 + p_1\right)\Delta p - c_\mathrm{p}\left(p_0 + p_1\right)\Delta n, \tag{3.b.1}$$

$$-\frac{\mathrm{d}\Delta n}{\mathrm{d}t} = c_\mathrm{n}\left(N_\mathrm{T}^0 + n_0 + n_1\right)\Delta n - c_\mathrm{n}\left(n_0 + n_1\right)\Delta p. \tag{3.b.2}$$

Next, take the derivative of (3.b.1) with respect to time and substitute (3.b.2) for $\frac{\mathrm{d}\Delta n}{\mathrm{d}t}$, and then substitute for Δn from (3.b.1). The result is:

$$\frac{\mathrm{d}^2\Delta p}{\mathrm{d}t^2} + G_1\frac{\mathrm{d}\Delta p}{\mathrm{d}t} + G_2\Delta p = 0, \tag{3.b.3}$$

where

$$\begin{aligned} G_1 &\equiv c_\mathrm{p}\left(N_\mathrm{T}^- + p_0 + p_1\right) + c_\mathrm{n}\left(N_\mathrm{T}^0 + n_0 + n_1\right) \text{ and} \\ G_2 &\equiv c_\mathrm{n}c_\mathrm{p}\left[\left(N_\mathrm{T}^- + p_0 + p_1\right)\left(N_\mathrm{T}^0 + n_0 + n_1\right) - \left(p_0 + p_1\right)\left(n_0 + n_1\right)\right]. \end{aligned} \tag{3.b.4}$$

Note (3.b.3) is homogeneous and has a solution:

$$\Delta p = Ae^{-H_+t} + Be^{-H_-t}, \tag{3.b.5}$$

in agreement with Eq. 4 of Sandiford (1957). The solutions for H are:

$$H_\pm = \frac{G_1 \pm \sqrt{G_1^2 - 4G_2}}{2}. \tag{3.b.6}$$

The corrected version of Sandiford's Eq. 5 (Sandiford 1957) is:

$$G_1^2 \gg 4G_2 \tag{3.b.7}$$

In this approximation, we find:

$$H_+ \cong G_1 \text{ and } H_- \cong \frac{G_2}{G_1}. \tag{3.b.8}$$

From the initial conditions, we have:

$$\begin{aligned} A + B &= \Delta p\,(0)\,, \text{ and} \\ \mathrm{H}_+A + H_-B &= G_3, \end{aligned} \tag{3.b.9}$$

where

$$G_3 \equiv c_\mathrm{p}[(p_o + p_1)\left(\Delta p\,(0) - \Delta n\,(0)\right) - N_\mathrm{T}^-\Delta p\,(0)]. \tag{3.b.10}$$

The solutions to (3.b.9) are:

$$A = \Delta p\,(0) - B \text{ and } B = \frac{G_3 - H_+\Delta p\,(0)}{H_- - H_+}. \tag{3.b.11}$$

A number of special cases can be treated. Suppose an n-type semiconductor with, $(E_{\mathrm{T}} - E_{\mathrm{F}})/k_{\mathrm{B}}T > 1$, so $n_1 > n_0 \gg p_0 > p_1$, and photo excitation so $\Delta p\,(0) = \Delta n\,(0)$. Then, we find:

$$A \cong 0 \text{ and } B \cong \Delta p\,(0)\,. \tag{3.b.12}$$

so

$$\Delta p\,(t) \cong \Delta p\,(0)\,e^{-t/\tau_-}, \tag{3.b.13}$$

where

$$\tau_-^{-1} \equiv c_p N^-. \tag{3.b.14}$$

Thus in this case $B >> A$ and the observed decay is the slow response corresponding to H_- not the fast decay due to H_+. This contrasts with the conclusion drawn by Sandiford.

This whole treatment only applies to the approximation where N_{T}^- and N_{T}^0 are taken to be independent of time, and hence does not correspond the real situation. However, it is a generalization of the treatment in Appendix 3.A where the populations of the trap states are taken to remain constant while the system relaxes. All three cases (a), (b), and (c) can be done. This is essentially case (a).

References

Adomaitis E, Grigoras K, Krotkus A (1990) Minority carrier recombination in p-type $Cd_xHg_{1-x}Te$. Semicond Sci Technol 5:836–841

Afromowitz MA, Didomenico M (1971) Measurement of free-carrier lifetimes in GaP by photoinduced modulation of infrared absorption. J Appl Phys 42:3205–3208

Baars J, Sorger F (1972) Reststrahlen spectra of HgTe and $Cd_xHg_{1-x}Te$. Solid State Commun 10:875–878

Baicker JA, Fang PH (1972) Transient conductivity of silicon. J Appl Phys 43:125–131

Bajaj J, Babulac LO, Newman PR, Tennant WE (1987) Spatial mapping of electrically active defects in HgCdTe using laser beam-induced current. J Vac Sci Technol A 5:3186–3189

Baker IM, Capocci FA, Charlton DE, Wotherspoon JTM (1978) Recombination in cadmium mercury telluride photodetectors. Solid State Electron 21:1475–1480

Bardeen J, Blatt FJ, Hall LH (1956) In: Breckenridge RC (ed) Photoconductivity Conference, Atlantic City 1954. Wiley, New York

Bartoli F, Allen R, Esterowitz L, Kruer M (1974) Auger-limited carrier lifetimes in HgCdTe at high excess carrier concentrations. J Appl Phys 45:2150–2154

Beattie AR (1962) Quantum efficiency in InSb. J Phys Chem Solids 23:1049–1056

Beattie AR, Landsberg PT (1959) Auger effect in semiconductors. Proc R Soc London Ser A 249:16–29

Beattie AR, Smith G (1967) Recombination in semiconductors by a light hole Auger transition. Phys Status Solidi 19:577–586

Blakemore JS (1962) Semiconductor Statistics. Pergamon, Oxford

Casselman TN (1981) Calculation of the Auger lifetime in *p*-type $Hg_{1-x}Cd_xTe$. J Appl Phys 52:848–854

Casselman TN, Petersen PE (1980) A comparison of the dominant Auger transitions in ptype (Hg,Cd)Te. Solid State Commun 33:615–619

Chen JS, Bajaj J, Tennant WE, Lo DS, Brown M (1987) On the Auger recombination process in p-type LPE HgCdTe. IN: Materials for infrared detectors and sources; Proceedings of the Symposium, Boston, MA, 1986. Materials Research Society, Pittsburgh, PA, pp 287–293

Chen MC, Colombo L (1992) Minority-carrier lifetime in indium-doped n-type $Hg_{0.78}Cd_{0.22}Te$ liquid-phase-epitaxial films. J Appl Phys 72:4761–4766

Chen MC, Colombo L, Dodge JA, Tregilgas JH (1995) The minority carrier lifetime in doped and undoped p-type $Hg_{0.78}Cd_{0.22}Te$ liquid phase epitaxy films. J Electronic Materials 24:539–544

Choo SC (1970) Carrier lifetimes in semiconductors with two interacting or two Independent recombination levels. Phys Rev B 1:687–696

Chu JH, Sher A (2007) Physics and Properties of Narrow Gap Semi-Conductors. Springer, New York

Chu JH, Xu SQ, Tang DY (1982) Chinese Sci. Bull 27:403

Chu JH, Xu SQ, Tang DY (1983) Energy gap versus alloy composition and temperature in $Hg_{1-x}Cd_xTe$. Appl Phys Lett 43:1064–1066

Chu JH, Li B, Liu K, Tang DY (1994) Empirical rule of intrinsic absorption spectroscopy in $Hg_{1-x}Cd_xTe$. J Appl Phys 75:1234–1235

Elliott CT (1977) Pat. 1488258

Elliott CT (1981) New detector for thermal imaging systems. Electron Lett 17:312–313

Fang FF, Howard WE (1966) Negative field-effect mobility on (100) Si surfaces. Phys Rev Lett 16:797–799

Gerhardts RR, Dornhaus R, Nimtz G (1978) The Auger-effect in $Hg_{1-x}Cd_xTe$. Solid State Electron 21:1467–1470

Gong HM (1993) PhD thesis, Shanghai Institute of Technical Physics

Gräfe W (1971) Extension of the surface recombination model and unique determination of the recombination energy by means of reverse current measurements. Phys Status Solidi A 4: 655–662

Hall RN (1952) Electron-hole recombination in Germanium. Phys Rev 87:387–387

Hall RN (1959) Proc Inst Electr Eng B Suppl 106:923

Hovel HJ (1992) Scanned photoluminescence of semiconductors. Semicond Sci Technol 7:A1–A9

Huang QS, Tang DY (1965) Recombination processes of carriers in Indium antimonide. Acta Phys Sin 21:1038

Jones CL, Capper P, Gosney JJ (1982) Thermal modelling of bridgman crystal growth. J Crystal Growth 56:581–590

Kinch MA (1981) In: Willardson RK, Beer AC (eds) Semiconductor and Semimetal, Vol. 18. Academic, New York, p 287

Kinch MA, Brau MJ, Simmons A (1973) Recombination mechanisms in 8–14-μ HgCdTe. J Appl Phys 44:1649–1663

Kopanski JJ, Lowney JR, Novotny DB, Seiler DG, Simmons A, Ramsey J (1992) High-spatial-resolution resistivity mapping applied to mercury cadmium telluride. J Vac Sci Technol B 10:1553–1559

Krishnamurthy S, Berding MA, Yu ZG (2006) Minority Carrier Lifetimes in HgCdTe Alloys J of Elect Materials 35:1369–1378

Kurnick SW, Powell JM (1959) Optical absorption in pure single crystal InSb at 298° and 78° K. Phys Rev 116:597–604

Lacklison DE, Capper P (1987) Minority carrier lifetime in doped and undoped ptype $Cd_xHg_{1-x}Te$. Semicond Sci Technol 2:33–43

Ling ZG, Lu PD (1984) Chinese J Semicond 5:144

Long D, Schmit JL (1970) In: Willardson RK, Beer AC (eds) Semiconductor and Semimetal, Vol. 5, Chapter 5. Academic, New York

Many A, Goldstein Y, Grover NB (1965) Semiconductor Surfaces. Willey, New York

Mao WY, Chu JH (1993) J Infrared MilliWaves 13:352

Milnes AG (1973) Deep impurities in semiconductors. Wiley, New York

Mooradian A, Fan HY (1966) Recombination Emission in InSb. Phys Rev 148:873–885

Moss TS, Burrell GJ, Ellis B (1973) Semiconductors Opto-electronics. Buttorworth, London, p 38

Mroczkowski JA, Shanley JF, Reine MB, LoVechio P, Polla DL (1981) Lifetime measurement in $Hg_{0.7}Cd_{0.3}Te$ by population modulation. Appl Phys Lett 38:261–263

Nemirovsky Y (1990) Passivation of II-IV compounds. J Vac Sci Technol A 8:1185–1187

Nemirovsky Y, Bahir G (1989) Passivation of mercury cadmium telluride surfaces. J Vac Sci Technol A 7:450–459

Petersen PE (1970) Auger Recombination in $Hg_{1-x}Cd_xTe$. J Appl Phys 41:3465–3467

Pines MY, Stafsudd OM (1980) Recombination process in intrinsic semiconductors using impact ionization capture cross sections in indium antimonide and mercury cadmium telluride. Infrared Phys 20:73

Polla DL, Tobin SP, Reine MB, Sood AK (1981a) Experimental determination of minority-carrier lifetime and recombination mechanisms in p-type $Hg_{1-x}Cd_xTe$. J Appl Phys 52:5182–5194

Polla DL, et al. (1981b) Proceeding of the 4th International Conference on the Physics of Narrow Gap Semiconductors. Linz, Austria, p 153

Pratt RG, Hewett J, Capper P, Jones CL, Quelch MJ (1983) Minority carrier lifetime in n-type Bridgman grown $Hg_{1-x}Cd_xTe$. J Appl Phys 54:5152–5157

Roosbroeck W, Shockley W (1954) Photon-Radiative Recombination of Electrons and Holes in Germanium. Phys Rev 94:1558–1560

Rosbeck JP, Starr RE, Price SL, Riley KJ (1982) Background and temperature dependent current-voltage characteristics of HgCdTe photodiodes. J Appl Phys 53:6430–6440

Sah CT, Tasch AF Jr (1967) Precise Determination of the Multiphonon and Photon Carrier Generation Properties using the Impurity Photovoltaic Effect in Semiconductors. Phys Rev Lett 19:69–71

Sandiford DJ (1957) Carrier Lifetime in Semiconductors for Transient Conditions. Phys Rev 105:524–524

Schacham SE, Finkman E (1985) Recombination mechanisms in p-type HgCdTe: Freezeout and background flux effects. J Appl Phys 57:2001–2009

Schiff L (1955) Quantum Mechanics. McGraw-Hill, New York

Schmit JL (1970) Intrinsic Carrier Concentration of $Hg_{1-x}Cd_xTe$ as a Function of x and T Using k·p Calculations. J Appl Phys 41:2876–2879

Shockley W (1950) Energy band structures in semiconductors. Phys Rev 78:173–174

Shockley W (1958) Electrons, holes, and traps. Proc IRE 46:973–990

Shockley W, Read WT (1952) Statistics of the recombinations of holes and electrons. Phys Rev 87:835–842

Souza ME de, Boukerche M, Faurie JP (1990). Minority-carrier lifetime in p-type $(111)B$ HgCdTe grown by molecular-beam epitaxy. J Appl Phys, 68:5195–5199

Srour JR, Curtis OL (1972) Techniques for obtaining recombination-center parameters from carrier lifetime studies. J Appl Phys 43:1779–1784

Staszewski GM von, Wollrab R (1989) Role of n+-accumulation layers on the carrier lifetime of n-$Hg_{1-x}Cd_xTe$. SPIE 1106:110

Tamm IE (1932) A possible kind of electron binding on crystal surfaces. Physikalische Zeitschrift der Sowjetunion 1:733–746

Tang DY (1974) Infrared Phys Technol 16:345

Tang DY (1976) Infrared Phys Technol 25–26:53

Tung T, Kalisher MH, Stevens AP, Herning PE (1987) Liquid-phase epitaxy of $Hg_{1-x}Cd_xTe$ from Hg solution – A route to infrared detector structures. Mat Res Soc Symp Proc Vol. 90. Mater Res Soc, Pittsburg, p 321

Wertheim GK (1958) Transient recombination of excess carriers in semiconductors. Phys Rev 109:1086–1091

White AM (1981) The influence of surface properties on minority carrier lifetime and sheet conductance in semiconductors. J Phys D: Appl Phys 14:L1–L3

Wijewarnasuriya PS, Lange MD, Sivananthan S, Faurie JP (1995) Minority carrier lifetime in Indium-doped HgCdTe (211)B epitaxial layers grown by molecular beam epitaxy. J Electronic Materials 24:545–549

Yadava RDS, Gupta AK, Warrior AVR (1994) Hole scattering mechanisms in $Hg_{1-x}Cd_xTe$. J Electron Mater 23:1359–1364

Chapter 4
Two-Dimensional Surface Electron Gas

4.1 MIS Structure

4.1.1 The Classical Theory of an MIS Device

The simplest metal-insulator-semiconductor device structure, shown in Fig. 4.1, includes a metal gate, an insulator layer on a semiconductor surface, and a back metal ohmic contact. t_{ox} is the thickness of insulator and ε_{ox} is its dielectric constant. The insulator layer lies between the metal gate and the top semiconductor surface. A bias voltage is applied to the metal gate to control the surface potential of the semiconductor surface. When the bias voltage reaches a threshold, the n-type semiconductor becomes strongly inverted. The resulting energy band bending is shown in Fig. 4.2. The potential Φ as a function of displacement x from semiconductor surface is obtained from the one-dimensional Poisson equation. Under nondegenerate and thermal equilibrium conditions, we have:

$$\frac{\mathrm{d}^2\phi}{\mathrm{d}x^2} = -\frac{q}{\varepsilon\varepsilon_0}\left[n_{\mathrm{n}0}(e^{q\phi/k_{\mathrm{B}}T} - 1) - p_{\mathrm{n}0}(e^{-q\phi/k_{\mathrm{B}}T} - 1)\right]. \tag{4.1}$$

The electrical field distribution is

$$E = -\frac{\mathrm{d}\phi}{\mathrm{d}x} = \pm\left(\frac{2k_{\mathrm{B}}T}{\varepsilon\varepsilon_0}\right)^{1/2} F(\phi), \tag{4.2}$$

where

$$F(\phi) = \left[n_{\mathrm{n}0}\left(e^{q\phi/k_{\mathrm{B}}T} - \frac{q\phi}{kT} - 1\right) + p_{\mathrm{n}0}\left(e^{-q\phi/k_{\mathrm{B}}T} + \frac{q\phi}{kT} - 1\right)\right]^{1/2}, \tag{4.3}$$

and $n_{\mathrm{n}0}$ and $p_{\mathrm{n}0}$ are the concentrations of majority and minority carriers in the semiconductor, ε is its permittivity, and q is the magnitude of the electron charge.

From Gauss's law, the space charge per unit area in a semiconductor required to produce this field is

$$Q_{\mathrm{s}} = \varepsilon\varepsilon_0 E_{\mathrm{s}} = \pm(2\varepsilon\varepsilon_0 k_{\mathrm{B}}T)^{1/2} F(\phi). \tag{4.4}$$

J. Chu and A. Sher, *Device Physics of Narrow Gap Semiconductors*, Microdevices,
DOI 10.1007/978-1-4419-1040-0_4, © Springer Science+Business Media, LLC 2010

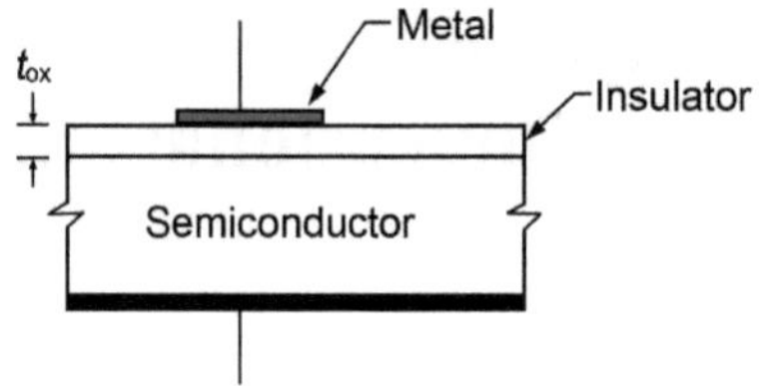

Fig. 4.1 The simplest metal-insulator-semiconductor device structure

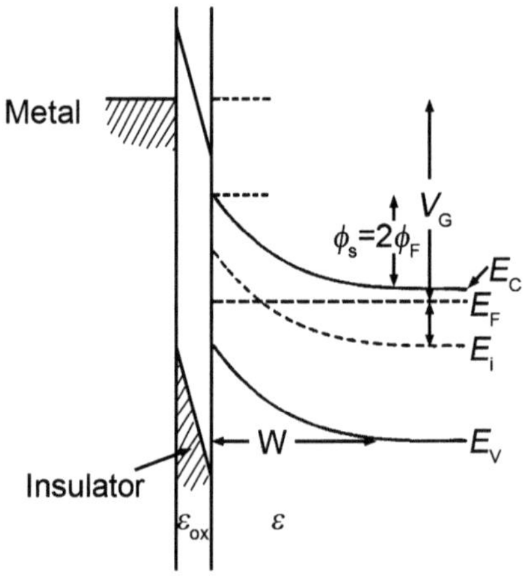

Fig. 4.2 The energy band diagram of an MIS device when the bias voltage reaches a threshold where the n-type semiconductor becomes strongly inverted

A typical variation of the space-charge density Q_s as a function of the surface potential φ_s is shown in Fig. 4.3 for a typical n-type HgCdTe detector ($n_0 = 2 \times 10^{15}\,\text{cm}^{-3}$) with a cutoff wavelength of $\lambda_c = 5\,\mu\text{m}$, at 77 K.

Note that for a positive φ_s, Q_s corresponds to accumulation. The function $F(\varphi)$ is dominated by the first term in (4.3), that is, $Q_s \sim \exp(q\phi_s/2k_BT)$. For $\varphi_s = 0$, we have the flat-band condition and $Q_s = 0$. For negative φ_s, the function $F(\varphi)$ is now dominated by the second term, that is, $Q_s \sim |\varphi_s|^{1/2}$. The space charge of a semiconductor is determined by ionized impurities in the depletion layer of the semiconductor. For $\varphi_s << 0$, the concentration of minority carriers on the semiconductor surface will be much larger than the concentration of bulk majority carriers; the function $F(\varphi)$ is now dominated by the fourth term, that is, $Q_s \sim \exp(q|\phi_s|/2k_BT)$. For $|\varphi_s| > |2\varphi_F|$, strong inversion begins, and the surface potential is:

$$\phi_s^{\text{inv}} = 2\phi_F = \left(-\frac{k_BT}{q}\right)\ln\left(\frac{n_{n0}}{p_{n0}}\right) \tag{4.5}$$

or

$$n_{n0} = p_{n0}\exp\left(-\frac{q\phi_s^{\text{inv}}}{k_BT}\right) = p_s,$$

where φ_F is the bulk Fermi potential shown in Fig. 4.2, and p_s is the surface concentration of minority carriers.

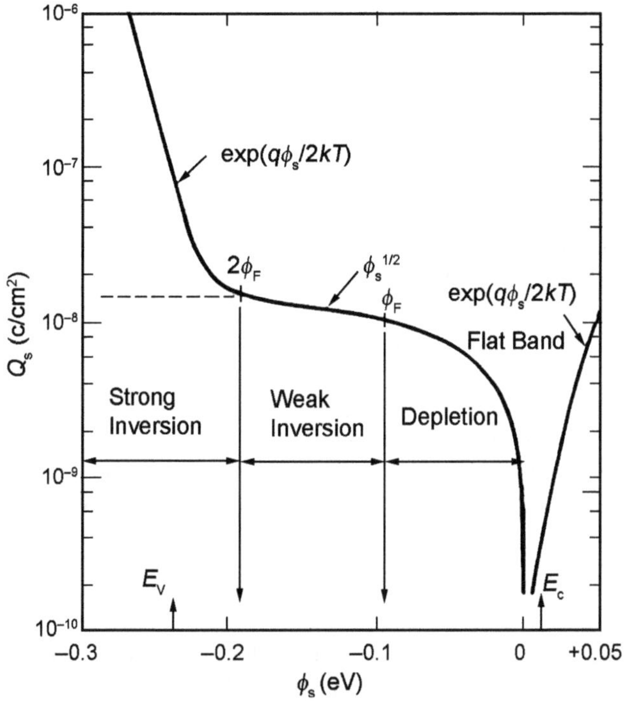

Fig. 4.3 A typical variation of the space-charge density Q_s as a function of the surface potential ϕ_s for an n-type semiconductor

The differential capacitance of the semiconductor space charge layer is given by:

$$C_d \equiv \frac{\partial Q_s}{\partial \phi_s} = \left[\frac{\varepsilon\varepsilon_0 q^2}{2k_B T}\right]^{1/2}\left[\frac{n_{n0}(e^{q\phi/k_B T}-1)-p_{n0}(e^{-q\phi/k_B T}-1)}{F(\phi)}\right]. \quad (4.6)$$

Similarly, for $\varphi_s > 0$, corresponding to the accumulation region,(4.6) can be approximated by $C_{acc} \propto \exp(q\phi_s/2k_B T)$. And then, for $\varphi_s >$ several $(k_B T/q)$, the capacitance of the semiconductor space charge layer will be large. In the strong inversion case, (4.6) can be approximated by $C_{inv} \propto \exp(q|\phi_s|/2k_B T)$. Similarly, for $|\varphi_s| >$ several$(k_B T/q)$, the capacitance of the strong inversion region will be large. For surface potentials φ_s in the depletion region and the weak inversion region, the space charge capacitance is determined by the depletion layer of the semiconductor. Equation (4.6) can be approximated by $C_d = (\varepsilon\varepsilon_0 qN/2\varphi_s)^{1/2}$. This is the depletion layer capacitance, $C_d = \varepsilon\varepsilon_0/W$, where W is width of the depletion layer, as shown in Fig. 4.2.

In an ideal strongly inverted MIS device, the inversion layer can be regarded as a layer with charge Q_{inv} and with a width approaching zero, and the depletion layer can be regarded as a region populated by intrinsic carriers. An applied voltage V_G

will partly appear across the insulator and partly across the space charge layer. Thus, by Gauss's law, we have:

$$\varepsilon_{ox}\varepsilon_0 \frac{V_G - \varphi_s}{t_{ox}} = Q_{inv} + qNW, \tag{4.7}$$

where N is net impurity concentration of the depletion layer with thickness W, and t_{ox} is the thickness of the oxide insulator layer. The surface potential related to the applied voltage is given by (Macdonald 1964):

$$\varphi_s = V_G' + V_0 - (2V_G'V_0 + V_G^2)^{1/2}, \tag{4.8}$$

where $V_G' = (V_G - V_{FB}) + (Q_{inv}/C_{ox})$, $V_0 = qN\varepsilon\varepsilon_0/C_{ox}^2$, $C_{ox} = \varepsilon_{ox}\varepsilon_0/t_{ox}$, and V_{FB} is the flat band voltage of the device.

The total capacitance of an ideal MIS device is a series combination of the insulator capacitance C_{ox} and the semiconductor space charge layer capacitance C_d:

$$C = \frac{C_{ox}C_d}{(C_{ox} + C_d)} \tag{4.9}$$

The capacitance C_d given by (4.6) depends on the voltage. Then from (4.9), we can find a relation between C and V_G. For an ideal MIS device ($V_{FB} = 0$), the theoretical variation of the C curve with V_G is shown in Fig. 4.4. The device parameters (n-HgCdTe) used are $E_g \approx 0.25$ eV, and $n_0 = 10^{15}$ cm^{-3}.

At a positive voltage, we have an accumulation of electrons and therefore a high differential capacitance ($C_d = C_{acc}$) of the semiconductor. As a result, the total capacitance is close to the insulator capacitance C_{ox}. When the voltage becomes sufficiently negative, $\varphi_s < 0$, and forms a depletion region acting as a dielectric in

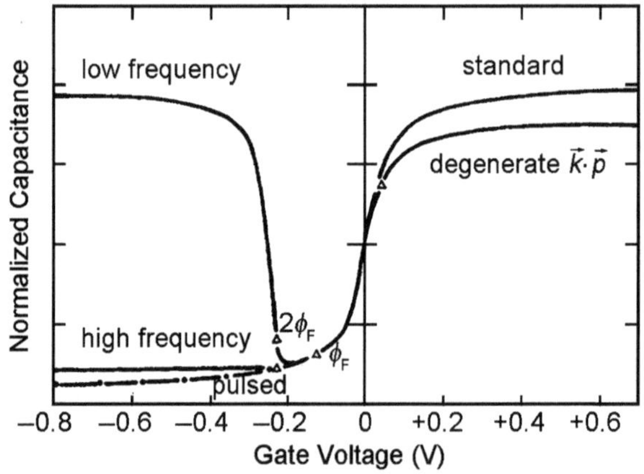

Fig. 4.4 Theoretical variation of low- and high-frequency capacitance curves with V_G for an n-HgCdTe ($E_g = 0.25$ eV) at 77 K (Chapman et al. 1978)

series with the insulator, the total capacitance ($C_d = (\varepsilon\varepsilon_0 qN/2\varphi_s)^{1/2}$) decreases. When the negative voltage is farther reduced, the capacitance goes through a minimum (C_{min}) and then increases again as an inversion layer of electrons forms at the surface. The differential capacitance of the semiconductor space charge layer ($C_d = C_{inv}$) again approaches C_{ox} with a sufficient reduction of the negative voltage. This curve is a low-frequency capacitance spectrum.

The above theory depends on the ability of the carriers to follow the applied ac voltage. It is possible in the accumulation layer and depletion layer. However, in the inversion layer, the minority carriers cannot keep up with a rapidly varying applied voltage. The carriers have a characteristic response time required to follow an applied ac voltage. The response time depends on the concentration of the effective minority carriers in the surface region. In the inversion layer, the response time of the minority carriers is inversely proportional to their induced dark current. At sufficiently high frequencies, the concentration of effective carriers in the inversion layer will change with the applied ac signal but cannot keep pace with it. The resulting curve is the high-frequency capacitance spectrum. The capacitance limit is $C_{ox}C_d/(C_{ox} + C_d)$. This constitutes a finite capacitance that joins the majority and the minority carriers region. In Fig. 4.4, the curve marked "pulsed" shows the capacitance curve of an MIS device response to high frequencies. When the frequency of a bias voltage is high enough, the minority carriers cannot respond at the semiconductor surface in time, and a deep depletion region is formed in the device. The total capacitance measured is equal to a series combination of C_{ox} and C_d, where $C_d = (\varepsilon\varepsilon_0 qN/2\varphi_s)^{1/2}$ is the capacitance of the depletion layer and φ_s is determined by (4.8). Figure 4.4 does take into account the effect of degenerate and nonparabolic bands, V_G is large and positive, so the value of the accumulation layer total capacitance C_{acc} is a little smaller than the standard value calculated by (4.6).

In above discussion, we ignored the surface states of the semiconductor. Figure 4.5 shows an oxide layer–semiconductor interface. There are three main types of states on a surface of HgCdTe: fixed oxide charges, slow-responding surface states, and fast-responding surface states. The number of fixed oxide charges plays an important role determining the flat band voltage of a device. In a typical

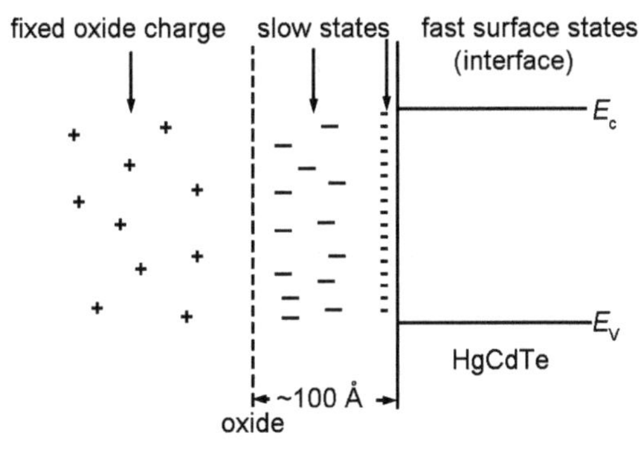

Fig. 4.5 HgCdTe surface states

slow surface state regime, near the interface within a tunnel distance of about 100 Å, the minority carrier interface-trap effect causes a shift of the ideal MOS curve along the voltage axis. The fast surface states cause the measured C-V characteristic curve to deviate from the theoretical prediction discussed above and causes an excess dark current. This deviation depends on the frequency of the bias voltage, the temperature, and the surface potential. The generation-recombination mechanism caused by fast interface states and that caused by bulk Shockley-Read centers follow the same dynamics. Figure 4.6 shows an n-type semiconductor in depletion, including the semiconductor surface states. The thermal generate rate N_{fs} can be modeled as effective resistances $R_{n,s}$ and $R_{p,s}$. These resistances act at the edge of energy band and will vary exponentially with the corresponding energy band in response to the gate potential. Figure 4.6 shows the contribution of the dark current due to the minority carriers. This dark current is generated by the generation-recombination mechanism that produces minority carriers in the neutral body and in the depletion layer.

The equivalent circuit displayed in Fig. 4.7a accurately models the behavior of an MIS device in depletion and/or inversion (Lehovec and Slobodskoy 1964). It includes the minority carrier response time in inversion and the surface state effects we discussed above. For the case of strong inversion, $R_{n,s} \to \infty$, $R_{p,s} \to 0$, and $(\omega C_{inv})^{-1} = (\omega C_s)^{-1}$, then the circuit in Fig. 4.7a simplifies to that in Fig. 4.7b.

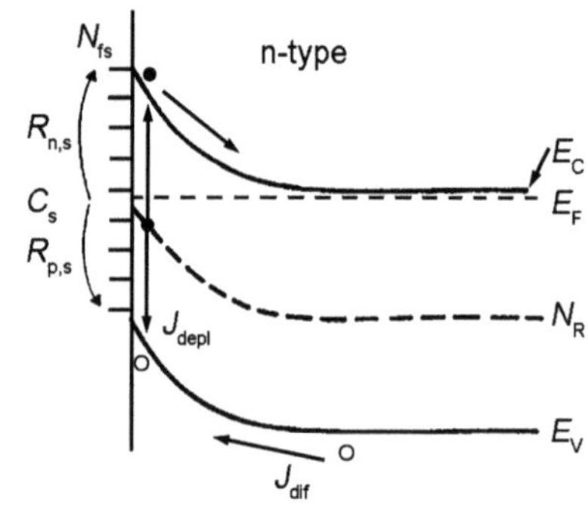

Fig. 4.6 A schematic of an n-type semiconductor in depletion, including its surface states

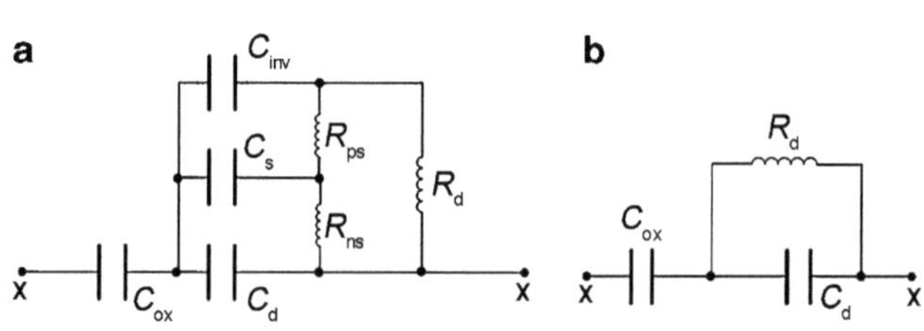

Fig. 4.7 The equivalent circuit model of an n-type MIS device. (**a**) in depletion/inversion and (**b**) in strong inversion

Analyzing the circuit in Fig. 4.7b results in an expression for the MIS device admittance:

$$Z^{-1} = \frac{\omega^2 R_d C_{ox}^2}{1 + \omega^2 R_d^2 (C_{ox} + C_d)^2} + \frac{j\omega[1 + \omega^2 R_d^2 C_d (C_{ox} + C_d)]}{1 + \omega^2 R_d^2 (C_{ox} + C_d)^2} = G_m + j\omega C_m \tag{4.10}$$

In the low-frequency range, $\omega R_d(C_{ox} + C_d) < 1$, (4.10) becomes $G_m = \omega^2 C_{ox}^2 R_d$, $C_m = C_{ox}$. In the high-frequency range, $\omega R_d(C_{ox} + C_d) > 1$, $C_m = C_{ox}C_d/(C_{ox} + C_d)$, and $G_m = R_d^{-1} C_{ox}^2/(C_{ox} + C_d)^2$. Therefore, much information can be gleaned by measuring the C-V curves of an MIS device under strong inversion. For example, we can learn the impurity energy level and the minority carrier-induced dark current. To estimate the fast surface state parameters C_s and R_s, it is necessary to measure the MIS device in the depletion/weak inversion regime and to extract information from the entire equivalent circuit in Fig. 4.7a. The extent of the information obtained from this kind of measurement technology can be found in the reference literature (Nicollian and Goetzberger 1967; Kuhn 1970; Amelio 1972; Tsau et al. 1986).

4.1.2 *Quantum Effects*

In the above analysis, when a positive voltage is applied to an MIS p-type semiconductor structure, the energy bands at semiconductor interface bend downward to form a potential well. Electrons accumulate in this inversion layer formed by the potential well. According to classical theory, electrons continuously gather in this potential well as the positive potential increases and the energy bands bend more. That is in agreement with the test result when the degree of energy band bending is small. For a large energy band bending, we must consider quantum effects. In this situation, the electron energy is not continuous but is quantized at an interface quantum well. Electrons are free to move in the plane parallel to the interface, but have quantized energy levels for the spatial dimension normal to the interface. Quantization into a set of discrete levels is expected that constrains the electron motion in the normal direction. In this potential well, a two-dimension electron gas is formed.

A two-dimensional electron gas (2DEG) is a system (Ando et al. 1982) in which electrons have quantized energy levels for one spatial dimension, but are free to move in two spatial dimensions. Since the 1970s, scientists have paid a high degree of attention to 2DEG systems because it not only engenders academic interest but also has attractive application prospects. Theoretical research on 2DEG systems focuses on the many peculiar phenomena (Dornhaus and Nimtz 1983) found there that do not exist in three-dimensional systems. For example, one finds integer and fractional quantum Hall effects, unusual phase change phenomena, and modified forms of superconductivity. Recently, people have found that oxide high-temperature superconductivity is also a two-dimensional phenomenon (Xiong 1987). The two-dimensional carriers in a Cu–O interface play a primary role in superconductivity. In an aspect of application research, Kinch (1981) recently developed a hetero-junction optical detector. But Hall devices, MIS units, multielement detectors etc. are all based on the special physical properties of two-dimension systems. With the advance of science, two-dimensional systems can be attained easily in the laboratory, such as an electron gas at the liquid helium and vacuum interface, the accumulation layer or inversion layer of metal-insulator-semiconductor structures, and electron and hole gasses at a hetero-junction interface. These systems are all two-dimensional well systems. However, in these systems, their electronic wave functions extend over a limited but finite range along in the quantized spatial dimension (about 10 nm), so we still cannot call these systems ideal two-dimensional systems. Even so, this kind of quasi two-dimensional system still basically reflects the special character of an ideal two-dimensional system. Most of the two-dimensional system photoelectrical research is all based on the quasi two-dimensional system discussed earlier.

There are by now many research methods developed to study two-dimensional systems. Theoretical research methods used include the Hartree effective mass approximation and the variational self-consistent method (Dornhaus and Nimtz 1983). Experiment methods used include interband optical transitions, magneto-optical transitions, electrical and magnetic transport, and cyclotron resonance (Dornhaus and Nimtz 1983). The research aim is to clarify the subband structure and properties of 2DEG systems, such as the subband energy dispersion relations, electron effective masses, and electron wave function distributions. So far, many researchers have addressed the physical properties of two-dimensional systems in wide bandgap semiconductor materials, such as Si, and GaAs. Few research projects on narrow gap semiconductor materials (such as MCT and InSb) have been reported. The reason is that the 2DEG narrow-band materials systems are harder to prepare. In recent years, because of the advancement of preparation technology of narrow-band semiconductor MIS devices, the research into the properties of narrow-band semiconductor 2DEG systems has made great progress.

Take inversion layer 2DEG of MIS devices as examples. 2DEG systems of narrow bandgap semiconductors (MCT and InSb) have several unique properties compared with those of 2DEGs of wide bandgap semiconductors (Si and GaAs):

1. Electron effective masses are small and state densities are low; there are many subbands filled with electrons in these 2DEG systems.
2. The band edges of both the conduction and valence bands of the narrow gap semiconductors MCT and InSb are located at the Γ point, so the energy-valley degeneracy is 1 ($g_v = 1$) and their subbands in 2DEG are nondegenerate. While the subbands in 2DEG Si are degenerate, so its degeneracy is decided by the crystal orientation.
3. The electron mobility is higher because their effective masses are small, and at low temperatures, the states occupied have long wavelengths, so interface scattering caused by roughness and other imperfections is weaker. In addition, the state density is small and many-body effects can be ignored because correlation and exchange energies are a small part of the subband electron energies.
4. The smaller band gap leads to nonparabolic energy bands, so the quantized part of its subband energy is influenced by the nonquantized part to produce unique properties.

5. At a critical applied electrical field for a given band gap, the conduction band and valence band wave functions will overlap, and then the 2DEG subband has completely different properties compared with those of the wide band gap semiconductors. Tunneling effects are examples.

This chapter emphasizes the inversion layer-induced 2DEG system in an MIS structure of p-type narrow bandgap semiconductors MCT and InSb. The issues treated include the subband structures, the ground state at the bottom of the subband, the excitation level of this ground state, the wave function distribution, the range of the quasi two-dimensional electron gases distribution, and the thickness of the depletion layer and its variation with the 2DEG electron area density. The special properties and new quantized character of these narrow gap semiconductors are treated in this discussion. In this chapter, we adopt a more convenient capacitance-voltage measurement method, and establish a quasi-quantized theory set based on the experimental technology. This combination provides a new approach to the study of two-dimensional subbands.

4.2 A Theory That Models Subband Structures

4.2.1 Introduction

We first discuss a theoretical model of an electron subband structure in an inversion layer. There are several existing theories that have been used to solve the hetero-junction 2DEG system problem. However, because a narrow bandgap semiconductor displays new properties, these theories cannot be taken over to solve the 2DEG system problem of a narrow bandgap semiconductor. This chapter formulates another view of the character of two-dimensional narrow bandgap systems, establishes a series of models to interpret experimental results, and then solves for some related parameters. This section and the next two following sections establish three models that predict experimental results: for a subband model of the ground state of a 2DEG, for a capacitance-voltage model of quantum-limited and nonquantum-limited situations, and for a subband potential model of a 2DEG. Finally, we test these predictions against a series of experimental results on the semiconductors MCT and InSb.

The subband electron behavior in the inversion layer of a p-type narrow bandgap MCT semiconductor provides an attractive way to confirm the properties of the subband electron ground-state energy E_0, deduced from studies of the electron Landau energy level and magneto-optic resonance (Chu et al. 1990a). From the quantum capacitance spectrum, the magnetic conductance oscillating spectrum, and the cyclotron resonance spectrum of a p-type MCT MIS structure, we can extract all the properties of the subband structure, including its ground-state energy E_0, its electron effective mass, the thicknesses of the inversion and the depletion layers, the Fermi level and its variation with the concentration of surface electrons. The calculation of the ground-state energy has already been developed for both a single band and multibands. It starts from the Poisson and Schrödinger equations and employs a self-consistent solution method. However, the calculational process is complex, the result does not agree well with experimental results (Sizmann et al. 1990). In this section, we will develop an improved calculation to one that accurately describes this subband structure. To do this, we add some influential factors, the nonparabolic property, a wave function averaging effect, the resonant defect state of p-type materials, and resonant tunneling of the inverted MIS structure. Then deduce a formula for the subband ground-state energy E_0 and its variation with the surface electron concentration that agrees well with experimental data.

Subband inversion layer studies of narrow bandgap semiconductor MIS structures have important consequences in the development of infrared detectors. For a p-type $Hg_{1-x}Cd_xTe$ semiconductor, the Fermi level lies close to the valence band top. When an applied bias voltage inverts the surface, and a quantum well forms to produce a distribution of quantized energy states normal to the surface and a 2DEG subband distribution parallel to the surface, it greatly affects detectors behavior. In an $Hg_{1-x}Cd_xTe$ semiconductor, inversion layer with subbands forms when the electrical field reaches $E \sim 10^5\,\mathrm{V/cm}$. The ground-state electron energy E_0, for an electron effective mass with an order of magnitude, $m^* \sim 0.01m$, and a thickness, z_0, of the inversion layer quantum well, are easily estimated from the uncertainty relation (Koch 1975):

$$p_0 z_0 = \sqrt{2m^* E_0} \cdot z_0 \approx \hbar. \tag{4.11}$$

For a triangular quantum well approximation: $E_0 \approx eEz$, then

$$\begin{cases} E_0 \approx \dfrac{(e\hbar E)^{2/3}}{(2m^*)^{1/3}} \approx 100\,\mathrm{meV}, \\ z_0 \approx \dfrac{\hbar^{2/3}}{(2m^* eE)^{1/3}} \approx 10\,\mathrm{nm} \end{cases} \tag{4.12}$$

In this triangular approximation, the thicknesses of the well for the higher excited-state energy levels $E_1, E_2, \ldots$, are larger. Their energies can also be estimated from the uncertainty relation.

Since the narrow bandgap semiconductors we plan to study have their valence band maximum and conduction band minimum lying at the Γ point, the electron band is highly nonparabolic and has a very low effective mass and therefore a low state density (about $10^{12}\,\mathrm{cm}^{-2}$), several surface quantum well states are always populated. Furthermore, the conduction band's strong nonparabolic behavior leads to a very large coupling between the 2DEG's motion parallel to surface with those in states normal to the surface. These complicate the dispersion relations. Another example is inversion layer electrons penetrate (tunnel) into continuous valence band states and cause a Fano resonance effect. (Fano 1961) Another notable characteristic of a narrow bandgap semiconductor inversion layer electron subband is that the spin degeneracy disappears because of the existence of the surface potential. For a bulk sample, the perturbation $(\nabla U \times \boldsymbol{k}) \cdot \boldsymbol{\sigma}$ term related to the spin–orbit interaction will lift the heavy hole band's twofold degeneracy. When a very large electrical field is applied perpendicular to the sample's surface, the perturbation term has a great influence on the subband structure. This leads to the subband having a larger electro-spin-splitting effect. This effect causes the subband's dispersion relation to split into two branches, even under a zero magnetic field, and leads to a series of new phenomenon: a Landau level shift, and an interface-induced mixing of the Landau level wave functions.

In first principle theories that do not include the exchange interaction, all semiconductors have predicted band gaps that are too small (Chen 1993; van Schilfgaarde 2006). Once the exchange interaction that goes beyond the Hartree approximation and other less important corrections are included, the band gaps are correctly predicted. This statement also applies to narrow gap semiconductors. The theories presented here depend on the Kane theory in which the band gap is a parameter that is taken to agree with experiments. Because the exchange interaction is not explicitly included here, there is a chance that it could make a meaningful correction to these results.

A narrow bandgap semiconductor electron energy band structure consists of a conduction band (a double degenerate Γ_6 band), two valence bands (quadruple degenerate Γ_8 bands), and a spin–orbit split band (a double degenerate Γ_7 band). The energy separation between the Γ_6 and the Γ_8 bands is very narrow, whereas the Γ_7 lies bellow Γ_6 and Γ_8 bands, $\Delta \gg E_g$. Near the Γ point, the energy band structure of narrow bandgap semiconductors can be well approximated by the Kane model, as long as the fundamental gap and P matrix elements are set equal to their experimentally observed values. In 1972, Stern (1972) deduced the kind of self-consistent calculational method that is used to deal with the subband structure of silicon a parabolic energy band semiconductor. This theory is one of the foundation methods for narrow bandgap semiconductor surface electron subband structures. However, complexities engendered by, for example, energy band mixing and nonparabolic behaviors cause the calculations to become difficult. Here we just briefly introduce an overview of the problems associated with this research.

Ohkawa and Uemura (1974), using a triangular quantum well surface potential, diagonalized the Kane Hamiltonian within the Wentzel-Kramers-Brillouin (WKB) approximation, and then calculated the $Hg_{0.79}Cd_{0.21}Te$ subband dispersion relation. They found a large spin-splitting effect in the dispersion relations, as well as an inversion phenomenon in the Landau levels. However, their calculation is not self-consistent, so it is hard to compare it quantitatively with experimental results. Takada et al. (Takada and Uemura 1977, Takada et al. 1980, 1982) have reported a Hartree self-consistent calculation. They developed a method that reduces the 6×6 Kane Hamiltonian matrix into a 2×2 matrix; thereby, simplifying the problem into a one-band approximation. In their calculation, the nonparabolic nature of the conduction band was included through the introduction of a k-dependent electron effective mass, but the interaction between the conduction and valence bands has been ignored. Starting from the 6×6 Kane Hamiltonian matrix, Marques and Sham (1982) calculated the InSb subband structure by a self-consistent calculational method. In principle, this method is also suitable for $Hg_{1-x}Cd_xTe$. Zawadzki (1983) formulated a $\boldsymbol{k} \cdot \boldsymbol{p}$ theory, deduced subband structures, and developed an optical transition theory of the subband electrons. He adopted the triangular quantum well approximation but did not consider spin splitting terms, so his calculation is based on a 4×4 Hamiltonian matrix. The dispersion relations obtained this way are different from the result based on the more precise 6×6 matrix. Merkt and Oelting (1987) adopted a three-level version of $\boldsymbol{k} \cdot \boldsymbol{p}$ theory along with a triangular quantum well approximation of the surface space charge layer, in a 6×6 matrix calculation of the subband structure. After this, Malcher and Oelting (1987) also extended this calculation to treat 2×2, 6×6, and 8×8 matrix versions. However, the Zener tunneling effect was not included in all of the above calculations. But, since narrow bandgap semiconductor subband energies can mix the states, this effect is very important.

Brenig and Kasai (1984) deal with the Zener tunneling effect by a Green function method, and calculated the width and shift of subbands caused by it. Their theory is based on a theory by Takada (1977, 1980, 1982). During calculations of narrow bandgap semiconductor subbands, the coupling between the valence and conduction bands plays a very important role. There are two coupling mechanisms: one is $\boldsymbol{k} \cdot \boldsymbol{p}$ coupling of bulk band structure; the other is the Zener coupling caused by the surface electric field. The former leads to nonparabolic energy band structures and a two-dimensional subband state density that depends on energy. This can be treated by a suitable kinetic energy operator in the subband calculation. The later leads to a degenerate subband state resonance character and a continuous bulk valence band state, which adds width to the subbands. Rössler et al. (Rössler et al. 1989; Ziegler and Rössler 1988) presented a detailed discussion of this treatment. However, it is not easy to deal with this problem accurately. This leads some researchers (Nachev 1988; Chu et al. 1991b) to study the influence of the k – linear term spin–orbit interaction to the subband structure. This new method and its results are presented in Sect. 4.2.2.

4.2.2 A Self-Consistent Calculational Model

First for an ideal situation, one must consider the interband interaction when a model by Stern (1972) is used to deal with MCT inversion layer subband structures. To a certain extent, the interband interaction can be embodied in a subband election effective mass $m^*(E)$ related to an energy found from the Kane model (Chu et al. 1991b; Nachev 1988). Figure 4.8 depicts the energy band bending of a p-type MCT MIS structure when it is inverted.

$E(z) = -e\Phi(z)$ is the energy at a distance z from the surface, where $\Phi(z)$ is the potential variation. The ground-state wave function parameter is calculated from the expression $j_0 = Z_{0,\mathrm{av}}/\langle Z\rangle_0$ [defined in (4.18) and (4.19)], then $E - E_0 j_0$ is the average ground-state subband electron energy referenced to the conduction band edge. According to the literature (Chu et al. 1991b) and considering interband interactions, the effective mass of electrons in the inversion layer-induced ground-state subband is given by:

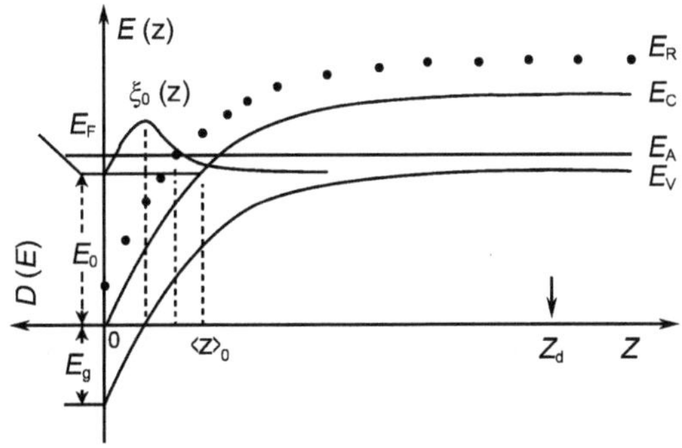

Fig. 4.8 Energy band bending diagram of a p-MCT MIS structure when it is inverted

$$m^*(E) = \left[1 + \frac{2(E - E_0 j_0)}{E_g}\right] m^*. \tag{4.13}$$

Here m_0^* is electron effective mass at the bottom of the conduction band.

In Fig. 4.8, the inversion layer subband system is deduced from the one-dimensional Poisson equation and the Schrödinger equation. Because the MCT inversion layer subband electron effective mass is small, and the state density is low, we can ignore electron–electron interactions in the subband. Also we suppose that (1) at $T = 4.2\,\mathrm{K}$, the system is in a quantum limited state with the Fermi level E_F pined at an acceptor level E_A, (2) there are no resonant defect states in the material, (3) the MIS structure depletion layer is thick enough so that Zener tunneling cannot occur, and (4) the depletion thickness drops quickly to zero in the transition region. So this subband system can be modeled by the following set of simplified equations:

$$\frac{d^2\Phi_z}{dz^2} = -\frac{[\rho_{dep}(z) + \rho_s]}{\varepsilon_s \varepsilon_0}, \tag{4.14}$$

$$\frac{d^2\zeta_0(z)}{dz^2} + \frac{2m^*(E)}{h^2}[E_0 + e\Phi(z)]\zeta_0(z) = 0, \tag{4.15}$$

$$e\Phi_s = (E_g - E_A) + E_F, \tag{4.16}$$

$$N_s = \int_{E_0}^{E_F} D(E)dE = \int_{E_0}^{E_F} \frac{m^*(E)}{\pi h^2} dE, \tag{4.17}$$

$$\langle Z \rangle_0 = \frac{\int_0^\infty z\zeta_0^2(z)dz}{\int_0^\infty \zeta_0^2(z)dz}, \tag{4.18}$$

$$Z_{0,aV} = \frac{\int_0^{\langle Z \rangle_0} z\zeta_0^2(z)dz}{\int_0^{\langle Z \rangle_0} \zeta_0^2(z)dz}, \tag{4.19}$$

and

$$Z_R = \frac{(E_F - E_R)}{E_0} \langle Z \rangle_0. \tag{4.20}$$

Here we have

$$\rho_{\text{dep}}(z) = 0, \quad (z > Z_{\text{d}}),$$
$$\rho_{\text{dep}}(z) = -eN_{\text{AD}} = -e(N_{\text{A}} - N_{\text{D}}), \quad (0 < z < Z_{\text{d}}),$$
$$\rho_{\text{s}}(z) = -\frac{eN_{\text{s}}}{\langle Z \rangle_0}, \quad (0 < z < \langle Z \rangle_0),$$

and $E_{\text{A}} = 0.0165 - 2.4 \times 10^{-8} \cdot N_{\text{A}}^{1/3}$. (Scott et al. 1976).

In (4.16), the first term on the right is the energy difference between the bulk conduction band edge, E_{c}, and the Fermi level, E_{F}. We assume that the Fermi level coincides approximately with the acceptor level in the bulk material, and the second term is the energy difference between the conduction band edge and the surface Fermi level. $\zeta_0(z) = (b^3/2)^{1/2} \cdot z \cdot \exp(-bz/2)$ is the zeroth order Fang-Howard function (Fang and Howard 1966), and b is a parameter. $\Psi_0(x, y, z) = \zeta_0(z) \cdot e^{i\theta z} \cdot e^{i(k_x x + k_y y)}$ is the ground-state subband electron wave function, and θz is a parameter related to the wave vectors k_x and k_y. Take $N_{\text{A}} = N_{\text{AD}}$ for highly doped samples.

The boundary conditions are

$$\Psi_0(x, y, z = 0) = 0, \tag{4.21a}$$

and

$$\Psi_0(x, y, z = Z_d) = 0. \tag{4.21b}$$

This is a good approximation for a p-type MCT MIS structure, since the bandgap of the ZnS insulator layer is 3.8 eV, which is far great than the MCT bandgap ($\sim$0.1 eV) with a composition $x \approx 0.20$. Therefore, we can ignore any wave function penetration into the medium. Considering that few electrons fill the subband ground state and counting only the average effect of the wave function, we can substitute the average effective mass m_{a}^* for $m^*(E)$. By using the boundary condition discussed above and solving (4.21a) and (4.21b), we obtain an expression for the subband energy ground state:

$$E_0 = \frac{h^2 b^2}{8m_{\text{a}}^*} + \left(\frac{3e^2}{\varepsilon_{\text{s}}\varepsilon_0 b}\right)\left[N_{\text{dep1}} + \frac{11 \cdot N_s}{16} - \frac{2N_{\text{AD}}}{b}\right]. \tag{4.22}$$

At the bottom of the ground-state band, $\partial E_0/\partial b = 0$, from which we can obtain the parameters:

$$E_0 = E_{00} + \delta E_0$$
$$E_{00} = (3/2)^{5/3}\left(e^2 h/\sqrt{m_{\text{a}}^* \varepsilon_{\text{s}}\varepsilon_0}\right)^{2/3},$$
$$(N_{\text{dep1}} + 55N_{\text{s}}/96)/(N_{\text{dep1}} + 11N_{\text{s}}/32)^{1/3} \tag{4.23}$$

$$\delta E_0 = (-2N_A e^2 \langle Z\rangle_0^2/3\varepsilon_s\varepsilon_0)(N_{\mathrm{dep1}} + 11N_s/96)/(N_{\mathrm{dep1}} + 11N_s/32)$$
$$\langle Z\rangle_0 = \frac{(N_s + 2N_{\mathrm{AD}}\cdot Z_d) + [(N_s + 2N_{\mathrm{AD}}\cdot Z_d)^2 - 8\cdot N_{\mathrm{AD}}\cdot \varepsilon_S\varepsilon_0 E_0/e]^{1/2}}{-2\cdot N_{\mathrm{AD}}}, \tag{4.24}$$

and

$$N_{\mathrm{dep1}} = N_{\mathrm{AD}}\cdot Z_d = (2\cdot\varepsilon_s\varepsilon_0 N_{\mathrm{AD}}\Phi_S/e)^{1/2}. \tag{4.25}$$

The surface potential Φ_s can be solved from (4.16). The Fermi level E_F can be obtained from the integral in (4.17):

$$E_F = -\frac{E_g - 2jE_0}{2} + \left[\frac{(E_g - 2jE_0)^2}{4} + E_0(E_g + E_0 - 2jE_0) + \frac{E_g\pi h^2 N_s}{m_0^*}\right]^{1/2} \tag{4.26}$$

Equations (4.24) and (4.27) are correlated, so $E_0(N_s = 0)$, $E_0(N_s)$ and $E_F(N_s)$ have to be calculated self-consistently.

Let's discuss a practical situation. In above discussion, we considered the influence of a narrow bandgap semiconductor interband interaction on E_0, then introduced the subband effective mass $m^*(E)$, and deduced a modified formula for E_0 still under ideal conditions. There are many other influential factors that need to consider in practical MIS devices. For example:

1. In depletion, the thickness region has N_{AD} basically constant. But in the transition region between the depleted region and the bulk, the charge density does not drop sharply to zero. Rather it exponentially decays to zero over the range of a shielding length. The influence of charge in the screening length to E_0 contributes to the surface energy band bending. The magnitude of this contribution is approximately, $e\Phi_s^{(1)} = -kT$ (Stern 1972). This influence of the term is small at low temperature.
2. For high doping ($N_{\mathrm{AD}} \approx 10^{17}\,\mathrm{cm}^{-3}$), composition $x = 0.21$, and a p-type MCT sample, there are theories indicating that positive ion vacancies are occupied by oxygen, aluminum, or silicon ions that leads to a series of resonance defect energy level states that lie above the conduction band edge. Recently, experiments have found that there exists an acceptor resonance defect state between the conduction band edge and the subband ground-state E_0. Its distance to the conduction band edge is about 45 meV, and its concentration N_R is about 30–40% N_A. In general, it does not bind electrons and is neutral (Chu et al. 1989). In the above ideal situation, consider that the subband ground state is filled with electrons when it is inverted. But in fact, when there is a resonance defect state, electrons also fill these energy levels prior to filling the ground state. So the presence of these resonant states also has an influence on E_0. One part of the effect of the existence of the defects states is equivalent to the thickness of the inversion layer having an additional term, $N_R Z_R$, another effect is an additional term contributing to the surface band bending, $e\Phi_s^{(2)} = -e^2 N_R Z_R^2/\varepsilon_s\varepsilon_0$.

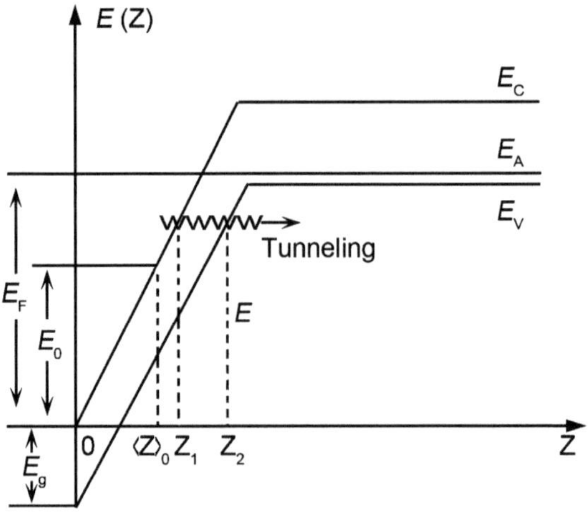

Fig. 4.9 A Zener resonant tunneling diagram for an inverted p-type MCT MIS structure

3. Since the subband energies, $E_{\mathrm{n}}(n = 0, 1, 2, \ldots)$ of a p-type narrow bandgap MCT semiconductor is close to the bandgap E_{g} and for highly doped sample, $(E_{\mathrm{F}}-E_0) \geq E_{\mathrm{A}}$ is easily satisfied, so the subband is located within the continuous valence band. In this case, electrons resonate in the valence band, and will tunnel to degenerated valence band states, leading hopping that broadens the inversion layer subband levels. This is the Zener effect. Brenig et al. (1984) and Zawadzki (1983) have introduced this Zener term and studied its influence on the inversion layer subband levels. Rössler et al. (1989) and Dornhaus and Nimtz (1983) have also undertaken a detailed discussion of this effect; however, their calculations are very complex. Here, we will deal with this problem by a simpler method. Figure 4.9 is diagram of Zener resonant tunneling.

The probability of electron resonant tunneling from an inversion layer subband level into a degenerate valence state energy E is

$$T \approx \exp\left\{\frac{2}{h}\int_{Z_1}^{Z_2} [2m^*(U(z) - E)]^{1/2}\mathrm{d}z\right\}. \tag{4.27}$$

Here we have:

$$\begin{aligned} Z_1 &= (\langle Z\rangle_0/E_0)\cdot E \\ Z_2 &= (\langle Z\rangle_0/E_0)\cdot(E + E_{\mathrm{g}}). \\ U(z) &= (E_0/\langle Z\rangle_0)\cdot z \end{aligned}$$

Then we find:

$$T_{\mathrm{inv}\to\mathrm{va1}} = \exp\left[-4\frac{(2m^*)^{1/2}E_{\mathrm{g}}^{3/2}\langle Z\rangle_0}{3hE_0}\right]. \tag{4.28}$$

From above equations, we see that the tunneling probability depends on only the band gap E_{g}, the inversion layer thickness $\langle Z\rangle_0$, the inversion layer subband electron ground-state energy E_0, and the subband electron average effective mass m_a^*. E_g and $\langle Z\rangle_0$ are small, so the Zener tunneling effect is more evident and $T_{inv\to val}$ approaches unity. The number of electrons that form the inversion layer and can tunnel into the valence band is:

$$N_{ST} = \int_{E_0}^{E_v} T(E)\cdot D(E)\mathrm{d}E \approx N_s \exp\left[-6.58\times 10^9 \left(\frac{m^*}{m_0}\right)^{1/2}\left(\frac{E_g}{E_0}\right)^{3/2}\langle Z\rangle_0\right]. \tag{4.29}$$

In this equation, the energy units are electron volts and the thickness units are meters. The influence of electrons tunneling on E_0 has two aspects: one aspect is its contribution to the total electron density in the inversion layer N_{ST}; the other is its contribution to the surface energy band bending, $e\Phi_s^{(3)} = e^2 N_{ST}\langle Z\rangle_0/\varepsilon_s\varepsilon_0$.

Consider the above factors, Φ_s and N_s in (4.24–4.26) can be amended to Φ_s^*, and N_s^*, and N_s in (4.27) can be amended to $N_s - N_{ST}$. Then we find:

$$\Phi_s^* = \Phi_s + \Phi_s^{(1)} + \Phi_s^{(2)} + \Phi_s^{(3)}, \tag{4.30}$$

$$N_s^* = N_s + N_R Z_R - N_{ST}. \tag{4.31}$$

Substitute the above additional factors into (4.31). Then we find:

$$e\Phi_s^* = \left(E_g - E_A + E_F - kT - \frac{e^2 N_R Z_R^2}{\varepsilon_s\varepsilon_0} + \frac{e^2 N_{ST}\langle Z\rangle_0}{\varepsilon_s\varepsilon_0}\right), \tag{4.32}$$

$$N_{dep1} = 1.110\times 10^4[\varepsilon_s N_{AD}\Phi_s^*]^{1/2}, \tag{4.33}$$

$$E_0 = E_{00} + \delta E_0. \tag{4.34}$$

Here we have defined new variables:

$$E_{00} \equiv 5.6\times 10^{-12}(m_a^*/m_0)^{-1/3}\varepsilon_s^{-3/2}(N_{dep1} + 55N_s^*/96)\cdot(N_{dep1} + 11N_s^*/32)^{-1/3}$$

$$\delta E_0 \equiv -1.2\times 10^{-8}\frac{N_{AD}\langle Z\rangle_0^2 (N_{dep1} + 11N_s^*/96)}{(N_{dep1} + 11N_s^*/32)\varepsilon_s}$$

In these equations, the energy units are electron volts, the length units are meters, and the area units are meters squared. Equations (4.33–4.35) are interrelated, so the inversion layer structure parameter must be obtained from a self-consistent calculation.

The calculational results will be discussed in the following paragraphs. The MCT inversion layer subband structure parameters can be obtained from (4.33–4.35), These include the ground-state subband energy E_0, the inversion layer and depletion layer thicknesses, the Fermi level, and their variation with the surface electron concentration N_s. From (4.18) and (4.19), we can also find the distribution parameter for the ground-state wave function: $j_0 = 0.162$.

Figure 4.10 is a calculated result for an $x = 0.21$ sample's variation of E_0 and E_F with N_s. Experimental points and related parameters come from the literature

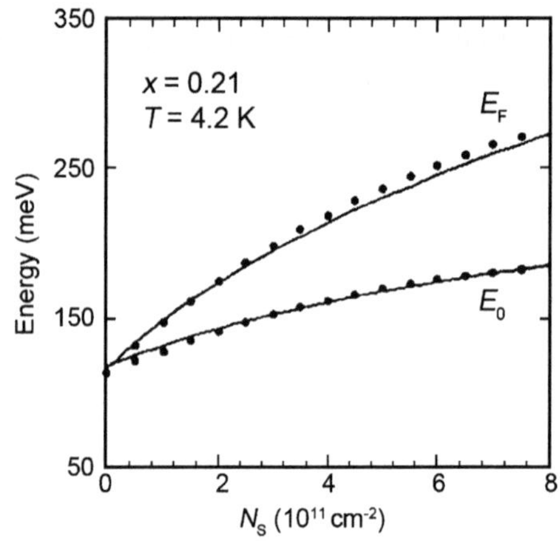

Fig. 4.10 The variation of an $x = 0.21$ sample's ($N_{AD} = 1.87 \times 10^{17}\,\text{cm}^{-3}$) variation of E_0 and E_F with N_s

Table 4.1 The calculational result for E_0 and E_F in different approximations

	N_S (cm^{-2})	E_0 (meV)	E_F (meV)
$T = 0$	0	114.02	114.00
$R = 0$	8×10^{11}	186.03	273.90
$T = 1$	0	114.05	114.00
$R = 0$	8×10^{11}	178.87	267.66
$T = 0$	0	117.20	117.00
$R = 1$	8×10^{11}	190.96	277.91
$T = 1$	0	117.24	117.00
$R = 1$	8×10^{11}	184.35	272.00

(Chu et al. 1989): with data taken at $T = 4.2\,K$, and parameters: $\varepsilon_s = 17$, $N_{AD} = 1.87\times10^{17}\,\text{cm}^{-3}$, $N_R = 9\times10^{16}\,\text{cm}^{-3}$, and $E_R = 45\,\text{meV}$. Electrons fill the levels between E_0 and E_F when the surface inversion layer electron concentration is N_s. Considering the electrons to be in a Fermi-Dirac distribution and the wave function averaging effect, we deduce an effective mass that differs from that at the conduction band edge, $(E_F - E_0)/4 + (E_0 - E_0{}^* j)$, as the electron average effective mass in our calculation.

Solid curves are calculated results including tunneling effects, and the round spot are experimental data.

In order to compare the results quantitatively, Table 4.1 show the influence of the resonant defect state and tunneling effect on E_0 and E_F when they exist separately, $R = 0$ denotes no resonant defect state, $R = 1$ denotes that there are resonance states, $T = 0$ denotes no tunneling effect, and $T = 1$ denotes that there is a tunneling effect. From the table, it is obvious that the calculated results for $E_0 \sim N_s$ and $E_F \sim N_s$ curves will be on the low side if one does not include resonance defect states, and will be on the high side if one does not include the tunneling effect. The larger the inversion layer electron concentration, the larger the influence of the two effects. These conclusions agree with those of Sizmann et al. (1990). So the influence of the two effects must be considered in any practical calculation.

We also calculated for an $x = 0.21$ sample, the variation of E_0 with the bulk doping concentration in two situations (Dornhaus and Nimtz 1983). $N_s \approx 8\times10^{11}\,\text{cm}^{-2}$

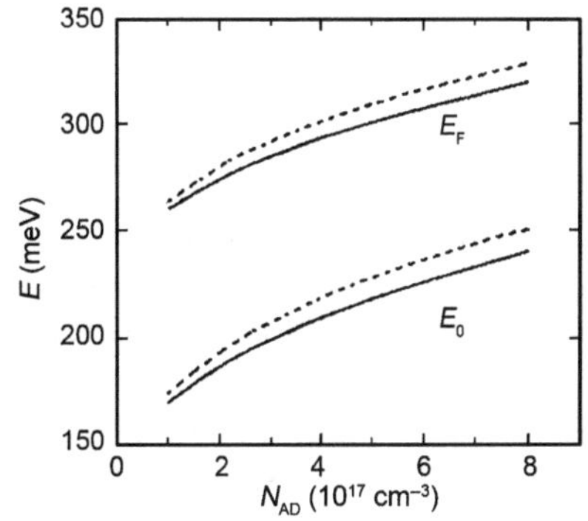

Fig. 4.11 The variation of E_0 with the bulk doping concentration when $N_s \approx 8 \times 10^{11}\,\text{cm}^{-2}$ and $x = 0.21$. The *dots* are experimental data and the *solid curve* is from theory calculation

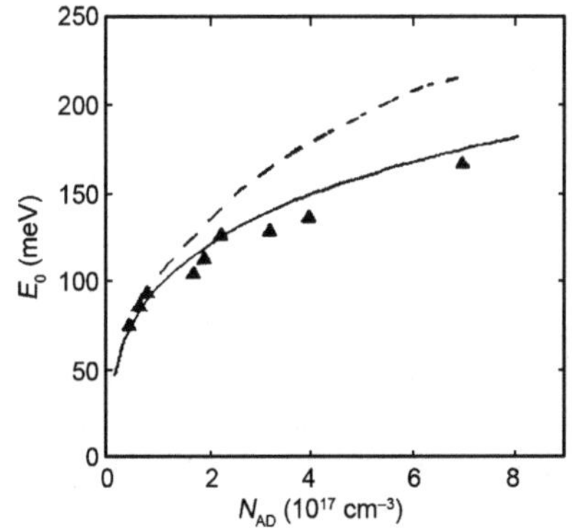

Fig. 4.12 The variation of E_0 with the bulk doping concentration when $N_s \approx 0\,\text{cm}^{-2}$ and $x = 0.21$. The *triangles* are experimental data and the *solid and dashed curves* are from theory calculation

(Fig. 4.11), the solid curve includes the Zener effect, and the dashed curve does not (Chu and Mi 1987). Next we take $N_s \approx 0$ (Fig. 4.11), where the tunneling effect is not included, though in this case, the doping concentration may be very high. However, few electrons tunnel to the valence band so the contribution to the band bending effect can be ignored. The parameters entering the calculation come from literature (Chu et al. 1989; Chu and Mi 1987; Chu 1985a): $T = 4.2\,\text{K}$, $N_R = 35\% N_A$, $E_R = 45\,\text{meV}$, and $\varepsilon_s = 17$. The triangles in Fig. 4.12 are experimental results from the literature (Sizmann et al. 1990).

In Fig. 4.11, the influence of tunneling on E_0 causing it to increase with an increase of the doping concentration is shown. This is caused by a thinning of the inversion layer with an increase of the doping concentration that leads to the material entering into a strongly degenerate state. Then the ground-state energy E_0 lies below the valence band edge and enhances the tunneling effect. In Fig. 4.12, the calculated result (solid curve) is more consistent with the experimental data than is a multiband model (Chu and Mi 1991) (dashed curve). One of the reasons for this deviation between multiband model and experiment is that this model does not account for the Zener effect. R. Sizmann thinks that E_0 will decrease when this effect is included (Sizmann et al. 1990). The other main reason for the multiband model's deviation probably is that it does not treat the averaging effects on the subband wave function or the subband electron effective mass.

As can be seen from above discussion, according to the Stern model, the introduction of a wave function distribution parameter, the subband electron average effective mass, considering the Zener tunneling effect, resonance defect states, and carrier decay within a shielding length, as influential factors, lead to a self-consistent calculational formula for the subband structure. The results calculated from this formula agree well with a p-type material's experimental results. This theory indicates that the presence of resonant defect states leads to a subband ground-state energy that increases, whereas the Zener effect leads to a subband ground-state energy that decreases; the net effect is augmented by an increase of the inversion layer electron concentration and the doping concentration.

4.3 Experimental Research on Subband Structures

4.3.1 Quantum Capacitance Subband Structure Spectrum Model

Since the charge density derived from the Poisson equation is coupled to the subband electron wave function distribution obtained by solving the Schrödinger equation and the surface potential function in the Schrödinger equation enters into the Poisson equation, a self-consistent theory is used to solve both equations. It begins by substituting $(1/i)\partial_z$ for k_z in the Hamiltonian for the subband and adding the surface potential:

$$H_{\text{subband}} = H_{8\times 8}\left(k_z \to \frac{1}{i}\partial_z\right) + V(z)I_{8\times 8}. \tag{4.35}$$

However, the usual self-consistent calculation depends on the selection of boundary conditions, so it does not always agree with the experimental results. For example, subband energy levels calculated for boundary conditions in which the wave functions exponential decay differ by 10–20 meV with results calculated for boundary conditions where the wave functions are taken to be zero in the middle of the band (Malcher et al. 1987). So we must not only improve the theory of subband energies but also establish a method that can directly obtain important properties from subband experimental data. That is, we must establish a phenomenological theory of the subband structure that can output subband structures from inputs derived from experimentally measured results. The result obtained from this method is subject to further testing by comparing its predictions against other experiments, for example, magneto-optical resonance phenomenon, Zener resonance, spin–orbit interactions, and contributions from k and k^3 terms. It also used to allow comparisons between different theory models.

This section chiefly discusses an n-type inversion layer subband behavior as a consequence of an inversion layer forming in a p-type $Hg_{1-x}Cd_xTe$ MIS structure within the electrical quantum limit condition. A focus is on the occupation of the ground-state subband, and once that is settled various physical properties

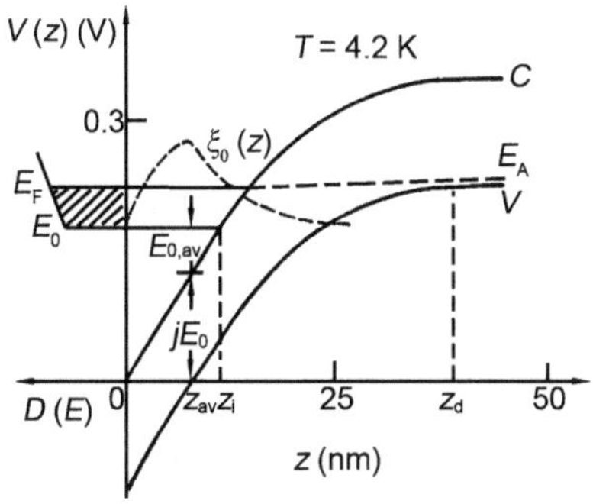

Fig. 4.13 The surface energy band bending diagram in the inversion layer of a p-type $Hg_{1-x}Cd_xTe$ MIS structured sample

of the subband can be displayed more clearly. Figure 4.13 shows the surface energy band bending diagram of the inversion layer of a p-type $Hg_{1-x}Cd_xTe$ MIS structured sample. In Fig. 4.13, $V(z) = -\phi(z)$ is the negative value of the surface potential, E_0 is the subband ground-state energy level at z_i, E_F is the Fermi energy level, $D(E)$ is the subband electron state density, z_i is the thickness of the inversion layer, z_d is the thickness of the depletion layer, z_{av} is the average distance between an electron inside inversion layer and the surface, and $E_{0,av}$ is the average value of the subband electron ground-state energy level inside inversion layer relative to the conduction band edge.

In order to approximately solve the Schrödinger equation, we assume that the subband energy E_0 formula can be developed in terms of a series expansion of the surface electron concentration N_s, and then introduce a first-order term E_{01}, a second-order term E_{02}, and a characteristic parameter $j \equiv z_{av}/z_i$ used to describe the wave function distribution (that is we approximate the subband electron to be distributed mainly in thickness z_{avt}). Adopting these simplified energy and wave function expressions, we solve the Schrödinger and Poisson equations to obtain the subband structure. In order to establish the physical parameters introduced in this theory, we need to fit measured C-V curves and an experimental value of the effective mass obtained from a cyclotron resonance experiment Also we need to fit the relation between N_s and the bias voltage obtained from a surface magnetic conductance (SdH: Shubnikov-de Haas)) oscillation experiment. Through these fittings, we can deduce a set of detailed quantitative properties of a subband structure as functions of N_s.

According to this physical idea, we can write the follow equation set (Chu et al. 1990b, 1991a):

$$\begin{cases} \dfrac{d^2\phi}{dz^2} = -\dfrac{\rho(z)}{\varepsilon\varepsilon_0} \\ E_0 = E_{00} + E_{01}N_s + E_{02}N_s^2 \\ j = \dfrac{z_{av}}{z_i} = \dfrac{E_0 - E_{0,av}}{E_0} \\ N_s = \int_{E_0}^{E_F} \dfrac{m^*(E)}{\pi\hbar^2} dE \\ e\phi_s = E_F + (E_g - E_A) \end{cases}, \tag{4.36a}$$

where

$$\rho(z) = \rho_s(z) + \rho_{\text{dep}}(z)$$
$$\rho_{\text{dep}}(z) = \begin{cases} -eN_A, (0 < z < z_\text{d}) \\ 0(z_\text{d} \leqslant z) \end{cases} \quad . \tag{4.36b}$$
$$\rho_\text{s}(z) = -e\sum_i N_i \xi_i^2(z)$$

The expression for E_0 and j substituted into the Schrödinger equation to a certain extent will be confirmed by fitting experimental results to E_{01} and E_{02} and finding that they are small. Analysis indicates that E_0 expanded to quadratic terms in N_s are sufficient to describe the relation between the subband energy E_0 and N_s in the range of $N_\text{s} \leqslant 8 \times 10^{11}\ \text{cm}^{-2}$. The value of N_s can be deduced from expressions in the literature (Ando et al. 1982) and considering that there is a nonparabolic band.

$\xi(z)$ is an envelope function of the subband electron wave function and takes on the common form of a Fang-Howard wave function (Fang and Howard 1966):

$$\xi_0(z) = \left(\frac{1}{2}b^3\right)^{1/2} ze^{-bz/2}. \tag{4.37}$$

In this wave function expression(4.39), the parameter b and the wave function distribution parameter j are interrelated, and j can be obtained by fitting experimental results.

If the area of an MIS structure is A, then according to (4.36), we have the expressions:

$$C_\text{s} = e\left(\frac{\partial N_\text{s}}{\partial \phi_\text{s}} + N_\text{A}\frac{\partial z_\text{d}}{\partial \phi_\text{s}}\right) \cdot A,$$
$$C = \frac{C_\text{i} C_\text{s}}{C_\text{i} + C_\text{s}} \tag{4.38}$$

where

$$N_\text{S} = \frac{C_\text{i}}{eA}(V_\text{g} - V_\text{t}) - N_\text{A} z_\text{d}, \tag{4.39}$$

and

$$\frac{m^*(E)}{m_0^*} = 1 + \frac{2(E - jE_0)}{E_\text{g}}, \tag{4.40}$$

enabling us to calculate the capacitance spectrum, effective mass, and the relation between N_s and V_g. By fitting the experimental results to these parameters, we can determine the physical parameters related to E_0 and j, and from them determine the subband structure. After we determine the effective mass, it allows us to specify the wave function. If we assume that the electrons are located at the average location in the plane separated at displacement z_av, from the surface, then the energy difference between the electron energy in this plane and the conduction band edge is

$$E_\text{av} = E - jE_0. \tag{4.41}$$

The formula for the subband electron effective mass is given by (4.42). In above expression, the calculated expression from (4.40) is fitted to the C-V experimental result. The calculated result from (4.41) must agree with the relation between N_s and V_g obtained from the SdH experiment, and the calculated result from (4.42) must coincide with the effective mass deduced from the cyclotron resonance experiment. In (4.36), E_{00} is given by:

$$E_{00} = \frac{C_i^2(V_t - V_{Fb})^2}{2\varepsilon\varepsilon_0 e N_A A} - (E_g - E_A). \tag{4.42}$$

Here V_t is the threshold voltage for the onset of inversion. Since the smooth part of a C-V curve in the depletion region goes through the (V_{Fb}, C_i) point, one of the parameters V_{Fb} or C_i is an undetermined coefficient. In the fitting procedure, j, $E_{01}(E_{02})$, and V_{Fb} are regarded as three adjustable parameters. The calculated result must agree with cyclotron resonance, SdH, and C-V curve fits, so it is unique. In this way, physical quantities related to N_s and the subband structure can be obtained. The detailed calculational process can be found in the reference (Chu et al. 1991b).

In order to implement the above model to get a subband structure, we need to measure a sample's C-V, SdH, and cyclotron resonance spectra. Experimental samples are selected to be MIS structures of p-type $Hg_{1-x}Cd_xTe$. After polishing and etching a p-type $Hg_{1-x}Cd_xTe$ sample, we grow either a layer of anodic oxide film on it to a thickness of about 90 nm, or a ~1-μm-thick film of ZnS is evaporated onto the substrate. Then about a 1 μm thick layer of lacquer us coated onto either type of the underlying passivation layers, and finally a last evaporate metal grid electrode is applied. Thus, there are two kinds of insulation layers on samples: (1) ZnS + lacquer and (2) SiO_2 + lacquer. We measure the C-V curve by a differential capacitance method to a precision of 0.005 pF/V (Chu et al. 1991b). To measure the surface conduction, magnetic field-induced oscillating conductance, and the cyclotron resonance of samples, a thin semitransparent Ni-Cr alloy metal grid electrode is evaporated on the surface followed by concentric circle of aluminum electrodes. A high-frequency voltage of about ~10^2 MHz is applied between concentric ring electrodes. This results in a high-frequency current across the sample's surface. Simultaneously, a low-frequency bias voltage is added, then surface conduction is measured by a lock-in-phase sensitive technique. If a longitudinal magnetic field is applied normal to the surface, then the SdH effect can be measured. Figure 4.14 shows the C-V curve, the surface conduction, and the SdH effect curve at $B = 7$ T. If one adds a farinfrared laser applied normal to sample's surface and parallel to a scanned magnetic field, then by measuring the flat band voltage V_{Fb} and reflection index difference ΔR with and without a bias voltage V_g, a cyclotron resonance signal can be measured. Figure 4.15 displays the surface cyclotron resonance curves for $x = 0.234$ p-type $Hg_{1-x}Cd_xTe$ samples prepared by liquid phase epitaxy (LPE) with different surface electron concentrations. From this figure, we see a cyclotron resonance peak and its movement with N_s. The effective mass of the subband electrons at the Fermi level can be calculated from the resonant peak's position.

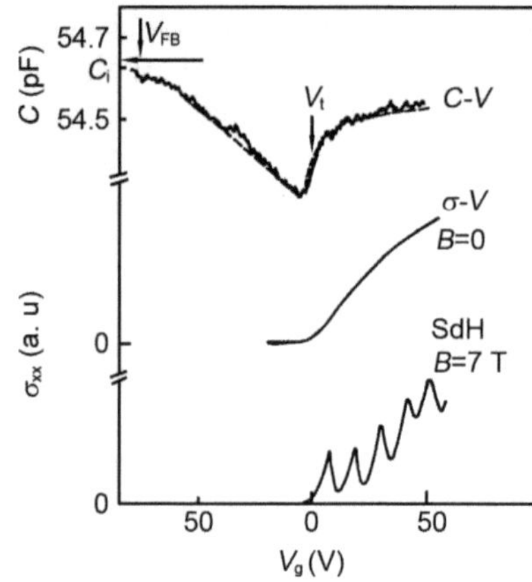

Fig. 4.14 The inversion layer's capacitance spectrum, the surface conduction, and the SdH effect curves at $B = 7\,\text{T}$ of a p-type $Hg_{1-x}Cd_xTe$ ($x = 0.234$, $N_A = 4.0 \times 10^{17}\,\text{cm}^{-3}$) sample

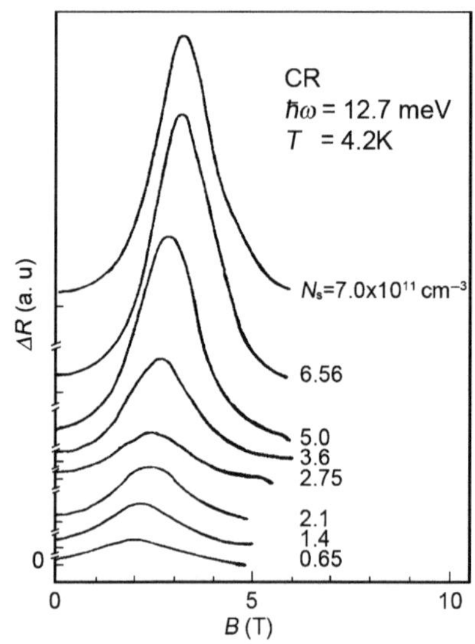

Fig. 4.15 The cyclotron resonance spectrum of a p-type $Hg_{1-x}Cd_xTe$ ($x = 0.234$, $N_A = 4.0 \times 10^{17}\,\text{cm}^{-3}$) sample at different surface electron concentrations N_S

For $x = 0.234$, $N_A = 4.0 \times 10^{17}\,\text{cm}^{-3}$ $Hg_{1-x}Cd_xTe$ MIS structured sample grown by LPE. The C-V curve fitted to a calculated expression is denoted by the dashed line in Fig. 4.14. Table 4.2 contains the fitted parameters and related physical quantities. In Fig. 4.16, "o" denotes the N_s-V_g relation obtained from the C-V fit, and "·" denotes that obtained from the SdH oscillation experimental data. In Fig. 4.17, the solid curves show the $m^*(E) - N_s$ relation obtained from the C-V fit, and "·" denotes cyclotron resonance experimental data. From Figs. 4.14, 4.16, and 4.17, the fitted curves deduced from the above model agree with the experimental C-V, SdH, and cyclotron resonance results. Then a sample's subband structure can be gleaned that agrees with an actual situation.

The surface energy band bending and the subband structure when the surface electron concentration is $N_s = 3 \times 10^{11}\,\text{cm}^2$ are shown quantitatively in Fig. 4.13. Figure 4.16 shows the variation of the subband ground-state energy level E_0 and

Table 4.2 Some relevant physical parameters for a liquid phase epitaxy sample

C_i (pF)	V_{Fb} (V)	V_i (V)	j (eV)	E_{00} (eV)	E_{01} (eV)	E_{02} (eV)	E_g (eV)	E_A
54.65	−75	0	0.84	0.146	1.2×10^{-12}	-5×10^{-26}	0.128	0

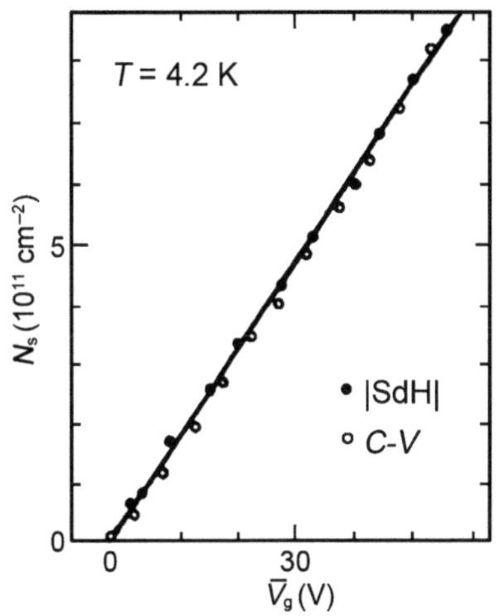

Fig. 4.16 The variation of a p-type $Hg_{1-x}Cd_xTe$ ($x = 0.234$. $N_A = 4.0 \times 10^{17}\,cm^{-3}$) sample's surface electron concentration N_S with bias voltage V_g (the "o" are C-V fitted to calculational results and the "·" are SdH experimental results)

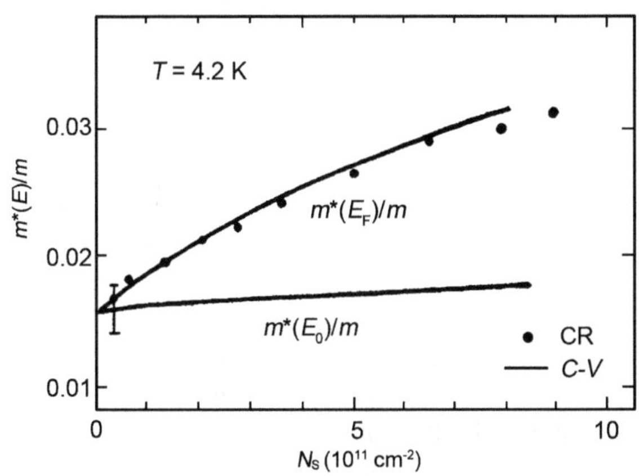

Fig. 4.17 The variation of a p-type $Hg_{1-x}Cd_xTe$ ($x = 0.234$, $N_A = 4.0 \times 10^{17}\,cm^{-3}$) sample's subband electron effective mass with the surface electron concentration N_S ("·" are SdH experimental results)

the Fermi energy level E_F with N_s. In Fig. 4.17, the curves are the variation of inversion layer average thickness z_0 and the depletion layer thickness z_d with N_s. From Figs. 4.18 and 4.19, we see that when the surface electron concentration N_s increases from 0 to $7 \times 10^{11}\,cm^{-2}$, E_0 from 0.146 to 0.206 eV, the Fermi level E_F from 0.146 to 0.276 eV, the depletion layer thickness from 36.6 to 41 nm, whereas the inversion layer thickness variation is small.

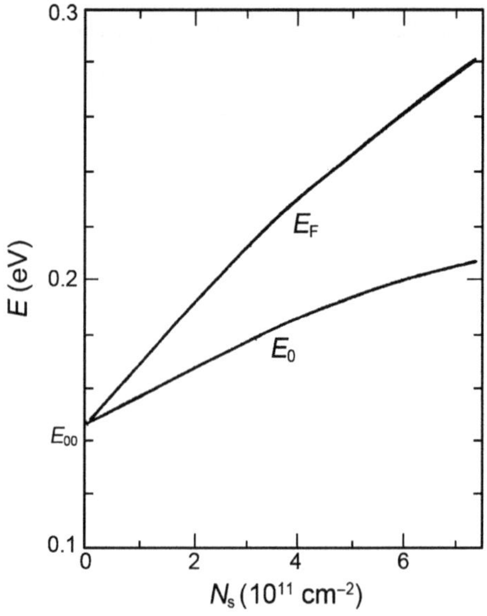

Fig. 4.18 The variation of a p-type $Hg_{1-x}Cd_xTe$ ($x = 0.234$, $N_A = 4.0 \times 10^{17}\,cm^{-3}$) sample's subband ground-state energy level E_0, and Fermi level E_F with the inversion layer electron concentration N_S

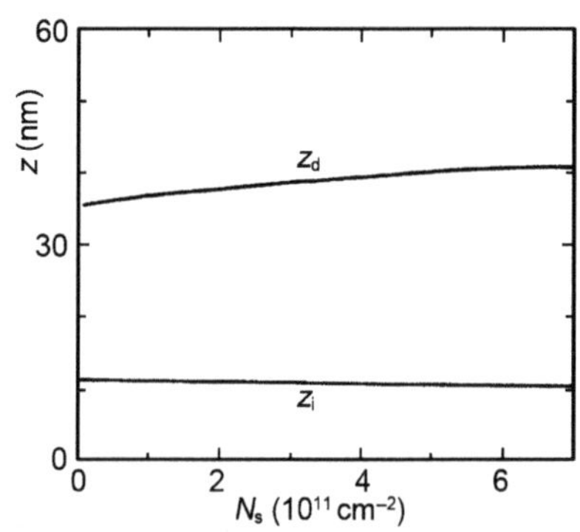

Fig. 4.19 The variation of a p-type $Hg_{1-x}Cd_xTe$ ($x = 0.234$, $N_A = 4.0 \times 10^{17}\,cm^{-3}$) sample's inversion layer average thickness z_i and depletion layer thickness z_d with the surface electron concentration N_S

We have adopted a physical parameter fitting model, and from the experimental measurements, it directly gives the subband structure's dependence on N_s. During the analysis, it specially calls attention to the influence of the resonant defect state on the subband structure. For a sample with a high p-type doping concentration of the $Hg_{1-x}Cd_xTe$ bulk material MIS structure, a peak has been observed in the C-V curve (Chu et al. 1989, 1992) that occurs before the inversion region is reached, which indicates that there is a resonant defect state that lies above the conduction band edge. The electron concentration dependence of this resonant defect state's peak energy level position and its line shape allow us to calculate its energy. Since an electron bound into a resonant defect state is localized, its fill and release process contributes to the differential capacitance (but not to the cyclotron resonance and SdH magnetic conductance oscillations). Therefore, its contribution must be considered when the C-V curve is fitted. Only then is the correct variation of the subband structure with the surface electron concentration N_s obtained.

4.3.2 Quantum Capacitance Spectrum in a Nonquantum Limit

In the inversion layer quantum limit, only the ground-state subband is filled with electrons. In this case, the variation of ground state's energy band with the electron concentration can be expressed by a second order function in N_s (Chu et al. 1990b, 1991b):

$$E_0 = E_{00} + E_{01}N_s + E_{02}N_s^2. \tag{4.43}$$

Here, E_{00} is the ground-state subband threshold energy, and E_{01} and E_{02} are ground-state subband energy parameters. According to semiconductor surface capacitance properties, it is easy to get the contribution of a p-type semiconductor surface 2DEG system to its MIS capacitance (Chu et al. 1990b):

$$C_{\mathrm{inv}} = e\left(\frac{\partial N_s}{\partial \Phi_s} + N_{\mathrm{AD}}\frac{\partial Z_{\mathrm{d}}}{\partial \Phi_s}\right). \tag{4.44}$$

Here, E_{00}, $\partial N_s/\partial\Phi_s$ and $\partial Z_d/\partial\Phi_s$ are deduced from (4.16), (4.18–4.21), and (4.23):

$$\frac{\partial Z_d}{\partial \Phi_s} = (2N_{AD}Z_d + N_s)^{-1}\left(\frac{2\varepsilon_0\varepsilon_s}{e} + N_s\frac{\partial(Z_d - \langle Z\rangle_0)}{\partial\Phi_s} - \langle Z\rangle_0\frac{\partial N_s}{\partial\Phi_s}\right). \tag{4.45}$$

$$\frac{\partial(Z_d - \langle Z\rangle_0)}{\partial\Phi_s} = \varepsilon_0\varepsilon_s\frac{(1 - \partial E/\partial N_s - \partial N_s/\partial\Phi_s)}{eN_{AD}(Z_d - \langle Z\rangle_0)}, \tag{4.46}$$

$$\partial N_s/\partial\Phi_s = \frac{e[1 + 2(E_F - j_0E_0)/E_g]}{[\pi h^2/m_0^* + (E_{01} + 2E_{02}N_s)(1 + 2(E_0 - 2j_0E_0 - j_0E_F)/E_g)]}, \tag{4.47}$$

and

$$E_{00} = \frac{C_i^2(V_t - V_{FB})^2}{2\varepsilon_s\varepsilon_0N_{AD} - (E_g - E_A)}. \tag{4.48}$$

The capacitance of an MIS device in depletion can be calculated from the following equation (Chu et al. 1990b):

$$C_{depl} = \frac{\varepsilon_0\varepsilon_S e}{[C_i \cdot (V_g - V_{FB})]}. \tag{4.49}$$

C_s is the semiconductor surface capacitance, representing C_{inv} and C_{depl} separately for different gate voltages. Then the subband structure in the quantum limit can be obtained by fitting to experimental C-V spectra.

There are four adjustable parameters to be fit: the inversion layer ground-state subband energy parameters E_{01} and E_{02}, the flat band voltage V_{FB} (or the insulating layer capacitance C_{i}), and the ground-state electron wave function distribution parameter j_0.

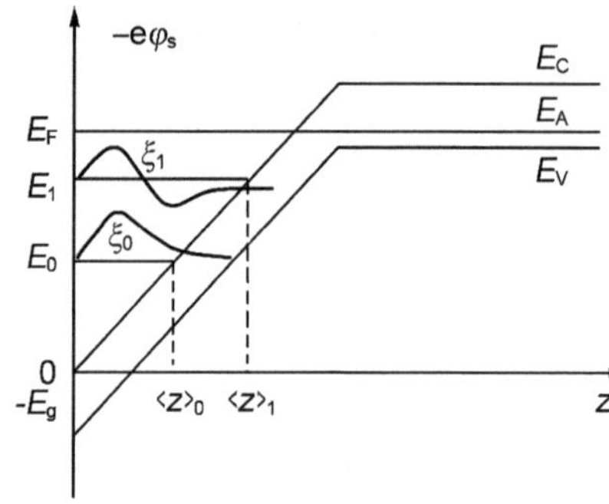

Fig. 4.20 The energy band bending diagram of an MIS structure in the nonquantum limit

In the nonquantum limit, there are at least two or even more than two subbands that are occupied by electrons. Figure 4.20 shows the energy band bending diagram for such a case, E_i $(i = 0, 1, 2, \ldots)$ are the ith subband energies, $\langle Z\rangle_i$ and $\zeta_i(z)$ are the penetration depths and wave functions of the ith subband, respectively. Considering interband interactions, the effective mass of each subband is:

$$m_i^*(E) = \left[1 + \frac{2(E - j_i \cdot E_i)}{E_g}\right] \cdot \frac{0.05966 \cdot E_g \cdot (E_g + 1)}{E_g + 0.667} \cdot m_0. \tag{4.50}$$

j_i is the ith subband wave function distribution parameter, and $E - E_0 \cdot j_i$ is the average energy of the ith subband referenced to conduction band edge. j_i is solved from the follow equations:

$$j_i = \frac{\langle Z\rangle_{i,\mathrm{av}}}{\langle Z\rangle_i}, \tag{4.51a}$$

$$\langle Z\rangle_i = \frac{\int_0^\infty z\zeta_i^2(z)\mathrm{d}z}{\int_0^\infty \zeta_i^2(z)\mathrm{d}z}, \tag{4.51b}$$

$$\langle Z\rangle_{i,\mathrm{av}} = \frac{\int_0^{\langle Z\rangle_i} z\zeta_i^2(z)\mathrm{d}z}{\int_0^{\langle Z\rangle_i} \zeta_i^2(z)\mathrm{d}z}. \tag{4.51c}$$

The density of electrons, N_{ai}, in each subband and the sum of the total electron density N_s in the inversion layer are obtained from the following equations:

$$N_{\mathrm{si}} = \int_{E_\mathrm{i}}^{E_\mathrm{F}} \frac{m_i^*(E)}{\pi h^2}\mathrm{d}E, \tag{4.52}$$

$$N_\mathrm{s} = \sum N_{\mathrm{si}}. \tag{4.53}$$

The subband electron effective mass in the inversion layer of MCT is small, and the state density is low, so the interactions within and among the subbands can be ignored. Thus, one can speculate that E_i can be expressed as a series expansion in the inversion layer electron concentration N_s, and that the quadratic term is sufficient to accurately represent the variation of each subband energy with N_s:

$$E_{\mathrm{i}} = E_{i0} + E_{i1} \cdot N_{\mathrm{s}} + E_{i2} \cdot N_{\mathrm{s}}^2. \tag{4.54}$$

Here, E_{i0} is the ith subband threshold energy, and E_{i1} and E_{i2} are the ith subband energy parameters.

In principle, the inversion layer subband system in Fig. 4.20 also can be solved from the coupled one-dimensional Poisson equation and Schrödinger equation. Following the same procedure as outlined in Sect. 4.2, the subband structure of a 2DEG system can still be solved through fitting theoretical expressions to an MIS device's capacitance spectrum. The contribution of a 2DEG system to a MIS device capacitance can still be written as (4.46), where E_{i0}, $\partial N_s/\partial \Phi_s$ and $\partial Z_d/\partial \Phi_s$ can also be deduced in combination with (4.16), (4.18–4.21), and (4.49–4.54).

In the following discussion as an example, examine a situation in which only two subbands are filled with electrons, and deduce concrete expressions to compare with experiments. The Fermi level can be solved from (4.56) and (4.57) when the total electron density in the inversion layer is N_{s}, and it is

$$E_F = \left[\frac{(E_g - j_0 \cdot E_0 - j_1 \cdot E_1)^2}{4} + E_0(E_g + E_0 - 2j_0 \cdot E_0) + E_1(E_g + E_1 - 2j_1 \cdot E_1) + \frac{E_g \pi h^2 N_s}{(2 \cdot m_{\mathrm{b}}^*)}\right]^{1/2} - \frac{E_g - j_0 \cdot E_0 - j_1 \cdot E_1}{2}. \tag{4.55}$$

For an MIS device, the measured capacitance C_{m} is the series combination of insulating layer capacitance C_{i} and semiconductor surface capacitance $\mathrm{C_s}$, namely $C_m = C_i \cdot C_s/(C_i + C_s)$. Here semiconductor surface capacitance C_{s} is C_{dep1} in depletion calculated from (4.51), and C_{s} is C_{inv} in inversion calculated from (4.46), namely $C_s = e \cdot (\partial N_s/\partial \Phi_s + N_{AD} \cdot \partial Z_d/\partial \Phi_s)$. Here, $\partial N_s/\partial \Phi_s = e/(\partial E_F/\partial N_s)$ is solved from (4.18) and (4.58), whereas $\partial Z_d/\partial \Phi_s$ is deduced from (4.16) and (4.18–4.21), the detailed deduction process follows. When the electron's area density in the inversion layer is N_{s}, the average electron energy E_{av} of two subbands and the surface potential are

$$\Phi_s = e \cdot \frac{(N_{AD} \cdot Z_d^2 + N_s \cdot Z_{av})}{(2\varepsilon_a \cdot \varepsilon_0)}, \tag{4.56}$$

and

$$E_{av} = e^2 \cdot \left[N_{AD} \cdot \frac{(2Z_d - Z_{av}) + N_s \cdot Z_{av})}{(2\varepsilon_a \cdot \varepsilon_0)}\right]. \tag{4.57}$$

In these equations, Z_{av} is the average inversion layer thickness. According to Fig. 4.20, an approximate solution for E_{av} is:

$$E_{av} = \frac{(N_{so} \cdot E_0 + N_{s1} \cdot E_1)}{N_s}. \tag{4.58}$$

Therefore, $\partial Z_d/\partial\Phi_s$ solved from (4.56–4.58) is:

$$\frac{\partial Z_d}{\partial\Phi_s}=\frac{[1.105\cdot 10^6\varepsilon_s+N_s\cdot\partial(Z_d-Z_{av})/\partial\Phi_s-Z_{av}\cdot\partial N_s/\partial\Phi_s]}{(2N_{AD}\cdot Z_d+N_s)} \tag{4.59}$$

In this equation, we have:

$$\frac{\partial(Z_d-Z_{av})}{\partial\Phi_s}=5.5\cdot 10^6\cdot\varepsilon_s\cdot\frac{1-\left(\frac{\partial E_{av}}{\partial N_s}\right)\cdot\left(\frac{\partial N_s}{\partial\Phi_s}\right)}{N_{AD}\cdot(Z_d-Z_{av})}. \tag{4.60}$$

$$\frac{\partial E_{av}}{\partial N_s}=-\frac{E_{av}}{N_s}+\frac{(E_0\cdot\partial N_{a0}/\partial N_s+N_{s0}\cdot\partial E_0/\partial N_s+E_1\cdot\partial N_{s1}/\partial N_s+N_{s1}\cdot\partial E_1/\partial N_s)}{N_s} \tag{4.61}$$

$$\frac{\partial N_{\mathrm{so}}}{\partial N_{\mathrm{s}}}=1-\frac{\partial N_{s1}}{\partial N_{\mathrm{s}}}, \tag{4.62}$$

$$\frac{\partial N_{s1}}{\partial N_s}=\left(\frac{\partial N_{s1}}{\partial\Phi_s}\right)\cdot\left(\frac{\partial\Phi_s}{\partial N_s}\right), \tag{4.63}$$

$$\frac{\partial N_{s1}}{\partial\Phi_s}=\frac{(\partial\Phi_s/\partial E_{\mathrm{F}})}{e}. \tag{4.64}$$

$\partial N_{s1}/\partial\Phi_s$ solved from (4.56), (4.57), and (4.59) is:

$$\begin{aligned}\frac{\partial N_{s1}}{\partial\Phi_s}&=k'\cdot\left[(2E_F+E_g-2j_1\cdot E_1)+(4j_1\cdot E_1-2j_1)\right.\\&\left.\quad\cdot(E_F-2E_1-E_g)\cdot\left(\frac{\partial E_1}{\partial N_s}\right)\cdot\left(\frac{\partial N_s}{\partial\Phi_s}\right)\right],\end{aligned} \tag{4.65}$$

with

$$k'=4.16667\cdot 10^{14}\cdot\frac{(m_0^*/m_0)}{E_g}. \tag{4.66}$$

Then $\partial Z_d/\partial\Phi_s$ can be written as a function that includes $\partial E_0/\partial N_s$, $\partial E_1/\partial N_s$. and $\partial N_s/\partial\Phi_s$, and be calculated from (4.58) and (4.59). Z_{av} and ∂Z_d are solved from (4.60) and (4.61). Therefore, the depletion region capacitance can be fitted by (4.51). The capacitance spectrum with only one subband filled is fitted by the quantum-limited experimental model (Xiong 1987); whereas when the second subband begins to fill, the nonquantum limit experimental model comes into play, which includes $E_0, E_1, E_F, Z_{av}, \partial Z_d$, and their variation with the electron concentration in the inversion layer N_s.

The fitting process of the two subband contribution to the inversion layer capacitance starts, given the initial values of the parameters E_{i0}, E_{i1}, and E_{i2}, at a certain bias voltage V, evaluate Ns from the C-V spectrum, then calculate E_0, E_1, and E_F; next calculate $\partial N_s/\partial\Phi_s, \partial N_{s1}/\partial\Phi_s$, $\partial N_{s0}/\partial\Phi_s, \partial E_0/\partial N_s, \partial E_1/\partial N_s, \partial E_{av1}/\partial N_s$, and $\partial Z_d/\partial \Phi_s$; and finally, the capacitance is found from (4.46). When adjusting parameters, E_1 is located at the second subband threshold voltage V_{t1} and is required to equal E_F since electrons are just filling the third subband at this voltage.

When the inversion layer has three or more subbands that are filled with electrons, by extrapolating this model one level at a time in the measured C-V spectrum, the multisubband system can be solved.

4.3.3 Experimental Research of Two-Dimensional Gases on the HgCdTe Surface

Samples of p-type MCT ($x = 0.24$) bulk materials were fabricated. They were ground, polished, and etched, and then had been evaporated layers of ZnS insulating layers of about 200 nm thickness deposited on the surface. Then through photolithography, these samples had been evaporated gold layers covering their surfaces, and had been made into MIS devices. The electrode areas were $4.91 \times 10^{-4}\,\mathrm{cm}^2$. The devices were fixed onto a sample holder with conductive silver glue. A gold thread was used as the lead wire to make the electrode into a good ohmic back contact.

A fabricated device is placed into a Dewar and its C-V spectrum is measured using a differential capacitance measurement device at 77 K. The AC frequency is 68.8 kHz and its amplitude is 10 mV (satisfying the small signal requirement). The scanning speed of the DC bias voltage is 120 mV/s. The measured result is a low-frequency spectrum (the solid curve in Fig. 4.21).

From this capacitance spectrum, we see that the saturation speed of the capacitance is low in the inversion layer, which is the expected C-V property of a narrow bandgap semiconductor MIS device. The quantum limit C-V experimental model is adopted to fit this experimental result (the dashed curve). The fitting parameters

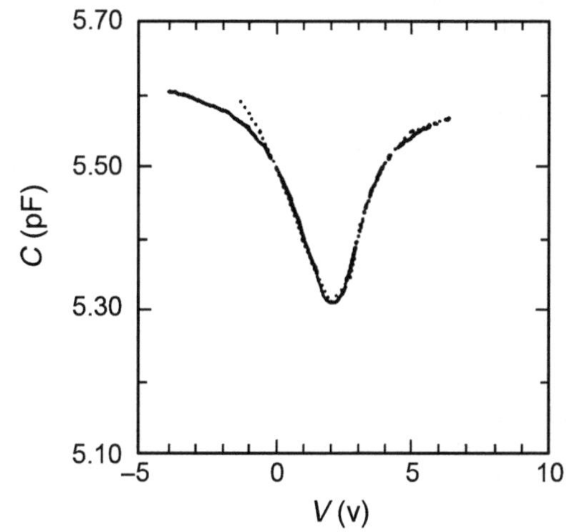

Fig. 4.21 The low-frequency capacitance spectrum of a p-type MCT($x = 0.24$) sample in an MIS structure

are $C_i = 8.66\,\text{pF}$, $V_{to} = 3.05\,\text{V}$, $N_{ad} = 3 \times 10^{16}\,\text{cm}^{-3}$, $j = 0.612$, $E_{00} = 68.04\,\text{meV}$, $E_{01} = 1.6 \times 10^{-13}\,\text{eV} \cdot \text{cm}^2$, and $E_{02} = -8.6 \times 10^{-26}\,\text{eV} \cdot \text{cm}^4$. From this fit, we obtain the subband structure parameters for the inversion layer, including the Fermi level E_F, the electronic ground-state energy E_0, the effective masses $m^*(E_0)/m_0$ and $m^*(E_F)/m_0$, the depletion layer thickness Z_d, the inversion layer's thickness $\langle Z \rangle_0$, and the variation of these parameters with the electron concentration in the inversion layer N_s. This variation is showed as solid curves in Figs. 4.22–4.25.

From Fig. 4.22, we see that the surface does not begin to invert and be filled with electrons until the energy band bending quantity gets to $E_{00} = 68.04\,\text{meV}$. When the electron concentration in the inversion layer N_S increases from 0 to $2 \times 10^{11}\,\text{cm}^{-2}$, E_0 and E_F increase from 88 to 87.40 and 121.70 meV, respectively. The effective masses at E_0 and E_F increase from 0.0186 to 0.020 and 0.024 m_0, respectively (Fig. 4.23). The depletion layer thickness increases from 107.85 to 122.4 nm (Fig. 4.24), whereas the inversion layer thickness decreases from 20.1 to 19.27 nm (Fig. 4.25) and remains almost constant. Since the sample's doping concentration is low, the depletion layer thickness is much larger than the inversion layer thickness. Dashed lines in the above figures are calculated results from a self-consistent variational theory. We see that they agree well with the experimental results.

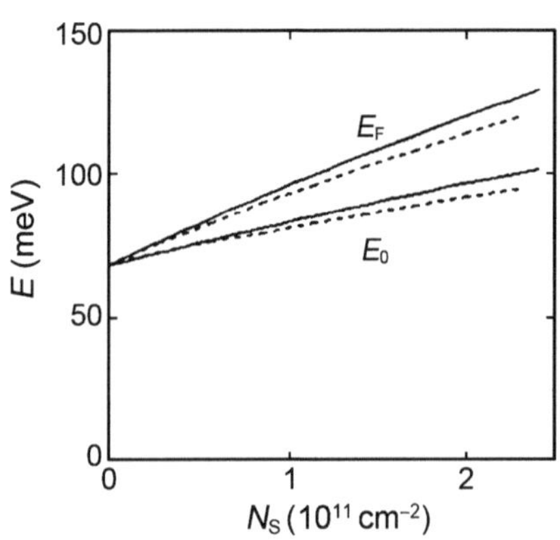

Fig. 4.22 The variation of ground-state energy level and the Fermi energy level in the inversion layer with the electron surface density

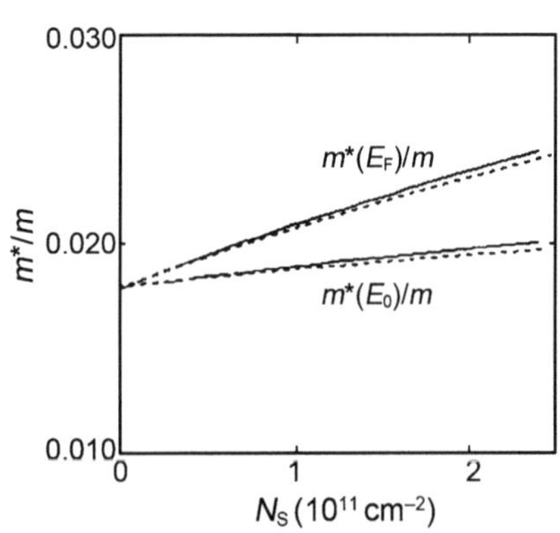

Fig. 4.23 The variation of the electron effective masses of the ground-state energy level and at the Fermi energy level in inversion layer with the electron surface density

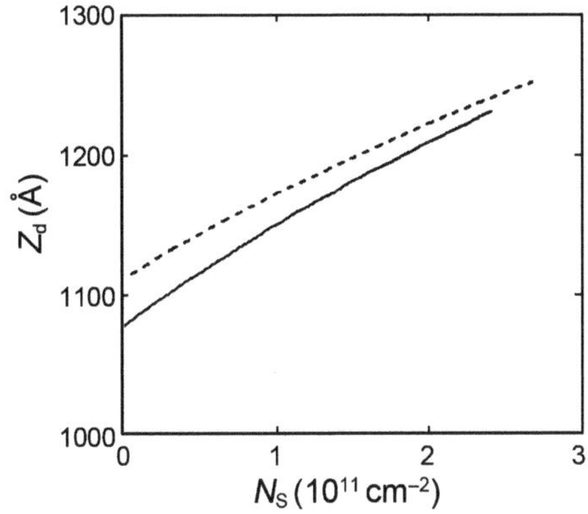

Fig. 4.24 The variation of the depletion layer thickness with the electron surface density in the inversion layer

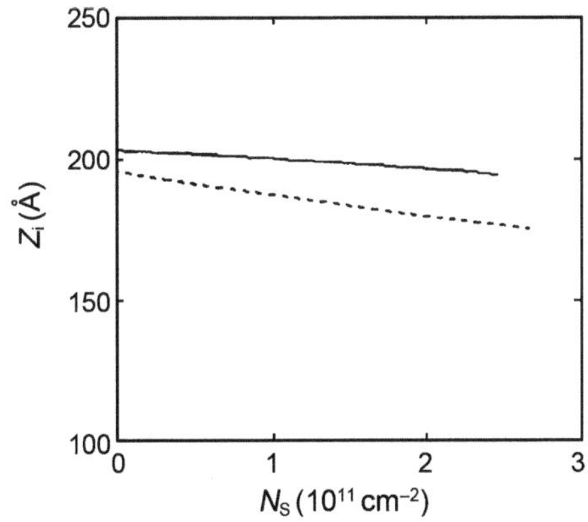

Fig. 4.25 The variation of inversion layer thickness with the electron surface density in the inversion layer

Figure 4.26 shows the C-V spectrum when the bias voltage is scanned forward and then reversed. The capacitance has a 0.3 V hysteresis along the voltage axes, and has no shift along the capacitance axes. A voltage hysteresis effect is caused by the response of slow interface states that are far from the ZnS–MCT interface. Ignoring the influence of other mechanisms, the slow interface state density in this device is estimated to be about $2.1 \times 10^{10}\,\text{cm}^{-2}$. The experimental value of the flat band voltage is -1.45 V that is obtained as a straight line extrapolation from the smooth part of the depletion region capacitance to the voltage axis. The voltage obtained from the flat band capacitance theory is -1.6 V ($C_{\text{FB}} = \sqrt{2} \cdot \varepsilon_0 \cdot \varepsilon_s / L_{\text{D}}$). The difference indicates that there exists some positive fixed charge in the insulator layer, and its density is $1.24 \times 10^{11}\,\text{cm}^{-2}$. It is obvious that the slowly responding interface state and fixed charge densities are both small.

The bulk material's doping concentration N_{AD} obtained from a Hall measurement is $4.5 \times 10^{16}\,\text{cm}^{-3}$. N_{AD} is $3 \times 10^{16}\,\text{cm}^{-3}$ obtained from the capacitance spectrum, which is obviously less than the Hall result. The same result is also found in other p-type MCT MIS devices. Since the C-V measurements are sensitive to the doping concentration in a thin layer surface (a few hundred nanometer thick), surface inversion and interface states both influence the doping concentration

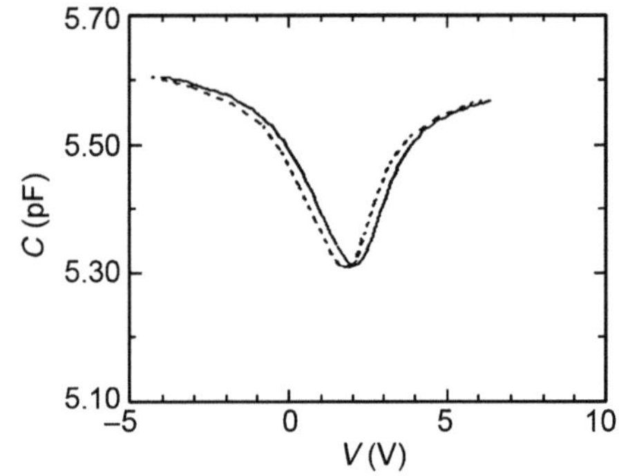

Fig. 4.26 The C-V spectrum when bias voltage scans forward and then is reversed (*solid curve*, forward scan; *dashed curve*, reverse scan)

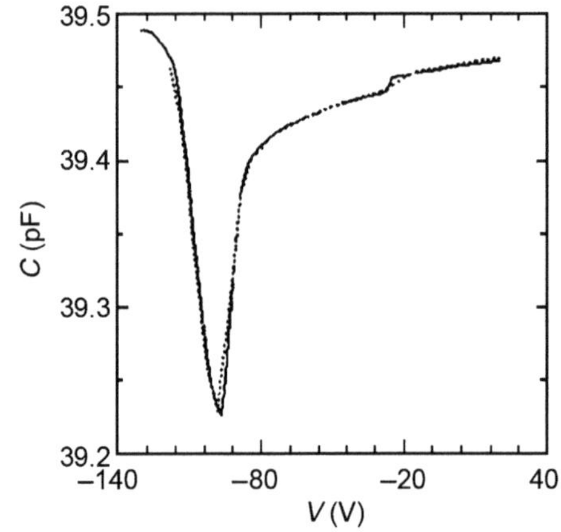

Fig. 4.27 The C-V spectrum of a p-type MCT ($x = 0.213, T = 4.2$ K) MIS device (*solid curve*: experimental measurements ($f = 238$ Hz); *dashed curve*: results fitted to the model)

measurements. The former mechanism leads to a result that is too small, and the latter leads to a result that is too large. Measurement results indicate that surface inversion cannot be ignored in MCT. Sometimes it will substantially modify the interface state properties and seriously degrade device performance, which is confirmed by thin film Hall measurements (Huang et al. 1993).

The capacitance spectrum in the nonquantum limit can also be observed in experiments. Figure 4.27 shows a low-frequency capacitance spectrum at 4.2 K of an $x = 0.213$ MCT MIS device whose electrode area is $A = 0.176\,\text{cm}^2$ and its doping concentration is $N_{AD} = 5.8 \times 10^{16}\,\text{cm}^{-3}$. From this figure, we see that the capacitance has two jumps in the inversion region; the first jump at the threshold voltage V_{t0} indicates that in the inversion layer, the first subband begin to be filled with electrons, whereas the second jump at threshold voltage V_{t1} then indicates that the second subband begins to be filled with electrons. The two threshold voltages read from this experiment are $V_{t0} = -90.2$ V and $V_{t1} = -25.5$ V, respectively.

The quantum- and nonquantum-limited capacitance spectrum model fit is shown as the dashed curve in Fig. 4.27. The fitting parameters are $\varepsilon_s = 16.5, C_\text{i} = 39.528$ pF, $V_\text{fb} = -120.2$ V; $j_0 = 0.8, j_1 = 0.7$; $E_{00} = 0.0787$ eV, $E_{01} = 1.1 \times 10^{-13}$ eV · cm, $E_{02} = -1.1 \times 10^{-26}$ eV · cm^2, $E_{10} = 0.172$ eV,

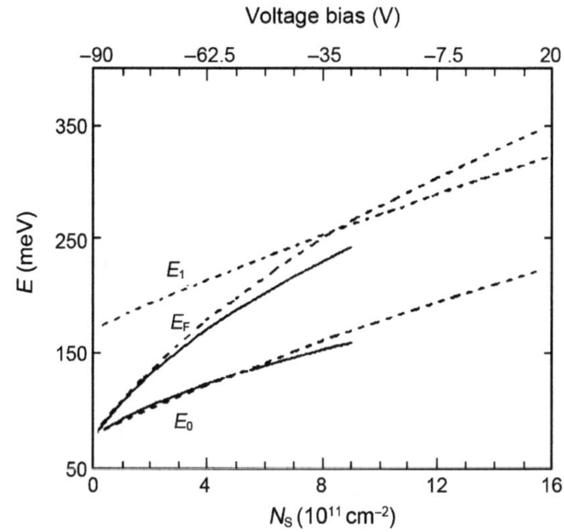

Fig. 4.28 E_0, E_1, and E_F verses $N_s(V_g)$

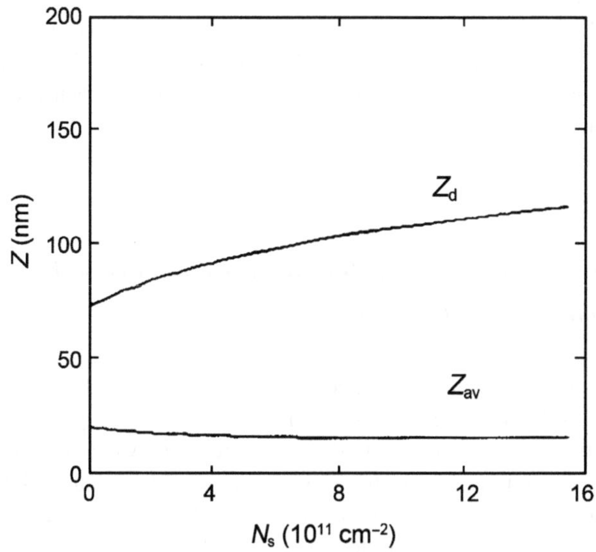

Fig. 4.29 Z_d, and Z_{av} verses N_s

$E_{11} = 1.1 \times 10^{-13}\,\text{eV} \cdot \text{cm}$, and $E_{12} = -1.0 \times 10^{-26}\,\text{eV} \cdot \text{cm}^2$. Also by fitting, we can obtained the inversion layer electron concentration variation of the ground-state subband energy level E_0, the first excited-state subband energy level E_1, the Fermi energy level, the depletion layer thickness, and the inversion layer average thickness shown in Figs. 4.28 and 4.29.

Figure 4.28 shows the variation of the subband energy levels (E_0 and E_1) and the Fermi energy level (E_F) with the inversion layer electron concentration and the bias voltage. The solid curves are the E_0 and E_F results calculated from the self-consistent theory model in the quantum limit, whereas the dashed curves are experimental results for E_0, E_1, and E_F obtained from fitting the capacitance spectrum. From this figure, we see that the first subband begins to fill with electrons when the bias voltage is −90.2 V, the second subband begin to fill with electrons when bias voltage increases to −25.5 V. Then the electron concentration in the inversion layer is about $9 \times 10^{11}\,\text{cm}^{-2}$. The E_0 and E_F theory and experimental results coincide before the second subband begins to fill, especially when the electron concentration in the inversion layer is low. While when inversion layer electron concentration increases, the agreement degrades because the self-consistent model holds only if the electron concentration in the inversion layer is low. For this measured MIS device, when the electron concentration in inversion layer increases from 0 to $1.58 \times 10^{12}\,\text{cm}^{-2}$, the first subband and second subband minimum energies increase from 80.14 and 172 to 223.38 and 318.36 meV, respectively, whereas E_F increases from 80.14 to 344.71 meV.

From Fig. 4.29, we see that when the electron concentration in the inversion layer increases from 0 to $1.58 \times 10^{12}\,\text{cm}^{-2}$, the depletion layer thickness increases from 72.4 to 116.14 nm, whereas the inversion layer average thickness decreases from 20.44 to 16.38 nm. We note that when the second subband begins to fill, the inversion layer average thickness first has only a feeble increase, then decreases with a continuing increase of the electron concentration. This occurs because the second subband begins to fill with electrons.

4.3.4 Experimental Research of a Two-Dimensional Electron Gas on an InSb Surface

MIS-structured samples are fabricated by first evaporating about a 170-nm-thick SiO_2 + SiO insulating layer on a p-type InSb substrate by a photo-CVD method. Then use a lithographic technique to define an area and evaporate gold onto the insulating layer to form a gate electrode. Use deposited In metal as the back ohmic contact electrode and a conductive silver glue as the contact to the gate electrode to finish the structure. The electrode area is $9 \times 10^{-4}\,\text{cm}^2$. N_{AD} from a Hall measurement is $2 \times 10^{16}\,\text{cm}^{-3}$. Figure 4.30 shows the capacitance-voltage spectrum measured at various frequencies.

It is seen from Fig. 4.30 that each capacitance spectrum curve has three rising regions as inversion occurs corresponding to onset voltages −2.4, −2.0, and 5.0 V marked V_0, V_T, and V_1, respectively at the top of the figure. When an MIS device transitions from depletion into inversion electrons begin to fill the ground-state subband and the capacitance increases rapidly, a situation which corresponds to the rising region at onset voltage V_0 in the figure, then the rate of rise of the capacitance begins to slow when the gate voltage increases to V_T. This rising behavior is

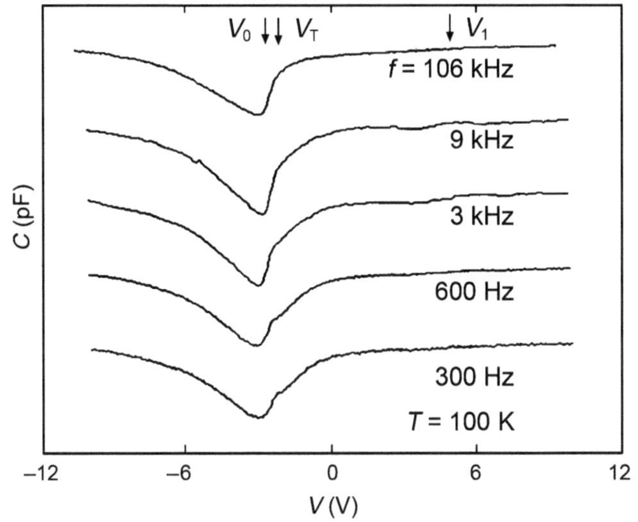

Fig. 4.30 Variable frequency C-V spectra of a p-type InSb MIS structure

Table 4.3 Fitted parameters

i	j_i	E_{i0} (eV)	E_{i1} (eV · cm^2)	E_{i0}(eV · cm^4)
0	0.610	0.046	1.50×10^{-14}	-3.0×10^{-27}
1	0.720	0.100	8.00×10^{-14}	-2.0×10^{-26}

interpreted as being caused by electrons beginning to fill the first excited state. So the structure at V_T is not caused by electrons filling only the ground-state subband. A more telling characteristic is that with an increase of the measurement frequency, the structure at V_T gradually disappears and the capacitance increase gradually becomes more rapid. So we think that there exists an acceptor-type resonance defect state in the conduction band that captures electrons and leads to a reduction of the number of electrons that participate and make contributions to the capacitance. As a result, this slows the rate at which the capacitance rises. With an increase of measurement frequency, the probability of a defect state capturing an electron decreases, and the rate at which the capacitance rises gradually recovers.

Fitting the low-frequency (300 Hz) experimental result to a nonquantum limited C-V model (using Table 4.3) is showed as the dashed curve in Fig. 4.31. In Fig. 4.31, the solid curve is an experimental result. The fit is not good near V_T because the trap effect is not included in the fitting procedure. The parameters extracted from the fitting procedure are $\varepsilon_s = 18$, $E_g = 0.206\,\text{eV}$, $m_0^* = 0.014\,\text{m}$, $C_i = 18.59\,\text{pF}$, $E_{10} = 0.172\,\text{eV}$, $V_{FB} = -7.8\,\text{V}$, $V_{t0} = -2.4\,\text{V}$, and $V_{t1} = 5.0\,\text{V}$.

Figure 4.32 shows the variations of the ground-state subband energy E_0, the first excited-state subband energy E_1, and the Fermi energy E_F as functions of the electron concentration in inversion layer. There are two subbands that are filled with electrons. The ground-state os filled first at the onset of inversion, then the first excited-state begins to fill when the electron concentration in the inversion layer

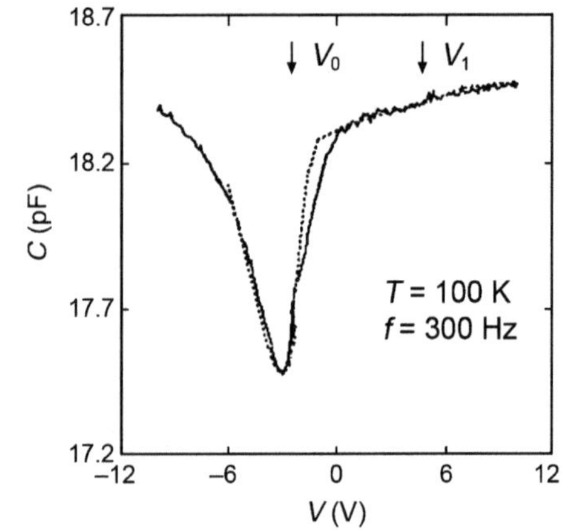

Fig. 4.31 Fitted curve (*solid curve*) and experimental result (*dashed curve*) of the low-frequency C-V spectrum of a p-InSb MIS device

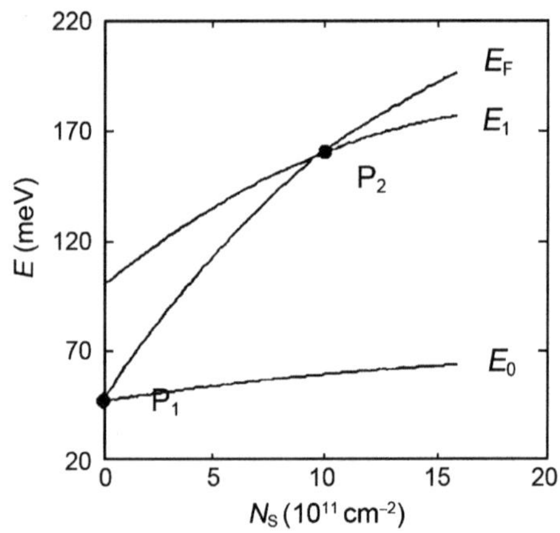

Fig. 4.32 The variations of the subband energy levels E_0, and E_i and the Fermi energy level as functions of the electron concentration in the inversion layer

increase to 9.4×10^{11} cm^{-2}. The corresponding bias voltages are -2.4 and 6.0 V, and are responsible for the energy crossovers at voltages V_0 and V_1 in the spectrum that are marked by P1 and P2 in Fig. 4.32. The rate at which the Fermi level increases begins to decrease when it crosses the first excited subband state as electrons fill this state. As the net electron density in the inversion layer continues to increase, the Fermi level is gradually pinned near the bottom of one of the subband excited-states. When the electron concentration in the inversion layer increases from 0 to 1.58×10^{12} cm^{-2}, the subband ground state and the first subband excited-state increase from 47.1 and 100.7 to 63.2 and 178.6 meV, respectively, whereas E_F increases from 47.2 to 196.0 meV.

Figures 4.33 and 4.34 show the variations of the subband bottom effective masses $m_i^*(E_i)/m$ $(i = 0, 2)$ and the subband electron penetration depths $\langle Z \rangle_i$ with the electron concentration in the inversion layer, respectively.

Figures 4.35 and 4.36 show the variation of the inversion layer average thickness z_{av} and the depletion layer thickness Z_d. P1 and P2 correspond to the places where the ground state and the first excited state begin to fill with electron. By careful observation, we find that, similar to the variations of Fermi level, the rate at which z_{av}

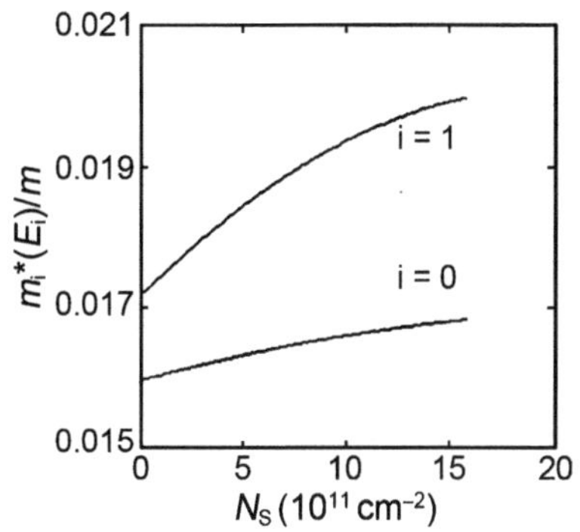

Fig. 4.33 The variations of the subband bottom effective masses as functions of the electron concentration in the inversion layer

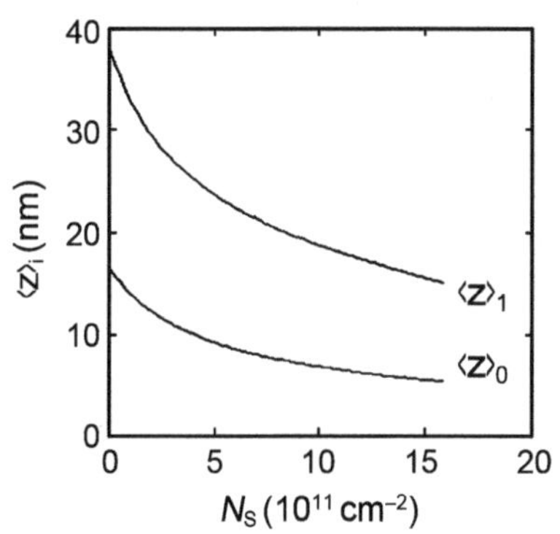

Fig. 4.34 The variations of the subband electron penetration depths as functions of the electron concentration in the inversion layer

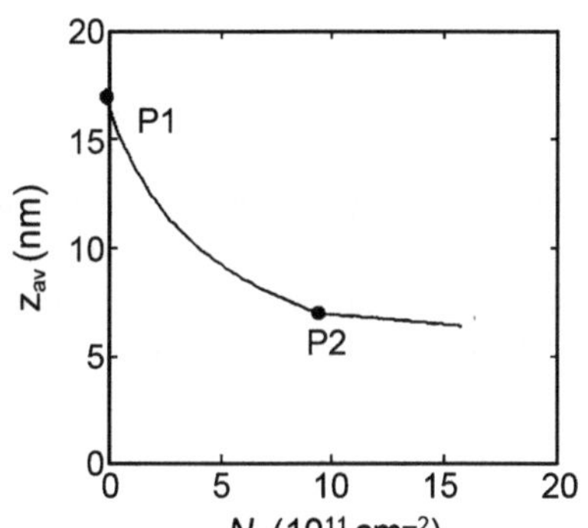

Fig. 4.35 The inversion layer's average thickness, z_{av}, variation as a function of the electron concentration in the inversion layer

and Z_d vary both slow when the first excited state begins to fill with electrons. These rates slow even more when electrons begin to fill higher level subbands. When the electron concentration in the inversion layer increases from 0 to $1.58 \times 10^{12}\,\text{cm}^{-2}$, the inversion layer average thickness decreases from 16.1 to 6.4 nm, whereas the depletion layer thickness still increases from 148.9 to 191.8 nm.

As a comparison, in Fig. 4.37, the solid curves are the variations of $E_{10}(= E_1 - E_0)$ and $E_{F0}(= E_F - E_0)$ as functions of the electron concentration in the inversion layer, and the dashed curves are the theoretical result of Malcher et al (1987). From Fig. 4.37, we see that the experimental data and the theory for E_{F0} agree well,

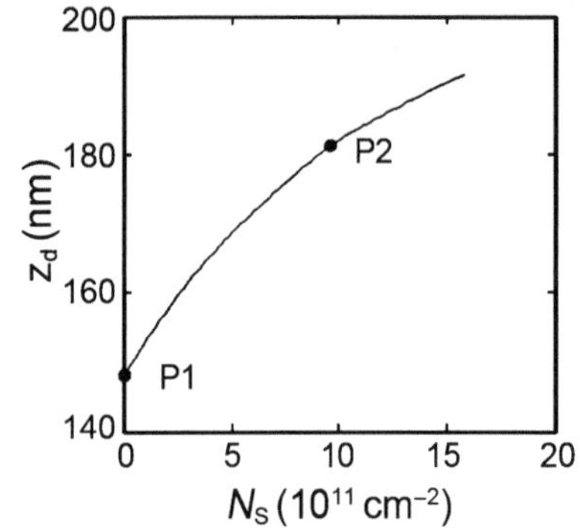

Fig. 4.36 The depletion layer thickness, Z_d, variation as a function of the electron concentration in the inversion layer

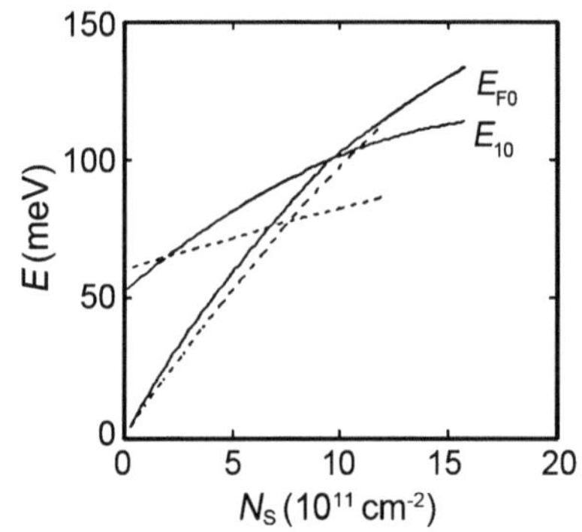

Fig. 4.37 A comparison between experimental (*solid curves*) and theoretical (*dashed curves*) for E_{F0} and E_{10} versus N_s (Malcher et al. 1987)

but the experimental data and the theory for E_{10} have a large discrepancy. The theory predicts that the third subband begins to fill when the electron concentration in inversion layer reaches $7 \times 10^{11}\,\text{cm}^{-2}$, but the threshold concentration measured by experiment is about $9.4 \times 10^{11}\,\text{cm}^{-2}$. The difference between the experimental data and the theory is probably caused by the adoption of a different parameter set. In their theory, they take some key parameters to be: $T = 0\,\text{K}$, $E_g = 0.235\,\text{eV}$, $m_0^* = 0.0136\,\text{m}$, and $N_{AD} = 5 \times 10^{16}$.

4.4 Dispersion Relations and Landau Levels

4.4.1 Expressions for Dispersion Relations and Landau Levels

Bychkov and Rössler have investigated theoretically the impact of the spin–orbit interaction on the surface electrons in narrow-gap semiconductor systems, in which the cubic term in k have also been taken in account (Bychkov and Rashba 1989; Rössler et al. 1989). The electron-subband dispersion relation and the main characteristics of the Landau levels were discussed qualitatively. The dispersion relation expressions and the energies of electron subband Landau levels in the inversion layer of p-type $Hg_{1-x}Cd_xTe$ have also been derived in subsequent papers

(Chu et al. 1990a, b). The physical quantities used in the expressions, such as the strength of the spin–orbit coupling, are obtained by fits to experiments. It is the first time these parameters have been determined quantitatively. The expressions can be used to describe qualitatively the zero-field spin splitting and the crossing of the Landau levels in the inversion layer of p-type $Hg_{1-x}Cd_xTe$.

From Kane's nonparabolic model, the dispersion relation of the electron subband is given by:

$$E_{/\!/} = -\frac{E_{g,eff}}{2} + \sqrt{\left(\frac{E_{g,eff}}{2}\right)^2 + E_g\frac{\hbar^2 k_{/\!/}^2}{2m_0^*}}, \tag{4.67}$$

or

$$E_{/\!/} = -\frac{E_{g,eff}}{2} + \sqrt{\left(\frac{E_{g,eff}}{2}\right)^2 + E_{g,eff}\frac{\hbar^2 k_{/\!/}^2}{2m^*(E_i)}}, \tag{4.68}$$

where $E_{g,eff}$ is the effective band gap energy, E_i is the minimum value of the subband energy, m_0^* and $m^*(E_i)$ are the effective masses at the conduction band edge and at E, respectively. The following relations are satisfied:

$$E_{g,eff} = E_g + 2E_i, \tag{4.69}$$

$$m^*(E_i) = \left(1 + \frac{2E_i}{E_g}\right) m_0^*. \tag{4.70}$$

When the z-direction dependence of E_i is taken into account, E_i will be approximated by $E_{i,av}$, which is the average of E_i in the z direction.

Equations (4.67) and (4.68) are derived from the eigenvalue equation without the spin–orbit coupling term, $(\nabla U \times \mathbf{k}) \cdot \boldsymbol{\sigma}$. But the spin–orbit interaction plays an important role in the band structures of many materials. Also the spin–orbit interaction causes the electron-subband dispersion relations to acquire an addition part related to a large spin splitting at zero-magnetic field (Bychkov and Rashba 1984). In the absence of a magnetic field, we define the Hamiltonian including the spin–orbit term to be:

$$\begin{cases} H_{/\!/} = H_0 + \hbar\alpha\boldsymbol{\sigma} \cdot (\mathbf{k} \times \boldsymbol{\varepsilon}), \\ \mathbf{k} = (k_x, k_y, 0), \text{ and} \\ \boldsymbol{\varepsilon} = (0, 0, \varepsilon_z). \end{cases} \tag{4.71}$$

Here the second term of $H_{/\!/}$ is the spin–orbit or Bychkov-Rashba term, and α is the spin–orbit coupling constant. The secular equation has the form:

$$\begin{vmatrix} E_{/\!/} - E & i\hbar\alpha\varepsilon_z k_{/\!/} e^{-i\phi} \\ -i\hbar\alpha\varepsilon_z k_{/\!/} e^{i\phi} & E_{/\!/} - E \end{vmatrix} = 0. \tag{4.72}$$

Then, with $E_{/\!/}$ given by (4.67) and (4.68), the solutions of (4.72) are:

$$E_{/\!/,\mp} = -\frac{E_{g,eff}}{2} + \sqrt{\left(\frac{E_{g,eff}}{2}\right)^2 + Eg\frac{\hbar^2 k_{/\!/}^2}{2m_0^*}} \pm \hbar\alpha\varepsilon_z k_{/\!/}. \tag{4.73}$$

Here the symbol$\pm$ denotes the different energies of the spin-split electron states ("$-$", "$+$"), and $\hbar\alpha\varepsilon_z$ is given by the relations:

$$\begin{cases} \hbar\alpha = \dfrac{2}{3}\dfrac{eP^2}{E_{\mathrm{g,eff}}}\left(\dfrac{1}{E_{\mathrm{g,eff}}} - \dfrac{1}{E_{\mathrm{g,eff}}+\Delta}\right), \text{ and} \\ \varepsilon_z(z) = -\dfrac{1}{e}\dfrac{\partial V(z)}{\partial z}. \end{cases} \tag{4.74}$$

which depends on the strength of the spin–orbit coupling. The accurate value of the spin–orbit coupling strength is approximated by an average over the electron distribution function. It follows from (4.73) that the energy dispersion in narrow-gap semiconductors splits into two branches even without a magnetic field present, and that the zero-field spin splitting energy is given by the strength of spin–orbit coupling.

If a magnetic field B is applied parallel to k_z, the contribution from the in-plane motion to the subband energy becomes quantized in units of the cyclotron energy and Landau levels are formed. In a system where the spin–orbit coupling is included, each Landau level splits into two nonspin-degenerate subbands with spin quantum number of $s = \pm 1/2$. The energy eigenvalues of the nth Landau level without a spin–orbit coupling are given by:

$$E'_{n,\pm} = -\frac{E_{\mathrm{g,eff}}}{2} + \sqrt{\left(\frac{E_{\mathrm{g,eff}}}{2}\right)^2 + E_{\mathrm{g}}\left[\hbar\omega_c\left(n+\frac{1}{2}\right) \pm \frac{1}{2}g^*\mu_{\mathrm{b}}B\right]} \tag{4.75}$$

If we treat the spin–orbit coupling term as a perturbation, then the Hamiltonian becomes

$$H_{/\!/} = \begin{pmatrix} H_1 & iA\sqrt{n} \\ -iA\sqrt{n} & H_2 \end{pmatrix}, \tag{4.76}$$

where $A = \sqrt{2}/\lambda_C\hbar\alpha\varepsilon_z$, λ_C is the cyclotron radius, H_1 and H_2 are ground-state Hamiltonians of the electron subband in a magnetic field but without a spin–orbit coupling term included. From this perturbation theory, the energy eigenvalues are

$$E^{n,-}_{n-1,+} = \frac{E'_{n-1,+} + E'_{n,-}}{2} \pm \sqrt{\left(\frac{E'_{n,-} - E'_{n-1,+}}{2}\right)^2 + An}. \tag{4.77}$$

The corresponding wave functions of the electronic states are given by:

$$\begin{cases} \psi_{n,-} = \sqrt{\dfrac{1+\sqrt{1+c}}{2\sqrt{1+c}}}\,|n,-\rangle + i\sqrt{\dfrac{-1+\sqrt{1+c}}{2\sqrt{1+c}}}\,|n-1,+\rangle. \\ \psi_{n-1,+} = \sqrt{\dfrac{1+\sqrt{1+c}}{2\sqrt{1+c}}}\,|n-1,+\rangle + i\sqrt{\dfrac{-1+\sqrt{1+c}}{2\sqrt{1+c}}}\,|n,-\rangle \end{cases} \tag{4.78}$$

Here we define:

$$c \equiv \frac{4An}{(E_{n,-} - E_{n-1,+})}. \tag{4.79}$$

Equations (4.77) and (4.78) give the energy of the nth spin-split Landau level and the corresponding wave functions of their ground-state subbands in the inversion layer of a narrow-gap semiconductor. From (4.77), a zero-field spin splitting can change the position of the nonspin-degenerate Landau level in a magnetic field. It causes the $E_{\mathrm{n},-}$ branch to move up, but the $E_{\mathrm{n}-1,+}$ branch to move down. Consequently, these two branches intersect one another at some magnetic field, which can be observed in magneto-transport measurements. The value of the magnetic field, at which the $E_{\mathrm{n},-}$ and $E_{\mathrm{n}-1,+}$ branches intersect, is determined by the strength of the spin–orbit coupling. So the spin–orbit coupling can be determined from the effect of the observed intersection.

Figure 4.38 shows SdH oscillation results measured at different magnetic fields. The oscillation maxima appear when the nth spin-split Landau level passes through the Fermi level. Figure 4.39 illustrates the magnetic-filed dependence of the gate-voltage spacing between the peaks 0^- to 1^+ and 1^- to 2^+ SdH oscillation curves. As the applied magnetic field decreases, the gate-voltage spacing ΔV_g also decreases (Fig. 4.39). The 0^- and 1^+ branches intersect at a critical magnetic field, which can be obtained by extrapolating the gate-voltage spacing curve to the magnetic field at which $\Delta V_g = 0$. The values of magnetic field obtained, at which the

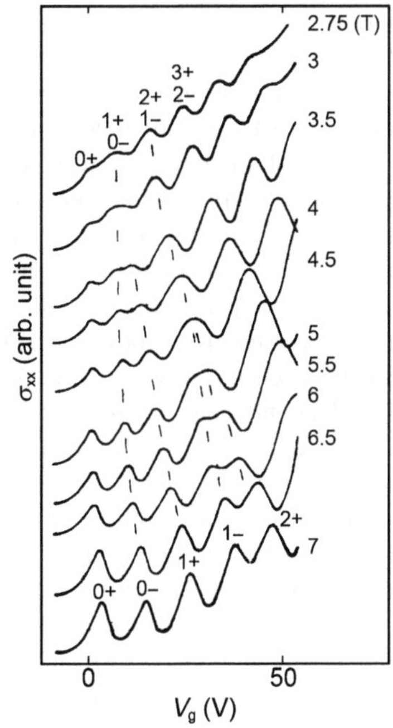

Fig. 4.38 The conductivity as a function of gate voltage in the inversion layer of p-type $Hg_{1-x}Cd_xTe$ ($x = 0.234$, $N_A = 4 \times 10^{17}\,\mathrm{cm}^{-3}$) at different magnetic fields

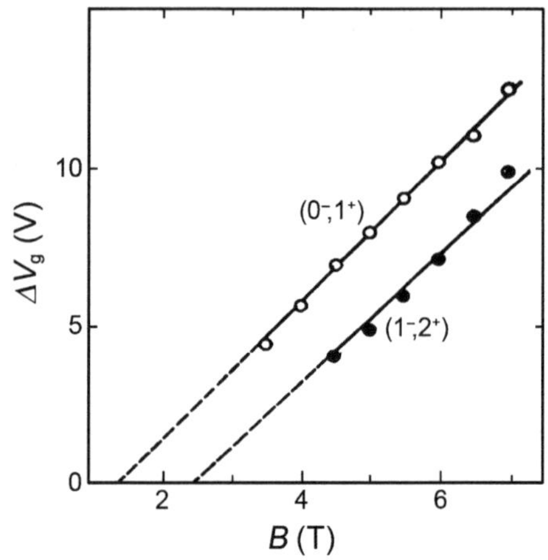

Fig. 4.39 The magnetic-filed dependence of the gate-voltage spacing between the peaks, 0^- to 1^+ and 1^- to 2^+

branches 0^- and 1^+, and 1^+ and 2^+ intersect, are 1.25 T and 2.35 T, respectively. The concentration, N_s, at a spin-split Landau level is given by the expression $2.4 \times 10^{10}\ B(T)\,\text{cm}^{-2}$. Therefore, the energy of the spin-split Landau level that passes through the Fermi level is obtained from the Fermi energy E_F versus N_s relation. On the contrary, we can calculate the energy of the nth spin-split Landau level from (4.77) if the parameter $\hbar\alpha\varepsilon_z$ is selected appropriately. This selection is made such that the magnetic fields, for which the intersection of the branches 0^- and 1^+, and 1^- and 2^+ occur, are the same as those from the measured SdH oscillations. As a result, the energy spacing of the Landau levels is also consistent with the results from cyclotron resonance measurements. Chu et al. (1990a) have measured the cyclotron-resonance spectra in the inversion layer of p-type $Hg_{1-x}Cd_xTe$ ($x = 0.234$) at different electron concentrations.

Figure 4.40 shows the results with photon energies of 12.7 and 17.5 meV. The calculated energy of the spin-split Landau levels is shown in Fig. 4.41 as a function of the magnetic field. The dotted-line and solid-line arrows show the electron transitions between the spin-split Landau levels with energy spacings of 12.7 and 17.5 meV, respectively. The photon energies for the indicated transitions are found to be the same in cyclotron resonance measurements. As shown in Fig. 4.42, as the electron density increases from 0 to $8 \times 10^{11}\,\text{cm}^{-2}$, the value of $\hbar\alpha\varepsilon_z$ required to fit this data only changes from 8×10^{-9} to 6×10^{-9} eV cm. The effective energy gap is also shown in Fig. 4.42 as a function of the electron concentration.

From the values deduced for $E_{g,\text{eff}}$ and $\hbar\alpha\varepsilon_z$, the dispersion relation of the spin-split subbands can be obtained from (4.73), and are shown in Fig. 4.43. The energy difference between E_- and E_+ is about 25 meV at $k_{/\!/} = 2 \times 10^6\,\text{cm}^{-1}$. The energy minimum of the branch "+" does not occur at $k_{/\!/} = 0$ but rather at $k_{/\!/} = 4 \times 10^5\,\text{cm}^{-1}$, and its minimum value is -1 meV. Because of the spin–orbit interaction, the dispersion relation splits into two branches in the inversion layer of narrow-gap semiconductors even without a magnetic field.

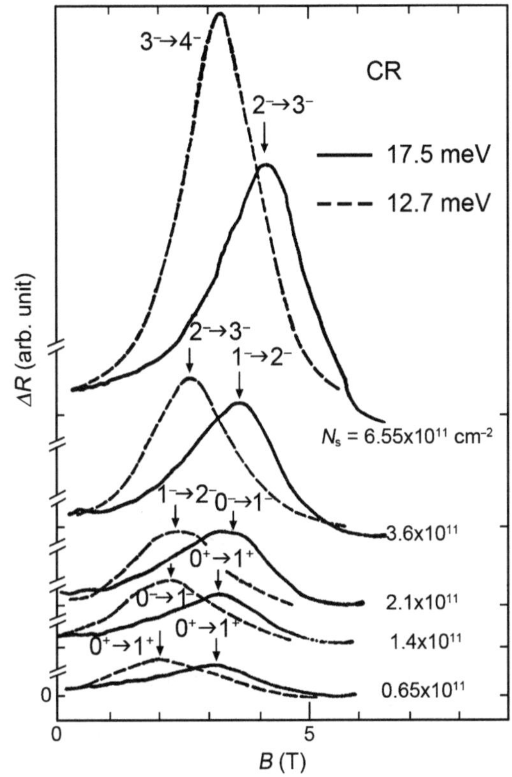

Fig. 4.40 Cyclotron-resonance spectra of a p-type $Hg_{1-x}Cd_xTe$ ($x = 0.234$, $N_A = 4 \times 10^{17}\,cm^{-3}$) at different surface electron concentrations with photon energies of 12.7 meV (*doted curves*) and 17.5 meV (*solid curves*)

4.4.2 Mixing of the Wave Functions and the Effective g Factor

From (4.78), the wave functions of the antiparallel-spin electrons belonging to neighboring Landau levels are mixed. So the wave function of $\psi_{n,-}$ includes two components $|n,->$ and $|n-1,+>$, and $\psi_{n-1,+}$ is a combination of the wave functions $|n-1,+>$ and $|n,->$.

According to (4.78), $\psi_{n,-}$ and $\psi_{n,+}$ can be expressed as:

$$\begin{cases} \psi_{n,-} = c_{n,-}\,|n,\ -\rangle + i\sqrt{1-c_{n,-}^2}\,|n-1,\ +\rangle \\ \psi_{n,+} = c_{n,+}\,|n,\ +\rangle + i\sqrt{1-c_{n,+}^2}\,|n+1,\ -\rangle \end{cases}. \tag{4.80}$$

Figure 4.44 is a plot of the mixing factors, $|C_{n,\pm}|^2$, as a function of the magnetic-field for several of the Landau spin-split levels in the inversion layer of p-type $Hg_{1-x}Cd_xTe$ ($x = 0.234$, $N_A = 4 \times 10^{17}\,cm^{-3}$). Except for the level 0^-, the wave functions $\psi_{n,\pm}$ are composed of two parts $|n,\ \pm\rangle$ and $|n \pm 1,\ \pm\rangle$. The rate of increase of the mixing factors slows as the order, n, of the Landau level increases, and it is weakened for all n as the magnetic field increases.

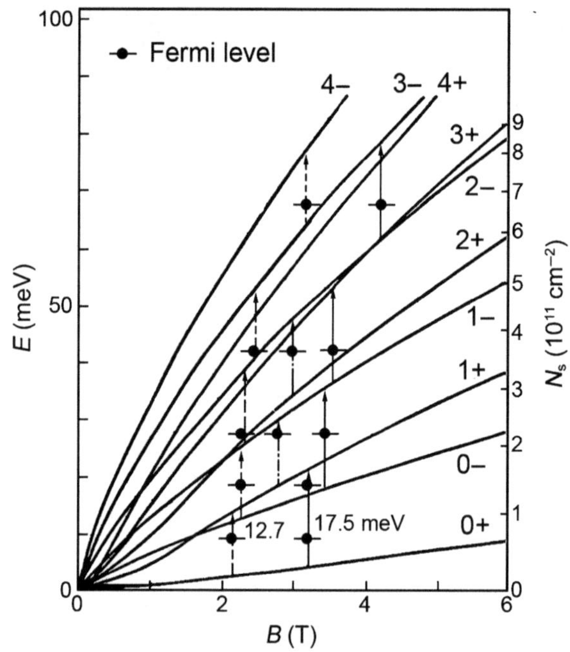

Fig. 4.41 Energy of the spin-split Landau levels in the inversion layer as a function of magnetic field of p-type $Hg_{1-x}Cd_xTe$ ($x = 0.234$, $N_A = 4 \times 10^{17}\,cm^{-3}$) (the arrows show the magneto-optical-induced electron transitions between different spin-split Landau levels)

Fig. 4.42 Effective band gap energy and spin–orbit coupling strength as a function of the electron concentration in the inversion layer of p-type $Hg_{1-x}Cd_xTe$ ($x = 0.234$, $N_A = 4 \times 10^{17}\,cm^{-3}$)

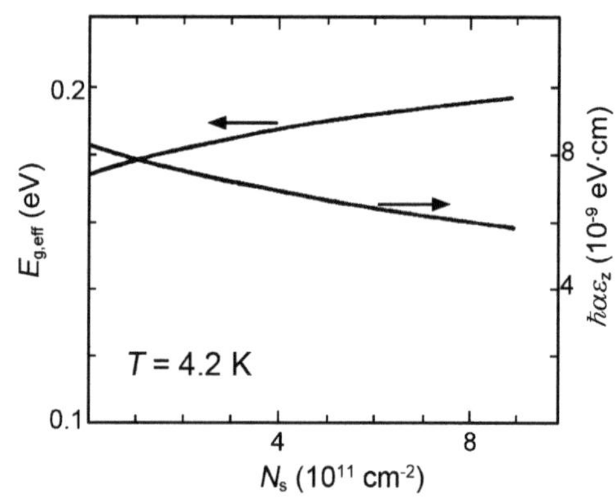

Equation (4.80) indicates that theoretically an electron can transition between the spin resonance states $\psi_{n,-}$ and $\psi_{n,+}$. There are two transition paths possible. One is a transition between the states $|n, +\rangle \rightarrow |n, -\rangle$, related to a normal spin resonance, and in this case results from mixing conduction band and valance band wave functions. It has the same origin as the spin resonance in bulk materials. The others are the transitions $|n, +\rangle \rightarrow |n-1, +\rangle$ or $|n+1, -\rangle \rightarrow |n, -\rangle$. In essence,

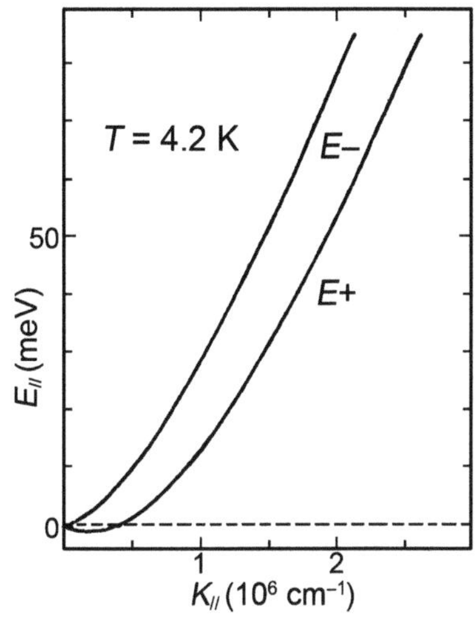

Fig. 4.43 The dispersion relations of the spin-split subbands in p-type $Hg_{1-x}Cd_xTe$ ($x = 0.234$, $N_A = 4 \times 10^{17}\,cm^{-3}$)

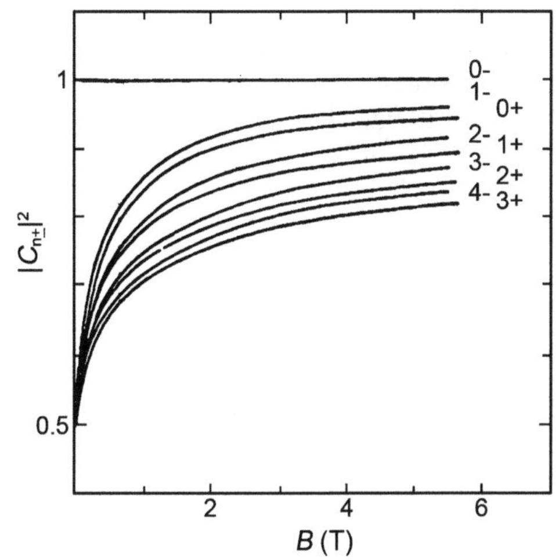

Fig. 4.44 The mixing factors in the inversion layer of p-type $Hg_{1-x}Cd_xT_e$ ($x = 0.234$, $N_A = 4 \times 10^{17}\,cm^{-3}$)

the last one is related to the cyclotron resonance with $\Delta n = -1$, and it is a new feature of spin resonance when the spin-split Landau levels form in semiconductor systems. Figure 4.45 shows how the mixing of the wave functions enables different transitions among the spin-split Landau levels.

Sizmann et al. (1988) were the first to report the spectra of electron-cyclotron resonance in the narrow-gap semiconductors. Since the signal of the spin resonance is 20 times weaker than that compared with cyclotron resonance, it is difficult to detect. But the spin resonance signal can be observed in the cyclotron-resonance-inactive mode if we change the direction of the magnetic field in a cyclotron resonance measurement. Figure 4.46 shows the cyclotron-resonance and spin-resonance spectra in the inversion layer p-type $Hg_{1-x}Cd_xTe$ with a photon energy of 17.6 meV. The resonance peak results from the electronic transition from state $2^+ \rightarrow 3^+$. With

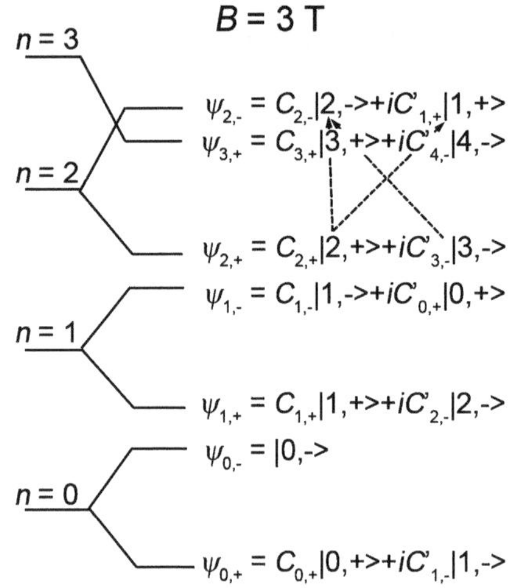

Fig. 4.45 The mixing of wave functions enables different transitions among spin-split Landau levels. Some permitted transitions are indicated by *dotted lines*

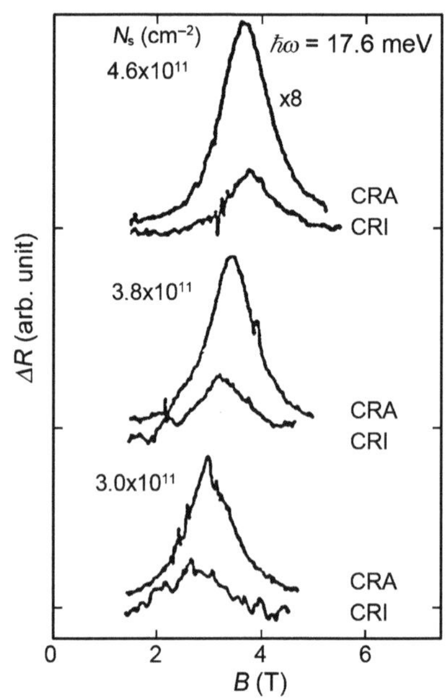

Fig. 4.46 Cyclotron-resonance and spin-resonance spectroscopy on the inversion layer of p-type $Hg_{1-x}Cd_x$ Te ($x = 0.234$, $N_A = 4 \times 10^{17}\,cm^{-3}$)

an electron concentration increase, the Fermi level moves to a higher energy and the peak moves to the high-magnetic-field side of the spectrum. As the electron concentration changes from 3×10^{11} to $4.6 \times 10^{11}\,cm^{-2}$, the spin-resonance peak approaches the cyclotron-resonance peak from the high-magnetic-field side, which results from the intersection of the Landau levels. Figure 4.47 shows the energies of the Landau levels as a function of the magnetic field. The levels 2^- and 3^+ intersect at 4 T. The spin-resonance peak of $2^+ \rightarrow 2^-$ and the cyclotron-resonance peak of

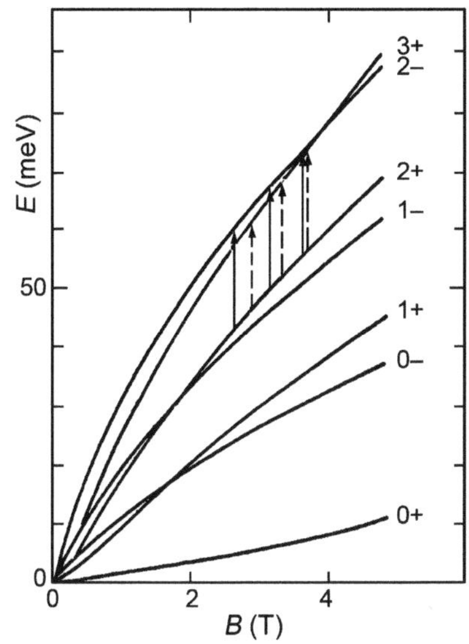

Fig. 4.47 Landau level energies in the inversion layer of p-type $Hg_{1-x}Cd_xTe$ ($x = 0.234$, $N_A = 4 \times 10^{17}\,cm^{-3}$) as a function of magnetic field

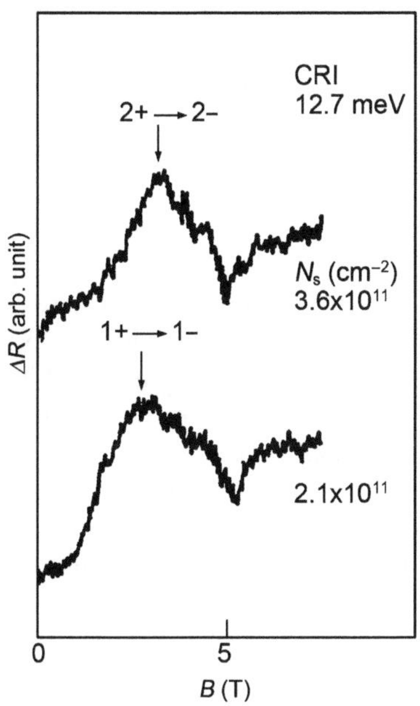

Fig. 4.48 Spin-resonance spectra for p-type $Hg_{1-x}Cd_xTe$ with $x = 0.234$ and $N_A = 4 \times 10^{17}\,cm^{-3}$ at different surface electron concentrations

$2^+ \rightarrow 2^+$ approach one another as the Fermi level moves to higher energy. The observed spin resonance confirms the wave function mixing and the intersection of the Landau levels.

Figure 4.48 shows the spin-resonance spectra in the inversion layer of p-type $Hg_{1-x}Cd_xTe$ ($x = 0.234$, $N_A = 4 \times 10^{17}\,cm^{-3}$) with the photon energy set at 12.7 meV. For an electron concentration of $N_S = 2.1 \times 10^{11}\,cm^{-2}$, the Fermi energy is 26.5 meV above the ground state, and the resonant peak is related to the transition,

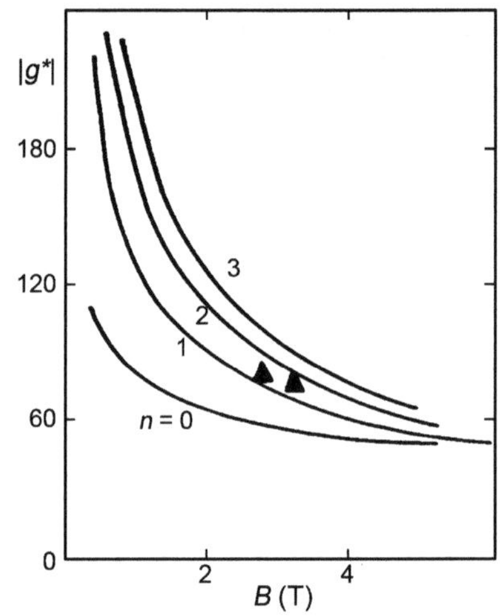

Fig. 4.49 Effective g factor of the nth Landau level in p-type $Hg_{1-x}Cd_xTe$ ($x = 0.234$, $N_A = 4 \times 10^{17}\,\text{cm}^{-3}$)

$1^+ \rightarrow 1^-$. For an electron concentration of $N_S = 3.6\times10^{11}\,\text{cm}^{-2}$, the Fermi energy is 42 meV above the ground state, and the resonant peak is relate to the transition, $2^+ \rightarrow 2^-$. These two kinds of transitions are illustrated in the diagram of the Landau level energies (the dotted curves in Fig. 4.41).

The effective g factor is defined by:

$$|g^*| \equiv \frac{(E_{n,-} - E_{n,+})}{\mu_b B}. \tag{4.81}$$

According to (4.77), due to the spin–orbit interaction, the energy level $E_{n,-}$ shifts up but $E_{n,+}$ shifts down. Consequently, the effective g factor is increased in semiconductor systems by the spin–orbit interaction. From (4.81), the effective g factor decreases as the magnetic field increases. Figure 4.49 shows the magnetic-field dependence of the effective g factor for the nth Landau level of the ground state in the inversion layer of p-type $Hg_{1-x}Cd_xTe$. The solid curves are calculated results for $n = 0, 1, 2, 3$, and the triangular points are the measured results for $n = 1, 2$. These results for the spin resonance clearly indicates that the spin–orbit interaction has an important effect on the subband structures of narrow-gap semiconductor surface layers.

4.5 Surface Accumulation Layer

In the preparation of an $Hg_{1-x}Cd_xTe$ photoconductive detector, surface passivation is usually needed. The passivation layer contains two parts: a prime insulating layer (anodic oxidation, anodic sulfideadation, or anodic fluorination) and then covered with a ZnS layer. A potential well and a large number of active defects are introduced into the surface in these complicated chemical processes (Nemirovsky and Kirdron 1979). Also band bending occurs on oxidation of an exposed surface of $Hg_{1-x}Cd_xTe$ (Nimtz et al. 1979). Because of these two effects, an accumulation layer or even an inversion layer is formed at the surface. The surface electron concentration of an $Hg_{1-x}Cd_xTe$ photoconductive detector is 10^{11}–10^{12} cm^{-2}, which often makes the surface potential larger than the bandgap of the material. In addition, a typical value of the thickness of a photoconductive detector is 10 μm, and the carrier concentration is 10^{14}–10^{15} cm^{-3} at 77 K. The net number of surface carriers is the same as that of the bulk or even larger, so the surface condition has an important influence on device performance.

The electron concentrations, effective masses, and energy levels in different subbands of an accumulation layer can be calculated from the WKB approximation. In a SdH oscillation experiment for an n-$Hg_{1-x}Cd_xTe$ accumulation layer, the electron concentrations for different subbands are gleaned from a Fourier analysis of the SdH oscillation curves. Furthermore, the electron concentration and mobility can be learned from the mobility spectrum.

Nemirosky and Kidron (1979) obtained electron concentrations and mobilities of an $Hg_{1-x}Cd_xTe$ accumulation layer from Hall Effect and capacitance-voltage measurements. After measuring the SdH oscillations of an accumulation layer, Nicholas et al. (1990) pointed out that a heavily accumulated surface layer of $Hg_{1-x}Cd_xTe$ could be described as a 2DEG containing several subbands. A theoretical calculation of the dispersion relations of these subbands depends on a self-consistent treatment of the surface potential and the surface electron concentration. Nachev (1988) and Lowney et al. (1993) applied an eight-band model to calculate the dispersion relations of the different subbands in a surface accumulation layer and an inversion layer, but the computational complexity of the model is quite high. Ando (1985) used a simpler semiclassical approximation to study the electronic properties of subbands in surface accumulation layers of narrow gap semiconductors.

4.5.1 Theoretical Model of n-HgCdTe Surface Accumulation Layer

The conduction band of $Hg_{1-x}Cd_xTe$ is nonparabolic. The Kane model can be rewritten as:

$$E_k\left(1+\frac{E_k}{E_g}\right)=\frac{\hbar^2k^2}{2m_0^*}, \tag{4.82}$$

where $k = (k_x, k_y, k_z)$ is the three-dimensional wave vector, E_k is the electron energy at $\boldsymbol{k}$, the wave vector, E_g and m_0^* are the bandgap of the material and the effective mass at the conduction band edge, respectively.

The electric potential distribution of a heavily accumulated surface layer of an $Hg_{1-x}Cd_xTe$ device follows the Poisson equation:

$$\frac{d^2V(z)}{dz^2}=\frac{\rho(z)}{\varepsilon_s\varepsilon_0}=-\frac{en(z)}{\varepsilon_s\varepsilon_0}, \tag{4.83}$$

where ϵ_s is the low-frequency dielectric constant, and $n(z)$ is the electron concentration, which is mainly due to the ionized active surface impurities. Bulk ionized impurities, and the image force caused by different dielectric constants of the layers between the bulk and the surface, have less influence on the electric potential than the surface impurities. Thus, their effect is neglected in (4.83).

For a low-temperature case, a semiclassical approximation, and spin degeneracy, the electron concentration can be written as:

$$n(z) = \frac{2}{(2\pi)^3}\frac{4\pi}{3}K_F(z)^3. \tag{4.84}$$

The Fermi wave vector K_F as a function of position z has the following relation:

$$\frac{\hbar^2 K_F^2}{2m_0^*} = \frac{\hbar^2(k^2+k_z^2)}{2m_0^*} = [E_F - eV(z)]\left[1 + \frac{E_F - eV(z)}{E_g}\right]. \tag{4.85}$$

where now $k^2 = k_x^2 + k_y^2$. In the limiting condition, $\hbar \to 0$, classical mechanics replaces quantum mechanics. The nature of the WKB approximation (Messiah 1959; Bohm 1954) is to introduce a series expansion in $\hbar$ and ignore the third and higher order terms. Therefore, we can replace the Schrödinger equation with its classical limit. Because this method is useful in a regime beyond the usual classical explanation (e.g., the regime where $E < V$), it is applicable over a wider scope than expected for a classical approximation. We use the WKB approximation to calculate the energy levels in a quantum well of $Hg_{1-x}Cd_xTe$ formed by a surface accumulation layer. The subband energy of surface accumulation layer $E_n(k)$ and the normal surface wave vector, k_z, follow from the expression:

$$\int_0^{z_n} k_z dz = \left(n + \frac{3}{4}\right)\pi, \tag{4.86}$$

where z_n is the turning point defined by $k_z[z_n, k, E_n(k)] = 0$. The only difference between (4.86) and Bohr-Sommerfeld (Eisberg and Resnick 1985) quantum law is that a fractional quantum number replaces an integer quantum number.

From (4.82–4.86), we obtain:

$$\int_{\phi^*(k)}^{E_F} \frac{d\phi}{[\alpha F(\phi, E_g)]^{1/2}} \left\{\beta[\phi - E_F + E_n(k)]\left[1 + \frac{\phi - E_F + E_n(k)}{E_g}\right] - k^2\right\}^{1/2} = \left(n + \frac{3}{4}\right)\pi, \tag{4.87}$$

where

$$\alpha = \frac{2e(2m_0^*)^{3/2}}{3\pi^2\varepsilon_s\varepsilon_0\hbar^3}, \tag{4.88}$$

$$\beta = \frac{2m_0^*}{\hbar^2}, \tag{4.89}$$

$$\phi = E_F - eV(z), \tag{4.90}$$

$$[\phi^* - E_F + E_n(k)]\left[1 + \frac{\phi^* - E_F + E_n(k)}{E_g}\right] = \frac{\hbar^2 k^2}{2m_0^*}, \tag{4.91}$$

and

$$F(\phi, E_g) = \int_0^\phi \mathrm{d}\zeta[\zeta(1 + \zeta/E_g]^{3/2}. \tag{4.92}$$

From (4.87), given $k = k_{\mathrm{Fn}}$ and $E_{\mathrm{n}}(k) = E_{\mathrm{F}}$, we can calculate the Fermi wave vector of the nth subband k_{Fn} and the electron concentration of the nth subband $N_{\mathrm{sn}} = k_{\mathrm{Fn}}^2/\pi$, where the sum over n of N_{sn} is N_{s}. Normally, according to the classical theory, the surface electron concentration N_{s}^0 is nearly the same as the sum of the subband electron concentrations N_{s}. N_{s}^0 can be calculated from the surface potential distribution:

$$N_s^0 = \frac{\varepsilon_s \varepsilon_0}{e} \left.\frac{\mathrm{d}V(z)}{\mathrm{d}z}\right|_{z=0} = \frac{\varepsilon_s \varepsilon_0}{e}[\alpha F(E_F, E_g)]^{1/2}. \tag{4.93}$$

The effective mass of the nth subband is defined as:

$$\frac{1}{m_n^*} = \frac{1}{\hbar^2 k} \frac{\partial E_n(k)}{\partial k}. \tag{4.94}$$

and it can be obtained by solving (4.87).

4.5.2 Theoretical Calculations for an n-HgCdTe Surface Accumulation Layer

The concentration x of an $\mathrm{Hg}_{1-x}\mathrm{Cd}_x\mathrm{Te}$ material, selected to produce an 8–14 μm wave band photoconductive detector, is approximately 0.2. From the CXT formula (Chu 1985b; Chu et al. 1982, 1983), we obtain $E_{\mathrm{g}} = 93.45\,\mathrm{meV}$, $m^* = 8.02 \times 10^{-3}\ m_0$, for $T = 1.2\,\mathrm{K}$, and the dielectric constant at low frequency is $\epsilon_{\mathrm{s}} = 17.6$ (Yadava et al. 1994).

Gui et al. (1997) calculated the surface accumulation layer subband parameters for two n-$\mathrm{Hg}_{1-x}\mathrm{Cd}_x\mathrm{Te}$ photoconductive detectors, with $x = 0.214$ and 0.191. Figure 4.50 shows the relation between N_{s} and energy at the bottom of the different subbands, $E_{\mathrm{n}}(k = 0)$. The Fermi level is pinned at the donor level. It is generally agreed that shallow donor levels in $\mathrm{Hg}_{1-x}\mathrm{Cd}_x\mathrm{Te}$ overlap the bottom of the conduction bands in the vicinity of $x = 0.2$. From Figure 4.50, we see that E_{n} decreases as N_{s} increases.

Figure 4.51 shows the electron concentration of the nth subband N_{sn} and the classical surface electron concentration N_{S}^0 as a function of N_{s}. These functions

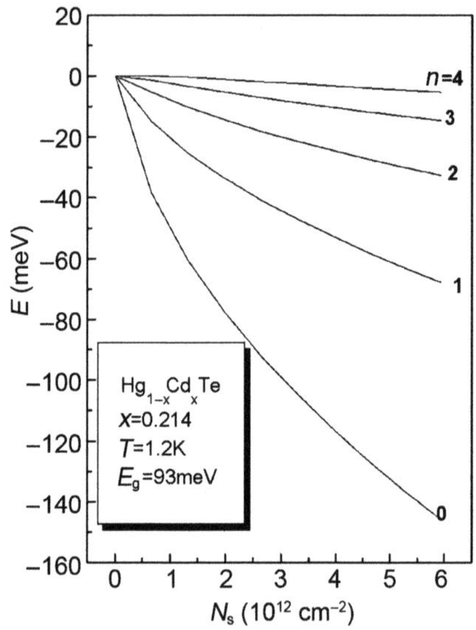

Fig. 4.50 Theoretically calculated relations between N_s and the energy E_n at the bottom of $Hg_{1-x}Cd_xTe$ surface subbands

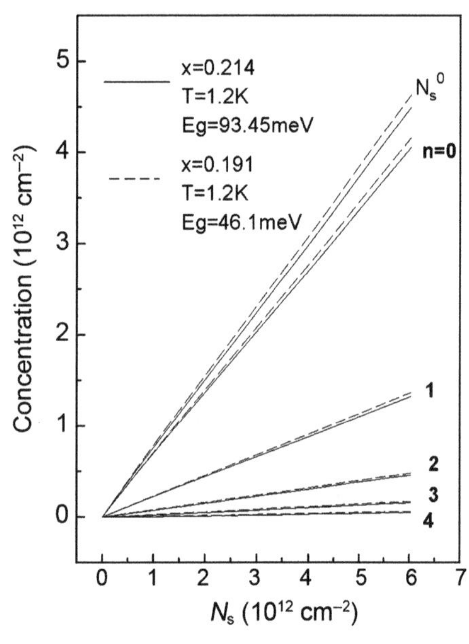

Fig. 4.51 Theoretically calculated electron concentrations of surface subbands for $Hg_{1-x}Cd_xTe$ detectors with different x values

are calculated in the heavily accumulated surface layers of long-wavelength photoconductive detectors with $x = 0.214$ and 0.191. In Fig. 4.51, it shows that N_s is about 20% higher than N_S^0. Calculations of heavily accumulated layers of $Hg_{1-x}Cd_xTe$ detectors having different compositions ($x = 0.191$–0.214), demonstrate that the relation between N_{sn} and N_s are linear and nearly independent of composition x. The slope changes from the ground state to the fourth excited state are 0.668, 0.219, 0.077, 0.027 and 0.009, respectively. Lowney et al. (1993) applied

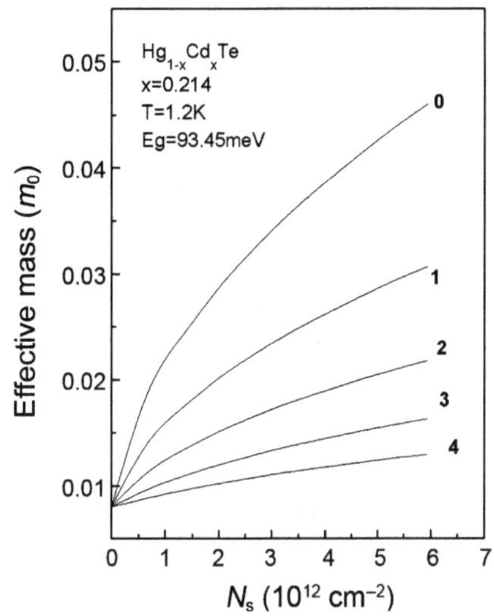

Fig. 4.52 Predicted relation between effective masses at the bottom of the different subbands m_n^* vs N_s for an $Hg_{1-x}Cd_xTe$ detector

the eight-band model to analyze the heavily accumulated layer of a $Hg_{1-x}Cd_xTe$ detector ($x = 0.191$). For $n = 0$–3, the slopes are 0.673, 0.223, 0.077, and 0.027, respectively. J. Singleton et al. (1986a) analyzed a lot of experimental results of n-$Hg_{1-x}Cd_xTe$ materials, for band gaps ranging from 30–90 meV, and obtained similar results. For $n = 0$–4, the slopes of $N_{sn}(N_s)$ in an accumulation layer are 0.6452, 0.2158, 0.097, 0.032, and 0.010. From these facts, we reach the conclusion that the gross carrier concentration of an n-$Hg_{1-x}Cd_xTe$ accumulation layer can be calculated based on the carrier concentration of any subband. Moreover, the carrier concentration of any subband can also be calculated knowing the gross carrier concentration of the accumulation layer.

Figure 4.52 shows the gross surface carrier density dependence of the effective masses at the bottom of the different subbands. From the figure, we see that the effective mass of the ground state is the largest, and the effective mass decreases with increasing subband index. A qualitative explanation is based on(4.11), and the physical reason is that the energy of the nth subband E_{0n} is higher than the energy of ground-state E_0, as well as wave function diffusion $Z_{0n} > Z_0$; therefore, m_n^* is smaller than the effective mass of the lowest subband.

4.5.3 Experimental Results for n-HgCdTe Surface Accumulation Layers

Gui et al. (1997) measured the transport properties of the two samples discussed above. The dimensions of the two samples were both 888 μm × 290 μm × 8 μm, the distance between the Hall electrodes was 336 μm, the top and bottom surfaces of samples were passivated by anodic oxidation, and extended In electrodes formed good ohmic contacts between the electrodes and the samples. At $T = 77\,\text{K}$, the carrier concentration and mobility of the bulk material are approximately $5.0 \times 10^{15}\,\text{cm}^3$ and $2.0 \times 10^5\,\text{cm}^2/\text{Vs}$. In order to accurately measure the SdH oscillations, the measurement system was automatic. It comprised a high precision digital current source and programmable digital voltmeter. In the range from 1.2–50 K, samples were measured by a DC method in a high magnetic field with tilt angles between the magnetic field direction and the sample's surface that were $0°, 15°, 90°\ldots$. (A 0° tilt angle means that the magnetic field direction is perpendicular to the sample's surface, whereas a 90° tilt angle means that the magnetic field direction is parallel to the sample's surface).

4.5.4 Results of an SdH Measurement

Figure 4.53 shows the dependence of the magneto-resistance at $T = 1.2\,\text{K}$ on the tilt angle of magnetic field. Samples exhibited complicated SdH oscillations at tilt angle, $\theta = 0°$, and SdH oscillations were found at low magnetic fields at tilt angle, $\theta = 90°$. Because the bulk electron concentration is approximately 10^{14}–$10^{15}\,\text{cm}^{-3}$ and the Fermi level is near the bottom of the conduction band, the SdH oscillations induced by the bulk carrier concentration could be found only at low magnetic fields. The two-dimensional property of the oscillations is proven by the fact that a peak marked by arrows in Fig. 4.53 drifts as the tilt angle varies. The SdH oscillations are complicated and are not a periodic function of $1/B$ with a fixed period, but

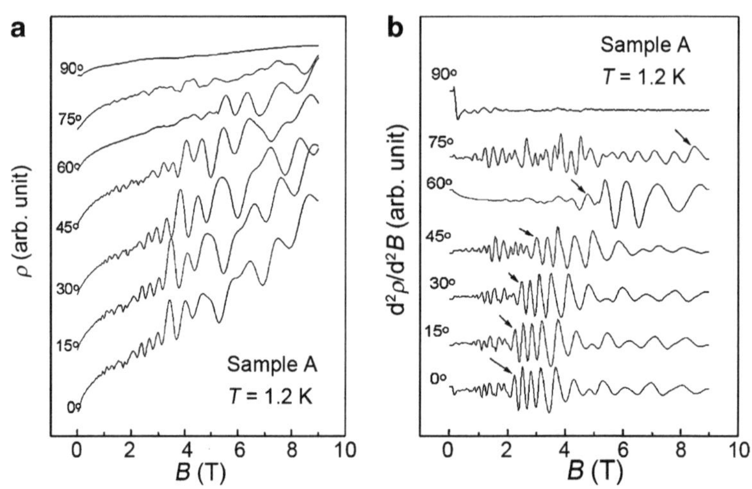

Fig. 4.53 SdH oscillations of sample A for different tilt angles. (**a**) 0° tilt angle means the magnetic field direction is perpendicular to the sample's surface; (**b**) the arrows denote corresponding peaks of the SdH oscillation that drift with a variation of the tilt angle, showing the two-dimensional property of the oscillations

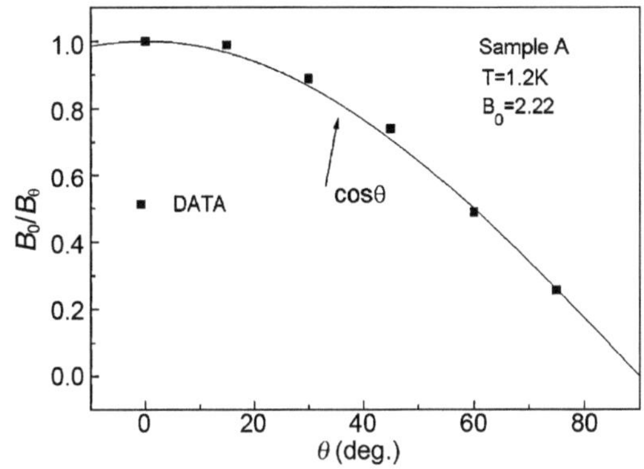

Fig. 4.54 The drift of a peak of the SdH oscillation with a variation of the tilt angle. The dependence of this peak on $\cos\theta$ proves that oscillations are caused by two-dimensional carriers

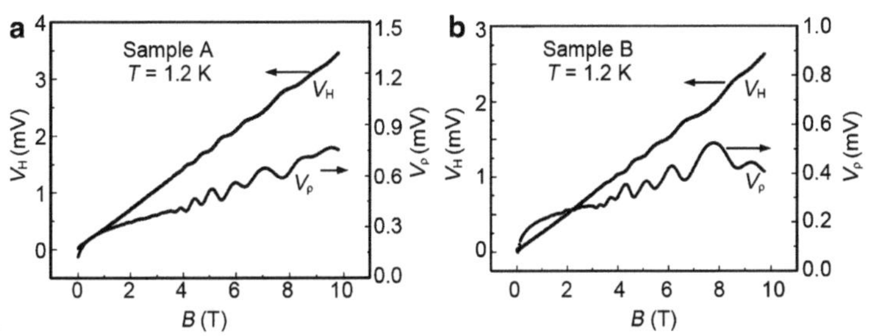

Fig. 4.55 The magnetic field intensity dependence of the Hall voltage and the magneto-resistance. (**a**) sample A; (**b**) sample B

rather a combination of several oscillation types with different periods. Because the oscillation period of a given carrier type is dependent on its concentration, the sample contains multiple carrier types.

Assuming that the Fermi level is not related to the tilt angle of the magnetic field, the tilt angle dependence of the SdH oscillatory peak's migration, marked by arrows in Fig. 4.53, is shown in Fig. 4.54. There is a cosine relation between the two parameters.

Figure 4.55a and b shows the magnetic field intensity dependence of the Hall voltage and the magneto-resistance. The tilt angle of the magnetic field was set at 0°, and the current applied to the samples was 2 mA. The oscillations in the figures weakened gradually and finally disappeared when the tilt angle changed from 0° to 90°. This fact shows that oscillations are caused by the surface accumulation layer and have a two-dimensional property. When the applied magnetic field is perpendicular to the device's surface accumulation layer, the subband splits into a series of Landau levels. By varying the magnetic field, the Landau levels move through the Fermi level that changes the density of electronic states and causes magneto-resistance oscillations. The minimum magneto-resistance oscillation corresponds to a Hall voltage plateau. Every subband causes a series of oscillations and their periods are related to $1/B$. The relation between n and P_n is $n = 4.82\times 10^{10} P_n (\mathrm{cm}^{-2})$, where P_n is the fundamental frequency of oscillation of the nth subband.

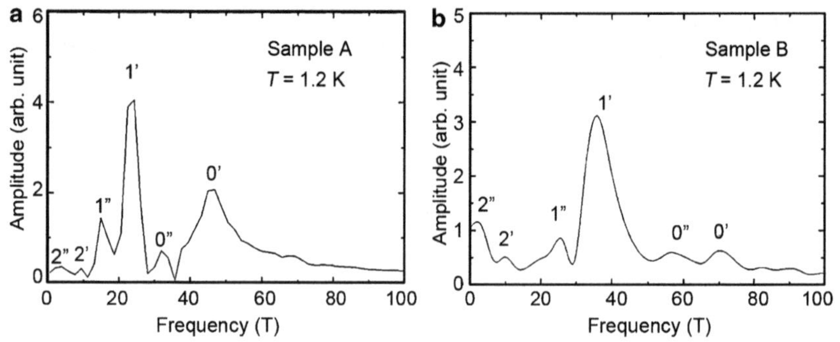

Fig. 4.56 Fourier transformation curves of SdH oscillations for samples A and B

Table 4.4 Surface electron concentrations for different subbands of samples A and B

		B_1(T)	B_2(T)	B_3(T)	N_{s0} (10^{12} cm^{-2})	N_{s1} (10^{12} cm^{-2})	N_{s2} (10^{12} cm^{-2})
A	Surface′	46.8	24.4	9.12	2.25	1.17	0.437
	Surface″	31.9	15.0	3.75	1.53	0.72	0.181
B	Surface′	68.9	36.1	10.1	3.31	1.73	0.485
	Surface″	55.8	26.4	3.28	2.68	1.27	0.157

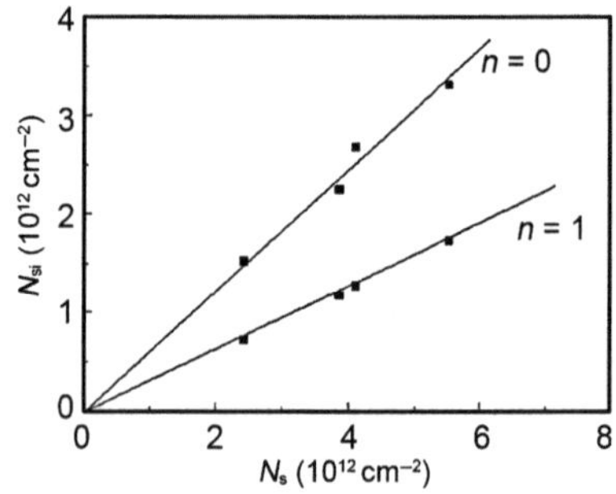

Fig. 4.57 Relations between the net electron concentration and the electron concentrations of the subbands (*dots* are plotted from the data in Table 4.4 and the *solid lines* are linear fitted results)

From Fourier transformations of the SdH oscillation curves, we can obtain the electron concentrations of the different subbands. Figure 4.56 shows the Fourier transform curves of SdH oscillations for samples A and B, The different surfaces are labeled by symbols (′) and (″) respectively. Table 4.4 shows the surface electron concentrations for different subbands calculated from Fig. 4.56.

The relations among the electron concentrations for different subbands are not predicted well by the theory of Sect. 4.2. The main problem is that value of N_{s1}/N_{s0} is less than its theoretical value. This is because SdH oscillations are complicated processes and(4.87) is just an approximation. Even so, according to the data in Table 4.4, the fitted lines for N_{s0}–N_s and N_{s1}–N_s (Fig. 4.57) are linear and pass through the origin, which proves that there is some merit to the approximation behind (4.87). Thus, approximate carrier concentrations for the different subbands in the two-dimensional system can be obtained from fits to the SdH oscillation measurements.

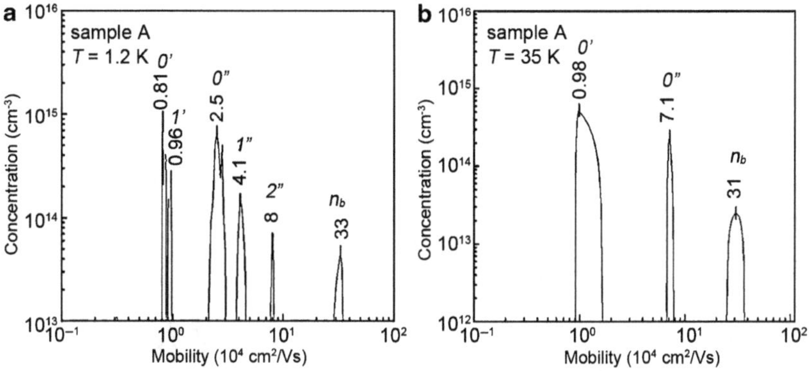

Fig. 4.58 QMS for sample A at different temperatures: (**a**) 1.2 K and (**b**) 35 K

Gui et al. (1998) studied these samples by using a quantitative mobility spectrum (QMS) analysis. Figure 4.58a and b shows the QMSs of sample A at $T = 1.2$ and 35 K. The exact carrier mobilities and the concentrations of different subbands in these samples can be obtained at $T = 35$ K because the SdH oscillations almost disappear at this temperature. Three peaks in Fig. 3.5.9b correspond to the mobilities of the bulk electrons, the two-dimensional electrons of the top, and bottom surface accumulation layers, respectively. The electron mobilities at 35 K for different subbands for the same surface are equal for sample A. The QMS is obviously more complicated at 1.2 K, compared with that at 35 K, because then the electron mobilities of different subbands in the surface two-dimensional electron systems are different. Ionized impurity scattering plays a predominant role at 1.2 K. The electron effective masses of different subbands are different so the corresponding mobilities are different too, and the peaks from the different subbands are distinguishable in the QMS. At the increased temperature, for example at 35 K, many scattering mechanisms become effective, for example, ionized impurity scattering, polar optical phonon scattering, alloy scattering, and dislocation scattering. Because of the different kinetic energies of the subbands, each scattering type plays a different role and the net effect is that the differences among the electron mobilities for different subbands tend to be reduced. Table 4.5 shows the concentrations of various electrons obtained from a QMS analysis, an SdH oscillation measurement, and theoretical calculations. Results of the QMS analysis at 35 K are the bulk electron concentration is $2.14 \times 10^{14}\,\text{cm}^{-3}$, the electron concentrations at the top and bottom surfaces are 4.31×10^{12} and $3.25 \times 10^{12}\,\text{cm}^{-2}$, respectively. The same results at 1.2 K are 2.20×10^{14}, 4.11×10^{12}, and $3.03 \times 10^{12}\,\text{cm}^{-2}$, which agree with the results at 35 K. In addition, the electron distributions of different subbands, obtained by QMS analysis, are in accord with the theoretically predicted distributions. So the results from the QMS measurements agree with theory better than those from SdH oscillation measurements.

The electron relaxation times for different subbands in an $Hg_{1-x}Cd_xTe$ surface accumulation layer differ due to the differences among their electron effective

Table 4.5 Magneto-resistance data fits for two kinds of carriers at 1.2 K

Sample	μ_1(cm^2/V s)	μ_2(cm^2/V s)	N_1(cm^{-3})	N_2(cm^{-3})	B_1(T)	B_2 (T)
D685–7	-2.65×10^5	-3.12×10^4	7.70×10^{14}	2.44×10^{15}	0.034	0.3
D685–5	-2.23×10^5	-2.18×10^4	8.10×10^{15}	5.08×10^{15}	0.06	0.54

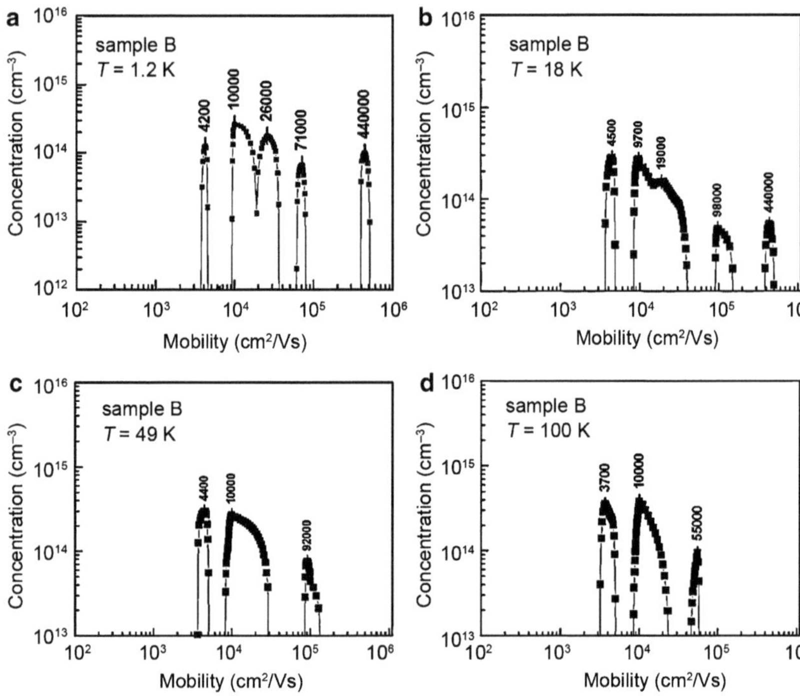

Fig. 4.59 The QMS of sample B at different temperatures

masses, kinetic energies, and subband wave functions. In order to study the temperature dependence of the subband electron mobilities in an $Hg_{1-x}Cd_xTe$ surface accumulation layer, we analyze the QMS experiment data in a variable magnetic field at different temperatures.

Figure 4.59 shows the QMS of sample B, from which we find the dependence of the subband mobility on temperature. At 1.2 K, there is a great difference between the electron mobilities of three subbands in surface ($'$), but the electron mobilities of the subbands are almost the same for surface ($''$). This is because the net electron concentration of surface ($'$) is 5×10^{15} cm^{-3}, but only 5×10^{14} cm^{-3} for surface ($''$). According to a theoretical calculation, there is little difference between the electron mobilities of subbands when the electron concentration is low and only the ground state has an appreciable occupancy; however, the effective mass of the ground state is more than two times that of the second excited state and it becomes occupied when the electron concentration reaches 5×10^{15} cm^{-3}. At low temperature, the electron mobility of the ground state ($n = 0$) is the lowest, and the mobilities of the excited

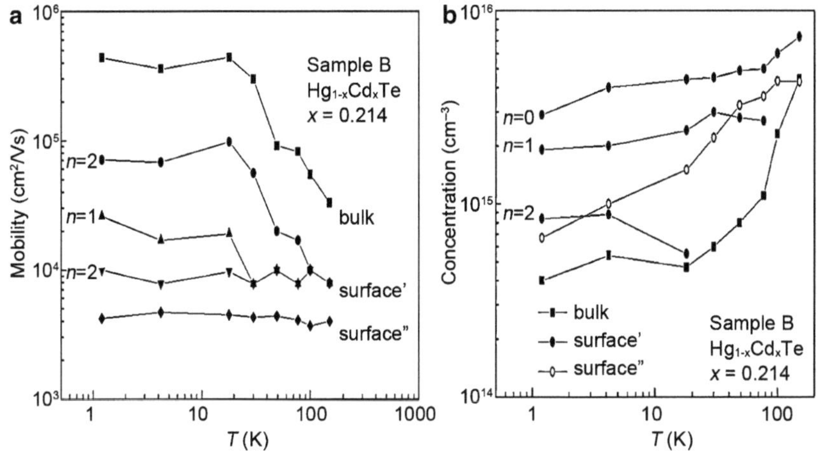

Fig. 4.60 (**a**) The temperature dependence of electron concentrations for sample B. (**b**) The temperature dependence of electron mobilities for sample B

states all progressively increase with increasing quantum index because their effective masses increase. With an increasing temperature, the electron mobilities of the excited states gradually decrease, and finally would be equal to the electron mobility of the ground state which is nearly independent of temperature. The temperature dependence of the electron concentrations and effective masses for sample B are shown in Fig. 4.60. From this figure, we see that the relationship between the temperature and net electron concentration are different for the two surfaces. The net electron concentration of surface (′) varies very slowly with a temperature variation; but the net electron concentration of surface (″) varies rapidly with a temperature variation. Physical mechanism differences between these two surfaces may explain these facts. One possible reason is that due to stress, defect states are introduced into the surface attached to the substrate. These donor defect states have a finite ionization energy, and ionize gradually as the temperature increases. From the relation: $n \propto \exp(-\Delta E_{\mathrm{D}}/k_{\mathrm{B}}T)$, we deduce an approximate ionization energy for these surface donors to be, $\Delta E_{\mathrm{D}} = 3.1\,\mathrm{meV}$.

4.6 Surfaces and Interfaces

4.6.1 The Influence of Surface States on the Performance of HgCdTe Photoconductive Detectors

For $Hg_{1-x}Cd_xTe$ photoconductive detectors, most of the passivation processes will introduce a surface accumulated layer to reduce surface recombination velocities and $1/f$ noise, as well as to reduce the intrinsic resistance of the device (Fig. 4.61).

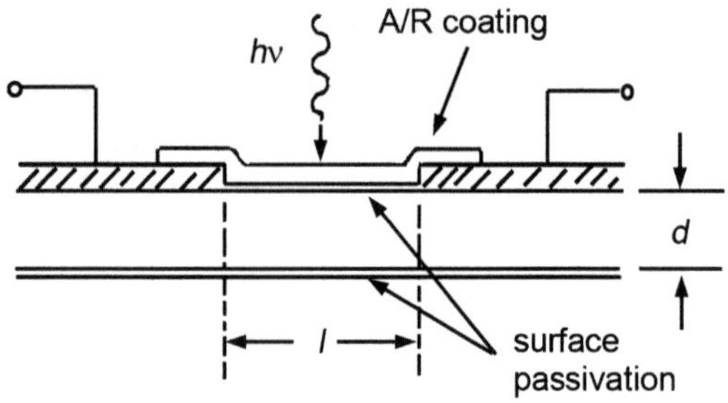

Fig. 4.61 Sectional view of a typical $Hg_{1-x}Cd_xTe$ photoconductive detector

The importance of passivation layers on device performance has attracted much previous consideration (Kinch 1981) and continuous studies (Nemirovsky and Bahir 1989; Nemirovsky 1990; Nemirovsky 1979). Singh pointed out that, for a detector, the effective minority carrier lifetime is tied to the surface states, especially, the fast surface states and the surface potential (Singh et al. 1991). However, all the studies mentioned above were mainly aimed at a qualitative analysis. Here the two-dimensional character of an accumulation layer due to passivation is taken into account (Singleton et al. 1986c; Lowney et al. 1993) when the distribution of surface potentials is calculated using Fang and Howard's variational method. To further illustrate the influence of passivation on device performance, the spacial distribution of photogenerated carriers and the voltage response of $Hg_{1-x}Cd_xTe$ photoconductive detectors are calculated using a one-dimensional model.

There exist large numbers of bound positive charges in the passivation layer of an $Hg_{1-x}Cd_xTe$ photoconductive detector. The equivalent concentration of electrons is induced at the surface of bulk materials due to the existence of the bound positive charges. The induced electrons are regarded as a quasi-2DEG since they exist in a very thin layer adjacent to the surface. There have been a series of theoretical models proposed, including a single-band model and multiband models to calculate the dispersive relation of the 2DEG. All the models, to be accurate, should be based on the self-consistent solution of the Schrödinger equation in conjunction with the Poisson equation. Among those models, Fang and Howard's variational method (Fang and Howard 1966) is relatively simple and practical. Since the electron concentration in the surface accumulation layer is much larger than that in the bulk, the surface potential V_H and the bound positive charge density based on the Fang and Howard's method obey the following correlation:

$$V_{\mathrm{H}} = \frac{3N_{\mathrm{I}}e^2}{2\varepsilon_{\mathrm{s}}\varepsilon_0 b}, \tag{4.95}$$

where b is defined as

$$b \equiv \left(\frac{33m_{\mathrm{n}}^{*}e^2N_{\mathrm{I}}}{8\varepsilon_s\varepsilon_0\hbar^2}\right)^{1/3}. \tag{4.96}$$

In this expression, m_n^* is the effective electron mass, e is the electric charge, ϵ_0 is the vacuum dielectric constant, and ϵ_s is the static dielectric constant. The static dielectric constant of $Hg_{1-x}Cd_xTe$ materials ϵ_s (Yadava et al. 1994) depends on the Cd concentration x:

$$\varepsilon_s = 20.8 - 16.8x + 10.6x^2 - 9.4x^3 + 5.3x^4. \tag{4.97}$$

When the device is irradiated, the photogenerated carriers in the bulk will move toward the surface where radiative recombination can take place. The surface potential of an $Hg_{1-x}Cd_xTe$ accumulated layer is a barrier for holes and this surface barrier will prevent holes from moving toward the surface. This causes the actual surface recombination velocity to be reduced. The effective surface recombination velocity is (White 1981):

$$S_{\text{eff}} = S_0 \exp\left(-\frac{V_{\text{H}}}{k_{\text{B}}T}\right), \tag{4.98}$$

where S_0 is the surface recombination velocity for a flat band, k_B is the Boltzmann constant, and T is the absolute temperature.

Accumulation layers exist at both the upper and lower surfaces of HgCdTe photoconductive detectors. Assuming the upper surface is the same as the lower surface, the resistance of a photoconductive detector can be expressed as:

$$R = \frac{1}{(N_b\mu_b + 2N_I\mu_s/d)e}\frac{l}{wd}, \tag{4.99}$$

where l, w, and d are the length, the width, and the height of the photoconductive detector, respectively. N_b is the bulk electron concentration, m_s is the mobility of the surface electrons, which is of the order of $10^4\ \text{cm}^2/\text{Vs}$ at 77 K (Gui et al. 1997), μ_b is the mobility of the bulk electrons, which can be determined by an expression proposed by Rosbeck et al. (1982). For an HgCdTe material with Cd content x, the mobility of the bulk electrons is about $1\text{–}3 \times 10^5\ \text{cm}^2/\text{Vs}$ at 77 K.

The minority carrier lifetime is significantly affected by surface recombination. The actual net lifetime due to surface recombination is determined as follows:

$$\frac{1}{\tau_{\text{net}}} = \frac{1}{\tau_b} + \frac{1}{\tau_s} = \frac{1}{\tau_b} + \frac{2S_{\text{eff}}}{d}. \tag{4.100}$$

In (4.100), the influence of the surface potential on the surface recombination velocity is included. The effective surface recombination velocity in(4.98) is adapted for the calculation of actual net lifetime in (4.100).

When a photoconductive detector operated in a constant current mode is irradiated by monochromatic light with wavelength λ, the voltage responsivity R_V is:

$$R_v = \left(\frac{\lambda}{hc}\right)\eta q R\left(\frac{\mu_b E \tau_{net}}{l}\right), \tag{4.101}$$

where E is the intensity of bias electric field, c is the velocity of light, h is Plank constant, and η is the quantum efficiency.

Next investigate the influence of the surface bound charge on the performance of n type photoconductive detector devices, operated at 77 K, with a Cd content x of 0.214, a bulk lifetime of 10 ns to 10 ms, a doping concentration of 5×10^{14} to $5 \times 10^{15}\,\text{cm}^{-3}$, and a mobility of surface electrons of $2 \times 10^4\,\text{cm}^2/\text{Vs}$. The photosensitive area of the detector is $50 \times 50\,\mu\text{m}^2$, and it has a thickness of 8 μm. The wavelength of the incident light is 10.6 μm. The quantum efficiency is 0.6. The magnitude of bias electric field is 20 V/cm.

The predicted dependence of the photoconductive detector resistance on surface bound charge is shown in Fig. 4.62. The surface accumulation layer induced by passivation is a high-conductivity region, which leads to a reduction of the device resistance. As shown in Fig. 4.62, the resistance of detectors changes with a surface-bound charge variation when the density of surface-bound charges is of the order of $N_\text{I} = 10^{11}$ to $10^{12}\,\text{cm}^{-2}$. At $N_\text{I} < 10^{11}\,\text{cm}^{-2}$, the resistance of detectors is almost independent of the surface passivation, and is determined primarily by the bulk electron concentration. While for $N_\text{I} > 10^{12}\,\text{cm}^{-2}$, the detector resistance is determined primarily by the concentration of surface electrons, and is almost independent of the bulk electron concentration.

Figure 4.63 shows the predicted influence of the surface bound charge on the net lifetime τ_net of detectors. When the value of N_I is relative small, τ_net increases rapidly with increasing N_I, finally saturates at value τ_b. The existence of surface-bound charge induces an electron accumulation layer at the surface of n type detectors. The accumulated layer leads to a band bending at the surface, resulting in electron traps, which serves as a potential barrier for minority carriers (holes). The barrier prevents the diffusion of photogenerated holes toward surface and slows the surface recombination velocity. As described in (4.96), the surface potential

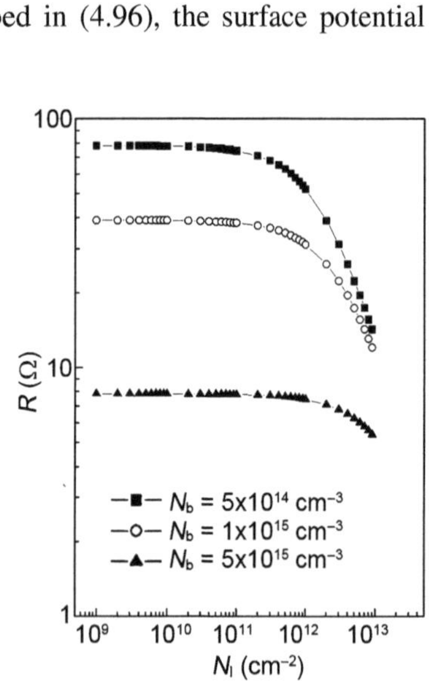

Fig. 4.62 The resistance is shown as a function of the surface bound charge density N_I, for photoconductive detectors with different doping concentrations N_b

Fig. 4.63 The net lifetime τ_{net} is shown as a function of the density of surface bound charge N_I, for devices with different surface recombination velocities S_0

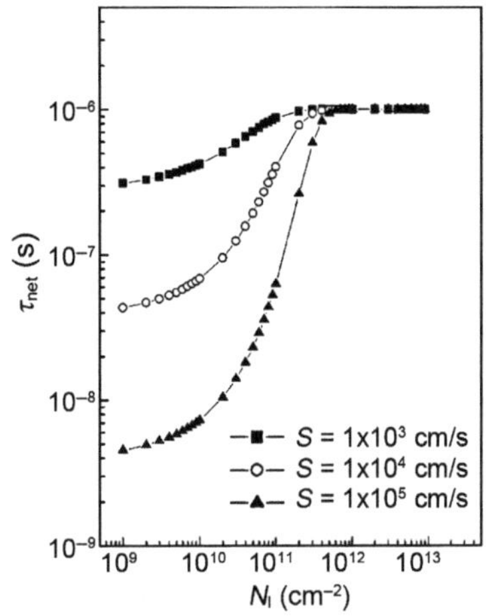

Fig. 4.64 The voltage responsivity R_v for devices with different dopant concentration N_b as a function of the concentration of the surface bound charge N_I. The values of other parameters are $S_0 = 1000\,\mathrm{cm/s}$ and $\tau_b = 1\,\mu\mathrm{s}$

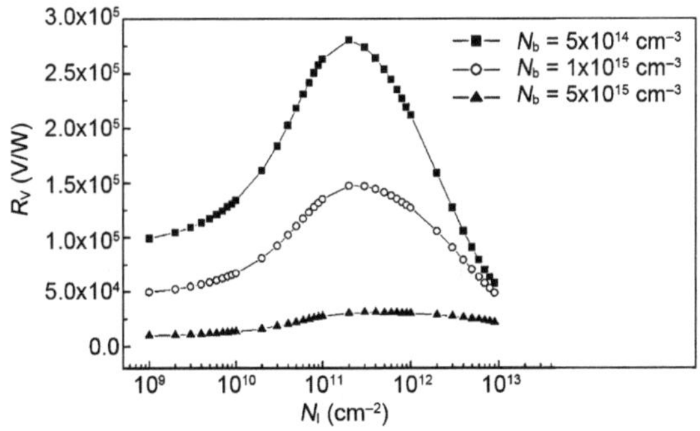

increases with an increase of the density of bound charges. When the density of bound charges N_I is larger than $3 \times 10^{11}\,\mathrm{cm}^{-2}$, due to the very high surface barrier for holes that they induce, the photogenerated holes can hardly overcome the barrier to reach the surface. This insight is useful for improving the performance of detectors.

Figure 4.64 shows the predicted dependence of the voltage responsivity on the concentration of bound charges with different bulk doping concentrations. The minority carrier lifetime is $1\,\mu\mathrm{s}$ and the surface recombination velocity is 1000 cm/s for this high detectivity device. When the concentration of bound charge in the passivation layer exceeds $10^{11}\,\mathrm{cm}^{-2}$, the surface potential reduces the surface recombination rate of photogenerated holes and improves the device performance. Once the concentration of bound charge exceeds $10^{12}\,\mathrm{cm}^{-2}$, the conductivity of the surface heavy accumulated layer is promoted, resulting in a reduction of the intrinsic resistance and device performance. With a further increase of the bound charge concentration, the device performance is reduced very fast, especially for the devices with low doping concentrations.

Figure 4.65 shows the predicted dependence of the voltage responsivity on the bound charge concentration with different surface recombination velocities.

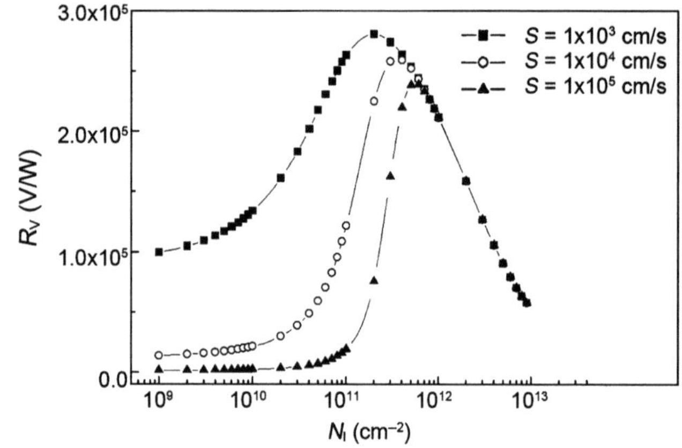

Fig. 4.65 The voltage responsivity R_v for devices with different surface recombination velocities S_0 as a function of the concentration of surface bound charges N_I. The values of other parameters are $N_b = 5 \times 10^{14}\,\mathrm{cm}^{-3}$ and $\tau_b = 1\,\mu\mathrm{s}$

The device minority carrier lifetime is $1\,\mu\mathrm{s}$ and the doping concentration is $5 \times 10^{14}\,\mathrm{cm}^{-3}$. As shown in Fig. 4.65, there exists a peak of the voltage responsivity curve when the concentration of bound charges is in the range of 10^{11}–$10^{12}\,\mathrm{cm}^{-2}$. The position of the voltage responsivity peak shifts to higher bound charge concentrations following an increase of the surface recombination velocity. The voltage responsivity maximum is directly determined by the surface recombination velocity. The smaller the surface recombination velocity is, the larger is the voltage responsivity maximum. When the bound charge concentration is larger than $6 \times 10^{11}\,\mathrm{cm}^{-2}$, the detector voltage responsivity is almost the same even if surface recombination velocity changes by two orders of magnitude. Indications are that device performance will be improved if the passivation process and the effective surface recombination velocity are selected properly.

Figure 4.66 shows the dependence of the voltage responsivity on the bound charge concentration for detectors with different minority carrier lifetimes, τ_b. The doping concentration is chosen to be $5 \times 10^{14}\,\mathrm{cm}^{-3}$ and the surface recombination velocity is $10^5\,\mathrm{cm/s}$ for this calculation. When the surface-bound charge is less than $10^{11}\,\mathrm{cm}^{-2}$, the performance of devices is predicted to be independent of the minority carrier lifetime. Thus, the effective lifetime τ_{net} of devices is determined by the surface recombination velocity. When the surface-bound charge concentration is in the range of 10^{11}–$10^{12}\,\mathrm{cm}^{-2}$, the variation of the voltage responsivity R_v is very large. For devices with a long lifetime ($\tau_b = 10\,\mu\mathrm{s}$), the voltage responsivity R_v varies with N_I by two orders of magnitude.

The discussion above indicates that the performance of HgCdTe photoconductive detectors depends sensitively on the concentration of surface-bound charges induced by the passivation process. On the one hand, the surface-bound charges reduces the recombination rate by preventing the photogenerated holes from reaching the

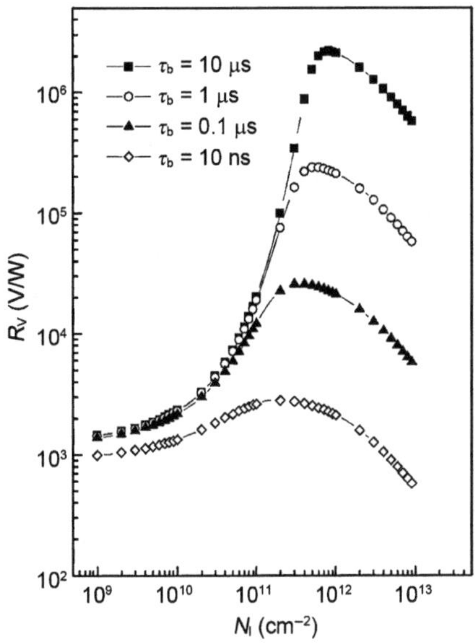

Fig. 4.66 The voltage responsivity R_v for devices with different minority carrier lifetimes τ_b as a function of the bound charge surface concentration N_I. The values of other parameters are $S_0 = 10^5\,\text{cm/s}$, $N_b = 5\times10^{14}\,\text{cm}^{-3}$

surface, resulting in an improvement of device performance; on the other hand, the surface-bound charges reduce the net device resistance, resulting in a reduction of device performance. The concentration of surface-bound charge can be changed by selecting a proper passivation process, so device performance can be optimized.

4.6.2 The Influence of the Surface on the Magneto-Resistance of HgCdTe Photoconductive Detectors

The technology related to photoconductive detectors is well established (Broudy 1981). However, the surface of HgCdTe oxidizes easily allowing an n^+ type accumulated or inversion layer with high conductivity to form, which affects device performance. An anodic oxidation process is adopted to reduce the surface recombination velocity and $1/f$ noise. However, as we have seen this n^+ type accumulated layer when not properly controlled also produces a high conductivity that degrades devices (Lowney et al. 1993).

A photoconductive detector is a two-terminal device and it cannot satisfy at least four electrode requirement for a Hall measurement. Most methods used to analyze multicarrier systems, such as a multicarrier fitting procedure, a mixed conduction analysis (Meyer 1993; Antoszewski and Faraone 1996), as well as a QMS analysis (Dziuba and Górska 1992; Antoszewski et al. 1995; Meyer et al. 1995), are incapable of investigating the electrical characteristic of surface and bulk electrons in photoconductive detectors. A magneto-resistance measurement is regarded as an effective means to investigate semiconductors with complicated energy bands. It can also be used to estimate carrier mobilities. Conventional methods can only be used to investigate single-carrier systems. However, prior to the work of Kim et al. (1993, 1995), they could not provide a solution to the multicarrier system due to the complexity of the magneto-resistance expressions. Kim et al. improved the magneto-resistance measurement technology, and developed a simplified model, named the reduced conductivity tensor (RCT) scheme, to reveal the transport properties of a multicarrier system.

This section describes the experiments on photoconductive detectors in which the magnetic field is varied and the RCT scheme is used in the data analysis. The theoretical result is consistent with that obtained from experiments. The electron concentrations and mobilities of both the bulk and surface electrons as a function of temperature are also given in this section.

Next the RCT scheme will be reviewed. The general procedure for the varying magnetic field data analysis using the RCT scheme is introduced in the papers by Kim et al. (1993, 1995). In the RCT method, only surface and bulk electrons that make the most significant contributions to the conductivity of photoconductive detectors are considered.

The magneto-resistance M is defined as a function of magnetic field B:

$$M(B) = \frac{\rho(B) - \rho(0)}{\rho(B)}, \tag{4.102}$$

where $\rho(B)$ is the resistivity of sample at magnetic field B, and $\rho(0)$ is the resistivity at $B = 0$. $\rho(B)$ is defined in terms of the longitudinal component σ_{xx} and the transverse component σ_{xy} of the conductivity tensor:

$$\rho(B) = \frac{\sigma_{xx}}{\sigma_{xx}^2 + \sigma_{xy}^2}. \tag{4.103}$$

The magneto-resistance M as a function of the conductivity tensor can be obtained by combining (4.102) with (4.103). Compared with other methods, RCT seems to be not very effective. However, the RCT method provides a unique way to obtain the carrier concentration and mobility for photoconductive detectors. This is because only the resistance as a function of magnetic field can be measured on these two-terminal photoconductive devices. Other potentially useful measurements, like the Hall voltage, cannot be accomplished on these structures.

A system with J types of carriers is assumed when the RCT analysis is used. $X(B)$ and $Y(B)$ are relative values of the longitudinal and the transverse components of the conductivity tensor, respectively. They are defined as:

$$\begin{aligned} X(B) &\equiv \frac{\sigma_{xx}(B)}{\sigma_{xx}(0)} = \sum_{j=1}^{J} X_j, \text{ and} \\ Y(B) &\equiv \frac{\sigma_{xy}(B)}{\sigma_{xx}(0)} = \sum_{j=1}^{J} Y_j. \end{aligned} \tag{4.104}$$

For a system with two types of carriers, $J = 2$. In the more general case, for materials with good homogeneity, the number of carrier species is rarely more than three. However, for a sample with very poor homogeneity, it is asserted that there are an infinite number of carrier types existing in the system.

The X_j and Y_j defined in (4.104) satisfy the following equations:

$$\begin{aligned} X_j &= \frac{f_j}{1+(\mu_j B)^2}, Y_j = \frac{f_j \mu_j B}{1+(\mu_j B)^2}, \text{ and} \\ f_j &= \frac{s_j \mu_j N_j}{\sum s_j \mu_j N_j} = \frac{q s_j \mu_j N_j}{\sigma_{xx}(0)}, \end{aligned} \tag{4.105}$$

where s_j denotes the sign of the charged carriers, and has the same sign as that of its mobility μ_j. $s_j = -1$ denotes electrons, and holes are denoted $s_j = 1$. f_j is a dimensionless parameter that denotes the fractional contribution to the conductance of the jth carrier type under the no magnetic field condition. N_j and μ_j are the concentrations and mobilities of the jth carrier types, respectively. q is the magnitude of electric charge. In an RCT process analysis, N_j is a dependent variable. When μ_j and f_j are given, the following expression is obtained:

$$N_j = \frac{f_j \sigma_{xx}(0)}{q s_j \mu_j}$$

For a system with only one carrier type, $f_j = 1$, $X_j = 1/[1 + (\mu_j B)^2]$, and $Y_j = \mu B/[1 + (\mu_j B)^2]$. For a system with multiple carrier types, the conditions $0 \leqslant f_j \leqslant 1$ and $\Sigma f_j = 1$ must be satisfied because all carrier types make positive contributions to the conductance.

From (4.102) to (4.105), we get:

$$M(B) = \frac{X}{X^2 + Y^2} - 1, \tag{4.106}$$

where the magneto-resistance is defined as a function of magnetic field, and depends on f_j and μ_j of each type of carrier. For a system with only two types of carriers, we have:

$$M(B) = \frac{(\alpha \Delta B)^2}{1 + (\beta \Delta B)^2}, \tag{4.107}$$

where $\alpha \equiv \sqrt{f_1(1-f_1)}$, $\beta \equiv \mu_1/(\Delta - f_1)$, and $\Delta \equiv (\mu_1 - \mu_2)/(\mu_1 + \mu_2)$.

When the magnetic field is low, one has $M(B) \cong (\alpha \Delta B)^2 \propto B^2$. For the case of a high magnetic field, $M(B) \cong (\alpha/\beta)^2$, so $M(B)$ is independent of the magnetic field B, and only depends on the mobility and carrier concentration. Here the transition magnetic field regime is defined as $B_1 < B < B_2$ ($B < B_1$ is the low magnetic field regime, and $B > B_2$ is the high magnetic field regime). The transition width is:

$$W = \log_{10}\left(\frac{B_2}{B_1}\right) = 0.954,$$

where $M(B_1) = 0.1(\alpha/\beta)^2$ and $M(B_2) = 0.9(\alpha/\beta)^2$.

Gui et al. (1997) measured the magneto-resistances of three HgCdTe photoconductive devices, and one HgCdTe film material grown by LPE. They analyzed the experimental data using a two-carrier -type RCT model. The fitted results are consistent with the experimental data.

Often two types of carriers exist in HgCdTe photoconductive detectors, that is, bulk electrons and surface electrons induced by the surface passivation. The operating temperature of these detectors is generally lower than 77 K. In this case, the electrical properties of these detectors are significantly affected by the surface electrons. The mobility of the surface electrons is of the order of $10^4\ \mathrm{cm}^2/\mathrm{Vs}$. The surface electron contribution to the conductance is comparable with that of the bulk electrons. This occurs because of the high surface electron concentration, up to 10^{11}–$10^{12}\ \mathrm{cm}^{-2}$, though the surface electron mobility is one order lower than that of the bulk electrons.

The $Hg_{1-x}Cd_xTe$ photoconductive detectors D685–6, D685–2 and D684–5 were prepared from bulk materials with a Cd concentration x of 0.214, dimensions of $888 \times 290 \times 8\ \mu\mathrm{m}^3$ and having passivated surfaces. The magneto-resistances of D685–7 and D684–5 as functions of the magnetic field at 1.2 K is shown in Fig. 4.67. The solid and open symbols are experimental data and the solid curves are fits to the experimental results using a two-carrier-type model. The curves fit the experimental results well. The parameters determined from the fitting procedure are shown in Table 4.6. In Table 4.6, μ_1 and N_1 are the bulk electron mobility and concentration, and, μ_2 and N_2 are the surface electron mobility and concentration, respectively.

In the RCT method, 2DEGs are considered to be homogeneously distributed across the whole thickness of the sample. Thus, the sheet concentration of surface electrons are found to be 1.92×10^{12} and $4.03 \times 10^{12}\ \mathrm{cm}^{-2}$ for sample D685–7 and

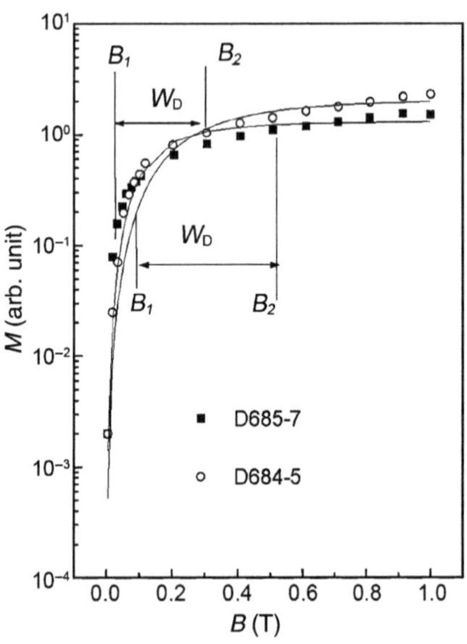

Fig. 4.67 The magneto-resistance of two HgCdTe photoconductive detectors as a function of the magnetic field at 1.2 K. The *solid and open symbols* are experimental data and the *solid curves* are fits to the data. For sample D685–7: $\mu_1 = -2.65 \times 10^5\ \mathrm{cm}^2/\mathrm{V\,s}$, $\mu_2 = -3.12 \times 10^4\ \mathrm{cm}^2/\mathrm{V\,s}$, $f_1 = 0.73$, and for sample D684–5: $\mu_1 = -2.23 \times 10^5\ \mathrm{cm}^2/\mathrm{V\,s}$, $\mu_2 = -2.18 \times 10^4\ \mathrm{cm}^2/\mathrm{V\,s}$, $f_1 = 0.86$. W_D is obtained from a calculation by inserting the parameters determined through the fitting process

Table 4.6 Parameters determined by fitting the magneto-resistance data taken at 1.2 K, using a two-carrier-type model

Sample	μ_1(cm^2/V s)	μ_2(cm^2/V s)	N_1(cm^{-3})	N_2(cm^{-3})	B_1(T)	B_2(T)
D685–7	2.65×10^5	3.12×10^4	7.70×10^{14}	2.44×10^{15}	0.034	0.3
D685–5	2.23×10^5	2.18×10^4	8.10×10^{15}	5.08×10^{15}	0.06	0.54

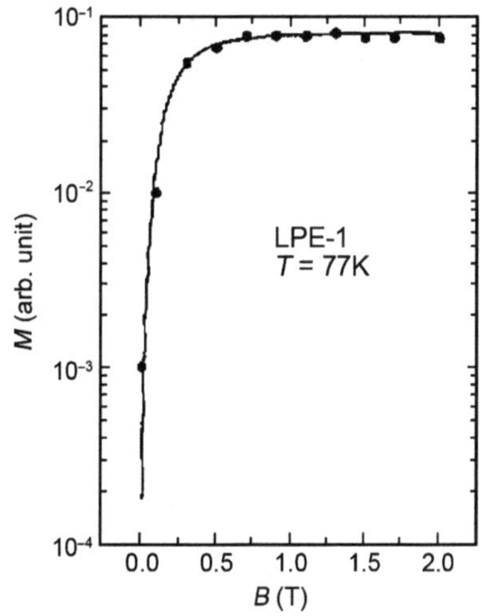

Fig. 4.68 The experimental magneto-resistance data (*solid symbols*) and the fitted curve (*dashed curve*) carried out by a two-carrier-type model versus magnetic field for the sample LPE-1 at 77 K

D684–5, respectively, when the net concentration of surface electrons in Table 4.6 is multiplied by the thicknesses of samples. The contributions to the conductance from surface electrons for samples D685–7 and D684–5 are 27 and 14 $(\Omega\text{cm})^{-1}$, respectively. Indications are that surface electrons have significant effects on the device performance.

In real HgCdTe detectors, there exist many other types of carriers, such as light holes and heavy holes, in addition to surface and bulk electrons. The mobilities of each carrier type are not unique, but can change with concentration and temperature. In particular, the nonparabolic energy bands encountered in HgCdTe also complicate its electrical properties. All the features mentioned above lead to some errors between the experimental results and the fitted curves. Even if the fits are nearly perfect the parameters extracted from them may be in error.

Magneto-transport measurements can be used to investigate not only photoconductive detectors but also p-type HgCdTe materials prepared as photovoltaic devices. There always exist 2DEGs at the surface of HgCdTe materials prepared as photovoltaic devices. In Hall measurements, p-type HgCdTe materials with potentially excellent performance grown by LPE, are sometimes mistakenly regarded as n-type materials with bad performance. Figure 4.68 shows the experimental magneto-resistance data (solid dots) and the fitted curve (dashed curve) carried out by a two-carrier-type RCT model calculation versus magnetic field for the sample LPE-1 at 77 K. The sample LPE-1 is a p-type material with excellent performance grown by LPE. But the sample acts n-type due to the existence of surface electrons

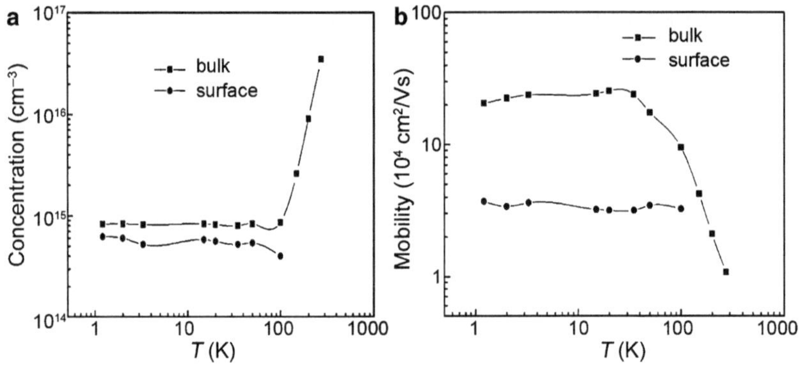

Fig. 4.69 For different types of carriers in the sample D685–2 at 77 K; (**a**) carrier concentration and (**b**) carrier mobility versus temperature

resulting in the hole contribution to the bulk conductance being only 7.5% of the total. However, the true information about the material's behavior can be obtained by an RCT analysis.

Transport measurements with a variable magnetic field were performed on sample D685–2 in the temperature range of 1.2–300 K. The carrier concentrations and mobilities of different types of carriers were investigated using an RCT analysis and the results determined are shown in Fig. 4.69. The curves in Fig. 4.69 were obtained after smoothing the experimental data. As shown in Fig. 4.69a and b, the bulk electron concentration almost does not vary as the temperature is elevated in the range from 1.2 to 100 K, while it rapidly increase with an increasing temperature beyond $T > 100$ K. The bulk electron mobility increases slowly with increasing temperature at low temperature, and reaches its maximum at about 35 K, and then decreases rapidly as the temperature continues to increase. The surface electron concentration and mobility are almost independent of temperature in the temperature range from 1.2 to 100 K, which is consistent with the behavior of the surface electrons in an n-type HgCdTe sample reported by Reine et al. (1993). The contribution to the net conductance from the surface electrons becomes progressively smaller compared with that from the bulk electrons as the temperature exceeds 100 K due to the rapid increase of the bulk electron concentration. Beyond 100 K, the surface electrons can be ignored. In that case ($T > 100$ K), it is fortunate that the surface electron contribution to the net conductance is small because their concentration and mobility determined by fitting experimental data using the RCT method are poorly given and are far from their actual values.

4.6.3 The Influence of Surfaces on the Magneto-Resistance Oscillations of HgCdTe Samples

The transport characteristics of two-dimensional systems formed in surface accumulation layers also can be investigated by Shubnikov–de Haas (SdH) measurements

(Justice 1988). For $Hg_{1-x}Cd_xTe$ narrow band gap semiconductors, the quantum effect is very significant due to its small effective electron mass resulting in a large separation of Landau energy levels. The SdH phenomenon quantum transport mainly takes place at low temperature where the magneto-resistance of the degenerate semiconductor will oscillate with a variation of the magnetic field. In contrast to other measurement methods, the behavior of two-dimensional and three-dimensional carriers can be distinguished by changing the angle between sample and magnetic field when the SdH measurement is performed. Besides, SdH measurements can be used to investigate the electron concentrations, effective electron masses, and scattering mechanisms (Chang et al. 1982; Koch 1982, 1984; Singleton et al. 1986a, b).

In this section, the transport characteristics of a surface accumulation layer of an $Hg_{1-x}Cd_xTe$ photoconductive detector is investigated using a SdH measurement. The photosensitive surface area of the investigated sample is 2.5×10^{-5}–1.4×10^{-4} cm^{-2}, and the sample thickness is 7–8 μm. The upper and lower surfaces of the sample are passivated using the same process. The resistivity of sample S9601 as a function of magnetic field perpendicular to the sample's surface at different temperatures are shown in Fig. 4.70. As shown in Fig. 4.70, the shape of the SdH oscillation varies little, but the magnitude of the oscillations decreases with increasing temperature. The SdH oscillations vanish when the magnetic field is parallel to the sample's surface (except for a partial oscillation for the magnetic field in the range of 0 to 1 T at 12.5 K). The oscillations are not simply periodic in $1/B$, which indicates that there exists more than one type of carrier. A fast Fourier transform (FFT) analysis can be used to extract the oscillation periods of the different carrier types. The results are shown in Fig. 4.71. Four peaks, 0, 1, 0′, and 1′ can be distinguished at all the measured temperatures. It indicates that there are two subbands in each surface for this $Hg_{1-x}Cd_xTe$ photoconductive detector (0 and 1 denote one surface, 0′ and 1′ denote the other surface). The multiple subbands occupied by electrons are due to the small effective electron masses of the subband electrons resulting in

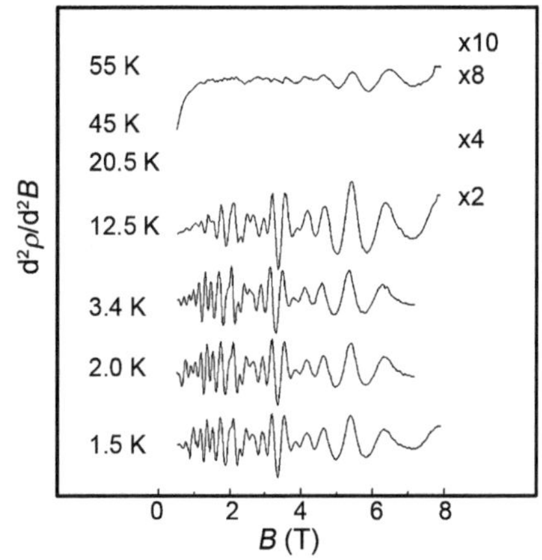

Fig. 4.70 The SdH oscillations of sample A with a thickness of 7 μm as a function of magnetic field at different temperatures

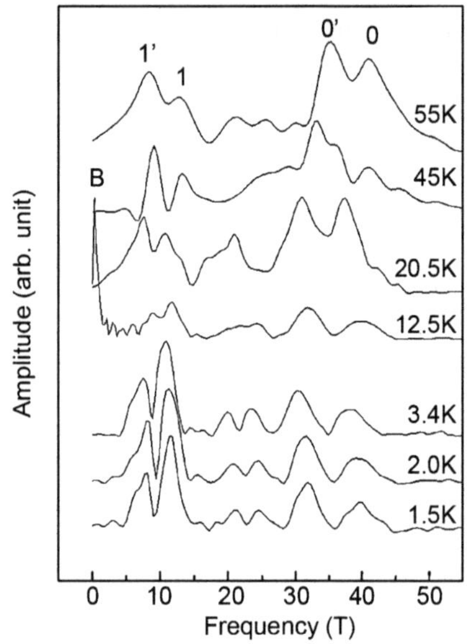

Fig. 4.71 The FFT of the SdH oscillations for sample S9601 at different temperatures

Table 4.7 The electron concentration of each subband for sample S9601 at different temperature

Electron concentration ($\times 10^{12}\,cm^{-2}$)				
T (K)	0	1	$0'$	$1'$
1.5	0.40	0.57	1.53	1.93
2.0	0.41	0.54	1.50	1.87
3.4	0.36	0.53	1.46	1.82
12.5	0.43	0.57	1.54	1.92
20.5	0.37	0.53	1.50	1.82
45	0.44	0.61	1.58	1.94
55	0.41	0.62	1.66	1.96

a relatively low density of states. The sharp peak in the FFT at 12.5 K corresponds to a feature from the bulk electrons with a concentration of $5.2 \times 10^{14}\,cm^{-3}$. The electron concentrations of different subbands extracted from the FFT analysis are shown in Table 4.7. It indicates that the electron concentration of each subband is independent of temperature.

4.6.4 The Influence of the Surface on the Correlation Between Resistivity and Temperature for an HgCdTe Photoconductive Detector

The carrier concentration of each subband being independent of temperature at low temperature was demonstrated in Sect. 4.6.3. However, the concentration of intrinsic carriers in bulk materials will increase exponentially with increasing temperature, whereas the mobility variation trends of the surface and bulk electrons are the same, so the contribution to the conductance from surface carriers will become progressively smaller as the temperature increases. Let's treat a simple case, in which it is assumed that the concentration of surface electrons is independent of temperature for both the upper and lower surfaces of an $Hg_{1-x}Cd_xTe$ photoconductive detector. Then a three-band model (bulk, surface ground state, and excited-state electrons) can be used to fit the correlation between the resistivity and temperature. A proposed model for an $Hg_{1-x}Cd_xTe$ photoconductive detector following these assumptions is:

$$\rho(T) = \frac{1}{e[n_\mathrm{b}(T)\mu_b(T) + 2n_{s0}(T)\mu_{s0}(T)/d + 2n_{s1}(T)\mu_{s1}(T)/d]}, \tag{4.108}$$

where $n_\mathrm{b}(T)$, $n_{s0}(T)$, $n_{s1}(T)$ and $\mu_b(T)$, $\mu_{s0}(T)$, $\mu_{s1}(T)$ are the concentrations and mobilities of the bulk electrons and the two types of surface electrons. The concentrations of the surface electrons are taken to be independent of temperature. The mobility of the bulk electrons is given by Parat et al. (1990) to be:

$$\mu_\mathrm{b} = \left\{\frac{1}{\mu_{300}[m_\mathrm{e}^*(300K)/m_\mathrm{e}^*(T)(300/T)^{1.9}]} + \frac{1}{\mu_{\mathrm{b}0}}\right\}^{-1}, \tag{4.109}$$

$m_e^*(300\,\mathrm{K})$ and μ_{300} are the effective mass and mobility of the bulk electrons at 300 K. $\mu_{\mathrm{b}0}$ is the asymptotic value of the mobility of the bulk electrons at very low temperatures. The mobilities of the surface electrons satisfy the following equations:

$$\mu_{s0} = \left(\frac{1}{\mu_b} + \frac{1}{\mu_0}\right)^{-1}, \tag{4.110}$$

and

$$\mu_{s1} = \left(\frac{1}{\mu_b} + \frac{1}{\mu_1}\right)^{-1}. \tag{4.111}$$

μ_0 and μ_1 are the asymptotic values of the surface electron mobilities at very low temperatures. The concentration of the bulk electrons satisfies the following equation:

$$n_\mathrm{b} = \frac{(N_\mathrm{D} - N_\mathrm{A}) + \sqrt{(N_\mathrm{D} - N_\mathrm{A})^2 + 4n_i^2(x,T)}}{2}. \tag{4.112}$$

μ_0, μ_1, $\mu_{\mathrm{b}0}$, $n_{\mathrm{s}0}$, $n_{\mathrm{s}1}$, and $N_\mathrm{D} - N_\mathrm{A}$ are the parameters found during the fitting process. The results fits the experimental data well if the subband surface electron concentrations extracted from the SdH measurement are selected as the initial values. Figure 4.72 shows comparisons of the fitted results to the experimental data for $Hg_{1-x}Cd_xTe$ photoconductive detector samples. As shown in Fig. 4.72, the theory fits the experimental results quite well. Table 4.8 collects the parameters extracted in the fitting procedure. In the surface accumulate layer, the mobility of ground-state electrons is smaller than that of the exited-state electrons. The magnitude of the ratio $(\mu_0/\mu_1) \leqslant 1$ reflects the ratio of the effective masses of the ground-state electrons

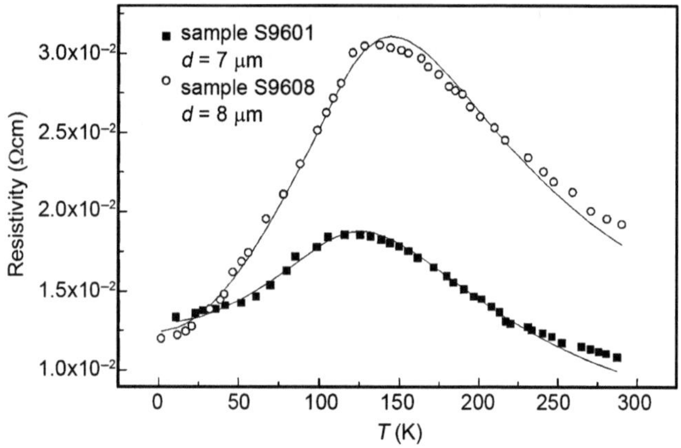

Fig. 4.72 The resistivity of samples S9601 and S9608 is as a function of temperature. The *symbols* are the experimental data, the *solid curves* are the fitted results. The thicknesses of samples S9601 and S9608 are 7 and 8 μm, respectively

Table 4.8 The concentration and mobility of different types of electrons at very low temperature

	$N_D - N_A$ (10^{14} cm^{-3})	n_{s0} (10^{12} cm^{-2})	n_{s1} (10^{12} cm^{-2})	μ_{b0} (10^5 cm^2/V s)	μ_0 (10^4 cm^2/V s)	μ_1 (10^4 cm^2/V s)
S9601	4.52	4.24	1.34	3.21	4.52	6.56
S9608	9.1	2.68	0.91	3.25	5.03	6.81

and the excited state electrons, if their relaxation times are considered to be the same. It also indicates that the effective mass of ground-state electrons is larger than that of the excited state electrons. The curves in Fig. 4.72 are quite smooth, and it would be amazing if they could not be fit with the six adjustable parameters available! With this much latitude, the accuracy of the six parameters must be treated with caution.

References

Amelio GF (1972) A new method of measuring interface state densities in mis devices. Surface Sci 29:125–143

Ando T (1985) Subbands in space-charge layers on narrow gap semiconductors: Validity of semiclassical approximation. J Phys Soc Jpn 54:2676–2681

Ando T, Fowler AB, Stern F (1982) Electronic properties of two-dimensional systems. Rev Mod Phys 54:437–672

Antoszewski J, Faraone L (1996) Analysis of magnetic field dependent Hall data in narrow bandgap $Hg_{1-x}Cd_xTe$ grown by molecular beam epitaxy. J Appl Phys 80:3881–3892

Antoszewski J, Seymour DJ, Faraone L, Meyer JR, Hoffman CA (1995) Magneto-transport characterization using quantitative mobility-spectrum analysis. J Electron Mater 24:1255–1262

Bohm D (1954) Quantum Theory. Constable, London

Brenig W, Kasai H (1984) Band mixing in narrow gap semiconductors. Z Phys B-condensed matter 54:191–194

Broudy RM (1981) In: Willardson RK, Beer AC (eds), Semiconductors and Semimetals, Vol 18. Academic, New York, p 157

Bychkov YA, Rashba EI (1984) Oscillatory effects and the magnetic susceptibility of carriers in inversion layers. J Phys C 17:6039–6045

Chang LL, Mendez EE, Kawai NJ, Esaki L (1982) Shubnikov-De Haas oscillations under tilted magnetic fields in InAs-GaSb superlattices. Surf Sci 113:306–312

Chapman RA, Kinch MA, Simmons A, Borrello SR, Morris HB, Wrobel JS, Buss DD (1978) $Hg_{0.7}Cd_{0.3}Te$ charge-coupled device shift registers. Appl Phys Lett 32:434–436

Chen YJ (1993) Quantum Hall Effect. Science Press, Beijing

Chu JH, Xu SQ, Ji HM (1985a) Determination of composition of $Hg_{1-x}Cd_xTe$ with the method of infrared absorption. Infrared Research 4:255

Chu JH (1985b) Effective mass of electrons in conduction band of $Hg_{1-x}Cd_xTe$ semiconductors. Infrared Research 4:439

Chu JH, Mi ZY (1987) Optical propertiers of narrow gap semiconductors. Prog Phys 7:311

Chu JH, Mi ZY (1991) Surface electron subband structures of $Hg_{1-x}Cd_xTe$. J Infrared Milli Waves 10:427–442

Chu JH, Xu SQ, Tang DY (1982) The energy band gap of ternary semiconductors $Hg_{1-x}Cd_xTe$. Chinese Sci Bull 27:403–406

Chu JH, Xu SQ, Tang DY (1983) Energy gap versus alloy composition and temperature in $Hg_{1-x}Cd_xTe$. Appl Phys Lett 43:1064

Chu JH, Sizmann R, Wollrab R, Koch F, Ziegler J, Maier H (1989) Study of resonant defect states in HgCdTe by capacitance spectroscopy. Infrared Research 8:327–334

Chu JH, Sizmann R, Koch F (1990a) Subband dispersion relations and Landau levels of the inversion layer in p-HgCdTe. Sci China A 33:515–521

Chu JH, Sizmann R, Koch F, Ziegler J, Maier H (1990b) Experimental study of inversion layer subband structure on HgCdTe. Chinese J Semicond 11:332–340

Chu JH, Shen SC, Sizmann R, Koch F (1991a) Magneto-optical resonance of subband electron for HgCdTe. J Infrared Milli Waves 10:44–50

Chu JH, Mi ZY, Sizmann R, Koch F (1991b) Subband structure in the electric quantum limit for $Hg_{1-x}Cd_xTe$. Phys Rev B 44:1717–1723

Chu JH, Mi ZY, Sizmann R, Koch F, Wollrab R, Ziegler J, Maier H (1992) Influence of resonant defect states on subband structures in $Hg_{1-x}Cd_xTe$. J Vac Sci Tech B 10:1569–1573

Dornhaus R, Nimtz G (1983) The properties and applications of the MCT alloy system in narrow-gap semiconductors. In: Springer Tracts in Modern Phys, Vol 98. Springer, Heidelberg, p 130

Dziuba Z, Górska M (1992) Analysis of the electrical conduction using an iterative method. J Phys III France 2:99–110

Eisberg R, Resnick R (1985) Sommerfeld's Model. In: Quantum Physics of Atoms, Molecules, Solids, Nuclei, and Particles, 2nd editon. Wiley, New York, pp 114–117

Fang FF, Howard WE (1966) Negative field-effect mobility on (100) Si surfaces. Phys Rev Lett 16:797–799

Gui YS, Zheng GZ, Chu JH, Guo SL, Tang DY, Cai Y (1997) Magneto-transport properties for accumulated layer on surface of $Hg_{1-x}Cd_xTe$ detector. Chinese J Semicond 18:667–673

Gui YS, Chu JH, Cai Y, Zheng GZ, Tang DY (1998) Study on the electron mobility for each subband on the n-HgCdTe accumulated layer. Acta Phys Sin 47:1354–1360

Huang H, Zheng GZ, Tong FM, Tang DY (1993) The effect of surface layer oxide on $Hg_{1-x}Cd_xTe$ Hall data. J Infrared Milli Waves 12:309–314

Justice RJ, Seiler DG, Zawadzki W, Koestner RJ, Goodwin MW, Kinch MA (1988) Characterization of molecular-beam epitaxially grown HgTe films by Shubnikov–de Haas measurements. J Vac Sci Technol A 6:2779–2784

Kim JS, Seiler DG, Tseng WF (1993) Multicarrier characterization method for extracting mobilities and carrier densities of semiconductors from variable magnetic field measurements. J Appl Phys 73:8324–8335

Kim JS, Seiler DG, Colombo L, Chen MC (1995) Characterization of liquid-phase epitaxially grown HgCdTe films by magnetoresistance measurements. J Electron Mater 24:1305–1310
Kinch MA (1981) In: Willardson RK, Beer AC (eds) Semiconductors and Semimetals, Vol. 18. Academic press, New York, pp 313–384
Koch F (1975) In: Zawadzki W (ed) Narrow Gap Semiconductors. Springer-Verlag, Berlin
Koch F (1982) In: Gornik E, Heinrich H, Palmetshafer L (eds) Physics of Narrow Gap Semiconductors, Vol. 152, Lecture Notes in Physics. Springer, New York, p 92
Koch F (1984) Two-dimensional systems, heterostructrues, and superlattics. In: Bauer G, Kuchar K, Heinrich H (eds) Solid State Science Series, Vol. 53. Springer, New York, p 20
Kuhn M (1970) A quasi-static technique for MOS C-V and surface state measurements. Solid State Electron 13:873–885
Lehovec K, Slobodskoy A (1964) Impedance of semiconductor-insulator-metal capacitors. Solid State Electron 7:59–79
Lowney JR, Seiler DG, Thurber WR, Yu Z, Song XN, Littler CL (1993) Heavily accumulated surfaces of mercury cadmium telluride detectors: Theory and experiment. J Electron Mater 22:985–991
Macdonald JR (1964) Distribution of space charge in homogeneous metal oxide films and semiconductors. J Chem Phys 40:3735–3737
Malcher F, Nachev I, Ziegler A, Rössler U (1987) Subband states in n-inversion layers on small-gap semiconductors. Z Phys B-condensed matter 68:437–444
Marques GE, Sham LJ (1982) Theory of space-charge layers in narrow-gap semiconductors. Surf Sci 113:131–136
Merkt U, Oelting S (1987) Simple description of nonparabolic two-dimensional subbands. Phys Rev B 35:2460–2462
Messiah A (1959) Mecanique quantique I. Dunod, Paris
Meyer JR, Hoffman CA, Bartoli FJ, Arnold DA, Sivananthan S, Fauri JP (1993) Methods for magneto transport characterization of IR detector materials. Semicond Sci Technol 8:805–823
Meyer JR, et al. (1995) Advanced Magneto-Transport Characterization of LPE-Grown $Hg_{1-x}Cd_xTe$ by QMSA. MCT Workshop
Nachev I (1988) Bound states in inversion layers on p-$Hg_{1-x}Cd_xTe$: self-consistent results. Semicond Sci Technol 3:29–34
Nemirovsky Y (1990) Passivation with II–VI compounds. J Vac Sci Technol A 8:1185–1187
Nemirovsky Y, Bahir G (1989) Passivation of mercury cadmium telluride surfaces. J Vac Sci Technol A 7:450–459
Nemirovsky Y, Kirdron I (1979) The interface between $Hg_{1-x}Cd_xTe$ and its native oxide. Solid State Electron 22:831–837
Nicholas RJ, Nasir F, Singleton J (1990) High magnetic field characterisation of (Hg,Cd)Te surface layers. J Cryst Growth 86:656–666
Nicollian EH, Goetzberger A (1967) The silicon-silica interface–electrical properties as determined by the metal-insulator-silicon conductance technique. Bell System Technical Journal 46:1055–1133
Nimtz G, Schlicht B, Dornhaus R (1979) Long-term Hall-type conversion by vacancy diffusion in $Hg_{1-x}Cd_xTe$ at room temperature. Appl Phys Lett 34:490–491
Ohkawa FJ, Uemura Y (1974) Quantized surface states of a narrow-gap semiconductor. J Phys Soc Jpn 37:1325–1333
Reine MB, Maschhoff KR, Tobin SP, Norton PW, Mroczkowski JA, Krueger EE (1993) The impact of characterization techniques on HgCdTe infrared detector technology. Semicond Sci Technol 8:788–804
Rosbeck JP, Starr RE, Price SL, Riley KJ (1982) Background and temperature dependent current-voltage characteristics of HgCdTe photodiodes. J Appl Phys 53:6430–6440
Rössler U, Malcher F, Lommer G (1989) In: Landwehr G (ed) High Magnetic Fields in Semiconductor Physics II. Springer, Berlin, p 157
Parat KK, Taskar NR, Bhat IB, Ghandhi SK (1990) Annealing and electrical properties of $Hg_{1-x}Cd_xTe$ grown by OMVPE. J Crystal Growth 106:513–523

Scott W, Stelzer EL, Hager RJ (1976) Electrical and far-infrared optical properties of p-type $Hg_{1-x}Cd_xTe$. J Appl Phys 47:1408–1414

Singh R, Gupta AK, Chhabra KC (1991) Surface passivation of mercury-cadmium-telluride infrared detectors. Def Sci J (India) 41:205–239

Singleton J, Nasir F, Nicholas RJ (1986a) High-magnetic field characterization of (Hg,Cd)Te detectors. Proc SPIE 659:99–108

Singleton J, Nicholas RJ, Nasir F, Sarkar CK (1986b) Quantum transport in accumulation layers on $Cd_{0.2}Hg_{0.8}Te$. J Phys C: Solid State Phys 19:35–42

Sizmann R, Chu JH, Wollrab R, Koch F, Ziegler J, Maier H (1988) In: Zawadzki W (ed) Proceeding of 19th International Conference on Physics of Semiconductors. Institute of physics, Warsaw, p 471

Sizmann R, Chu JH, Koch F, Ziegler J, Maier H (1990) Sub-band and resonant level spectroscopy from capacitance measurements. Semicond Sci Technol 5:S111–S114

Stern F (1972) Self-consistent results for n-type Si inversion layers. Phys Rev B 5:4891–4899

Takada Y, Uemura Y (1977) Subband structures of N-channel inversion layers on III–V compounds – possibility of the gate controlled Gunn effect. J Phys Soc Jpn 43:139–150

Takada Y, Arai K, Uchimura N, Uemura Y (1980) Theory of electronic properties in N-channel inversion layers on narrow-gap semiconductors I. Subband structure of InSb. J Phys Soc Jpn 49:1851–1858

Takada Y, Arai K, Uemura Y (1982) In: Gornik E (ed) Physics of Narrow Gap Semiconductors, Lecture Notes in Physics, Vol. 152. Springer, Berlin, p 101

Tsau GH, Sher A, Madou M, Wilson JA, Cotton VA, Jones CE (1986) Low-frequency admittance measurements on the HgCdTe/Photox SiO_2 interface. J Appl Phys 59:1238–1244

van Schilfgaarde M, Kotami T, Faleev SV (2006) Adequacy of approximations in GW theory. Phys Rev B 74:245125

White AM (1981) The influence of surface properties on minority carrier lifetime and sheet conductance in semiconductors. J Phys D: Appl Phys 14:L1–L3

Xiong XM (1987) PhD thesis, Shanghai Institute of Technical Physics

Yadava RDS, Sundersheshu BS, Anandan M, Bagai RK, Borle WN (1994) Precipitation in CdTe crystals studied through mie scattering. J Electron Mater 23:1349–1357

Zawadzki W (1983) Theory of optical transitions in inversion layers of narrow-gap semiconductors. J Phys C: Solid State Phys 16:229–240

Ziegler A, Rössler U (1988) Zener-tunnelling of inversion layer electrons on small gap semiconductors. Solid State Commun 65:805–808

Chapter 5
Superlattice and Quantum Well

5.1 Semiconductor Low-Dimensional Structures

5.1.1 Band Dispersion Relation

In three-dimensional crystals, the band edge carrier motion can be described by a quasi-particle model. The interaction of the particle with the periodic crystal field is included in the effective mass, m^*. To first order, the electrons in the conduction band of a crystal with inversion symmetry have m^* independent of crystal direction, and the quasi-particle energy, E^{3D}, is in an isotropic distribution in $\boldsymbol{k}$-space,

$$E^{3D}(k) = \frac{\hbar^2}{2m^*}(k_x^2 + k_y^2 + k_z^2), \tag{5.1}$$

where k_x, k_y, k_z denote the wave numbers in the x, y, z directions.

If a carrier is confined in low-dimensional structures, such as a quantum well, quantum wire or dot, whose size is comparable with the De Broglie wave length of an electron, the properties of such electrons, such as their density of states (DOS), will display quantum mechanical effects. In comparison to three-dimensional bulk materials, quantum well structures have quantum confinement effects in the growth direction. In a parabolic approximation, the energy in a quantum well changes from (5.1) to:

$$E^{2D} = E_{z,n_z} + \hbar^2/2m_{xy}^*(k_x^2 + k_y^2). \tag{5.2}$$

E_{z,n_z} can be solved from the Schrödinger equation in the framework of an effective mass approximation (EMA) (Weisbuch and Vinter 1991):

$$\left[-\frac{\hbar^2}{2m^*(z)}\frac{\partial^2}{\partial z^2} + V(z)\right]\phi(z) = E_{z,n_z}\phi(z). \tag{5.3}$$

Here, $m^*(z)$ is the carrier effective mass in the well or barrier layer, $V(z)$ the potential profile, and $Z(z)$ is the envelope wave function of the confined quantum state.

J. Chu and A. Sher, *Device Physics of Narrow Gap Semiconductors*, Microdevices, DOI 10.1007/978-1-4419-1040-0_5, © Springer Science+Business Media, LLC 2010

For an extreme case, i.e. an infinitely deep quantum well, $V(z) \to \infty$ at the well boundaries, E_{z,n_z} in (5.2) and (5.3) has an analytical solution:

$$E_{z,n_z} = \frac{\hbar^2 \pi^2}{2m_z^*}\left(\frac{n_z^2}{L^2}\right). \tag{5.4}$$

Consequently, the eigen energy of a carrier in the quantum well is comprised of a continuous part and a discrete part. Going from three to two dimensions, the band structure evolves from a continuous distribution to a series of discrete subbands.

In reality, the barrier height of the quantum well is finite. For the case of a finite but deep quantum well, if the potential has even symmetry, i.e., $V(z) = V(-z)$ (see Fig. 5.1), the solutions of (5.3) are odd or even states. The even parity solutions are (Haug and Koch 1993; Bastard et al. 1991):

$$\phi(z) = \begin{cases} \cos kz, & |z| < L/2 \\ B\, e^{-\kappa(|z|-L/2)}, & |z| > L/2 \end{cases}. \tag{5.5}$$

The odd wave functions are:

$$\phi(z) = \begin{cases} \sin kz, & |z| < L/2 \\ B\, e^{-\kappa(|z|-L/2)}, & |z| > L/2 \end{cases}. \tag{5.6}$$

In the above equations, L is the well width. The eigenvalues are determined by: $E_{z,n_z} = \hbar^2 k_{z,n_z}^2/2m_{z,w}^*$ or $E_{z,n_z} = V - \hbar^2 k_{z,n_z}^2/2m_{z,b}^*$ and $0 < E_{z,n_z} < V_{z,0}$. Because most often the properties (e.g., effective mass, dielectric constant, and bond length) of the well material and the barrier materials are different, the boundary conditions are set by matching envelope functions at the boundary producing what are called the BenDaniel–Duke boundary conditions (BenDaniel and Duke 1966). Rather than requiring that the wave functions and their derivatives match at the boundary the new conditions require ϕ and $\frac{1}{m^*}\frac{\mathrm{d}\phi}{\mathrm{d}z}$ both be continuous. At $z = \pm \mathrm{L}/2$, the even parity solutions of (5.5) thus satisfying the boundary conditions:

$$(k_{z,n_z}/m_{z,\mathrm{w}}^*)\tan(k_{z,n_z}L/2) = \kappa_{z,n_z}/m_{z,\mathrm{b}}^*, \tag{5.7}$$

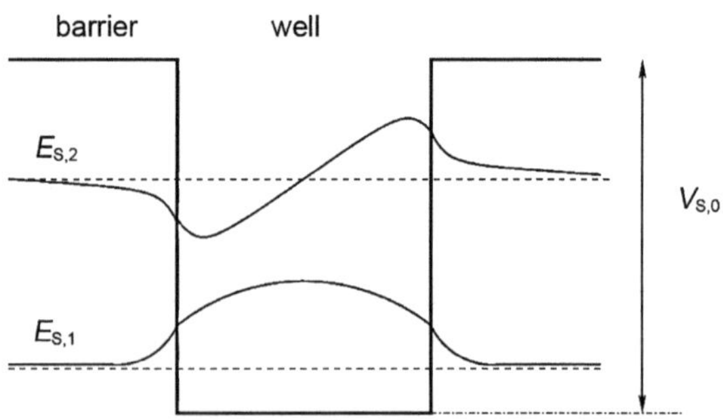

Fig. 5.1 The energy levels and wave functions of the first and second subbands in a finite-barrier quantum well. The inflection points of the wave functions originate from differences of the effective masses in the quantum well

Similarly, for odd parity solutions of (5.7), to satisfy the boundary conditions requires:

$$(k_{z,n_z}/m^*_{z,\mathrm{w}})\tan(k_{z,n_z}L/2) = \kappa_{z,n_z}/m^*_{z,\mathrm{b}}. \tag{5.8}$$

Summarizing the above equations, k_{z,n_z} can be obtained from:

$$\kappa_{z,n_z} = \begin{cases} (m^*_{z,\mathrm{b}}/m^*_{z,\mathrm{w}})k_{z,n_z}\tan(k_{z,n_z}L/2), & \tan(k_{z,n_z}L/2) > 0 \\ -(m^*_{z,\mathrm{b}}/m^*_{z,\mathrm{w}})k_{z,n_z}\cot(k_{z,n_z}L/2), & \tan(k_{z,n_z}L/2) < 0 \end{cases} \tag{5.9}$$

or

$$k_{z,n_z} = \begin{cases} k_{z,0}\left(\dfrac{m^*_{z,\mathrm{w}}}{m^*_{z,\mathrm{w}} + m^*_{z,\mathrm{b}}\tan^2(k_{z,n_z}L/2)}\right)^{1/2}, & \tan(k_{z,n_z}L/2) > 0, \\ k_{z,0}\left(\dfrac{m^*_{z,\mathrm{w}}}{m^*_{z,\mathrm{w}} + m^*_{z,\mathrm{b}}\cot^2(k_{z,n_z}L/2)}\right)^{1/2}, & \tan(k_{z,n_z}L/2) < 0, \end{cases} \tag{5.10}$$

$k_{z,0}$ in (5.10) is:

$$k_{z,0} = \left(\frac{2m^*_{z,\mathrm{w}}V_{z,0}}{\hbar^2}\right)^{1/2}. \tag{5.11}$$

To fully describe the subbands in stained heterojunctions, the influence of strain on the properties of the well material must be included. Strain can change the band structure, and also cause energy level splittings by reducing the band structure symmetry.

The adjustment of strain at the band edge can be described by "Model-Solid" theory. It was originally proposed by Van de Walle and Martin (Van de Walle and Martin 1986, 1987; Van de Walle 1989), lately developed by many others (Satpathy et al. 1988; Wang and Stringfellow 1990; Krijin 1991; Shao et al. 2002), and used to calculate strain-induced band edge shifts. If the strained layer is a pseudo-morphic growth on a (001) substrate (Fig. 5.2), the strain can be modeled by a biaxial strain

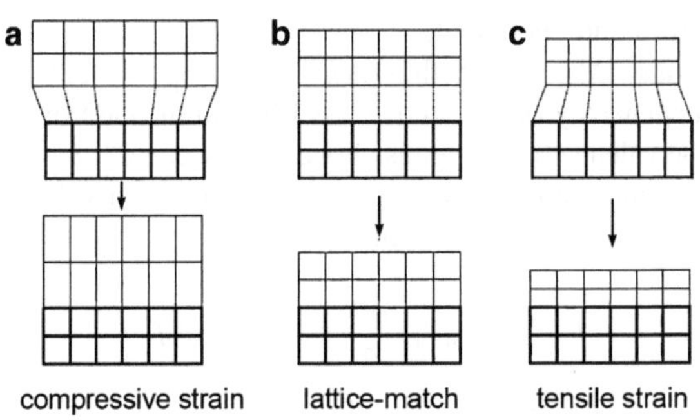

Fig. 5.2 The schematic diagrams of the strained layers grown on a (001) substrate (**a**) compressive stress, (**b**) lattice-matched, (**c**) tensile stress

coefficient parallel to the growth interface, $\varepsilon_{/\!/}$, and a nonaxial strain coefficient, $\varepsilon_{\perp}$, perpendicular to the growth interface, which satisfy:

$$\varepsilon_{\parallel} = \frac{a_0}{a_f} - 1 \text{ and } \varepsilon_{\perp} = \left(-\frac{2C_{12}}{C_{11}}\right)\varepsilon_{\parallel}. \tag{5.12}$$

Here, a_0 is the substrate lattice constant, a_f is its equilibrium (without strain) lattice constant. For a zinc-blende semiconductor, the strain at the Γ point is equivalent to the sum of a hydrostatic pressure and a uniaxial stress along the z (shear) direction. The hydrostatic pressure causes a shift of the conduction band edge E_c. The average valence band energy is $E_{v,av} = (E_{hh} + E_{lh} + E_{so})/3$:

$$\Delta E_c^{hy} = a_c\left(2\varepsilon_{\parallel} + \varepsilon_{\perp}\right) \text{ and } \Delta E_{v,av}^{hy} = a_v\left(2\varepsilon_{\parallel} + \varepsilon_{\perp}\right) \tag{5.13}$$

where a_v and a_c are hydrostatic deformation potentials of the valence and the conduction bands, respectively. Although the uniaxial shear stress doesn't affect the conduction band's shape at the Γ point, it can interact with the spin–orbit effect and cause an additional splitting in the valence band. The $|J, m_J>=|3/2, 3/2>$ state (E_{hh}), the $|J, m_J>=|3/2, 1/2>$ state (E_{lh}), and the $|J, m_J>=|1/2, 1/2>$ state (E_{so}), can all be expressed relative to the average valence band energy E_{av},

$$\begin{aligned}
\Delta E_{hh}^{sh} &= \frac{\Delta^{so}}{3} + \frac{\Delta_{001}^{sh}}{3},\\
\Delta E_{lh}^{sh} &= -\frac{\Delta^{so}}{6} - \frac{\Delta_{001}^{sh}}{6} + \frac{1}{2}\left[(\Delta^{so})^2 - \frac{2}{3}\Delta^{so}\Delta_{001}^{sh} + (\Delta_{001}^{sh})^2\right]^{1/2},\\
\Delta E_{so}^{sh} &= -\frac{\Delta^{so}}{6} - \frac{\Delta_{001}^{sh}}{6} - \frac{1}{2}\left[(\Delta^{so})^2 - \frac{2}{3}\Delta^{so}\Delta_{001}^{sh} + (\Delta_{001}^{sh})^2\right]^{1/2},
\end{aligned} \tag{5.14}$$

where Δ_{so} is the spin–orbit splitting energy without strain. In the (001) direction, Δ_{001}^{sh}, is:

$$\Delta_{001}^{sh} = 3b\frac{C_{11} + 2C_{12}}{C_{11}}\varepsilon, \tag{5.15}$$

b is the shear deformation potential.

For a zero strain, $\Delta_{001}^{sh} = 0$, thus $\Delta E_{hh}^{sh} = \Delta E_{lh}^{sh} = \Delta^{so}/3$, and $\Delta E_{so}^{sh} = -2\Delta^{so}/3$. The degeneracy of E_{hh} and E_{lh} is lifted in the case of a lattice matched interface, now the valence band edge is above $E_{v,av}$, i.e., $\Delta^{so}/3$.

In the Kelvin's thermometric scale, the conduction and valence bands energy levels are expressed by following formulas:

$$\begin{aligned}
E_c &= E_{v,av} + \frac{\Delta^{so}}{3} + E_{g,0} + \Delta E_c^{hy},\\
E_{hh} &= E_{v,av} + \Delta E_{v,av}^{hy} + \Delta E_{hh}^{sh},\\
E_{lh} &= E_{v,av} + \Delta E_{v,av}^{hy} + \Delta E_{lh}^{sh}.
\end{aligned} \tag{5.16}$$

$E_{\text{v,av}}$, Δ_{so}, and $E_{\text{g,0}}$ are material parameters in the absence of strain. The band edge discontinuity between the strained well layer and the lattice matched strain free barrier layer is:

$$\begin{aligned} V_{\text{e,0}} &= E_{\text{c}}(\text{barrier layer}) - E_{\text{c}}, \\ V_{\text{hh,0}} &= E_{\text{hh}} - E_{\text{v}}(\text{barrier layer}), \\ V_{\text{lh,0}} &= E_{\text{lh}} - E_{\text{v}}(\text{barrier layer}), \end{aligned} \tag{5.17}$$

and

$$\begin{aligned} E_{\text{v}}(\text{barrier layer}) &= E^{\text{b}}_{\text{v,av}} + \frac{\Delta^{\text{so,b}}}{3}, \\ E_{\text{c}}(\text{barrier layer}) &= E_{\text{v}}(\text{barrier layer}) + E^{\text{b}}_{\text{g,0}}. \end{aligned} \tag{5.18}$$

In fact the parabolic band dispersion relation is only valid close to the Brillouin center, $k_{xy} = 0$. As k_{xy} increases the effective mass becomes larger. To account for this, Kane proposed a model that treats the interaction between different bands; it is especially useful for narrow gap semiconductors (Kane 1957).

In a second-order approximation, the dispersion of the conduction band is:

$$\begin{aligned} E^{\text{2D}}_{z,n}(k_{xy}) &= E_{z,n} + \frac{\hbar^2 k^2_{xy}}{2m^*_{xy}(E)}, \\ m^*_{xy}(E) &= \frac{m^*_{xy}}{1 + \sqrt{4\alpha(E - E_{z,n})/E_{\text{g}}}}, \\ \alpha &= -\frac{(1 - m^*_{xy}/m_0)^2(3E^2_{\text{g}} + 4E_{\text{g}}\Delta^{\text{so}} + 2(\Delta^{\text{so}})^2)}{(E_{\text{g}} + \Delta^{\text{so}})(3E_{\text{g}} + 2\Delta^{\text{so}})}. \end{aligned} \tag{5.19}$$

Similar expressions are also valid for the light-hole valence band. For the heavy-hole band, because of the high DOS at finite energy, the nonparabolic band effect is always neglected.

The dispersion relation of quantum wires and dots can be obtained in a similar way. There will be new quantum effects if the carrier motion is confined in space. For a quantum wire, if the barrier is infinitely high in the z, and x directions, the energy eigenvalues become:

$$E^{\text{1D}}_{n_x,n_z} = \frac{\hbar^2\pi^2}{2}\left(\frac{n^2_x}{m^*_x L^2_x} + \frac{n^2_z}{m^*_z L^2_z}\right) + \frac{\hbar^2 k^2_y}{2m^*_y}. \tag{5.20}$$

The carrier energy eigenvalues also consists of a continuous and a discrete part, the difference is that two-dimensional confinement leads to a subband dispersion in a single direction.

A quantum dot is a zero-dimensional system. If the barrier is assumed to be infinitely high, the 0 energy eigenvalues are:

$$E^{\text{0D}}_{n_x,n_y,n_z} = \frac{\hbar^2\pi^2}{2}\left(\frac{n^2_x}{m^*_x L^2_x} + \frac{n^2_y}{m^*_y L^2_y} + \frac{n^2_z}{m^*_z L^2_z}\right), \tag{5.21}$$

n_x, n_y, n_z are quantum numbers, they are integers but are not simultaneously equal to zero. L_x, L_y, L_z are sizes of the dot in the x, y, z directions. m_x^*, m_y^*, m_z^* are effective masses in the three different direction. A carrier is totally localized in a quantum dot, hence the energy levels are discrete.

In reality the quantum dots are not rectangles, and the barrier height is finite. In this case, the energy eigenvalues are obtained by numerically solving the Schrödinger Equation (Zunger 1998; Grundmann et al. 1995).

5.1.2 Density of States

In relation to the dimensional dependence of the band dispersion relations, the DOS function also varies with quantum confinement in real space. The DOS is defined as follows:

$$\rho(E) = 2\sum_{n,k} \delta(E - E_n(k)), \tag{5.22}$$

where δ denotes a delta function, $E_n(k)$ is the eigenenergy, for systems with different dimensionalities. The DOS functions are (Arakawa and Sakaki 1982):

$$\begin{aligned}
\rho^{3D}(E) &= \frac{(2m^*/\hbar^2)^{3/2}}{2\pi^2}\sqrt{E}, \\
\rho^{2D}(E) &= \sum_{n_z} \frac{m^*}{\pi\hbar^2 L_z} H(E - E_{n_z}), \\
\rho^{1D}(E) &= \sum_{n_x,n_z} \frac{(m^*/2\hbar^2)^{1/2}}{\pi L_x L_z} \frac{1}{\sqrt{E - E_{n_x} - E_{n_z}}}, \\
\rho^{0D}(E) &= \sum_{n_x,n_y,n_z} \frac{1}{L_x L_y L_z} \delta\left(E - E_{n_x} - E_{n_y} - E_{n_z}\right),
\end{aligned} \tag{5.23}$$

m^* is the carrier effective mass, here assumed to be isotropic in $\boldsymbol{k}$-space, E is the energy relative to the band edge, and $H(E)$ is the unit step function, $H(E \geq 0) = 1$ and $H(E < 0) = 0$.

Figure 5.3 shows the DOS functions corresponding to different dimensional materials. In a three-dimensional system, the carrier eigenenergies have a continuous distribution, and the density of function increases as the square root of energy. In a two-dimensional quantum well, the first subband shifts to higher energy owing to the quantum confinement effect, consequently the DOS remains zero below the first subband, jumps to a finite constant value until the next subband is reached. Overall it is a series of step functions as a function of energy. For a one-dimensional quantum wire, similarly, the DOS also displays a step transit behavior at the onset of each subband, but after the step it drops as $E^{-1/2}$ until the next step is reached.

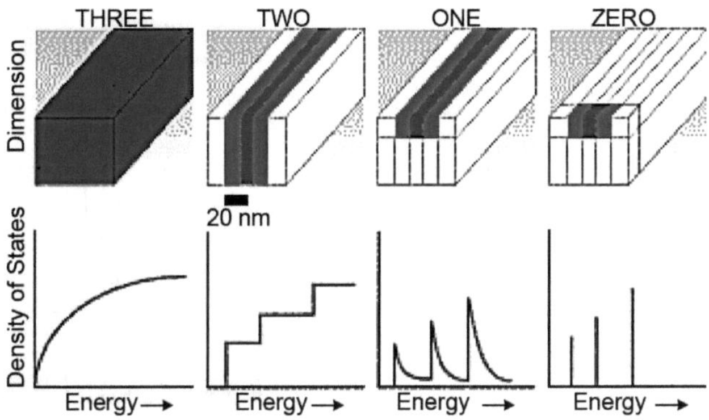

Fig. 5.3 The density states of materials with different dimensions

For a quantum dot, the energy levels are discrete, and the DOS is a series of delta functions centered at each level. For each level, there is a maximum population of only two electrons with opposite spins.

5.1.3 Optical Transitions and Selection Rules

Optical transitions related to semiconductor low-dimensional structures can be classified into interband transitions and intraband transitions. The latter are also called intersubband transitions.

An interband transition occurs between conduction and valence bands. It is related to two kinds of carriers, namely electrons and holes, hence it is a dipolar transition. The transition energy is:

$$E_{\text{inter}} = E_{\text{g}} + E_{\text{c,n}_\text{c}} + E_{\text{v,n}_\text{v}} - E_{\text{ex}}, \tag{5.24}$$

$E_{\text{c,nc}}$ and $E_{\text{v,nv}}$ are energy levels of electrons and holes, respectively, and E_{ex} is the exiton binding energy. The intraband transitions occur between two conduction or valence subbands, it relates to only one type of carrier, electrons or holes, thus it is a unipolar transition. There is another type of intrasubband transition, the absorption of a photon the emission of one phonon.

Figure 5.4 shows schematically the interband and intraband absorption transitions. "I" designates the interband transition from the valence to the conduction band. "II" designates the intraband transition occurring within the conduction and the valence subbands, including the transitions between two bound energy levels, and from one bound energy to the continuum states.

The existence of optical intraband transitions was discovered by Kamgar on a 2DEG in an Si inversion layer (Kamgar et al. 1974). Since the absorption efficiency of single inversion layer is rather low, less than 1%, it is difficult to use this for

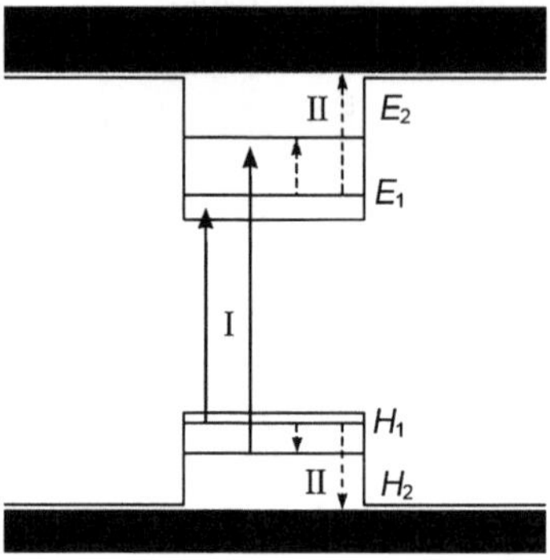

Fig. 5.4 The interband and intraband absorption in a quantum well

infrared detectors. Esaki and Sakaki proposed to use the intrasubband transition in GaAs/AlGaAs multiple quantum wells (Esaki and Sakaki 1977).

In symmetric quantum wells, the optical transitions can be described by effective mass theory. In an electric dipole approximation, an optical transition matrix element is expressed as:

$$p_{ij} = \langle \Psi_j | \vec{\varepsilon} \cdot \mathbf{p} | \Psi_i \rangle, \tag{5.25}$$

$\vec{\varepsilon}$ is the polarized vector of optical field, $\mathbf{p}$ the momentum operator, Ψ_j and Ψ_i are wave functions of the initial and the final states. In envelope function theory, wave function can be expressed as a product of a band edge function $u_i(\mathbf{r})$ and one slowly varying spatial function $F_i(\mathbf{r})$,

$$\Psi_i = u_i(\mathbf{r}) \cdot F_i(\mathbf{r}) = u_i(\mathbf{r}) \cdot e^{(\mathrm{i}\mathbf{k}_{/\!/} \cdot \mathbf{r}_{/\!/})} \cdot \phi_i(z), \tag{5.26}$$

where $u_i(\mathbf{r})$ is a Bloch wave function component at the band edge, $\mathbf{k}_{/\!/}$ a wave vector at the quantum well interface, $\phi_i(z)$ the envelope wave function of the ith subband in the z direction. The optical transition matrix element can be expressed as:

$$p_{ij} = \varepsilon \cdot \langle u_j | p | u_i \rangle \cdot \langle F_i | F_j \rangle + \langle u_j | u_i \rangle \cdot \langle F_j | \varepsilon \cdot p | F_i \rangle . \tag{5.27}$$

The first term on right side is the interband optical transition matrix element, and the second term sets the intraband transition selection rule. It is worthwhile to point out that, interband optical transitions occur between different Bloch states at the band edge, but intraband transitions occur between different envelope functions. For an intraband transition, the Bloch component is the same for i and j states, hence the optical transition element, because $\langle u_j | u_i \rangle = \delta_{ij}$, can be expressed as:

$$\langle F_j | \vec{\varepsilon} \cdot \mathbf{p} | F_i \rangle = \frac{i\sqrt{2(E_j - E_i)m_0}}{e\hbar} \mu_{ij},$$
$$\mu_{ij} = e \langle \phi_j(z) | z | \phi_i(z) \rangle \cdot \vec{\varepsilon} \cdot \hat{z}. \tag{5.28}$$

μ_{ij} is the intraband transition dipole coupled element, $\hat{z}$ the unit vector in the z direction, e the electron charge, E_i and E_j are eigenenergies of the i and j states, and m_0 is the free electron mass. It can be seen that μ_{ij} is related to polarization vector $\vec{\varepsilon}$ and envelope functions of the initial and final states. Since **z** is an odd function, for a symmetric quantum well, the intraband transitions can only happen between two envelope functions with different parities. But for asymmetric quantum wells, such as a coupling quantum well, a ladder quantum well, and a DC-biased quantum well, principally all transitions are allowed (Pan et al. 1990; Mii et al. 1990).

For quantum well with an infinite barrier, the intraband transition dipole element is:

$$\mu_{ij} = \frac{8e}{\pi^2} \cdot \frac{i \cdot j}{(j^2 - i^2)^2} \cdot L_{\mathrm{w}} \cdot \sin\theta, \tag{5.29}$$

where L_{w} is the well width and θ is the incidence angle. It is obvious that, if $\theta = 0$, i.e., the incident light is perpendicular to the sample surface or equally if the polarization direction of the light is perpendicular to the growth direction, the intraband transition element will be zero. This is the so-called polarization selection rule (Yang et al. 1994; Liu et al. 1998). Chu (2001) directly verified this selection rule by measuring the absorption spectrum of InGaAs/GaAs multiple quantum wells with a Fourier transform infrared spectrometer. The sample arrangement adopted has a waveguide form, as shown in Fig. 5.5. The incident light is perpendicular to one side of the sample which has a 45° angle relative to the quantum well growth direction. The polarization direction of the light can be adjusted by a polarizer. In Fig. 5.5, the TE component is parallel to quantum well plane, and the TM has a component perpendicular to quantum well plane.

When $\theta = 90°$, μ_{ij} has maximum. For this case, it is not difficult to derive that the intraband dipole transition element can be expressed as:

$$\mu_{12} = \frac{16}{9\pi^2} \cdot e \cdot L_{\mathrm{w}} \simeq 0.18eL_{\mathrm{w}}. \tag{5.30}$$

According to the definition of the oscillator strength, f_{ij}, for infinitely deep quantum well is:

$$f_{12} = |p_{12}|^2 = \frac{2m_0(E_2 - E_1)}{e^2\hbar^2} \cdot \mu_{12}^2 = 0.96 \cdot \frac{m_0}{m^*}. \tag{5.31}$$

Obviously the oscillator strength is independent of the transition energy, and inversely proportional to the carrier effective mass.

A similar analysis can be done for quantum wires and quantum dots. It can be proven that the optical transition with polarization parallel to the quantum wire direction is forbidden. For quantum dots, principally there is no forbidden direction

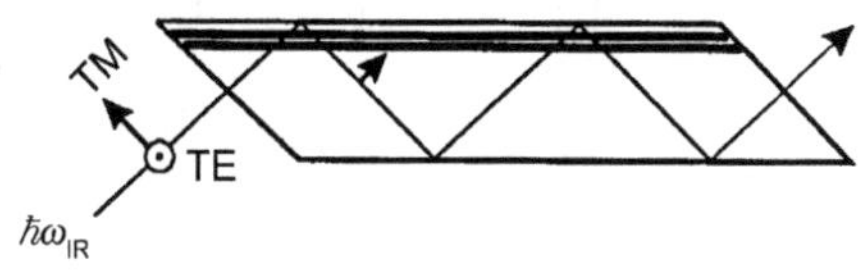

Fig. 5.5 The intraband absorption of a waveguide-form sample

for an optical transition. The polarization direction of an optical transition between two subbands is determined by the spatial symmetry of the relevant wave functions. Thus by analyzing the symmetry, the polarization properties of the intraband dipole transition can be predicted. It is worth noting that for perpendicular incidence, intraband absorption is usually prohibited in quantum wells, but it is allowed in quantum dots and quantum wires. This is significant, meaning it is possible to develop perpendicular incidence infrared detectors.

In quantum dot structures, the absorption coefficient between subbands 1 and 2 can be expressed as:

$$\alpha(\omega) = \frac{\pi E_{21} e^2 (n_1 - n_2)}{2\varepsilon_0 c\tilde{n} m_0 \omega \Omega} \cdot f \cdot g(E_{21} - \hbar\omega), \tag{5.32}$$

where $n_1 - n_2$ is the number of carriers that can absorb photons in a volume element Ω, ε_0 the vacuum dielectric constant, $\tilde{n}$ the refractive index, f the oscillator strength, and $g(\omega)$ is the spectral line shape.

5.2 Band Structure Theory of Low-Dimensional Structures

5.2.1 Band Structure Theory of Bulk Semiconductors

In this section, a brief introduction to band structure theory based on the $\mathbf{k}\cdot\mathbf{p}$ method is given. Much more detail is provided in Chapter 3 of "Physics and Properties of Narrow Gap Semiconductors" (Chu and Sher 2007). First a qualitative picture of the band structure formation in bulk materials is presented, followed by the $\mathbf{k}\cdot\mathbf{p}$ model for narrow gap materials originally proposed by Kane. The next subsection describes the principles underlying the envelope function model for the subband structure of hetero-structures. From a model based on an eight-band $\mathbf{k}\cdot\mathbf{p}$ method, the last subsection demonstrates the specific features of subband structures of HgTe/HgCdTe quantum wells.

For semiconductor materials such as Ge or GaAs, there are eight outer electrons per unit cell that contribute to the chemical bonds. The other electrons of the constituent atoms are localized in closed shell configurations and their wave functions are tightly bound to their atomic nuclei. They do not contribute appreciably to electronic properties observable in electric transport experiments or optical properties near the band gap. In Fig. 5.6, Ge is used, as an example, to show the formation of the band structure. The eight outermost electrons (four from each Ge atom and four from its four neighboring atoms) hybridize to form tetrahedral bonds between the Ge atom and its four nearest neighbors. In basic terms, one could say that the orbitals of every atom (s-like or p-like) hybridize with an orbital of their neighboring atoms, to produce two levels; one bonding and the other antibonding. The bonding states are stable and lower in energy. Because there are a large number of unit cells in a solid, the hopping integrals between the bonding and antibonding levels broaden

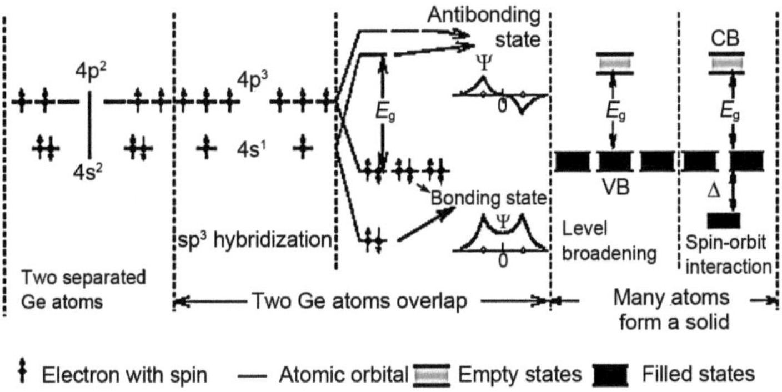

Fig. 5.6 Formation process of the Ge energy bands

Table 5.1 The spin–orbit splitting energy for important semiconductors

Mater.	HgTe	CdTe	InAs	GaAs	AlAs	GaSb	Ge	Si
Δ (eV)	1.08	1.08	0.38	0.341	0.275	0.752	0.29	0.044

them into bands. The bonding s levels are deeply bound and always occupied by two electrons per unit cell. The remaining six electrons per unit cell completely fill the three bonding p orbitals. Thus, the bonding states are completely filled and form the valence bands. The bands originating from the antibonding orbitals are all empty, the lowest lying (often originating from an s band) form the conduction band of the material. Obviously, the band gap emerges between the antibonding and bonding states (Bastard 1988).

In most semiconductors the top of the valence band occurs at the center of the Brillouin zone (BZ), namely the Γ point. In the absence of a spin–orbit coupling, the three valence bands (which originate from bonding p orbits) are degenerate at the Γ point. The spin–orbit coupling lifts the sixfold degeneracy and gives rise to a quadruplet (symmetry Γ_8) that corresponds to $J = 3/2$, and to a doublet (symmetry Γ_7) that corresponds to $J = 1/2$, where J is the total angular momentum, i.e., the sum of the spin and the orbit terms. Since the spin–orbit coupling is essentially an atomic property, it should scale with the atomic number. Table 5.1 summarizes the spin–orbit splitting energy for important semiconductors.

The main problem of band structure theory is the determination of the energy E values of interest for any k inside or on the surface of the BZ, and particularly near those points at which E is an extremum. In semiconductors, electrons and holes are located at these points and the nature of the dispersion relations at these points determines many of the properties of these substances.

From the dispersion relation, $E(\mathbf{k})$, the cyclotron effective mass can be deduced. Often the motion of an electron in a magnetic field can be treated in a semi-classical way (Omar 1975):

$$\hbar \frac{\mathrm{d}\boldsymbol{k}}{\mathrm{d}t} = -e\left[\boldsymbol{v}(k) \times \boldsymbol{B}\right]. \tag{5.33}$$

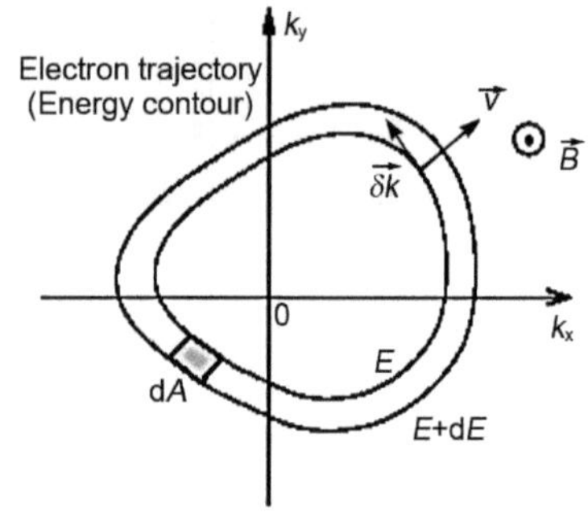

Fig. 5.7 The electron trajectory in $\boldsymbol{k}$ space in a magnetic field

From (5.33), it can be seen that $\boldsymbol{k}$ is perpendicular to both $\boldsymbol{v}$ and $\boldsymbol{B}$, and the electron velocity $\boldsymbol{v} = 1/\hbar \cdot \nabla_k E(\boldsymbol{k})$. Thus, the electrons move along the equal potential curves (or surfaces for 3D) at E_F. And the cyclotron mass is (Pincherle 1971):

$$m_\mathrm{c}^* = \frac{\hbar^2}{2\pi} \cdot \frac{\mathrm{d}A}{\mathrm{d}E}, \tag{5.34}$$

where $\mathrm{d}A$ is the area element between two equal energy curves (shown in Fig. 5.7. For the case of an isotropic, though nonparabolic band, we can write:

$$\frac{1}{m_\mathrm{c}^*} = \frac{1}{\hbar^2 k} \cdot \left.\frac{\mathrm{d}E}{\mathrm{d}k}\right|_{E_\mathrm{F}}. \tag{5.35}$$

By comparing the theoretical and the cyclotron resonance experimental value for m_c^*, the band structure can be verified. An excellent example is the cyclotron resonance absorption experiment in germanium (Kittel 1968). Also the band dispersion is essential to determine the DOS $D(E)$ and hence the optical and transport properties. $D(E)$ for a two-dimensional system is given by:

$$D(E) = \frac{1}{(2\pi)^2} \int_\mathrm{L} \frac{\mathrm{d}l}{|\nabla_k E|}, \tag{5.36}$$

where $\mathrm{d}l$ is the length element of the constant energy curve, and the integration is along the equal potential curve with energy E.

Below is a brief summary of $\boldsymbol{k} \cdot \mathbf{p}$ theory. In a bulk crystal, the one-electron Schrödinger equation has the following form:

$$\left[\frac{(\boldsymbol{p})^2}{2m_0} + V(\boldsymbol{r}) + \frac{\hbar}{4m_0^2c^2}(\boldsymbol{\sigma} \times \nabla V) \cdot \boldsymbol{p}\right] \psi(\boldsymbol{r}) = E\psi(\boldsymbol{r}), \tag{5.37}$$

where $\boldsymbol{p}$ is the momentum operator, the third term is the spin–orbit coupling, and $V(\boldsymbol{r})$ is the crystal field, which has translation symmetry. The electron wave functions are Bloch functions:

$$\psi_{nk}(\boldsymbol{r}) = u_{nk}(\boldsymbol{r}) \cdot \exp(\mathrm{i}\boldsymbol{k} \cdot \boldsymbol{r}), \tag{5.38}$$

where $u_{nk}(\boldsymbol{r})$ is a periodic function of the lattice translation operator and n is the band index. With periodic boundary conditions $\boldsymbol{k}$ forms a quasi-continuum in the first BZ. The periodic part of the Bloch functions, $u_{nk}(\boldsymbol{r})$, are solutions of the equation:

$$\left[\frac{(\boldsymbol{p})^2}{2m_0} + V + \frac{\hbar^2(\boldsymbol{k})^2}{2m_0} + \frac{\hbar \boldsymbol{k}}{m_0} \cdot \left(\boldsymbol{p} + \frac{\hbar}{4m_0c^2}\boldsymbol{\sigma} \times \nabla V\right) + \frac{\hbar}{4m_0^2c^2}(\boldsymbol{\sigma} \times \nabla V) \cdot \boldsymbol{p}\right] u_{nk}(\boldsymbol{r}) = E u_{nk}(\boldsymbol{r}). \quad (5.39)$$

The Bloch functions at k_0, $u_{nk_0}(\boldsymbol{r})$, form a complete and orthogonal basis set, where k_0 is the k value at a band extremum. The Bloch function $u_{nk}(\boldsymbol{r})$ can be expanded in terms of $u_{nk_0}(\boldsymbol{r})$:

$$u_{nk}(\boldsymbol{r}) = \sum_{n'} c_{nn'}(\boldsymbol{k}) u_{n'k_0}(\boldsymbol{r}), \quad (5.40)$$

if $k \approx k_0$. After some algebraic simplifications, the following eigenvalue problem for the coefficients $c_{nn'}(\boldsymbol{k})$ is finally obtained:

$$\sum_{n'} \left\{ \left[E_n(\vec{k}_0) + \frac{\hbar k^2}{2m_0} \right] \delta_{nn'} + \frac{\hbar}{m_0} \vec{k} \cdot \pi_{nn'} \right\} c_{nn'}(\vec{k}) = E_n(\vec{k}) c_{nn}(\vec{k}), \quad (5.41)$$

where $\pi_{nn'}$ is expressed as:

$$\pi_{nn'} = \int \mathrm{d}^3 r u^*_{nk_0}(\boldsymbol{r}) \left(\boldsymbol{p} + \frac{\hbar}{4m_0c^2} \boldsymbol{\sigma} \times \nabla V \right) u_{n'k_0}(\boldsymbol{r}). \quad (5.42)$$

The integral is over the unit cell of the lattice.

In a one-conduction-band model, one needs to consider, for example, only the coupling between the Γ_6 conduction band and the other n bands. As long as $\boldsymbol{k}$ is small (i.e., $\hbar^2k^2/2m_0 \ll |E_c - E_n|$), then from second-order perturbation theory, the dispersion relation of the nondegenerate conduction band is parabolic in the vicinity of the Γ point with an effective mass:

$$\frac{1}{m^*} = \frac{1}{m_0} + \frac{2}{m_0^2} \sum_{n \neq c} \frac{|\pi_{cn}|^2}{E_c - E_n}, \quad (5.43)$$

where π_{cn} is a generalized momentum matrix element between states c and n. The degenerate valence band structure can be described by a Luttinger model (Luttinger and Kohn 1955; Luttinger 1956) considering only the fourfold degenerate Γ_8 bands at $k = 0$ or together with the Γ_7 spin–orbit split off bands, with or without the incorporation of an external magnetic field. For small band gap semiconductors, the coupling between the conduction and valence bands must be included. The original work on this problem was done by Kane for InSb, later named the Kane model (Kane 1957). Pidgeon and Brown (1966) extended the Kane model to include an

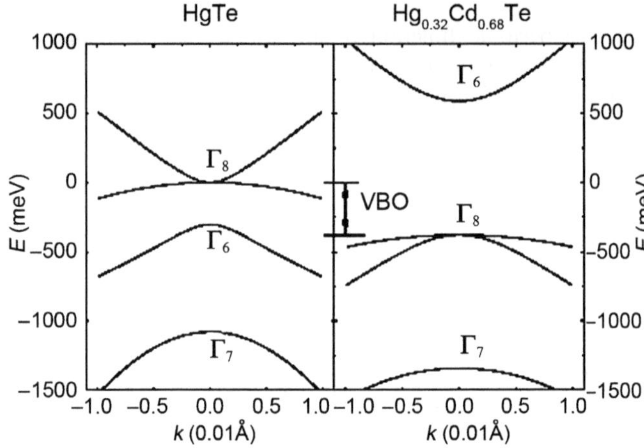

Fig. 5.8 The band structures of HgTe and $Hg_{1-x}Cd_xTe$ calculated in the $\mathbf{k}\cdot\mathbf{p}$ model (Pfeuffer 1998)

external magnetic field. Groves et al. (1967) showed that this model is also suitable to describe the band structure of the semimetal HgTe.

Figure 5.8 shows the band structures of $Hg_{0.32}Cd_{0.69}Te$ and HgTe calculated following the $\mathbf{k}\cdot\mathbf{p}$ method (Pfeuffer et al. 1998). They are the barrier and well materials of our QW structures. This $Hg_{0.32}Cd_{0.69}Te$ alloy has a normal band structure with an open gap. The lowest conduction band is Γ_6^c with spin, $s = 1/2$. Due to the spin–orbit interaction, the uppermost valence band is split into two bands, Γ_8^v and Γ_7^v, with $J = 3/2$ and $J = 1/2$, respectively. The Γ_8^v band is further split into two branches. The $J_z = \pm 3/2$ branches with a larger effective mass are called the heavy-hole (HH) bands. The other, $J_z = \pm 1/2$ branches with a smaller effective mass are called the light-hole (LH) bands. The HH and LH bands are degenerate at $k = 0$.

HgTe has a drastically different band alignment. The s-like bands move downward in energy, and now the Γ_6 band lies between the Γ_8 and Γ_7 bands. Because of the $\mathbf{k}\cdot\mathbf{p}$ interaction between the Γ_6 and Γ_8 bands, the light Γ_8 band and Γ_6 band repel each other at $k = 0$, and they bend upward or downward, respectively. The two Γ_8 bands are degenerate at $k = 0$, the light-hole band acts as the conduction band edge and the heavy-hole band acts as the valence band edge, and the thermal bandgap is zero. Thus, at a finite temperature HgTe is practically a semimetal. The special band alignment of HgTe is called an inverted band structure, because the Γ_6 band lies 0.3 eV below the Γ_8 band.

5.2.2 *Envelope Function Theory for Heterostructures*

Heterostructures can be classified into three types according to their different band edge alignments, as shown in Fig. 5.9. In type I systems, the electrons and

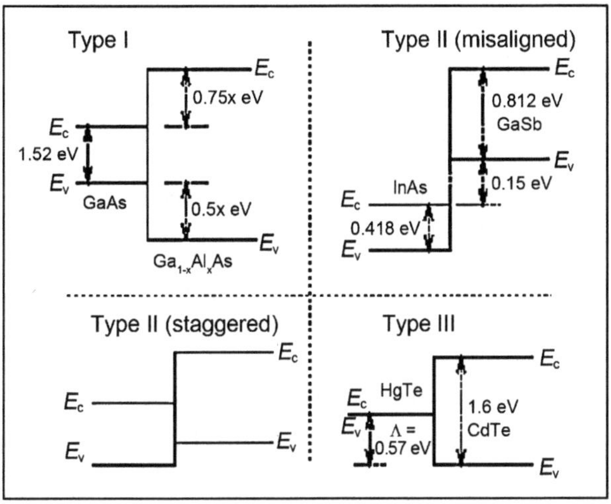

Fig. 5.9 Band edge alignment of three types of heterostructures

holes are confined in the same layer, e.g., a typical example is the GaAs/GaAlAs heterojunction. But in type II hetero-structures, the electrons and holes are separated and located in different materials. Type II structures can be further divided into two subgroups, namely "staggered" and "misaligned." The HgTe-based heterostructures belong to the type III group that is formed by a semimetal and a semiconductor. The band offset of the heterostructure, shown in Fig. 5.9, is an important parameter that helps to determine its electrical properties. The valence band offset (VBO) of the type III Hg/Cd heterostructure has been determined to be 570 ± 60 meV at 5 K (Truchseβ et al. 1995; Becker et al. 2000).

Next, a brief introduction to the envelope function method is presented. Although other approaches, such as the empirical tight-binding and pseudo-potential methods, have already been done for subband structure calculations, only the envelope function method results in a wave function and a carrier effective mass that have a clear physical meaning and are suitable for self-consistent calculations in cases for which a substantial charge transfer across the interface takes place. Finally, it allows the highly desirable incorporation of an external magnetic field.

Two key assumptions are made in the envelope function model.

1. Inside each layer the wave function is expanded in terms of the periodic part of the band edge Bloch functions:

$$\psi(\boldsymbol{r}) = \sum_n f_n^{\mathrm{A}}(\boldsymbol{r}) u_{n,k_0}^{\mathrm{A}}(\boldsymbol{r}), \tag{5.44}$$

where $\boldsymbol{r}$ lies in layer A and $f_n^{\mathrm{A}}(\boldsymbol{r})$ are envelope functions. In the above relation, it has been tacitly assumed that the heterostructure states are constructed using the host wavevectors k_{A}, which relative to k_0, are small. Or equivalently, the

$\mathbf{k} \cdot \mathbf{p}$ method only applies to states near the band edge. Periodicity parallel to the interface always allows us to write:

$$f_n^{\mathrm{A}}(\boldsymbol{r}) \rightarrow \exp(\mathrm{i}\boldsymbol{k}_{/\!/} \cdot \boldsymbol{r}_{/\!/}) f_n^{\mathrm{A}}(z). \tag{5.45}$$

2. The periodic part of the Bloch functions are assumed to be the same in both layer types that constitute the heterostructure:

$$u_{n,k_0}^{\mathrm{A}}(\boldsymbol{r}) \equiv u_{n,k_0}^{\mathrm{B}}(\boldsymbol{r}). \tag{5.46}$$

This implies that the interband matrix element $\langle S \,|p_x|\, X\rangle$ is the same in the A and B layers, and there is a lattice constant match at the interface.

The Bloch functions are seldom known explicitly. However, at the high-symmetry points of the first BZ (notably the Γ point), the way in which the Bloch functions change under the symmetry operations belonging to the crystal point group can be analyzed using group theory arguments. The four Bloch functions at the Γ point are labeled S, X, Y, Z, whose associated wave functions transform in the same way as the atomic s, x, y, z functions under the symmetry operations that map the local tetrahedron onto itself. The Γ_6^c band edge wave function with a spin degeneracy of 2 takes the following form:

$$\Gamma_6^c, \left[s = \frac{1}{2}\right] \begin{cases} u_{c0}^{1/2} = S\uparrow \\ u_{c0}^{-1/2} = S\downarrow \end{cases}, \tag{5.47}$$

where S is a function invariant under all symmetry operations belonging to the crystal point group. The valence band edge wave functions are:

$$\Gamma_8^{\mathrm{v}}, \left[J = \frac{3}{2}\right] \begin{cases} u_{\mathrm{v}0}^{3/2} &= \frac{1}{\sqrt{2}}(X + \mathrm{i}Y)\uparrow \\ u_{\mathrm{v}0}^{1/2} &= -\frac{1}{\sqrt{6}}[(X + \mathrm{i}Y)\downarrow - 2Z\uparrow] \\ u_{\mathrm{v}0}^{-1/2} &= -\frac{1}{\sqrt{6}}[(X - \mathrm{i}Y)\uparrow + 2Z\downarrow] \\ u_{\mathrm{v}0}^{-3/2} &= \frac{1}{\sqrt{2}}(X - \mathrm{i}Y)\downarrow \end{cases}, \tag{5.48}$$

and

$$\Gamma_7^{\mathrm{v}}, \left[J = \frac{1}{2}\right] \begin{cases} u_{\mathrm{v}0}^{1/2} = \frac{1}{\sqrt{3}}[(X + \mathrm{i}Y)\downarrow + Z\uparrow] \\ u_{\mathrm{v}0}^{-1/2} = \frac{1}{\sqrt{3}}[(X - \mathrm{i}Y)\uparrow - Z\downarrow] \end{cases}. \tag{5.49}$$

The next question is how to construct a suitable Hamiltonian for the investigated system; the topic of Sect. 5.2.2.1.

The principle driving the Hamiltonian construction is based on quantum mechanical perturbation theory. The basic idea is very simple: first a complete orthogonal basis set is chosen. The choice of the basis depends on the system being investigated. For the narrow gap system studied here, the basis functions given in (5.47)–(5.49) are suitable, because both the coupling between the conduction and valence band, and the spin–orbit coupling of the valence bands, are included. Then, the Hamiltonian can be constructed according to:

$$\sum_{s'\in S}\left(\sum_{\alpha,\beta}^{x,y,z} k_\partial D_{ss'}^{\alpha\beta} k_\beta + \sum_{\alpha}^{x,y,z} P_{ss'}^{\alpha} k_\alpha\right) f_{s'}(\boldsymbol{r}) + E_s(\boldsymbol{r}) f_s(\boldsymbol{r}) = E f_s(\boldsymbol{r}),\ \forall s \in S, \tag{5.50}$$

where $\boldsymbol{k} = -\mathrm{i}\nabla$, the indices s and s' sum over the dimensionality S of the chosen basis set, $E_s(\boldsymbol{r})$ are the respective band edge potentials. $P_{ss'}^{\alpha}$ describes the coupling between the two bands in set S in a first-order perturbation approximation.

$D_{ss'}^{\alpha\beta}$ describes the interaction between a band in set S and a remote band not in S in second-order approximations.

According to perturbation theory, they are expressed by:

$$P_{ss'}^{\alpha} = \frac{\hbar}{m_0}\left\langle s\left|p_\alpha\right|s'\right\rangle \tag{5.51}$$

and

$$D_{ss'}^{\alpha\beta} = \frac{\hbar^2}{2m_0}\left(\delta_{ss'}\delta_{\alpha\beta} + \frac{2}{m_0}\sum_{r\notin S}\frac{\left\langle s\left|p_\alpha\right|r\right\rangle\left\langle r\left|p_\beta\right|s'\right\rangle}{E-E_r}\right). \tag{5.52}$$

5.2.2.1 The Luttinger Model

The valence band structure of bulk semiconductors was solved by Luttinger in the 1950s (Luttinger and Kohn 1955; Luttinger 1956). The basis functions in (5.48) and (5.49) are a representation in which the spin–orbit coupling has been diagonalized. The representation is $\left|j, m_j\right\rangle$, equivalent to that in atomic physics, where j and m_j are the quantum numbers for the total angular momentum $J = L + S$ and its z component. After a tedious calculation the Hamiltonian can be written as:

$$H_{\mathrm{L}} = \begin{pmatrix} U+V & -S_- & R & 0 & \frac{1}{\sqrt{2}}S_- & -\sqrt{2}R \\ -S_+ & U-V & 0 & R & \sqrt{2}V & -\sqrt{\frac{3}{2}}S_- \\ R^+ & 0 & U-V & S_- & -\sqrt{\frac{3}{2}}S_+ & -\sqrt{2}V \\ 0 & R^+ & S_+ & U+V & \sqrt{2}R^+ & \frac{1}{\sqrt{2}}S_+ \\ \frac{1}{\sqrt{2}}S_+ & \sqrt{2}V & -\sqrt{\frac{3}{2}}S_- & \sqrt{2}R & U-\Delta & 0 \\ -\sqrt{2}R^+ & -\sqrt{\frac{3}{2}}S_+ & -\sqrt{2}V & \frac{1}{\sqrt{2}}S_- & 0 & U-\Delta \end{pmatrix}, \tag{5.53}$$

where the matrix elements in (5.53) are:

$$
\begin{aligned}
U &= E_{\mathrm{v}}(x,y,z) - \frac{\hbar^2}{2m_0}\sum_{\alpha}^{x,y,z}(k_\alpha \gamma_1^{\mathrm{L}} k_\alpha),\\
V &= \frac{\hbar^2}{2m_0}\sum_{\alpha}^{x,y,z}(k_\alpha \gamma_2^{\mathrm{L}} k_\alpha),\\
R &= -\frac{\hbar^2}{2m_0}\sqrt{3}(k_+\mu k_+ - k_-\overline{\gamma} k_-),\\
S_\pm &= -\frac{\hbar^2}{2m_0}\sqrt{3}(k_\pm \gamma_3^{\mathrm{L}} k_z + k_z \gamma_3^{\mathrm{L}} k_\pm),\\
k_\pm &= k_x \pm \mathrm{i}k_y,\\
\boldsymbol{k} &= -\mathrm{i}\nabla,
\end{aligned}
\tag{5.54}
$$

and γ_1^L, γ_2^L, γ_3^L, μ, and $\overline{\gamma}$ are parameters describing the coupling between the p-like valence band states and the remote bands with Γ_1, Γ_2, Γ_3, and Γ_4 symmetries. For a 2DEG, k_x, k_y are still good quantum numbers, but k_z must be replaced by its operator form, $k_z = -\mathrm{i}\partial/\partial z$.

5.2.2.2 The Kane Model

The Luttinger model discussed above is suitable to describe the valence band structure of a semiconductor with a relatively large band gap. But for a narrow gap semiconductor or semimetal, the coupling between the conduction and valence band is fundamentally important in order to account for the nonparabolicity of both bands. It is essential to include the Γ_6 states into the basis set at the very beginning. The $\mathbf{k}\cdot\mathbf{p}$ coupling term in (5.52) leads to a Hamiltonian term linear in k coupling the Γ_6 to the Γ_8, and Γ_7 states. Kane first treated this problem for InSb within the $\mathbf{k}\cdot\mathbf{p}$ method (Kane 1957). The following basis set was used:

$$
\begin{aligned}
|1\rangle &= |\Gamma_6, +1/2\rangle = S\uparrow\\
|2\rangle &= |\Gamma_6, -1/2\rangle = S\downarrow\\
|3\rangle &= |\Gamma_8, +3/2\rangle\\
|4\rangle &= |\Gamma_8, +1/2\rangle\\
|5\rangle &= |\Gamma_8, -1/2\rangle\\
|6\rangle &= |\Gamma_8, -3/2\rangle\\
|7\rangle &= |\Gamma_7, +1/2\rangle\\
|8\rangle &= |\Gamma_7, -1/2\rangle
\end{aligned}
\tag{5.55}
$$

The Hamiltonian of a 2DEG in a (001) orientation is:

$$H = \begin{pmatrix} T & 0 & -\frac{Pk_+}{\sqrt{2}} & \frac{Pk_z}{\sqrt{3/2}} & \frac{Pk_-}{\sqrt{6}} & 0 & -\frac{Pk_z}{\sqrt{3}} & -\frac{Pk_-}{\sqrt{3}} \\ 0 & T & 0 & -\frac{Pk_+}{\sqrt{6}} & \frac{Pk_z}{\sqrt{3/2}} & \frac{Pk_-}{\sqrt{2}} & -\frac{Pk_+}{\sqrt{3}} & \frac{Pk_z}{\sqrt{3}} \\ -\frac{Pk_-}{\sqrt{2}} & 0 & \ddots & & & & & \\ \frac{Pk_z}{\sqrt{3/2}} & -\frac{Pk_-}{\sqrt{6}} & & \ddots & & & & \\ \frac{Pk_+}{\sqrt{6}} & \frac{Pk_z}{\sqrt{3/2}} & & & \ddots & & & \\ 0 & \frac{Pk_+}{\sqrt{2}} & & & & H_{\mathrm{L}}' & & \\ -\frac{Pk_z}{\sqrt{3}} & -\frac{Pk_-}{\sqrt{3}} & & & & & \ddots & \\ -\frac{Pk_+}{\sqrt{3}} & \frac{Pk_z}{\sqrt{3}} & & & & & & \ddots \end{pmatrix}. \tag{5.56}$$

$P = -\dfrac{\hbar}{m_0}\langle S\,|p_x|\,X\rangle$ is the Kane matrix element. T is given by:

$$T = E_c(z) + \frac{\hbar^2}{2m_0}[(2A+1)k_{/\!/}^2 + k_z(2A+1)k_z], \tag{5.57}$$

where A is

$$A = \frac{1}{m_0}\sum_j^{\Gamma_5} \frac{\left|\langle S\,|p_x|\,u_j\rangle\right|^2}{E_c - E_j}. \tag{5.58}$$

Here, the summation is over the Γ_5 states, apart from those of valence bands. H_{L}' is the corrected Luttinger Hamiltonian.

The wave functions of (5.56) can be written in the following form:

$$\psi_{k_x,k_y}(z) = e^{\mathrm{i}}(k_x x + k_y y) \begin{pmatrix} f_{k_x,k_y}^{(1)}(z) \\ f_{k_x,k_y}^{(2)}(z) \\ f_{k_x,k_y}^{(3)}(z) \\ f_{k_x,k_y}^{(4)}(z) \\ f_{k_x,k_y}^{(5)}(z) \\ f_{k_x,k_y}^{(6)}(z) \\ f_{k_x,k_y}^{(7)}(z) \\ f_{k_x,k_y}^{(8)}(z) \end{pmatrix} \begin{matrix} \Gamma_6, +1/2 \\ \Gamma_6, -1/2 \\ \Gamma_8, +3/2 \\ \Gamma_8, +1/2 \\ \Gamma_8, -1/2 \\ \Gamma_8, -3/2 \\ \Gamma_7, +1/2 \\ \Gamma_7, -1/2 \end{matrix} \tag{5.59}$$

The eight envelope functions f_i can be obtained by the solving the eigen-value problem consisting of eight coupled, second-order differential equations.

The carrier density distribution is given by:

$$\left|\psi_{k_x,k_y}(z)\right|^2 = \sum_{i=1}^{8}\left|f_{k_x,k_y}^{(i)}\right|^2. \tag{5.60}$$

5.2.2.3 The Kane Model in an External Magnetic Field

For a magnetic field applied normal to the 2DEG, the motion parallel to the heterostructure interface is quantized into Landau levels. The gauge invariance requires that k be replaced by:

$$\boldsymbol{k} = -\mathrm{i}\nabla + \frac{e}{\hbar}\boldsymbol{A}. \tag{5.61}$$

$\boldsymbol{A}$ is the vector potential and $\boldsymbol{B} = \nabla \times \boldsymbol{A}$, where $\boldsymbol{k}$ satisfies the following relation:

$$\boldsymbol{k} \times \boldsymbol{k} = \frac{e}{\mathrm{i}\hbar}\boldsymbol{B}. \tag{5.62}$$

We introduce the following operators (annihilator and creator, respectively):

$$\begin{aligned} a &= \frac{l}{\sqrt{2}}k_- = \frac{l}{\sqrt{2}}(k_x - \mathrm{i}k_y) \\ a^+ &= \frac{l}{\sqrt{2}}k_+ = \frac{l}{\sqrt{2}}(k_x + \mathrm{i}k_y), \end{aligned} \tag{5.63}$$

with

$$\left[a, a^+\right] = 1. \tag{5.64}$$

For harmonic oscillator eigenstates, the following relations hold:

$$\begin{aligned} a^+ a\,|n\rangle &= n\,|n\rangle\,, \\ a\,|n\rangle &= \sqrt{n}\,|n-1\rangle\,, \\ a^+\,|n\rangle &= \sqrt{n+1}\,|n+1\rangle\,. \end{aligned} \tag{5.65}$$

Now, the Hamiltonian in (5.56) is a function of a^+, a, $k_z = -\mathrm{i}\partial/\partial z$ and z-dependent band structure parameters.

Additionally, new terms should be added to the Hamiltonian arising from the electron or hole spin coupling to the magnetic field, i.e., the Zeeman spin splitting terms. Altarelli (1985) has considered this problem within a 6×6 band model (Γ_6 and Γ_8 bands). For the conduction band this means that $g^*\mu_\mathrm{B}B$ should be added to the diagonal terms, here

$$\frac{1}{m^*} = \frac{1}{m_0}\left[1 + \frac{2P^2}{3(E_g + \Delta)}\right], g^* = 2\frac{m_0}{m^*}. \tag{5.66}$$

The above equation embodies the spin–orbit split Γ_7 band contribution to the conduction band effective g-factor. For the valence band, this means adding the term (Luttinger and Kohn 1955; Luttinger 1956):

$$(e/c)\kappa J_z B + (e/c) q J_z^3 B, \tag{5.67}$$

where J_z is the spin 3/2 matrix and κ and q are material parameters. Actually q is usually very small for the semiconductors of interest and the second term can be neglected.

The Hamiltonian including the Zeeman term for the eight-band Kane model has been treated by Weiler et al. (1978). A solution of the Hamiltonian in an external magnetic field has the following form:

$$\psi_N(z) = \begin{pmatrix} f_1(z)\phi_{n1} \\ f_2(z)\phi_{n2} \\ f_3(z)\phi_{n3} \\ f_4(z)\phi_{n4} \\ f_5(z)\phi_{n5} \\ f_6(z)\phi_{n6} \\ f_7(z)\phi_{n7} \\ f_8(z)\phi_{n8} \end{pmatrix} = \begin{pmatrix} f_1^N(z)\phi_N \\ f_2^N(z)\phi_{N+1} \\ f_3^N(z)\phi_{N-1} \\ f_4^N(z)\phi_N \\ f_5^N(z)\phi_{N+1} \\ f_6^N(z)\phi_{N+2} \\ f_7^N(z)\phi_N \\ f_8^N(z)\phi_{N+1} \end{pmatrix}, \tag{5.68}$$

in the axial approximation. For every $N = -2, -1, 0, 1, 2, \ldots (\varphi \equiv 0$, if $N < 0)$, eight coupled, second-order real differential equations for f_i are obtained.

5.2.3 *Specific Features of Type III Heterostructures*

With the theoretical model developed earlier, the subband dispersion of HgTe QWs can be calculated. Its specific features compared to a type I system are summarized below.

5.2.3.1 Three Different Band Structure Regimes

One specific feature of HgTe QWs is that it has three different band structure regimes corresponding to different well widths. Figure 5.10 shows the subband dispersion. Panel (b) shows the subband edge energies as a function of the well width. It can be understood in terms of two effects. One is the quantum confinement effect and its variation with the well width. The other is the inverted band structure of bulk HgTe. With increasing well width, the quantum confinement effect decreases, thus the electron-like (hole-like) subband will decrease (increase) its energy. The heavy-hole states have a weaker confinement effect due to their larger effective masses,

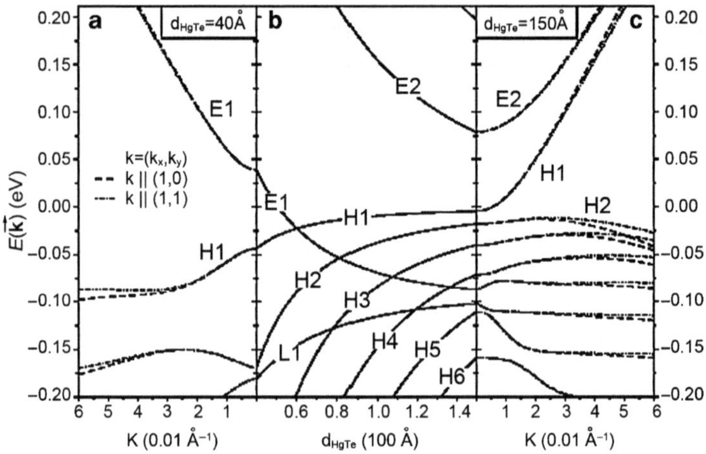

Fig. 5.10 Three different band structure regimes (**a**) The subband structure of a $d_w = 4$ nm wide HgTe/$Hg_{0.32}Cd_{0.68}$Te quantum well. (**b**) The variation of the subband edges with d^w. (**c**) The subband structure of a $d_w = 15$ nm wide HgTe/$Hg_{0.32}Cd_{0.68}$Te quantum well. In panel (**a**) and (**c**) the subband dispersions for two different directions in k-space, $k//(1,0)$ and $k//(1,1)$ respectively, are shown

consequently the first valence subband is $H1$. When, $d_w < 6$ nm, the QW is in the normal band regime, where the $E1$ subband lies above the $H1$ valence subband, see panel (a). If, $d_w = 6$ nm, the $E1$ and $H1$ subbands coincide, and a semimetal is realized. If, $d_w = 6$ nm, the $E1$ and $H1$ subbands change roles, the QW is in the so-called inverted band regime, now the $H1$ subband becomes the first conduction subband, and $E1$ becomes one of the valence subbands, see panel (c). The inverted band alignment is a unique feature of type III heterostructures, which is a direct consequence of the inverted band structure of bulk HgTe.

5.2.3.2 Interface States

HgTe has an unusual inverted band structure, in contrast to that of HgCdTe, as shown in Fig. 5.8. The effective masses of the light particle bands, i.e., the Γ_6 and Γ_8 light-hole bands in both materials, have opposite signs. When these two materials are combined to form a QW, the maximum of the carrier density distribution of the light particles will be located near the interface. The interface states are illustrated in Fig. 5.11.

For the purpose of comparison, the situation of a type I GaAs/AlGaAs QW is sketched in Fig. 5.12.

The formation of the interface state in type III systems can be understood in a simple one-band model. The continuity of the charge density and the divergence-free character of the probability current lead to the following boundary conditions at the interface:

$$\psi_1(z)\,\big|_{z0} = \psi_2(z)\,\big|_{z0}$$

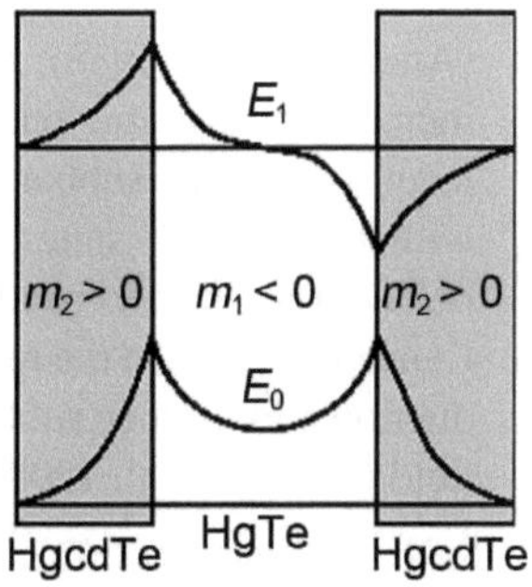

Fig. 5.11 Carrier density distribution of the electron (or light-hole) state for a HgTe/HgCdTe QW

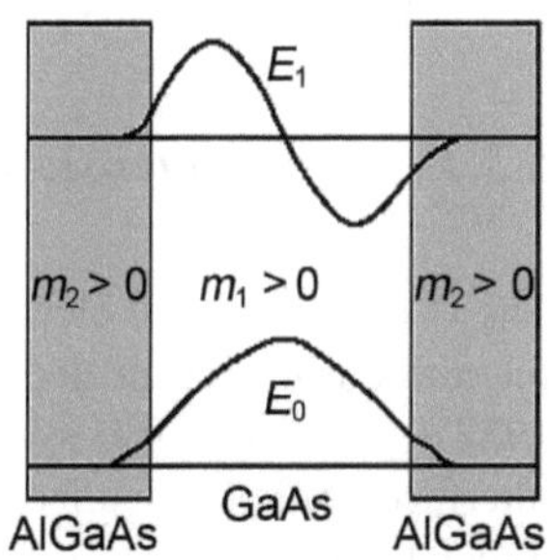

Fig. 5.12 Carrier density distribution of the electron (or light-hole) state for a GaAs/AlGaAs QW

and

$$\frac{1}{m_1}\frac{\mathrm{d}}{\mathrm{d}z}\psi_1(z)\,\Big|_{z0} = \frac{1}{m_2}\frac{\mathrm{d}}{\mathrm{d}z}\psi_2(z)\,\Big|_{z0}. \tag{5.69}$$

If $m_1 \cdot m_2 < 0$, as in HgTe QWs, the slope of the wave function in the well and barrier will have opposite signs, which lead to the formation of the interface states.

5.2.3.3 Consequences of an Inverted Band Structure

The main feature of the inverted band structure is the heavy-hole nature of the first conduction subband. As a result, it may have the following consequences.

1. Electron density probability distribution
   The maximum of the electron density distribution is not located near the minimum of the confinement potential as is true in type I heterojunctions, but is shifted significantly to the opposite side of the QW.
2. Crossing of conduction and valence subband Landau levels
   For HgTe QW's in the inverted band regime, the first conduction and valence subbands are $H1$ and $H2$, respectively. They are a mixture of heavy-hole and light-particle states. A quantizing magnetic field applied along the growth direction of the QW structure uncovers the mixed nature of the heavy-hole subbands in the inverted band regime. The lowest Landau level $n = 0$ of the $H1$ subband contains pure heavy-hole states which do not mix with the light particle states

(Ancilotto et al. 1988). Accordingly, this level lowers its energy linearly with an increasing magnetic field and shows a hole-like behavior, while all other Landau levels of the $H1$ subband rise in energy with magnetic field due to the coupling with light particle states. Thus, they show an electron-like behavior. The unusual behavior in inverted band QWs together with the peculiar dispersion of the $n = 2$ Landau level from the topmost valence subband, $H2$, leads to a crossing of conduction and valence subband Landau levels at a critical value B_c of the magnetic field (Schultz et al. 1998).

5.3 Magnetotransport Theory of Two-Dimensional Systems

5.3.1 Two-Dimensional Electron Gas

The two-dimensional electron gas (2DEG) has been the subject of broad research interest during recent decades (for a review of the early work see (Ando et al. 1982)). The first 2DEG system ever studied was the in-version layer at the Si–SiO_2 interface, which is realized in the Si-MOSFET (metal oxide semiconductor field effect transistor). In this system the quantum Hall effect (QHE) was originally discovered (Klitzing et al. 1980). It soon became known that the large effective mass of electrons ($m^* = 0.19m_0$) and the inherent strong interface scattering give an upper limit of the mobility, of about 30,000 cm^2/Vs in an Si-MOSFET. In the 1970s, with the development of MBE growth techniques, a high-mobility 2DEG was realized at the interface of $GaAs/Al_{1-x}Ga_xAs$ heterojunctions (Esaki and Tsu 1970; Chang et al. 1974; Cho and Arthur 1975). The lattice mismatch between GaAs and AlAs is only 0.12%. This is a very suitable arrangement to achieve an atomically abrupt interface, which is essential for a high-mobility 2DEG. The introduction of the modulation doping technique has also turned out to be a very powerful tool to achieve a high mobility. It was first applied by Dingle et al. (1978) to GaAs/AlGaAs heterojunctions. Its principle is that the electrons in the well are spatially separated from their parent donors in the barrier, as shown schematically in Fig. 5.13. The spatial separation can be further enhanced by inserting a spacer, which is a nominally undoped part of the barrier, between the donors and the well. Hence, the scattering rate from the ionized donors will be drastically reduced, resulting in a large enhancement of the carrier mobility. The highest reported mobility is $2 \times 10^7\,cm^2/Vs$ for this type of system (Störmer 1999). The utilization of the high mobility led to the discovery of the fractional quantum Hall effect (FQHE) by Tsui et al. (1982). The HgTe/HgCdTe heterostructure, as a representative of type III systems, on the other hand, is in its preliminary stage and still needs much research.

Below is a discussion of subband energies and carrier distributions for 2DEG in heterostructures.

As shown in Fig. 5.13, after the charge transfer process induced by the modulation doping, the band bending at the interface forms an almost triangular well that confines the electrons. Consequently, the motion of the electrons in the z direction

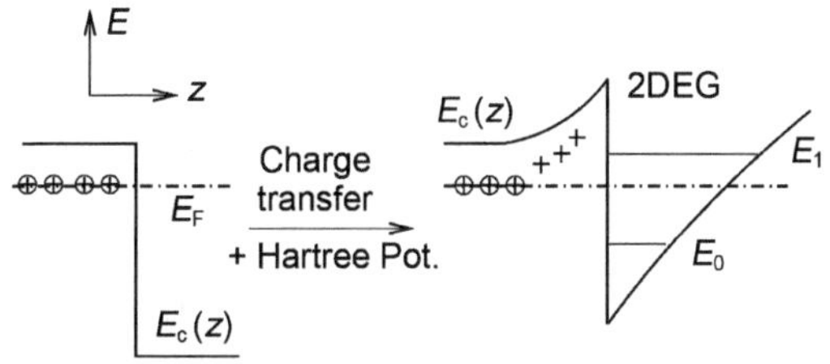

Fig. 5.13 Schematic of the conduction band edge profile of a modulation doped heterostructure before (*left side*), and after (*right side*) the charge transfer. The lowest two subbands with energy levels E_0 and E_1 and the Fermi energy E_F are indicated. *Circles* with crosses: neutral donors; *Crosses*: ionized donors

will be quantized into discrete energy levels. The charge distribution determines the confining potential, which in turn determines the charge distribution. This imposes a stringent self-consistency requirement on the solution of the Poisson and the Schrödinger equations. The electron–electron interaction, a many body problem, has been treated within the simplest scheme: the Hartree approximation. It amounts to replacing the exact many-electron potential by an average one. Each electron is assumed to move in a self-consistent potential $V_H(z)$ and the coupled Poisson and Schrödinger equations must be solved self-consistently:

$$\frac{d}{dz}\left[\varepsilon(z)\frac{dV_H}{dz}\right] = -4\pi e^2\rho(z), \tag{5.70}$$

where $\rho(z) = -en(z) + e\left[N_D^+(z) - N_A^-(z)\right]$ is the density of free charges. In this case, the Schrödinger equation is:

$$\left[-\frac{\hbar^2}{2}\frac{d}{dz}\frac{1}{m(z)}\frac{d}{dz} + V(z)\right]\varphi_i(z) = E_i\varphi_i(z), \tag{5.71}$$

where $V(z) = V_0\theta(z) + V_H(z)$ is the total potential, $\theta(z)$ a step function, and E_i is an eigen-state energy. The net electron density $n(z)$ is:

$$n(z) = \sum_i n_i\left|\varphi_i(z)\right|^2, \tag{5.72}$$

where n_i are the electron densities in the separate subbands. Self-consistency can be achieved by solving the equations in the following sequence: (5.72)→(5.70) →(5.71)→(5.72). The wave functions of the electrons are:

$$\varphi_{i,k_x,k_y} = \frac{1}{\sqrt{L_xL_y}}e^{i(k_xx+k_yy)}\varphi_i(z), \tag{5.73}$$

where the electrons can move freely in the x–y plane. Here, L_x, L_y gives the extent of the 2DEG in the x and y directions, respectively. The corresponding energy values are given by:

$$E_{i,k_x,k_y} = E_i + \frac{\hbar^2(k_x^2 + k_y^2)}{2m^*}, \tag{5.74}$$

where m^* is the effective mass of the electrons. The DOS of each subband takes on the constant value, $m^*/\pi\hbar^2$. The total DOS is the sum over all subbands and is given by:

$$D(E) = \sum_i \theta(E - E_i)\frac{m^*}{\pi\hbar^2}. \tag{5.75}$$

The carrier densities in different subbands at $T = 0\,\mathrm{K}$ are:

$$n_i = (E_\mathrm{F} - E_i)\theta(E_\mathrm{F} - E_i)\frac{m^*}{\pi\hbar^2}. \tag{5.76}$$

At finite temperatures, this becomes:

$$n_i = \frac{m^*}{\pi\hbar^2}k_\mathrm{B}T\ln\left[1 + e^{-\frac{E_i - E_\mathrm{F}}{k_\mathrm{B}T}}\right]. \tag{5.78}$$

5.3.2 *Classical Transport Theory: The Drude Model*

In 1900 Drude used an ideal molecular gas model to describe the free electrons in metals (Drude 1900a, b). His model contains the following additional assumptions:

1. The electrons move freely between collisions (drift transport). The interactions of an electron with other electrons and neutral atomic cores are neglected.
2. The collisions are instantaneous and change the electron velocity abruptly. These collisions are mainly with the ion cores.
3. The probability of a collision in the time interval $\mathrm{d}t$ is $\mathrm{d}t/\tau$, where τ is the average time between two collisions, also called momentum relaxation time (see the following item).
4. The electrons reach equilibrium through these collisions. After each collision, the electron completely loses the memory of its previous velocity and is randomly oriented.

The equation of motion for the drift velocity $\boldsymbol{v}$ in the presence of a crossed magnetic field $\boldsymbol{B}$ and an electric field $\boldsymbol{E}$ is:

$$m^*\frac{\mathrm{d}\boldsymbol{v}}{\mathrm{d}t} = -e(\boldsymbol{E} + \boldsymbol{v} \times \boldsymbol{B}) - \frac{m^*}{\tau}\boldsymbol{v}. \tag{5.79}$$

The last term on the right side of (5.79) represents the frictional force. Under stationary conditions, i.e., $\dot{\boldsymbol{v}} = 0$, when $\boldsymbol{j} = -ne\boldsymbol{v}$ and $\boldsymbol{B} = 0$, we obtain:

$$j = \sigma_0 E, \quad \sigma_0 = \frac{e^2 n\tau}{m^*}, \quad E = \rho_0 j, \quad \text{and} \quad \rho_0 = \frac{1}{\sigma_0}. \tag{5.80}$$

In an external electric field $\boldsymbol{E}$, the electrons acquire a final constant average velocity $\boldsymbol{v} = \mu \boldsymbol{E}$, here μ is the mobility, a measure of how easily electrons can move in crystals. From the relations given above, one can easily show that:

$$\mu = \frac{\boldsymbol{v}}{\boldsymbol{E}} = \frac{e\tau}{m^*}. \tag{5.81}$$

But for $\boldsymbol{B} \neq 0$, $\overleftrightarrow{\rho}$ becomes a resistivity tensor with $\boldsymbol{E} = \overleftrightarrow{\rho}\boldsymbol{j}$. Using (5.79), we obtain:

$$\overleftrightarrow{\rho} = \rho_0 \begin{pmatrix} 1 & \omega_c\tau \\ -\omega_c\tau & 1 \end{pmatrix}. \tag{5.82}$$

The conductivity tensor $\overleftrightarrow{\sigma}$ is defined as:

$$\overleftrightarrow{\sigma} \equiv \overleftrightarrow{\rho}^{-1} = \frac{\sigma_0}{1 + (\omega_c\tau)^2} \begin{pmatrix} 1 & -\omega_c\tau \\ \omega_c\tau & 1 \end{pmatrix}. \tag{5.83}$$

From (5.83), the Hall angle $\boldsymbol{\varphi}$, which is defined as the angle between $\boldsymbol{E}_x$ and $\boldsymbol{j}$, is $\phi = \arctan(\omega_c\tau)$. From (5.82) and (5.83), we can obtain the following relations:

$$\rho_{xx} = \frac{\sigma_{xx}}{\sigma_{xx}^2 + \sigma_{xy}^2}, \quad \rho_{xy} = \frac{\sigma_{xy}}{\sigma_{xx}^2 + \sigma_{xy}^2}, \tag{5.84}$$

$$\rho_{xx} = \frac{1}{\sigma_0} = \frac{m^*}{e^2 n\tau} = \frac{1}{ne\mu}, \tag{5.85}$$

and

$$\rho_{xy} = \frac{\omega_c\tau}{\sigma_0} = R_H B = \pm\frac{B}{ne}. \tag{5.86}$$

From (5.85), ρ_{xx} is independent of B. This means a 2DEG with one occupied subband, ρ_{xx} should be a constant at low magnetic fields (if τ is energy independent, i.e., for a nondegenerate system or for an anisotropic Fermi surface, or due to a quantum interference effect (weak-localization)) In the nondissipative limit, i.e., $\tau \to \infty$, $\rho_{xx} = (\sigma_{xx})^{-1} = 0$, then the Hall resistivity is independent of τ, and linearly proportional to B. This means in the QHE there is no dissipation in the quantum Hall state, because $\rho_{xx} = 0$. But obviously ρ_{xy} in the Drude model cannot explain the plateaus observed in the QHE.

5.3.3 *Landau Levels in a Perpendicular Magnetic Field*

When a 2DEG is subjected to a perpendicular magnetic field B, and the cyclotron radius of the electron is comparable to or smaller than the Fermi wavelength, the motion of the electrons in the x–y plane will be quantized into Landau levels (Landau 1930). The orbital movement of the electron is described by the Schrödinger equation:

$$\frac{(-\mathrm{i}\hbar\nabla - e\boldsymbol{A})^2}{2m^*}\phi(x,y) = E\phi(x,y), \tag{5.87}$$

where $\boldsymbol{A}$ is the vector potential of the magnetic field, with $\boldsymbol{B} = \nabla \times \boldsymbol{A}$. Choosing the Landau gauge, $\boldsymbol{A}(x,y,Z) = (0, B_x, 0)$, and because $[p_y, H] = 0$, the wave function can be written as:

$$\phi(x,y) = \frac{e^{\mathrm{i}k_y y}}{\sqrt{L_y}}\varphi(x). \tag{5.88}$$

Substituting (5.88) into (5.87), we get an effective one-dimensional Schrödinger equation:

$$H_{x0}\phi_{x0}(x) = \left[-\frac{\hbar^2}{2m^*}\frac{\mathrm{d}^2}{\mathrm{d}^2x} + \frac{1}{2}m^*\omega_{\mathrm{c}}^2(x-x_0)^2 + V(x)\right]\phi_{x0}(x) = E\phi_{x0}(x), \tag{5.89}$$

where ω_{c} is the cyclotron resonance frequency, $\omega_{\mathrm{c}} = eB/m^*$. x_0 is the center coordinate, $x_0 = -\boldsymbol{l}^2\,k$. $\boldsymbol{l}$ is the magnetic length, $l = (\hbar/eB)^{1/2}$. The eigenvalues of the above equation are given by:

$$E_n = \left(n + \frac{1}{2}\right)\hbar\omega_{\mathrm{c}}, \quad n = 0, 1, 2, \ldots \tag{5.90}$$

where n denotes the index of the Landau levels. Each state is degenerate with respect to x_0. $\varphi_{x_0,n}(x)$ is the eigenfunction of a harmonic oscillator centered at x_0 and with level index n. The number of states (per spin) with energy E_n per unit area is:

$$\frac{1}{L_xL_y}\sum_k = \frac{1}{2\pi L_x}\int \mathrm{d}k = \frac{1}{2\pi l^2}\int \frac{\mathrm{d}x_0}{L_x} = \frac{1}{2\pi l^2}. \tag{5.91}$$

In the first equality, the periodic condition along the y direction, i.e., $k = (2\pi/L_y)\xi$, where ξ is an integer, has been applied. Thus, the degeneracy of each Landau level is:

$$n_{\mathrm{L}} = \frac{eB}{h}. \tag{5.92}$$

Given the flux quantum defined as, $\phi_0 \equiv h/e$, the degeneracy of the Landau level is equal to the number of flux quanta per unit area. If n_{s} denotes the total carrier density, the fill factor is defined as:

$$v = \frac{n_{\mathrm{s}}}{n_{\mathrm{L}}} = \frac{h}{eB}n_{\mathrm{s}}. \tag{5.93}$$

The density of Landau level states is a summation over the Delta-functions:

$$D(E) = \frac{\mathrm{d}N}{\mathrm{d}E} = \frac{g_\mathrm{s}}{2\pi l^2} \sum_{n=0}^{\infty} \delta(E - E_n). \tag{5.94}$$

In the above discussion, the electron spin has been neglected. In magnetic fields, due to the Zeeman effect, the twofold spin degeneracy is lifted. This leads to an additional term in the Hamiltonian:

$$H = \frac{1}{2m^*}(\mathbf{p} + e\mathbf{A})^2 + g^* \mu_\mathrm{B} \vec{\sigma} \cdot \mathbf{B} + V(z), \tag{5.95}$$

where $\mu_\mathrm{B} = e\hbar/2m_0$ is the Bohr magneton and g^* is the effective Landau g factor of the carriers ($g^* = 2$ for electrons in a vacuum, $g^* < 0$ in many semiconductors and is a function of band structure parameters). The operator $\vec{\sigma}$ has eigenvalues, $\pm 1/2$. The energy spectrum can be written as:

$$E_{i,n,s} = E_i + \left(n + \frac{1}{2}\right)\hbar\omega_\mathrm{c} + s g^* \mu_\mathrm{B} B. \tag{5.96}$$

The DOS is a summation over the Landau and spin levels:

$$D(E) = \frac{g_\mathrm{s}}{2\pi l^2} \sum_{i,n,s} \delta(E - E_{i,n,s}). \tag{5.97}$$

An additional feature of the 2DEG is as follows. Contrary to the Landau level splitting which is dependent upon only the perpendicular component of B, the Zeeman spin splitting is isotropic and depends on the total field B. If the sample's surface normal is tilted by an angle θ with respect to the magnetic field, then:

$$E_{i,n,s} = E_i + \left(n + \frac{1}{2}\right) B \cos\theta \pm \frac{1}{2} g^* \mu_\mathrm{B} B. \tag{5.98}$$

This forms the basis of the "coincidence method" in which by changing the angle, θ, g^* can be determined.

Next the temperature broadening of the DOS will be discussed. The electron density at a finite temperature T is:

$$n(E) = \frac{g_s}{2\pi l^2} \sum_{n=0}^{\infty} f(E - E_n), \tag{5.99}$$

where $f(E)$ is Fermi–Dirac distribution, $f(E) = (1 + e^{\beta\varepsilon})^{-1}, \beta = 1/k_\mathrm{B}T$ and $\varepsilon = E - E_\mathrm{F}$. The DOS at finite temperature is:

$$D(E) = \frac{\mathrm{d}n}{\mathrm{d}E} = \frac{g_\mathrm{s}}{2\pi l^2} \sum_{n=0}^{\infty} \frac{\beta}{4\cosh^2\left[\beta\left(E - E_\mathrm{F}\right)/2\right]}. \tag{5.100}$$

The temperature broadening of the Landau level DOS is, $4k_BT/\hbar\omega_c$. For narrow gap materials, due to their small electron effective masses, $\hbar\omega_c$ is large, and thus the temperature will have less influence on the broadening of the Landau levels. Consequently, SdH oscillations can be more easily observed in narrow gap materials.

5.3.4 The Broadening of the Landau Levels

In the above sections the effect of imperfections has not been considered. The scattering by imperfections will give rise to a lifetime τ and hence level broadening $\Gamma = \hbar/\tau$ of the electron states. Thus, scattering will alter the singularities of, $D(E)$, rounding off the peaks, adding states in the gaps, etc.

A simple model of a broadened DOS has been derived by Ando and Uemura (1974). It is based on a second-order self-consistent Born approximation (SCBA) induced by the electron-disorder potential interaction. The disorder is assumed to originate from randomly distributed impurities. The scattering potential of a given impurity is taken to have a Gaussian spatial shape. Thus, effects associated with the short or long range nature of the impurity potential can be studied in this model. Finally the magnetic field is assumed to be strong enough to be able to neglect the disorder induced mixing of the Landau levels. The calculations result in replacing the delta functions in (5.97) by semi-ellipses:

$$D(E) = \frac{1}{2\pi l^2}\sum_{i,n,s}\frac{2}{\pi\Gamma_{n,s}}\left[1-\left(\frac{E-E_{i,n.s}}{\Gamma_{n,s}}\right)^2\right]^{1/2}, \tag{5.101}$$

where, if scattering is short ranged, the Landau level broadening $2\Gamma_n$ is expressed as:

$$\Gamma_{n,s} = \sqrt{\frac{2}{\pi}\hbar\omega_c\frac{\hbar}{\tau}} \propto \sqrt{\frac{B}{\mu}}. \tag{5.102}$$

The broadening is independent of the Landau level index n and proportional to $(B/\mu)^{1/2}$. It has been shown that the SCBA results in distinguishable errors, especially for Landau levels with small quantum numbers (Gerhardts 1975b). The obvious drawback of SCBA is the abrupt cut-off at $E = E_{i,n,s} \pm \Gamma_{n,s}$, which is unphysical. Experimental results (Weiss and Klitzing 1987) indicate that the DOS between two adjacent Landau levels is not zero. Even the method of higher order approximations, such as single-site approximation (SSA) (Ando 1974a) or many-site approximation (MSA) (Ando 1974b), also lead to a sharp cut-off overestimate for small Landau quantum numbers, but for large quantum numbers they yield approximately the same result as that given by SCBA.

Gerhardts chose the method of the lowest-order cumulate approximation (or path-integral method) to calculate the DOS (Gerhardts 1975a, b, 1976). The resultant DOS takes on a Gaussian form:

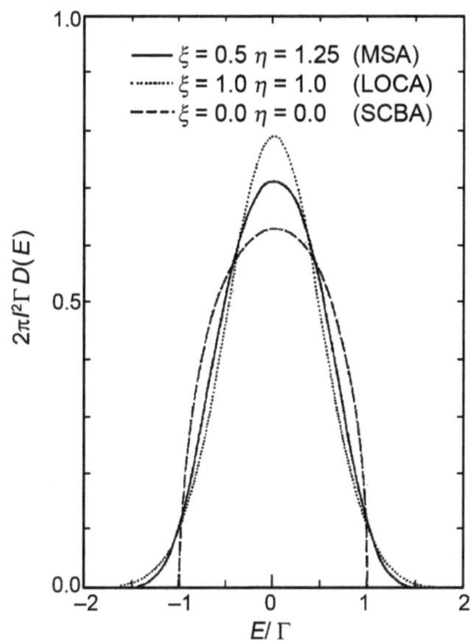

Fig. 5.14 The density-of-states profile of the ground Landau level calculated by the self-consistent Born approximation (SCBA, *dashed*), many-site approximation (MSA, *solid*) and the lowest-order cumulate approximation (LOCA, *dotted*)

$$D(E) = \frac{1}{2\pi l^2} \sum_n \left(\frac{\pi}{2}\Gamma_n^2\right)^{-1/2} \exp\left[-2\frac{(E - E_n)^2}{\Gamma_n^2}\right]^{1/2}. \tag{5.103}$$

The advantage is that it is analytical, however, one does not know how to relate Γ_n to B and n. Weiss and Klitzing (1987) have shown that at least between two adjacent Landau levels the DOS can be described by (5.103), and the broadening agrees well with the prediction of (5.102) from the SCBA theory. Here, the DOS is taken to be given in Gaussian form expressed by (5.103) and the broadening by (5.102). Figure 5.14 summarizes the DOS of the Landau level calculated by different theoretical models.

5.3.5 *Shubnikov-de Haas Oscillations of a 2DEG*

In 1930, oscillations in the magnetoresistance of bismuth single crystals were first observed (Shubnikov and De Haas 1930), they were later named the Shubnikov–de Haas (SdH) oscillations in honor of the discovers. They arise due to quantum effects and reflect the oscillations of the DOS of the Landau levels. The following conditions must be fulfilled in order to observe SdH oscillations:

1. The thermal broadening of the Fermi level, $k_B T$, must be smaller than the energy separation of the Laudau levels, $\hbar\omega_c$. And $E_F > \hbar\omega_c$ should be satisfied, in order that more than one Landau level can be occupied:

$$k_B T \ll \hbar\omega_c < E_F, \tag{5.104}$$

2. The broadening of the Landau levels must be smaller than $\hbar\omega_c$,

$$\Gamma_{n,s} \ll \hbar\omega_c, \tag{5.105}$$

i.e., with $\tau = \hbar/\Gamma_{n,s}$ and $\mu = e\tau/m^*$,

$$\omega_c\tau \gg 1, \ \mu B \gg 1. \tag{5.106}$$

Equation (5.106) means that an electron should complete several cycles of its cyclotron motion before it is scattered by an impurity in time interval τ.

Several authors have calculated the magnetoresistance of two-dimensional systems. Ohta(1971a, b) expressed the magneto-conductivity in terms of Green's functions and derived a formula which shows that the conductivity maximum increases in proportion to the Landau level index. A simpler approach was given by Ando (1974c) based on the SCBA. He arrived at the same increase in the conductivity maximum as Ohta. But Isihara and Smrčka (1986) pointed out that the SCBA is inadequate even for small impurity concentrations, and is more suitable for strong magnetic fields. For low and intermediate magnetic fields, multiple scatterings must be taken into account. Their results can be expressed by:

$$\rho_{xx} = \rho_0 \left(1 + 2\frac{\Delta g\,(T)}{g_0}\right) \tag{5.107}$$

and

$$\rho_{xy} = \frac{\omega_c\tau_0}{\sigma_0}\left(1 - \frac{1}{(\omega_c\tau_0)^2}\frac{\Delta g\,(T)}{g_0}\right), \tag{5.108}$$

where g_0 and Δg are the DOS at zero magnetic field and its oscillatory part. $\Delta g/g_0$ is expressed as:

$$\frac{\Delta g\,(T)}{g_0} = 2\sum_{s=1}^{\infty} \underbrace{\exp(-\pi s/\omega_c\tau)}_{(1)} \underbrace{\frac{\left(\dfrac{2\pi^2 s k_B T}{\hbar\omega_c}\right)}{\sinh\left(\dfrac{2\pi^2 s k_B T}{\hbar\omega_c}\right)}}_{(2)} \times \underbrace{\cos\left[\frac{2\pi s(E_F - E_0)}{\hbar\omega_c} - s\pi\right]}_{(3)} \underbrace{\cos\left(\frac{s\pi g^* m^*}{2}\right)}_{(4)} \tag{5.109}$$

The designated components are:

1. This term describes the magnetic field dependence of the SdH oscillations. It enables the determination of the scattering rate, $1/\tau$, and the Dingle temperature, $T_D = \hbar/2\pi k_B\tau$.

2. This is a temperature-dependent term. From this term, the effective mass of electrons can be deduced from the temperature dependence of the amplitudes of SdH oscillations.
3. This is the oscillation term. The carrier concentrations n_s in separate subbands can be determined from the oscillation frequencies f by means of the relation, $n_s = (e/h)f$.
4. The spin splitting is taken into account in this term. In some cases the effective g factor can be obtained, e.g., if it is temperature- and magnetic-field dependent in dilute semiconductors.

5.3.6 Quantum Hall Effect

The QHE was discovered by Klitzing et al. in 1980 (Klitzing et al. 1980). In the last 2 decades, extensive investigations have been devoted to this subject. And comprehensive reviews are also available (Prange and Girvin 1990; Janssen et al. 1994; Chakraborty and Pietiläinen 1995; Sarma 2001). The original measurement by Klitzing et al. was made on n-type inversion layers of p-type Si. Well-defined plateaus in Hall resistance and a simultaneously vanishing magnetoresistance were found. The surprising result of precision measurements was that the Hall resistance is quantized in integer multiples of h/e^2.

In a p-type Si MOSFET structure, a positive gate voltage makes an n-type inversion layer present in the surface of the structure. The carrier motion normal to the surface is quantized. When a magnetic field is applied normal to the surface, the degenerate 2DEG gas parallel to the surface is quantized into Landau levels. The experiments of Klitzing were performed at 1.5 K and in magnetic fields of 13.9 and 18 T. Figure 5.15 shows the Hall voltage V_y and the voltage drop V_p in the x direction as a function of the gate voltage. Well-defined plateaus in Hall resistance and a simultaneously vanishing magnetoresistance were found. If the Fermi level falls in the middle of two Landau levels, the effective DOS is zero, and the carriers in the inversion layer cannot be scattered, namely, $\tau \to \infty$. So, the longitudinal conductivity is infinite, and the longitudinal resistivity vanishes.

According to (5.84) and (5.86), it follows that:

$$\frac{V_y}{I_x} = \rho_{xy} = \frac{\sigma_{xy}}{\sigma_{xx}^2 + \sigma_{xy}^2} = \frac{1}{\sigma_{xy}} = \rho_{xy} = \pm\frac{B}{ne}, \tag{5.110}$$

where n is the total carrier density,

$$n = (i+1)\,n_L. \tag{5.111}$$

From (5.92), the Hall effect:

$$V_y = \frac{1}{i+1}\frac{h}{e^2}I_x, \tag{5.112}$$

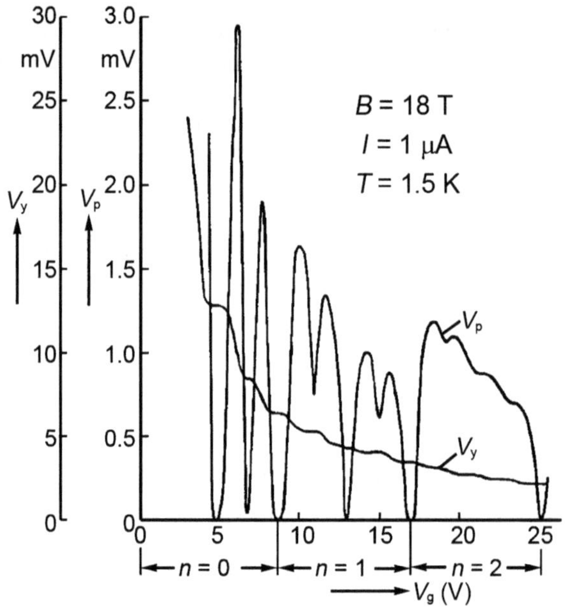

Fig. 5.15 The Hall voltage V_y and the voltage drop V_p in the x direction as a function of the gate voltage

in which I_x is the source–drain current. The Hall voltage is insensitive to the magnetic field, the dimensions of the MOS device, and the material properties. Figure 5.16 shows the experimental results on two MOS devices with length-to-width ratios of, $L/W = 25$ and $L/W = 0.65$. In Fig. 5.16, the fill factor is 4, and $h/(4e^2) = 6453.2\,\Omega$. The mean value of 120-repete measurements is 6453.198 Ω. A fine-structure constant of $\alpha = 2\pi e^2/ch = 1/137.03604$ is obtained.

In 1983, Klitzing found the QHE in a GaAs/AlGaAs heterostructure, as shown in Fig. 5.17 (Klitzing and Ebert 1983). And Sizmann (1986) also found the QHE in a p-type HgCdTe MOSFET.

The surprising result of precision measurements was that the Hall resistance is quantized in integer units of h/e^2:

$$\rho_{xy} = \frac{1}{\nu}\frac{h}{e^2}, \tag{5.113}$$

with an accuracy on the order of 10^{-6}. Another interesting feature of the above finding is that the quantization condition of the conductivity is insensitive to the details of the sample (geometry, amount of disorder, etc.). Due to its high precision and reproducibility, the QHE was adopted in defining the ohm (Ω) unit in 1990 by the International Bureau of Weights and Measures (BIPM) in Paris. The ohm was defined in terms of the Klitzing constant, $R_{K-90} = 25812.807$. More recent measurements have achieved an accuracy of 2–5 times 10^{-9} (Braun et al. 1997).

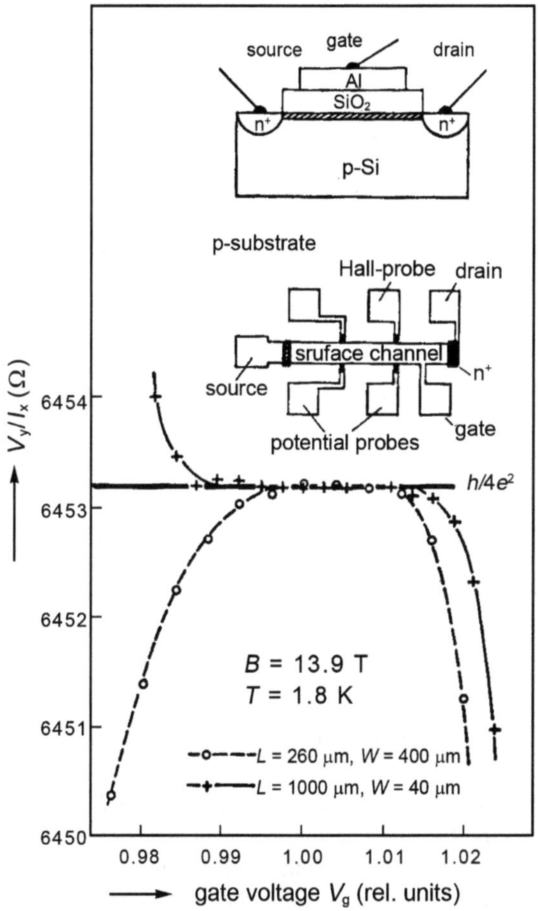

Fig. 5.16 The experimental results for two MOS devices with length-to-width ratios of $L/W = 25$ and $L/W = 0.65$

K.V. Klitzing first related the values of the ρ_{xy} plateaus with the natural constants h/e^2 after performing precision measurements (Klitzing et al. 1980). The effect was named after Klitzing even though the plateaus had been observed before by other authors (Englert and Klitzing 1978). The quantum Hall plateaus can be understood as a localization and delocalization effect of the Landau levels. Anderson first predicted the localization of electrons in the presence of disorder (Anderson 1958). Mott developed this idea and proposed the concept of the mobility edge (Mott 1966). A coherent potential approximation (CPA) treatment of $s-d$ scattering induced localization in CuNi alloys was done by Chen et al. (1972). For some concentrated alloy concentrations this theory predicts the transport transitions from drift (delocalized) to hopping (localized) conductivity in agreement with experiments in which the temperature variation of the conductivity switches sign from positive to negative (Touloukian 1967; Meaden 1965). Another systematic theory about localization is a

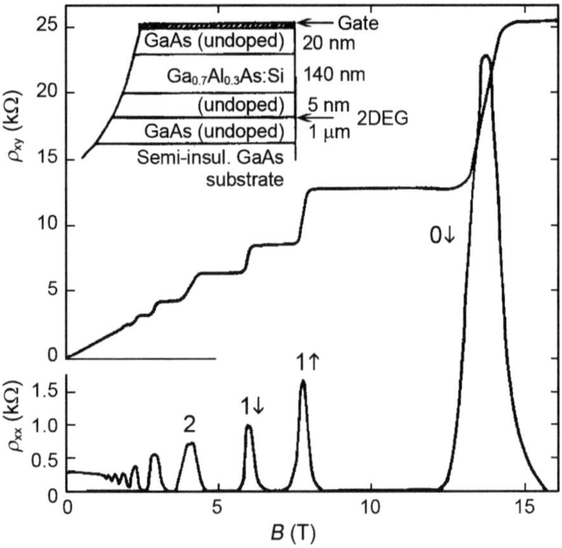

Fig. 5.17 The Hall resistivity (ρ_{xy}) and longitudinal resistivity (ρ_{xx}) in GaAs/AlGaAs heterostructure at 0.35 K

scaling theory given by Abrahams et al. (1979). According to this theory, electrons moving in a two-dimensional random potential are always localized in the absence of a magnetic field B at $T = 0\,\text{K}$. But if $B \neq 0$, the time-symmetry of the system will be broken, causing extended states to exist in the Landau levels. Ono (1982) was the first to suggest that only in the vicinity of the center of each Landau level do extended states exist, while the other states are exponentially localized. The inverse localization length is $\alpha(E) \propto |E - E_n|^s$, where E_n is the energy at the center of each Landau level. The behavior of the localization length in strong magnetic fields has been an active area of research (Ando 1983; Aoki and Ando 1985, 1986; Chalker and Coddington 1988; Huckestein and Kramer 1990). Finite-size scaling studies for the lowest Landau band revealed a universal exponent of $s = 2.34 \pm 0.04$, which has been experimentally confirmed by Koch et al. (1991).

The QHE can be understood as a phase transition between localized and delocalized transport, as shown in Fig. 5.18. If E_F is located in the localized states of the Landau level band tails, they make no contribution to σ_{xx}. Thus, $\sigma_{xx} = 0$, because σ_{xx} is only related to the states near E_F. At the same time, the number of extended states below E_F is also constant, and because σ_{xy} is determined by all the states below E_F, thus σ_{xy} shows a plateau. The exact quantization of ρ_{xy} was explained by Prange (1981) by studying a model of free electrons interacting with impurities that broaden the δ-function potentials. He concluded that the Hall current at the integral quantization is exactly the same as that for free electrons because the loss of Hall current, due to the formation of one localized state, is exactly compensated by an appropriate increase in the Hall current carried by the remaining extended states.

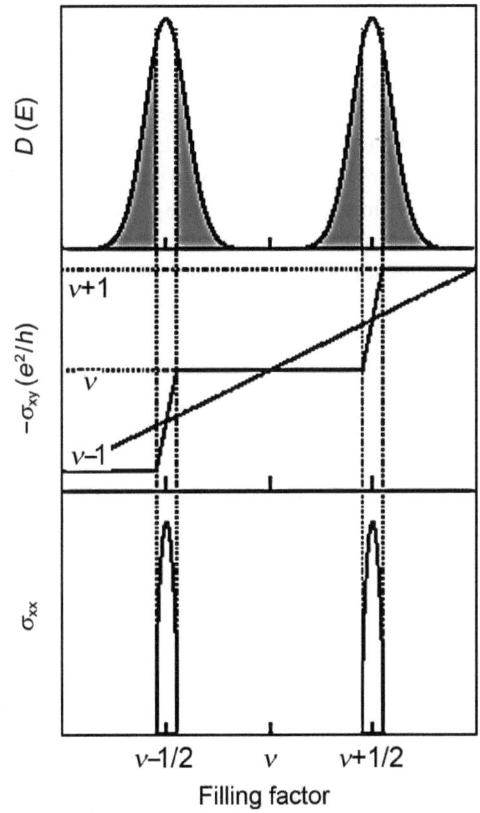

Fig. 5.18 The diagonal conductivity σ_{xx}, the Hall conductivity σ_{xy} and the density of states $D(E)$ as a function of the Landau level filling factor. The dashed diagonal line corresponds to the classical value ne/B

On the other hand, if E_F lies in the extended states region, a dissipative current will exist. Because, $\sigma_{xx} \propto D^2(E)$, σ_{xx} will have a maximum when E_F lies at the centre of each Landau level, and σ_{xy} will increase with B.

Laughlin's gauge invariance theory (Laughlin 1981) gives an elegant explanation of the QHE. The universal character of the QHE suggests that the effect is due to a fundamental principle. According to Laughlin's approach, the quantization is so accurate because it is based on two very general principles: gauge invariance and the existence of a mobility gap. Another theoretical approach by Kubo et al. (1965) is a formulation that also provides an insight into the QHE (Aoki and Ando 1981; Ando 1984a, b). From this formulation, an important conclusion is that if all the states below E_F are localized, $\sigma_{xx} = 0$, or equivalently, $\sigma_{xy} \neq 0$ suggests that there are extended states below E_F. This implies as discussed above, the presence of a magnetic field provides a situation different from that predicted by scaling theory (Abrahams et al. 1979).

In the above discussions, the system is considered to be infinitely large. But the real samples often have edges and ohmic contacts; in this case, the QHE can be explained in terms of the Landauer–Buettiker formalism (Landauer 1957, 1988; Büttiker 1986, 1988a, b). Some authors have pointed that in the quantum Hall regime, the current flows along the one-dimensional channel at the edge of the layer (Halperin 1982; Ando 1992), which is called an "edge state." Classically, these edge states correspond to hopping orbits. The current in each edge state is, $I_n = (\mu^1 - \mu^2)e/h$, where μ^1 and μ^2 are the chemical potentials at both edges. Edge channels on opposite sides of the sample carry current in opposite directions. A net current is established if there is a difference in the magnitudes of those oppositely flowing currents. The existence of edge states has been proven by several experimental studies (Haug et al. 1988; Müller et al. 1990, 1992; Wees et al. 1989).

Edge states in the integer QHE are shown in Fig. 5.19. In the single-electron model, the position of the edge state is determined by the intersection of a Landau level's energy and the Fermi energy. Its width is of the order of the magnetic length, l. In an actual sample, the density decreases gradually near the boundary at $B = 0$. Thus, there will be an extra electrostatic energy cost to change the gradual

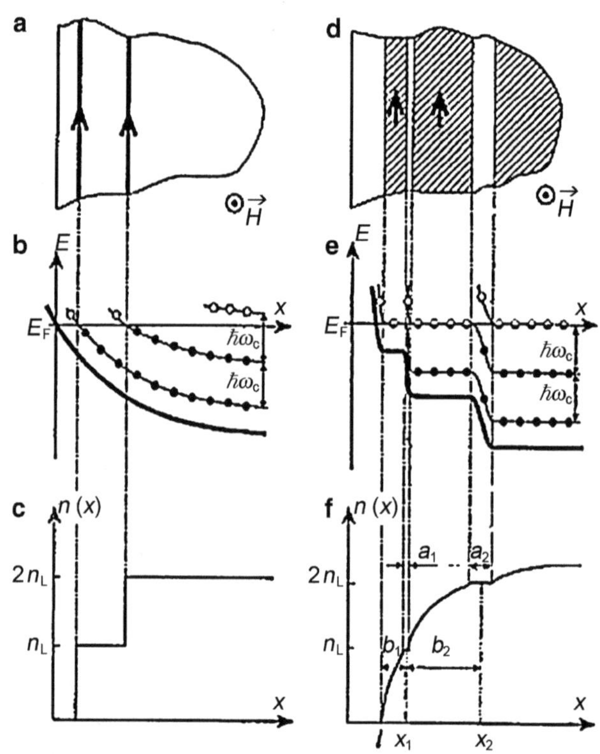

Fig. 5.19 Edge states in the QHE regime (Chklovskii et al. 1992). In the single-electron scheme: (**a**) Top view of the 2DEG near the edge. Arrows indicate the direction of the electron flow in the two edge channels. (**b**) Landau level bending near the edge. Filled circles indicate occupied Landau levels. (**c**) Electron density as a function of distance to the boundary. In the self-consistent electrostatic picture: (**d**) Top view of the 2DEG near the edge with compressible (*unshaded* regions) and incompressible strips (*shaded* regions), (**e**) bending of the electrostatic potential and the Landau levels, and (**f**) electron density as a function of distance to the middle of the depletion region

density profile at $B = 0$ into a step-like one at $B \neq 0$. On the other hand, because of its degeneracy, a Landau level can accommodate additional electrons without changing its energy. Hence, the Landau levels at E_F can be flattened to avoid an electrostatic energy cost. Only when E_F resides between two adjacent Landau levels, will adding an electron cost extra energy. Therefore, as the sample boundary is approached, there will be a series of alternating compressible and incompressible regions. Chklovskii et al. (1992) and Lier and Gerhardts (1994) quantitatively calculated the width of the compressible and the incompressible strips. It is necessary to point out that the QHE is insensitive to the current distribution and is not exclusively an edge effect but for Hall devices with small currents, $I < 0.1\mu A$, the edge current dominates the electronic properties of the system (Klitzing 1990, 1993).

After the discovery of the QHE in III–V compounds, the QHE was also observed in HgCdTe MIS structures by Kirk et al. (1986) and Sizman (1986).

5.4 Experimental Results on HgTe/HgCdTe Superlattices and QWs

5.4.1 Optical Transitions of HgTe/HgCdTe Superlattices and Quantum Wells

The early stage research on HgTe/HgCdTe superlattices and quantum wells is summarized in several reviews (Faurie et al. 1982; Hetzler et al. 1985; Jones et al. 1985; Harris et al. 1986). The foremost goal in these earlier studies is to verify that the grown HgTe/HgCdTe samples were real heterostructures, not HgCdTe alloys owing to interface diffusion. The photoluminescence and transmission spectroscopy indicated the heterostructures had indeed been formed, furthermore the band gap could be adjusted by changing the well and barrier widths.

Hetzler et al. (1985) first observed the infrared photoluminescence of HgTe/HgCdTe superlattices. His first HgTe-CdTd sample, with a 38–40 Å well width and a 18–20 Å thick CdTe barrier, shows a PL peak at 105 meV at 140 K, equivalent to a 11.8 μm wavelength cutoff. From the interface diffusion model, the above mentioned well and barrier thicknesses correspond to a $Hg_{1-x}Cd_xTe$ alloy with $x = 0.33$, which has a band gap of 320 meV at 140 K, and an equivalent cutoff wavelength of 3.9 μm. Consequently, the PL peak energy is much lower than the band gap of an equivalent HgCdTe alloy. Hetzler also measured the PL spectra of a $Hg_{0.71}Cd_{0.29}Te$ alloy at 135 K, which has a peak at 260 meV, i.e. a wavelength of 4.8 μm. Therefore, the PL peak energy indicates that the grown structure is indeed a superlattice not an alloy formed by interface diffusion. Figure 5.20 shows the PL spectra of the first HgTe(40 Å) – CdTe(20 Å) sample. The top curve shows the PL spectra of a $Hg_{0.71}Cd_{0.29}Te$ alloy at 135 K. The bottom curve is the PL spectra at 170 K for the second HgTe(50 Å) – CdTe(50 Å) sample, with a peak centered at

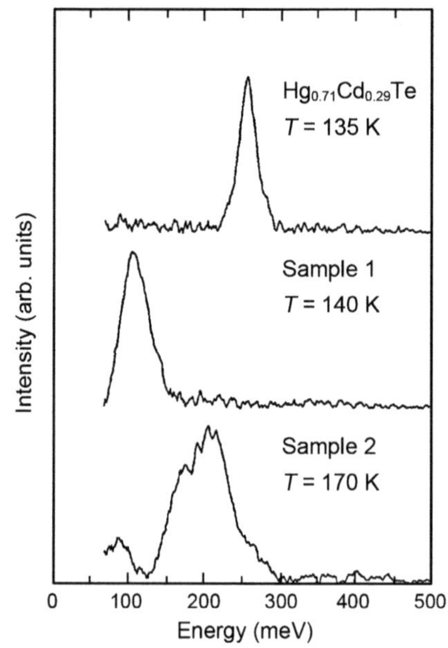

Fig. 5.20 Photoluminescence spectrum of HgTe-CdTd superlattices (sample 1: has a 40 Å HgTe well and 20 Å thick CdTe barrier; sample 2: has a 50 Å well and 50 Å barriers)

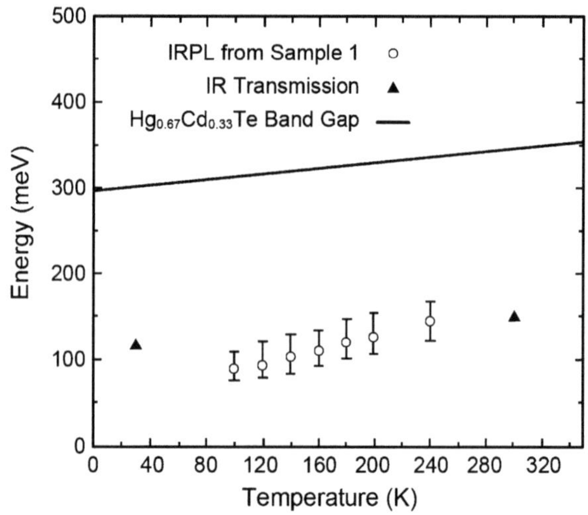

Fig. 5.21 The PL peak positions as a function of temperature of sample 1 and a $Hg_{0.67}Cd_{0.33}Te$ alloy sample

205 meV corresponding a wavelength of 6.0 μm. Its energy is also lower than the band gap of an equivalent $Hg_{0.50}Cd_{0.50}Te$ alloy which is 588 meV (2.1 μm).

The PL peak positions as a function of temperature are shown in Figs. 5.21 and 5.22. The solid line is the energy gap of an equivalent HgCdTe alloy, calculated from the CXT empirical formula. The triangles in Fig. 5.21 are the energy gap determined

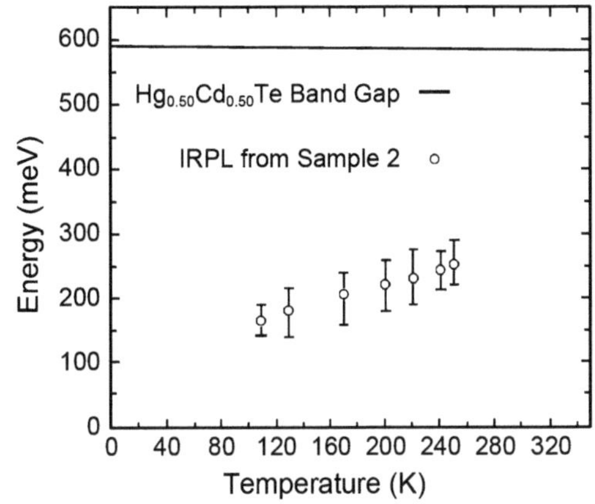

Fig. 5.22 The PL peak positions as a function of temperature of sample 2 and a $Hg_{0.50}Cd_{0.50}Te$ alloy sample

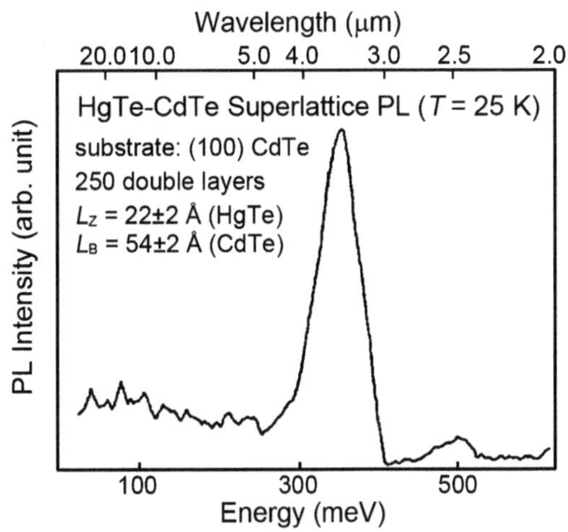

Fig. 5.23 The PL spectra for a HgTe (22 ± 2Å) – CdTe (54 ± 2Å) multi-layer structure

by transmission spectroscopy. It can be seen in the temperature range investigated, the energy gap of a HgTe–CdTe superlattice is smaller than that of its equivalent HgCdTe alloy.

Similar results have also been obtained by Harris et al. (1986). Figure 5.23 shows the PL spectra measured for a HgTe(22 ± 2 Å) – CdTe(54 ± 2 Å) multi-layer structure. The peak centers at 357 meV(3.47 μm). Its equivalent alloy, $Hg_{1-x}Cd_xTe$ alloy with $x = 0.71$, has an energy gap of 974 meV(1.3 μm). Consequently, the PL peak energy is also smaller than the energy gap of the equivalent HgCdTe alloy.

From the results presented above, one can easily see that the PL peak is related to the HgTe well width, the smaller the well width, the larger the PL peak's energy. The quantitative relation can only be obtained by solving the band structure problem of a HgCdTe superlattice.

Experimental determination of the VBO of the HgTe–CdTe superlattice is a vital issue. One method to deduce the VBO is to measure the optical transition from the

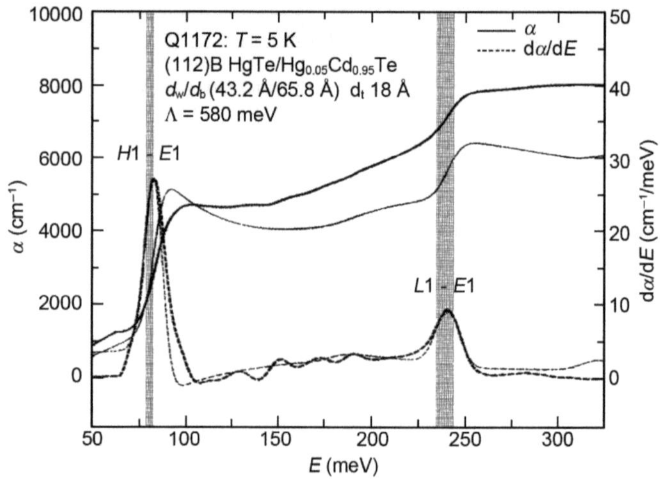

Fig. 5.24 The absorption spectrum of $HgTe/Hg_{1-x}Cd_xTe$ superlattice at $T = 5$ K. The measured absorption coefficient and its differential results are shown by a thick solid curve and dashed curve, the calculated ones are shown by a thinner solid curve and a dashed curve

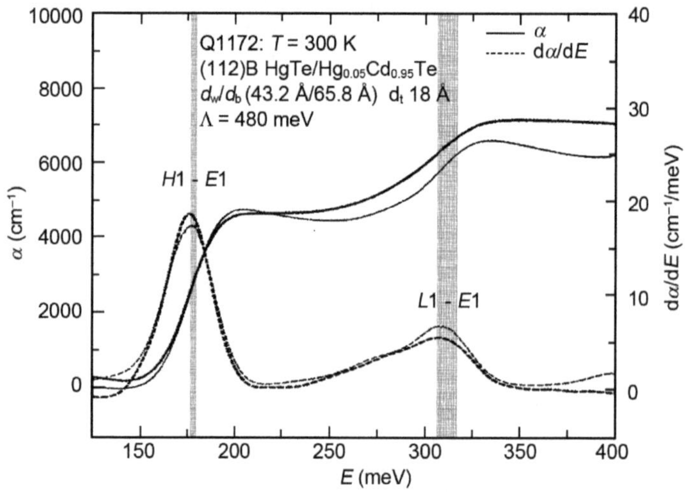

Fig. 5.25 The absorption spectrum of a $HgTe/Hg_{1-x}Cd_xTe$ superlattice at $T = 300$ K

first heavy-hole subband $H1$ and first light-hole subband $L1$, to the first conduction subband $E1$, respectively, and then compare them with a band structure calculation. The $HgTe(43\,\text{Å})/Hg_{1-x}Cd_xTe(66\,\text{Å})$ superlattice with $x = 0.95$ used by Becker et al. (1999) were grown by MBE on $Cd_{0.96}Zn_{0.04}Te$ substrate, topped with 600 Å CdTe buffer layer. The transmission spectra are depicted in Figs. 5.24 and 5.25, and the temperature dependence of $H1$-$E1$, $L1$-$E1$, and $L1$-$H1$ transitions are

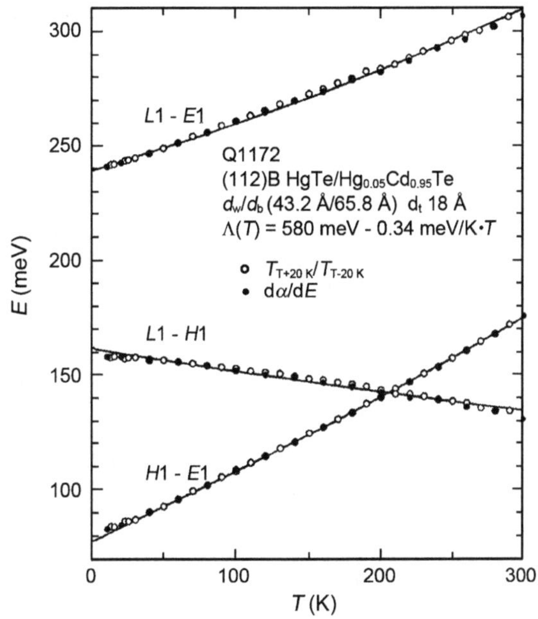

Fig. 5.26 The temperature dependence of the $H1$-$E1$, $L1$-$E1$ and $L1$-$H1$ transitions

shown in Fig. 5.26. The experimental VBO values determined are 580 $\pm$ 40 meV at 5 K, and 480 $\pm$ 40 meV at 300 K. The temperature dependence is: $\Lambda(T)(\text{meV}) =$ 580–0.34 T. The VBO determined by UPS and XPS measurements is 350$\pm$60 meV (Sporken et al. 1989).

5.4.2 *Typical SdH Oscillations and the Quantum Hall Effect*

This section briefly describes SdH oscillations and an integer QHE observed in n-type modulation doped HgTe/HgCdTe quantum wells. Similar to GaAs/AlGaAs heterojunctions, SdH oscillations and the QHE are readily seen in HgTe quantum wells. The SdH oscillations persist at relatively high temperatures because of the small effective mass of HgTe. Hoffman et al.(1991, 1993) first studied the transport properties of unintentionally doped HgTe quantum wells. Lately, the Würzburg group (Goschenhofer et al. 1998; Pfeuffer-Jeschke et al. 1998; Landwehr et al. 2000; Zhang et al. 2001) systematically studied the transport properties of both n- and p-type modulation doped HgTe quantum wells. Typical SdH oscillations and QHE measurements are illustrated in Fig. 5.27 on a $HgTe/Hg_{0.32}Cd_{0.68}Te$ quantum well between 1.6 and 60 K (Zhang 2001). This sample has a carrier concentration of $5.1 \times 10^{11}\,cm^{-3}$, and a mobility of $6.3 \times 10^{4}\,cm^{2}/V \cdot s$. As depicted in Fig. 5.27, the onset of SdH oscillations occurs at about 1 T, the spin splitting first appears at 2 T because of the large effective g-factor.

The effective mass, an important physical parameter to describe the carrier motion, can be deduced from the temperature dependence of the SdH oscillation

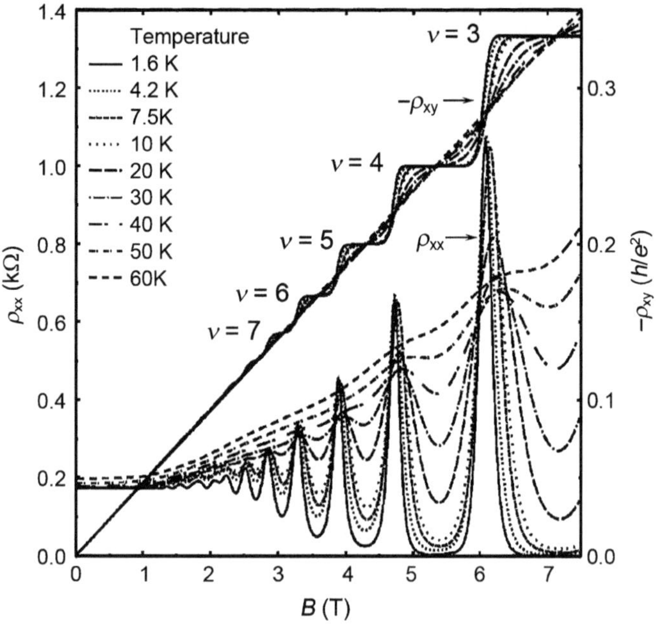

Fig. 5.27 Typical SdH oscillations and quantum Hall effect measurements in n-type a modulated $HgTe/Hg_{0.32}Cd_{0.68}Te$ quantum well

amplitudes. The second term in the formula, (5.109), indicates the decay of the SdH oscillation amplitudes with temperature is related to the effective mass. It is described by the Ando formula:

$$\frac{A(T_1, B)}{A(T_2, B)} = \frac{T_1 \sinh(\beta T_2 m^*/m_0 B)}{T_2 \sinh(\beta T_1 m^*/m_0 B)}, \tag{5.114}$$

where $\beta = 2\pi^2 k_B m_0/\hbar e = 14.707\,(\mathrm{T/K})$. Figure 5.28 shows the temperature dependence of SdH oscillation amplitudes; the nonoscillatory background has been subtracted from ρ_{xx}.

The temperature dependence of SdH oscillation amplitude ratios are shown in Fig. 5.29. From this data, the effective mass can be easily derived from the formula in (5.114). The resulting effective mass is in rough agreement with values obtained by other methods.

In the formula, (5.109), the factor containing the Dingle temperature T_D is caused by collision broadening. The damping of the SdH oscillation amplitudes with $1/B$ can be caused both by T and T_D, but the influence of T_D can be separated. The SdH oscillation amplitude, $A(T, B)$, satisfies (Coleridge 1991):

$$A(T, B) = 4\rho_0 X(T) \exp(-2\pi^2 k_B T_D/\hbar\omega_c), \tag{5.115}$$

where ρ_0 is the zero magnetic field resistivity, $X(T) = \frac{2\pi^2 k_B T/\hbar\omega_c}{\sinh(2\pi^2 k_B T/\hbar\omega_c)}$ is a thermal damping factor, and ω_c is the cyclotron resonance frequency. If

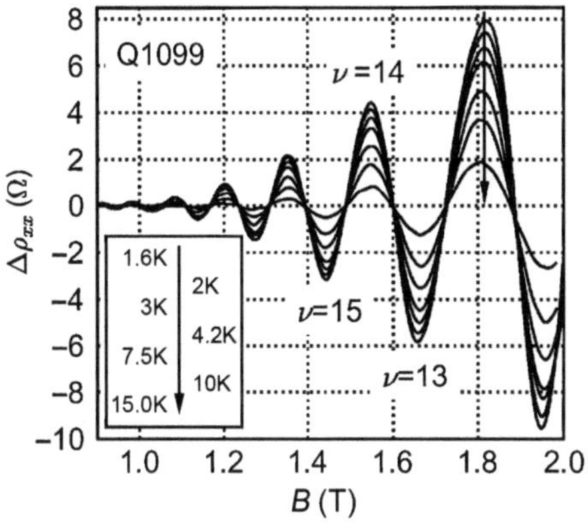

Fig. 5.28 The temperature dependence of SdH oscillation amplitudes

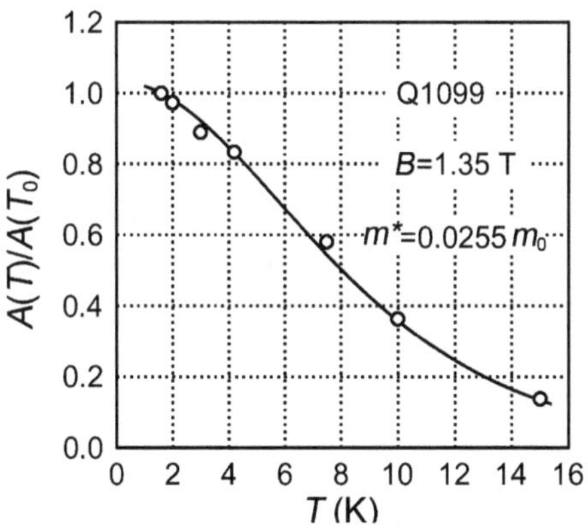

Fig. 5.29 The temperature dependence of SdH oscillation amplitude ratio. The effective mass has been deduced to be $m^* = 0.0255m_0$

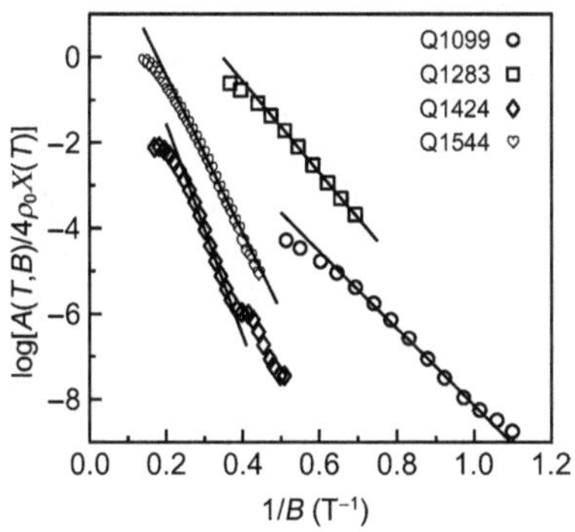

Fig. 5.30 The magnetic-field dependence of SdH amplitudes for several HgTe QWs with different structures

$\ln[A(T, B)/4\rho_0 X(T)]$ is plotted versus $1/B$, its slope is related to T_D. Figure 5.30 shows this kind of plot for several HgTe QWs with different structures. $\ln[A(T, B)/4\rho_0 X(T)]$ varies approximately linearly with $1/B$, but sample Q1283 shows an obvious deviation. This is caused by the Rashba spin–orbit interaction which will be discussed in Sect. 5.4.3.

5.4.3 Rashba Spin–Orbit Interaction in n-Type HgTe Quantum Wells

This section will describe the Rashba spin–orbit interaction induced by symmetry annihilation. Recently, spin effects in semiconductor hetero-structures has aroused a lot of attention, and begins to form a totally new subject: spintronics (spin transport electronics or spin-based electronics). In correspondence with the conventional electronics (Wolf et al. 2001), the spin degree of freedom of an electron is used to design electronics. In this kind of device, the information carrier is the spin of an electron, unlike the charge in conventional electronics. In comparison with charge, the spin of an electron possesses the following advantages in addition to its non-volatility:

(1) a spin can be easily manipulated by an external magnetic field, (2) spins can have a very long coherence length or relaxation time, the phase coherence of its wave function can be maintained for a long time, while the phase coherence of a charge state can be easily destroyed by defect, and impurity scattering as well as collisions from other charges. Compared to devices based on electron charge transport, the spin properties render the possibility to develop even smaller sized, less power consuming, and faster electronics. By implementing spintronics, it is possible to realize the long sought device that integrates electronics, optoelectronics, and magnetoelectronics.

It is well known that in semiconductors bulk inversion asymmetry can lift the spin degeneracy even without an external magnetic field (Dresselhaus 1955). In semiconductor quantum wells or heterostructures, because of the structure lacks inversion symmetry, the spin degeneracy is also lifted. This is usually called the Rashba spin–orbit splitting or the Rashba effect (Rashba 1960; Bychkov and Rashba 1984). It was first observed by Störmer et al. in GaAs/AlGaAs heterojunctions (Störmer et al. 1983). The spin–orbit splitting in the inversion layer of p-type HgCdTe was discussed in Chapter 4. Owing to its controllability by gate voltage, the Rashba effect possibly can be used to make the so called Datta & Das spin FET (Datta and Das 1990).

Figure 5.31 shows the variation of SdH oscillations with gate voltage for a symmetric HgTe quantum well. From a fast Fourier transform spectrum the electron density in two split branches of first $H1$ conduction subband shows a clear evolution with gate voltage. Zhang et al. (2001) pointed out the Rashba effect is indeed negligible in the case of a symmetric quantum well. This is clearly evidenced by the carrier density difference in two split subbands as a function of gate voltage shown in Figure 5.32. When $V_g = 0.2\,\mathrm{V}$, the quantum well confinement potential is exactly symmetric, and no splitting in the $H1$ subband is visible, but if the gate voltage is varied in either direction, a clear Rashba splitting is observed in the $H1$ subband.

Chou et al. (2004) also studied the Rashba spin splitting in a HgTe/HgCdTe quantum well with an inverted band structure by analyzing the beating pattern in the SdH oscillations. A strong Rashba splitting is observed. Three different analysis methods,

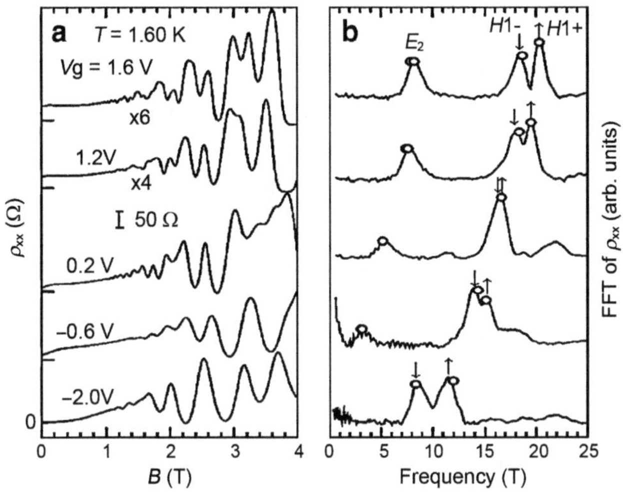

Fig. 5.31 The variation of SdH oscillations and its FFT resulting variation with gate voltage for a symmetric HgTe quantum well

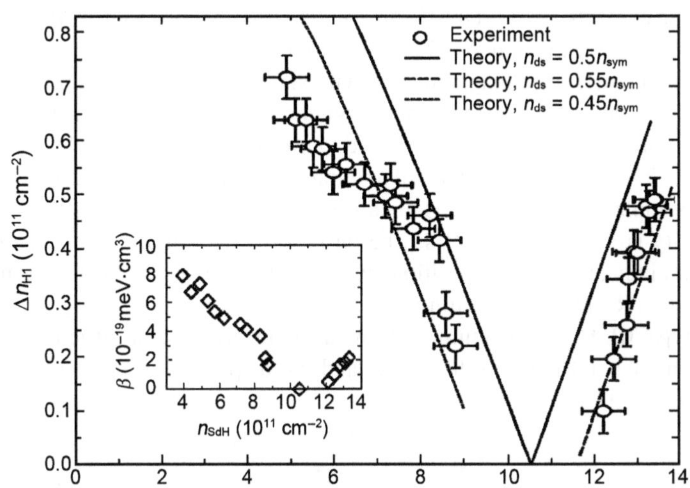

Fig. 5.32 The measured and calculated carrier density difference in two split subbands as a function of total density (Zhang 2001)

a FFT of SdH oscillation versus $1/B$, an analysis of node positions, and a numerical fit to a SdH beat pattern, give the same Rashba spin splitting energy, 28–36 meV.

Chau studied two n-type modulation doped HgTe/HgCdTe quantum well sample, A and B. They were grown by MBE under the same condition. The well width is 11 nm, the substrate is (001) CdZnTe, with a CdTe-doped layer as the upper barrier. The HgCdTe barrier includes a 5.5 nm spacer and 9 nm doped layer. A Hall bar

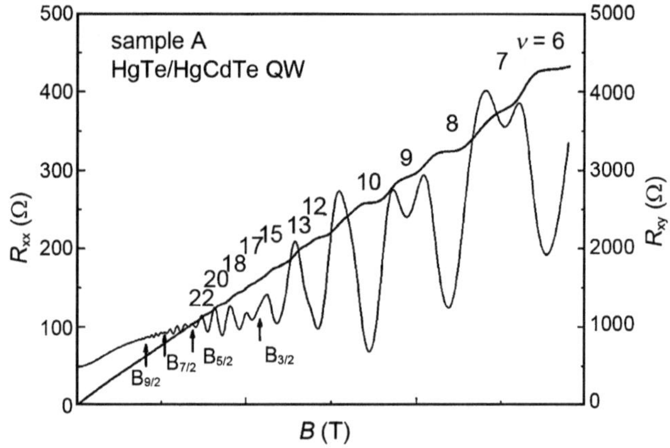

Fig. 5.33 The longitudinal and transverse magneto-resistance, R_{xx} and R_{xy}, for sample A at 1.4 K. The arrows indicate the node positions in the beat pattern

mesa was made by a standard photolithography and chemical etching process, and good ohmic contacts were established by indium thermal bonding. A gate structure which consists of a 200 nm Al_2O_3 layer and an Al contact were made on top of sample B. The transverse magneto-resistivity and Hall resistivity were measured in a temperature range of 1.4–35 K with magnetic fields ranging from 0 to 15 T. An excitation current less than 1 μA was used to avoid overheating the electrons.

Figure 5.33 shows the longitudinal and transverse magneto-resistances, R_{xx} and R_{xy}, for sample A at 1.4 K. The Hall carrier concentration is $2.0 \times 10^{12}\,\mathrm{cm}^{-2}$, and the mobility is $9.5 \times 10^4\,\mathrm{cm}^2/\mathrm{Vs}$. The onset of SdH oscillations in R_{xx} is 0.8 T (the arrows indicate the node positions in the beat pattern). The beat pattern can be attributed to a Rashba spin splitting.

In a type III HgTe quantum well with an inverted band structure, the first conduction subband is heavy-hole like, and the dispersion of spin split band can be approximated by:

$$E_{\pm}(k_{/\!/}) = \frac{\hbar^2 k_{/\!/}^2}{2m^*} \pm \beta k_{/\!/}^3, \tag{5.116}$$

where $k_{/\!/}$ is the parallel wave vector. The spin splitting at the Fermi energy is $\Delta_{\mathrm{R}} = 2\beta k_{\mathrm{F}}^3$, β is the spin–orbit coupling constant:

$$\beta = \frac{\hbar^2}{2m^*}\sqrt{\frac{X(2-X)}{4\pi n}}, \tag{5.117}$$

where

$$X = \frac{2(2+\sqrt{1-a^2})}{a^2+3}, \tag{5.118}$$

$a = (n_+ - n_-)/n, n = n_+ + n_-$, and $n_\pm$ are the carrier densities in two spin split subbands.

The amplitude modulation of SdH oscillations because of the carrier density difference between the spin split subbands is expressed by (Das et al. 1989):

$$A \propto \cos(\pi \nu), \tag{5.119}$$

where $\nu = \delta / h\omega_c$, δ is the spin splitting energy and $h\omega_c$ is the cyclotron energy. When ν is a half integer (1/2, 3/2,. . .), the corresponding B value is at the node position of the SdH oscillations. The last node is $\delta = 3/2h\omega_c$ (Teran et al. 2002). The number of oscillation periods between two neighboring nodes is approximately inversely proportional to the spin splitting energy. Compared to InGaAs quantum wells with a similar carrier density, the number of oscillation periods is far less than that in HgTe quantum wells. This implies there is a much larger Rashba spin splitting in HgTe quantum wells.

A HgTe quantum well has a very strong nonparabolic band effect, the effective mass at the Fermi energy differs form the band edge effective mass. The effective mass at the Fermi energy, $m^* = (0.044 \pm 0.005)m_0$, can be deduced from the temperature dependence of SdH oscillations which is shown in the inset to Fig. 5.34. From the Fourier transformation of the SdH oscillations at various temperatures depicted in Fig. 5.35, the carrier density in the two spin split subbands are found to be 0.8×10^{12} and $1.06 \times 10^{12}\,\text{cm}^{-2}$, respectively. From this, the spin splitting energy is deduced to be 28.2 meV. The FFT peaks don't shift when the temperature is raised up to 35 K.

Sample B was covered by an insulator and a gate electrode, the electron concentration was found to increase a lot because of the work function difference between gate electrode and semiconductor. Figure 5.36 shows measured SdH oscillations and node positions of the beat pattern indicated by the arrows. From the FFT

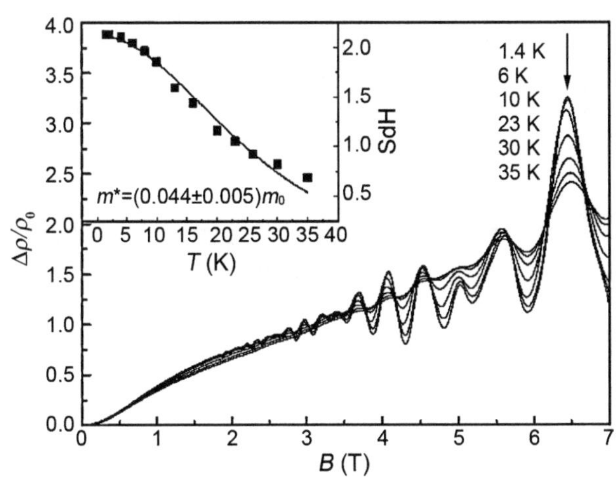

Fig. 5.34 The temperature dependence of the SdH oscillations (the insert shows the temperature dependence of the SdH amplitudes)

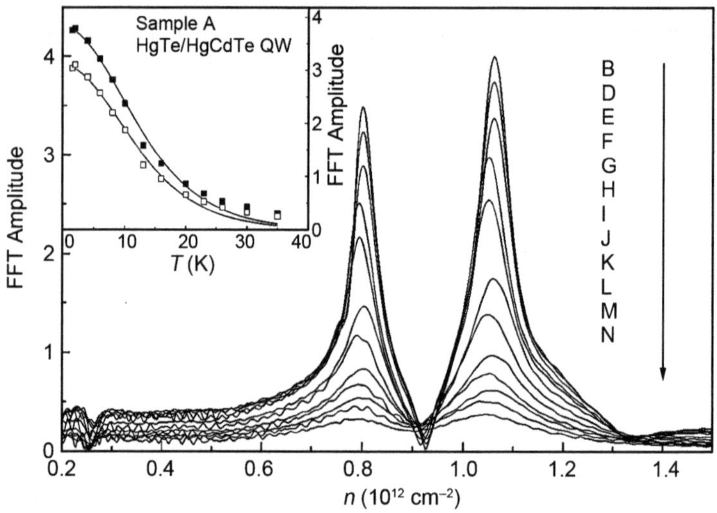

Fig. 5.35 The Fourier transformation of the SdH oscillations at various temperatures (the insert shows the temperature dependence of the FFT amplitudes)

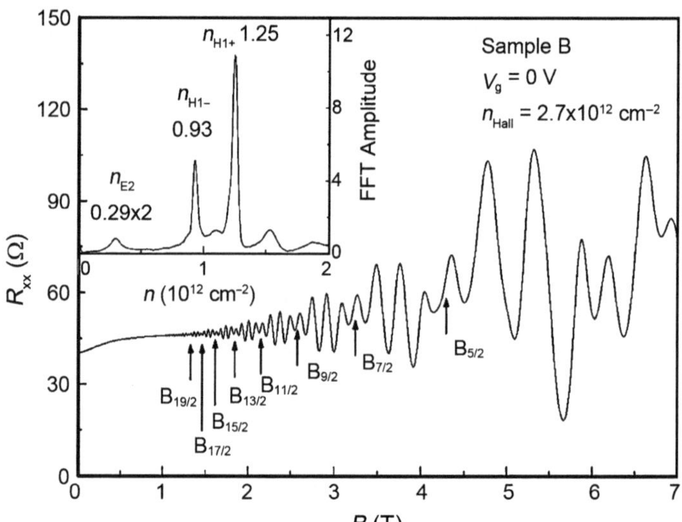

Fig. 5.36 The measured SdH oscillations R_{xx} of sample B. The arrows show the node positions of the beat pattern. The insert shows the FFT results

shown in the inset, the carrier densities in the first spin split subbands are 1.25 and 0.93×10^{12} cm^{-2}, respectively. The second subband is also found to be populated. The even higher population difference found in sample B leads to a larger Rashba spin splitting, 34.7 meV.

The spin splitting energy in 2DEG can be expanded in a series in powers of $\hbar\omega_c$ (Grundler 2000):

$$\delta = \delta_0 + \delta_1\hbar\omega_c + \delta_2(\hbar\omega_c)^2 + \cdots, \tag{5.120}$$

where δ_0 is the zero magnetic field spin splitting, and δ_1 is the linear term. Only at very high magnetic fields are the parabolic and higher terms important. If δ_0 dominates, only the first two terms need be considered. Accordingly, from (5.119) and (5.120), the Rashba spin splitting δ_0 can be deduced by linearly fitting the total spin splitting versus Landau level splitting. The Rashba spin splittings obtained are 28.5 and 35.5 meV for sample A and B, respectively, in good agreement with FFT method results (Fig. 5.37).

To further confirm the beating patterns are caused by Rashba spin splittings, SdH oscillations were numerically simulated. According to Gerhardts (1976), considering the spin splitting in a low-order approximation, the density of Landau level states is:

$$D(E) = \frac{1}{2\pi l^2}\sum_{N\pm}\left[\frac{\pi}{2}\Gamma^2\right]^{-1/2}\exp\left[-2\frac{(E-E_{N\pm})^2}{\Gamma^2}\right], \tag{5.121}$$

where $l = (h/eB)^{1/2}$, is the magnetic length, and $E_{N\pm}$ is Nth spin up a down Landau level energy. To simplify, the Landau level broadening, Γ, is assumed to be a constant, and the spin split Landau level is expressed by:

$$E_{N\pm} = \hbar\omega_c\left[N \pm \frac{1}{2}\sqrt{(1-m^*g^*/2)^2 + N\frac{\Delta_R^2}{E_F\hbar\omega_c}}\right], \tag{5.122}$$

where g^* is an effective g-factor. Figure 5.38 shows the SdH oscillations and numerical results. By adjusting Δ_R, Γ and g^* to minimize the deviation between theory and experiment, the node positions coincide. The results obtained are: sample A,

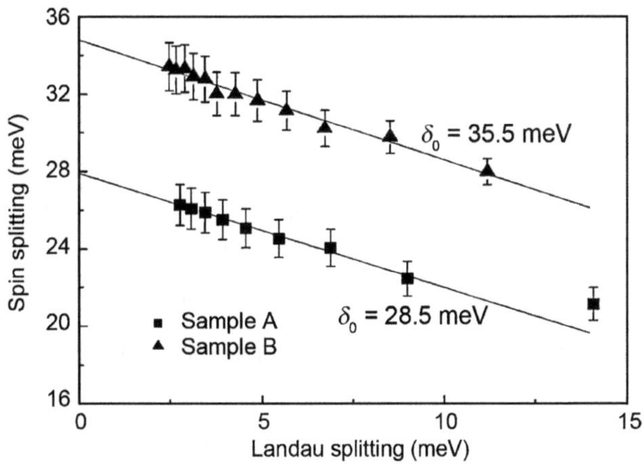

Fig. 5.37 The total spin splitting energy as a function of the Landau splitting energy

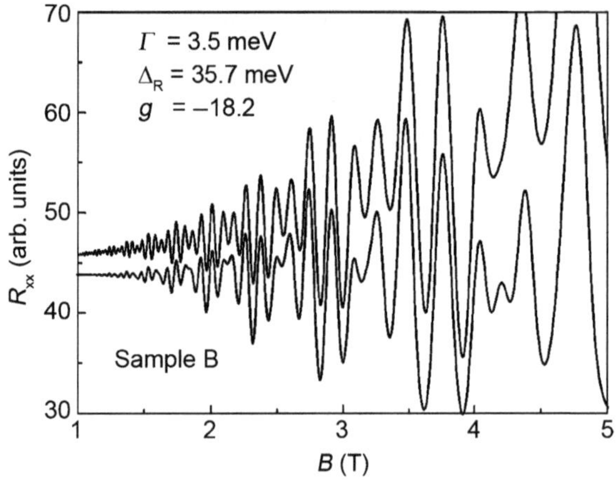

Fig. 5.38 The measured and fitted SdH oscillation results (the upper curves are the measured results, the lower ones are the fits)

$\Delta_R = 28.8\,\text{meV}$, $\Gamma = 3.0\,\text{meV}$ and $g^* = -18.2$; sample B, $\Delta_R = 35.7\,\text{meV}$, $\Gamma = 3.5\,\text{meV}$, and $g^* = -18.3$. Consequently, the three different methods, FFT of SdH oscillation, analysis of node positions, and the numerical fit to SdH beat pattern, give results in complete agreement. This confirms the strong Rashba spin splitting in HgTe quantum wells.

By analyzing SdH oscillation beat patterns in HgTe quantum wells, Chou et al. found a strong Rashba spin–orbit interaction ($\sim$35 meV) which is one order of magnitude larger than that is an InGaAs quantum well with the same carrier concentration, and an even larger room temperature thermal broadening ($\sim$26 meV). The strong spin–orbit interaction between the Γ_7 and Γ_8 bands in HgTe bulk material and the heavy-hole nature of the first conduction subband in a HgTe quantum well, enhances the Rashba spin–orbit interaction. These results are significant for future designs of electron spin-based electronic devices.

References

Abrahams E, Anderson PW, Licciardello DC, Ramakrishnan TV (1979) Scaling theory of localization: Absence of quantum diffusion in two dimensions. Phys Rev Lett 42:673–676

Altarelli M (1985) Electronic structure of two-dimensional semiconductor systems. J Lumines 30:472–487

Ancilotto F, Fasolino A, Maan JC (1988) Hole-subband mixing in quantum wells: A magnetooptic study. Phys Rev B 38:1788–1799

Anderson PW (1958) Absence of diffusion in certain random lattices. Phys Rev 109:1492–1505

Ando T (1974a) Theory of quantum transport in a two-dimensional electron system under magnetic fields II: Single-site approximation under strong fields. Jpn J Phys Soc 36:1521–1529

Ando T (1974b) Theory of quantum transport in a two-dimensional electron system under magnetic fields III: Many-site approximation. Jpn J Phys Soc 37:622–630

Ando T (1974c) Theory of quantum transport in a two-dimensional electron system under magnetic fields IV: Oscillatory conductivity. Jpn J Phys Soc 37:1233–1237

Ando T (1983) Electron localization in a two-dimensional system in strong magnetic fields I: Case of short-range scatterers. Jpn J Phys Soc 52:1740–1749

Ando T (1984a) Electron localization in a two-dimensional system in strong magnetic fields II: Long-range scatterers and response functions. Jpn J Phys Soc 53:3101–3111

Ando T (1984b) Electron localization in a two-dimensional system in strong magnetic fields III: Impurity-concentration dependence and level-mixing effects. Jpn J Phys Soc 53:3126–3135

Ando T (1992) In: Fukuyama H, Ando T (eds) Transport Phenomena in Mesoscopic systems. Springer, Berlin, p 185

Ando T, Uemura Y (1974) Theory of quantum transport in a two-dimensional electron system under magnetic fields I: Characteristics of level broadening and transport under strong fields. Jpn J Phys Soc 36:959–967

Ando T, Fowler AB, Stern F (1982) Electronic properties of two-dimensional systems. Rev Mod Phys 54:437–672

Aoki H, Ando T (1981) Effect of localization on the hall conductivity in the two-dimensional system in strong magnetic fields. Solid State Commun 38:1079–1082

Aoki H, Ando T (1985) Critical localization in two-dimensional Landau quantization. Phys Rev Lett 54:831–834

Aoki H, Ando T (1986) Critical localization and low-temperature transport in two-dimensional Landau quantization. Surf Sci 170:249–255

Arakawa Y, Sakaki H (1982) Multidimensional quantum well laser and temperature dependence of its threshold current. Appl Phys Lett 40:939

Bastard G (1988) Wave Mechanics Applied to Semiconductor Heterostructures. Halsted, New York

Bastard G, Brum JA, Ferreira R (1991) In: Ehrenreich, Turnbull D (eds) Solid State Physics, Vol. 44. Academic, New York, p 229

Becker CR, Latussek V, Li M, Pfeuffer-Jeschke A, Landwehr G (1999) Valence band offset in $HgTe/Hg_{1-x}Cd_xTe$ superlattices. J Electron Mater 28:826–829

Becker CR, Latussek V, Pfeuffer-Jeschke A, Landwehr G, Molenkamp LW (2000) Band structure and its temperature dependence for type-III $HgTe/Hg_{1-x}Cd_xTe$ superlattices and their semimetal constituent. Phys Rev B 62:10353–10363

BenDaniel DJ, Duke CB (1966) Space-charge effects on electron tunneling. Phys Rev 152: 683–692

Braun E, Schumacher B, Warnecke P (1997) In: Landwehr G, Ossau W (eds) High magnetic fields in the physics of semiconductors II. World Scientific, Singapore, p 1005

Büttiker M (1986) Four-terminal phase-coherent conductance. Phys Rev Lett 57:1761–1764

Büttiker M (1988a) Absence of backscattering in the quantum Hall effect in multiprobe conductors. Phys Rev B 38:9375–9389

Büttiker M (1988b) Symmetry of electrical conduction. IBM J Res Dev 32:317–334

Bychkov YA, Rashba EI (1984) Oscillatory effects and the magnetic susceptibility of carriers in inversion layers. J Phys C 17:6039–6045

Chakraborty T, Pietiläinen P (1995) The quantum hall effect. In: Cardona M, Fulde P, Klitzing KV, et al. (eds) Springer Series in Solid-State Sciences Vol 85. Springer, Berlin

Chalker JT, Coddington PD (1988) Percolation, quantum tunnelling and the integer Hall effect. J Phys C 21:2665–2679

Chang LL, Esaki L, Tsau R (1974) Resonant tunneling in semiconductor double barriers. Appl Phys Lett 24:593

Chen AB, Weisz G, Sher A (1972) Temperature dependence of the electron density of states and dc electrical resistivity of disordered binary alloys. Phys Rev B 5:2897–2924

Chklovskii DB, Shklovskii BI, Glazman LI (1992) Electrostatics of edge channels. Phys Rev B 46:4026–4034

Cho AY, Arthur JR (1975) In: Somerjai G, McCaldin J (eds) Progress in Solid State Chemistry. Vol 10. Pergamon, New York, p 175

Chou ZJ, Gui YS, Su XZ, Dai N, Guo SL, Chu JH (2004) Giant Rashba spin splitting in HgTe/HgCdTe quantum wells. Acta Physica Sinica 53:1186–1190
Chu JH, Sher A (2007) Physics and Properties of Narrow Gap Semi-Conductors. Springer, New York
Chu L (2001) PhD thesis. Technische Universität München
Coleridge PT (1991) Small-angle scattering in two-dimensional electron gases. Phys Rev B 44: 3793–3801
Das B, Miller DC, Datta S, Reifenberger R, Hong WP, Bhattacharya PK, Singh J, Jaffe M (1989) Evidence for spin splitting in InxGa1-xAs/In0.52Al0.48As heterostructures as B $\rightarrow$ 0. Phys Rev B 39:1411–1414
Datta S, Das B (1990) Electronic analog of the electro-optic modulator. Appl Phys Lett 56:665
Dingle R, Störmer HL, Gossard AC, Wiegmann W (1978) Electron mobilities in modulation-doped semiconductor heterojunction superlattices. Appl Phys Lett 33:665
Dresselhaus G (1955) Spin-Orbit Coupling Effects in Zinc Blende Structures. Phys Rev 100: 580–586
Drude P (1900a) Zur Elektronentheorie der Metalle I. Teil, Annalen der Physik 1:566–613
Drude P (1900b) Zur Elektronentheorie der Metalle: II. Teil, Annalen der Physik 3:369–402
Englert T, Klitzing KV (1978) Analysis of ρ_{xx} minima in surface quantum oscillations on (100) n-type silicon inversion layers. Surf Sci 73:70–80
Esaki L, Sakaki H (1977) New photoconductor. IBM Tech Disc Bull 20:2456–2457
Esaki L, Tsu R (1970) Superlattice and negative differential conductivity in semiconductors. IBM J Res Dev 14:61–65
Faurie JP, Million A, Piaguet J (1982) CdTe-HgTe multilayers grown by molecular beam epitaxy. Appl Phys Lett 41:713
Gerhardts RR (1975a) Path-integral approach to the two-dimensional magneto-conductivity problem I: General formulation of the approach. Z Physik B 21:275–283
Gerhardts RR (1975b) Path-integral approach to the two-dimensional magneto-conductivity problem II: Application to n-type (100)-surface inversion layers of p-silicon. Z Physik B 21:285–294
Gerhardts RR (1976) Cumulant approach to the two-dimensional magneto-conductivity problem. Surf Sci 58:227–234
Goschenhofer F, Gerschütz J, Pfeuffer-Jeschke1 A, Hellmig R, Becker CR, Landwehr G (1998) Investigation of iodine as a donor in MBE grown Hg1–xCdxTe. J Electron Mater 27:532–535
Groves SH, Brown RN, Pidgeon CR (1967) Interband Magnetoreflection and Band Structure of HgTe. Phys Rev 161:779–793
Grundler D (2000) Large Rashba splitting in InAs quantum wells due to electron wave function penetration into the barrier layers. Phys Rev Lett 84:6074–6077
Grundmann M, Stier O, Bimberg D (1995) InAs/GaAs pyramidal quantum dots: Strain distribution, optical phonons, and electronic structure. Phys Rev B 52:11969–11981
Halperin BI (1982) Quantized Hall conductance, current-carrying edge states, and the existence of extended states in a two-dimensional disordered potential. Phys Rev B 25:2185–2190
Harris KA, Hwang S, Blanks DK, Cook JW, Schetzina JF, Otsuka N, Baukus JP, Hunter AT (1986) Characterization study of a HgTe-CdTe superlattice by means of transmission electron microscopy and infrared photoluminescence. Appl Phys Lett 48:396
Haug H, Koch SW (1993) Quantum Theory of the Optical and Electronic Properties of Semiconductors. World Scientific, Singapore
Haug RJ, MacDonald AH, Streda P, Klitzing KV (1988) Quantized multichannel magnetotransport through a Barrier in two dimensions. Phys Rev Lett 61:2797–2800
Hetzler SR, Baukus JP, Hunter AT, Faurie JP (1985) Infrared photoluminescence spectra from HgTe-CdTe superlattices. Appl Phys Lett 47:260
Hoffman CA, Meyer JR, Bartolil FJ, Lansari Y, Cook JW, Schetzina JF (1991) Electron mobilities and quantum Hall effect in modulation-doped HgTe-CdTe superlattices. Phys Rev B 44: 8376–8379

Hoffman CA, Meyer JR, Bartoil FJ (1993) Novel magnetotransport and magneto-optical processes in semimetallic HgTe-CdTe superlattices. Semicond Sci Technol 8:S48–S57
Huckestein B, Kramer B (1990) One-parameter scaling in the lowest Landau band: Precise determination of the critical behavior of the localization length. Phys Rev Lett 64:1437–1440
Isihara V, Smrčka L (1986) Density and magnetic field dependences of the conductivity of two-dimensional electron systems. J Phys C: Solid State Phys 19:6777–6789
Janssen M, Hajdu J, Viehweg O (1994) In: Hajdu J (ed) Introduction to theory of the integer quantum hall effect. VCH Verlagsgesellshaft GmbH, Weiheim
Jones CE, Casselman TN, Faurie JP, Perkowitz S, Schulman JN (1985) Infrared properties and band gaps of HgTe/CdTe superlattices. Appl Phys Lett 47:140
Kamgar A, Kneschaurek P, Dorda G, Koch JF (1974) Resonance spectroscopy of electronic levels in a surface accumulation layer. Phys Rev Lett 32:1251–1254
Kane EO (1957) Band structure of indium antimonide. J Phys Chem Solids 1:249–261
Kirk WP, Kobiela PS, Schiebel RA, Reed MA (1986) Investigation of the two-dimensional electron gas in HgCdTe by quantum Hall effect measurements. J Vac Sci Technol A 4:2132–2136
Kittel C (1968) Introduction to Solid State Physics, 3rd Ed. Wiley, New York, p 320
Klitzing KV (1990) Ten years quantum hall effect. Adv Solid State Phys 30:25–39
Klitzing KV (1993) The quantum hall effect - An edge phenomenon? Physica B 184:1–6
Klitzing KV, Ebert G (1983) The quantum hall effect. Physica B + C 117–118 Part 2:682–687
Klitzing KV, Dorda G, Pepper M (1980) New method for high-accuracy determination of the fine-structure constant based on quantized hall resistance. Phys Rev Lett 45:494–497
Koch S, Haug RJ, Klitzing KV, Ploog K (1991) Size-dependent analysis of the metal-insulator transition in the integral quantum Hall effect. Phys Rev Lett 67:883–886
Krijin MPCM (1991) Heterojunction band offsets and effective masses in III-V quaternary alloys. Semicond Sci Technol 6:27–31
Kubo R, Miyake SJ, Hashitsume N (1965) In: Seitz F, Turnbell D (eds) Solid State Physics Vol 17. Academic, New York, p 269
Landau L (1930) Diamagnetism in metals. Z Physik 64:629–637
Landauer R (1957) Spatial variation of currents and fields due to localized scatterers in metallic conduction. IBM J Res Dev 1:223–238
Landauer R (1988) Spatial variation of currents and fields due to localized scatterers in metallic conduction. IBM J Res Dev 32:306–316
Landwehr G, Gerschütz J, Oehling S, Pfeuffer-Jeschke A, Latussek V, Becker CR (2000) Quantum transport in n-type and p-type modulation-doped mercury telluride quantum wells. Physica E 6:713–717
Laughlin RB (1981) Quantized hall conductivity in two dimensions. Phys Rev B 23:5632–5633
Lier K, Gerhardts RR (1994) Self-consistent calculations of edge channels in laterally confined two-dimensional electron systems. Phys Rev B 50:7757–7767
Liu H, Buchanan M, Wasilewski Z (1998) How good is the polarization selection rule for intersubband transitions? Appl Phys Lett 72:1682
Luttinger JM (1956) Quantum theory of cyclotron resonance in semi-conductors: General theory. Phys Rev 102:1030–1041
Luttinger JM, Kohn W (1955) Motion of electrons and holes in perturbed periodic fields. Phys Rev 97:869–883
Meaden GT (1965) Electrical Resistance of Metals. Plenum, New York
Mii YJ, Wang KL, Karunasiri RPG, Yuh PF (1990) Observation of large oscillator strengths for both $1 \rightarrow 2$ and $1 \rightarrow 3$ intersubband transitions of step quantum wells. Appl Phys Lett 56:1046
Mott NF (1966) The electrical properties of liquid mercury. Philos Mag 13:989–1014
Müller G, Weiss D, Koch S, Klitzing KV, Nickel H, Schlapp W, Lösch R (1990) Edge channels and the role of contacts in the quantum Hall regime. Phys Rev B 42:7633
Müller G, Weiss D, Khaetskii AV, Klitzing KV, Koch S, Nickel H, Schlapp W, Lösch R (1992) Equilibration length of electrons in spin-polarized edge channels. Phys Rev B 45:3932–3935
Ohta K (1971a) Broadening of Landau levels in two-dimensional electron gas: Its effect on surface capacitance. Jpn J Appl Phys 10:850–863

Ohta K (1971b) Broadening of Landau levels in two-dimensional electron gas II: Transverse magnetoconductance. Jpn J Phys Soc 31:1627–1638

Omar MA (1975) Elementary Solid State Physics: Principles and Applications. Addison-Wesley, Michigan, p 238

Ono Y (1982) Localization of electrons under strong magnetic fields in a two-dimensional system. Jpn J Phys Soc 51:237–243

Pan JL, West L C, Walker S J, Malik RJ, Walker JF (1990) Inducing normally forbidden transitions within the conduction band of GaAs quantum wells. Appl Phys Lett 57:366

Pfeuffer-Jeschke A, Goschenhofer F, Cheng SJ, Latussek V, Gerschutz J, Becker CR, Gerhardts RR, Landwehr G (1998) Cyclotron masses of asymmetrically doped HgTe quantum wells. Physica B-Condensed Matter 256–258:486–489

Pidgeon CR, Brown RN (1966) Interband magneto-absorption and faraday rotation in InSb. Phys Rev 146:575–583

Pincherle L (1971) Electron energy bands in solids. MacDonald, p 98

Prange RE (1981) Quantized hall resistance and the measurement of the fine-structure constant. Phys Rev B 23:4802–4805

Prange RE, Girvin SM (1990) The Quantum Hall Effect. Springer, New York

Rashba EI (1960) FiZ Tverd Tela (Leningrad) 2:1224; Sov Phys Solid State 2:1109

Sarma SD (2001) Spintronics. American Scientist 89:516

Satpathy S, Martin RM, Van de Walle CG (1988) Electronic properties of the (100) (Si)/(Ge) strained-layer superlattices. Phys Rev B 38:13237–13245

Schultz M, Merkt U, Sonntag A, Rössler U, Winkler R, Colin T, Helgesen P, Skauli T, Løvold S (1998) Crossing of conduction- and valence-subband Landau levels in an inverted HgTe/CdTe quantum well. Phys Rev B 57:14772–14775

Shao J, Dörnen A, Winterhoff R, Scholz F (2002) Ordering parameter and band-offset determination for ordered $Ga_xIn_{1-x}P/(Al_{0.66}Ga_{0.34})_yIn_{1-y}P$ quantum wells. Phys Rev B 66:035109

Shubnikov L, DeHaas WJ (1930) Leiden Commun 207c

Sizmann R (1986) Diploma Arbeiter. T U München, München

Sporken R, Sivananthan S, Faurie JP, Ehlers DH, Fraxedas J, Ley L, Pireaux JJ, Caudano R (1989) Temperature-dependent photoemission study of the HgTe–CdTe valence-band discontinuity. J Vac Sci Technol A 7:427–430

Störmer HL (1999) Nobel Lecture: The fractional quantum Hall effect. Rev Mod Phys 71:875–889

Störmer HL, Schlesinger Z, Chang A, Tsui DC, Gossard AC, Wiegmann W (1983) Energy structure and quantized hall effect of two-dimensional holes. Phys Rev Lett 51:126–129

Teran FJ, Potemski M, Maude DK, Andrearczyk T, Jaroszynski J, Karczewski G (2002) Pauli paramagnetism and Landau level crossing in a modulation doped CdMnTe/CdMgTe quantum well. Phys Rev Lett 88:186803

Touloukian J (1967) Thermophysical properties research literature retrieval guide Vols II and III. Purdue University Press, La Fayette

Truchseβ von M, Latussek V, Goschenhofer F, Becker CR, Landwehr G, Batke E, Sizmann R, Helgesen P (1995) Magneto-optics and valence-band discontinuity in a HgTe-Hg1-xCdxTe superlattice. Phys Rev B 51:17618–17623

Tsui DC, Störmer HL, Gossard AC (1982) Two-dimensional magnetotransport in the extreme quantum limit. Phys Rev Lett 48:1559–1562

Van de Walle CG (1989) Band lineups and deformation potentials in the model-solid theory. Phys Rev B 39:1871–1883

Van de Walle CG, Martin RM (1986) Theoretical calculations of heterojunction discontinuities in the Si/Ge system. Phys Rev B 34:5621–5634

Van de Walle CG, Martin RM (1987) Theoretical study of band offsets at semiconductor interfaces. Phys Rev B 35:8154–8165

Wang TY, Stringfellow GB (1990) Strain effects on GaxIn1–xAs/InP single quantum wells grown by organometallic vapor-phase epitaxy with $0 \leqslant x \leqslant 1$. J Appl Phys 67:344–352

Wees BJV, Willems EMM, Harmans CJPM, Beenakker CWJ, Houten HV, Williamson JG, Foxon CT, Harris JJ (1989) Anomalous integer quantum Hall effect in the ballistic regime with quantum point contacts. Phys Rev Lett 62:1181–1184

Weiler MH, Aggarwal RL, Lax B (1978) Warping- and inversion-asymmetry-induced cyclotron-harmonic transitions in InSb. Phys Rev B 17:3269–3283

Weisbuch C, Vinter B (1991) Quantum Semiconductor Structures. Academic, San Diego, CA

Weiss D, Klitzing KV (1987) In: Landwehr G (ed) Springer Series in Solid-State Sciences Vol 71. Springer, Berlin, p 57

Wolf SA, Awschalom DD, Buhrman RA, Daughton JM, Molnár SV, Roukes ML, Chtchelkanova AY, Treger DM (2001) Spintronics: A spin-based electronics vision for the future. Science 294:1488–1495

Yang RQ, Xu JM, Sweeny M (1994) Selection rules of intersubband transitions in conduction-band quantum wells. Phys Rev B 50:7474–7482

Zhang XC (2001) Dissertation Zu Erlangung des naturwissenschaftlichen Doktorgrades der Bayerischen Julius-Maximilians. Universität Würzburg, Würzburg

Zhang XC, Peuffer-Jeschke A, Ortner K, Hock V, Buhmann H, Becker CR, Landwehr G (2001) Phys Rev B 63:245305

Zunger A (1998) Electronic-structure theory of semiconductor quantum dots. MRS Bulletin 23:35

Chapter 6
Devices Physics

6.1 HgCdTe Photoconductive Detector

6.1.1 Brief Introduction to Photoconductive Device Theory

A series of excellent fundamental properties of HgCdTe make it the almost ideal material for infrared detectors (Long 1970; Levinstein 1970; Dornhaus and Nimtz 1976). There are two photon detector classification categories: photovoltaic and photoconductive. The semiconductor properties of HgCdTe make it suitable for fabricating both detector kinds (Long 1970; Kruse 1979; Kingston 1978; Eisenman et al. 1977; Broudy and Mazurczyk 1981; Tang and Tong 1991; Xu and Fang 1996).

Let us first discuss device theory and performance characteristics of photoconductive detectors. Consider a slab of a photoconductive device of length l, width w and thickness d connected in a circuit shown in Fig. 6.1. For a case where there is no radiation incident on the device, with an applied voltage V_0:

$$V_0 = I_0(R_0 + R_L), \tag{6.1}$$

where R_0 and R_{L} are the device and load resistances. R_0 is given by:

$$R_0 = \frac{l}{\sigma_0 w d} \tag{6.2}$$

where the conductivity σ_0 includes both the dark-conductivity, σ_{d}, and any background radiation-induced conductivity, σ_{B}:

$$\sigma_0 = \sigma_{\mathrm{d}} + \sigma_{\mathrm{B}}. \tag{6.3}$$

For a case where there is also an incident signal radiation present on the device, the conductivity is given by:

$$\sigma = \sigma_{\mathrm{d}} + \sigma_{\mathrm{B}} + \sigma_{\mathrm{s}}, \tag{6.4}$$

J. Chu and A. Sher, *Device Physics of Narrow Gap Semiconductors*, Microdevices,
DOI 10.1007/978-1-4419-1040-0_6, © Springer Science+Business Media, LLC 2010

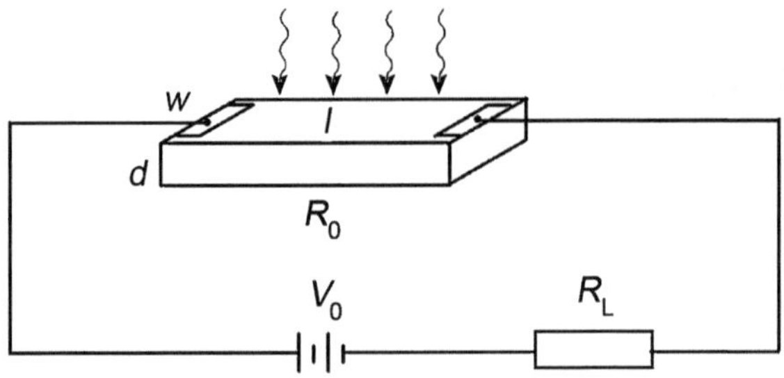

Fig. 6.1 Sketch of the working process of a photoconductive detector

where σ_s arises from the signal radiation. The net resistance of the device is therefore:

$$R = \frac{l}{\sigma w d}. \tag{6.5}$$

For the case where an incident signal radiation is present, the voltage difference between two ends of the device (or load-resistor) differs from that when there is no incident signal radiation. The difference between the signal and no signal voltages equals the signal voltage V_s, which is given by:

$$V_s = \frac{V_b R_0 R_L}{(R_0 + R_L)^2} \cdot \frac{\sigma_s}{\sigma} = \frac{V_b R_0 R_L}{(R_0 + R_L)^2} \cdot \frac{\Delta n}{(n_0 + n_b)}, \tag{6.6}$$

where $\Delta n = n_s - (n_b + n_0)$ is the difference between the photogenerated carrier density from the signal, n_s, and the sum of the thermal generated carrier density, n_0, plus the background radiation-induced carrier density, n_B. Usually $R_L \gg R_0$, so the coefficient term in (6.6) simplifies to:

$$\frac{V_b R_0 R_L}{(R_0 + R_L)^2} = \frac{V_b R_0}{R_0 + R_L} \cdot \frac{R_L}{R_0 + R_L} \cong \frac{V_b R_0}{R_0 + R_L} = V_b, \tag{6.7}$$

where V_b is the bias voltage applied across the device. Hence we have:

$$V_s = V_b \cdot \frac{\sigma_s}{\sigma} = V_b \cdot \frac{\Delta n}{(n_0 + n_B)} \tag{6.8}$$

Electrons and holes both contribute to the total conductivity. Then the conductivity from the dark current is:

$$\sigma_d = e(n_o \mu_n + p_0 \mu_p), \tag{6.9}$$

where n_0, and p_0 are electron and hole densities and μ_n, and μ_p are electron and hole motilities in thermal equilibrium. Also, the background radiation-induced conductivity σ_B is:

$$\sigma_B = e n_B(\mu_n + \mu_p), \tag{6.10}$$

where n_B is the excess e–h pair density generated by the background radiation.

For the case where the incident signal radiation is modulated, the electron and hole life times τ_n and τ_p must be considered in a conductivity analysis. Given a quantum efficiency, η, defined as the ratio of the number of electron–hole pairs generated per incident photon, and the incident photon flux, J_s, defined as the number of photons per unit time and unit area striking the detector, the steady state photo-generated carrier density is given by:

$$\Delta n = \eta J_s.\frac{wl}{wld} \cdot \tau_n = \eta J_s \frac{\tau_n}{d}, \text{ and } \Delta p = \eta J_s \frac{\tau_p}{d}.$$

Thus, the photo signal-induced conductivity is:

$$\sigma_s = e(\mu_n \Delta n + \mu_p \Delta p) = \frac{\eta J_s e}{d}(\mu_n \tau_n + \mu_p \tau_p). \tag{6.11}$$

Combining (6.9)–(6.11), with (6.8), yields an expression for the signal voltage:

$$V_s \approx \frac{\eta J_s(\mu_n \tau_n + \mu_p \tau_p) V_b}{d[(n_0 \mu_n + p_0 \mu_p) + n_B(\mu_n + \mu_B)]}. \tag{6.12}$$

The rate of signal photons incident at the photosensitive surface per unit time is, $J_s \cdot lw$. The total incident energy is, $J_s \cdot lw \cdot E_\lambda$, where E_λ is the energy ($h\nu = hc/\lambda$) of each incident photon with wavelength λ. Then for a detector with a photosensitive area, $A = lw$, the incident power giving rise to the signal is:

$$P_\lambda = J_s \cdot AE_\lambda. \tag{6.13}$$

The voltage responsibility is defined as the ratio of the photo-induced signal voltage V_s to the incident photopower, P_λ. From (6.12) and (6.13), the voltage responsibility of a photoconductive device is given by expression:

$$R_\lambda = \frac{V_s}{P_\lambda} = \frac{\eta(\mu_n \tau_n + \mu_p \tau_p) V_b}{AdE_\lambda[(n_0 \mu_n + p_0 \mu_p) + n_B(\mu_n + \mu_p)]}. \tag{6.14}$$

The units of the responsivity are usually taken to be: volts/watt. Actually a more rational measure of a device is the normalized responsivity defined as: $R_\lambda^* \equiv \frac{V_s A}{P_s}$. This normalized responsivity depends on, P_s / A, which is independent of the geometry of the device and therefore a more accurate measure of its material properties.

Generally photoconductive devices are fabricated from n-type material. For an n-type material with, $n_0 \gg p_0$, the background radiation-induced conductivity being smaller than the dark-current-induced conductivity, $\sigma_B < \sigma_d$, and supposing, $\tau_n \approx \tau_p$, (6.14) becomes:

$$R_\lambda = \frac{\eta \tau V_b}{AdE_\lambda n_0}, \tag{6.15}$$

or

$$R_\lambda = \frac{\eta(\lambda)}{lw} \cdot \frac{\lambda}{hc} \cdot \frac{V_b \tau}{n_0}. \tag{6.16}$$

Assume there are no drift, or diffusion, processes then the relation between the average excess carrier density and time is given by:

$$\frac{\partial \Delta p}{\partial t} = J_{\mathrm{s}} A \eta = -\Delta p / \tau, \tag{6.17}$$

where Δp is the average excess carrier density and τ is the recombination life time. Then we have:

$$\Delta p(t) = J_s A \eta \tau \mathrm{e}^{-t/\tau}, \tag{6.18}$$

or after a Fourier transform, in terms of frequency:

$$\Delta p(f) = J_{\mathrm{s}} A \eta \tau [1 + (2\pi f \tau)^2]^{-1/2}. \tag{6.19}$$

Thus the responsibility's dependence on frequency is given by:

$$R_\lambda(f) = R_\lambda(0)[1 + (2\pi f \tau)^2]^{-1/2}, \tag{6.20}$$

where $R_\lambda(0)$ is the static value given by (6.16).

The responsibility, R_λ, is an important parameter to characterize the performance of a device. However, a full characterization of a device cannot be attained from only the responsibility. Another important parameter is noise. If a small noise power, $P_{\mathrm{N}} = V_{\mathrm{N}}^2 / R_{\mathrm{L}}$, is generated or received by a device, it will induce an rms noise voltage, V_{N}, across the load resistance. The smaller V_{N} is the more sensitive the device is regarded to be. $P_{\mathrm{N}\lambda} \equiv (P_\lambda P_{\mathrm{N}})^{1/2}$ is the so-called noise-equivalent-power. It is related to the signal power needed to produce a signal voltage equal to the rms noise voltage. Obviously, the smaller $P_{\mathrm{N}\lambda}$ is, the more sensitive the device will be. The detectivity is a parameter that accounts for the noise as well as the responsivity. It is defined as the reciprocal of noise-equivalent-power multiplied by the square root of the product of the bandwidth, Δf, times the device area, A.

$$D_\lambda^* \equiv \frac{\sqrt{A \Delta f}}{P_{\mathrm{N}\lambda}}. \tag{6.21}$$

The units of the detectivity are usually taken to be: $\mathrm{cm} \cdot (\mathrm{Hz})^{1/2}/\mathrm{watt}$.

The relation between the responsivity and the detectivity is:

$$\begin{aligned} D_\lambda^* &= \frac{(A\Delta f)^{1/2}}{P_{\mathrm{N}\lambda}} = \frac{(A\Delta f)^{1/2}}{(P_\lambda P_{\mathrm{N}})^{1/2}} = \frac{V_{\mathrm{s}}(A\Delta f)^{1/2}}{V_{\mathrm{N}} P_\lambda} = \frac{R_\lambda (A\Delta f)^{1/2}}{V_{\mathrm{N}}}. \\ &\approx \frac{\eta \tau V_{\mathrm{b}} \Delta f^{1/2}}{A^{1/2} d E_\lambda n_0 V_{\mathrm{N}}}. \end{aligned} \tag{6.22}$$

Note that, P_λ / A, depends only on material properties, not the geometry of a detector, and, P_{N}, may contain geometric terms but they are often essential to the device's material properties, so D_λ^* is also properly normalized.

Suppose only thermal noise is included, the so-called Johnson noise limit, so that: $V_N = \left(\bar{V}^2\right)^{1/2} = (4k_\mathrm{B}TR\Delta f)^{1/2}$. The device resistance is, $R = \dfrac{l}{(\sigma_\mathrm{d} + \sigma_\mathrm{B})\, wd} \approx \dfrac{l}{en_0\mu_\mathrm{n}wd}$, so the detectivity is given by:

$$D_\lambda^* \cong \frac{\eta\tau V_\mathrm{b}}{2E_\lambda l}\left(\frac{e\mu_\mathrm{n}}{dn_0k_\mathrm{B}T}\right)^{1/2}. \tag{6.23}$$

From (6.23), generally speaking, a high detectivity is achieved with a large bias field V_b/l applied across the sample, with a small thickness d, a low operating temperature T, a long life-time τ, a large mobility μ_n, and a small carrier density n_0. Surely when an applied bias field increases, Joule-heating and generation–recombination noise will become more important. Thus the above analyses are correct only in the Johnson noise limited condition.

6.1.2 Device Performance Characterization Parameters

In the section above the principles and general description of photoconductive device performance are briefly introduced. In this section, important parameters that characterize device performance will be treated. The spectral detectivity, $D_\lambda^*(\lambda, f, \Delta f)$, and responsibility, $R_\lambda(f)$, (6.20) are given by:

$$D_\lambda{}^* \equiv D^*(\lambda, f, \Delta f) = [R_\lambda(f)/V_\mathrm{N}(f)](A\Delta f)^{1/2}, \tag{6.24}$$

and

$$R_\lambda = \frac{V_\mathrm{S}(f)}{P_\lambda}, \tag{6.25}$$

or

$$D_\lambda{}^* = \frac{V_\mathrm{S}}{V_\mathrm{N}} \cdot \frac{(A\Delta f)^{1/2}}{P_\lambda}, \tag{6.26}$$

where V_S is root-mean-square signal voltage, V_N is root-mean-square noise voltage, and all quantities are measured in an electronic frequency bandwidth, Δf. P_λ is the optical power in watts, incident on the photosensitive area of the detector within the spectral range, λ to $\lambda + \Delta\lambda$.

P_λ can be obtained by using a calibrated blackbody light source and a narrow band filter. The spectral detectivity $D_\lambda{}^*$ can be found by first measuring blackbody detectivity, $D_\mathrm{B}^*(T_\mathrm{B}, f, \Delta f)$. D_B^* is the detectivity of a detector exposed to the entire blackbody spectrum at temperature T_b,

$$D_\mathrm{B}^* = \frac{V_\mathrm{S}/V_\mathrm{N}}{P_\mathrm{B}(T_\mathrm{b})}[A\Delta f]^{1/2}. \tag{6.27}$$

Because $P_\mathrm{B}(T_\mathrm{b})$ is total optical power emitted from a blackbody of temperature T_b, its net emitted power can be calculated accurately. As long as the detector's spectral response encompasses the entire spectral range emitted by the blackbody, and its emissivity is known, this method is useful. If the spectral range of the detector is too narrow, then a calibrated filter can be used to determine P_{Beff}.

Since the radiation spectrum from a blackbody is known, if the spectral distribution of the quantum efficiency is also known, then the detectivity spectrum and blackbody detectivity can be linked by the expression:

$$g \equiv {}^{D^*_\lambda}/_{D^*_\text{B}}, \tag{6.28}$$

$$g = \frac{\eta(\nu_\text{s})}{\nu_\text{S}} \left[\int_{\nu\text{c}}^{\infty} \frac{\eta(\nu) M(\nu, T_\text{B}) d\nu}{P_\text{B} \nu} \right]^{-1}, \tag{6.29}$$

where ν_S is the signal frequency. $M(\nu, T_\text{B})$ is the radiated power per unit frequency interval at frequency ν from a blackbody of temperature T_B. P_B is the net power emitted from an ideal blackbody of temperature T_B. $\eta(\nu)$ is the quantum efficiency spectrum; the number of e–h pairs generated by incident photons of frequency ν. For a photon detector, the signal is proportional only to the number of absorbed photons. For those photons that are not absorbed, or are reflected from the detector's surface, they do not contribute to the response. For short wavelength photons, D^* increases with an increase of their wavelength, because, E_λ, the energy per photon becomes smaller. When the wavelength corresponding to the bandgap of direct bandgap semiconductors is reached, then D^* drops rapidly to zero with a further wavelength increases, because then the photons no longer have sufficient energy to excite an electron–hole pair.

The peak responsivity appears at wavelength, λ_p, and D^* at this wavelength is labeled, $D^*_{\lambda\text{p}}$. The cut-off wavelength, λ_c, is usually defined as wavelength at which responsivity is 50% of its peak on a responsivity-wavelength curve. The frequency corresponding to λ_c is ν_c. For HgCdTe detectors, $\lambda_\text{c}/\lambda_\text{p}$, is approximately 1.1. A rough estimate of, λ_c, is $\lambda_\text{c} = 1.24/E_\text{g}$. In practice, because HgCdTe is a direct gap material its cut off is abrupt, so the wavelength corresponding to the forbidden bandwidth is near or slightly smaller than, λ_p, depending on the thickness of the sensing material.

For wavelengths larger than λ_c ($h\nu < E_\text{g}$), the quantum efficiency η drops rapidly to zero due to rapid drop of the sensing material's light absorption coefficient. One simple assumption is that η remains constant at wavelength smaller than λ_c and zero at other wavelengths. From (6.29), g depends not only on T_B but also on λ_c. Take as an example, $T_\text{B} = 500\,\text{K}$, $\lambda_\text{c} = 12\,\mu\text{m}$, $g = 3.5$, then the spectral dependence of $\eta(\lambda)$ can be measured by fitting D^*_λ data to (6.29).

It is a little rough to calculate λ_c from $1.24/E_\text{g}$. A more precise calculation for λ_c must consider the spatial distribution of the photogenerated carriers. In practice, the cut-off wavelength, λ_c, depends on sample thickness; the minority carrier diffusion length, and device design. λ_c should be acquired from an analyses of the device spectral response. The expression is given by:

$$R_\lambda = \frac{\lambda}{hc} \frac{V_\text{b}}{J_\text{s} A} \frac{\Delta \bar{p}(\mu_\text{n} + \mu_\text{p})}{(n\mu_\text{n} + p\mu_\text{p})}, \tag{6.30}$$

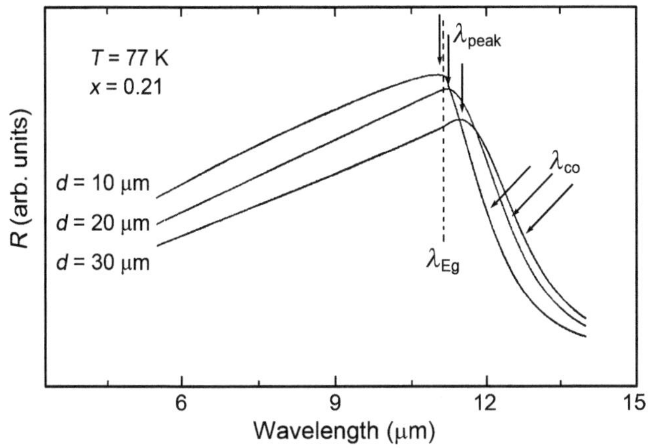

Fig. 6.2 Responsibility of $Hg_{1-x}Cd_xTe$, $x = 0.21$, with thicknesses $d = 10$, 20, and 30 μm

where V_b is bias voltage, J_s the incident photon flux, A the device area, n and p are the electron and hole densities in thermal equilibrium; μ_n and μ_p are electron and hole mobilities. $\Delta\bar{p}$ is the minority carrier density that depends on sample thickness, absorbance and the diffusion length of the carriers. The spatial distribution of minority carriers should be included in the calculation. The wavelength, λ_{peak}, at which the responsivity has its peak, R_{peak}, and that for which it reaches its half-maximum, the cut-off, λ_{co}, R_{co}, are obtained from the calculated $R-\lambda$ curve shown in Fig. 6.2.

In Fig. 6.2, the calculated responsibility of $Hg_{1-x}Cd_xTe$ with $x = 0.21$ at temperature of 77 K is presented, supposing thicknesses of the specimens are $d = 10$, 20, and 30 μm. As indicated by the dotted line, λ_{E_g} stays fixed no matter how the thickness varies, but λ_{peak} and λ_{co} vary with sample thickness (see Fig. 6.2). Relations between the cut-off wavelength, λ_{co}, on sample thickness, for HgCdTe photoconductive devices with different compositions, at 77 K, are shown in Fig. 6.3. In Fig. 6.3, for sample thicknesses in the range 5–50 μm, λ_{co} increases strongly with sample thickness increases. For sample thicknesses over 50 μm, λ_{co} increases only weakly with sample thickness.

From the voltage responsibility spectrum calculation for HgCdTe photoconductive devices with different compositions and different thicknesses, we can get a set of λ_{peak} and λ_{co} data. Expressions for the cut-off wavelength of detectors versus the peak-response wavelengths, which include the spatial distribution of photogenerated carriers involved, can be deduced from these data (Chu et al. 1998):

$$\lambda_{co} = \frac{a(T)}{x - b(T) - c(T)\log(d)}, \tag{6.31}$$

$$\lambda_{peak} = \frac{A(T)}{x - B(T) - C(T)\log(d)}, \tag{6.32}$$

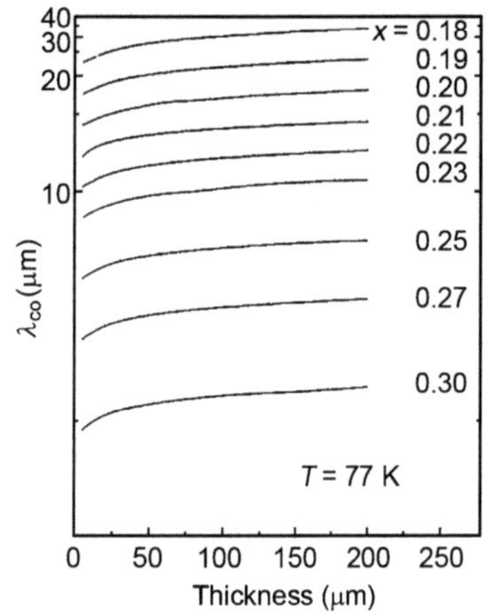

Fig. 6.3 Cut-off wavelength, λ_{co}, of HgCdTe photoconductive detectors with different compositions at 77 K versus the thickness of these samples

where

$$
\begin{aligned}
a(T) &= 0.7 + 6.7 \times 10^{-4}T + 7.28 \times 10^{-8}T^2 \\
b(T) &= 0.162 - 2.6 \times 10^{-4}T - 1.37 \times 10^{-7}T^2 \\
c(T) &= 4.9 \times 10^{-4} + 3.0 \times 10^{-5}T + 3.51 \times 10^{-8}T^2 \\
A(T) &= 0.7 + 2.0 \times 10^{-4}T + 1.66 \times 10^{-8}T^2 \\
B(T) &= 0.162 - 2.8 \times 10^{-4}T - 2.29 \times 10^{-7}T^2 \\
C(T) &= 3.5 \times 10^{-3} - 3.0 \times 10^{-5}T - 5.85 \times 10^{-8}T^2
\end{aligned}
$$

The applicable range of the above formula is $0.16 < x < 0.60$, $4.2 < T < 300$ K, and $5 < d < 200\,\mu$m. In this formula, λ_{peak}, λ_{co}, and d are all in units of μm. Comparison of the calculated λ_{co} from this formula with experimental data is shown in Fig. 6.4. The experimental data cited are from different laboratories (Schmit and Stelzer 1969; Hansen et al. 1982; Price and Royd 1993).

The influence of photogenerated carrier distributions need to be considered especially when very-long cut-off wavelength HgCdTe infrared detectors are designed (Phillips et al. 2001, 2002).

6.1.3 Noise

In detectivity expressions, noise which affects the performance of detectors is a very important factor. Noise originates from many sources, not only those related to the photon absorption process, but also those involved in the frequency and temperature dependences. The photoconductive detector noise catalog includes thermal, g–r, $1/f$, and amplifier noise. Besides these there is also a noise source arising from the background radiation.

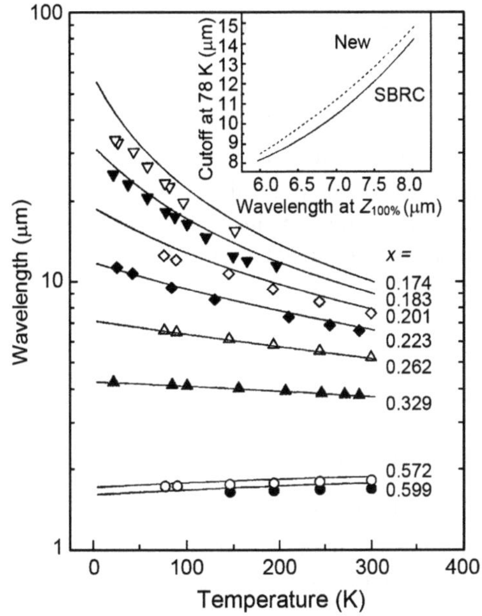

Fig. 6.4 Comparison of calculated λ_{co} from (6.31) with corresponding experimental data, cited from Schmit and Stelzer (1969), Hansen et al. (1982), and Price and Royd (1993)

6.1.3.1 Thermal Noise

Thermal noise voltage V_j, also named Johnson noise, comes from temperature-induced voltage fluctuations across the resistive component of the sensing material with resistance R_j. The rms (root-mean-square) fluctuations generated by the temperature induces a noise per unit band-pass Δf that does not depend on properties of material other than its net resistance the Johnson noise can be shown to be:

$$V_j^2 = (4kTR_j)\Delta f. \tag{6.33}$$

Supposing one is sensing the signal across the detector, i.e., the resistance R_0, then the noise generated by R_L adds in quadrature with that from R_0, but R_0 is in parallel with R_L and so shunts it. Thus the expression is:

$$V_{N_{\text{eff}}}^2 = 4k_B T\Delta f\left[R_0^2 + \left(\frac{R_0 R_L}{R_0 + R_L}\right)^2\right]^{1/2},$$

or

$$V_{N_{\text{eff}}} = \left[1 + \left(\frac{R_L}{R_L + R_0}\right)^2\right]^{1/4}\sqrt{4k_B T R_0 \Delta f} \tag{6.34}$$

6.1.3.2 Generation–Recombination Noise

Generation–Recombination (g–r) noise comes from generation and then recombination processes of carriers in the sensing material. The average number of carriers in a detector is determined by a balance between these two processes. Fluctuations of this carrier number results in another noise, called the "generation–recombination" noise. For a semiconductor dominated by a direct recombination process, the noise voltage per unit band–pass Δf, is (Van Vliet 1958, 1967, 1970):

$$V_{\mathrm{g-r}}^2 = \frac{4V_{\mathrm{b}}^2}{(lwd)^2}\left\langle \Delta N^2 \right\rangle \frac{\tau}{n_0^2}\frac{\Delta f}{1+\omega^2\tau^2}. \tag{6.35}$$

where $\left\langle \Delta N^2 \right\rangle$ is the average fluctuation of the majority carrier number, and n_0 is the carrier concentration. For a two-level system, $\left\langle \Delta N^2 \right\rangle = g\tau$, where g is generation probability per unit time, and τ is minority carrier lifetime. $\tau = \dfrac{\Delta N}{J_{\mathrm{s}}(\lambda)\,\eta(\lambda)\,A}$, where $J_{\mathrm{s}}(\lambda)$ is the signal photon flux. Suppose the thermally induced g–r process is independent of the photon-induced g–r process, then the total average majority carrier fluctuation equals the sum of these two g–r processes:

$$\left\langle \Delta N^2 \right\rangle = (g\tau)_{\mathrm{thermal}} + (g\tau)_{\mathrm{photon}}, \tag{6.36}$$

where:

$$\begin{aligned} (g\tau)_{\mathrm{thermal}} &= \frac{n_0 p_0}{n_0 + p_0} lwd \\ (g\tau)_{\mathrm{photon}} &= p_{\mathrm{b}} lwd, \end{aligned} \tag{6.37}$$

where p_{b} is the hole density excited by background radiation. Thus we have:

$$V_{\mathrm{g-r}} = \frac{2V_{\mathrm{b}}}{(lwd)^{1/2} n_0}\left[\left(1 + \frac{p_0}{p_{\mathrm{b}}}\frac{n_0}{n_0 + p_0}\right)\left(\frac{p_{\mathrm{b}}\tau\Delta f}{1+\omega^2\tau^2}\right)\right]^{1/2}. \tag{6.38}$$

At low temperature, the material is extrinsic. p_0 is very small and p_{b} is dominates. This is the so-called background-limited inferred photon noise, which is determined by fluctuations of the incident photon flux. The background-limited infrared photon-detector (BLIP) detectivity D_{BLIP}^* is determined by this limitation.

6.1.3.3 1/*f* Noise

There is another noise source, named 1/*f* noise, in detectors related to practical device structures. Improving device design and fabrication technology can reduce this noise. The low-frequency response of a device is mainly limited by 1/*f* noise. In photoconductive theory, 1/*f* noise can be studied by classical methods. It is independent of all other noise sources and its magnitude is inversely proportion to frequency. This noise is assumed to be coming from every part of detector. A noise expression given by Kruse et al. (1962) is:

$$V_{1/f} = \frac{C_1}{d}\frac{l}{w}E^2\frac{\Delta f}{f}, \tag{6.39}$$

where l, w, d is length, width and thickness of the detector, E the dc bias electric field strength, Δf the band-width of the noise, f the frequency, and C_1 is a coefficient fixing the intensity of the $1/f$ noise. C_1 depends on the carrier density rather than the detector dimensions. There exists a frequency f_0 at which the $1/f$ noise power and the g–r noise power are equal:

$$V_{1/f_0}^2 = V_{\text{g–r}}^2(0). \tag{6.40}$$

Thus the $1/f$ noise at frequency f can be expressed as:

$$V_{1/f}^2 = (f_0/f)V_{\text{g–r}}^2(0), \tag{6.41}$$

where f_0 may be a function of bias, temperature, and background photon flux.

Broudy (1974) developed a $1/f$ noise theory for HgCdTe photoconductive devices by investigating a large number of HgCdTe photoconductive device working parameters. In his empirical-based theory, the $1/f$ noise voltage $V_{1/f}$ has a simple relation with the g–r voltage, $V_{\text{g–r}}$, captured in the formula:

$$V_{1/f}^2 = (k_1/f)V_{\text{g–r}}^3, \tag{6.42}$$

where k_1 is a constant. The $1/f$ noise is a "current noise" in this theory. Since $V_{\text{g–r}}$ varies with current, $1/f$ noise decreases with an increase of the detector resistance. The transition frequency f_0 in the $V_{\text{g–r}}$ expression is given by:

$$f_0 = k_1 V_{\text{g–r}}. \tag{6.43}$$

A classical transition frequency can be calculated from (6.38)–(6.40). For an n-type HgCdTe, $n_0 \gg p_b$, then f_0 is:

$$f_0 = \frac{C_1 n_0^2}{4(p_b + p_0)\tau}, \tag{6.44}$$

where, $(\omega\tau)^2 \ll 1$, has been assumed.

If the temperature is low enough, then thermally generated carriers can be ignored, and we can write an f_0 in the "background limited infrared photodetector" case:

$$f_0^{\text{BLIP}} = C_1 n_0^2 d \,/\, 4\tau^2 \eta J_B, \tag{6.45}$$

so, f_0 depends on photosensitive material thickness rather than its area.

In a classical theory, C_1, can be calculated without consideration of the noise coming from the surface or inside of the bulk material. If there are traps at the semiconductor surface, fluctuations of electrons trapped or released by the traps will result in fluctuations of the bulk electron density, therefore fluctuations of the conductance (Van der Ziel 1959). The $1/f$ frequency dependence of the noise power arises from these fluctuations. Suppose electrons go into the bulk from these surface traps by tunneling, then from this process, we can deduce a carrier life time probability distribution. The tunneling probability is exponentially proportion to the tunneling distance. A C_1 expression given by this model is:

$$C_1 \approx \left(N_t/4n_0^2\right) l/\alpha d, \tag{6.46}$$

where N_t is the trap density, and α is the tunneling characteristic length.

The Hooge model (Hooge 1969) which assumes the $1/f$ noise totally originates from a bulk process gives:

$$C_1 = 2 \times 10^{-3}/n_0. \tag{6.47}$$

However, it lacks support from experiments.

Equations (6.42) and (6.43) offer a reason, $1/f$ noise increases as the g–r noise increases. Therefore, to reduce the $1/f$ noise one needs to select a low g–r noise material, if all other parameters remain the same. This conclusion is significant. For example, one should use the lowest applicable bias current, keep the detector temperature lower than the thermal carrier generation range, try as best one can to lower background radiation, select a semiconductor material with a high donor concentration, and adopt improved fabrication technology to reduce k_1.

Figure 6.5 illustrates a typical HgCdTe detector's measured f_0 versus the background photon flux. When the background photon flux is high, f_0 of the detector follows a $J_B^{-1/2}$ behavior. When the background photon flux is smaller than 10^{17} photons/cm$^2 \cdot$ s, the dependency of f_0 on Q_B will be weak, then the thermal g-r noise will become more important. (Borrello et al. 1977).

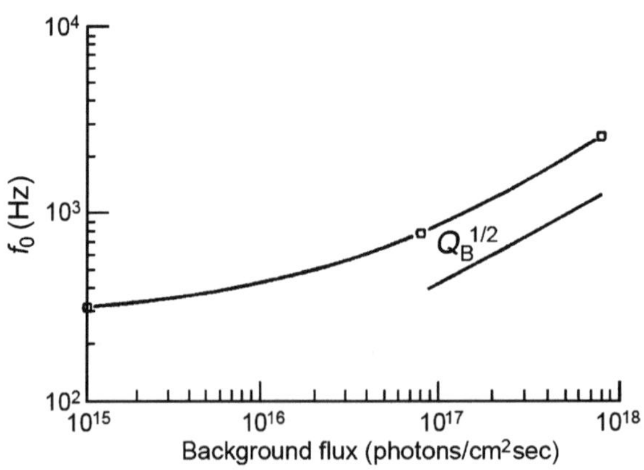

Fig. 6.5 A HgCdTe detector's measured f_0 versus the background photon flux

6.1.3.4 Amplifier Noise

Furthermore, there are also a voltage noise source, e_a, and a current noise source, i_a, usually called the white noise source at the input of the amplifier. If a detector with resistance of r_d is connected to the amplifier's input, it introduces noise, also called the amplifier noise, which is:

$$V_a^2 = e_a^2 + i_a^2 r_d^2.$$

For a HgCdTe photoconductive detector, r_d is usually smaller than 100Ω. Since a commonly used upper frequency limitation is 10 MHz, the input capacitance contribution to the impedance usually can be omitted.

6.1.3.5 Total Noise

From above analyses, the total noise of a detector is:

$$V_t^2 = V_j^2 + V_{g-r}^2 + V_{1/f}^2 + V_a^2. \tag{6.48}$$

Figure 6.6 shows a typical noise spectrum and indicates contributions from each source.

Combining the above equations, D_B^* under illumination by blackbody radiation is:

$$D_B^* = \frac{V_s(A\Delta f)^{1/2}}{P_B(T_b)} \cdot \frac{1}{V_N} = \frac{V_s(A\Delta f)^{1/2}}{P_B(T_b)} \cdot \left(\frac{1}{V_j^2 + V_{1/f}^2 + V_{g-r}^2 + V_a^2}\right)^{1/2}, \tag{6.49}$$

or

$$D_B^* = D_{BLIP}^* \left(1 + \frac{p_0}{p_b} \frac{n_0}{n_0 + p_0}\right)^{-1/2} \left(1 + \frac{f_0}{f} + \frac{V_j^2 + V_a^2}{V_{g-r}^2(f)}\right)^{-1/2} \tag{6.50}$$

in this formula, D_{BLIP}^* is the detectivity of the best performance of a detector.

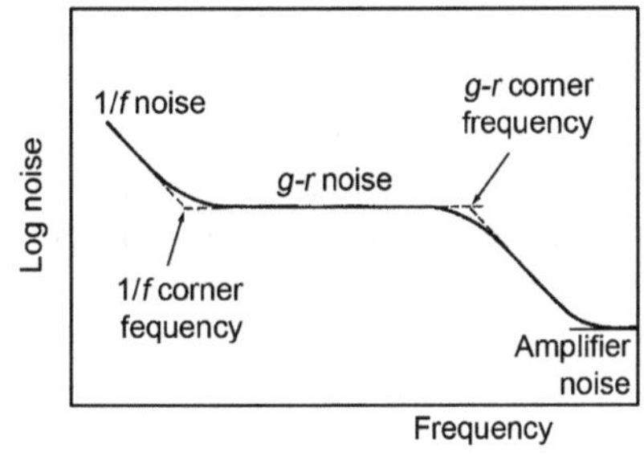

Fig. 6.6 A typical noise spectrum of a photoconductive detector

At higher temperatures, the detector material is intrinsic, $n_0 = p_0$, so:

$$\left(1+\frac{p_0}{p_b}\frac{n_0}{n_0+p_0}\right)^{1/2} \approx \left(\frac{p_0}{2p_b}\right)^{1/2} \gg 1. \tag{6.51}$$

From the D_B^* expression, (6.49), the performance of the detector is degraded. At lower temperatures, n_0, is approximately a constant, and p_0 is very small compared with n_0. Then the detector material becomes extrinsic and D_B^* is given by:

$$D_B^* = D_{BLIP}^*\left(1+\frac{p_0}{p_b}\right)^{-1/2}\left[1+\frac{f_0}{f}+\frac{V_j^2+V_a^2}{V_{g-r}^2(f)}\right]^{-1/2}. \tag{6.52}$$

Obviously, when we have:

$$\frac{V_j^2+V_a^2}{V_{g-r}^2(f)} \ll 1,\ \frac{p_0}{p_b} \ll 1,\ \text{and}\ \frac{f_0}{f} \ll 1, \tag{6.53}$$

then background limited performance is regained.

When amplifier and Johnson noise dominate, that is $V_j^2 + V_a^2 \gg V_{1/f}^2 + V_{g-r}^2$, then from (6.49), the detectivity is given by:

$$D_B^* = R_\lambda (A\Delta f)^{1/2} / (V_j^2 + V_a^2)^{1/2}. \tag{6.54}$$

The temperature and background dependence of D^*, the responsibility and the g–r noise are determined by the density increases and shorter majority and minority carrier time constants (Kinch and Borrello 1975).

For an n-HgCdTe semiconductor, the donator activation energy can be ignored and n_0 as well as its temperature dependence are:

$$n_0 = \frac{N_D - N_A}{2} + [(N_D - N_A)^2 / 4 + n_i^2]^{1/2}, \tag{6.55}$$

and the minority carrier density is:

$$p_0 = \frac{n_i^2}{n_0}. \tag{6.56}$$

For Auger recombination, the carrier life-time is (Kinch et al. 1973), $\tau = 2\tau_1 n_i^2/(n_0 + p_0)(n_0)$, where τ_1 is $\tau_1 = C_0(E_g/kT)^{3/2}\exp(E_g/kT)$, E_g is the forbidden bandgap, and C_0 is a constant. n_i can be calculated from one of several empirical formulas (Mazurczyk et al. 1974; Finkman and Nemirovsky 1979; Schmit and Stelzer 1969; Schmit 1970; Chu et al. 1983, 1991). In general we have, $n_i = n_i(E_g, T, x)$, where $E_g = E_g(x, T)$ related to the concentration x and temperature T. When these formulas are combined with that for the detectivity, its dependence on temperature is given.

In above noise analysis, strictly speaking, once the minority carrier life time is involved, surface recombination must be considered, since surface recombination affects the performance of photoconductive devices. The recombination probability per unit time at a semiconductor surface is often bigger than that in the bulk. It lowers the life-time of minority carriers.

6.1.3.6 Background Noise and Background Limited Detectivity

When thermal noise and g–r detector noise are very small, the noise from the background radiation plays a leading role. The detectivity under this condition is called background limited. The density fluctuation of the carriers excited by this background radiation produces a noise. The opto-induced carries exited by the background radiation also generate a signal. The ratio of the signal to noise is $V_s / V_N = \frac{(\eta P_s / h\nu)}{\sqrt{2\eta Q_B A \Delta f}}$. The background radiation limited detectivity is (Broudy and Mazurczyk 1981; Kruse et al. 1962; Tang and Mi 1989):

$$D^*_{\mathrm{BLIP}} = \frac{V_s}{V_N} \cdot \frac{\sqrt{A \cdot \Delta f}}{P_s} = \frac{1}{h\nu}\sqrt{\frac{\eta}{2Q_B}}. \tag{6.57}$$

For photoconductive detector, a fluctuation of the recombination rate introduces a fluctuation in the occupation probability of the carriers at this same rate. Therefore, the noise should be $\sqrt{2}$ times of V_N. The background radiation limited operation is:

$$D^*_{\mathrm{BLIP}} = \frac{1}{2h\nu}\sqrt{\frac{\eta}{Q_B}}, \tag{6.58}$$

where Q_B is background radiation photon flux.

$$Q_B = \int_{\nu_c}^{\infty} \frac{2\pi}{c^2} \frac{\nu^2 e^{h\nu/k_B T}}{(e^{h\nu/k_B T} - 1)} d\nu, \tag{6.59}$$

and ν_c is the frequency corresponding to the cut-off wavelength. Q_B represents the photon flux from the background radiation available to excite minority carriers.

From the above discussion, it can be seen from the Planck formula that the background radiation falls when the background temperature decreases. Therefore, the number of carriers excited by the background radiation also decreases. Then the background-induced noise weakens and the background limited performance (BLIP) detectivity will increase.

6.1.4 The Impact of Carrier Drift and Diffusion on Photoconductive Devices

In an actual device, drift and diffusion of carriers play important roles and must be treated carefully. On the one hand, there is a spatial density gradient adjacent to the electrical contacts, and it drives a distinct diffusion current. On the other hand, if the electric field is strong enough, carriers once generated will be swept to the electrical contacts without any recombination occurring. This is the so-called sweep-out limit. Rittner (1956) presented the earliest report on drift and diffusion in photoconductive devices. From the continuity equation and the Poisson equation, Pittmer deduced the basic photoconductive theory equations. In his derivation, a trap-effect is included, ion current is ignored, and the existence of space charge neutrality is assumed. If defects in the sensing material are ignored, then basic equation is:

$$\frac{\partial \Delta p}{\partial t} = g - \frac{\Delta p}{\tau_{\mathrm{g}}} + D\nabla \cdot \nabla \Delta p + \mu E \nabla(\Delta p), \tag{6.60}$$

where g is the generation probability per unit time, $g \equiv \eta J_{\mathrm{s}}/d(\mathrm{cm}^{-3}\mathrm{s}^{-1})$, and τ_{g} is the recombination life-time of the carriers. The appropriate diffusion constant and carrier mobility are:

$$D = \frac{n+p}{\dfrac{n}{D_{\mathrm{h}}} + \dfrac{p}{D_{\mathrm{e}}}}, \tag{6.61a}$$

$$\mu = \frac{p-n}{\dfrac{n}{\mu_{\mathrm{h}}} + \dfrac{p}{\mu_{\mathrm{e}}}}, \tag{6.61b}$$

where D_{h} and D_{e} are hole and electron diffusion constants, and μ_{h} and μ_{e} are their mobilities. Usually (6.60) is nonlinear because the coefficients D and $\mu\tau_{\mathrm{g}}$ depend on n. For a low light intensity case (low injection) $\Delta p \ll n$, so:

$$\frac{\partial \Delta p}{\partial t} = g - \frac{\Delta p}{\tau} + D_0\nabla \cdot \nabla \Delta p + \mu_0 E \nabla(\Delta p), \tag{6.62}$$

where τ is the life-time with a low excitation level, and D_0 and μ_0 are:

$$D_0 = \frac{n_0+p_0}{\dfrac{n_0}{D_{\mathrm{h}}} + \dfrac{p_0}{D_{\mathrm{e}}}}, \tag{6.63a}$$

$$\mu_0 = \frac{p_0-n_0}{\dfrac{n_0}{\mu_{\mathrm{h}}} + \dfrac{p_0}{\mu_{\mathrm{e}}}}. \tag{6.63b}$$

From the Einstein relation, $D = kT\mu/e$, the diffusion coefficient and mobility are given by:

$$D_0 = (kT\,/\,e)\mu_e\mu_h(n_0 + p_0)/(\mu_e n_0 + \mu_h p_0), \tag{6.63c}$$

$$\mu_0 = (p_0 - n_0)\mu_e\mu_h\,/\,(\mu_e n_0 + \mu_n p_0), \tag{6.63d}$$

where now τ, D_0, and μ_0 are all constant coefficients.

Let the electric field lie along the x direction. Then for, $[-(L/2) < x < (L/2)]$, and in the low light intensity approximation, $n = n_0 + \Delta n$ and $p = p_0 + \Delta p$; with boundary conditions $\Delta n = \Delta p = 0$ at $x = -L/2$ and $x = L/2$. From (6.62) the excess minority carrier density at x is given by:

$$\Delta p = \frac{\eta J_s}{d}\tau_p\left[1 + \frac{e^{\alpha_1 x}\sinh(\alpha_2 L/2) - e^{\alpha_2 x}\sinh(\alpha_1 L/2)}{\sinh(\alpha_1 - \alpha_2)L/2}\right], \tag{6.64}$$

where

$$\alpha_{1,2} = \frac{\mu_0 E}{2D_0} \pm \left[\left(\frac{\mu_0 E}{2D_0}\right) + \frac{1}{D_0\tau}\right]^{1/2}, \tag{6.65}$$

and

$$D_0 = (kT/q)\mu_0. \tag{6.66}$$

If the minority carrier drift length l_1 is $\mu_0 E\tau$, and the diffusion length l_2 is $\sqrt{D_0\tau}$, for convenience, $\alpha_{1,2}$ are given in term of l_1 and l_2 by:

$$\alpha_{1,2} = -\frac{l_1}{2l_2^2} \pm \left[\left(\frac{l_1}{2l_2^2}\right)^2 + \frac{1}{l_2^2}\right]^{1/2}. \tag{6.67}$$

Integrating the expression, (6.64), from $x = -L/2$ to $x = L/2$, the net hole density is:

$$\Delta p = \frac{\eta J_s}{dw}\tau_p\left[1 + \frac{(\alpha_2 - \alpha_1)\sinh(\alpha_1 L/2)\sinh(\alpha_2 L/2)}{\alpha_1\alpha_2(L/2)\sinh(\alpha_1 - \alpha_2)L/2}\right]. \tag{6.68}$$

From the above results, Rittner (1956) calculated a steady state photo current:

$$\Delta J = e\mu_n(b+1)\eta J_s\tau E\xi/d, \tag{6.69}$$

where $b = \mu_e/\mu_h$, and

$$\xi = 1 + \frac{(\alpha_2 - \alpha_1)\sinh(\alpha_1 L/2)\sinh(\alpha_2 L/2)}{\alpha_1\alpha_2(L/2)\sinh(\alpha_1 - \alpha_2)L/2}. \tag{6.70}$$

For $b \gg 1$, the voltage responsivity is:

$$R_\lambda \cong \lambda\eta e R_d\mu_e E\tau\xi/h_c d, \tag{6.71}$$

where R_d is the resistance of the detector. In a strong electric field, the drift length l_1 is much longer than detector length L, or the diffusion length l_2. To a first-order approximation we have:

$$\alpha_1 \approx \frac{1}{l_1} \ll 1, \quad \text{then} \quad \sinh\frac{\alpha_1 L}{2} \cong \frac{L}{2l_1} \tag{6.72a}$$

and

$$-\alpha_2 \approx \frac{l_1}{2{l_2}^2} \gg 1, \quad \text{then} \quad \frac{\sinh(\alpha_2 L/2)}{\sinh(\alpha_1 - \alpha_2) L/2} \cong -1 + \frac{L}{2l_1}. \tag{6.72b}$$

Applying these approximations to (6.70), ξ in a strong field becomes:

$$\xi_{\text{hf}} \to \frac{L}{2l_1}, \quad l_1 > l_2,\ l_1 > L.$$

Hence the responsivity in a strong field reduces to the expression:

$$R_{\text{hf}} = (\lambda/hc)(\eta e \mu_{\text{e}}/2\mu_0) R_{\text{d}}. \tag{6.72c}$$

Since in HgCdTe materials, the electron mobility depends on the electric field, for a fully accurate calculation, the dependence of the resistance on the electric field should be included.

Also, the drift length l_1 depends on the mobility, μ_0. For an extrinsic n-type material, μ_0 can be simplified to the hole mobility. For an intrinsic semiconductor, μ_0 is ~ 0 and therefore no sweep-out effect is present.

The effect of carrier drift and diffusion must be included in a g–r noise calculation (Williams 1968; Kinch et al. 1977). According to the discussion of g–r noise in last section, the g–r noise voltage is given below when drift and diffusion effects are included:

$$V_{\text{g-r}} = \frac{2V_{\text{b}}F}{n_0(Lwd)^{1/2}}\left[\left(1 + \frac{p_0}{p_{\text{b}}}\frac{n_0}{n_0 + p_0}\right)(p_{\text{b}}\tau\xi)\Delta f\right]^{1/2}. \tag{6.73}$$

Then the "sweep-out" limited detectivity is:

$$D_\lambda{}^* = \frac{1}{2}\frac{(\lambda e/hc)(\mu_{\text{e}}/\mu_0)\eta R_{\text{d}}}{(V_{\text{j}}^2 + V_{\text{a}}^2)^{1/2}}(A\Delta f)^{1/2}. \tag{6.74}$$

This extreme value of the detectivity is independent of the time constant, and consequently is independent of surface recombination. Only in a case where surface recombination is present, should the electric field be increased to enable the detectivity to reach its extreme value.

Smith (see Broudy and Mazurczyk 1981) derived the g–r noise expression for an n-type material by again considering carrier drift and diffusion effects. For an n-type material, the noise voltage is:

$$V_{\text{N}} = \frac{2(b+1)V}{(n_0 b + p_0)(lwd)^{1/2}}\left[p_0 + \langle p_{\text{b}}\rangle F(\omega)\phi(\omega)\right]^{1/2}\sqrt{\Delta f}, \tag{6.75}$$

where the effective-time dependent function is: $\phi(\omega) = \frac{\tau}{1+\omega^2\tau^2}\xi = \frac{\tau}{1+\omega^2\tau^2}\frac{1}{F(\omega)}$, and the sweep-out factor, related to the spatial variation of the electric field is $F(\omega) = \xi^{-1}$. At low frequency, $\omega^2\tau^2 \ll 1$, these expressions reduce to:

$$\langle p_b \rangle = \frac{\eta J_b}{d}\tau\xi, \tag{6.76}$$

so that:

$$\begin{aligned} D_\lambda^* &= \frac{\lambda}{2hc}\left(\frac{\eta}{J_b}\right)^{1/2}\left[\frac{\langle p_b \rangle}{p_0 + \langle p_0 \rangle F(0)}\right]^{1/2}, \\ &= D_{\mathrm{BLIP}}^*\left[\frac{\langle p_b \rangle}{p_0 + \langle p_0 \rangle F(0)}\right]^{1/2}. \end{aligned} \tag{6.77}$$

For an n-type semiconductor, at low temperature, p_0 is very small. Then the above formula reduces to:

$$D_\lambda^* = D_{\mathrm{BLIP}}^*(F(0))^{-1/2}. \tag{6.78}$$

$F(0)$ has a value between $1/\sqrt{2}$ and 1 (Broudy and Mazurczyk 1981).

Figures 6.7 and 6.8 show a comparison of experimental data with a theoretical calculation (Broudy et al. 1975, 1976). Some experimental results do not coincide

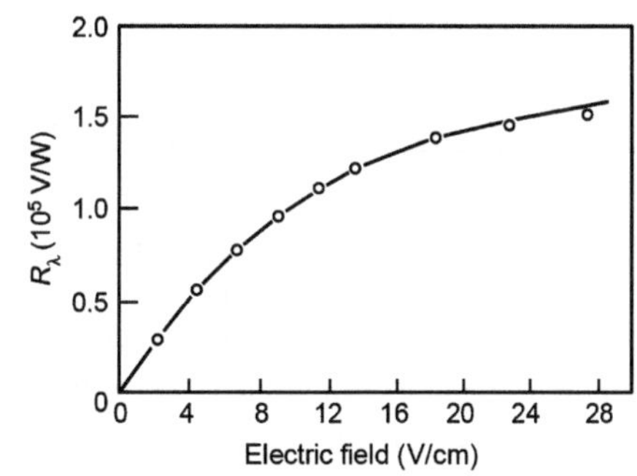

Fig. 6.7 The detectivity of a HgCdTe detector's dependence on the electric field. *Dots* represent experimental data, while the *curve* is theory

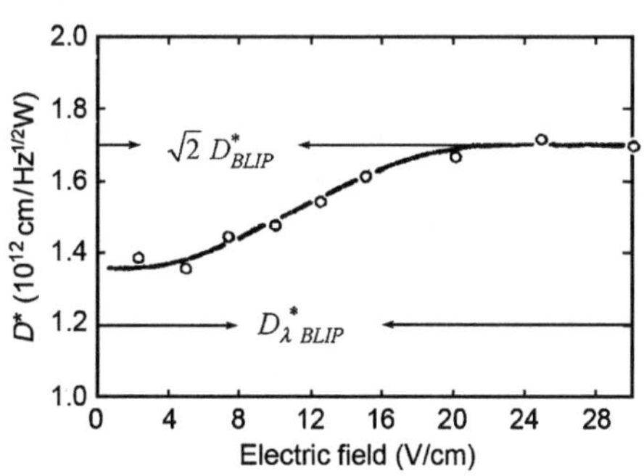

Fig. 6.8 The detectivity D^* of a HgCdTe detector's dependence on the electric field. *Dots* are experimental data and the *curve* is theory

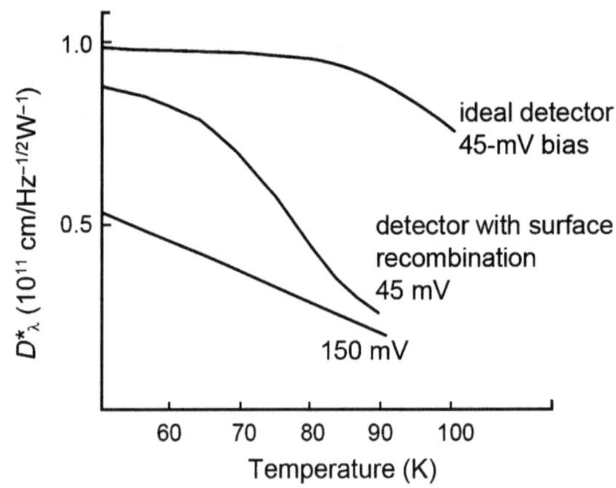

Fig. 6.9 Theoretical curves of the detector's detectivity dependence on temperature for different bias voltages

with theoretically calculated results (Kinch et al. 1977). Figure 6.9 shows the detectivity as a function of temperature. For an ideal detector, according to (6.53), a 45 mV bias voltage is required. However, if surface recombination is present, it is more difficult to reach the sweep-out condition, and device performance degrades rapidly. If the bias voltage is increased from 45 to 150 mV, the performance will be degraded even more.

6.2 Photovoltaic Infrared Detectors

6.2.1 Introduction to Photovoltaic Devices

Pseudobinary HgCdTe photovoltaic devices are composed of reverse biased p–n junctions (Fig. 6.10). The incident radiation is absorbed within several microns below the surface of the detector to produce electron–hole pairs, one of which transfers to the p–n junction and the other to the back contact by a combination of diffusion and drift under the influence of the space charge field produced by the reverse bias. This modifies the voltage drop across an open circuited device. If the diode terminals are shorted, a current will be generated in the circuit to restore the minority carrier concentration in the p–n junction to its equilibrium value. An operating point can be set on a current–voltage characteristic curve by a choice of bias voltage and load resistance. If the incident radiation is modulated, there will be an AC signal voltage generated in response. Photovoltaic devices generally have faster responses than those of photoconductive devices.

One form of device is an n-on-p, i.e., the n-type layer is grown on the p-type layer. The other one is a p-on-n device, i.e., the p-type layer is grown on the n-type layer. Two parameters are used to describe the sensitivity of a photovoltaic device. One is the quantum efficiency η, which is the number of carriers in the junction

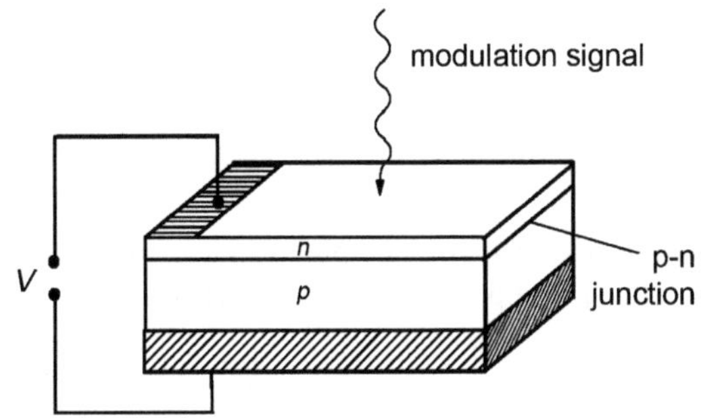

Fig. 6.10 Schematic representation of an open circuited photovoltaic device

generated per incident photon. The other is the resistance at zero-bias voltage, $R_0 = \left(\frac{\partial V}{\partial I}\right)_{V=0}$. When the device is irradiated by light the total current I_t of the diode is composed of the photocurrent I_p and the dark current $I_d(V)$. The ideal dark current is:

$$I_d(V) = I_0\left(\exp\frac{eV}{k_B T} - 1\right), \tag{6.79}$$

where I_0 is the saturation current of the diode. The direction of the photocurrent is the reverse of the forward-bias-current:

$$I_t = -I_P + I_d(V). \tag{6.80}$$

The photocurrent is expressed by:

$$I_P = e\eta Q, \tag{6.81}$$

where Q is the number of the photons arriving per second. For a small bias voltage V, and assuming a dark current which is linearly dependent on the bias voltage, we have:

$$I_t = -I_P + \frac{V}{R_0}. \tag{6.82}$$

The total current is zero if a bias voltage is taken to make the dark current equal to the photocurrent, then:

$$V = R_0 I_P = R_0 e\eta Q. \tag{6.83}$$

This operating mode has the advantage that it functions around a zero signal, but it is not the way devices typically are run. The normal mode is one that operates in reverse bias.

The power of the incident radiation P_λ is given by the number of the incident photons arriving per second, Q, times the energy per photon, $E_\lambda = h\nu$, so we have:

$$V = \frac{e\eta R_0 P_\lambda}{E_\lambda}. \tag{6.84}$$

Then the voltage responsivity is:

$$R_{V\lambda} = \frac{V}{P_\lambda} = \frac{e\eta R_0}{E_\lambda}. \tag{6.85}$$

In this atypical case, the photovoltage equals to the rms noise voltage in the bandwidth Δf. Suppose Johnson noise is the primary noise source, then in this case:

$$\sqrt{\bar{V}_\mathrm{N}^2} = \sqrt{4k_\mathrm{B}TR_0\Delta f}, \tag{6.86}$$

If $\Delta f = 1\,\mathrm{Hz}$ and the photovoltage equals the noise voltage, the incident radiation power equals the noise-equivalent power, $P_{\mathrm{N}\lambda}$. Then from (6.84) and (6.85), we have:

$$\left(\frac{e\eta R_0}{E_\lambda} \cdot P_{\mathrm{N}\lambda}\right)^2 = 4k_\mathrm{B}TR_0, \tag{6.87}$$

or

$$P_{\mathrm{N}\lambda} = \frac{2E_\lambda(k_\mathrm{B}T)^{1/2}}{e\eta R_0^{1/2}}. \tag{6.88}$$

Let A be the area of the detector, the detectivity is the reciprocal of the noise-equivalent power after removing the effect of the area of the detector and the bandwidth (selected to be 1 Hz in this case), i.e.,:

$$D_\lambda^* = \frac{A^{1/2}}{P_{\mathrm{N}\lambda}} = \frac{e\eta(R_0A)^{1/2}}{2E_\lambda(k_\mathrm{B}T)^{1/2}}. \tag{6.89}$$

This shows that the detectivity is proportional to the quantum efficiency η, and the square root of the product of the area and the resistance of the detector, $\sqrt{R_0A}$. For a small V, from (6.1), we have:

$$I(V) = I_0 \cdot \frac{eV}{k_\mathrm{B}T} - I_0. \tag{6.90}$$

Then we have $R_0 = \left(\frac{\partial V}{\partial I}\right)_{V=0} = \frac{k_\mathrm{B}T}{eI_0}$ and

$$D_\lambda^* = \frac{e^{1/2}\eta A^{1/2}}{2E_x {I_0}^{1/2}} = \frac{e^{1/2}\eta}{2E_x J_0^{1/2}}, \tag{6.91}$$

where J_0 is the saturation current density. $J_0^{1/2}$ can also be express in terms of the mobility and the lifetimes of the electrons and holes, as well as their concentrations. The majority carrier concentration must be increased to decrease the minority carrier concentration, if we want a small saturation current.

A similar derivation can be carried out in the reverse-bias case where a multiplicative factor of $\sqrt{2}$ is present. In practice the shot-noise also must be taken into account.

To increase the quantum efficiency, η, a small reflectivity for the incident radiation is required at the detector surface, a small surface recombination velocity, and a junction whose depth is smaller than the hole diffusion length. The response speed is limited by the lifetime of the photogenerated carriers and circuit parameters such as the junction capacitance.

6.2.2 *Current-Voltage Characteristic for p–n Junction Photodiodes*

Reine et al. (1981) has analyzed in his review article, the current–voltage characteristic (*I–V*)for p–n junction photodiodes, that determines the dynamic resistance and the thermal noise of these devices.

From the current–voltage characteristic $I(V)$, the dynamic resistance of a photodiode at zero bias voltage, denoted by R_0, is given by:

$$R_0^{-1} = \left.\frac{\mathrm{d}I}{\mathrm{d}V}\right|_{V=0}. \tag{6.92}$$

A frequently used figure of merit for a photodiode is the R_0A product, i.e., the product of the R_0 given by (6.92) and the area of the p–n junction. Because $J = I/A$ is the current density, the product R_0A is:

$$(R_0A)^{-1} = \left.\frac{\mathrm{d}J}{\mathrm{d}V}\right|_{V=0}. \tag{6.93}$$

Equation (6.93) shows that R_0A represents the variation of the current density caused by a small variation of the voltage at zero bias, and it is one way to characterize the performance of a device. Obviously, R_0A is independent of the area of the junction, and thus it is used widely as an important device figure of merit.

Various current mechanisms of a p–n junction photodiode will be discussed later. The diffusion current, which is a fundamental dark current mechanism in a p–n junction photodiode, originates from the random thermal generation and recombination of electron–hole pairs within a diffusion length of the minority carrier space charge region. The diffusion current is the primary junction current of the HgCdTe photodiode at high temperature, and it sets the dark current in devices, especially those that operate at temperatures higher than 77 K. The main contribution to the dark current at low temperatures is a tunneling current across the space charge region.

Figure 6.11 is a simplified sectional view of an n-on-p junction photodiode. It can be divided into three regions: (1) an electrical quasineutral region with a thickness,

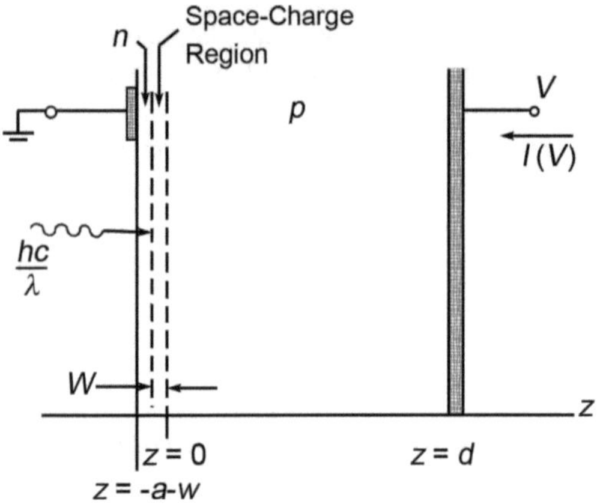

Fig. 6.11 Sectional view of a simple n-on-p photodiode with a back ohmic contact

"a", in the heavily doped n-layer; (2) a space charge region with a thickness, "w", in the lightly doped p-layer; and (3) a electrical quasineutral region with a thickness "d", in the p-layer. Assume that the transition region between the n-type layer and the p-type layer is thin enough so that both the n and p layers are homogeneous and all the applied voltage is supported by the space charge region; the low injection case, that is, the injected minority carrier concentration is small compared with the majority carrier concentration; and the distribution of the carriers is nondegenerate, so the carrier concentrations at thermal equilibrium in all the regions will follow the relation:

$$n_0(z)p_0(z) = n_i^2. \tag{6.94}$$

The minority carrier concentration in the regions fulfils the following boundary conditions (Hauser 1971):

$$p(-w) = p_{n0} \exp\left(\frac{eV}{k_B T}\right), \tag{6.95a}$$

$$n(0) = n_{p0} \exp\left(\frac{eV}{k_B T}\right), \tag{6.95b}$$

where p_{n0} is the minority carrier concentration in the n-layer in thermal equilibrium, n_{p0} the minority carrier concentration in p-layer in thermal equilibrium, e, the charge, k_B the Boltzmann constant, and T is the temperature of the diode. The nonequilibrium carrier concentration in the space charge region follows the relation:

$$n(z)p(z) = n_i^2 \exp\left(\frac{eV}{k_B T}\right). \tag{6.96}$$

Obviously, (6.96) becomes (6.94) in the case where $V = 0$. Consider the z location in the p region, when the thermal equilibrium condition is broken by the application of an external potential, then the carrier concentrations will be:

$$n(z,t) = n_{p0} + \Delta n(z,t), \tag{6.97}$$

and

$$p(z,t) = p_{p0} + \Delta p(z,t). \tag{6.98}$$

Assume the p region is electrically neutral, so:

$$\Delta n(z,t) = \Delta p(z,t). \tag{6.99}$$

Then the excess minority carrier concentration is the solution of the equation:

$$D_e \frac{d^2 \Delta n}{dz^2} - \frac{\Delta n}{\tau_e} = 0, \tag{6.100}$$

where D_e is the minority carrier diffusion coefficient in the p region, and τ_e is the minority carrier lifetime in p region. The boundary condition is (6.95b) at $z = 0$, and the boundary condition at $z = d$ is equivalent to that at $z = \infty$:

$$\Delta n(z \to d) \cong \Delta n(z \to \infty) \to 0. \tag{6.101}$$

So the solution of (6.100) is:

$$\Delta n(z) = n_{p0} \left[\exp\left(\frac{eV}{k_B T} \right) - 1 \right] \exp\left(-\frac{z}{L_e} \right), \tag{6.102}$$

where L_e is the minority carrier diffusion length given by:

$$L_e = \sqrt{D_e \tau_e}. \tag{6.103}$$

The boundary condition approximation, (6.101), is equivalent to taking the thickness of the p region effect on the diffusion current of the space charge region in the p layer. This is equivalent to assuming the back contact is ohmic. If there were a barrier at the back contact then an accumulation layer would be formed there.

Thus from the diffusion current density expression, $J_e = eD_e \frac{\partial n}{\partial z}$, the diffusion current density derived at $z = 0$ is:

$$J_{e\infty} = en_{p0} \frac{D_e}{L_e} \left[\exp\left(\frac{eV}{k_B T} \right) - 1 \right], \tag{6.104}$$

where the subscript, ∞, is added to indicate the condition, $d \gg L_e$, has been assumed. The above relation was obtained by Shockley in 1949 (Shockley 1949).

From (6.93) and (6.103), the contribution of the diffusion current in the p region to the product $R_0 A$ is:

$$(R_0 A)_{p\infty} = \frac{k_B T}{e^2} \frac{1}{n_{p0}} \frac{\tau_e}{L_e}. \tag{6.105}$$

From (6.94), $n_0(z)p_0(z) = n_i^2$, if $p_{p0} = N_A$, the acceptor concentration, and with the Einstein relation:

$$D_\mathrm{e} = (k_\mathrm{B}T/e)\mu_\mathrm{e}, \tag{6.106}$$

Equation (6.105) becomes:

$$(R_0A)_{\mathrm{p}\infty} = \frac{1}{e}\frac{N_A}{n_i^2}\sqrt{\frac{k_\mathrm{B}T}{e}\frac{\tau_\mathrm{e}}{\mu_\mathrm{e}}}. \tag{6.107}$$

The temperature dependence of $(R_0A)_{\mathrm{p}\infty}$ is primarily determined by n_i^2.

Now consider the diffusion current in the n region. With the boundary condition (6.95a); and assuming that the thickness of n region, a, is far larger than the minority carrier diffusion length L_h:

$$L_\mathrm{h} = \sqrt{D_\mathrm{h}\tau_\mathrm{h}}, \tag{6.108}$$

where D_h and τ_h are the diffusion coefficient and the lifetime of the minority carriers in the n region, respectively; and with a similar derivation to that of (6.105), we find the contribution to R_0A from the diffusion current in the n region to be:

$$(R_0A)_{\mathrm{n}\infty} = \frac{k_\mathrm{B}T}{e^2}\frac{1}{p_{\mathrm{n}0}}\frac{\tau_\mathrm{h}}{L_\mathrm{h}}. \tag{6.109}$$

This equation can be converted to:

$$(R_0A)_{\mathrm{n}\infty} = \frac{1}{e}\frac{N_\mathrm{D}}{n_i^2}\sqrt{\frac{k_\mathrm{B}T}{e}\frac{\tau_\mathrm{h}}{\mu_\mathrm{h}}}, \tag{6.110}$$

where we made the assumption that, $n_{\mathrm{n}0} = N_\mathrm{D}$, and N_D is the donor concentration in the region, as well as the Einstein relation:

$$D_\mathrm{h} = (k_\mathrm{B}T/e)\mu_\mathrm{h}. \tag{6.111}$$

The above discussion has supposed that the interface is far away from the space charge layer, a distance that is far larger than the minority carrier diffusion length. In fact, this distance is frequently smaller than the minority carrier diffusion length, so the two surfaces at $z = -a-w$ and $z = d$ will have an effect on the diffusion current and consequentially R_0A. The steady-state minority carrier concentration in the p region, $\Delta n(z)$, is given by the solution to the continuity equation. The boundary condition at $z = 0$ is given by (6.95b), and the boundary condition at $z = d$ can be expressed in terms of the surface recombination velocity, $S_\mathrm{p} = -D_e \left.\frac{1}{\Delta n}\frac{\partial \Delta n}{\partial z}\right|_{z=d}$:

$$J_\mathrm{e}(d) = eD_\mathrm{e}\left.\frac{\partial \Delta n}{\partial z}\right|_{z=d} = -eS_\mathrm{p}\Delta n(d). \tag{6.112}$$

The solution, $\Delta n(z)$, is given by:

$$\Delta n(z) = n_{p0}\left[\exp\left(\frac{eV}{k_B T}\right) - 1\right]\left[\frac{\cosh\left(\frac{z-d}{L_e}\right) - \beta\sinh\left(\frac{z-d}{L_e}\right)}{\cosh\left(\frac{d}{L_e}\right) + \beta\sinh\left(\frac{d}{L_e}\right)}\right], \tag{6.113}$$

where β is defined as:

$$\beta \equiv S_p L_e / D_e = S_p/(L_e/\tau_e). \tag{6.114}$$

Thus β is exactly the ratio of the surface recombination velocity to the diffusion velocity. The result for the product R_0A is:

$$(R_0A)_p = (R_0A)_{p\infty}\left[\frac{1 + \beta\tanh\left(\frac{d}{L_e}\right)}{\beta + \tanh\left(\frac{d}{L_e}\right)}\right], \tag{6.115}$$

where $(R_0A)_{p\infty}$ is given by (6.109). The results from (6.115) are plotted in the Fig. 6.12, which shows the dependence of the ratio of $(R_0A)_p$ to $(R_0A)_{p\infty}$ on the ratio d/L_e for various β values. The figure shows that R_0A can increase or decrease, depending on the value of β. Because one wants $(R_0A)_p$ to be as large as possible, there is a benefit to having β be positive or even zero, which occurs when $\left.\frac{\partial \Delta n}{\partial z}\right|_{z=d}$ is negative.

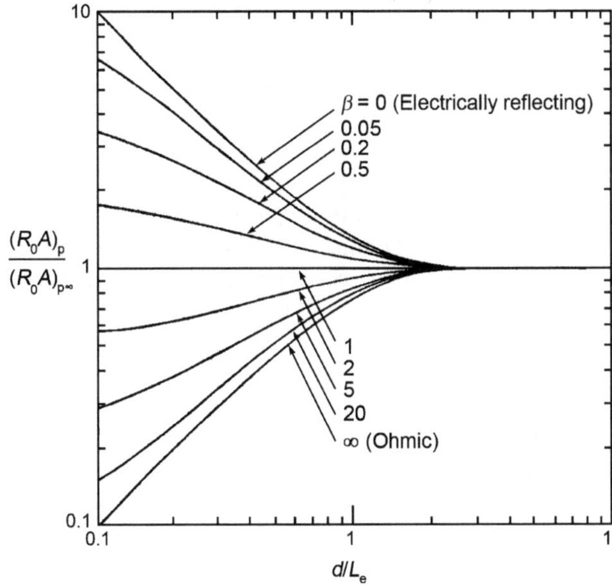

Fig. 6.12 Relationships between $(R_0A)_p$ for various boundary conditions, and the ratios of the distance d to the space charge region to the diffusion length

A similar method can be used to deal with effects occurring at the n region interface, i.e., the interface at $z = -a - w$. Then we get an expression similar to (6.115):

$$(R_0A)_{\mathrm{n}} = (R_0A)_{\mathrm{n}\infty}\left[\frac{1+\beta\tanh\left(\frac{d}{L_{\mathrm{h}}}\right)}{\beta+\tanh\left(\frac{d}{L_{\mathrm{h}}}\right)}\right], \quad \text{with } \beta = S_{\mathrm{h}}/(L_{\mathrm{h}}/\tau_{\mathrm{h}}). \tag{6.116}$$

For a sample with a low carrier concentration, the minority carrier diffusion length is about 45 μm in a p-type $Hg_{0.8}Cd_{0.2}Te$ alloy, while it is about 100 μm in a p-type $Hg_{0.7}Cd_{0.3}Te$ alloy. Because the diffusion length is larger than the thickness of the p region in a typical focal plane array device, i.e., $d < L_{\mathrm{e}}$, the above discussion is especially important. It is shown in Fig. 6.12 that when $d \ll L_{\mathrm{e}}$, which means the thinner the p region is, the better, and if the surface recombination velocity is negative and is small compared to the diffusion velocity, then $(R_0A)_{\mathrm{p}}$ is large and can be ignored, so (6.112) becomes:

$$J_{\mathrm{e}}(d) \cong 0. \tag{6.117}$$

This means there is little recombination at the back interface and no minority carriers flow into or out of the boundary at $z = d$ when $\beta = 0$ and $L_{\mathrm{e}} \gg d$. It is effectively a reflecting surface for minority carriers. However, it can still be an ohmic contact for the majority carriers. Then the product R_0A, which is determined by the diffusion current in the p region, can be calculated from:

$$(R_0A)_{\mathrm{p}} = \frac{k_{\mathrm{B}}T}{e^2}\frac{N_{\mathrm{A}}}{n_i^2}\frac{\tau_{\mathrm{e}}}{d}, \tag{6.118}$$

which means a decrease in the thickness d will increase the product R_0A. This equation still holds when taking into account the diffusion current in the p region and radiative recombination.

There are three ways suggested by Long (1977) to make $J_{\mathrm{e}}(d) = 0$, the first one is forming a high concentration p^+ region at the end of the p region by ion implantation or acceptor diffusion. A p^+–p junction, formed in this way, can block the minority carriers while it is an ohmic contact for the majority carriers at the back contact of the p region. The n^+–p–p^+ structure has been discussed in detail by Long (1977) Long et al. (1978), and Sood et al. (1979a–c). The second way is a proper surface treatment that can adjust the surface potential and consequentially decrease the boundary recombination velocity S_{p}. The third one is a p region of HgCdTe grown by LPE on the substrate of CdTe, which has a wide forbidden bandgap. With this kind of buffering transition, the electric field is present producing a barrier to both minority and majority carriers, causing an accumulation layer to form at this boundary. It is then an n–p–i structured device. This kind of boundary condition has been discussed in detail by Lanir et al. (1979a, b), Lanir and Shin (1980). It does not lead to a high quality device and requires that the incident radiation be modulated in order for it to produce a signal.

The contribution of the diffusion current to the product, R_0A, can be derived from the total diffusion current, which is given by adding an n region and a p region so the diffusion currents add together (the resistances add in parallel):

$$\frac{1}{R_0A} = \frac{1}{(R_0A)_n} + \frac{1}{(R_0A)_p}. \tag{6.119}$$

We can compare the contribution of the diffusion current in the n region to that in the p region. For an ideal case, i.e., $\beta = 0$ and $L_e \gg d$, and $L_n \gg a$, we have:

$$\frac{(R_0A)_n}{(R_0A)_p} = \frac{N_D}{N_A}\frac{\tau_e}{\tau_h}\frac{d}{a}. \tag{6.120}$$

The product R_0A is determined by the contribution to the p region diffusion current if the ratio is far larger than 1.

The theory of the contribution to the R_0A product upper limit for a HgCdTe photodiode from diffusion current can now be calculated. From the above analysis, we conclude that the diffusion current in the n region can be ignored compared to that in the p region. Also we assume that the primary recombination mechanism in the in p region is radiative recombination, that is, there is no Shockley–Read recombination center in the p region, and in the case considered, Auger recombination is slower than radiative.

The τ_e is taken to be the radiative lifetime, τ_{rad}, which is given by (Blackemore 1962):

$$\tau_{rad} \approx \frac{1}{B(p_{p0} + n_{p0})} \approx \frac{1}{BN_A}. \tag{6.121}$$

An assumption, $p_{p0} = N_A \gg n_{p0}$, has been taken in the above equation. The radiative recombination coefficient B is given by (Long 1977):

$$B = 5.8 \times 10^{-13}\sqrt{\varepsilon_\infty}\left(\frac{1}{m_c + m_v}\right)^{3/2}\left(1 + \frac{1}{m_c} + \frac{1}{m_v}\right)\left(\frac{300}{T}\right)^{3/2} E_g^2, \tag{6.122}$$

where in this equation E_g is in eV, B is in cm^3/s, and T is in K. ε_∞ is the dielectric constant at high frequency, and m_c and m_v are the specific effective mass ratios relative to the free electron mass, of the conduction band and valence band, respectively. The specific effective mass of holes in the p region, m_v, is 0.5, and m_c is given by the Weiler equation (Weiler 1981):

$$\frac{1}{m_c} = 1 + 2F + \frac{E_p}{3}\left(\frac{2}{E_g} + \frac{1}{E_g + \Delta}\right), \tag{6.123}$$

where $F = 1.6$, $E_p = 19\,\text{eV}$, and $\Delta = 1\,\text{eV}$. $\varepsilon_\infty(x)$ is given by the Baars and Sorger (1972) equation:

$$\varepsilon_\infty(x) = 9.5 + 3.5[(0.6 - x)/0.43]. \tag{6.124}$$

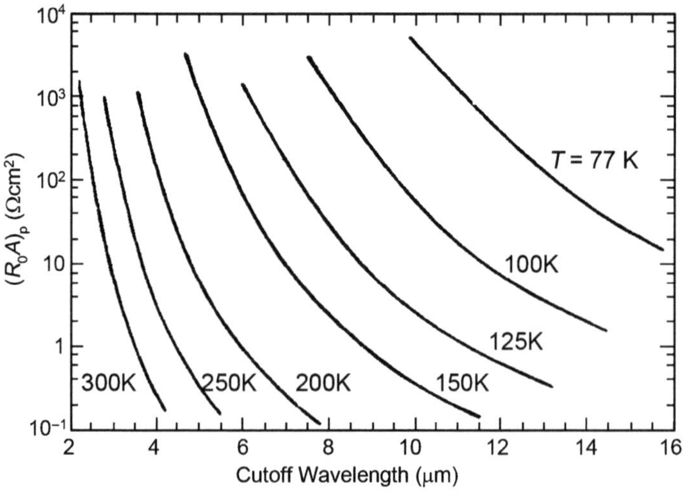

Fig. 6.13 The relations between $(R_0A)_p$ of the HgCdTe photodiodes and their cut-off wavelengths at various temperatures

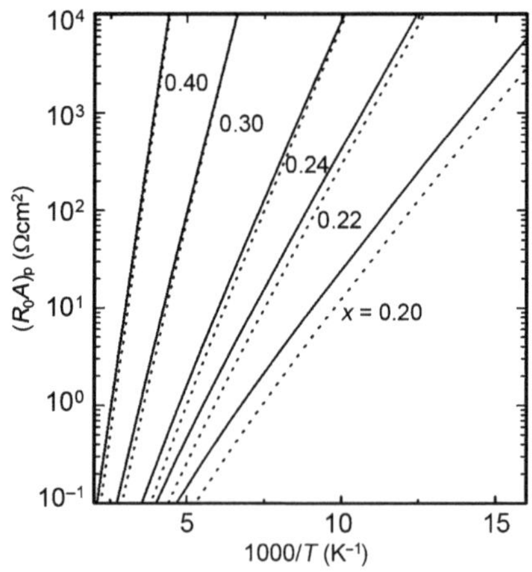

Fig. 6.14 A comparison of the theoretical calculations of the relation between $(R_0A)_p$ and $1/T$. The *dashed curves* are results with E_g and n_i taken from the Schmit method, and the solid ones denote the results with the new parameters

If E_g is taken from the reference of Schmit and Stelzer (1969), and n_i is from the reference of Schmit (1970), the result for $(R_0A)_p$ is shown in Fig. 6.13, assuming the p region thickness $d = 10\,\mu\text{m}$.

If E_g and n_i are taken from newer research, see "Physics and Properties of Narrow Gap Semiconductors" (3.108) and (5.28) (Chu and Sher 2007), the results are shown in Fig. 6.14. Attention should be paid to the upper limit of the product R_0A, which is independent of the acceptor concentration N_A.

Recombination current in the space charge region was not taken into account in the above discussion, but in fact, impurities or the energy levels of native point defects in the space charge region may act as Shockley-Read generation and recombination centers (g–r) that will cause a junction current. The importance of this current mechanism was first pointed out by Sah et al. (1957), who proved that at low temperature the g–r current is more important than the diffusion current in the space charge region. This condition still holds when the thickness of the space charge region is far less than the minority carrier diffusion length. The variation of the g–r current with temperature is proportional to n_i, while the variation of the diffusion current is proportional to ${n_i}^2$. The diffusion current becomes the primary contributor at relatively higher temperatures, and decreases with the temperature while the g–r current in space charge region decreases more slowly, and then finally a temperature is reached at which the diffusion current equals the g–r current. The diffusion current dominates above this temperature. There are some other current mechanism, such as the surface generation–recombination current, and the band-to-band tunneling current, that decrease with temperature like the g–r current, slower than n_i^2. One needs to figure out which minority current mechanism of a photodiode dominates the noise at low temperature before a successful analysis of the problems of a photodiode can be accomplished.

The net steady-state recombination rate, $U(z)$, occurring at g–r centers having an energy E_t relative to the valence band edge, is given by (3.3.18):

$$U(z) = -\frac{dn}{dt} = \frac{np - n_i^2}{\tau_{p0}(n + n_1) + \tau_{n0}(p + p_1)}, \tag{6.125}$$

where $U(z)$ is the number of the carriers per unit volume that recombine per unit time, and $n = n(z)$ and $p = p(z)$ are the non-equilibrium electron and hole concentrations in the space charge region, respectively. We have from Sect. 3.4 the definitions:

$$n_1 = N_c \exp\left(\frac{E_t - E_g}{k_B T}\right), \tag{6.126}$$

$$p_1 = N_v \exp\left(\frac{-E_t}{k_B T}\right), \tag{6.127}$$

$$\tau_{n0} = \frac{1}{C_n N_t}, \tag{6.128}$$

$$\tau_{p0} = \frac{1}{C_p N_t}, \tag{6.129}$$

where N_c and N_v are the effective density of states of the conduction and valence bands, respectively, C_n and C_p are the electron and hole capture coefficients, and N_t is the number of g–r centers per unit volume. The product of $n(z)$ and $p(z)$ is independent of the position, z, in the space charge region and approximately follows the Shockley relation, (6.96). So the g–r centers causes a net recombination at a positive bias, $U(z) > 0$, and a net generation in a reverse bias, $U(z) < 0$. Taking an integral of (6.125) over the whole space charge region, the junction current density J_{g-r}, generated by the centers, is given by:

$$J_{g-r} = e \int_{w}^{0} U(z) dz. \tag{6.130}$$

The $n(z)$ and $p(z)$ must be obtained before calculating the integral.

Sah et al. (1957) assumed $n(z)$ and $p(z)$ vary linearly with distance, z, in the space charge region, and got the result:

$$J_{\text{g-r}} = \frac{en_i w}{\sqrt{\tau_{\text{n0}}\tau_{\text{p0}}}} \frac{\sinh\left(\dfrac{-eV}{2k_{\text{B}}T}\right)}{e\left[\dfrac{V_{\text{bi}}-V}{2k_{\text{B}}T}\right]} f(b), \tag{6.131}$$

where V_{bi} is the built-in electric potential difference of the p–n junction, thus eV_{bi} is the difference between the quasi-Fermi energy levels in the n region and in the p region. The function $f(b)$ is given approximately by:

$$f(b) = \int_0^\infty \frac{\text{d}u}{u^2+2bu+1}, \tag{6.132}$$

where

$$b = \exp\left(\frac{-eV}{2k_{\text{B}}T}\right)\cosh\left[\frac{E_{\text{t}}-E_{\text{i}}}{k_{\text{B}}T} + \frac{1}{2}\ln\left(\frac{\tau_{\text{p0}}}{\tau_{\text{n0}}}\right)\right], \tag{6.133}$$

and E_{i} is the intrinsic energy level of a state relative to the top of valence band. E_{i} is defined as being equal to the Fermi energy level ($E_{\text{i}} = E_{\text{F}}$) when $n = p$.

$$E_{\text{i}} = \frac{1}{2}(E_{\text{C}} + E_{\text{V}}) + \frac{1}{2}k_{\text{B}}T\ln\left(\frac{N_{\text{V}}}{N_{\text{C}}}\right).$$

Equation (6.125) reaches its maximum and the recombination center will display its maximum effect at voltage V, where $E_{\text{i}} = E_{\text{t}}$ and $\tau_{\text{p0}} = \tau_{\text{n0}}$.

The contribution to the product R_0A from the g–r current in the depletion layer is obtained from (6.131) and is given by:

$$(R_0A)_{\text{g-r}} = \frac{\sqrt{\tau_{\text{n0}}\tau_{\text{p0}}}V_{\text{bi}}}{en_{\text{i}}wf(b)}. \tag{6.134}$$

For the most effective g–r centers, i.e., $E_{\text{t}} = E_{\text{i}}$, $\tau_{\text{p0}} = \tau_{\text{n0}}$, $b = 1$, and $f(0) = 1$ when $V = 0$, thus $(R_0A)_{\text{g-r}}$ varies with temperature proportional to n_{i}^{-1}, but differs from the contribution from the diffusion current that is also proportional to n_{i}^{-1}. The dashed curve in Fig. 6.15 shows the temperature dependence of $(R_0A)_{\text{g-r}}$ for $\text{Hg}_{1-x}\text{Cd}_x\text{Te}$ with various compositions, compared with the solid curves for $(R_0A)_{\text{p}}$. This calculation is performed following (6.134), assuming $f(b) = 1$, $\tau_{\text{p0}} = \tau_{\text{n0}} = 0.1\,\mu\text{s}$, $eV_{\text{bi}} = E_{\text{g}}$, and $w = 0.1\,\mu\text{m}$, which is equivalent to taking an effective space charge concentration, N_{B}, to be $\sim 1\times 10^{16}\,\text{cm}^{-3}$. In fact, τ_{p0} and τ_{n0}, which depend on the concentration of Shockley–Read centers, will vary in different crystals grown using different fabrication methods.

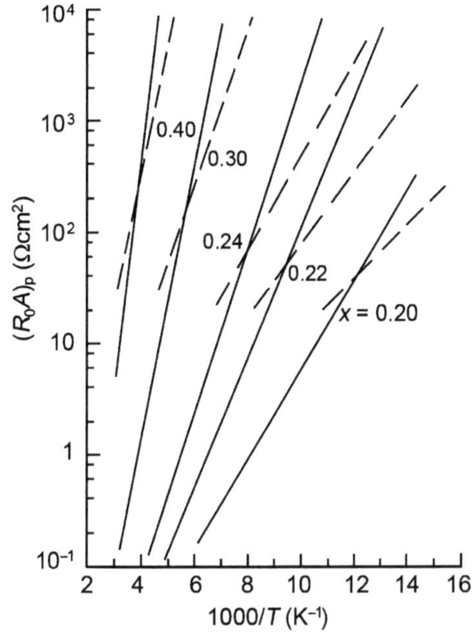

Fig. 6.15 Comparison of the temperature dependences of $(R_0A)_p$ (*solid curves*) and $(R_0A)_{g-r}$ (*dashed curves*) of HgCdTe with various compositions

The ratio of the R_0A contributed by the g–r current in the depletion layer to that contributed by the diffusion current in the p region is given by:

$$\frac{(R_0A)_{g-r}}{(R_0A)_{p\infty}} = \frac{n}{N_A}\frac{eV_{bi}}{k_BT}\frac{L_e}{w}\frac{\sqrt{\tau_{n0}\tau_{p0}}}{\tau_e}, \tag{6.135}$$

where $f(b)$ has been taken as 1. $\frac{eV_{bi}}{k_BT}$ is also large, and if L_e/w is the order of 100, then in the depletion layer, the g–r current prevails over the diffusion current only at the temperatures for which, $n_i < N_A \times 10^{-3}$ or smaller.

Surface leakage current frequently needs to be taken into account. For an ideal p–n junction, the dark current originates from the generation and recombination of the carriers in the quasineutral region, i.e., the diffusion current, and that in the space charge region, i.e., the g–r current. In fact, there are often other dark current mechanisms in devices; especially at low temperature, there are dark currents related to surfaces. There are, in the surface oxide layer and overlaying insulation layer, fixed charges and fast interface states that act as g–r centers and alter the surface potentials on all sides of the device. These factors induce many dark current mechanisms related to the surfaces.

For research into the current mechanisms derived from the surfaces, a gate electrode isolated by an insulating layer is frequently used to control the surface externally. As shown in Fig. 6.16, the conditions to establish an accumulation, a depletion and an inversion are, $V_G < V_{Fb}$, $V_G > V_{Fb}$, and $V_G \gg V_{Fb}$, respectively, where V_{Fb} is the flat band voltage.

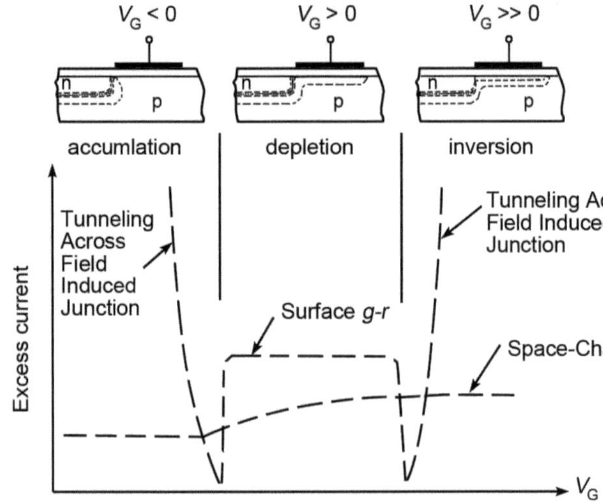

Fig. 6.16 Various current mechanisms established in a gate-controlled diode fabricated from a narrow bandgap material

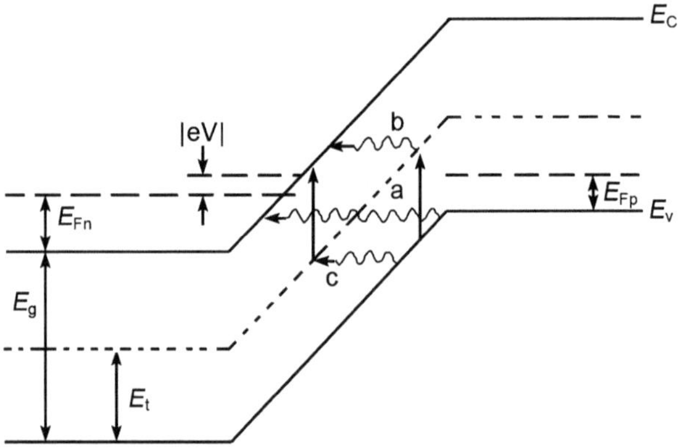

Fig. 6.17 A diagrammatic sketch of the two basic tunneling processes in the HgCdTe: (**a**) direct tunneling and (**b**) and (**c**) trap-assisted tunneling

Band-to-band tunneling is an important junction current mechanism in addition to the various current mechanisms discussed earlier. Next we discuss the effect of a band-to-band tunneling current in a junction with a zero-bias resistance R_0. As we know, tunneling affects the current–voltage characteristics of a p–n junction in reverse-bias.

There are two basic tunneling transitions in HgCdTe, as shown in Fig. 6.17. The transition labeled, a, denotes the direct tunneling, i.e., an electron transits from one side of the space charge region to the other side with the energy conserved. The transitions b and c denote defect-assisted tunneling, i.e., impurities and defect-induced states in the space charge region act as intermediate transition states. For a theoretical calculation of direct tunneling refer to Anderson (1977); for a calculation of defect-assisted tunneling in an MIS structure of HgCdTe refer to Chapman et al. (1978); and for a calculation of tunneling assisted by p–n junction center defects in long wavelength HgCdTe refer to Wong (1980). Many of the approximations made in these treatments as well as that of Nemirovsky et al. (1991), have been eliminated in the more recent work of Krishnamuthy et al. (2006).

Kane (1961) calculated, $J_{\mathrm{t}}(V)$, the junction current density induced by direct tunneling between the conduction band of the n region and the valence band of the p region, and found the form:

$$J_{\mathrm{t}} = Be^{-a}D(V), \tag{6.136}$$

where

$$B = \frac{4\pi em^*}{h^3}E_{\perp}, \quad \text{and} \quad a = \frac{\pi}{4}\left(\frac{E_{\mathrm{g}}}{\theta}\right)^{3/2}, \tag{6.137}$$

$$\theta^{3/2} = \frac{eFh}{2\pi\sqrt{2m^*}}, \tag{6.138}$$

where F is the average field in the space charge region, assumed to be uniform, m^* the effective mass at the conduction band-edge, h the Planck constant, and $E_{\perp}$ is the kinetic energy of particles which move in the planes perpendicular to the tunneling direction:

$$E_{\perp} = \theta\sqrt{\frac{\theta}{E_{\mathrm{g}}}}. \tag{6.139}$$

$D(V)$ in (6.136) is a quantity related to the probability that a tunneling transition joins an initial state to a final state with the same energy on each side of the space charge region. $D(0)$ is 0 at zero-bias, that is, no net junction current occurs. For a tunnel diode, in which the energy levels of both sides of the junction are highly degenerate, $D(V)$ near zero-bias is approximately given by:

$$D(V) \propto eV = e(V_{\mathrm{b}} + V_{\mathrm{bi}})y, \tag{6.140}$$

where V_{b} is the external bias voltage and V_{bi} is the built-in voltage. It is only at low temperature where thermal excitation over the barrier becomes small that direct tunneling becomes important. The additional contribution to R_0A from the tunneling current, (6.93) and (6.136), is:

$$\frac{1}{(R_0A)_{\mathrm{t}}} = eB_0\exp(-a_0). \tag{6.141}$$

where a_0 and B_0 are the values of a and B, respectively, when $V_{\mathrm{b}} = 0$. Note, when V_{b} goes to zero, the built-in potential is still present. The primary contribution to the temperature dependence of the tunneling current arises from E_{g}, contained in a_0. For wide gap semiconductors, E_{g} decreases with the temperature, so $(R_0A)_t$

decreases with temperature, but for sufficiently narrow gap semiconductors E_g increases with temperature (Krishnamurthy et al. 1995), thus reversing this trend.

This derivation was done for a constant average F in the space charge layer, but in fact F varies as z^2 in this layer. This is one of the important approximations removed from the tunneling analysis in the work by Krishnamurhty et al. (1995).

If the freeze-out at low temperature of free holes at the top of the p region valence band is taken into account, then $D(V)$ near $V_b = 0$ can be expressed approximately by:

$$D(V) \approx eV\left(\frac{k_B T}{k_B T + E_\perp}\right)\frac{p_0}{N_v}, \tag{6.142}$$

where p_0 is the free hole concentration in the p region, and N_v the effective density of states in the valence band. Then (6.141), which was used to calculate, $(R_0A)_t$, changes to:

$$\left(\frac{1}{R_0A}\right)_t = eB_0\exp(-a_0)\left(\frac{k_B T}{k_B T + E_\perp}\right)\frac{p_0}{N_v}. \tag{6.143}$$

This equation reflects the effect of the exponential dependence of p_0 on temperature. The impaction of the holes becomes more significant at low temperature, and makes $(R_0A)_t$ increase with decreasing temperatures.

Figure 6.18 illustrates the temperature dependence of $(R_0A)_t$ contributed by the direct band-to-band tunneling current in an n^+-on-p junction of $Hg_{0.8}Cd_{0.2}Te$ with various acceptor energies. Despite the use of arbitrary units the tendency of the variation with the temperature is proper. Direct tunneling becomes difficult when the temperature increases, so $(R_0A)_t$ increases steeply. $(R_0A)_t$ increases once again when the temperature is decreased for a sample with $E_A \geq 0.003$, because of the hole freeze-out in the p region. Defect-assisted tunneling was first calculated by

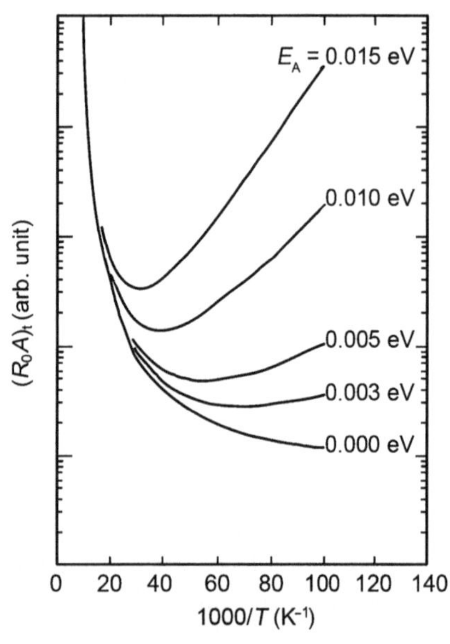

Fig. 6.18 Temperature dependence of the $(R_0A)_t$ contributed by the direct band-to-band tunneling current of a $Hg_{0.8}Cd_{0.2}Te$ junction as a function of temperature with various doping acceptor energy levels

Wong (1980) for the n-on-p junction of long wavelength HgCdTe, and the results were qualitative similar to those reported in Fig. 6.18.

The tunneling current depends significantly on the electric field F in the space charge region, which is contained in the parameter a_0 with its exponential form. For usual values of F, a_0 is small so direct tunneling transitions would not have a great effect on the properties of HgCdTe photodiodes even at low temperature, but more significantly, there is a field-induced junction caused by the high field in the surface. It is necessary to suppress the field-induced junction as much as possible by controlling the surface potential. However, even though direct tunneling is insignificant, trap-assisted tunneling may still play an important role in setting the low temperature dark current (Krishnamurthy et al. 2006).

6.2.3 The Photocurrent in a p–n Junction

An infrared photon with energy larger than the band-gap energy will generate an electron–hole pair when it is absorbed by the photodiode. The electron–hole pair will be separated by the high electric field and contributes to the photocurrent in the external circuit if the absorption occurs in the space charge region. If the absorption occurs in either the n or p neutral regions, but is within a diffusion length distance from the space charge region, the photo-induced electron or hole will transfer by diffusion to the space charge region and then be separated by the high electric field to contribute the photocurrent in the external circuit. Thus a voltage is produced between the two sides of the p–n junction when the photodiode is irradiated. A photo-induced open-circuit voltage will appear if the terminals of the n and p regions are open-circuited, while a current will flow through the photodiode if a conducting wire is placed across these terminals. This is the photovoltaic effect of a p–n junction.

If photons with a nonequilibrium stable flux, Q, the number of photons per $cm^2 \cdot s$, are injected into the photodiode, a stable photocurrent density $J_{ph}(Q)$ will be generated:

$$J_{ph}(Q) = \eta\, eQ, \tag{6.144}$$

where η is the quantum efficiency of the photodiode, defined as the number of the electron–hole pairs generated per incident absorbed photon, and it has a maximum value of 1. The quantum efficiency η is a function of the wavelength of the incident radiation, and depends on the geometry of the photodiode, its reflection coefficient, and on the diffusion length of minority carriers in the quasineutral region.

The current–voltage characteristic, $J(V, Q)$, of a photodiode that is irradiated and has an applied bias voltage V_b, is usually given by:

$$J(V, Q) = J_d(V) - J_{ph}(V, Q) \tag{6.145}$$

where $J_d(V_b)$ is the characteristic dark current of a photodiode in the absence of radiation, and it only depends on V_b. Equation (6.145) indicates that the current–voltage characteristic of the photocurrent of an irradiated photodiode is exactly the result given by the difference between the photocurrent and the dark current.

The quantum efficiency can be directly calculated as a function of the wavelength, the absorption coefficient, and the minority carrier lifetime, if the dark current and photocurrent are linearly independent.

In the following paragraph we will discuss the effect on the quantum efficiency of geometric configurations and of the material properties of an ion-implanted HgCdTe n-on-p photodiode. Assuming the thickness of the p region is semi-infinite, as shown in Fig. 6.19, the concentration of the stable photo-induced minority carrier in the p region, $\Delta n(z)$, is given by Van der Wiele (1976):

$$\Delta n(z) = \left(\frac{\alpha Q \tau_e}{\alpha L_e + 1}\right)\left[\frac{\exp(\frac{-z}{L_e}) - \exp(-\alpha z)}{\alpha L_e - 1}\right], \quad (\alpha L_e \neq 1) \tag{6.146a}$$

$$\Delta n(z) = \left(\frac{\alpha Q \tau_e}{\alpha L_e + 1}\right)\left(\frac{z}{L_e}\right)\exp\left(\frac{-z}{L_e}\right), \quad (\alpha L_e = 1) \tag{6.146b}$$

If the reflectivity of the front surface and the absorption in the n region and in the space charge region are ignored, the quantum efficiency can be calculated from: $J_d = \eta e Q = e D_e \cdot \left.\frac{\partial n}{\partial z}\right|_{z=0}$, i.e.

$$\eta = \frac{\alpha L_e}{\alpha L_e + 1}. \tag{6.147}$$

The quantum efficiency depends on the wavelength, in part because the absorption coefficient depends on the wavelength, thus it is important to obtain, $\alpha(\lambda)$. As

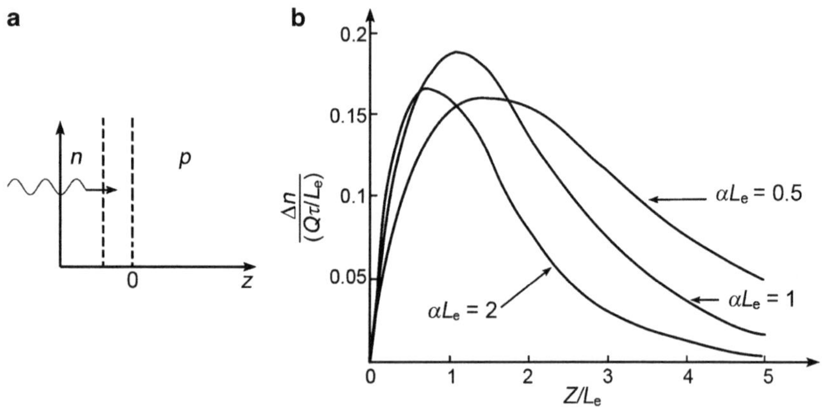

Fig. 6.19 (**a**) Schematic diagram of a p–n junction with irradiation, (**b**) Relationship curves between the spatial concentrations, $\Delta n(z)$, of stable photo-generated minority carriers and the normalized distance to the front surface

indicated in the (6.147) the larger the absorption coefficient is, and the larger the minority carrier diffusion length L_e is, the larger η is. When $\alpha L_e \gg 1$, then we have $\eta \to 1$; and when $\alpha L_e = 1$, $\eta = 1/2$, so the quantum efficiency decreases to its half of its maximum value, and then:

$$\alpha(\lambda_{co}) = \frac{1}{L_e}, \tag{6.148}$$

a peculiar special case. In general the quantum efficiency depends on the wavelength, the composition, the temperature, and the minority carrier diffusion length of carriers in the p region because the absorption coefficient depends on the wavelength, the composition, and the temperature.

The behavior of $\Delta n(z)$ is indicated in (6.146a) and (6.146b); see Fig. 6.19. Also from (6.2.146), the maximum of $\Delta n(z)$ occurs at:

$$z_{\max} = \frac{L_e \ln(\alpha L_e)}{\alpha L_e - 1} \cong \frac{\ln(\alpha L_e)}{\alpha}, \quad (\alpha L_e \gg 1) \tag{6.149}$$

and the maximum value of Δn is:

$$\Delta n(z_{\max}) = \left(\frac{\alpha Q \tau_e}{\alpha L_e + 1}\right) \exp(-\alpha z_{\max}) \approx \frac{1}{\alpha L_e} \frac{Q \tau_e}{L_e}, (\alpha L_e \gg 1) \tag{6.150}$$

Thus, $\Delta n(z_{\max}) = 1.6 \times 10^{12} \mathrm{cm}^{-3}$ when the photon flux is 1×10^{17} photons/cm$^2 \cdot$s, the lifetime is 0.5 μs, the absorption coefficient is $5 \times 10^3\ \mathrm{cm}^{-1}$, and the diffusion length is 25 μm. The absorption coefficients of the n region and the space charge region become progressively more important, the shorter the wavelength becomes compared to the cut-off wavelength, despite the thicknesses of these regions being far smaller than L_e. The penetrate depth decreases because the absorption coefficient increases at shorter wavelengths. The Moss–Burstein (M–B) effect (Moss 1954; Burstein 1954) must be taken into account if the n region is heavily doped so the Fermi level lies in the conduction band, thus increasing the effective bandgap, so the absorption edge is shifted to shorter wavelengths.

There are two modes of a general p–n junction photodiode, the front irradiated n-on-p photodiode and back irradiated photodiode (Fig. 6.20a, b). In the front irradiated mode, it requires that the thickness of the p region be approximately equal to or smaller than the diffusion length, to allow the photo-generated carriers to reach the space charge region before recombination occurs. If, $d \ll L_e$, the cut-off wavelength of the two modes is determined by the thickness of the p region, not by the diffusion and absorption in the n region or in the space charge region, where the surface at $z = d$ is assumed to be a perfectly electrical reflective surface. Thus the quantum efficiency of the front irradiated mode is (Van der Wiele 1976):

$$\eta = \left(\frac{\alpha L_e}{\alpha^2 L_e^2 - 1}\right) \left[\alpha L_e - \frac{\sinh\left(\frac{d}{L_e}\right) + \alpha L_e e^{-\alpha d}}{\cosh\left(\frac{d}{L_e}\right)}\right]. \tag{6.151}$$

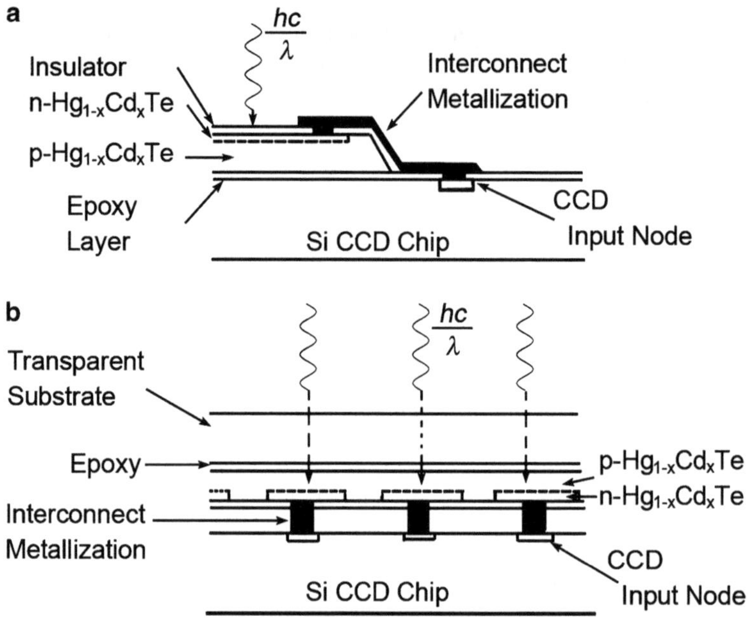

Fig. 6.20 The two modes of p–n junction photodiodes: (**a**) the front irradiated mode and (**b**) the back irradiated mode

And the quantum efficiency of the back irradiated mode is:

$$\eta = \left(\frac{\alpha L_e}{\alpha^2 L_e^2 - 1}\right)\left[\frac{\alpha L_e - \sinh\left(\frac{d}{L_e}\right) e^{-\alpha d}}{\cosh\left(\frac{d}{L_e}\right)} - \alpha L_e e^{-\alpha d}\right]. \tag{6.152}$$

If $d \ll L_e$ and the wavelength is such that, $\lambda \ll \lambda_{co}$, $\alpha L_e > 1$, then the two equations simplify to:

$$\eta \approx 1 - e^{-\alpha d}. \tag{6.153}$$

The absorption coefficient that establishes the cut-off wavelength at which the quantum efficiency decreases to 1/2 of its maximum value is determined by the thickness of the p region:

$$\alpha(\lambda_{co}) = \frac{2.7}{d}. \tag{6.154}$$

The absorption coefficient at the cut-off wavelength λ_{co} can be roughly evaluated from the thickness of the p region: d. Then the cut-off wavelength can be determined by the expression that determines the absorption coefficient. For example, if $d = 10\,\mu\text{m}$, $\alpha(\lambda_{co}) = 690\,\text{cm}^{-1}$, then the cut-off wavelength is 12.4 μm for an MCT alloy with $x = 0.210$ at 80 K. The cut-off wavelengths are 12.7 μm for $L_e = 25$, and 13.1 μm for $L_e = 50\,\mu\text{m}$, respectively, from results for a semi-infinite thickness. The above analysis indicates that to increase the quantum efficiency requires decreasing the surface reflection loss and the surface recombination velocity, as well as a longer diffusion length of p-type HgCdTe.

6.2.4 Noise Mechanisms in Photovoltaic Infrared Detectors

Many authors have discussed the noise in the photovoltaic infrared detectors. Pruett and Petritz (1959) studied the noise in InSb photodiodes. Tredwell and Long (1977) proved the noise mechanisms in HgCdTe photodiodes is the same as those in HgCdTe photoconductive detectors. Studies of the noise in the infrared detectors can be found in Kruse et al. (1962), Kingston (1978), Van Vliet (1967), and Van der Ziel and Chenette (1978).

When a photodiode is in thermal equilibrium, i.e., no external bias and no external photon flux, the mean square noise current $\bar{I}_{\mathrm{n}}^2$ equals the Johnson noise of the photodiode with a zero-bias resistance R_0.

$$\bar{I}_{\mathrm{n}}^2 = \left(\frac{4k_{\mathrm{B}}T}{R_0}\right)\Delta f. \tag{6.155}$$

The noise voltage is:

$$V_{\mathrm{N}} = R_0\sqrt{\bar{I}_{\mathrm{n}}^2} = \sqrt{4k_{\mathrm{B}}TR_0\Delta f}, \tag{6.156}$$

where Δf is the bandwidth of the noise. So the mechanism responsible for the junction current determines not only R_0, but also the noise in the photodiode.

In the following paragraphs, we will discuss the noise in a photodiode that is not in thermal equilibrium, for two special cases, diffusion current and g–r current in the space charge region. And then we will discuss $1/f$ noise in the p–n junction.

We can analyze the noise as follows, if only a diffusion current and a photocurrent are present. The noise in the infrared photodiode that is not in thermal equilibrium can be treated as a shot noise. The junction current is the sum of the diffusion current and the background-induced photocurrent from a flux Q_{B}. For the ideal case we have:

$$I(V) = I_0\left[\exp\left(\frac{eV}{k_{\mathrm{B}}T}\right) - 1\right] - I_{\mathrm{ph}}, \tag{6.157}$$

where

$$I_{\mathrm{ph}} = \eta e Q_{\mathrm{B}} A \tag{6.158}$$

and A is the surface area of the photosensitive element of the photodiode. In fact, the diffusion current in (6.157) is the sum of two currents, a forward current, $I_0\exp\left(\frac{eV}{k_{\mathrm{B}}T}\right)$, that depends on the voltage, and a reverse current I_0, that is a constant. The two diffusion currents and the background photocurrent fluctuate independently, and contribute to the shot noise in quadrature. Thus the mean square of the shot noise current is:

$$\bar{I}_{\mathrm{n}}^2 = 2e\left\{I_0\left[\exp\left(\frac{eV}{k_{\mathrm{B}}T}\right) + 1\right] + I_{\mathrm{ph}}\right\}\Delta f, \tag{6.159}$$

where Δf is the noise bandwidth.

When the photodiode works at zero bias, the resistance, R_0, is:

$$\frac{1}{R_0} = \left.\frac{\mathrm{d}I}{\mathrm{d}V}\right|_{V=0} = \frac{eI_0}{k_{\mathrm{B}}T}. \tag{6.160}$$

Substitute the I_0 from (6.160) into (6.159), yields the noise current at zero bias:

$$\bar{I}^2_{\mathrm{n}(V=0)} = \left(\frac{4k_{\mathrm{B}}T}{R_0} + 2\eta e^2 Q_{\mathrm{B}} A\right)\Delta f. \tag{6.161}$$

The first term is the Johnson noise at zero bias, and the second term is the photocurrent shot noise generated by the background radiation. The second term will play the dominant role if R_0 is large enough, i.e.:

$$R_0 \gg \frac{2k_{\mathrm{B}}T}{\eta e^2 Q_{\mathrm{B}} A}. \tag{6.162}$$

When the reverse bias is large, $|eV| \gg k_{\mathrm{B}}T$, then from (6.159) and (6.160), we have a new expression for the noise current:

$$\bar{I}^2_{\mathrm{n}(V<0)} = \left(\frac{2k_{\mathrm{B}}T}{R_0} + 2\eta e^2 Q_{\mathrm{B}} A\right)\Delta f, \tag{6.163}$$

where now the shot noise is reduced from its valve at zero bias. Equation (6.159) is simpler than the more general one deduced by Van der Ziel and Chenette (1978).

Next we will analyze the noise related to the g–r current in the space charge region. Many studies have been carried out in cases where the noise from the g–r current in the space charge region dominates (Lauritzen 1968; Van Vliet 1976; Van Vliet and Van der Ziel 1977; Van der Ziel and Chenette 1978). The noise current at zero bias can be obtained by substituting the $(R_0A)_{\mathrm{g-r}}$ expression from (6.133) into (6.155). The g–r current in the space charge region is a sum of two currents:

$$I_{\mathrm{g-r}} = I_{\mathrm{r}} - I_{\mathrm{g}}. \tag{6.164}$$

The current component, I_{r}, can be ignored when the reverse bias is large, $|eV| \gg k_{\mathrm{B}}T$. For a single trap energy level $E_{\mathrm{t}} = E_{\mathrm{i}}$ and $\tau_{\mathrm{n}0} = \tau_{\mathrm{p}0} = \tau_0$, the g–r current in the space charge region is:

$$I_{\mathrm{g-r}}(V) = -I_{\mathrm{g}}(V) = \frac{en_i Aw}{2\tau_0}. \tag{6.165}$$

Van der Ziel and Chenette (1978) proved that the mean square noise current at low frequency is given by:

$$\bar{I}^2_{\mathrm{n}} = 2eI_{\mathrm{g}}\Delta f, \tag{6.166}$$

which is exactly the shot noise of the photocurrent I_g. At higher frequency the mean square noise current is slightly smaller:

$$\bar{I}_n^2 = \frac{2}{3}(2eI_g)\Delta f. \tag{6.167}$$

Tobin et al. (1980) studied the $1/f$ noise in an ion–implanted n^+-on-p HgCdTe photodiode. It was discovered that $1/f$ noise does not depend on the photocurrent or the diffusion current, but is linearly proportional to the surface leakage current. This was accomplished by measuring the $1/f$ noise, while varying the background photon flux, the temperature, the reverse bias, and the gate bias. It was also proved that the rms. $1/f$ noise current $I_{n,ex}$, is a function of the measuring frequency, the reverse bias V, the gate bias, and the temperature, and is given by:

$$I_{n,ex}(f, V, V_g, T) = \left[\frac{\alpha_{ex} I_s(V, V_g, T)}{\sqrt{f}}\right]\Delta f, \tag{6.168}$$

where $I_s(V, V_g, T)$ is the surface leakage current of the photodiode, which significantly depends on the reverse gate bias and the temperature; the dimensionless absorption coefficient, α_{ex}, is about 1×10^{-3} for all photodiodes; and Δf is the noise bandwidth and is assumed to be smaller than f.

The dependence of the $1/f$ noise current on the dark current is shown in Fig. 6.21 (Tobin et al. 1980). The two currents were measured at a reverse bias voltage $\sim$50 mV and at 83–160 K. The sample is a photovoltaic detector array made of ion-implant n^+-on-p $Hg_{0.7}Cd_{0.3}Te$. The area of the junction is $1.3 \times 10^{-5} - 4.8 \times 10^{-4}\,cm^2$. The primary dark current in this temperature range is the g–r current. The solid line was obtained from (6.168) with $\alpha_{ex} = 1 \times 10^{-3}$. The temperature

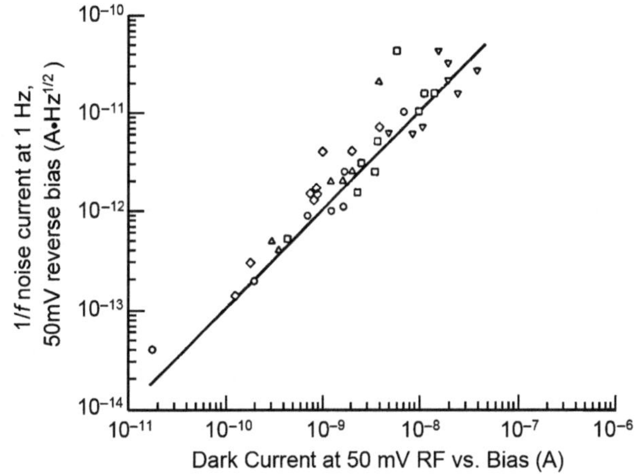

Fig. 6.21 Relation between the $1/f$ noise current and the dark current

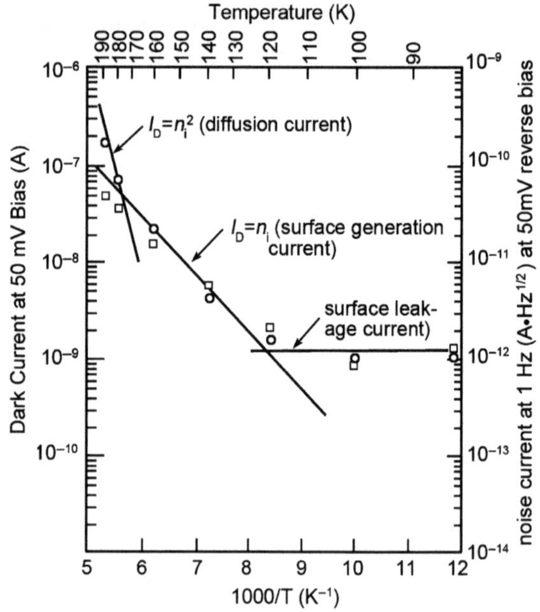

Fig. 6.22 Temperature dependence of the dark current and the $1/f$ noise current of a $Hg_{0.7}Cd_{0.3}Te$ photodiode

dependence of the dark current and of the $1/f$ noise current of this $Hg_{0.7}Cd_{0.3}Te$ photodiode is shown in Fig. 6.22. Both currents were measured at a reverse bias voltage, -50 mV. The primary dark current at the temperatures above 180 K, is the diffusion current, which varies with temperature proportional to, n_i^2, while the $1/f$ noise current exhibits the characteristic surface generation–recombination behavior, which varies with temperature as n_i until the temperature decreases to 110 K. While below 110 K, the $1/f$ noise current and dark current do not depend on the temperature, and the primary dark current is the surface leakage current.

6.2.5 *Responsivity, Noise Equivalent Power and Detectivity*

The physical quantities that are usually used to characterize the sensitivity of an infrared photodiode are detectivity, D_λ^*, and noise equivalent power, NEP_λ. These two quantities have been discussed previously for specific cases, but now will be treated more generally.

Assume a photodiode is uniformly irradiated by a monochromatic light with a signal photon flux Q_s; then the r.m.s. signal photocurrent I_s is:

$$I_s = \eta e Q_s A, \tag{6.169}$$

where A is the area of the photosensitive element. The root-mean-square irradiant power P_λ received by the detector is given by:

$$P_\lambda = \left(\frac{hc}{\lambda}\right) Q_s A. \tag{6.170}$$

The current responsivity $R_{I\lambda}$ is exactly the ratio of I_s to P_λ, and has the units of ampere/watt.

$$R_{I\lambda} = \left(\frac{\lambda}{hc}\right) \eta e. \tag{6.171}$$

The noise equivalent power NEP_λ is the power of the incident radiation with a wavelength λ that equals the noise power in a bandwidth, Δf, that is, such that the ratio of the signal to the noise is 1. The signal-to-noise ratio is:

$$\frac{S}{N} = \frac{I_s}{\sqrt{\bar{I}_n^2}} = \frac{R_{I\lambda} P_\lambda}{\sqrt{\bar{I}_n^2}}, \tag{6.172}$$

where $\bar{I}_n^2$ is the mean square noise current within a bandwidth Δf. Taking the $S/N = 1$ and specifying the noise current for a unit bandwidth, we have the noise equivalent power:

$$\mathrm{NEP}_\lambda = \frac{\sqrt{\bar{I}_n^2}}{R_{I\lambda}\sqrt{\Delta f}}. \tag{6.173}$$

Apparently, the smaller the noise equivalent power is, the greater is the detector's ability. The detectivity is related to the reciprocal of the noise equivalent power. If the root-mean-square noise current is specified for unit area of the detector's sensitive element, the detectivity D_λ^* is given by:

$$D_\lambda^* = \frac{\sqrt{A}}{\mathrm{NEP}_\lambda} = \frac{R_{I\lambda}\sqrt{A\Delta f}}{\sqrt{\bar{I}_n^2}}, \tag{6.174}$$

where D_λ^* is in $\mathrm{cm \cdot Hz^{1/2}/W}$.

The noise current is given by (6.161) at zero bias, and from (6.171) and (6.173), the detectivity is given by:

$$D_\lambda^* = \frac{\lambda}{hc}\eta e \left[\frac{4k_B T}{R_0 A} + 2\eta e^2 Q_B\right]^{-1/2}. \tag{6.175}$$

When thermal noise plays the primary role in a photodiode, (6.174) can be simplified to:

$$\left(D_\lambda^*\right)_{th} = \frac{\lambda}{hc}\eta e \sqrt{\frac{R_0 A}{4k_B T}}, \tag{6.176}$$

where the subscript "th" denotes the thermal noise limited case. When the noise from the background photon flux plays the primary role, (6.175) simplifies to:

$$\left(D_{\lambda}^{*}\right)_{\mathrm{BLIP}} = \frac{\lambda}{hc}\sqrt{\frac{\eta}{2Q_{\mathrm{B}}}}, \tag{6.177}$$

which is the well-known detectivity for the background limited performance (BLIP) of infrared detectors. It requires a very large R_0A to achieve the background limited condition, which makes $\frac{4k_{\mathrm{B}}T}{R_0A} \ll 2\eta e^2 Q_{\mathrm{B}}$, that is, $R_0A \gg \frac{4k_{\mathrm{B}}T}{2\eta e^2 Q_{\mathrm{B}}}$.

Next the response time will be discussed. The main factors that determine the response time are: the time that the photo-generated electron–hole pair in the quasineutral n or p region needs to transfer to the space charge region by diffusion; the time that the photo-generated carrier needs to transmit through the space charge region by drift; the RC time constant related to the junction capacitance and the junction impedance as well as the external circuit impedance connected to the junction.

One contribution to the response time relates to the diffusion rate in the quasineutral region. Taking as an example the n-on-p photodiode in the Fig. 6.11 and assuming that all the incident radiation is absorbed in the quasineutral p region, the alternating generation rate per unit volume at point z, produced by the signal radiation modulated at frequency ω is:

$$G(z,t) = \alpha Q \exp(-\alpha z + \mathrm{i}\omega t). \tag{6.178}$$

The excess photo-generated carrier concentration, $\Delta n(z,t)$, is a solution of the continuity equation with the boundary condition:

$$\Delta n(0,t) = 0, \tag{6.179}$$

and $\Delta n(z,t) \to 0$ as $(z \to \infty)$. The $\Delta n(z,t)$ will produce a photocurrent density:

$$J_{\mathrm{ph}}(t) = e\eta(\omega)Q\exp[\mathrm{i}(\omega t - \phi)] \tag{6.180}$$

where the phase angle ϕ is given by the following equation:

$$\phi = \tan^{-1}\left[\frac{b(\omega)}{\alpha L_{\mathrm{e}} + a(\omega)}\right]. \tag{6.181}$$

The AC quantum efficiency $\eta(\omega)$ is:

$$\eta(\omega) = \left[\left(1 + \frac{a(\omega)}{\alpha L_{\mathrm{e}}}\right)^2 + \left(\frac{b(\omega)}{\alpha L_{\mathrm{e}}}\right)^2\right]^{-1/2}, \tag{6.182}$$

where $a(\omega)$ and $b(\omega)$ are dimensionless functions of the frequency:

$$a(\omega) = \left[\frac{\sqrt{1+\omega^2\tau_e^2}+1}{2}\right]^{1/2},$$
$$b(\omega) = \left[\frac{\sqrt{1+\omega^2\tau_e^2}-1}{2}\right]^{1/2}. \tag{6.183}$$

In the low frequency limit, $a \to 1,\ b \to 0$, and $\eta(\omega)$ simplifies to the dc quantum efficiency given by (6.147), i.e., $\eta = \dfrac{\alpha L_e}{(\alpha L_e + 1)}$. At high frequency such that, $\omega\tau_e \gg 1,\ \eta(\omega)$ simplifies to:

$$\eta(\omega) \to \frac{\alpha L_e}{\sqrt{\omega\tau_e}}. \tag{6.184}$$

A mid range frequency f_0 above which the quantum efficiency decreases rapidly with a frequency increase is:

$$f_0 = \frac{(\alpha L_e)^2}{2\pi\tau_e} = \frac{\alpha^2 D_e}{2\pi}. \tag{6.185}$$

f_0 can be estimated as follows. The minority carrier mobility is about $3 \times 10^5\ \mathrm{cm^2/V\ s}$ for a p-type $Hg_{1-x}Cd_xTe$ with $x = 0.2$ at $T = 77\ \mathrm{K}$. $D_e = 2 \times 10^4\ \mathrm{cm^2/s}$ is obtained from (6.106). The frequency f_0 depends greatly on the wavelength arising from $\alpha(\lambda)$. For a wavelength that is not much smaller than the cut-off wavelength, let $\alpha = 500\ \mathrm{cm^{-1}}$, then, $f_0 = 800\ \mathrm{MHz}$, which is equivalent to a response time of 1.25 ns. A similar calculation can be performed in the n region, and it is discovered that the high frequency limit due to the diffusion effect in the n region of HgCdTe is far larger than that in the p region, because the hole mobility in the n region is so small (smaller than the mobility of the minority carriers (electrons) in the *p* region by about two orders of magnitude) that it will produce a smaller cut-off frequency f_0 and consequentially a longer response time.

The diffusion effect in the space charge region has a small affects on the device response time. The effect of the photo-generated carrier's finite transit time on the frequency of the photodiode was analyzed early by Gärtner (1959), and the readers can also refer to later work by Sze (1969). The transit time, that the carriers with a drift velocity v_d need to transit through the space charge region with a thickness w, is given by:

$$t_r = \frac{w}{v_d}. \tag{6.186}$$

This transit time is about 1×10^{-11} s, if $w = 1\ \mu\mathrm{m}$, and v_d is about $1 \times 10^7\ \mathrm{cm/s}$ limited by lattice scattering. It is shown that the transit time will affect the frequency response in the range $\sim(2\pi t_r)^{-1}$, i.e., 16 GHz.

The junction capacitance effect will be discussed next. The junction capacitance times the dynamic resistance, the device's series resistance plus that of the external circuit, have an effect on the high frequency response of a $Hg_{0.8}Cd_{0.2}Te$ photodiode (Peyton et al. 1972; Shanley and Perry 1978). A simple description will be given here. Assume that the capacitance C_j of the space charge region and the load resistance R_L of the external circuit play the primary role, thus the high frequency limit f_0 is:

$$f_0 = \frac{1}{2\pi R_L C_j}, \tag{6.187}$$

where

$$C_j = \frac{\varepsilon_s \varepsilon_0 A_j}{w} \tag{6.188}$$

ε_s is the static dielectric constant, ε_0 the permittivity of the free space, A_j the area of the junction, and w is the width of the space charge region. w is usually given as:

$$w = \sqrt{\frac{\varepsilon_s \varepsilon_0 (V_{bi} - V)}{e N_B}}, \tag{6.189}$$

where $N_B = N_A N_D/(N_A + N_D)$ is the effective doping concentration in the space charge region, and V_{bi} the built-in junction voltage:

$$V_{bi} = \frac{k_B T}{e} \ln\left(\frac{N_A N_D}{n_i^2}\right). \tag{6.190}$$

For a $Hg_{1-x}Cd_xTe$ photodiode of with composition $x = 0.206$, the cut-off wavelength is about 12 μm at 77 K, $n_i = 5 \times 10^{13}\,\text{cm}^{-3}$, ε_s is 17, $N_D = 2 \times 10^{14}\,\text{cm}^{-3}$, and $N_A = 1 \times 10^{17}\,\text{cm}^{-3}$. $eV_{bi} = 0.059\,\text{eV}$, is given by (6.190), and w, C_j, and f_0 can be calculated. Relations between the cut-off frequency and the reverse bias voltage is shown in Fig. 6.23. The cut-off frequencies for various N_D are calculated from (6.187)–(6.190), and are shown in the figure assuming that the load resistance is 50 Ω, the area of the junction A_j is $1 \times 10^{-4}\,\text{cm}^2$, and $N_A = 1 \times 10^{17}\,\text{cm}^{-3}$. The width, w, of the space charge region is also shown in the figure. We see that the

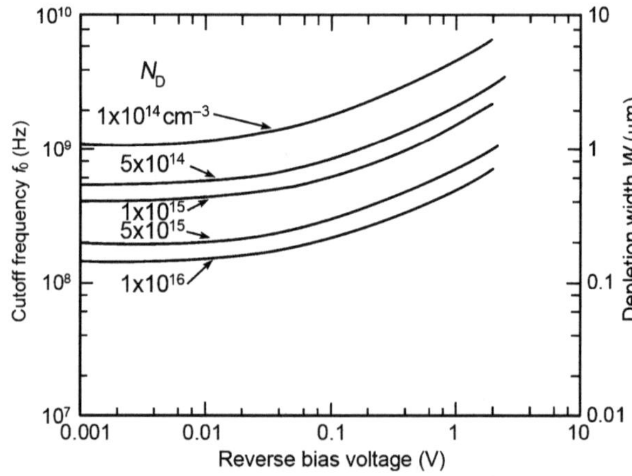

Fig. 6.23 The dependence of the photodiode cut-off frequency f_0 on the reverse bias voltage

cut-off frequency can reach 2 GHz for a low n-type doping concentration at some biases. Certainly, the presence of the other distributed capacitances and series resistances from the electrodes will increase these response times.

6.3 Metal-Insulator-Semiconductor Infrared Detectors

6.3.1 MIS Infrared Detector Principles

The dependence of the surface potential on the bias voltage in an MIS device has been described in the first section of Chap. 4. Kinch (1981) presented a well rounded analysis of the MIS structured infrared detectors. Here we will give only a brief introduction into their operating principles. For an n-type semiconductor MIS device, a negative bias voltage makes the surface energy bands bend upward. The surface valence band is totally filled with the holes when the top of the valence band near the surface reaches or exceeds the Fermi level E_F, and consequentially at this point an inversion layer begins to form. However, the inversion condition at thermal equilibrium differs from that where the surface is irradiated, which is depicted in Fig. 6.24. Figure 6.24a shows the inverted band structure in thermal equilibrium, i.e., inversion occurs when the surface energy band is bent to $\phi_s(\text{inv}) = 2\phi_F$.[1] Whereas Fig. 6.24b shows the case of in the presence of optical radiation, i.e., inversion occurs when the surface energy band is bent to $\phi_s'(\text{inv}) = 2\phi_F - V_\phi$.

$$\begin{aligned} &\phi_s'(\text{inv}) < \phi_s(\text{inv}) \\ &\Delta\phi(\text{inv}) = \phi_s(\text{inv}) - \phi_s'(\text{inv}). \end{aligned} \tag{6.191}$$

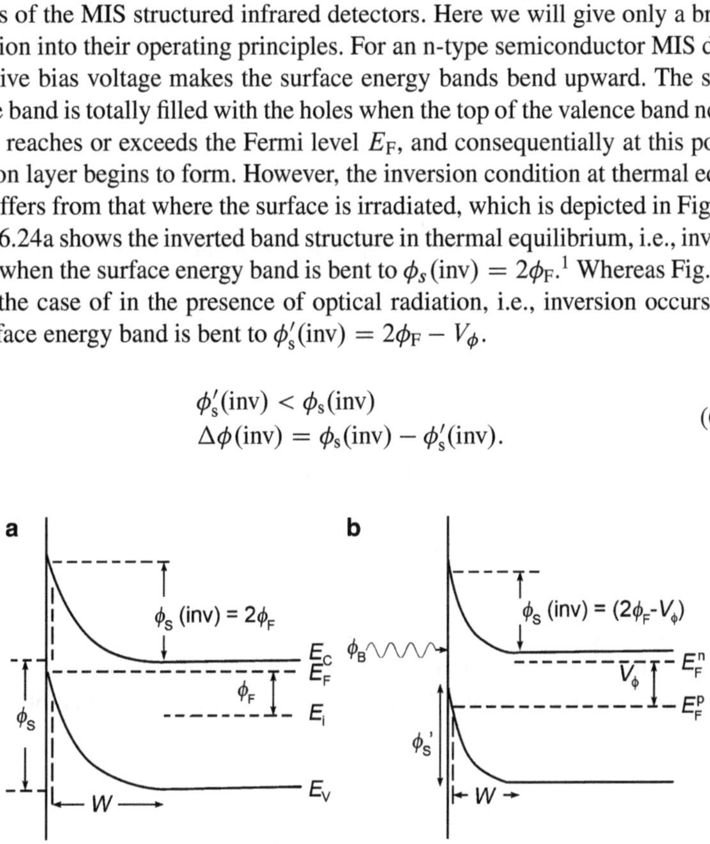

Fig. 6.24 The diagrams for an n-type MIS devices in inversion (**a**) in thermal equilibrium; (**b**) with irradiation present

[1] This is an approximation satisfied well in materials where the ratio of the hole to electron effective mass is near unity, but fails somewhat for narrow gap materials where the ratio can reach values as high as 10^5. However, the approximation gets progressively better as the temperature is lowered.

The condition for strong inversion is that the surface hole concentration is larger than the bulk electron concentration, i.e., $p_s = p_0 \exp(-e\phi_s(\text{inv})/k_B T) > n_0$. The light radiation generates electron–hole pairs so under a negative bias, the optically generated electrons move into the bulk while the holes congregate at the surface, which make it easier to produce an inverted surface. Taking the Fermi level as a reference, the hole quasi-Fermi level, E_F^p, has a difference of eV_ϕ from the electron quasi-Fermi level, E_F^n, and the width of the depletion layer w decreases. The hole concentration, p_0 varies by Δp_0 under irradiation, and consequentially causes a variation $\Delta\phi_s(\text{inv})$ in the surface potential. From the analysis of MIS structures in Sect. 4.1, we have:

$$\Delta\phi_s(\text{inv}) = \left(\frac{k_B T}{e}\right)\frac{\Delta p_0}{p_0}. \tag{6.192}$$

Thus, the bias voltage needed to maintain the inversion in MIS devices with light radiation present differs from that without light radiation; the difference is:

$$\Delta V = \Delta\phi_s(\text{inv}). \tag{6.193}$$

If diffusion plays the primary current mechanism role in an MIS device, then we have:

$$R_d A_d = \frac{k_B T}{e} \cdot J_{\text{dif}} = k_B T \tau_p / e^2 p_0 L_p, \tag{6.194}$$

where L_p and τ_p are the diffusion length and the lifetime of the minority carriers, respectively, while R_d and A_d are the impedance and the area of the device, respectively. Assuming that Δp_0 is caused by the incident signal photon flux, ΔQ_s, then we have:

$$\Delta p_0 = \eta \Delta Q_s \tau_p / L_p. \tag{6.195}$$

Substituting this expression into (6.192) and (6.194), yields:

$$\Delta\phi_s = \eta e \Delta Q_s R_d A_d, \tag{6.196}$$

which shows that the variation of the surface potential caused by the signal photon flux, ΔQ_s, is proportional to the product of the impedance and the area of the device.

The operational circuit diagram of an MIS device is shown in Fig. 6.25. A voltage is produced at the device terminals of the circuit when external signal photons are

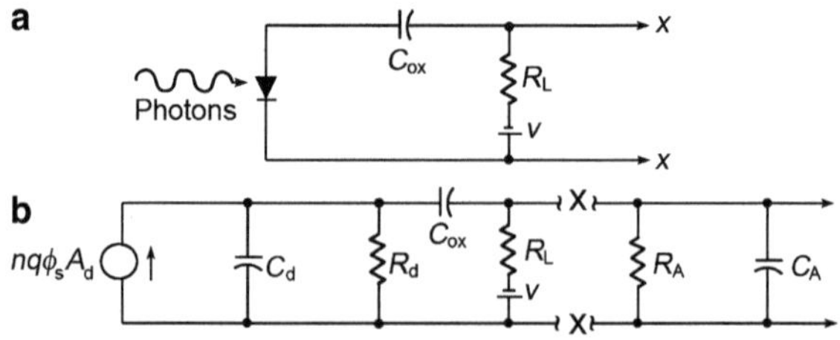

Fig. 6.25 (**a**) An operational diagram and (**b**) an equivalent circuit diagram of an MIS device in which the signal is modeled as a current source and the input impedance of the amplifier is modeled as the parallel combination of a resistance and a capacitance

injected, if an n-type HgCdTe MIS structure has a DC reverse bias on the gate electrodes to invert it. The device equivalent circuit, shown in Fig. 6.25, is the parallel circuit of a photo-generated current signal source with a capacitance, $C_{\rm d}$, and a resistance, $R_{\rm d}$.

When a signal radiation is present, the voltage across $R_{\rm d}$ is:

$$\Delta V = \eta e \Delta Q_{\rm s} R_{\rm d} A_{\rm d}, \tag{6.197}$$

where $R_{\rm d}C_{\rm d}$ is a time constant. The variation of the voltage is reflected in the response across the two terminals of the load resistance, $R_{\rm L}$, when $R_{\rm L}C_{\rm ox} \gg R_{\rm d}C_{\rm d}$, and consequentially will serve to drive the preamplifier.

If the incident light is modulated with a frequency $\omega_{\rm m}$, there is an AC signal at the input of the preamplifier:

$$\Delta V = \frac{\eta e \Delta Q_{\rm s} R_{\rm d} A_{\rm d}}{\left[1+\omega_{\rm m}^2 R_{\rm d}^2 C_{\rm d}^2\right]^{1/2}} \cdot \frac{\omega_{\rm m} R_{\rm L} C_{\rm ox}}{\left[1+\omega_{\rm m}^2 R_{\rm L}^2 C_{\rm ox}^2\right]^{1/2}} \tag{6.198}$$

When we have, $\omega_{\rm m} R_{\rm d} C_{\rm d} \ll 1$ and $\omega_{\rm m} R_{\rm L} C_{\rm ox} \gg 1$, (6.198) transforms into:

$$\Delta V = \eta e \Delta Q_{\rm s} R_{\rm d} A_{\rm d},$$

which is the same as (6.197).

At low temperature, the device impedance is also determined by the background radiant flux $Q_{\rm B}$, if the device is only irradiated by it. When the diffusion current and the background radiation photocurrent play the primary roles in the device; for the open circuit case:

$$I_{\rm t} = I_0 \left[\exp\left(\frac{eV}{k_{\rm B}T}\right) - 1\right] - \eta e Q_{\rm B} A_{\rm d} = 0.$$

If $V > k_{\rm B}T/e$ then we have:

$$R_{\rm d} = \left(\frac{{\rm d}I}{{\rm d}V}\right)^{-1} = \frac{k_{\rm B}T}{e^2 \eta Q_{\rm B} A_{\rm d}}. \tag{6.199}$$

From (6.196), (6.199), and (6.193), it follows that:

$$\Delta V = \left(\frac{k_{\rm B}T}{e}\right) \frac{\Delta Q_{\rm s}}{Q_{\rm B}}, \tag{6.200}$$

which shows that in this special case the signal voltage is independent of the quantum efficiency; where $\Delta Q_{\rm s}$ is the sum of the signal photon flux and the background photon flux $Q_{\rm B}$. Equation (6.200) transforms back into (6.197), if $Q_{\rm B}$ obtained by solving (6.199) is substituted into (6.200).

The noise voltage at the output contacts in Fig. 6.25 is approximately equal to:

$$\bar{V}_{\mathrm{n}}^{2} \approx 4k_{\mathrm{B}}TR_{\mathrm{d}}\Delta f + 4k_{\mathrm{B}}T'R_{\mathrm{L}}\Delta f \cdot \frac{\left|R_{\mathrm{d}} + \frac{1}{\mathrm{i}\omega C_{\mathrm{ox}}}\right|^{2}}{\left|R_{\mathrm{L}} + R_{\mathrm{d}} + \frac{1}{\mathrm{i}\omega C_{\mathrm{ox}}}\right|^{2}}. \tag{6.201}$$

For a case where, $\omega_{\mathrm{m}} C_{\mathrm{ox}} R_{\mathrm{L}} \gg 1$, this equation reduces to the expression:

$$\bar{V}_{\mathrm{n}}^{2} \approx 4k_{\mathrm{B}}TR_{\mathrm{d}}\Delta f + \frac{4k_{\mathrm{B}}T'R_{\mathrm{L}}\Delta f}{R_{\mathrm{L}}\omega^{2}C_{\mathrm{ox}}^{2}}. \tag{6.202}$$

The noise voltage at the output contacts generated by the device resistance, R_{d}, is larger than the noise voltage from the load resistance, if $R_{\mathrm{L}} > \frac{T'}{T} \cdot \frac{1}{R_{\mathrm{d}}\omega^{2}C_{\mathrm{ox}}^{2}}$. In order to obtain a better performing device, the resistance, R_{A}, and the capacitance, C_{A}, of the preamplifier should satisfy the conditions, $R_{\mathrm{A}} > R_{\mathrm{L}}$ and $C_{\mathrm{A}} > C_{\mathrm{ox}}$.

The detectivity is defined by:

$$D^{*} = \frac{V_{\mathrm{s}}}{V_{\mathrm{N}}} \cdot \frac{\sqrt{A\Delta f}}{P_{\mathrm{in}}}, \tag{6.203}$$

where $V_{\mathrm{S}} = \Delta V$ is the signal voltage, which is given by (6.197); $V_{\mathrm{N}} = \sqrt{\bar{V}_{\mathrm{n}}^{2}}$ is the noise voltage, which is given by the first term on the right-hand side of (6.202), and P_{in} the incident radiation power. Thus we have:

$$D^{*} = \frac{\Delta V}{\sqrt{\bar{V}_{\mathrm{n}}^{2}}} \cdot \frac{\sqrt{A\Delta f}}{\Delta\phi_{\mathrm{s}} h\nu A_{\mathrm{d}}} = \frac{\eta e}{2h\nu} \cdot \left(\frac{R_{\mathrm{d}}A_{\mathrm{d}}}{k_{\mathrm{B}}T}\right)^{1/2}. \tag{6.204}$$

When diffusion current is the dominant current mechanism (refer to 6.194), then:

$$R_{\mathrm{d}}A_{\mathrm{d}} = k_{\mathrm{B}}T\tau_{\mathrm{p}}/e^{2}p_{0}L_{\mathrm{p}}.$$

If the minority carrier lifetime, τ_{p}, is determined by the Auger band-to-band recombination lifetime:

$$\tau_{\mathrm{p}} = 2\tau_{\mathrm{Ai}}\frac{n_{\mathrm{i}}^{2}}{n_{0}(n_{0} + p_{0})}, \tag{6.205}$$

where, τ_{Ai}, is the Auger lifetime of the intrinsic material. With the Einstein relation:

$$L_{\mathrm{p}} = \sqrt{D_{\mathrm{p}}\tau_{\mathrm{p}}} = \sqrt{(k_{\mathrm{B}}T\mu_{\mathrm{p}}/e) \cdot \tau_{\mathrm{p}}}, \tag{6.206}$$

we have:

$$R_{\mathrm{d}}A_{\mathrm{d}} = \frac{k_{\mathrm{B}}T}{e^{2}n_{\mathrm{i}}}\left(\frac{2e\tau_{\mathrm{Ai}}}{k_{\mathrm{B}}T\mu_{\mathrm{p}}}\right)^{1/2}. \tag{6.207}$$

Fig. 6.26 A structural sketch of an MIS infrared detector

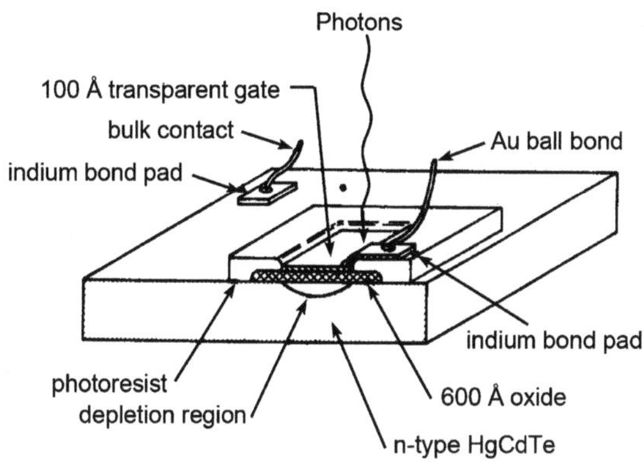

The dependence of R_dA_d and D^* on the temperature and the alloy concentration, can be obtained from (6.207) and (6.204), respectively.

We will next present a description of the structure and major parameters of an MIS infrared detector. The structure of these devices is showed in Fig. 6.26, where each part is labeled. For example, for a cut-off wavelength of, $\lambda_c = 5\,\mu m$, (equivalent to a 0.25 eV bandgap), and if an n-type HgCdTe alloy is covered with a ten nm thick insulator layer; and assuming that the device operates at 190 K, where diffusion current plays the primary role; and $n_i = 2 \times 10^{14}\,cm^{-3}$, $\mu_p = 200\,cm^2/Vs$, $\tau_{Ai} = 4 \times 10^{-4}$ s, then from (6.207), we have $R_dA_d = 8\,\Omega cm^2$; so if the area of the device is $1.5 \times 10^{-4}\,cm^2$, $R_d = 5.3 \times 10^4\,\Omega$. Taking $\eta = 1$, it follows from (6.204), that $D^* = 10^{11}$ cm Hz$^{1/2}$ W^{-1}. Furthermore, we have:

$$C_d = \frac{\varepsilon_0 \varepsilon A_d}{w}, \tag{6.208}$$

where w is the width of the depletion layer, and is expressed by:

$$w^2 = 4\varepsilon_0 \varepsilon \phi_F / n_0 e, \tag{6.209}$$

and ϕ_F, is the Fermi potential. For an n-type HgCdTe alloy with a cut-off wavelength of, $\lambda_c = 5\,\mu m$, ε is ~19.5, and if n_0 is ~$10^{15}\,cm^{-3}$, $\phi_F \sim 25$ mV, and then, $w = 3.4 \times 10^{-5}$ cm. As has been evaluated above for, $A_d = 1.5 \times 10^{-4}\,cm^2$, and $C_d = 8$ pF, consequentially the corresponding time constant is, $R_dC_d = 4.25 \times 10^{-7}$ s. The capacitance of the insulator layer, which covers the HgCdTe surface, can be calculated from its thickness and the dielectric constant of the insulator layer. $C_{ox} = 34.5$ pF, if the thickness is, 70 nm, and the dielectric constant, ε_{ox}, is ~18.3. (This is not SiO_2 which has a much smaller dielectric constant, but some combination of Hg, Cd, and Te oxides.) The output signal voltage across the load resistance is calculated from (6.198). The noise voltage at the two terminals of R_L is determined by R_d, which is found from (6.197), if the R_L is large enough. Substituting for the quantities such as, C_{ox}, and taking, $T' = 300$ K, if $f = 10^3$ Hz we have, $R_L \cong 6.1 \times 10^{14}\,F^{-1}\,Hz$; or, $R_L \cong 6.1 \times 10^8\,\Omega$.

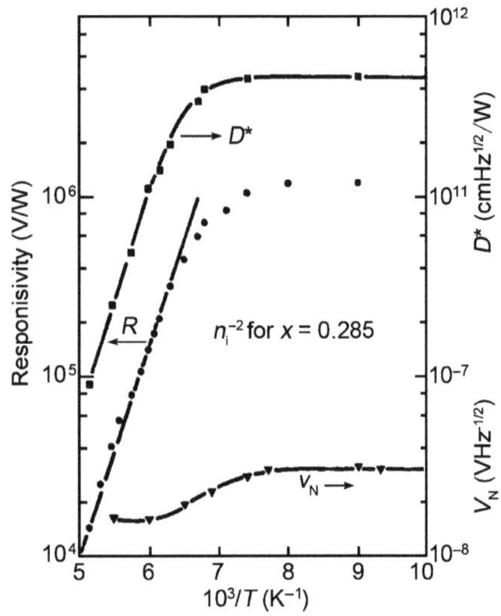

Fig. 6.27 The temperature dependence of the detectivity D^*, the responsivity R_λ, and the noise voltage of an MIS device made from p-type HgCdTe with $x = 0.285$

Figure 6.27 shows the temperature dependences of the detectivity, D^*, the voltage responsivity, R_λ, and the noise voltage of an MIS device made from p-type HgCdTe with $x = 0.285$ (Kinch et al. 1980). The symbols are experimental results and the solid curves are results of calculations with $\eta = 1$.

6.3.2 The Dark Current in MIS Devices

Next a study of the dark current in devices will be undertaken. When an n-type HgCdTe MIS device is biased into deep depletion, the dark current density is given by:

$$J_d = en_i\left[\frac{n_i L_p}{n_0\tau_p} + \frac{w}{2\tau_0} + \frac{s}{2}\right] + e\eta Q_B + J_t, \tag{6.210}$$

where the first term is the minority carrier current that is generated in the electrically neutral region and transfers to the depletion layer by diffusion; the second term is the minority carrier current generated in the depletion layer of width, w; the third term is the current produced by the fast interfaces states related to the surface recombination velocity s; the fourth term is the current generated by the incident background photon flux; and the fifth term is the current generated by direct carrier tunneling from the valence band to the conduction band. This expression is applicable to the zero bias case. The noise property of MIS devices is frequently characterized with the R_0A product of the device. Next, we will analyze each term contributing to the dark current in (6.210) one by one.

The first term in (6.210) is the diffusion current. The minority carrier diffusion current of an n-type semiconductor is given by:

$$J_{dif} = en_i^2 L_p/n_0\tau_p, \tag{6.211}$$

where in the bulk, L_p, is the minority carrier diffusion length, and τ_p is their lifetime. For a good-quality material, the lifetime τ_p is determined by Auger band-to-band recombination. The Auger lifetime τ_A is given by:

$$\tau_A = 2n_i^2\tau_{Ai}/n_0(n_0 + p_0), \tag{6.212}$$

where τ_{Ai} is the Auger lifetime of the intrinsic material. For $n_0 \gg p_0$, from (6.211) and (6.212), it follows that:

$$J_{dif} = en_i\left[k_B T\mu_p/2e\tau_{Ai}\right]^{1/2}, \tag{6.213}$$

where μ_p is the hole mobility. (The Einstein relation $L_p = \sqrt{D_p\tau_p} = \sqrt{(k_B T\mu_p/e)\cdot\tau_p}$ has been used in this derivation.)

The diffusion current of an n-type HgCdTe alloy with various concentrations and temperatures can be calculated from (6.213). The results for HgCdTe with $0.20 < x < 0.35$, for which the cut-off wavelengths are in the range of about 4–13.5 μm, are shown in the Fig. 6.28. The calculated currents are lower limits for n-type material and they would be somewhat different if the thickness of the material were small compared to L_p. A typical value of L_p in n-type material is in the range of 25–75 μm.

For p-type material, the minority carrier lifetime for band-to-band recombination is given by:

$$\tau_R = 2\tau_{Ri}\cdot n_i/p_0, \tag{6.214}$$

where τ_{Ri} is the radiative lifetime of an intrinsic material. The diffusion current is given by:

$$J_{dif} = \frac{en_i^{3/2}}{p_0^{1/2}}\left[\frac{k_B T\mu_n}{2e\tau_{Ri}}\right]^{1/2}, \tag{6.215}$$

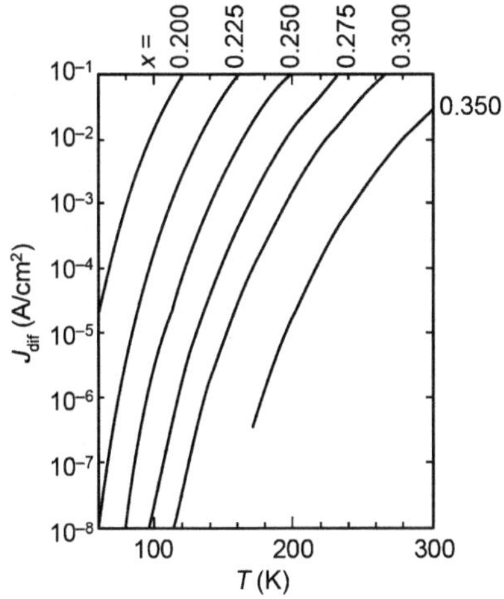

Fig. 6.28 The dependence on the temperature and on the concentration of the diffusion current limited by Auger recombination in n-type HgCdTe

Fig. 6.29 The temperature dependence of the diffusion current in p-type HgCdTe

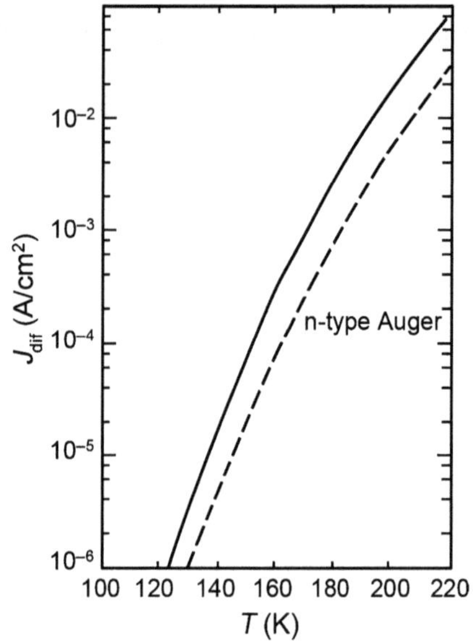

where μ_n is the electron mobility. In this case, $J_{dif} \propto p_0^{-1/2}$, so a high doping concentration can be used to limit the dark current and consequentially to improve the device performance. But too large a p_0 will causes a large depletion layer capacitance, decreasing the diffusion length and the breakdown voltage when operating with reverse bias. Figure 6.29 shows the temperature dependence of J_{dif} for a p-type HgCdTe with a cut-off wavelength, $\lambda_c = 5\,\mu\text{m}$ and $p_0 = 10^{15}\,\text{cm}^{-3}$, where the τ_{Ri} used is the its measured value; the one used above when discussing the minor carrier lifetime. The diffusion current of an n-type MIS structure is also shown in the figure as reference. Apparently, the diffusion current of a p-type material is three times that of n-type material for $p_0 = 10^{15}\,\text{cm}^{-3}$, because the high minority carrier mobility causes a large diffusion length. But the diffusion current can be decreased if the thickness of the n-type substrate is decreased.

The second term of (6.210) is the depletion current. This current is generated by a Shockley–Read center in the MIS structure with a strong reverse bias, and is given by:

$$J_{dep} = ewn_i^2 / \left(\tau_{p0} n_1 + \tau_{n0} p_1\right), \tag{6.216}$$

where w is the thickness of the depletion layer; $n_1 = N_C \exp\left[-E_t/k_B T\right]$, $p_1 = N_V \exp\left[(-E_g + E_t)/k_B T\right]$, $\tau_{p0} = \left(c_p N_R\right)^{-1}$, $\tau_{n0} = \left(c_n N_R\right)^{-1}$; N_C and N_V are the density of states of the conduction band and valence band, respectively; c_p and c_n are the hole and electron capture coefficients of the Shockley–Read center, respectively; N_t is the density of the traps, with energy E_t referenced to the conduction band. $n_1 = p_1 = n_i$, when in an operating mode with, $V = 0$. If the trap energy level is located at the intrinsic energy level; and if $c_n = c_p$, it follows that:

$$J_{dep} = ewn_i/2\tau_0, \tag{6.217}$$

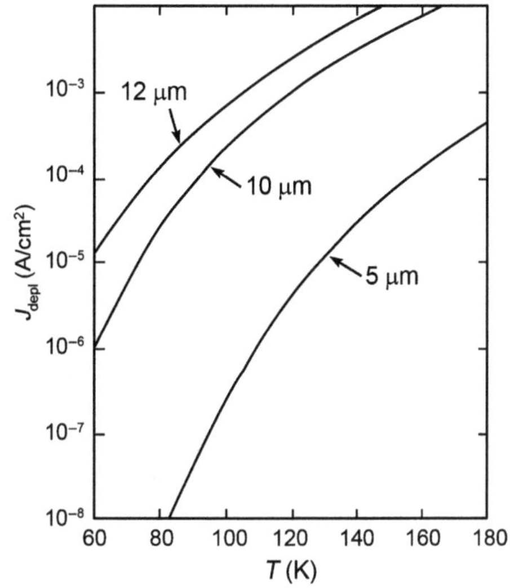

Fig. 6.30 The temperature dependence of the depletion current of devices with three different cut-off wavelengths

where $\tau_0 = \left(c_p N_t\right)^{-1}$. Equation (6.217) is exactly the second term of (6.210). The depletion current varies with the thickness of the depletion layer that is given by:

$$w = (2\varepsilon\varepsilon_0\phi_s/qn_0)^{1/2}, \tag{6.218}$$

where ϕ_S is the surface potential of the MIS structure at a given gate voltage. Figure 6.30 shows the temperature dependence of the depletion current of three devices with various cut-off wavelengths, where parameters, $\tau_0 = 10^{-5}$ s, $n_0 = 5 \times 10^{14}\,\mathrm{cm}^{-3}$, $\phi_S = 3.5\,\mathrm{V}(\lambda_c = 5\,\mu\mathrm{m},\ 10\,\mu\mathrm{m},\ 12\,\mu\mathrm{m})$ were taken.

The current produced by the fast surface states is similar to the depletion current produced by the bulk Schockley–Read center, and is given by:

$$j_s = \frac{1}{2} e n_i s, \tag{6.219}$$

where s is the maximum surface recombination velocity. At 77 K, for $\lambda_c = 10\,\mu\mathrm{m}$ and $n_i = 6 \times 10^{12}\,\mathrm{cm}^{-3}$, $j_s = 5 \times 10^{-7}$ s A/cm^2 from (6.219). The surface recombination velocity, S, can be less than 10^2 cm/s for a better performing device, and then j_s is small, and consequentially is unimportant at 77 K.

The fourth term is the background current. The current density produced by the incident background photon flux is given by:

$$J = \eta e Q_B, \tag{6.220}$$

where η is the quantum efficiency and, Q_B, is the background photon flux density. Equation (6.220) provides a lower limit to the infrared device dark current. For a background limited infrared photodetector (BLIP), this current must be the primary contributor to the dark current.

The fifth term in (6.202) is the tunneling current. The tunneling current through a delta function barrier is given by (Sze 1969):

$$J_t = \frac{e^3 E \phi_s}{4\pi^3 \hbar^2} \left(\frac{2m^*}{E_g}\right)^{1/2} \exp\left(\frac{4\,(2m^*)^{1/2}\, E_g^{3/2}}{3e\hbar E}\right), \tag{6.221}$$

where E is the electric field at the center of the barrier. The electric field at the MIS semiconductor device surface is given by:

$$E_s = (2en_0\phi_s/\varepsilon\varepsilon_0)^{1/2}, \tag{6.222}$$

where ϕ_s is the surface potential of the empty well. From (6.222) and (6.223), and taking $m^*/m_0 = 7 \times 10^{-2} E_g$ (Kinch and Buss 1971), it follows that:

$$J_t = 10^{-2} n_0^{1/2} \phi_s^{3/2} \exp\left[-\frac{4.3 \times 10^{10} E_g^2}{(n_0\phi_s)^{1/2}}\right] \text{A/cm}^2, \tag{6.223}$$

where n_0 is in cm^{-3}, E_g is in eV, and ϕ_s is in V. The tunneling current has a strong dependence on the bandgap of the semiconductor, while it has a weaker dependence on the doping concentration and on the surface potential. The dependence of the tunneling current on the surface potential of an empty well is showed in Figs. 6.31 and 6.32, where the calculation was done from (6.223). There are different carrier concentrations for the two HgCdTe sample with $E_g = 0.25$ eV and $E_g = 0.1$ eV. The current generated by the background radiation is also shown in Figs. 6.31 and 6.32, where the calculations correspond to cut-off wavelengths for the two sample, taking the quantum efficiency to be, $\eta = 0.5$, and the infrared system field of view f in a typical IR system taken as 2.5.

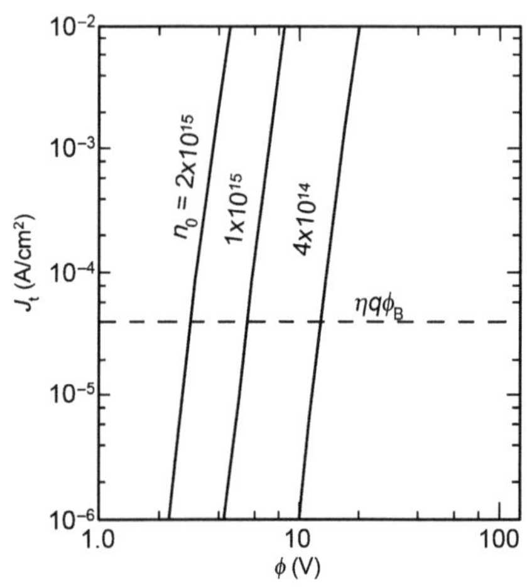

Fig. 6.31 The dependence of the tunneling current J_t and the background current J_B on the surface potential of a sample with $E_g = 0.25$ eV and with different doping concentrations

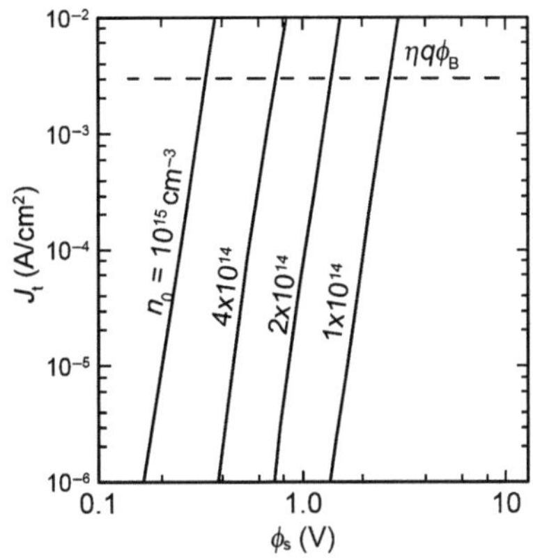

Fig. 6.32 The dependence of the tunneling current J_t and the background current J_B on the surface potential of a sample with $E_g = 0.1$ eV and with different doping concentrations

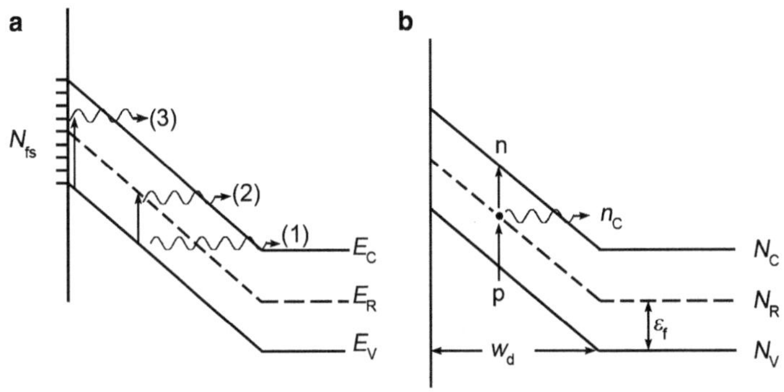

Fig. 6.33 The different tunneling mechanisms in HgCdTe

What has been discussed above is the direct band-to-band tunneling, the transition (1) shown in the Fig. 6.33a. But tunneling from the semiconductor valence band can occur by other processes as well. Other permitted processes are trap-assisted tunneling. In transition (2) showed in Fig. 6.33, the electrons first transfer from the valence band to the Shockley–Read recombination center N_t by thermal excitation, and then transfer from N_t to the conduction band by tunneling; or as depicted in the transition (3), the electrons transfer from the valence band to the surface states N_{fs} by the thermal excitation, and then transfer from the surface states N_{fs} to the conduction band by tunneling. Calculations of these tunneling current mechanisms can be found in Anderson (1977), Sah (1961), Kinch et al. (1980), and Krishnamurthy et al. (2006).

Besides the metal–insulator–semiconductor (MIS) structures, a metal/semiconductor structure can also be applied as an infrared detector. To make a metal/semiconductor structure, a metal is deposited on the surface of the semiconductor that forms a barrier at the interface, called a Schottky Barrier. Many metal types form Schottky barriers rather than Ohmic contacts. Schottky barrier photodiode detectors are treated in Tang and Mi (1989), Pellegrini and Shepherd (1983), Shepherd (1983).

6.4 Low-Dimensional Infrared Detectors

6.4.1 Introduction

Although HgCdTe is still the most frequently employed semiconductor with a tunable bandgap, it is now facing more and more challenges. There is a major effort to devise technologies to replace HgCdTe. On one hand, a fragile Hg–Te bond causes instabilities in bulk material as well as at surfaces and interfaces. And the progress has become somewhat sluggish in the fabrication of large-scale photovoltaic HgCdTe infrared focal plane arrays. On the other hand, research is making flying progress on new semiconductor heterostructure systems. Hence attempts continue to explore other infrared detecting methods besides HgCdTe.

A quantum well infrared photodetector (QWIP) is one of these options. Typically, a QW is made of one thin semiconductor film (QW layer, which is normally narrower than 15 nm) sandwiched by two other semiconductor films (barrier layers) with relatively large bandgap energies. In contrast to the continuous electronic states in bulk semiconductors, the electronic states in a QW are separated into several distinct levels. The number, the energies as well as the separations of these levels are all tunable. When the energy of an absorbed photon corresponds to the separation of two QW levels, it is feasible to realize infrared detection with photon energies ranging from the mid-infrared to the far-infrared region.

When an electronic transition takes place between two levels both in the conduction band of a material system, it is called intersubband or intraband transition. Currently, available intersubband transition QWIPs are mostly fabricated from GaAs/AlGaAs QWs. Because the conduction band bandgap discontinuity of these materials is less than 210 meV, such a QWIP normally functions in the LWIR air-window at, 8–10 μm. The relatively small value of the discontinuity also causes difficulty in control of the device dark current, because the electrons in the QW can more easily be thermally excited to jump over the barrier. This forces the device to operate only at quite low temperatures.

To solve this issue, a good candidate is the $In_{1-x}Ga_xAs/AlGaAs$ material system, which has a large conduction-band discontinuity relative to the GaAs/AlGaAs system. And further, the discontinuity is tunable by varying the composition x. Also a deeper QW has more permitted electronic transitions, so it is easier to realize multicolor detection with a single material system. But InGaAs is lattice-mismatched with AlGaAs. When InGaAs is epitaxily grown on AlGaAs, strain will inevitably occur, which plays a key role in determining the material's optical and electrical properties. Strain may also introduce defects, which will in some situations capture excited carriers induced by intraband-absorption.

To detect infrared emissions with an interband transition, the bandgap energy of semiconductor bulk and QW material as well, must be larger than the radiant energy to be detected. In sharp contrast to wide-band QW materials, narrow-bandgap semiconductor materials are significantly more difficult, not only in growth, but also in device fabrication (Levine 1993; Sher et al. 1991). Meanwhile, narrow-bandgap materials normally display lower electrical performance. For example, their dark currents are in general larger because of the carrier thermal excitation. Therefore, it is an obvious choice to employ widely studied III–V low-dimensional semiconductor structures. In contrast to HgCdTe, III–V semiconductor material systems are superior in that their growth techniques are not only well developed but also highly controllable. Table 6.1 provides a performance comparison of different types of infrared detectors. In the following, major interest will be focused on QW and quantum-dot (QD) infrared detectors.

A more detailed characterization of low-dimensional semiconductor infrared detectors is in general similar to forthcoming treatment. Here, a brief introduction is presented to ease into the more detailed discussion to follow.

A direct response of a photodetector to infrared emission is a current responsivity, R_i, defined as (Rogalski 2003):

$$R_i \equiv \frac{\lambda\eta}{hc} eg, \tag{6.224}$$

where λ is the wavelength of the infrared radiation, h the Plank constant, c the light velocity, and e is the electron charge. η is the quantum efficiency, defined as the number of electron–hole pairs produced per incident photon. g is the photoelectric gain, which characterizes the rate at which a photo-induced pair of carriers finally reaches the contact electrodes, and is defined as the ratio of the photo-induced electron life-time (τ_L) to the time required for an electron to pass through an electrode (τ_T). Note R_i has dimensions of (voltage)$^{-1}$.

Current noise always exists, due to generation and recombination rate processes (Rogalski 2003):

$$I_N^2 = 2(G + R)A_e t \Delta f e^2 g^2, \tag{6.225}$$

where G and R represent the thermal carrier generation and recombination rates per unit volume, respectively, Δf is the frequency bandwidth, t is the detector thickness, and A_e is the "electrical" area of the device.[2]

From (6.224) and (6.225), a principal parameter characterizing a detector, the detectivity D^*, can be deduced. From its definition we have:

$$D^* = \frac{R_\mathrm{i}(A_o \Delta f)^{1/2}}{I_\mathrm{n}}, \tag{6.226}$$

where A_0 is the optical area of the device, i.e., the photosensitive area. Thus we find:

$$D^* = \frac{\lambda}{hc}\,\eta\,[2(R+G)t]^{-1/2}\left(\frac{A_o}{A_\mathrm{e}}\right)^{1/2}. \tag{6.227}$$

[2] Since G and R are treated as separate independent noise sources they should add in quadrature. Thus the $G + R$ from the literature should be replaced by $\sqrt{G^2 + R^2}$ in (6.225), (6.227), and (6.228). Then for $G = R$ the factor 0.31 in (6.230) should be replaced by 0.44.

Table 6.1 Performance comparison of different types of infrared detectors

		Detector type	Advantages	Disadvantages
Thermal detector		IV–VI (PbS, PbSe, PbSnTe)	Light, reliable, low-cost, room-temperature operational, reliable material easily fabricated	Low detectivity in high frequency ranges, slow response (on the order of magnitude of ms); a high thermal expansion factor, large permittivity
Photodetector	Intrinsic semi-conductor	II–VI (HgCdTe)	Bandgap feasibly adjustable, well-developed theories and experiments, multicolor detector attainable	Nonuniformity for large-scale material, high-cost growth/process, bad surface stability
		III–V (InGaAs, InAs, InSb, InAsSb)	Good quality, mature technology, single integrated circuit desirable	High lattice-mismatch heteroepitaxy, long-wavelength limit at 7 μm (77 K)
	Nonintrinsic semi-conductor	Si:Ga, Si:As, Ge:Cu, Ge:Hg	Extra-long wavelength, relatively simple technology	High thermal production, extremely low temperature
	Free carrier	PtSi, Pt_2Si, IrSi	Low cost, high output, large-scale 2D arrays	Low-quantum efficiency, low-temperature operation
	QW	Type-I: GaAs/AlGaAs, InGaAs/AlGaAs	Well-developed growth techniques, good large-scale uniformity, multi-color detection	High heat productive, design/growth complicated
		Type-II: InAs/InGaSb, InAs/InAsSb	Low Auger recombination rate, wavelength easily controled	Design/growth complicated, interface sensitive
	Quantum dot	InAs/GaAs, InGaAs/InGaP, Ge/Si	Normal incidence possible, low heat produced	Design/growth complicated

If assumptions can be made that $A_e/A_o = 1$, the radiation only passes through the active layer once, and the reflectivities are negligible, then the quantum efficiency and detectivity can, respectively, be written as:

$$\eta = 1 - e^{-\alpha t}, \\ D^* = \frac{\lambda}{hc}\eta\,[2(G+R)t]^{-1/2}, \quad (6.228)$$

where α is the absorption coefficient. It can be easily deduced that D^* reaches its maximum when $t = 1.26/\alpha$. Therefore, at equilibrium, i.e., $G = R$, there is:

$$D^* = 0.31\frac{\lambda}{hc}\left(\frac{\alpha}{G}\right)^{1/2}. \quad (6.229)$$

As (6.229) ignores the contribution of the reflectivity at the backside of the detector, the real value of the detectivity will be higher than that predicted by (6.229) (Piotrowski and Gawron 1997):

$$D^* = 0.31\frac{\lambda}{hc}k\left(\frac{\alpha}{G}\right)^{1/2}, \quad (6.230)$$

where $1 < k < 2$.

Besides the detectivity, the noise equivalent difference temperature (NEDT) is also an important criterion characterizing device performance. It represents the temperature difference needed to produce an electrical signal equal to the root-mean-square noise voltage, and is defined as:

$$\mathrm{NEDT} = V_n\frac{(\partial T/\partial Q)}{(\partial V_s/\partial Q)} = V_n\frac{\Delta T}{\Delta V_s}, \quad (6.231)$$

where V_n is the root-mean-square noise voltage, V_s the voltage signal corresponding to the temperature difference ΔT.

6.4.2 Basic Principles of QW Infrared Photodetectors

Levine et al. demonstrated the first QW infrared photodetector with a GaAs/AlGaAs QW structure (Levine 1993; Levine et al. 1987). It was based on the absorption transition between two confined conduction-band levels. The operating process is illustrated in Fig. 6.34. Photon absorption between the subbands produces photoexcited electrons, which then tunnel out of the QW and go into the continuum states at the top of the barriers.

Then driven by an applied external electric field, the electrons are transported for a particular distance L (i.e., the mean-free-path before the electron are recaptured into a QW) within the average lifetime of the electron in the excited-state level, and hence produce a photocurrent.

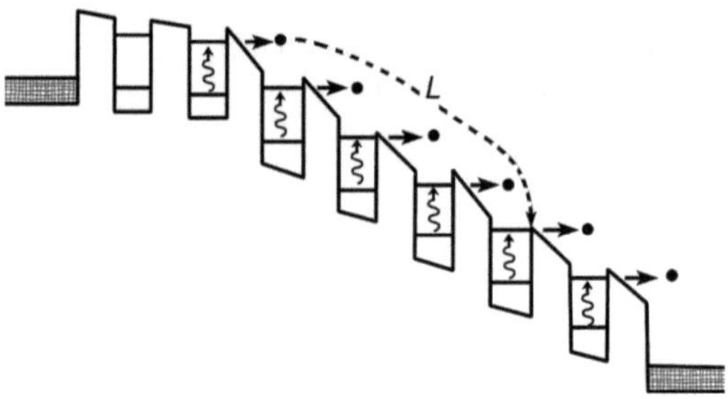

Fig. 6.34 Process for the photo-induced carriers and photocurrent induced by intersubband absorption

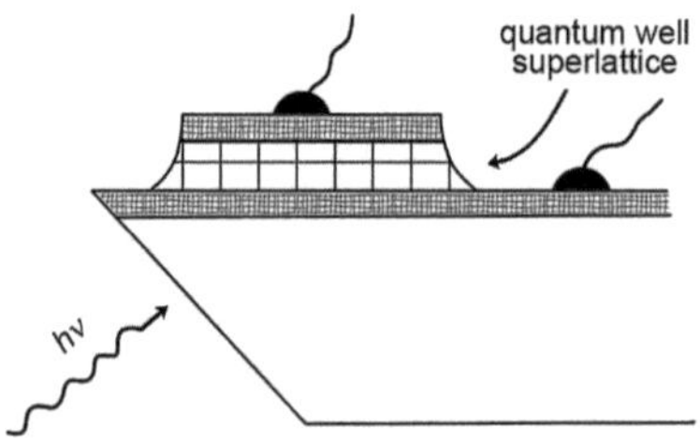

Fig. 6.35 Schema for the sample's geometric structure, the detector is irradiated by infrared emission that impinges from sample's backside

It is noteworthy that though the photocurrent depends on an applied bias, it can follow either along the QW plane, or in the direction perpendicular to the plane. However, from the point of view of detection, it is significantly superior for the transport of carriers to occur along the QW plane. On one hand, the mobilities are higher for ground- and excited-state carriers in the perpendicular direction, and hence a higher photo-response can be established. On the other hand, because along this direction hetero-barriers restrict the transport of ground-state carriers in the doped region of the QW, the dark current is very low.

QW infrared detectors have been a hot subject of research in the last few decades. In what follows, a brief discussion will be directed to its mechanisms and fundamental structures. For more details please refer to Levine's review article (Levine 1993).

The first detector considered has 49 periods of $GaAs/Al_{0.27}Ga_{0.73}As$ QWs, for which the well width is $L_w = 7.6\,nm$, and the barrier width is $L_b = 8.8\,nm$. To avoid possible interface states, Si doping is made just in the center of well with a width of 5.6 nm (doping density is $N_D = 3 \times 10^{17}\,cm^{-3}$). Meanwhile, both the top- (0.5 μm-thick) and bottom- (1 μm-thick) contact layers are doped to realize current injection (doping density is about $10^{18}\,cm^{-3}$). Figure 6.35 schematically illustrates the geometric structure of the sample. To achieve intersubband electronic transitions, selection rules dictate that the incident light must possess an electric field component that is normal to the QW. Therefore, the incident light should enter in the QW plane from the well side. To increase light-collecting efficiency, the substrate is polished into a wedge with a particular angle θ, e.g., 45°. With this configuration, infrared emission can effectively reach the detector from the backside.

In Fig. 6.36, an experimental result is illustrated for the conductance–voltage relation under forward bias. The temperature is $T = 20\,K$. The conductance shows a sequence of 48 negative peaks, as the bias voltage increases starting from 0, the

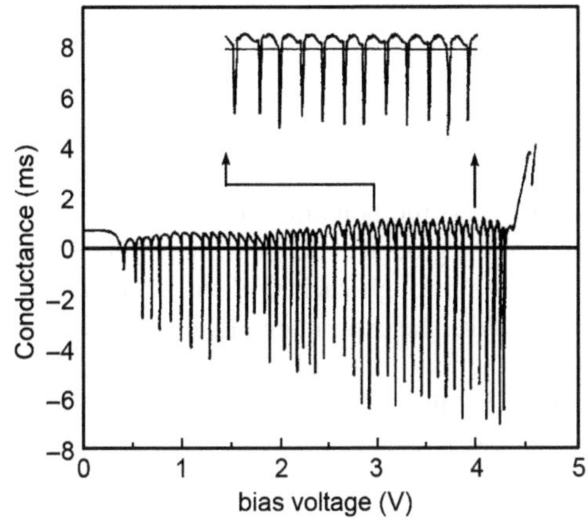

Fig. 6.36 Conductance as a function of bias voltage measured for a superlattice with 49 $GaAs/Al_{0.27}Ga_{0.73}As$ periods

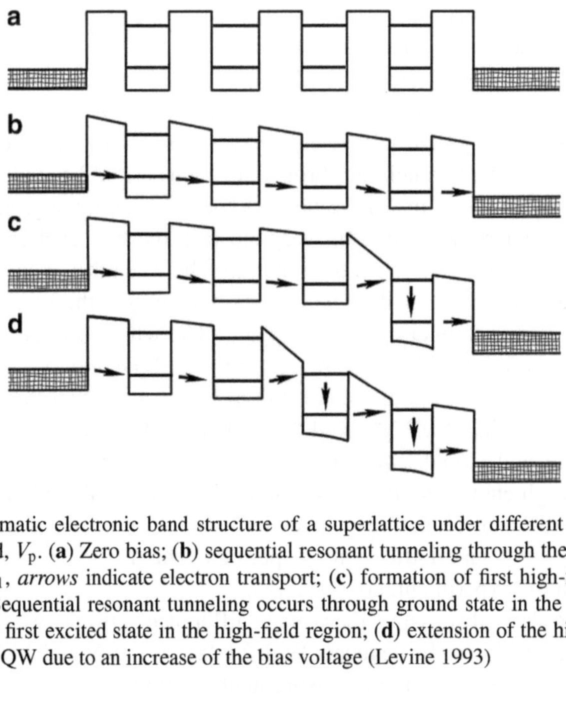

Fig. 6.37 Schematic electronic band structure of a superlattice under different average voltage drops per period, V_p. (**a**) Zero bias; (**b**) sequential resonant tunneling through the ground state E_1 for $eV_p < 2h/\tau_1$, *arrows* indicate electron transport; (**c**) formation of first high-field domain for $eV_p \geq 2h/\tau_2$. Sequential resonant tunneling occurs through ground state in the low-field region and through the first excited state in the high-field region; (**d**) extension of the high-field domain to an additional QW due to an increase of the bias voltage (Levine 1993)

first negative peak occurs at $V_b = 0.35$ V, and the voltage separation between two adjacent peaks is nearly identical, about 85 ± 18 meV. It will be shown in the following discussion that such a periodic behavior is determined by continuous resonant tunneling.

In principle, due to the two-dimensional character of the electron gas in QWs, resonant tunneling occurs only when the corresponding levels in different QWs are coincide (see Fig. 6.37a). As an external electric field is applied, however, the condition is normally not fulfilled. Kazarinov et al. reported that as soon as acoustic phonon scattering and impurity scattering exist in QWs, the requirement on energy as well as momentum conservation is decreased. If further, the relation can be ensured that:

$$eV_p < \hbar/\tau_1, \tag{6.232}$$

where V_p is the potential difference between two adjacent QWs, τ_1 the scattering time of the ground state, then the resonant tunneling remains possible, and electrons may pass through each QW via ground-state resonant tunneling, as illustrated in Fig. 6.37b. The experimental value of conductance is therefore finite.

As the bias voltage V_b increases to 0.35 V, the first negative peak occurs. This means that the resonant tunneling condition is broken. Now there is a relation (Choi et al. 1987):

$$eV_p = 2\hbar/\tau_1. \tag{6.233}$$

If the voltage drop per period is assumed to be identical, then $V_p = V_b/(N + 1)$, where N is the number of periods of the superlattice. As it is here $V_b = 0.35\,\text{V}$, $N = 49$, so $V_p = 7\,\text{mV}$, and hence $\tau_1 \sim 10^{-13}\,\text{s}$.

When the bias voltage increases further, $eV_p > 2h/\tau_1$, the resonant tunneling through the ground state is no longer possible, and a negative differential resistance occurs. The resistance is significantly enhanced to form a high-field domain, as the precondition for resonant tunneling is broken for each period. The voltage increment fully drops across this domain. Such a process continues until the ground-state level in the next adjacent well is lifted up very close to the first excited-state level of this domain (see Fig. 6.37c). When the separation between the ground state and the first excited state becomes smaller than $2h/\tau_2$, where τ_2 is the lifetime of the first excited state, resonant tunneling appears again. The whole process repeats as the bias voltage increases steadily, as shown in Fig. 6.37d. Due to the screening effect due to accumulated space charge under a high bias voltage and the consequential electronic band bending, the high-field domains do not randomly occur, but occur first at the anode, and extend sequentially to cathode as the bias increases. When the bias becomes high enough so the high-field domain covers the whole superlattice, the sequential resonant tunneling reaches its limit. As a result, there will be p-1 negative peaks for a superlattice with p periods. This theoretical result is in full agreement with the experimental observation.

Based on the above discussion, the voltage difference between two adjacent negative conductance peaks can be deduced. With $2h/\tau_1 = 7\,\text{meV}$ and $2h/\tau_2 = 11\,\text{meV}$ determined by infrared absorption measurement, we find:

$$\Delta V = E_2 - E_1 - 2\hbar/\tau_1 - 2\hbar/\tau_2 = 85 \pm 18\ \text{meV}, \tag{6.234}$$

so the energy separation between two subbands is hence derived to be, $E_2 - E_1 = 103 \pm 18\,\text{meV}$. This value is much closed to that predicted by a theoretical analysis (Choi et al. 1987).

If the bias voltage is increased further, another set of negative conductance peaks will occur. In contrast to the first set, this one is introduced by carriers sequentially resonant tunneling into the continuum states at the top of barriers.

For such a detector, only the photo-induced carriers in the high-field domains can tunnel out of a QW, and run away by effective transport (escape probability p_e). Therefore, the photocurrent is solely determined by a high-field domain,

$$I_p = n_p e v\,, \qquad n_p = (\alpha P \cos\theta / h\nu) p_e \tau_L\,, \tag{6.235}$$

where v is the transport velocity of the carriers in the superlattice direction, n_p the number density of photo-induced carriers, $P\cos\theta$ is the light power on active region with an incident angle θ, τ_L the re-captured life-time of the escaped (thermal) carriers, $\tau_L = L/v$, and L is the mean-free-path of the thermal carriers. From this relation, the R_i peak value R_p, and g can be derived:

$$R_p = \frac{\lambda\, \eta_a p_e}{hc} eg\,, \qquad g = \left(\frac{v\tau_L}{l}\right) = \left(\frac{\tau_L}{\tau_T}\right) = \left(\frac{L}{l}\right). \tag{6.236}$$

It is noteworthy that, in comparison to (6.225), here η is separated into two parts: one represents the absorption contribution, η_a, and the other is the QW escape factor, p_e (Martinet et al. 1992).

Levine et al. (1987) have measured the first bound-bound QWIP grown by MBE. The detector structure consists of 50 periods of 6.5-nm-GaAs/9.5-nm-$Al_{0.25}Ga_{0.75}As$ QW/barrier sandwiched by top and bottom heavily n^+ doped GaAs contact layers. The doping density of the QW is $N_D = 1.4 \times 10^{18}\,cm^{-3}$. The conduction band contains two bound states. For such a system, the probability of a bound–bound transition is large. Meanwhile, that of the transition from the ground to a continuum state at the top edge of barrier is very low (West and Eglash 1985). The absorption transition between the ground and excited state is shown in Fig. 6.38.

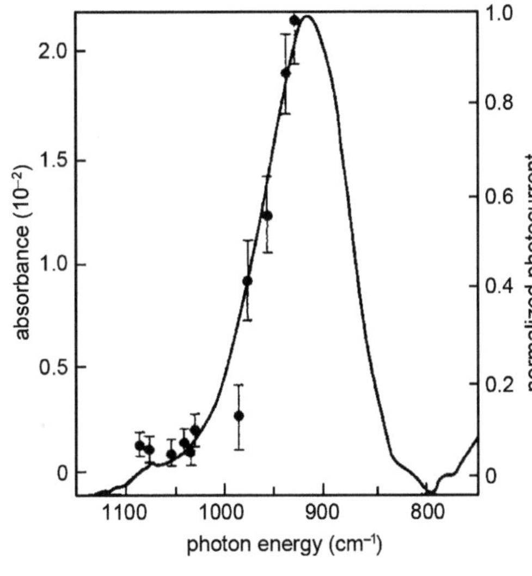

Fig. 6.38 Absorption transition between bound states in the conduction band. The *curve* is an experimental result, while the *dots* are from a photocurrent-photon energy measurement

The absorption peak occurs at $\lambda_p = 10.9\,\mu m$, with a full-width at half-maximum (FWHM) of $\Delta\nu = 97\,cm^{-1}$, corresponding to an excited-state life time of $\tau_2 = (\pi\Delta\nu)^{-1} = 1.1 \times 10^{-13}$ s. The absorption coefficient of the detector is experimentally determined to be, $\alpha = 600\,cm^{-1}$, with $R_p = 0.52$ A/W, $L = 250$ nm, and $p_e = 60\%$.

This type of bound-to-bound transition QWIP is the earliest attempt to make a semiconductor low-dimensional structured infrared detector. However, it has the disadvantages of a large dark current, a low detectivity, along with a low quantum efficiency.

6.4.3 Bound-to-Continuum State Transition QW Infrared Detector

As mentioned earlier, the transition probability of bound-to-bound states is much higher than that of bound-to-continuum states for a QW comprising two bound levels. However, the strong oscillator strength of the excited bound state can be pushed up into the continuum by decreasing the thickness of the QW, resulting in a strong bound-to-continuum state absorption (Coon and Karunasiri 1984). The corresponding conduction band structure is as illustrated in Fig. 6.39.

The major advantage of this structure is that the photoexcited electrons can escape from the QW without tunneling through an energy barrier and thus, the bias voltage required for the photoelectron to efficiently escape from the well is dramatically reduced, strongly lowering the dark current. In addition, the thickness of the barrier can be significantly increased, leading to a reduction of the ground state sequential tunneling by many orders of magnitude. Taking advantage of these improvements, Levine et al. demonstrated the first bound-to-continuum QWIP and achieved a dramatic increase in its performance, i.e., the detectivity, by several orders of magnitude (Levine et al. 1989, 1990).

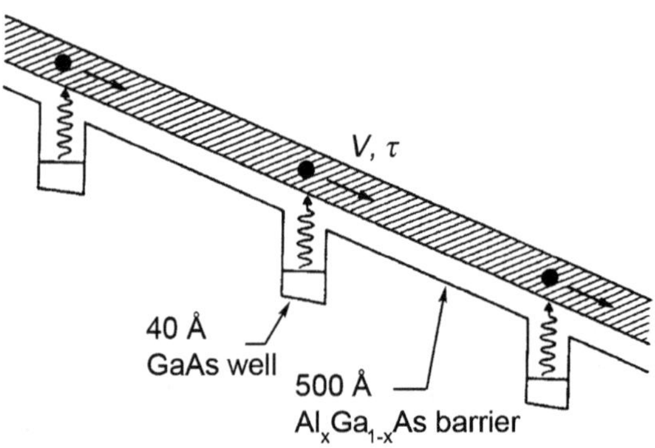

Fig. 6.39 Conduction-band structure for a bound-to-continuum QWIP, showing the photoexcitation and hot-electron transport processes

Quantitatively, bias-dependent dark current can be expressed as:

$$I_d(V) = n^*(V)ev(V)A\,,$$
$$n^*(V) = \left(\frac{m^*}{\pi\hbar^2 L_p}\right)\int_{E_1}^{\infty} f(E)T(E,V)dE, \qquad (6.237)$$
$$f(E) = \left[1 + e^{(E-E_1-E_F)/kT}\right]^{-1},$$

where $n^*(V)$ is the effective density of electrons thermally excited out of the well into the continuum transport states, e the electronic charge, A the device area, v the average transport velocity given by $v = \mu F[1 + (\mu F/v_s)^2]^{-1/2}$, μ the mobility, F the average electric field, v_s the saturated drift velocity, m^* electron effective mass, L_p the period of the superlattice, $f(E)$ the Fermi factor, E_1 the bound ground-state level, E_F the two-dimensional Fermi level measured relative to E_1, and $T(E,V)$ is the bias-dependent tunneling current transmission factor for a single barrier. This equation simultaneously accounts for both the thermal emission over the energy barrier E_b (for $E > E_b$) and the thermally assisted tunneling (for $E < E_b$).

Figure 6.40 shows the dark current as a function of bias voltage at various temperatures for a $\lambda_c = 8.4\,\mu$m bound-to-continuum QWIP, in which solid curves represent experimental data, while dashed curves are theoretical results derived from (6.238). Obviously, there is excellent agreement between the experiments and the theory (Levine et al. 1990). This indicated that tunneling through the top of the barrier makes a major contribution to the dark current.

The dark-current noise I_N, on the other hand, shows different features under positive and negative bias: it is much higher for positive bias than that for negative bias, and shows a sudden increase at rather high positive biases due to an avalanche gain process (Levine et al. 1992b). Analytically, I_N can be expressed as Hasnain et al. (1990); Janousek et al. (1990):

$$I_N \equiv \sqrt{I_N^2} = \sqrt{4eI_d g\Delta f} \qquad (6.238)$$

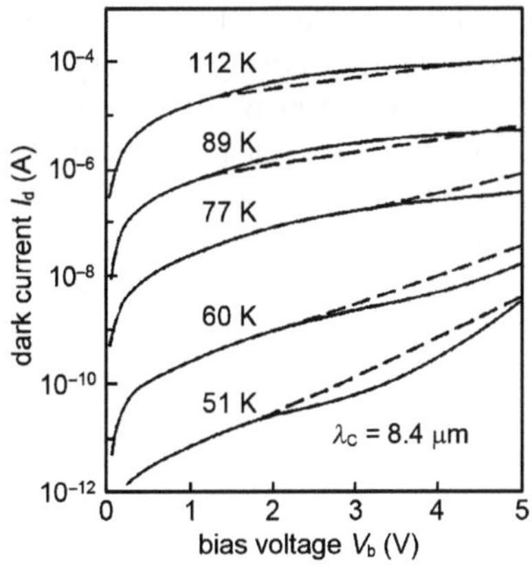

Fig. 6.40 The dark current versus voltage at various temperatures for a $\lambda_c = 8.4\,\mu$m QWIP. Positive bias means mesa top is positive. *Solid curves* represent the experimental data and the *dashed* the theoretical results

Table 6.2 Structural parameters of the measured $GaAs/Al_xGa_{1-x}As$ QWIP detectors (here L_w and L_b are the thicknesses of the well and barrier layers, respectively, in units of Å; N_D is the doping density, in units of $10^{18}\,cm^{-1}$, B-C, B-B and B-QC stand for the bound-to-continuum, bound-to-bound and bound-to-quasi-continuum transitions, respectively)

Sample	L_w	L_b	x	N_D	Type of doping	Period	Subband
A	40	500	0.26	1	N	50	B–C
B	40	500	0.25	1.6	N	50	B–C
C	60	500	0.15	0.5	N	50	B–C
D	70	500	0.10	0.3	N	50	B–C
E	50	500	0.26	0.42	N	25	B–B
F	50	50/500	0.30/0.26	0.42	N	25	B–QC

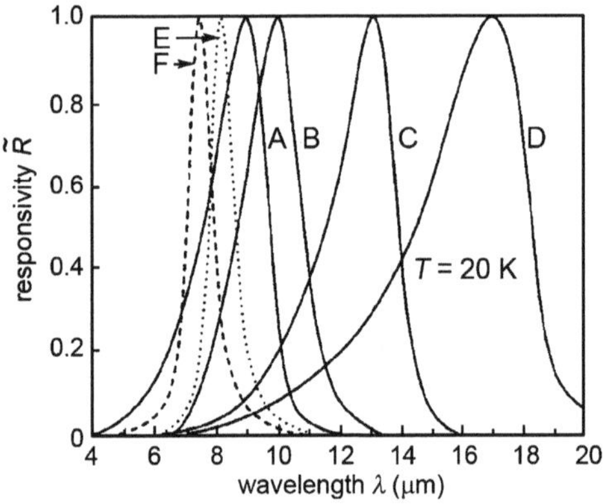

Fig. 6.41 Normalized responsivities for the bound-to-continuum (*solid curves*) and bound-to-bound (*dashed curves*) QWIP detectors, A–F correspond to the samples listed in Table 6.2

where I_N^2 is mean-squared noise current, and Δf is the bandwidth. From this relation, the optical gain, g, can be derived.

Levine et al. have measured the responsivity spectra of bound-to-continuum and bound-to-bound QWIP detectors with the structural parameters listed in Table 6.2.

Sample F has a bound-to-quasicontinuum transition and has a double-barrier structure with the thickness and Al fraction of the two barriers being 50/500 Å, and 0.30/0.26, respectively. The responsivity of each sample is illustrated in Fig. 6.41 (Levine et al. 1992b), its absolute amplitude is determined by measuring the photocurrent I_p. It is clear that the linewidths of the bound-to-bound QWIP detectors (dashed curves), $(\Delta\lambda/\lambda)_n = 10$–$11\%$, is much narrower than that of the bound-to-continuum ones (solid curves), $(\Delta\lambda/\lambda)_w = 19$–$28\%$.

According to the Zussman et al. (1991) theory, the photocurrent I_p is determined by:

$$\begin{aligned} I_p &= \int_{\lambda_1}^{\lambda_2} R(\lambda)P(\lambda)\mathrm{d}\lambda, \\ P(\lambda) &= W(\lambda)\sin^2(\Omega/2)AC_F\cos\theta\,, \\ W(\lambda) &= (2\pi c^2 h/\lambda^5)(\mathrm{e}^{hc/\lambda k_B T_B} - 1)^{-1}. \end{aligned} \tag{6.239}$$

where λ_1 and λ_2 define the spectral range of the responsivity, $R(\lambda) = R_p^0 \tilde{R}(\lambda)$, R_p^0 is the peak responsivity, $\tilde{R}(\lambda)$ the normalized spectral responsivity, $P(\lambda)$ the power incident on the detector from blackbody emission per unit wavelength at a wavelength λ, $W(\lambda)$ the blackbody spectral density, A the active area of the detector, Ω the solid angle of optical field, θ the incident angle, C_F the lumped coupling factor, $C_F = T_f(1 - r)M$, T_f the net transmission coefficients of the filters and windows, r the reflectivity of the detector surface, M the optical-beam modulation factor, and T_B is the blackbody temperature. With this relation, R_p^0 can be accurately determined by measuring the blackbody photocurrent at a temperature $T_B = 1{,}000$ K.

Figure 6.42 illustrates the bias-dependence of the responsivity for bound-to-continuum and bound-to-bound QWIP detectors, respectively (Levine et al. 1992b). It is obvious in Fig. 6.42a that the responsivity is approximately linear at low bias and saturates at high bias for the bound-to-continuum samples. For the bound-to-quasicontinuum sample (sample F), there is a similar responsivity-bias relation as illustrated in Fig. 6.42b. Meanwhile, obvious shape differences of the responsivity curves are seen for the fully bound sample (sample E). The responsivity does not start out linearly with bias, but starts close to zero for a finite bias, because the photoexcited carriers need field assistant to tunnel out and escape from the well. Because sample E has a bound excited state close to the top of the well, only a 1 V bias is required to get the photoexcited carrier to efficiently tunnel out of the well.

Figure 6.43 shows the experimentally determined optical gain and hot-electron mean free path as a function of bias voltage for sample B listed in Table 6.2 (Levine et al. 1992b) at a temperature of 77 K. Both positive (open circles) and negative

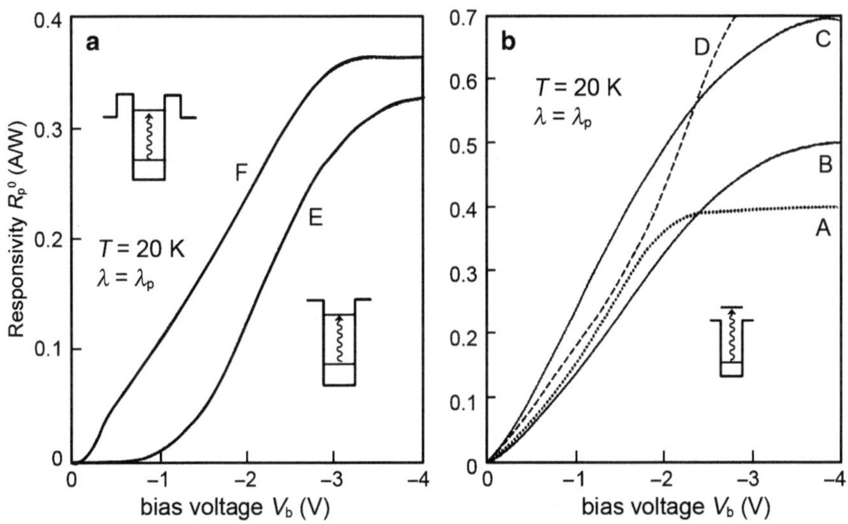

Fig. 6.42 Bias-dependent peak responsivity $R_p{}^0(\lambda = \lambda_p)$ at $T = 20$ K for (**a**) bound-to-continuum and (**b**) bound-to-quasicontinuum QWIP detectors. The *insets* illustrate the corresponding conduction-band electronic structures

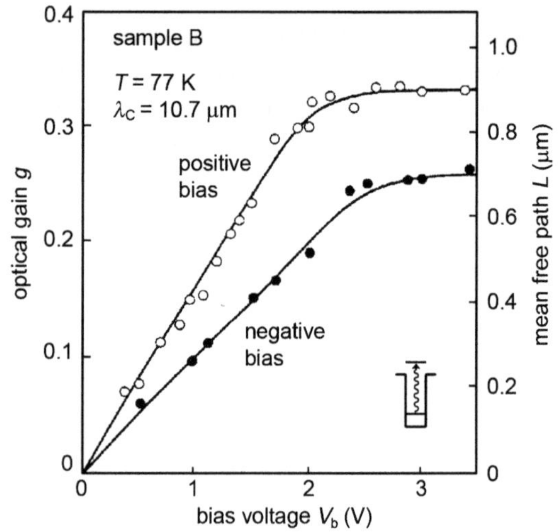

Fig. 6.43 Bias-dependent optical gain g (*left-hand-side scale*) and hot-electron mean free path L (*right-hand-side scale*) for the sample B at $T = 77$ K. Both positive (*open circles*) and negative (*solid circles*) bias are shown. Smooth curves are drawn through the measured data. The inset shows the conduction-band diagram

(solid circles) bias are shown. With the aid of (6.236), the hot-electron mean free path L is also determined, as illustrated on the right-hand-side scale of Fig. 6.43. It is clear that the optical gain manifests a nearly a linear increase at a low-bias, and saturates in the high-bias region. For sample B, it is clear that $g \sim 0.3$ when $V \geq 2$ V. It is noteworthy that, in contrast to the dark-current noise, the optical gain does not change obviously when changing the sign of the bias, indicating that although the number of electrons escaping out of the QW into the continuum states strongly depends on the bias direction. The continuum-state transport-dependent optical gain is insensitive to the motion direction of the carriers. Levine et al. found that the optical gain is very similar for all three different types of QWIP detectors. This means as soon as carriers escape out of QW into continuum states, their transport is generally identical. Furthermore, since $l = 2.7\,\mu$m for this sample, L is about 1 μm. The total quantum efficiency η is also accessible from (6.236) and the bias-dependent optical gain g.

Figure 6.44 shows the results of bound-to-continuum transition samples A–D (Levine et al. 1992b), in which η_0 and $\eta_{\max}$ represent the zero-biased quantum efficiency and maximum quantum efficiency, respectively. It is clear that even under zero bias, the quantum efficiency is not zero but falls in the range of 3.2–13%. As the bias increases, the quantum efficiency η increases at first linearly, and finally saturates at $\eta_{\max} = 8$–25%.

The bound-to-quasicontinuum QWIP detectors have quite similar zero-biased quantum efficiencies η_0. But the bound-to-bound QWIP detectors manifest an obviously different feature. Its zero-biased quantum efficiency is an order of magnitude lower than that of the bound-to-continuum ones, because the bound-state photoelectron need field-assisted tunneling to escape from the QW (Levine et al. 1988a).

The $\eta - V_b$ relation is similar for all the three types of QWIP detectors. This means that once the photoexcited electrons escape from the QW the hot-carrier transport process is nearly identical. From the relation $\eta = \eta_{\max} p_e$ (6.4.236), the

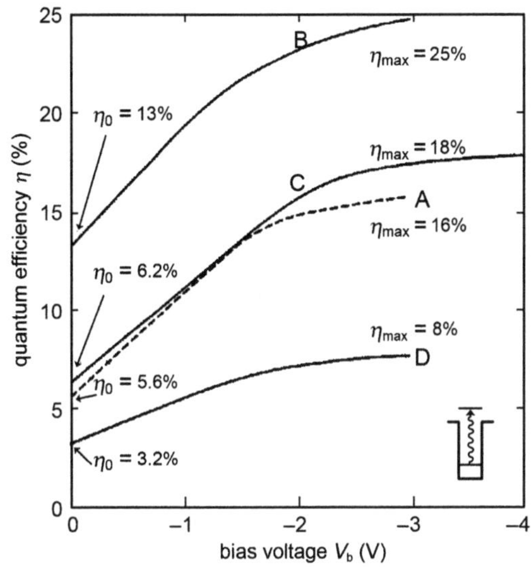

Fig. 6.44 Quantum efficiency η versus bias V_b for bound-to-continuum QWIP detectors, η_0 and η_{max} are the zero-biased and maximum quantum efficiencies, respectively. The inset illustrates the conduction-band electronic band structure. A–D are the corresponding samples listed in Table 6.2

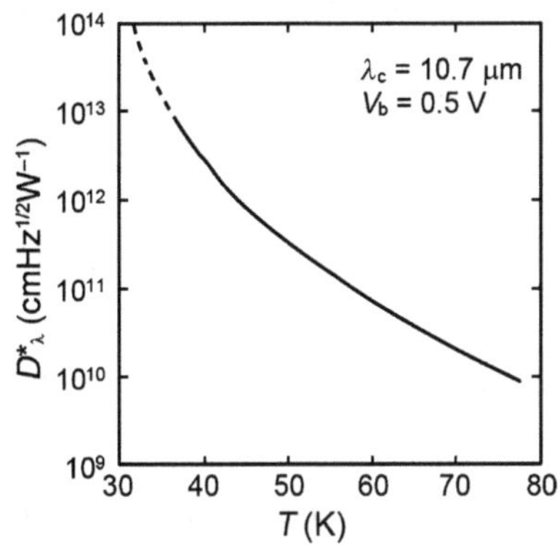

Fig. 6.45 The detectivity versus temperature for a bound-to-continuum QWIP having $\lambda_c = 10.7\,\mu$m, $V_b = 0.5$ V

escape probability p_e is also determined. For all three different types of detectors, p_e shows a similar bias dependence. But the bound-to-quasi continuum ones have a significantly different zero-biased value, because the escape time for an electron in the continuum states near the top edge of the barrier is smaller than the time for the electron to be re-captured by the same QW (Levine et al. 1992a). For the bound-to-bound detector, the zero-bias escape time is more than one order of magnitude longer than any of the other QWIPs.

From the definition of the detectivity (6.236), the peak detectivity $D_\lambda{}^*$ can be derived:

$$D^*_\lambda = R_p\sqrt{A\Delta f}/I_N, \tag{6.240}$$

where A is the detector area and $\Delta f = 1$ Hz. Figure 6.45 depicts the temperature-dependent detectivity for a bound-to-continuum QWIP having $\lambda_c = 10.7\,\mu$m. It is obvious that as the temperature increases the detectivity decreases drastically.

Combing (6.238)–(6.240) the generation–recombination-limited detectivity can be expressed as (Rosencher et al. 1992):

$$D_\lambda^* = \left(\frac{\eta_a p_e}{2h\nu}\right)\left(\frac{\tau_L}{n^* l}\right)^{1/2}. \tag{6.241}$$

Obviously, to obtain a high detectivity η_a, p_e and τ_L must be as large as possible, while n^* is as small as possible.

For a high incident infrared power the detectivity reaches background-limited performance determined by the background photon noise, $D_{BLIP}{}^*$, (Rosencher et al. 1992; Levine et al. 1988b):

$$D_{BLIP}^* = \frac{1}{2}\left(\frac{\eta}{h\nu I_B}\right)^{1/2}, \tag{6.242}$$

and thus for BLIP operation the only important parameter is the net quantum efficiency $\eta = \eta_a p_e$. A numerical example of QWIP performance is illustrated in Fig. 6.45. Taking a net quantum efficiency $\eta = 10\%$ gives a background-limited detectivity of $D_{BLIP}^* = 2.5\times10^{10}$ cm Hz$^{1/2}$/W, which occurs at $T = 68$ K. If the optical flux is reduced by one order of magnitude, then $D_{BLIP}^* = 7.9\times10^{10}$ cm Hz$^{1/2}$/W.

Furthermore, the noise equivalent temperature difference can be derived from (6.231) (Zussman et al. 1991; Rosencher et al. 1992) to be:

$$\text{NEDT} = \frac{(A\Delta f)^{1/2}}{D_B^*(\mathrm{d}P_B/\mathrm{d}T)}, \tag{6.243}$$

where D_B^* is the blackbody detectivity defined by:

$$D_B^* = R_B\sqrt{A\Delta f}/I_n, \quad R_B = \frac{\int_{\lambda_1}^{\lambda_2} R(\lambda)W(\lambda)d\lambda}{\int_{\lambda_1}^{\lambda_2} w(\lambda)d\lambda}. \tag{6.244}$$

In most cases, R_B is reduced by only a relatively small amount from its peak value R_p. dP_B/dT is the change with temperature of the incident integrated blackbody power in the spectral range of the detector. From (6.240), an approximation can be obtained by setting $C_F = 1$:

$$\frac{\mathrm{d}P_B}{\mathrm{d}T} = \left(\frac{P_B}{T_B}\right)\left(\frac{h\nu}{k_B T_B}\right), \tag{6.245}$$

by combine (6.243)–(6.245), it is derivable that (Zussman et al. 1991):

$$\text{NEDT} = (I_n/I_p)(kT_B/h\nu)T_B. \tag{6.246}$$

It is obvious that NEDT is proportional to the ratio of the noise current I_n and photocurrent I_p.

6.4.4 Miniband Superlattice QWIPs

The pioneering work on miniband superlattice QWIPs was carried out by Kastalsky et al. (1988) and Byungsung et al. (1990), who employed QW structures containing two bound states. To improve their performance, Gunapala et al. (1991) proposed a bound-to- continuum miniband detector structure. Theoretically, the physical process of optical absorption and electron transport can be described with a Kronig–Penney model (Choi et al. 1987). The wavefunction of the allowed and forbidden miniband $E(k)$ satisfies the Schrödinger equation, and is determined by:

$$\begin{aligned} \Psi_1(z) &= a_n e^{k_1(z-n\Lambda)} + b_n e^{-k_1(z-n\Lambda)}, \\ \Psi_2(z) &= a'_n e^{-ik_2(z-n\Lambda)} + b'_n e^{-ik_2(z-n\Lambda)}, \end{aligned} \tag{6.247}$$

where $\hbar k_1 = \sqrt{2m_1^*(V-E_z)}$, $\hbar k_2 = \sqrt{2m_2^* E_z}$, Λ is the period of the superlattice, n represents the nth QW, subscripts 1 and 2 denote the barrier and well layers, respectively, $m_1{}^*$, $m_2{}^*$ are the z direction effective masses, V the barrier height, and E_z is the z-direction component of the kinetic energy. To satisfy: (1) the continuous boundary conditions that Ψ and $(1/m^*)\mathrm{d}\Psi/\mathrm{d}z$ must be continuous at $z = n\Lambda$ and $z = n\Lambda + L_b$, respectively, (2) the periodic character of the Bloch functions, and (3) the normalization condition, the coefficients are derivable. All can be determined to yield the band dispersion $E_z(k)$, the z-direction group velocity $v_g = (1/\hbar)\mathrm{d}E_z/\mathrm{d}k$ as well as optical absorption probability from the first bound miniband to the second continuum miniband above the barrier. To make the analysis of the superlattice states more reliable, a necessary improvement is made by taking into consideration the thickness fluctuations of the barrier and well layers (Gunapala et al. 1991). The thickness fluctuations of the barrier/well layer can be described with a Gaussian function:

$$G(E) = \frac{1}{\sqrt{2\pi\Delta E^2}} e^{-(E-E_0)^2/2\Delta E^2}, \tag{6.248}$$

where a monolayer fluctuation corresponds to an energy width $\Delta E = 10\,\mathrm{meV}$.

Figure 6.46 shows normalized room-temperature absorption spectra for two miniband superlattice QWIPs with different barrier thicknesses. The solid curves are experimental values, while the dashed and dotted ones are from theoretical calculations. A geometry is used in which an angle of 45° is present between the incident light and the normal to the sample's surface.

The inset schematically illustrates the miniband structure of the detector. The two samples were prepared by an MBE technique. One consists of a 40-Å-thick GaAs QW layer doped with $n = 1 \times 10^{18}\,\mathrm{cm}^{-3}$ and a 30-Å-thick undoped $Al_{0.28}Ga_{0.72}As$ barrier layer. The top- and bottom-contact GaAs layers are doped ($n = 1 \times 10^{18}\,\mathrm{cm}^{-3}$) with thicknesses of 0.5 and 1 μm, respectively. The other sample has a similar structure with only one exception, the barrier thickness is $L_b = 45$ Å.

It can be seen from Fig. 6.46 that the energy position of the peak absorption coefficient is independent of the barrier thickness but is mainly determined by the

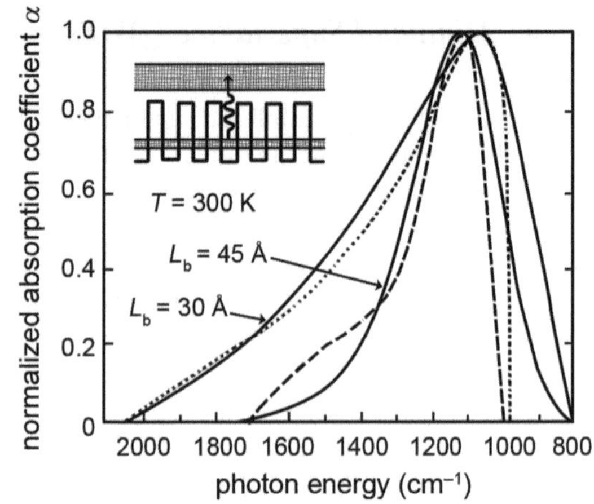

Fig. 6.46 Normalized room-temperature absorption coefficient spectra for two miniband superlattice QWIPs with barrier thickness of 30 and 45Å, respectively. The *solid curves* are experimental values and the *dashed and dotted curves* are from theoretical calculations. The inset schematically illustrates the miniband structure

well thickness, which is about $\lambda_p \sim 9\,\mu$m. Meanwhile, the widths of the lowest and the first excited minibands are obviously barrier-thickness dependent. Theoretical calculations indicate that for $L_b = 30$ Å, $\Delta E_I = 41$ meV, $\Delta E_{II} = 210$ meV, and for $L_b = 45$ Å, $\Delta E_I = 17$ meV, $\Delta E_{II} = 128$ meV. Correspondingly, the width $\Delta \nu$ of the absorption spectrum is also barrier-thickness dependent: for $L_b = 30$ Å, $\Delta\nu = 530\,\text{cm}^{-1}$, and for $L_b = 45$ Å, $\Delta\nu = 300\,\text{cm}^{-1}$. If a similar dependence is taken into account for the absolute absorption coefficient peak value ($\alpha = 3{,}100\,\text{cm}^{-1}$ for $L_b = 30$ Å, and $\alpha = 1{,}800\,\text{cm}^{-1}$ for $L_b = 45$ Å), it is conclusive that a narrower barrier corresponds to a stronger integrated absorption in these miniband structures. Furthermore, the aforementioned theoretical model incorporates all the principal physical mechanisms of the optoelectronic process in these miniband structures, and predicts good agreement with the measurement on the peak positions, the linewidth, as well as the general shapes of the absorption coefficient spectra.

In Fig. 6.47, experimental responsivities are illustrated for the two samples, and the reflectance loss is already corrected. The measurement is performed with a 45° polished incident surface at a temperature of 20 K. It is noteworthy that the responsivity is generally temperature-independent in the range of $T \leq 77$ K.

Here the peak values of the responsivity is 1–2 orders of magnitude higher than the Byungsung et al. early result (Byungsung et al. 1990). Though the peak position of absorption spectra are nearly barrier-thickness independent, the responsivity peaks for the $L_b = 30$ and 45 Å samples occur at $\lambda = 5.4$ and $7.3\,\mu$m, respectively. The dashed curves in Fig. 6.47 are fits based on (6.224) and (6.236) with $\eta = (1 - e^{-2\alpha l})/2$, and choosing τ_L as the only fitting parameter. Obviously, the calculated peak position fits the experimental result. The fitting procedure yields $\tau = 5$ ps, which is very close to the measured value of $\tau = 4$ ps. This means the reason the spectral peaks in these responsivity curves do not exactly coincide with the absorption peaks is that the group velocity v_g peaks near the center of the band and vanishes at its edges, whereas the absorption curves peak near the low energy edge of the band.

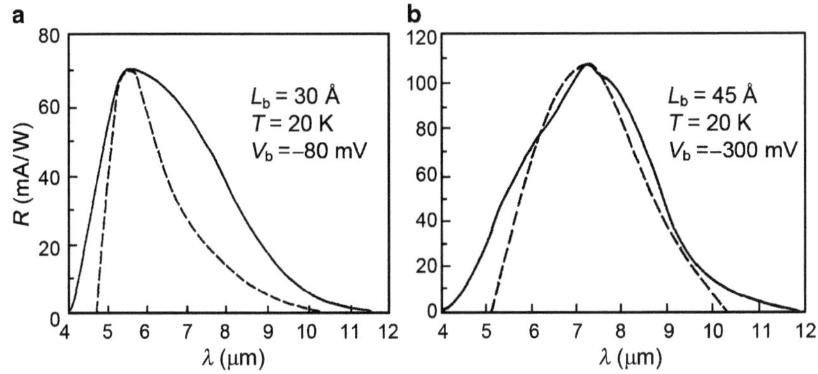

Fig. 6.47 Responsivity at $T = 20$ K for the sample with (**a**) $L_b = 30$ Å, $V_b = 80$ mV, and (**b**) $L_b = 45$ Å, $V_b = -300$ mV. The *solid curves* are experimental data and *dashed curves* are theoretical values

The dark current measurement indicates on two 200-μm-diameter detectors with different barrier widths indicate that the narrower-barrier detector obviously has a higher dark current (Gunapala et al. 1991). This is because a narrower barrier corresponds to a wider miniband, and hence a stronger conduction. On the other hand as expected, the measurements at two different temperatures 77 and 4 K show that a higher temperature leads to a higher dark current, due to different mechanisms being active at different temperatures: at 77 K, the main dark current contribution arises from thermionic emission from the first miniband below the top of the barriers to the second miniband above the barriers, while at 4 K the dark current is due to conduction within the first miniband.

Gunapala et al. also measured the change of responsivity with bias voltage. For an.$L_b = 45$ Å miniband detector, the responsivity is zero at $V_b = 0$ and increases monotonically with positive or negative bias. The same behavior is observed in wider barrier ($L_b = 300$–500 Å) multiple QW detectors. In strong contrast, the $L_b = 30$ Å detector has a zero bias response as shown in Fig. 6.48. The responsivity increases slightly with negative bias up to $V_b = -100$ mV, but decreases strongly with positive bias and in fact vanishes at $V_b = +110$ mV.

From the responsivities and dark currents, the peak detectivity D_λ^* can be calculated using (6.240). For the $L_b = 30$ Å structure the result is $D_\lambda{}^* = 2.5 \times 10^9$ and 5.4×10^{11} cm Hz$^{1/2}$W for $T = 77$ and 4 K at $V_b = -80$ mV. For $L_b = 45$ Å the result is $D_\lambda{}^* = 2.0 \times 10^9$ cm Hz$^{1/2}$W and 2.0×10^{10} cm Hz$^{1/2}$W for 77 and 4 K at $V_b = -300$ mV. These values are several orders of magnitude larger than the previous bound-to-bound miniband results reported by Byungsung et al. (1990).

6.4.5 Multiwavelength QW Infrared Detectors

One of the most distinct advantages of the QWIP technology is that the responsivity peak can be varied from ~2 to 20 μm while maintaining a lattice match between the

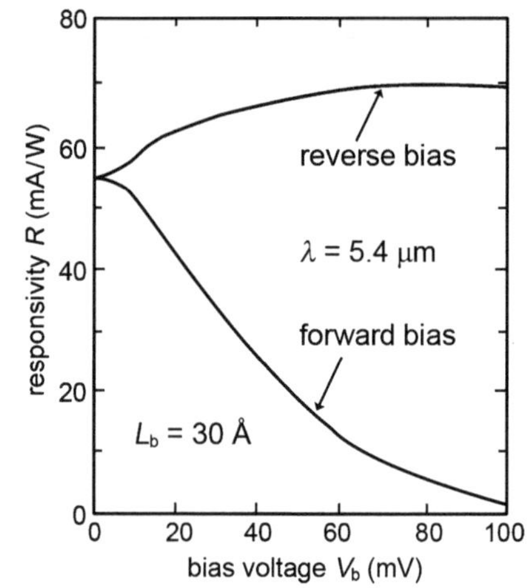

Fig. 6.48 Responsivity vs V_b for both forward and reverse bias, measured at $\lambda = 5.4\,\mu$m and $T = 20$ K for an $L_b = 30$ Å detector (Gunapala et al. 1991)

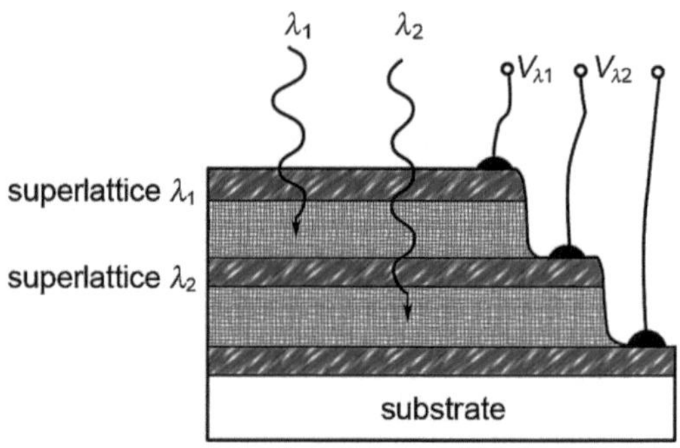

Fig. 6.49 A schematic structure of a three-electrodes vertically integrated two color QWIP allowing each wavelength to be individually detected (Levine 1993)

epilayers and the substrate, by controlling the QW width, the barrier composition, and the structure. It is thus possible to monolithically stack several lattice-matched QWIPs during crystal growth (as shown in Fig. 6.49 to achieve a desired multi-spectral response. This multi-QWIP stack can be designed to add together all the photocurrents and thus achieve a very wide spectral bandwidth or to independently measure each spectral component enabling its use as a spectrometer or a multiwavelength imager.

Köck et al. have demonstrated for the first time two color QWIPs with the structure illustrated in Fig. 6.49 (Köck et al. 1992) by stacking two series of multi-QWs. One of the QWIPs has a 50-Å-GaAs-well/$Al_{0.34}Ga_{0.66}As$-barrier structure, with its peak responsivity at $\lambda_p = 7.4\,\mu$m. The other has an 80-Å-GaAs-well/$Al_{0.29}Ga_{0.71}As$-barrier structure, with its peak responsivity at $\lambda_p = 11.1\,\mu$m. The QWIPs are separated by an intermediate highly doped contact layer, and thus by properly adjusting the bias on this three-terminal detector the photoresponse from either the 80 or the 50 Å wells could be selected. Both QWIPs exhibit extremely high detectivities at $T = 77$ K, i.e., $D^* = 5.5 \times 10^{10}$ cm Hz$^{1/2}$W for the shorter wavelength, and $D^* = 3.88 \times 10^{10}$ cm Hz$^{1/2}$W for the longer wavelength.

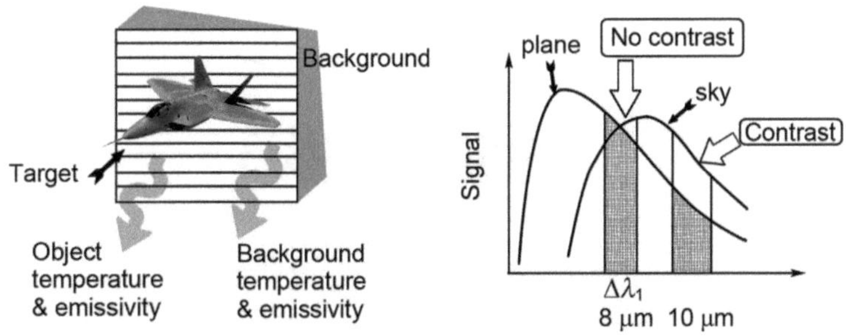

Fig. 6.50 Target recognition: is one application of multiwavelength QWIPs

Kheng et al. has used another approach to demonstrate a QWIP operating in both the midwave ($\lambda = 3$–$5\,\mu$m) as well as the long-wave ($\lambda = 8$–$10\,\mu$m) spectral regions (Kheng et al. 1992). They used a wide well $L_{\mathrm{w}} = 95$ Å containing two bound levels and doped it very heavily $N_{\mathrm{D}} = 6.8 \times 10^{18}\,\mathrm{cm}^{-3}$ so that the Fermi level was above the second bound state. Due to this high doping, both the E_1 to E_2 bound-to-bound intersubband absorption (at $\lambda \sim 10\,\mu$m) and the bound-to-continuum absorption ($\lambda \sim 4\,\mu$m) are observed. Because the photoexcited electrons must tunnel out from the E_2 bound level, the relative intensities of these two bands can be varied by over two orders of magnitude with bias voltage.

One application of such detectors is target recognition. Each object has its unique infrared radiation pattern or "signature." This pattern consists of different intensities at different wavelengths and hence multispectral infrared imaging is the only method to distinguish the signature of particular object from its background. As illustrated in Fig. 6.50, the infrared radiation of an airplane is identical to that of the sky at an $8\,\mu$m wavelength, while there is a high contrast at $10\,\mu$m. Though in this case having a $10\,\mu$m single wavelength detector would do equally well.

A multispectral QWIPs is a new type device. According to the same principle, bicolor, tricolor, and even spectroscopic type QWIPs are achievable. In addition, the QWIPs with QCL structures are also novel devices, which are desirable for room-temperature operation (Hofstetter et al. 2002).

6.4.6 Quantum-Dots Infrared Detectors

The QWIPs discussed above rely on a unipolar intersubband absorption transition. In quantum dots, there also exists the intersubband transition within the same band, and quantum dot detectors with these types of transition are expected to play an important role in the mid- and far-infrared spectral regions (Jelen et al. 1997). Compared to the QWIPs, the QD detector has the advantage of a long intersubband relaxation time due to a reduced electron–phonon scattering rate (Benisty et al. 1991). In addition a QD detector is also sensitive to normally incident photons

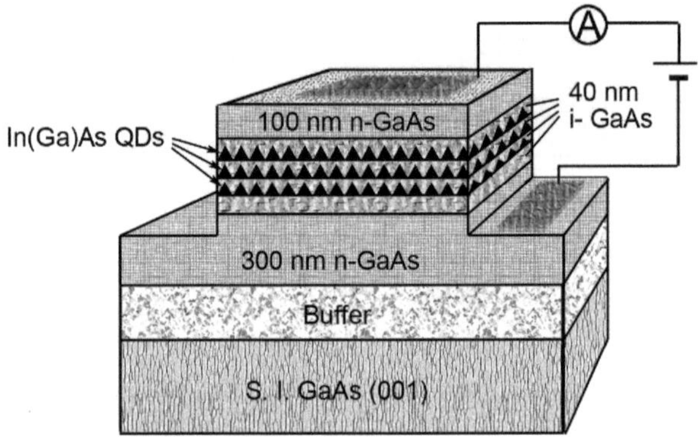

Fig. 6.51 A schematic structure for a QD detector, with the electrical connection set for measuring a vertical photocurrent spectrum (Chu et al. 1999b)

because the intersubband transition in a QD is not restricted by the polarization selection rule. There are two types of operational modes for QD detectors, namely, a longitudinal photocurrent mode and a transverse photocurrent mode.

First, we consider the longitudinal photocurrent QD detector. Similar to the QWIPs, an n-i-n structure can be used to fabricate QD detectors. In Fig. 6.51, the In(Ga)As/GaAs QD detector structure studied by Chu et al. (1999a–c) is schematically illustrated. The active region contains three In(Ga)As QD layers. They are grown in GaAs infiltrating layers (referred to as wetting layers) and are thin enough to maintain their QW feature. Using an atomic force microscope it is observed that the QD density is about $1.2 \times 10^{10}\,\mathrm{cm}^{-2}$ (Chu et al. 1999c). Separating layers between the GaAs layers are rather thick (40 nm) to avoid any coupling between the QDs in different layers. The growth temperature is 530 °C (Chu et al. 1999a).

The QDs are Si doped with a nominal doping density of $7.2 \times 10^{10}\,\mathrm{cm}^{-2}$. The doping layer lies 2 nm below the corresponding QD layer to enable electrons to effectively transfer from the dopants to the quantum dots. It is estimated that each quantum dot binds about six electrons. The active region is sandwiched between two n-type doped contact layers. Photolithography and lift-off techniques are employed to fabricate these mesa structures, as illustrated in Fig. 6.51. Also, a $400 \times 400\,\mu\mathrm{m}^2$ optical window is made in the top contact layer to realize a normal incidence photocurrent measurement. The electrical connection is also schematically illustrated. Because the bias voltage is applied along the sample's growth direction, the photocurrent is known as a vertical photocurrent (Chu et al. 1999b).

Figure 6.52 schematically depicts the QD conduction band structure. Obviously, when a vertical bias voltage is applied, the excited electrons can transport between the top and bottom contact electrodes, and therefore leads to a photocurrent. A typical photoluminescence spectrum is plotted in Fig. 6.53 (Chu et al. 1999b). It manifests obvious peaks at 1.08 and 1.14 eV, respectively, which correspond to transitions in the QDs between their conduction and valence bands with an identical principal quantum number. In addition, a weak peak at 1.43 eV is also

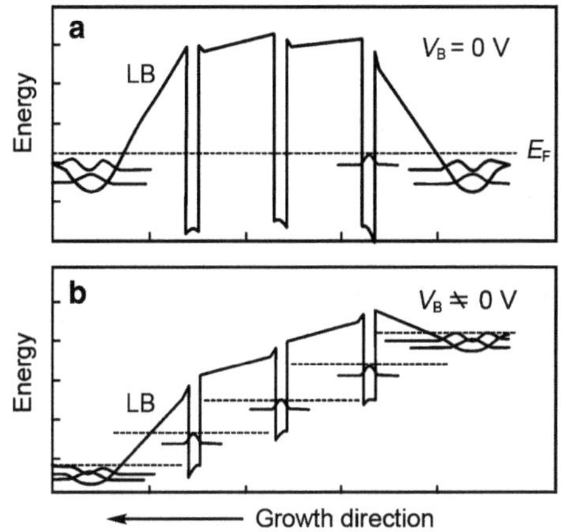

Fig. 6.52 A schematic conduction-band structure for a QD detector (**a**) the thermal equilibrium state and (**b**) with a bias voltage V_B applied

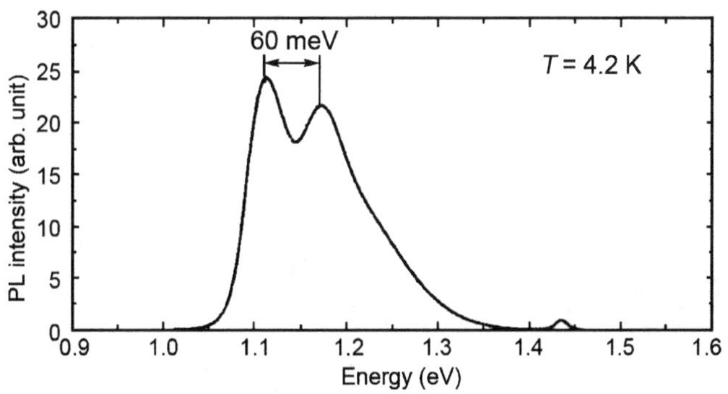

Fig. 6.53 Photoluminescence spectrum for an In(Ga)As/InGaAs QD detector

observable, which is related to the two-dimensional GaAs wetting layers. The measured conduction-band intersubband photocurrent spectra are depicted in Fig. 6.54 for a temperature of 10 K.

It can be seen from Fig. 6.54a that for the normal incident geometry the photocurrent obviously has two peaks, one at 240 and the other at 280 meV, both of which are due to conduction-band intersubband transitions. This indicates the polarization selection rule, valid for QWs, is not applicable to the QD system with its three-dimensional confinement. The doping density ensures the first excited level in the QD is partially filled, abd the two peaks are identified to be a ground state-to-continuum and a first-excited state-to-continuum transitions. The 40-meV energy separation is close to the energy splitting between the ground and the first excited state levels in the QD conduction band. Also the photocurrent occurs even under zero bias, and then increases with bias voltage. This can be understood from the asymmetry of the dopent's location.

In Fig. 6.54b, polarization-dependent measurements are plotted for a QD detector with a multi-pass waveguide geometry. One lateral side of the sample is polished

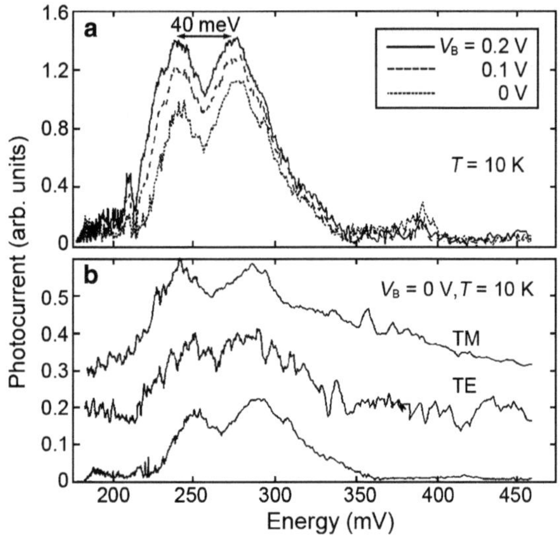

Fig. 6.54 A normal incidence photocurrent spectra of an In(Ga)As/InGaAs QD detector: (**a**) measured at different bias voltages, (**b**) the polarization-dependent photocurrent for a waveguide geometry (Chu et al. 1999b)

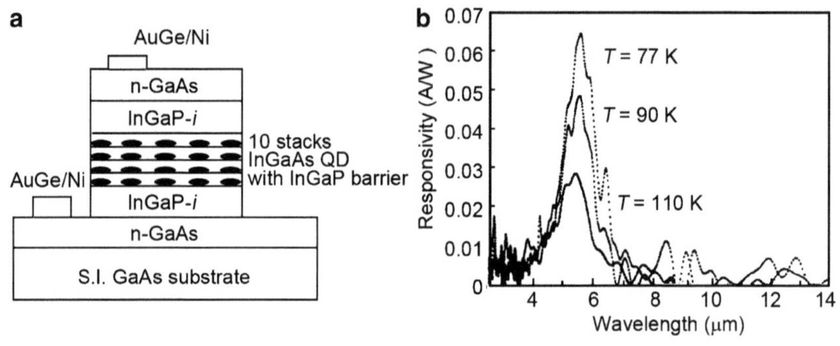

Fig. 6.55 (**a**) A schematic structure, and (**b**) responsivities at different temperatures for an InGaAs/InGaP QD photoconductive detector (Kim et al. 1998)

into a 45° angle to receive the incident light. The TE polarized mode denoted in the figure is parallel to, while the TM contains a component perpendicular to the QD layer. Clearly, the two polarized modes correspond to a nearly identical photocurrent, indicating that the conduction-band bound-to-continuum transitions in QDs are basically polarization independent. Note, however, the bound-to-bound intersubband transitions in a QD depend on the symmetries of the bound states, and hence may exhibit some polarization dependence.

Figure 6.55a illustrates, as an example, a schematic structure for an InGaAs/InGaP QD infrared detector (Kim et al. 1998).

This QD sample was grown on semi-insulating (100) GaAs substrate by a low pressure MOCVD technique at a temperature of 480 °C. First, a 5,000-Å-thick Si-doped ($n = 1 \times 10^{18}\,\mathrm{cm}^{-3}$) GaAs back-contact layer was deposited, followed by a 1,000-Å-thick lattice-matched InGaP layer. Then an active region was grown consisting of 10 periods of InGaAs-QD/350-Å-InGaP-barrier layers. The structure was finished with a 1,500-Å undoped InGaP layer and an n-doped ($n = 8 \times 10^{17}\,\text{cm}^{-3}$) top-contact GaAs layer. A $400 \times 400\,\mu\text{m}^2$ detector structure was fabricated by a chemical wet etching technique. The detector responsivity was measured using a blackbody emission source and a Fourier transform infrared spectrometer. With the help of the following equations (Kim et al. 1998; Mohseni et al. 1997), the responsivity is given by:

$$R_1 = k\frac{I_p}{\frac{(D_A^2 A_d)}{4r^2}\sigma T_{BB}^4},\qquad k = \frac{\sigma T_{BB}^4}{\int_0^\infty M(T_{BB},\lambda)S(\lambda)d\lambda},\tag{6.249}$$

where R_1 represents the absolute responsivity, in units of A/W, I_p the QD detector current, A_d the detector area, D_A the aperture diameter of the blackbody emission source, r the distance from the blackbody aperture to the detector, σ the Stefan–Boltzmann constant, T_{BB} the blackbody temperature, and k is a calibration factor caused by the overlap of detector relative responsivity $S(\lambda)$ and the blackbody emission spectrum $M(T_{BB}, \lambda)$.

Figure 6.55b shows the detector responsivity measured at different temperatures. At $T = 77\,\text{K}$, the QD intersubband transition occurs at $5.5\,\mu\text{m}$, while the cut-off wavelength of the responsivity is $6.5\,\mu\text{m}$. For a bias voltage $V_B = -2\,\text{V}$, the peak responsivity is 0.067 A/W. As the temperature increases, the responsivity decreases. If $T > 130\,\text{K}$, photoconductive signal is nearly buried in the noise. And if the temperature falls into range of $T \leq 110\,\text{K}$, the responsivity's linewidth is about 48 meV, i.e., $\Delta\lambda/\lambda \approx 20\%$, and is independent of temperature. This is significantly narrower than that reported by Benisty et al. (1991). The difference is mainly due to QD size fluctuation in the sample. In Fig. 6.56, the bias-dependent responsivity is depicted for an InGaAs/InGaP QD photoconductive detector measured at a temperature of 77 K. Obviously, as the bias voltage increases, the responsivity increases linearly at first, and then saturates at a voltage of about $-4.5\,\text{V}$. This is ascribed to the saturation of the drift velocity of photo-induced carriers in a high external electric field.

It is experimentally proved that at 77 K and a $-2\,\text{V}$ bias voltage the detectivity is about $4.74 \times 10^7\,\text{cm}\,\text{Hz}^{1/2}\,\text{W}$ at a detection wavelength of $5.5\,\mu\text{m}$. The corresponding dark current is $1.49 \times 10^{-10}\,\text{A/Hz}^{1/2}$, which is two orders of magnitude

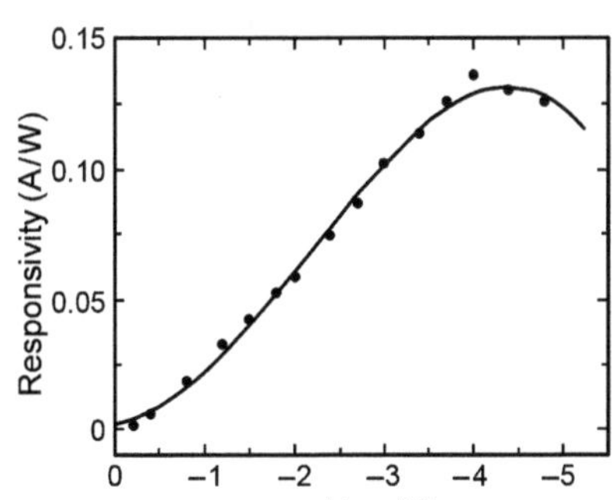

Fig. 6.56 The bias voltage dependence of the responsivity at 77 K for an InGaAs/InGaP QD photoconductive detector (Kim et al. 1998)

lower than that of a GaAs/AlGaAs QWIPs at the same wavelength and temperature. The sensitivity to a transverse photon flux is helpful for enhancing the photocurrent intensity, and hence is hopeful for achieving higher detectivities for QD infrared detectors.

As aforementioned, bound electrons in QD samples can either transport in the sample's growth direction, forming a longitudinal photocurrent, or be excited into the wetting layer and conduct along the layer plane driven by a transverse electric field, forming a transverse photocurrent (Chu et al. 2000). By measuring the transverse photocurrent, the transition between bound states of the QD and the wetting layer can be identified. The measurement of QD transverse photocurrent is easier than that of QWs, as QDs are a three-dimensional confinement system.

Figure 6.57a illustrates the electrode configuration of the QD sample, the width of each finger-type electrode is 20 μm, with a separation of 40 μm between them. The electrodes are made from an Ni/Ge/Au alloy, and treated by a quick thermal anneal. The sample contains of 15 InGaAs QD layers embedded in intrinsic GaAs layers. Figure 6.57b depicts the transverse photocurrent spectra for the In(Ga)As/GaAs QD sample. A wide and weak peak occurs in the energy range of 230–330 meV, which roughly corresponds to that of the aforementioned longitudinal photocurrent, and is therefore ascribed to the transition between the bound and continuum states in QDs. Meanwhile, a strong peak with a weak shoulder occurs at an energy of 180 meV with a linewidth of 20 meV, and is asymmetric. It is caused by transitions between filled states in the QDs and subbands in the wetting layers. By measuring a photoluminescence spectrum, it can be further identified that the transition for the strong peak is mostly between the first-excited state in the QDs and the subbands in the wetting layers (a type II transition), while the transition for the shoulder is from the ground state in the QDs to the subbands in wetting layers (a type I transition)

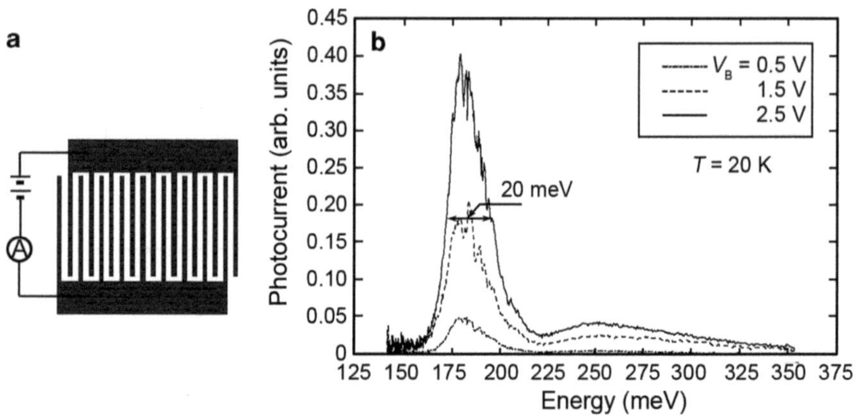

Fig. 6.57 (**a**) Electrode configuration used for measuring photocurrent spectra. (**b**) Transverse photocurrent spectra for a QD sample measured at $T = 20$ K, the stronger peak is due to the transition between the first-excited state in QDs and a subband in the wetting layer (Chu et al. 2000)

(Chu 2001). Because of the doping density, the QD ground level is filled with two electrons while the first excited level can accommodate four electrons. This leads to the intensity of the transition II being stronger than that of the transition I. The wavefunction symmetry also plays an important role in the transition intensity. The ground state of QDs has *s* (even) symmetry, and the first-excited state *p* (odd) symmetry (Stier et al. 1999). The state in the wetting layer intersubband is bound along the *z*-direction and has even symmetry. Furthermore, in the *x* and *y* directions, the wavefunction in the wetting layer close to the QDs is also quasibound and shows even symmetry. This is due to a combination of effects. The QDs are negatively charged and thickness fluctuations of the wetting layer caused by the QDs form a lateral QW. This leads transition I to be nearly forbidden, and hence significantly weakens its photocurrent (Chu et al. 2000; Chu 2001).

It is indicated by an experimental study that the transverse mode photocurrent in the conduction band of QD detectors is much stronger than the longitudinal one. For the transverse geometry, the transport of carriers occurs in the wetting layers, and the corresponding photocurrent has therefore a longer lifetime than the longitudinal photocurrent. A longer carrier lifetime gives a higher photoconductive gain (Liu 1992), and therefore a stronger photocurrent (Chu et al. 2001).

However, for the sake of just carrier transport, a wetting layer is indeed not an ideal medium. All the thickness, composition, and strain of the wetting layer have negative effects on the photo-induced carrier transport. If the photo-induced carriers can be transferred into transport channels with higher carrier mobility, and be transported there, the transverse photocurrent can be dramatically enhanced. The conduction-band structure designed with this consideration is illustrated in Fig. 6.58 for a transverse-photocurrent QD detector. The detector contains three In(Ga)As/GaAs self-assembled QD layers. A 40-nm-thick separation layer is inserted between two adjacent QD layers. The growth temperature is 530 °C. The doping is similar to that in the above discussion, ensuring some six electrons in each QD. An extra InGaAs QW layer is grown above the top QD layer. The photo-induced carriers will be transferred into the QW layer for transport. In addition, a δ-doped layer is inserted at 100 nm below the sample's surface to saturate the GaAs surface states and form a weak electric field in the QD region so as to drive the

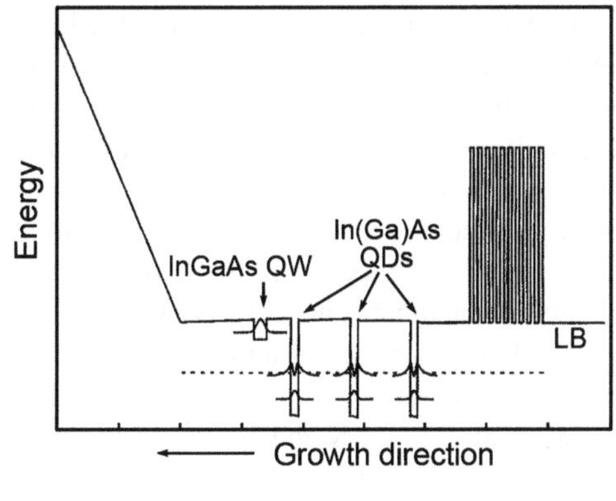

Fig. 6.58 Conduction band structure for a transverse-photocurrent QD detector. Photo-induced carriers are transferred into QWs for transport (Chu et al. 2001)

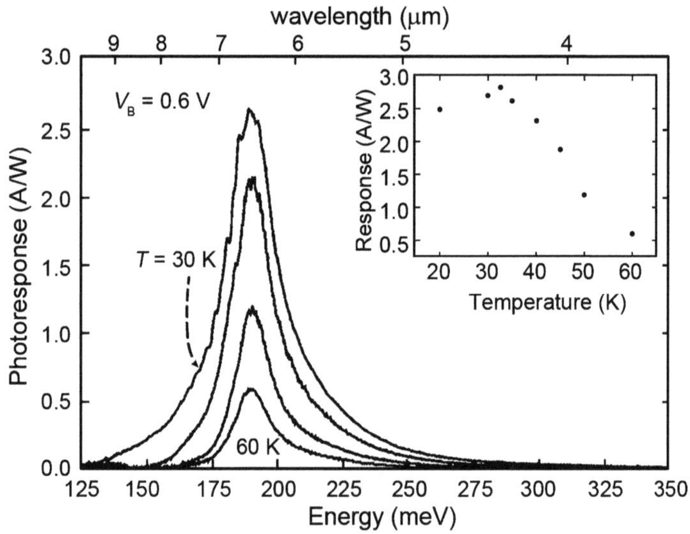

Fig. 6.59 Photo-responsivity for a transverse-photocurrent QD detector measured at different temperatures. The peak responsivity versus temperature is illustrated in the insert. The bias voltage is $V_B = 0.6$ V, and the modulation frequency is 44 Hz (Chu et al. 2001)

photo-induced carriers into the QW layer. The transfer of electrons in real space between the QDs and the two-dimensional channel was first observed by Lee et al. (1999). The electron transfer time acts as an important time constant determining the detector performance.

Figure 6.59 depicts the photo-responsivity of a transverse-photocurrent QD detector measured at different temperatures. The bias voltage is $V_B = 0.6$ V, and the modulation frequency is 44 Hz. The responsivity's peak energy is 185 meV, similar to that illustrated in Fig. 6.57. However, the responsivity is much higher than that in Fig. 6.57, indicating there is a very high photoconductive gain. The inset shows the temperature-dependent peak responsivity. Below 33 K, the responsivity is slightly enhanced as the temperature increases. For temperatures higher than 33 K, the responsivity begins to be drastically reduced.

In general, an increase in temperature has a direct influence on the electron transfer between the QDs and QW layer, and on the detector dark current as well. When temperature is slightly increased from a rather low value, the electron transfer is thermally enhanced between the wetting layer and the transport channel, and the peak photocurrent increases slightly. As the temperature increases further, the probability of the bound electrons in QDs being thermally excited grows, which in turn causes a drastic increase in the dark current (Ryzhii 1996). Meanwhile, due to the reduction of the bound electron number in the QDs, intersubband absorption is reduced. To a first-order approximation, the transition matrix element is independent of temperature. It is experimentally found that the peak responsivity reaches 10 W/A at $T = 30$ K and $V_B = 1.5$ V, and still has a value of the order of magnitude of 1 W/A at 60 K and only 60 mA/W at 77 K.

6.5 Low-Dimensional Semiconductor Infrared Lasers

6.5.1 Introduction

Conventional semiconductor lasers, including type-I electronic band structure quantum-well lasers, rely on a dipole emitting transition between the conduction and valence bands in a forward-biased PN junction (Wang 2001). They are used mostly in optical communication and data reading and writing. As the transition energy is mainly determined by the fundamental properties of the material and hence falls in the near-infrared and visible regions, their emitting wavelengths are not longer than 4 μm. Different materials have to be chosen to significantly change the emitting wavelength.

On the other hand, quantum cascade semiconductor lasers manifest obvious advantages in the mid-infrared and far-infrared regions, which rely on unipolar transitions between intersubbands.

A laser has to fulfill the following prerequisite conditions:

1. Excess high density carriers should be provided externally by heavy electrical injection.
2. The gain must be larger than the photon loss. The gain is proportional to the carrier density, so the electrical injection must be higher than a particular critical value, i.e., the threshold value.
3. A strong enough positive feedback must be assured for the photons. This can be realized by setting two reflecting mirrors at each end of the optical path to form a waveguide cavity, as illustrated in Fig. 6.60a. When the wavelength λ and the cavity length L satisfy:

$$\lambda = \frac{2n_1 L}{N}, \tag{6.250}$$

then constructive interference occurs in the cavity, leading to a positive feedback. Here n_1 is refractive index and N is an integer number. The reflecting mirrors also serve to restrict the light propagation direction and determine the light wavelength.

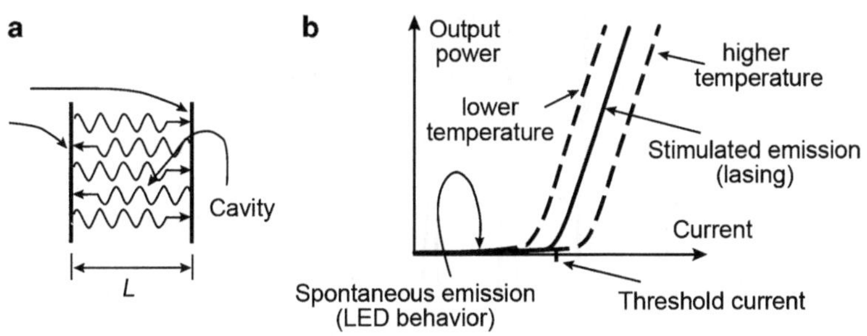

Fig. 6.60 (**a**) Positive-feedback conditions, and (**b**) output power as a function of electrical current for a semiconductor laser

In Fig. 6.60b, a typical power-current characteristic is schematically illustrated for a semiconductor laser. It is obvious that the power-current relation is strongly affected by temperature. It is the current required for a given power increases with temperature because carrier diffusion and nonradiative recombination are enhanced when the temperature gets higher. Because the active region of a semiconductor laser is commonly very small, a high injection current may easily cause it over-heat, therefore, cooling is necessary, and a reduction of the threshold current is always a crucial target. It is also clear that at a particular temperature, the laser output power exhibits light-emitting-diode features at low currents, and drastically increases when current exceeds threshold, showing a stimulated emission character. Figure 6.61 depicts an emission spectrum of a semiconductor laser. Obviously, a set of discrete narrow laser lines (modes) is selected from the continuum emission spectrum, due to the periodic constructive interference in the optical waveguide cavity.

In 1994, the quantum cascade laser (QCL) was invented at the Bell Labs, USA. Before its emergence, progress in developments was very slow for mid- and far-infrared lasers. In this period, homojunction lasers invented in 1962, GaAs/A1GaAs heterojunction lasers emerged in 1970's, and the quantum well lasers a bit later. The lasing of all these devices rely on the stimulated recombination of conduction-band electrons with valence-band holes, with the subsequent emission of photon energy. The lasing wavelength was fully determined by the semiconductor bandgaps, while the desirable semiconductor materials in nature are very rare with band-gaps suitable for mid- and far-infrared lasers. Since the 1970s, tremendous efforts have gone into the establishment of a new semiconductor laser emission theory. For example, in 1971, Kazatinov et al. proposed a new concept of a "photon-assisted" tunneling transition, i.e., intersubband optical transition energy being equal to the energy difference between two quantized energy states within the same energy band of a semiconductor. In 1986, Capasso at the Bell Labs came up with a new idea of an intersubband light that are emitted by electron tunneling in one QW region. In 1988, Liu suggested using a three-well structure for mid- and far-infrared light emission. However, it was later recognized that to realize a population inversion between two

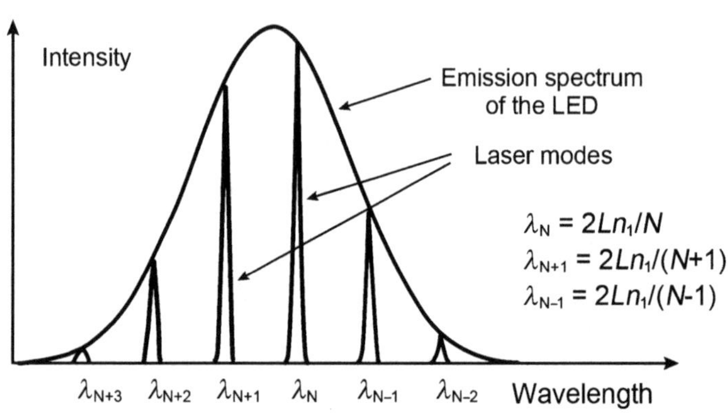

Fig. 6.61 A schematic emission spectrum from a semiconductor laser

intersubbands separated by an energy larger than that of an optical phonon (36 meV) is rather difficult, because then carrier lifetimes are of the order of nanoseconds (10^{-9} s), much longer than that induced by optical phonons (10^{-12} s). In the beginning 1990's, the Bell Labs group employed an InGaAs/InA1As material's system and designed a three-well structure. The width of injector well was reduced to 0.8–1 nm. The photon transition lifetime from the injector well level to active region well level reduces to the same order as an optical phonon-induced lifetime, making population inversion between intersubbands in the active region possible. In 1994, Faist et al. (1994) announced the first 4.3 μm mid-infrared quantum cascade laser.

6.5.2 *Basics of Intersubband Cascade Lasers*

QCLs are the first semiconductor laser types with wavelength that are arbitrarily tunable over a very wide range. It is easy to achieve different operational wavelengths with a single material system by changing the QW width and/or barrier height. In fact, the Bell Labs and other research institutions have employed an InP/GaInAs/AlInAs material system to realize lasers with wavelengths ranging from 4 to 11 μm.

The materials used in fabricating QCLs are alloys grown by MBE on the substrates GaAs and InP, which have the advantages of a relatively large bandgap energy and a mature growth technique. This makes QCLs more manageable and reliable relative to conventional infrared diode lasers made from narrow bandgap semiconductors such as InAs-, or InSb-based alloys.

The structure of QCLs is similar to the four-level system for many atomic and solid-state lasers. Stimulated emission happens between two excited states, in which the higher energy level is full of electrons while the lower energy level is empty, i.e., the system has a population inversion. This inversion occurs because of electron tunneling.

The structure of a QCL is made of multistages of active regions; each active region contains three parts, an injector region, a coupled QW active region, and a relaxation region. The relaxation region also serves as the injector region for the next stage. The injector/relaxation region is designed as a graded bandgap superlattice structure. In its operation a QCL is similar to a waterfall. When electrical current runs through a QCL, electrons drop sequentially from higher stage to adjacent lower stage, giving off at each drop one photon. At each stage, an electronic transition happens between two bound levels. The emitted photon is reflected back and forth between two internal reflecting mirrors stimulating other quantum transitions and hence emitting more photons. Such an amplification process ensures a high output power.

Figure 6.62 shows conduction band electronic levels for a $Ga_{0.47}In_{0.53}As$/$Al_{0.48}In_{0.52}As$ QCL, that contains two active regions and an intermediate injector region (Gmachl et al. 2001). An applied external electric field is visually illustrated by the linear slope of the electrical potential. The conduction band-offset discontinuity

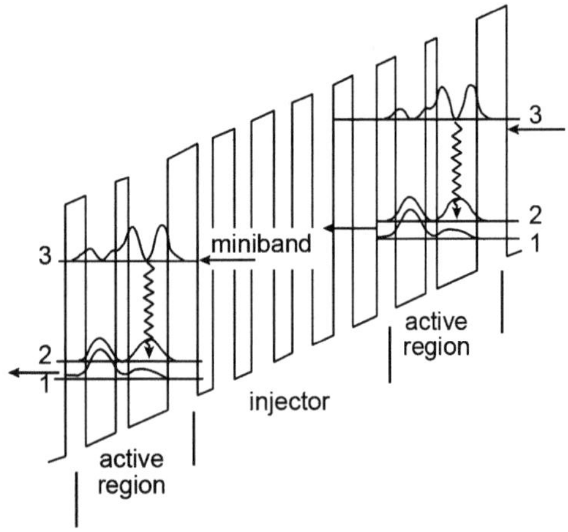

Fig. 6.62 Conduction-band structure for a $Ga_{0.47}In_{0.53}As/Al_{0.48}In_{0.52}As$ QCL. Two active regions and an intermediate injector region are illustrated. The moduli squared of the wavefunctions are also illustrated involved in the laser transition (labeled 1, 2, and 3). The optical transition is illustrated by wavy *arrows*, and the electron flow by the straight

is about 520 meV between the QW and the barrier. Solving the Schrödinger equation for this potential results in the energy levels drawn, and the square of the wavefunctions as illustrated. Also illustrated are the emissions related to the first three confined states of each active region. Extrinsic electrons are supplied to the injector region through doping with silicon to a sheet density of typically $1–5 \times 10^{11}\,\text{cm}^{-2}$ per period of active region and injector. This is roughly equivalent to a bulk doping density of $3–5 \times 10^{16}\,\text{cm}^{-3}$. This low carrier density justifies solving only the Schrödinger equation using the QW and external bias potential, while neglecting additional potentials resulting from the ionized donors and extrinsic electrons.

For this particular example, the QW thicknesses are chosen, in the active region, to be 6.0 and 4.7 nm, and the thickness of the intermediate barrier is 1.6 nm. This results in energy separations between levels 3 and 2 of 207 meV (corresponding to a wavelength of 6.0 μm) and between levels 2 and 1 of 37 meV. The latter is deliberately chosen to be very close to the energy of the LO phonon modes of the InGaAs/AlInAs/active region materials. By applying an appropriate bias (in this case, the applied electric field is $62\,\text{kV cm}^{-1}$, 0.29 V per stage of active region and injector), electrons tunnel from the injector region into energy level 3, the upper laser state, of the active region. Electrons scatter from this level into both lower-lying levels 2 and 1, by very rapidly emitting LO phonons. Using a Froehlich interaction model, the scattering times can be calculated to be $\tau_{32} = 2.2$ ps and $\tau_{31} = 2.1$ ps. This results in a total upper-state lifetime of:

$$\tau_3 = \frac{1}{1/\tau_{32} + 1/\tau_{31}} = 1.1\,\text{ps}. \tag{6.251}$$

Similarly, a scattering time can also be determined between levels 2 (the lower laser state) and 1:

$$\tau_{21} = \tau_2 = 0.3\,\text{ps}. \tag{6.252}$$

This scattering time is ultra-short due to the resonant nature of this transition with the LO phonons. With $\tau_{32} \gg \tau_2$, it is obvious that population inversion occurs between the laser levels 3 and 2, as soon as electrons are supplied fast enough into level 3 by tunneling from the preceding injector region, and exit from level 2 and more so from level 1 into the following downstream injector region at a high rate, again by tunneling. Inside this injector region and brought about by the externally applied electric field, electrons gain energy once again and are injected into the following downstream active region. Although laser action has been achieved for a single active region and has been tested for as many as 100 active regions (Gmachl et al. 1998b), typical QC lasers have a cascade containing 20–30 active regions alternated with injector regions.

In what follows the gains and losses will be briefly introduced; details can be found in related references (Gmachl et al. 2001; Faist et al. 1996).

Based on the usual rate equation model with the inclusion of a phenomenological broadening of the states between which transitions occur (which includes the Fermi-distributions in states 3 and 2), a gain coefficient g, defined as the gain per unit current density, can be deduced (Gmachl et al. 2001; Faist et al. 1996):

$$g = \tau_3 \left(1 - \frac{\tau_2}{\tau_{32}}\right) \frac{4\pi e z_{32}^2}{\lambda_0 \varepsilon_0 n_{\text{eff}} L_{\text{p}}} \frac{1}{2\gamma_{32}}. \tag{6.253}$$

where λ_0 is the wavelength in vacuum, ε_0 the vacuum dielectric constant, e the elementary charge, n_{eff} the effective refractive index of the mode, L_{p} the thickness of one period of the active and injector regions, $2\gamma_{32}$ the full width at half maximum of the luminescence spectrum, and z_{32} is the optical dipole matrix element. For the aforementioned example, the parameters are $\lambda_0 = 6\,\mu\text{m}$, $n_{\text{eff}} = 3.25$, $L_{\text{p}} = 47\,\text{nm}$, $2\gamma_{32} \sim 20\,\text{meV}$, $z_{32} = 2.0\,\text{nm}$, and therefore $g = 30\,\text{cm} \cdot \text{kA}^{-1}$.

It is worth noting that (6.253) was derived on the basis of the gain of a QW laser, but taking the total thickness of one period of active and injector regions as the net active-region thickness.

The loss is the second import factor in determining a threshold current, which has three main contributing parts. The first part is the so-called mirror loss. The laser resonator is formed by two semitransparent mirrors, typically uncoated, as-cleaved, and parallel semiconductor facets. Each of them provides a reflectivity:

$$R = \left(\frac{n_{\text{eff}} - 1}{n_{\text{eff}} + 1}\right)^2, \tag{6.254}$$

that results in a mirror- or outcoupling loss:

$$\alpha_{\text{m}} = \frac{1}{L} \ln(R). \tag{6.255}$$

Clearly, this part of loss can be reduced/enhanced by using high-reflection/antireflection coating at the mirror facets. Additional outcoupling losses include the scattering losses due to imperfections and interfaces. However, quantifying these losses is difficult.

The second part of the contribution to the loss is known as waveguide loss, which is caused by free-carrier absorption in the doped semiconductor regions and in the metallic contact layers. While the influence of the metallic layers can be suppressed, the waveguide loss, α_w, is unavoidable. It increases approximately quadratically with the wavelength (Yu and Cardona 1999), i.e., $\alpha_w \propto \lambda^2$.

The third part is introduced by resonant intersubband transitions. Due to the considerable optical dipole matrix element of interminiband transitions, it is in general conceivable that the extrinsic electrons in the injector region can cause significant absorption if optical transitions are present that resonant with the laser wavelength. It is therefore necessary to avoid such resonantly absorbing intersubband or interminiband transitions in the design of the QCL active regions and injectors.

Based on these loss mechanisms, an equation for the threshold current density J_{th} can be established (Gmachl et al. 2001):

$$J_{th} = \frac{\alpha_w + \alpha_m}{g\Gamma}, \tag{6.256}$$

where Γ represents the overlap of the guided mode with the stack of active regions and injectors; called the "confinement factor". The number of stages in the cascade enters the into threshold current density via this quantity. Taking the temperature-dependence of J_{th} into account, the α_m, α_w and Γ can be assumed to be temperature independent, which is valid only if thermally generated free carriers and changes in the electron band structure are negligible. However, the gain coefficient is clearly temperature dependent. Electron scattering times for LO phonon emission are reduced by the Bose factor through the larger phonon population found at higher temperatures. Meanwhile, due to the thermal spread of carrier populations, enhancement of nonparabolicity effects become more visible, leading to enhancement of other electron scattering mechanisms, such as impurity and interface roughness scattering. In addition, the width of the gain spectrum is broadened at higher temperatures. Finally, extrinsic electrons from the injector region can be thermally back excited into the lowest laser level, where they reduce the population inversion. A temperature-dependent threshold current density was given by Faist et al. (1998b):

$$J_{th} = \frac{1}{\tau_3(T)\left(1 - \frac{\tau_2(T)}{\tau_{32}(T)}\right)} \left[\frac{\varepsilon_0 n_{eff} L_p \lambda_0 (2\gamma_{32}(T))}{4\pi e z_{32}^2} \cdot \frac{\alpha_w + \alpha_m}{\Gamma} + e n_g e^{-\frac{\Delta}{kT}}\right], \tag{6.257a}$$

$$\tau_i = \tau_{i0} \frac{1}{1 + \frac{2}{e^{(E_{LO}/k_B T)} - 1}}, \quad i = 1, 2, 3, \tag{6.257b}$$

where τ_{i0} is the scattering time at low temperature, E_{LO} the LO phonon energy, T the temperature, n_g the carrier sheet density in the injector ground state, and Δ is the energy separation between the injector ground state and level 2 of the preceding active region. This relation has been observed to provide a reasonably accurate prediction of experimental results.

Nevertheless, just as for conventional semiconductor lasers, a so-called "characteristic temperature," T_0, has proven to be useful for QCLs as well:

$$J_{\text{th}}(T) = J_0 e^{(T/T_0)}. \tag{6.258}$$

It ranges for different QCL's and laser types from 100 to 200 K, which is a large value compared to all other semiconductor lasers with typical T_0 values found to lie below 100 K.

From the discussion of the loss mechanisms, the slope efficiency, i.e., the rate of increase in optical power (per outcoupling facet) per unit current, is also derivable:

$$\frac{\partial P}{\partial I} = \frac{1}{2}\frac{h\nu}{e} N_{\text{p}} \frac{\alpha_{\text{m}}}{\alpha_{\text{m}} + \alpha_{\text{w}}} \left(1 - \frac{\tau_2}{\tau_{32}}\right), \tag{6.259}$$

where $h\nu$ is the photon energy and N_{p} is the number of stages of active regions and injectors. It is obvious that a large number of stages result in a large slope efficiency, and consequently in general, a large optical output power. In practice, values in the range of W/A and W, respectively, are common.

6.5.3 Basic Structures of Intersubband Cascade Lasers

Design of a QCL consists of two tasks, that of designing active regions and the waveguide. Since the first demonstration of the QCL in 1994 by Faist et al. (1994), three typical designs of active regions have been established, namely, a "vertical transition" design, a "diagonal transition" design, and a superlattice active region design. Many different types of QCLs have been fabricated by combining active region designs with conventional dielectric waveguides or surface-plasmon waveguides.

Our first concern is the structure of the active regions. In the so-called three-well vertical transition design, a thin quantum well layer is inserted between the injector region and the active region, as illustrated in Fig. 6.63. This additional QW leads to a significant reduction of electron scattering from the injector directly into the laser ground state 2 and also level 1. This improvement gave birth to the first high-performance, room-temperature operation QC laser (Faist et al. 1996).

Figure 6.63 shows a portion of the conduction-band structure for a InGaAs/InAlAs QCL (Gmachl et al. 1999b). The active region consists of three InGaAs QWs closely coupled by thin AlInAs barriers. At an electric field of 45 kV/cm, corresponding to the measured laser threshold, the upper laser level (level 3) is separated from the lower laser level (level 2) by $E_{32} = 153.6$ meV ($\lambda_0 \approx 8\,\mu$m). The LO-phonon scattering time between the two levels is calculated to be $\tau_{32} = 3.1$ ps, and the dipole matrix element of the optical transition is $z_{32} = 1.9$ nm. The lower laser level is closely coupled to the ground state (level 1) of the active material, such that the two levels are strongly anticorrelated. This increases the matrix element z_{32} with respect to z_{31} and maximizes τ_{32} with respect

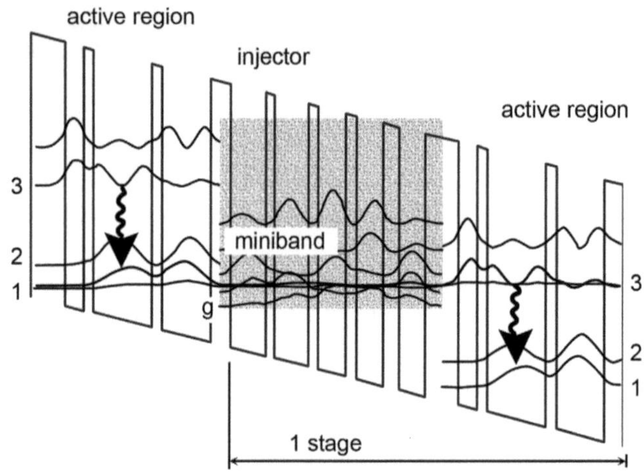

Fig. 6.63 Conduction band diagram of two active regions with the intermediate injector region and the absolute value of the wavefunctions squared that are involved in the laser transition. The wavy *arrows* indicate the laser transitions. The layer thicknesses in nanometres of one period of active material and injector are, from *left* to *right* starting from the injection barrier (*left-most* layer): 3.8/2.1/1.2/6.5/1.2/5.3/2.3/4.0/1.1/3.6/1.2/3.2/1.2/3.0/1.6/3.0 nm. The *underlined* layers are n-type doped. The barriers are indicated in *bold symbols; italic style* shows the injector region. The applied electric field is 45 kV/cm (Gmachl et al. 2001)

to τ_{31}. The energy difference is $E_{21} = 38.3\,\text{meV}$, designed to efficiently deplete the lower laser level of electrons via resonant LO-phonon scattering ($E_{\text{LO}} \sim 34\,\text{meV}$). The corresponding lifetime is calculated to be τ_2–$\tau_{21} = 0.3\,\text{ps}$, which is much smaller than $\tau_{32} = 3.1\,\text{ps}$. The LO-phonon scattering time from level 3 to level 1 is $\tau_{31} = 3.6\,\text{ps}$, and the scattering time into all states of the following injector region is estimated to be $\tau_{3\text{i}} \sim 14.6\,\text{ps}$, significantly longer than intra-active region scattering times.

With (6.254) and taking into account (1) the scattering times determined above, (2) the measured luminescence linewidth $2\gamma_{32} \sim 10\,\text{meV}$, (3) the length of active/injector region period, $L_{\text{p}} = 44.3\,\text{nm}$, (4) the effective refractive index of the waveguide $n_{\text{eff}} = 3.27$, and (5) the lasing wavelength $\lambda_0 = 8\,\mu\text{m}$, the gain coefficient can be derived to be $g \approx 60\,\text{cmkA}^{-1}$ (Gmachl et al. 1998a).

The mirror-losses α_{m} for an $L = 2.25\,\text{mm}$ long laser with as-cleaved facets are calculated to be $\alpha_{\text{m}} = 5.59\,\text{cm}^{-1}$. The waveguide losses at threshold have been measured from subthreshold spectra using the Hakki–Paoli method (Hakki and Paoli 1975) to be $\alpha_{\text{w}} \approx 24\,\text{cm}^{-1}$.

In addition, for the entire stack of 30 active region and injector periods, it is obtained from (6.256) that the confinement factor $\Gamma = 0.5$ and the threshold current density $J_{\text{th}} = 0.98\,\text{kA cm}^{-2}$. The latter is in excellent agreement with the experimental value of $J_{\text{th}} \approx 1\,\text{kA cm}^{-2}$. Similarly, the slope efficiency is $\partial P/\partial I = 394\,\text{mW/A}$ per facet for $N_{\text{p}} = 30$. This is also in very good agreement with the experimental value of $\partial P/\partial I \approx 400\,\text{mW/A}$. This good agreement between calculated and experimentally measured values implies that besides excellent growth quality, the major physical effects are understood to a satisfactory degree. It is observed that as N_p increases, the confinement factor is enhanced. Normally, for $N_p \leq 30$, Γ grows linearly with N_p, while for higher N_p, Γ tends to saturate, e.g., $\Gamma(N_p = 30) = 0.49$, $\Gamma(75) = 0.81$, $\Gamma(100) = 0.88$, and $\Gamma(200) = 0.97$.

In Fig. 6.64, both the theoretical (curves) and experimental (dots) results for the N_p-dependence of the QCL threshold current densities are depicted. The theoretical calculation is based on (6.258). The experimental values are from pulsed- and continuous-operation modes, respectively. For $N_p \geq 15$, a threshold current density $J_{th} \leq 2\,\text{kA cm}^{-2}$ is always obtainable at low temperature. For a small N_p, J_{th} increases significantly with N_p, and is nearly proportional to N_p^{-1}. For $N_p \leq 6$, the laser does not work at room temperature, and for $N_p = 3$–30, the laser can operate in a continuous-wave mode.

Figure 6.64b shows an experimental current–voltage relation at low temperature. The voltage drop per stage is $E_{32} + E_{21} = E_{\text{photon}} + E_{\text{LO-phonon}} = 193\,\text{meV}$, and the voltage drop across the entire N_p-stage cascade is given by:

$$V_S(N_p) = \frac{(E_{32} + E_{21})}{e} \cdot N_p \propto N_p. \tag{6.260}$$

In accordance with conventional semiconductor diodes, here the turn-on voltage V_{to} is introduced, as the bias voltage at which current starts to flow through the device. The measured values of V_{to} as a function of N_p is depicted as inset in Fig. 6.64b, which can be well fit by:

$$V_{to}(N_p) = V_S(N_p) + V_{\text{offset}} = e^{-1} \cdot (E_{32} + E_{21}) \cdot N_p + V_{\text{offset}}, \tag{6.261}$$

V_{offset} is the only fitting parameter, and is determined from the fit to be, $V_{\text{offset}} = 0.33\,\text{V}$. Such a residual voltage can be well explained by the voltage drops due to (1) non-alloyed contacts and graded-gap regions ($\sim$0.1 V), (2) the connection between the first injector region and bulklike InGaAs layer ($\sim$0.05 V), and (3) the connection between the last injector region and the bulklike InGaAs layer ($\sim$0.2 V) of the waveguide. On the other hand, the voltage at laser threshold $V_{th}(N_p)$ can be calculated as:

$$V_{th}(N_p) = V_{to}(N_p) + \frac{\partial V}{\partial J}(N_p) \cdot J_{th}(N_p), \tag{6.262}$$

where the last term accounts for the excess voltage above V_{to} needed to sustain the threshold current density $J_{th}(N_p)$. The theoretical results are depicted in the inset to Fig. 6.64b, with the corresponding experimental values plotted as dots. Obviously, the theory is in good agreement with the experiment.

In Fig. 6.64c, the measured slope efficiency per facet at low temperature is plotted as a function of the number of stages (Gmachl et al. 1999b). The inset gives the corresponding slope efficiency per stage. Once corrected for the different collection efficiencies, pulsed and, where available, cw data are in good agreement. The best fit to the data results in a slope efficiency per facet and per stage of 13.24 mW/A.

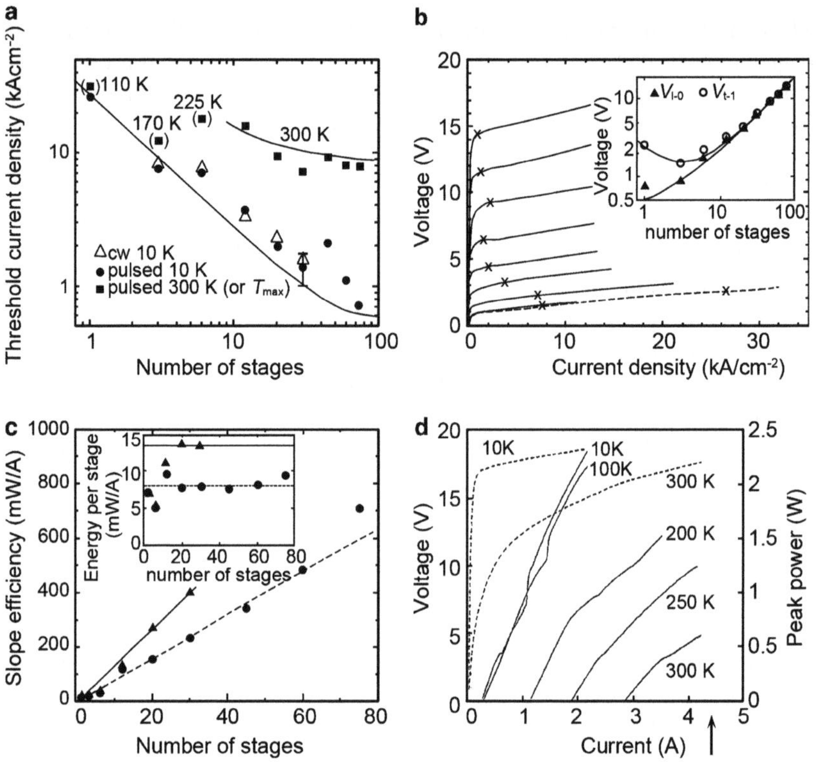

Fig. 6.64 QCL characteristic versus the number of stages: (**a**) threshold current density measured at cryogenic temperatures in pulsed mode (*solid circles*) and in cw mode (*open triangles*), and in pulsed mode at RT (*solid squares*). The respective highest operating temperature is displayed. The *solid curves* are calculated for cryogenic temperature and RT. The error-bar shown for the $N_p = 30$-stage device indicates the range of values that are obtained for nominally identical devices from a variety of wafers and processing cycles. (**b**) Voltage–current characteristics of various QCL devices with different N_p, the uppermost trace corresponds to the sample with $N_p = 75$ stages, the lowermost to $N_p = 3$. The trace of $N_p = 1$ is drawn as a *dashed curve*. The times on each curve denotes the voltage at which the laser threshold is reached. Inset, turn-on voltage (V_{to}, *solid triangles*) and threshold voltage (V_{th}, *open circles*) as a function of N_p. The symbols show experimental data, the *solid curves* the model calculations. (**c**) Measured slope efficiency per facet. The triangles correspond to data obtained in cw operation, the circles to pulsed data. The curves are the best fit to the data. The data have been obtained at cryogenic temperatures. However, the slope efficiency degrades only marginally with temperature. *Inset*, slope efficiency per facet per stage as a function of N_p. (**d**) Light output versus current (*solid*) and voltage versus current (*dashed*) characteristics at various heat-sink temperatures of the laser containing $N_p = 75$ active region and injector stages. The collection efficiency is ∼90%, and the power is collected from one facet

From the slope efficiency the total external differential quantum efficiency per facet is determined as:

$$\eta_D(N_p) = \frac{e}{E_{32}} \frac{\partial P}{\partial I}(N_p), \tag{6.263}$$

resulting in 8.5% and 8.3% per stage for the theoretical and experimental values, respectively. For $N_p \geq 12$, $\eta_D(N_p)$ excess unity, e.g., η_D (75) $\approx$ 6.4.

Figure 6.64d shows the pulsed light output versus current (solid) and voltage versus current (dashed) characteristics at various heat-sink temperatures of the laser containing $N_p = 75$ stages. The collection efficiency is ~90%, and the power is collected from one facet. Peak output powers of 1.4, 1.1, and 0.54 W are obtained at 50, 200, and 300 K (RT), respectively. In general, for small N_p, the peak light output power is nearly proportional to N_p, meanwhile, for a large N_p, the output tends to saturate.

With the benefit of the combination of a very short length of the active region with large flexibility to design injector regions, the three-well vertical active regions have been successful for wavelengths as long as 13 μm and have been particularly successful for shorter wavelengths. For example, Faist et al. first succeeded in demonstrating a 3.5 μm QCL in 1998 (Faist et al. 1998a).

The second type of active region is known as superlattice active region. Superlattice (SL) active region QC lasers were first introduced in 1997 (Scamarico et al. 1997b). Laser action takes place between minibands, rather than between the subbands of quantum well active regions QCLs. SL-QC lasers excel through their high gain, large current carrying capabilities, and weaker temperature sensitivity. However, they can generally only be applied in the longer wavelength range (i.e., $\lambda_0 \geq 7\,\mu$m, for lattice-matched InP-based SL-QC lasers) as the wider minibands take up more energy space than simple subbands.

In general, a semiconductor SL consists of a periodic stack of nanometre-thick layers of QWs and barriers. The period of this artificial crystal is typically much larger than the lattice constant of the bulk crystalline constituents. This superimposed crystal potential splits the conduction band (and the valence bands as well) along the direction normal to the layers into a series of narrow (typically, tens to a few hundreds of meV wide) minibands separated by energy gaps ("minigaps") in the strong-tunnel-coupled regime. For a given set of materials, miniband and minigap widths can be engineered by suitable choices of the layer thicknesses.

Figure 6.65 illustrates the original interminiband SL-QC laser concept and its first realization in an InGaAs/InAlAs structure (Scamarico et al. 1997b). The SL consists of eight strongly coupled QWs, i.e., the intermediate barriers are very thin (typically 1–2 nm for conduction band discontinuities of several hundreds of meV). This results in two minibands substantially confined within the SL with its 520 meV band offset. The SL is homogeneously doped to 6–7 $\times$ 10^{16} cm^{-3} in order for the extrinsic charges to screen the electric field applied externally during laser operation. The ensuing quasi-Fermi level (~12 meV) is still located well below the top of the first miniband, which has a width ~100 meV, so that even at elevated temperatures the states near the top of the latter band can be assumed to be essentially empty. Electrons are injected electrically from a preceding, equally doped, injector region directly into a state near the bottom of the second "upper" miniband, as was the case with the simple three-well active region discussed in earlier sections. From there, electrons can make an optical transition to the top of the lower miniband. In **k**-space, this transition occurs at the so-called mini-Brillouin-zone boundary.

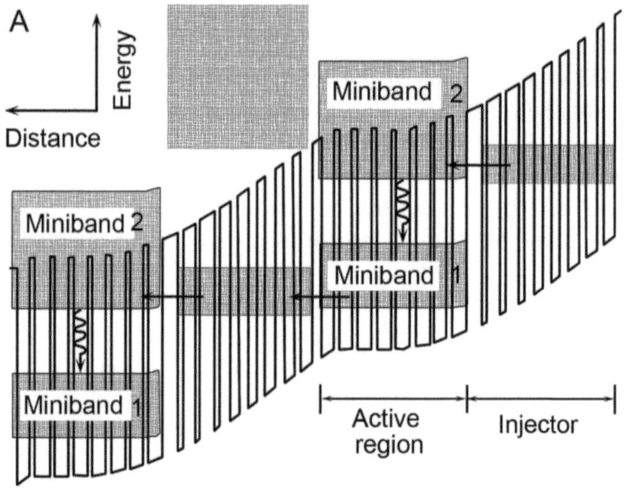

Fig. 6.65 A schematic conduction-band structure of the original InGaAs/InAlAs SL-QC laser. The *shaded* regions indicate the minibands. Laser action, denoted by the wavy *arrows*, takes place across the first minigap 2 → 1. The applied electric field is screened from the flat SLs through extrinsic electrons

In general, the interminiband scattering processes are dominated by LO phonon emission as it is in other QC lasers. Specifically, an electron thermalized near the bottom of the higher miniband can relax down to states near the top of the lower miniband by scattering via LO phonon emission employing a high momentum transfer. This process is characterized by a comparatively long scattering time of $\sim$10 ps. Within each miniband, it involves the emission of small-wavevector optical phonons, and hence the electrons relax much faster, in a few tenths of a picosecond, by intraminiband scattering. The resulting large relaxation time ratio of inter- versus intra-miniband scattering events ensures an intrinsic population inversion across the minigap. A second unique design feature of the SL-QC laser is the high oscillator strength for the optical transition at the mini-Brillouin zone boundary of the SL. According to Helm's research (Helm 1995), the oscillator strength of radiative transitions between the first two minibands of an SL increases with wavevector and is a maximum at the zone boundary, and in particular, it strongly increases with decreasing barrier thickness.

In the original SL-QC laser structure illustrated in Fig. 6.65, the wavelength was selected to be 8 μm, using 1-nm-thick AlInAs barriers and 4.3-nm-thick InGaAs QWs in a lattice-matched composition. The structure was optimized for maximum oscillator strength with an optical dipole matrix element of 3.6 nm, which is approximately twice as large as that of similar-wavelength QC lasers featuring few-well active regions.

To properly compute the energy levels, optical dipole matrix element of the laser transition, as well as the scattering times between the relevant levels, the actual band profile of the structure has to be calculated first. To accomplish this, knowledge of both the applied external bias and the electric field generated by the charge distribution of the donor ions plus the extrinsic electrons is required. The latter, in turn, is determined by the wavefunctions of the electronic levels. Therefore, a self-consistent solution of the Schrödinger and Poisson equations have to be done numerically, which relates the charge distribution to the electrostatic potential. Once energy levels and wavefunctions are properly calculated most parameters of the laser operation can just be estimated in the same fashion as before for the few-well QC lasers based on intersubband transitions. These calculations are always based on potentials derived from the virtual crystal approximation, which holds reasonably well for these alloys. However because of the large strains associated with the lattice mismatches among AlAs, GaAs, and InAs these alloys are highly correlated and not well, approximated as random (Chen and Sher 1995), but this is also an approximation always used in this theory. For example, for the laser illustrated in Fig. 6.65, the gain coefficient can be determined to be 25 cm/kA. If it is used with conventional waveguide ($\alpha_w \sim 30\,\text{cm}^{-1}$), a threshold current density of $8.5\,\text{kA/cm}^2$, and a peak output power of 850 mW was achieved in pulsed operation at a 5 K heatsink temperature. It is clear that already these first SL-QC lasers excelled through their high optical output power, which at that time was primarily a result of their much larger current carrying capabilities compared with conventional QC lasers. In a comparison of their threshold current densities, high-temperature, cw operation, and slope efficiency, however, these first SL-QC lasers were inferior to other QC lasers. The most important reasons are: (1) the large waveguide loss brought about by the generally high doping level, and (2) the broad luminescence linewidth as a result of ionized impurity scattering and interface roughness scattering because of the increased number of interfaces in the active region, and (3) a reduced population inversion at elevated temperatures, caused by high electron densities in the active regions. Unfortunately, such substantial extrinsic doping levels were originally needed to screen the externally applied electric field from penetrating into the SLs. Hence, subsequent efforts to improve SL-QC laser performance focused on retaining the advantages of the SL active region while reducing the required doping levels.

The first major performance improvement came with an SL-QC laser design that places the dopants only in the injector regions. The underlying idea of the design is to significantly spatially separate the extrinsic electrons from the donor ions in such a way that the field generated by these charges exactly compensates the external applied electric field across the SLs. Tredicucci et al. have ensured the SL regions will be almost field free with the first miniband of each period aligned in energy with the second miniband of the next period to facilitate efficient carrier transport between the stages. This is accomplished by applying proper n-type doping in the layers near the adjacent SL "downhill" in the energy diagram (Tredicucci et al. 1998). In this way they obtained scattering times and a gain coefficient very similar to those of doped SL-QC lasers. Nevertheless, even for a waveguide design routinely employed in QC lasers, which now displays a clearly lower waveguide loss due to the reduced effective bulk doping of the active waveguide core, both RT pulsed and low-temperature cw operation were now easily achieved. With this improvement, an RT threshold current density of 10 kA/cm, and peak optical power of 175 mW collected from one facet with a 50% efficiency have been obtained.

An even further improved design of SL-QC lasers was found in the so-called chirped SL QC lasers, that features an alternative SL-QC laser design to achieve flat minibands with undoped active regions. It is based on SL active regions with gradually decreasing well thicknesses in the direction of the electron motion. At zero bias, the states of the QWs are localized due to the variation in the thicknesses of adjacent layers and the corresponding energy levels of the isolated wells are out of resonance. As an electric field of the appropriate value and polarity is applied, these states are brought into resonance so that their wave functions anticorrelate forming manifolds of closely spaced extended states, i.e., "minibands". Pioneering work was done and initial results were obtained (Capasso et al. 1999; Tredicucci et al. 2000a). In Fig. 6.66, the calculated conduction band diagram is illustrated for a structure with a so-called funnel-type injector. Two well-defined minibands are created in the chirped SLs, matched to a single injection miniband that bridges them together across the cascading stages. The states involved in the interminiband laser transition are uniformly delocalized over at least six periods, which ensures a large optical dipole matrix element. With optimization, peak powers of almost 900 mW at 5 K and nearly 300 mW at RT were obtained for devices processed with a broad-area waveguide configuration. Also a maximum cw optical power of 300 mW at 5 K and 200 mW at 80 K was measured from one facet.

It is worth stressing that the "chirped" SL active region design has been subsequently demonstrated in the long-wavelength region with laser operation at 17, 19, 21.5, and 24 μm, respectively (Tredicucci et al. 1999, 2000b; Gianordoli et al. 2000; Colombelli et al. 2001). These lasers were the first demonstration of semiconductor injection lasers based on intraband transitions at wavelengths beyond

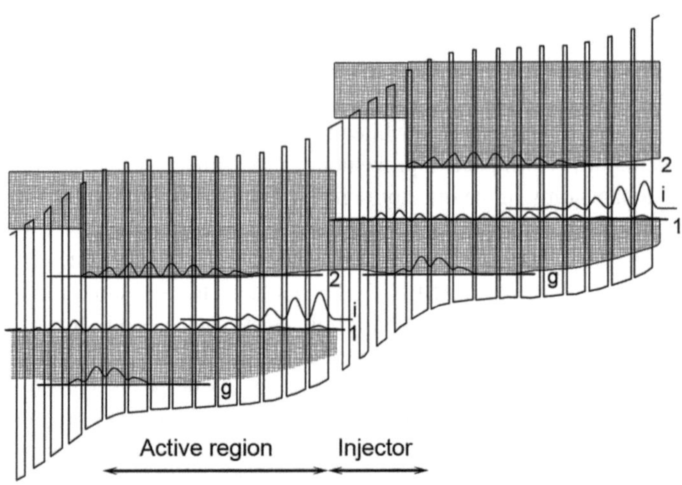

Fig. 6.66 The calculated energy diagram of an SL-QC laser with undoped SL active regions. The laser transition 2 → 1 is indicated by the *wavy arrow*, and the absolute value squared of the respective wavefunctions is also shown. The *straight arrow* indicates the injection barrier. The shadowed regions represent electronic minibands

the atmospheric windows (3–5 μm and 8–13 μm). In general, above the optical phonon energy, i.e., $\lambda \leq 30\,\mu$m, intersubband transitions are characterized by an increasingly smaller radiative efficiency with decreasing transition energy (Ferreira and Bastard 1989). The intraband free-carrier absorption coefficient in doped semiconductors is roughly proportional to λ_0^2, indicating increasingly larger optical waveguide losses at longer wavelengths. Furthermore, two-phonon absorption becomes a noticeable factor.

All these aspects lead to prohibitively high threshold current densities for conventional intersubband QC laser structures. In contrast, QC lasers based on interminiband transitions in semiconductor SLs can be driven with very large current densities (up to 30 kA/cm^2 and above) without compromising laser performance.

The QC and SL-QC laser designs described above have one identical feature, the laser-action involves wavefunctions of the upper and lower states that are essentially located in the same region of real space, hence the title "vertical" transition. There is a third type called a "diagonal" transition design. In contrast to "vertical" transitions, the "diagonal" transition QC lasers show a strong dependence of the wavelength on the applied electric field through a voltage-induced Stark effect. The design parameters are furthermore characterized by a smaller dipole matrix element but also clearly longer scattering times, thus allowing again for sufficient gain to achieve laser action.

In the design shown in Fig. 6.67, laser action takes place from level G_+ to level **1** via photon-assisted tunneling or a diagonal transition denoted by the wavy arrow, since the two states have reduced spatial overlap under an appropriate applied positive bias (Faist et al. 1997a). Here G_+ is the ground state of the miniband in the superlattice injector, while level **1** is the ground state of the active region QW. This miniband design ensures that under application of a suitable electric field,

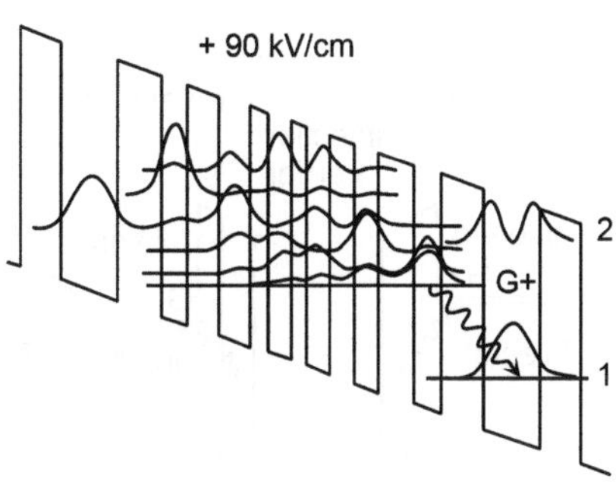

Fig. 6.67 Conduction band diagram of two active regions with the intermediate injector region of a diagonal transition active region QC laser under a positive applied electric field of 90 kV/cm approximately corresponding to laser threshold. The layer thicknesses in nanometres are from left to right starting from the first injection barrier: 3.5/4.8/3.5/2.4/2.5/2.6/1.5/2.2/1.0/2.2/1.9/2.2/2.9/2.2/3.5/4.8/3.5. The $In_{0.52}Al_{0.48}As$ layers (energy barriers) are in *bold symbols* alternated with the $In_{0.53}Ga_{0.47}As$ wells. The *underlined layers* are doped to $n = 4 \times 10^{17}$ cm^{-3}; *italic style* indicates the injector region. The absolute value squared of the wavefunctions in the active and injector regions are shown. The *wavy arrow* indicates the laser transition $G_+ \rightarrow 1$

G_+ is spatially localized close to the injection barrier. For a given active region QW, the emitted photon energy is controlled by the thickness of the injector layers in the immediate vicinity of the injection barrier and by the applied electric field. For the illustrated structure under a laser threshold electric field of $+90\,\text{kV/cm}$, the optical matrix element, the lifetime of the G_+ level, and the transition energy are calculated to be $z_+ = 0.35\,\text{nm}$, $\tau_+ = 46\,\text{ps}$, and $E_+ = 198\,\text{meV}$, respectively. Due to the relatively large spatial separation of the states (G_+ and **1**) of the diagonal laser transition, enabling the long lifetimes of the G_+ level, τ_+, the population inversion necessary for laser action is guaranteed. The low-temperature gain coefficient and optical peak output power are obtained to be about $32\,\text{cm/kA}$ and $>100\,\text{mW}$, respectively, which are of the same order of magnitude as those of optimized vertical-transition QC and SL-QC lasers.

Just from the performance point of view, the "diagonal" transition seems to be inferior to the "vertical" one. However, the "diagonal" transition laser has a distinct advantage that in contrast to all other QC and SL-QC lasers; it operates as a light source under both positive and negative bias voltage. With it, dual-wavelength lasers can be achieved. Figure 6.68 schematically illustrates the basic principle of operation of the bi-directional QC laser. Under an appropriate applied positive bias (Fig. 6.68a), laser action takes place from level G_+ to level 1, via photon-assisted tunneling, a diagonal transition (wavy arrow). A reversal of the bias polarity (Fig. 6.68b) will localize the injector ground state G_- at the opposite end of the injector miniband. With an injector entirely symmetric around its center, G_+ and G_- would be equivalent and the laser wavelength would be the same in both polarities. If, however, the injector regions are designed to be asymmetric, the energy position of G_- will be different from that of G_+ leading to different laser wavelengths λ_+ and λ_- (Gmachl et al. 1999a).

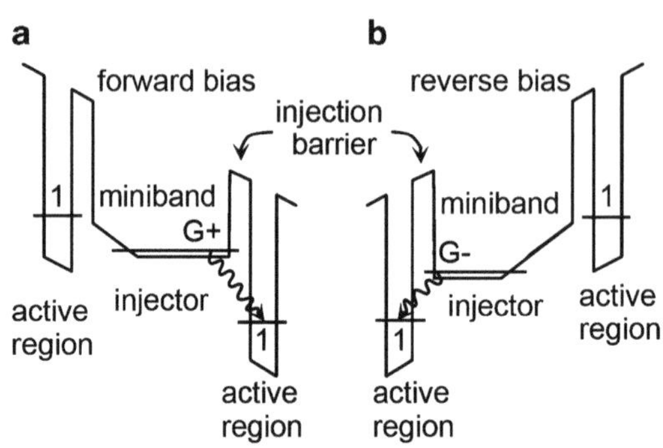

Fig. 6.68 Schematic conduction-band structure for a bi-directional QC laser: (**a**) a portion of the conduction-band structure (two active regions with the intermediate injector region) is shown under a positive applied bias, in which electrons traverse the structure from left to right. (**b**) Schematic band structure of the same QC laser under negative bias, in which electrons traverse the structure in the opposite direction. Depending on bias polarity, laser action takes place between different pairs of energy levels ($G_+ - 1$ and $G_- - 1$), leading to a bias-polarity-dependent laser characteristic

In earlier sections focus has been directed to the influence of the gain coefficients by various designs of active regions and injectors currently in use with QC lasers. Total losses are the second important contribution to laser performance. As was briefly discussed, a low waveguide loss is crucial to high-quality QCL performance. Taking into account the biggest impact in the mid-infrared wavelength range, that of free carrier absorption in the doped semiconductor, it is obvious that the question of waveguide design is more complex than it is at near-infrared wavelengths, where the obvious choices are semiconductor dielectric waveguides. In the following several advantageous QCL waveguide designs, e.g., dielectric waveguides, will be discussed.

First, a dielectric waveguide is a straightforward choice if the material system employed, provides appropriate and sufficiently large refractive index variations to build a high refractive index waveguide core; one that contains the active material and a lower-refractive-index cladding layer. The confinement factor, i.e., the overlap of the optical guided mode with the active regions and injectors, should be as large as possible to obtain a low-laser threshold. This can be achieved with an as large as possible refractive index step between the core and the cladding layer.

For the most commonly used InGaAs/InAlAs material system on an InP substrate, the InP substrate (with a refractive index $n_{\mathrm{InP}} \sim 3.10$) and AlInAs ($n_{\mathrm{InAlAs}} \sim 3.20$) is a natural cladding layer for the waveguide core. The latter consists of the stack of 500 or more alternating ultrathin layers of InGaAs and AlInAs, with the average refractive index of the stack being normally evaluated through a linear interpolation between the refractive indices of AlInAs and InGaAs ($n_{\mathrm{InGaAs}} \sim 3.49$) according to the volume fraction of the two constituents. It is therefore obviously higher, about $n_{\mathrm{QC}} \sim 3.35$ in general, than the refractive indices of the cladding layers. To increase the average refractive index even more, in particular when the actual cascade stack is chosen to be rather thin, the active region and injector stack is sandwiched between two several hundred nanometres thick InGaAs layers.

It is noteworthy that such an estimate does not include the doping effects. In fact, all or at least a significant portion, of the waveguide layers have to be doped to enable carrier transport in current injection lasers, the free-carrier absorption in these layers usually reduces the refractive index by a few percent relative to their purely dielectric value. In addition, a noticeable attenuation coefficient is introduced, which is ultimately responsible for the waveguide loss α_{w}. There is another additional waveguide loss caused by the coupling of the mid-infrared wavelength light to the surface plasmon at the semiconductor–top-metal interface contact. A pathway to alleviate this unwanted coupling is to make the dielectric layers thick enough to separate the top metal from the waveguide core. However, this approach conflicts with the desirable requirement of thin epilayers for reduced MBE growth time. Sirtori et al. has proposed a solution of a "plasmon-enhanced" waveguide design (Sirtori et al. 1995b).

In Fig. 6.69, a cross section of an optimized plasmon-enhanced waveguide is schematically illustrated for a $\lambda = 8.5\,\mu$m wavelength. The low-doped ($\sim 2 \times 10^{17}\,\mathrm{cm}^{-3}$) InP substrate is the lower waveguide cladding layer. The top AlInAs cladding region consists of a thick low-doped ($1 \sim 3 \times 10^{17}\,\mathrm{cm}^{-3}$) layer. This reduces

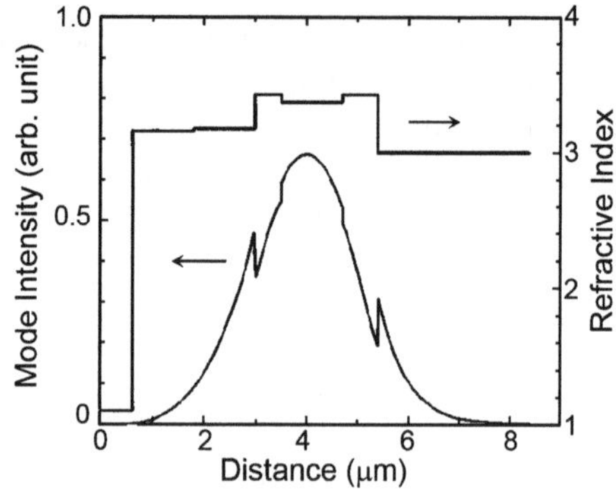

Fig. 6.69 Calculated intensity mode profile and profile of the real part of the refractive index in the growth direction of an optimized dielectric QCL waveguide incorporating plasmon enhanced confinement, employing a highly doped, very low-refractive-index semiconductor layer. This waveguide is optimized for a $\lambda = 8.5\,\mu$m wavelength

the waveguide losses and provides low-loss optical confinement. The following topmost highly doped InGaAs layer plays a crucial role in suppressing the coupling between the fundamental waveguide mode and the surface plasmon mode propagating along the metal contact–semiconductor interface. Its doping level is high enough ($7 \times 10^{18}\,\mathrm{cm}^{-3}$) that the plasma frequency approaches but does not exceed that of the waveguide mode at $8.5\,\mu$m. At such a high doping level the advantage of the anomalous dielectric dispersion near the plasma frequency can be fully explored. The resulting decrease of the refractive index from $n_{\mathrm{InGaAs}} \sim 3.49$ to 1.26 enhances the confinement factor of the mode. The confinement factor is calculated to be 0.51, the effective refractive index of the mode is $n_{\mathrm{eff}} \approx 3.27$ and the calculated waveguide loss is $\alpha_{\mathrm{w}} \approx 8.7\,\mathrm{cm}^{-1}$. Calculations show that the loss would be significantly increased ($15.1\,\mathrm{cm}^{-1}$) and the confinement factor would be reduced to 0.44, if the plasmon-enhanced confinement layer were eliminated.

So far, the discussion has been focused on the layer structure in the growth direction. In this direction the waveguide is usually in a single mode because even if a higher-order transverse mode exists for shorter wavelengths, it has such a high loss and low confinement factor that it will not be excited during laser operation.

In what follows the in-plane direction waveguide is examined. Conventional semiconductor lasers are often processed as entirely planar structures, where waveguiding is provided only by current flow from a top stripe contact into the large-area bottom contact. Such "gain-guiding" works, in principle, also for QCLs. However, such a waveguide is very inefficient due to a very large in-plane current spreading aided by a high turn-on bias and large differential resistance (about few 10Ω to few $10\,\mathrm{k}\Omega$) below its turn-on bias. Hence, the preferred QCL waveguides are etched to various depths to encourage efficient current and light confinement. The resulting stripe widths can range from a few to several tens of micrometers depending on the application. Qualitatively, the wider the stripe is the larger the output power. Nevertheless, too large stripes often allow excitation of higher-order transverse modes.

To achieve the best performance, QC lasers are commonly processed into narrow, deep-etched ridges for strong current confinement and efficient heat removal as a result for a larger surface-to-volume ratio. The narrow ridges are first passivated by thin (few 100 nm thick) layers of SiN, SiO or ZnSe, and then overlaid with a contact metallization (Faist et al. 2000). Another approach is using $Ge_{0.25}Se_{0.75}$ chalcogenide glass as a thick sidewall coating and burying layer for the narrow laser ridges (Gmachl et al. 2001). Chalcogenide glasses are well known to be low-loss mid-IR materials with their refractive indices varying around 2.3, and hence they allow strong optical confinement.

Distributed feedback (DFB) waveguide lasers have found important application in particular cases, e.g., trace gases detection, due to its advantage of having a narrow-linewidth, a tunable wavelength range, a high sensitivity and specificity, usually combined with the advantages derived from its compactness and the robustness of a packaged semiconductor laser device. To provide the required narrow-band emission, however, a simple Fabry–Perot (FP) type resonator described above is not sufficient. The typically 1–3 mm long cavity leads to a spacing of the FP modes which are narrow compared with the width of the gain spectrum. This causes lasers to operate in multiple modes under ordinary conditions, certainly in pulsed and also in cw operation at high current levels. To sustain a single-mode output with a good side-mode suppression ratio and a controllable and tunable wavelength over a certain range, QC lasers are fabricated as DFB lasers incorporating a first-order Bragg grating. They were first demonstrated in 1996, and have very rapidly evolved and have already shown great promise in many different gas-sensing applications (Faist et al. 1997b). In QC lasers, repeated scattering from a Bragg grating, incorporated into the waveguide, favours one wavelength (the Bragg wavelength) and therefore it is the grating period rather than the laser gain spectrum that determines the single-mode emission wavelength.

In fact, the waveguide of a conventional QCL provides several possible ways to produce a grating modulation strong enough to produce the necessary wavelength selective feedback. One grating type is etched into the surface of the waveguide. It is relatively easy to fabricate gratings in the top-cladding layer, and it shows good performance, however, further improvement of the performance is ultimately restricted because the gratings are located at the exponential tail of waveguide mode. To overcome this insufficiency, another grating type is proposed positioned close to the active waveguide core (Gmachl et al. 1997). In a first growth cycle by MBE, the active waveguide core is grown, which consists of a several 100 nm thick InGaAs layers followed by the stack of many ($\sim$30) periods of alternating active regions and injectors and capped by another $\sim$500 nm thick InGaAs layer. The wafer is then removed from the growth chamber, and the Bragg grating is fabricated into the upper InGaAs layer using a conventional technique. The wafer is then transferred back into another growth chamber, where an InP top cladding is grown on top of the Bragg grating using solid-source MBE. Qualitatively, a large coupling coefficient and a strong modulation of the effective refractive index of the waveguide is due to the refractive index contrast between InGaAs and InP and the strong overlap of the grating constraint with the cavity mode.

Although this procedure that involves two growth cycles increases the fabrication effort, Gmachl et al. have successfully demonstrated QC-DFB lasers with buried gratings make the best single-mode laser devices fabricated to date (Gmachl et al. 2000).

6.5.4 Antimony Based Semiconductor Mid-Infrared Lasers

For the wavelength region over $2\,\mu$m, it is possible to use antimony-based III–V semiconductor alloys in fabricating conventional diode lasers. Other material systems-based semiconductor lasers commercially available operate normally in the wavelength range of 0.6–$2.0\,\mu$m. This wavelength limitation can be overcome by using the aforementioned novel QC designs.

Before the 1990s, the development of antimony-based lasers relied mainly on an LPE technique (Choi and Turner 2000). Bochkarev et al. had demonstrated using this technique to fabricate a double heterostructure (DH) laser, that had GaInAsSb active regions and AlGaAsSb cladding layers and operated at $2.3\,\mu$m (Bochkarev et al. 1988). At RT and in a cw operating mode, its threshold current density was as low as $1.5\,\mathrm{kA/cm^2}$. To improve further its performance and extend its wavelength to longer regions, it is crucially important to significantly reduce the Auger recombination rate. This requires a drastic change to the device structure, e.g., the use of a quantum well structure. However, the LPE technique cannot fulfill this requirement with ease.

On the other hand, MBE and MOVPE techniques have evolved since 1990s to enable a solid foundation for promoting the performance of Sb-based lasers. In 1992, Choi et al. demonstrated the first type-I GaInAsSb/AlGaAsSb QW laser (Choi and Eglash 1992), operating at $2.1\,\mu$m and having a threshold current density of only $260\,\mathrm{A/cm^2}$ at RT. After that, different QW laser structures covering a wavelength range of 1.9–$4.5\,\mu$m have been subsequently developed with InAsSb/InAlAsSb and InAsSb/InAsP material systems by using MBE or MOVPE techniques. In addition, type-II lasers have also made rapid progress. Comparatively, type-II lasers (Zhang 1995; Meyer et al. 1995), especially cascade lasers (Youngdale et al. 1994; Zhang et al. 1997), have two distinct advantages, their working wavelengths easily overcome bandgap limitations to function at longer wavelengths and the Auger recombination can be effectively diminished. As already demonstrated, for the wavelength range of 1.9–$3\,\mu$m, GaInAsSb/AlGaAsSb QW lasers manifest the best performance (Garbuzov et al. 1997), and good performance is also obtained by replacing the AlGaAsSb barrier with GaSb (Baranov et al. 1996). For wavelengths longer than $3\,\mu$m, InAsSb is used for the active region; however, its operational temperature is far below RT due to fast Auger recombination (Choi et al. 1994; Lane et al. 1997). In addition, for the wavelength range of 2.7–$5.2\,\mu$m, type-II GaInSb/InAs superlattice or quantum well laser is also a choice, It shows excellent pulse-mode performance with light pumping. Its highest operational temperature is $350/310\,\mathrm{K}$ for wavelengths of $3.2/4.5\,\mu$m, respectively. Nevertheless, a long-existing problem

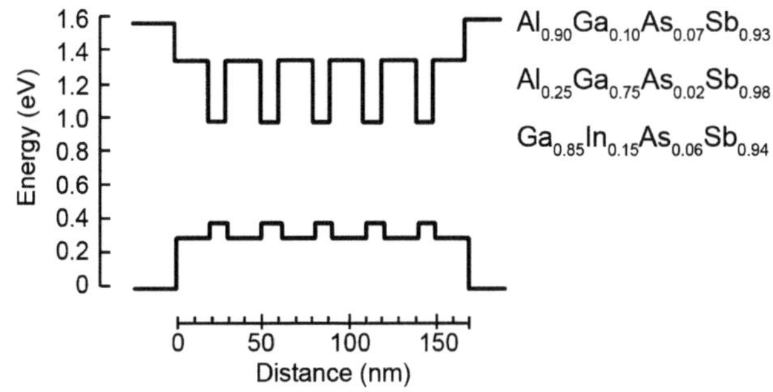

Fig. 6.70 The active region band structure for a $Ga_{0.86}In_{0.14}As_{0.05}Sb_{0.95}/Al_{0.25}Ga_{0.75}As_{0.02}Sb_{0.98}$ QW laser. The lasing wavelength is 1.9 μm (Choi and Turner 2000)

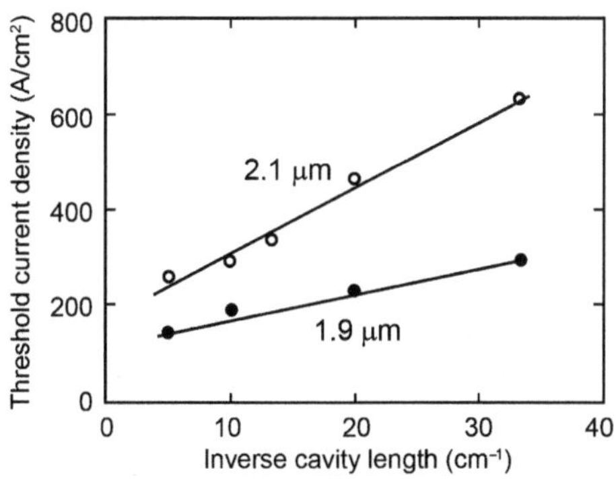

Fig. 6.71 Threshold current density as a function of reciprocal cavity length for a pulsed-mode laser working at room temperature. The *solid dots* are experimental results for the laser illustrated in Fig. 6.70; *circles* are results for a laser operating at 2.1 μm reported by Choi and Eglash (1992)

with the type-II diode lasers is that its real performance is much lower than that with light pumping (Choi and Turner 2000).

In what follows, a brief introduction is directed to GaInAsSb/AlGaAsSb QW lasers.

Figure 6.70 plots a theoretical calculational result for the near active region bandedge structure for a $Ga_{0.86}In_{0.14}As_{0.05}Sb_{0.95}/Al_{0.25}Ga_{0.75}As_{0.02}Sb_{0.98}$ QW laser grown by MBE on a GaSb substrate (Choi et al. 1994). First, an n^+-GaSb buffer layer is grown, followed by a 2-μm n-$Al_{0.9}Ga_{0.1}As_{0.02}Sb_{0.98}$ cladding layer, then an active region consisting of 5-, 10-nm-$Ga_{0.85}In_{0.16}As_{0.06}Sb_{0.96}$ QW layers, and 6-, 20-nm-$Al_{0.25}Ga_{0.75}As_{0.02}Sb_{0.98}$ barrier layers, is grown, followed by a 2-μm p-$Al_{0.9}Ga_{0.1}As_{0.02}Sb_{0.98}$ cladding layer, and a 0.05-μm p^+-GaSb electric contact layer. Calculations reveal that a valence-band discontinuity of about 90 meV exists between the QWs and barriers in the active region. The corresponding conduction-band discontinuity is about 350 meV.

It is noteworthy that this structure is similar to the first type-I GaInAsSb/AlGaAsSb QW laser demonstrated by Choi and Eglash (1992), with the only exception being that the composition of QW/barrier layers for the two structures is slightly different. In Fig. 6.71, solid dots show the cavity-length-dependent threshold current density $J_{\rm th}$ of a pulsed mode at room temperature. Circles on the other

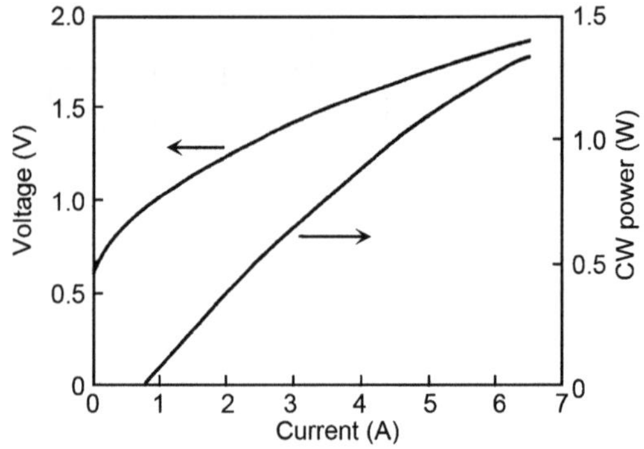

Fig. 6.72 The output power/voltage versus current for a 1,000-μm-long, 200-μm-wide cw-mode laser with a heat sink temperature set at 12 K (Choi and Turner 2000)

hand, represent the results of the first type-I GaInAsSb/AlGaAsSb QW laser by Choi and Eglash (1992). It is clear that $J_{\text{th}} = 143\,\text{A/cm}^2$ at $L = 2\,\text{mm}$, which corresponds to $29\,\text{A/cm}^2$ per QW layer. This is equivalent to just half of the minimum value of the InGaAs/AlGaAs laser operating at $\sim 1\,\mu\text{m}$ (Choi and Turner 2000). As L is reduced, J_{th} increases gradually. For $L = 300\,\mu\text{m}$, J_{th} reaches a value of $280\,\text{A/cm}^2$.

Figure 6.72 plots output power/voltage change with current for a $1,000$-μm-long, 200-μm-wide cw-mode laser, with its heat sink temperature set at 12 K. The front side and back side of the laser are processed by coatings to avoid oxidation of the Al. The corresponding reflectivities of the two sides are 4 and 95%, respectively. The threshold current is about 650 mA. The initial slope efficiency is about 0.3 W/A, corresponding to a differential quantum efficiency of 47%. The maximum output power is 1.3 W. The turn-on voltage of the laser is about 0.6 V, nearly equal to the bandgap energy of the active-region quantum wells. The effective series resistance is about 0.3 Ω at a low driving current and decreases to 0.1 Ω at a high driving current.

Based on type-II structures, many improvements have been made to significantly enhance laser performance. For example, Garbuzov et al. have employed a lightly doped wide-waveguide layer and an enhanced doping density p-type cladding layer (Garbuzov et al. 1996), resulting in a reduction of innerband loss and effective series resistance. By using a single quantum well structure (Garbuzov et al. 1997), the threshold current is further reduced. Lee et al. have successfully extended the operating wavelength to 2.78 μm by increasing In and Sb compositions of the GaInAsSb quantum wells. To achieve a low threshold current and good quality light beam, Choi et al. have facilitated a ridge-waveguide laser structure (Choi et al. 1993). The pulsed threshold current at RT is as low as 29 mA for the demonstrated 8-μm-wide ridge-waveguide laser, operating at 2.1 μm, and with a cw maximum output power iof 28 mW.

To realize single-frequency operation, York et al. have fabricated a ridge-waveguide distributed Bragg reflection laser (Choi and Turner 2000), as illustrated in Fig. 6.73. In the gain region, a 5-μm-wide ridge is made, while in the grating region, the ridge and first-order gratings are fabricated at the same time. The highest

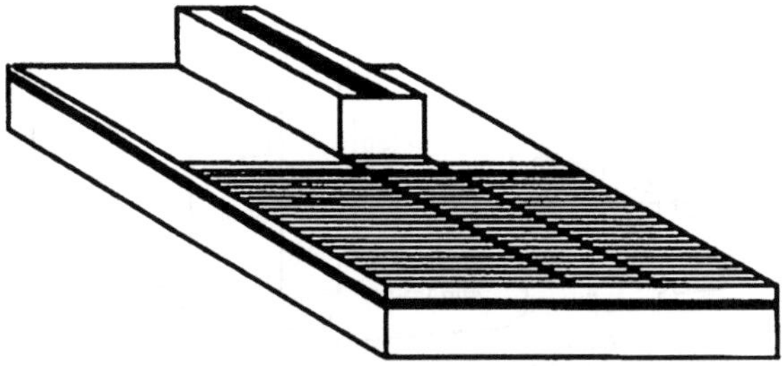

Fig. 6.73 Schematic of a ridge-waveguide distributed Bragg reflection laser. The portion of the structure with gratings is not pumped (Choi and Turner 2000)

operating temperature in a pulsed mode is 37 °C. The threshold current is about 270 mA. Note that, it can work in single longitudinal mode, and the output wavelength is tunable in the range of 1.966–1.972 μm, with a temperature dependence of 0.16 nm/K.

6.5.5 Interband Cascade Lasers

In contrast to intersubband cascade lasers, interband cascade lasers are based on interband optical transitions between conduction-band electrons and valence-band holes. Its cascade feature is realized by using type-II heterostructures. With the aid of strong real-space inter- and intra-band tunneling, electrons entering the conduction band of an initial injector stage, tunnel into the conduction band of the next QW, then make a radiative intraband transition to a state in the valence band, and finally intraband tunnel into the following adjacent stage conduction band. Theoretically, the emitted wavelength of an interband cascade laser can be much longer than that of type-I lasers and hence extend into the far-infrared region, thanks to their type-II heterostructure. For example, Zhang et al. have reported electro-luminescence at 15 μm for an interband cascade laser at room temperature (Zhang et al. 1997).

Figure 6.74a schematically illustrates the electronic band structure of a type-II InAs/AlSb/GaSb QW interband transition QCL (Yang 2000). Each active region consists of one InAs/AlSb QW and one GaSb/AlSb QW. The band-structure design of the two QWs ensures simultaneously that: (1) the ground level of the InAs QW's conduction band is energetically higher than the top of the GaSb QW's valence band, and (2) the ground level of the GaSb QW's valence band is energetically higher than the bottom of the InAs QW's conduction band. As heavy holes are localized in the GaSb QW, the GaSb layer must be thin enough to warrant a sufficient overlap of the electron and heavy-hole states wave functions. On the other hand, thin GaSb layers cannot effectively stop electron leakage.

To solve this conflict, the E_{h} in Fig. 6.74a can be designed to be a light-hole level. The light-hole states can easily tunnel into InAs layers and form a strong coupling to the conduction band there. This provides a possibility to optimize simultaneously

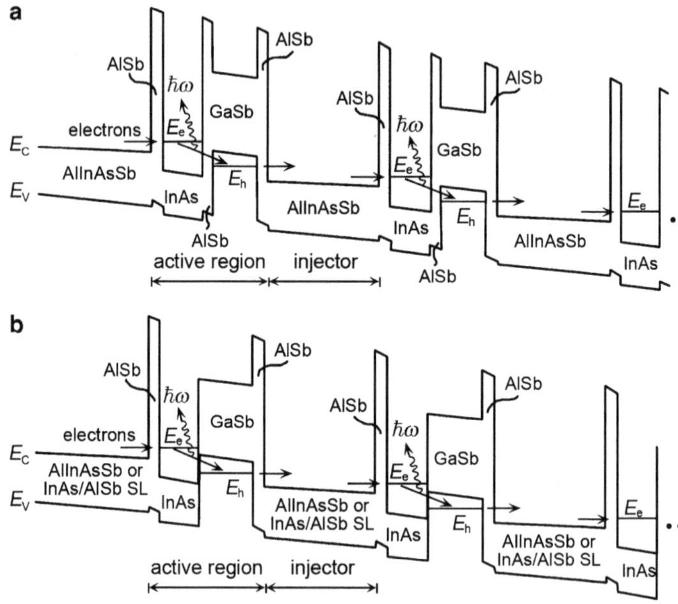

Fig. 6.74 (**a**) The band structure of a type-II InAs/AlSb/GaSb interband-transition QW cascade laser; (**b**) strong real-space interband coupling in type-II QW structure (Yang 2000)

the two aspects of stopping electron leakage and enhancing the efficiency of electron injection. A high efficiency of the production of a population inversion is accessible due to the high intraband-tunneling rate of light holes. Such an interband-transition design effectively eliminates nonradiation relaxation process caused by intersub-band phonon scattering. At the same time, as a heavy-hole state is localized in a relatively thick GaSb layer, it only slightly overlaps with an electron state, so the transition probability between them is very low. In turn, the Auger recombination is effectively diminished, which contributes a dominant portion to the nonradiative loss of conventional interband-transition infrared lasers. Grein et al. (1994) and Flatte et al. (1995) theoretically indicate that two choices may lead to a reduction of the Auger recombination in Sb-based type-II QW structures.

One is band engineering in energy-wavevector space, with which the emission efficiency can also be enhanced. The other is real-space wavefunction engineering, with which the performance can be optimized for type-II electron–light-hole interband transition QCL.

Optical transitions are mainly determined by real-space interband coupling between two adjacent QWs. By adjusting the coupling, i.e., changing the thickness of AlSb barrier between the InAs and GaSb QWs, the gain spectrum can also be controlled. In fact, this barrier layer can be canceled, as illustrated in Fig. 6.74b, to maximize the real-space overlap of the wavefunctions. The emitted wavelength determined by E_e and E_h can be tuned from the mid-infrared to the far-infrared ($\sim$100 μm), by adjusting InAs and GaSb QW layer thicknesses.

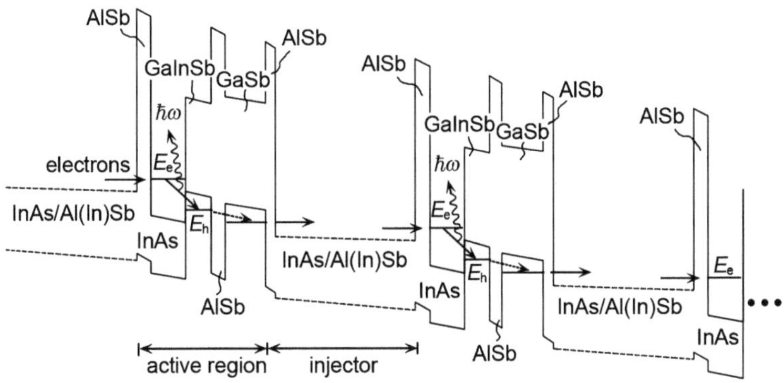

Fig. 6.75 The band structures for a type-II InAs/AlSb/GaInSb QCL. Inter-well electron–heavy-hole transitions are illustrated (Yang 2000)

Another choice is retaining the electron–heavy-hole recombination mode, but employing a valence-band-coupled double QW structure, as depicted in Fig. 6.75. The first QW layer is thin enough GaInSb, to establish a ground heavy-hole E_h state in the QW that is sufficiently close to one in the real-space InAs QW layer, and at the same time, increasing the separation between the valence band sublevels. An electron–heavy-hole optical transition efficiency is thereby obtained, and the Auger recombination loss is reduced (Youngdale et al. 1994). The second QW layer is a relatively thick GaSb, which is adjacent to the next stage injector. Its light-hole ground state level can be energetically higher than the bottom of the injector conduction band by selecting suitable thicknesses. The advantages gained are: (1) the electron leakage through the InAs QW layer can be well blocked, leading to a nearly 100% injection efficiency; and (2) the light-hole state provides carriers a fast channel for transporting via interband tunneling from a GaSb QW layer to the next stage injector. As the interband relaxation time is much longer than the time for carriers to pass through the active region, a population inversion is therefore quickly established.

It is noteworthy that the aforementioned device structure is an electron injection type, its lasing action is realized through injection of electrons into its n-doped injector. In practice, several structural variations, including a p-doped injector and an n-doped and a p-doped coexisting injector, have been demonstrated (Yang 2000).

Figure 6.76 plots emission spectra measured at 80 and 120 K, respectively, of the type-II InAs/AlSb/GaInSb interband cascade laser schematically illustrated in Fig. 6.76. The laser is prepared by MBE on a p-doped GaSb substrate, which consists of 20 periods of active/injector regions. The injector is InAs/Al(In)Sb multilayer structure, with the InAs being n-type Si-doped (6×10^{17} cm^{-3}), and an undoped Al(In)Sb. The whole multilayer structure is strain balanced, and lattice-matched to the substrate. The active region, on the other hand, has a double QW structure, the thickness of each AlSb/InAs(QW)/$Ga_{0.7}In_{0.3}Sb$/AlSb/GaSb(QW) layer is sequentially 23, 25.5, 34, 15, and 53 Å. It is clear that the peak of

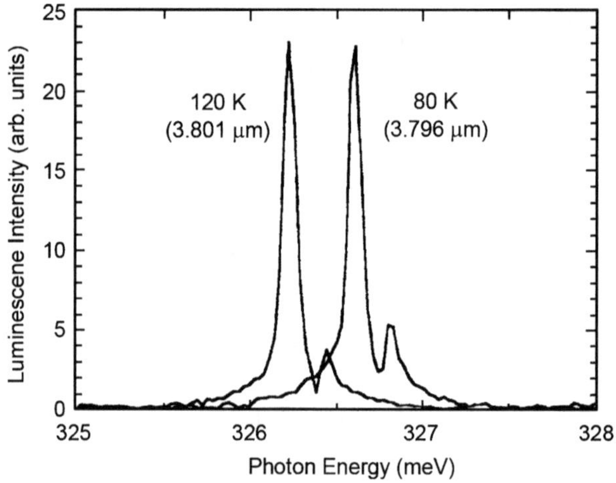

Fig. 6.76 The emission spectra of a type-II InAs/AlSb/GaInSb interband-transition QCL (Yang 2000)

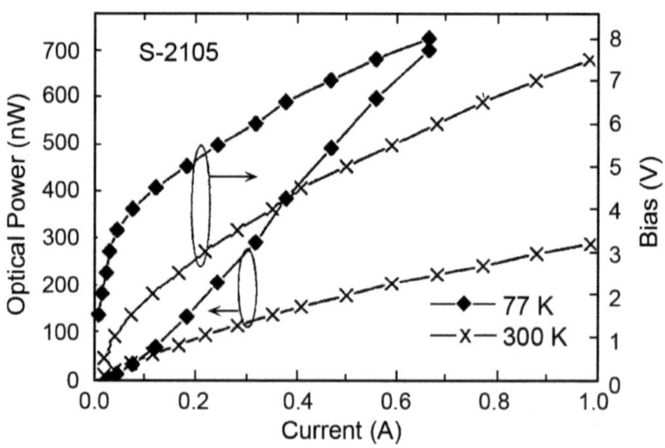

Fig. 6.77 Output optical powers/voltages versus injection current for a type-II InAs/AlSb/GaInSb QW interband transition quantum cascade emitting diode, *horizontal arrows* point to the appropriate vertical axes (Yang 2000)

the emission spectrum occurs at 326.2 meV ($\lambda = 3.801\,\mu$m) at 120 K, and at 326.6 meV ($\lambda = 3.796\,\mu$m) at 80 K. This is because as the temperature increases, the bandgap decreases, and the coupling window between the InAs and GaSb QWs is widened. The real-space overlap between transition-involved two energetic states' wavefunctions is therefore enhanced (Lin et al. 1997).

For emitting wavelengths exceeding 5 μm, it is beneficial to use an LED structure to study carrier transport and emission spectra, so as to avoid long material preparation times and complicated laser structure fabrication. In Fig. 6.77, the output optical power as a function of injection current is illustrated for a type-II InAs/AlSb/GaInSb interband cascade LED with 15 periods of active/injector regions (Yang et al. 1997), measured at 77 and 300 K, respectively. For the 77 K and a 5 V forward bias, the electro-luminescence (EL) peak occurs at 6.25 μm, and its FWHM is about 48 meV. For a forward bias voltage of 6.5 V at room temperature, the peak occurs at 7.3 μm, with an FWHM of 49 meV. At 77 K, the output optical power is nearly a linear function of the drive current. The onset of EL occurs at 3–4 V, thereafter, the current increases more rapidly with bias voltage, as does the EL. As $T = 300$ K, the output power shows an obvious tendency to saturate as the current is increased. This can be ascribed to a heating effect, and to a thermally induced leakage enhancement.

Note that for a particular current, the output power manifests a weak temperature dependence, e.g., for a drive current of 600 mA, the output power only decreases a factor of about 3 as the temperature increases from 77 to 300 K. This is a common phenomenon for type-II interband cascade LEDs, and is obviously superior to the conventional interband transition LED. The physics behind this is that Auger recombination, which exponentially increases with temperature in conventional interband transition mid-IR lasers, may be suppressed in the type-II QW interband transition QCLs (Zegrya and Andreev 1995). On the other hand, the output power of this cascade LED is significantly higher than that of intersubband cascade LEDs (Sirtori et al. 1995a; Scamarcio et al. 1997a), indicating the fact that its radiation efficiency is improved.

Figure 6.78 shows a schematic band structure of a type-II InAs/GaInSb/Al(In)Sb QW interband transition QCL. The numbers near the bottom of the figure denote layer thicknesses in Å. The composition of the ternary alloy layers are $Ga_{0.7}In_{0.3}Sb$ and $Al_{0.7}In_{0.3}Sb$, respectively. The laser consists of 23 periods of active/injector regions with a total thickness of about 1.6 μm, in which the injectors are n-doped

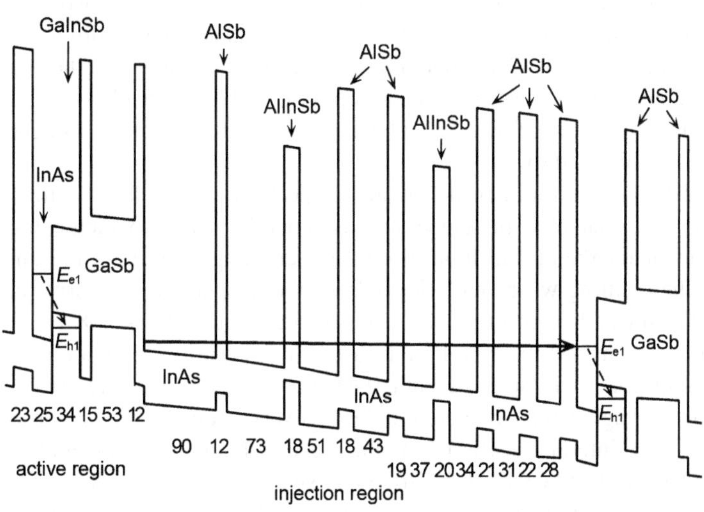

Fig. 6.78 A schematic electronic band structure for a type-II InAs/GaInSb/Al(In)Sb QW interband transition QCL. The numbers near the bottom of the figure indicate layer thicknesses (in Å). The compositions of the ternary alloys are $Ga_{0.7}In_{0.3}Sb$ and $Al_{0.7}In_{0.3}Sb$, respectively (Yang 2000)

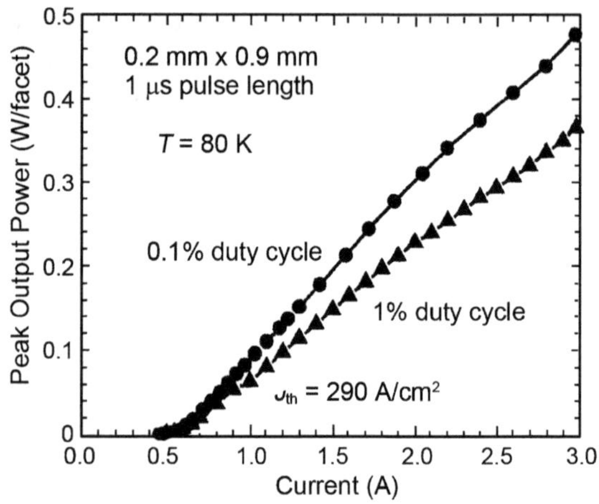

Fig. 6.79 Peak output optical power versus current for a gain-guided InAs/GaInSb/Al(In)Sb QW interband transition QCL driven by 1 and 10 kHz pulse currents, respectively, at 80 K. The laser is 0.9-mm long and 0.2-mm wide. *Dots* are for a 1 kHz pulse current, and triangles for a 10 kHz pulse current (Yang 2000)

multiple QWs. Its top and bottom waveguide cladding layers are 1.6 and 2.0-μm-thick n-doped 24.3-Å-InAs/23-Å-AlSb superlattices, respectively. As a transition region, a digitally graded multi-QW region is inserted into the interspaces between the waveguide core and the cladding layer and between the cladding layer and the electrical contact layer. The theoretical confinement factor of the waveguide is 0.73.

In Fig. 6.79, the output optical power as a function of drive current is depicted for a 0.9-mm-long and 0.2-mm-wide gain-guided laser in pulsed mode at a temperature of 80 K. The repeat frequency of the drive current is 1 and 10 kHz, respectively, with corresponding duty cycles of 0.1 and 1%.

It is clear that for the duty cycle of 0.1%, the peak output power can be as high as 0.5 W per facet, with a slope of $\mathrm{d}P/\mathrm{d}I = 311\,\mathrm{mW/A}$. If the two facets of the laser can be assumed to be identical, the corresponding external differential quantum efficiency is 131%. Correspondingly, for the duty cycle of 1%, the peak output power is about 0.37 W per facet, with a slope of $\mathrm{d}P/\mathrm{d}I = 155\,\mathrm{mW/A}$. The external differential efficiency is 96%. This phenomenon is attributed to a stronger thermal effect for the higher duty cycle.

The threshold current density versus temperature of this laser is illustrated in Fig. 6.80. The characteristic temperature is about $T_0 = 81$ K, which is higher than conventional interband transition diode laser emitting at 3 μm (Choi and Turner 1995). The emitting wavelength depends on the temperature, the drive current as well as the threshold current density, and is in the range of 3.85–3.98 μm. However, in contrast to conventional type-I diode lasers, the threshold current of a type-II QW interband QCLs is not improved, but is even worse than the type-I lasers. The reason is that the leakage current related to defects is very strong, reflecting stronger contributions from interface scattering, tunneling, and thermal electron emitting for the type-II QW interband QCLs. It is difficult to significantly reduce these mechanisms by promoting material quality and/or improving device design. In response to this influence, an applicable measure may be to combine the concept of a cascade with a conventional structure (Yang 2000).

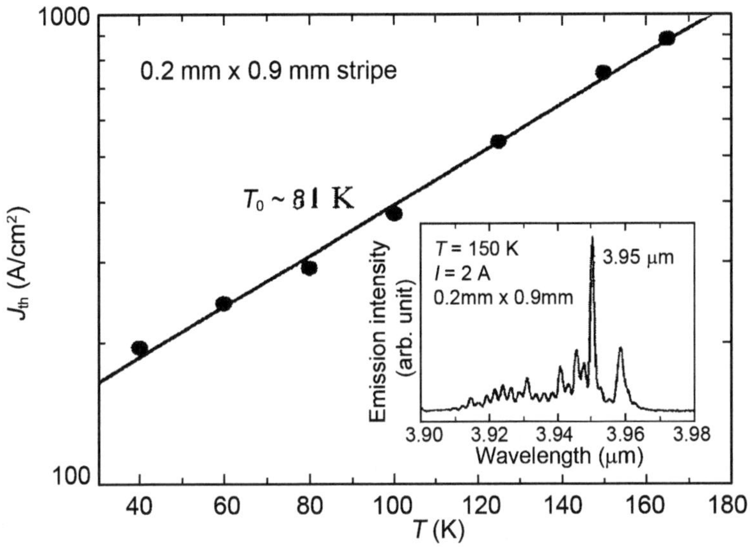

Fig. 6.80 The threshold current density versus temperature of the heat sink for a gain-guided InAs/GaInSb/Al(In)Sb QW interband transition QCL. The inset shows a highly resolved emission spectrum at 150 K (Yang 2000)

6.5.6 Applications of Quantum Cascade Lasers

In recent years, the research and development of 3–5 μm and 8–13 μm novel lasers have drawn worldwide interest. The lasers possess crucial application potentials in the areas of environment pollution monitoring and environment security control of industrial processes as well as construction of free-space wireless optical communication networks. Because the QCLs based on low-dimensional semiconductor structures operate in the mid- and far-infrared regions, they have found successful applications in special areas, such as optical detection of trace gases.

Infrared laser spectroscopy is a powerful tool in monitoring trace gases in air. QCLs are recognized as desirable light sources for this spectroscopy. Because trace gas characteristic features in their absorption spectra are caused by molecular rotation and vibration transitions, they are located in the mid-infrared region, in general $\lambda \sim 3.5$–$13\,\mu$m. They can be qualitatively or semi-quantitatively studied by using narrow linewidth tunable semiconductor lasers. The lasers, on the other hand, can operate both at in a cw mode at liquid-nitrogen temperature, and in a pulsed mode near room temperature.

Namjou et al. (1998) have for the first time measured mid-infrared ($\lambda \sim 7.8\,\mu$m) absorption spectra of nitrogen samples containing N_2O and CH_4 at a temperature close to room temperature. The noise equivalent sensitivity is about 5×10^{-5}.

Sharpe et al. (1998) have performed direct measurements of NO and NH_3 absorption at low temperature with 5.2 and 8.5-μm cw DFB QCLs by using a high-resolution rapid scan technique. The drive current of the lasers is a saw-shaped impulse with a repeat frequency in range of 6–11 kHz. With such a drive current, the laser linewidth can be tuned to $2.5\,cm^{-1}$, and hence covers about 11 characteristic spectroscopic lines. The noise equivalent absorption is as low as 3×10^{-6}.

Kosterev et al. (1999) have detected different isotopes of methane (CH_4, CH_3D, $^{13}CH_4$) at liquid-nitrogen temperature with an 8.1-μm cw DFB QCL. The measurement accuracy is ± 0.5 ppm for a density of 15.6 ppm, and the minimal detectable absorption is about 1×10^{-4}.

Williams et al. (1999) have measured the intrinsic linewidth of several QC DFB lasers operating at about 8 μm, by identifying the fluctuations of laser intensity while traveling through an N_2O gas cell. The results indicate that the linewidth is smaller than 1 MHz in an integration time of 1 ms. By controlling the drive current with a feedback detector signal, the laser electrical status can be further stabilized, and the linewidth is therefore reduced to below 20 kHz.

Paldus et al. (1999) have employed an 8.5 μm cw quantum cascade DFB laser as the source in measuring photo-acoustic (PA) spectra of dilute NH_3 and H_2O in nitrogen. The PA spectrum is based on the absorption of incident light by trace gases, and therefore can achieve very high sensitivity when a high power source is used. In fact, Paldus et al. have obtained a sensitivity of 100 ppbv in detecting NH_3 for a integration time of 1s. Paldus and Oomens et al. (Paldus et al. 2000) have further employed a QC DFB laser in a cavity-ring-down (CRD) spectroscopic measurement of dilute NH_3 in nitrogen, and reached a noise equivalent sensitivity of $3.4 \times 10^{-9}\,cm^{-1}\,Hz^{-1/2}$, which is equivalent to a detection limit of 0.25 ppbv.

6.6 Single-Photon Infrared Detectors

6.6.1 Introduction

In the preceding sections, we have discussed conventional infrared detectors and low-dimensional infrared detectors. However, neither of these device classes is capable of measuring extremely low light intensities, such as a single photon. Sticking to the traditional photon description of light, a single photon creates an e–h pair in a photodiode. This induces a current that lies well below the thermal noise of the diode, and thus is not measurable. In fact, typical signal values for a modest bandwidth are of the order of 1 pA, which corresponds to approximately 6.5 million photons per second at a 633 nm wavelength – quite far from single photon resolution.

A single photon detector requires some amplification mechanism, as noiseless as possible, which per primary generated electron or e–h pair, generates a secondary pulse of electrons consisting of at least 10^5 electrons per pulse. If the secondary pulse is abrupt, the resulting current can be quite large and the pulse becomes easily detectable by additional electronics.

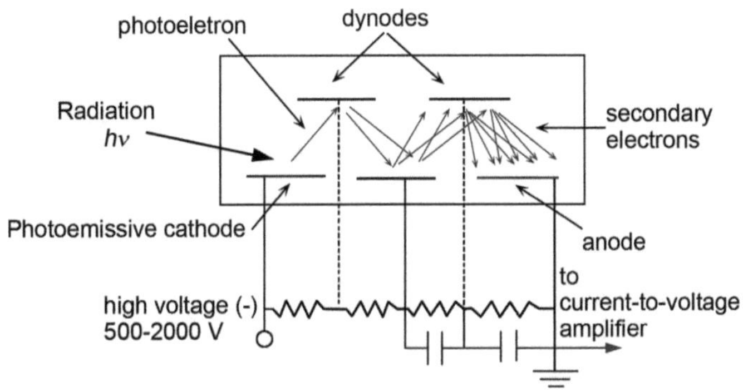

Fig. 6.81 Schematic structure of a photomultiplier tube

A typical member of the family of internal gain photodetectors is the photomultiplier tube (PMT). Its structure is schematically illustrated in Fig. 6.81. It is a vacuum device which has a photosensitive area, the photocathode, and a number of "dynode stages" that multiplies the number of electrons generated by the initial photon. This is accomplished when the incident photon has enough energy to excite an electron from the photocathode, then this electron is accelerated by a rather strong electric field to gain sufficient energy so when it collides with the first dynode it releases on average 3–6 secondary electrons. The number of secondary electrons emitted depends mainly on the dynode material and electric field strength used. These secondary electrons will be accelerated in turn do the same thing on successive dynode stages. It does not require much algebra to prove that an average amplification on the order of 10^6 is easily achieved after 8–14 stages. Since the flight time of the electrons within the tube is very short, as is the electron release time from the photocathode, which can be viewed as being instantaneous, a pulse width of the order of a few 10 ns is generated. This current pulse is relatively high (a few 10 μA) and thus is easy to post-amplify.

However, as soon as it is used the drawbacks of this single-photon detector PMT mode become clear. First, its quantum efficiency is reasonably small. Since the primary electron must be physically released from the photocathode, the photocathode cannot be made arbitrarily thick – however, a thin layer does not allow a high photon capture probability – the photocathode is semitransparent. Usually the quantum efficiency is no better than 20% in the UV/blue wavelength regime, and is below 5% in the red wavelength regime. Additional problems with all PMTs arise from two effects, a recharging lag of the last few dynodes in the chain, which are the ones emitting quite a high number of electrons in a short time, and after pulses, "phantom" pulses generated some time after the primary photon-induced pulse was generated. Usually these are caused by residual gas atoms that get ionized when traveling through the electron "cloud" between the last dynode in the chain and the anode of the PMT, then recombine and release photons in doings so.

An avalanche photodiode (APD) is another member of the family of internal gain photodetectors. It is a PIN diode operated in reverse bias. In an ideal PIN diode, each

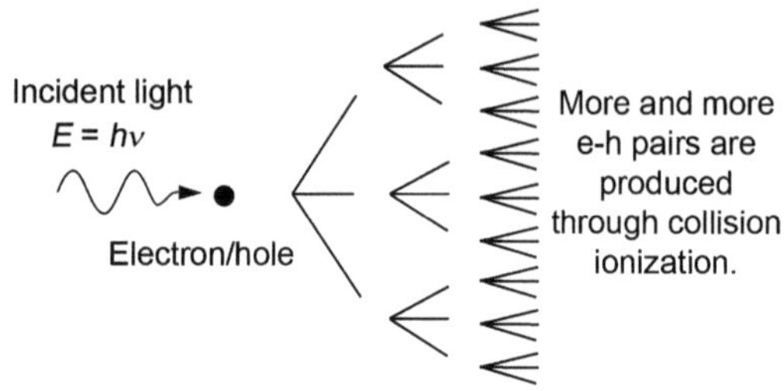

Fig. 6.82 A schematic of an avalanche induced by collision ionization by high kinetic energy electrons

incident photon creates an e–h pair. Through a considerable increase of the reverse voltage across the p–n junction, the field strength can become so strong that the charge carriers accelerated in the field gain sufficient kinetic energy to excite another electron from the valence band into the conduction band and still pass through the depletion region. The new e–h pair thus produced can in turn excite still more e–h pairs. In this way, a carrier avalanche effect occurs. The process is schematically illustrated in Fig. 6.82. The photo-induced signal is therefore amplified.

An APD has two operational modes, an analog mode (AM) and a Geiger mode (GM), in which the internal gain is significantly different. For the AM, the gain is normally only of the order of magnitude of 10^2 and hence cannot be used for single-photon detection. The GM is reached when the diode is operated well above its breakdown voltage. The distinctive feature of the GM mode is that nothing happens as long as there is no free electron, while an avalanche of any free electrons will be quickly generated, to produce a large current pulse, even larger than that of a dynode chain generated electron pulse of a PMT. As soon as the first electron appears, and once the avalanche is launched, the APD stays in a conducting mode forever. Given this feature, the GM APD is literally a single photon detector, but it can only work to detect another photon after the avalanche is quenched in some manner.

In contrast to a PMT, the quantum efficiency of an APD depends on its electron multiplication behavior, and is very high. An overall quantum efficiency of 70% at 633 nm is reachable for a modern APD, and it still shows more than 20–30% in the near IR and in the blue wavelength regime.

In this section, APDs and GM APDs will be briefly discussed.

6.6.2 Fundamentals of an APD

The generation of e–h pairs as well as the current gain of an APD relies on the impact ionization coefficient, which is defined as the average number of e–h pairs created by a carrier per unit distance traveled. It has units of cm^{-1}. Generally, the ionization coefficient for electrons, α_n, is different from that for holes, β_p, their ratio is called the ionization coefficient ratio, $\kappa = \beta_p/\alpha_n$. Its value as well as electric field dependence is determined by carrier scattering and the p–n junction's structure, and has an obvious influence on low-frequency avalanche gain and gain-bandwidth product.

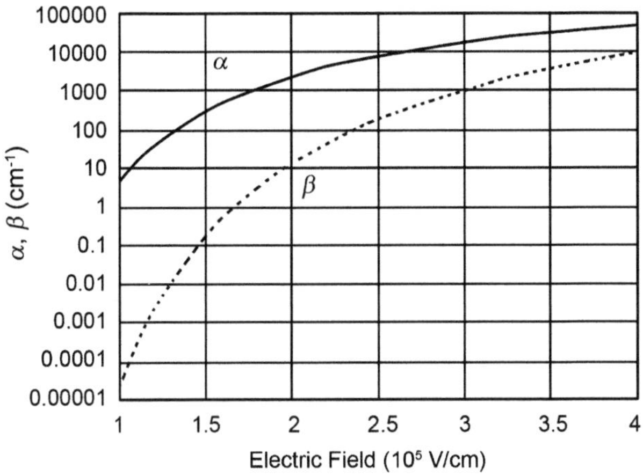

Fig. 6.83 Carrier ionization coefficients as a function of electric field for a silicon APD, α is for electrons, and β is for holes

Figure 6.83 shows the electric field dependence of the ionization coefficients for a silicon APD, where α is the value for electrons and β for holes (Musienko et al. 2000).

The avalanche gain process for APDs can be quantitatively described by analyzing the high-field depletion region. In Fig. 6.84a the direction of the electric field and the electron- and hole-current ($J_n(x)$, $J_p(x)$) are illustrated. In this figure W is the width of depletion region, and $\alpha_n(x)$ and $\beta_p(x)$ are the ionization coefficients at x for electrons and holes, respectively. The spatial distribution of the electric field is not considered. The $J_n(x)$ and $J_p(x)$ as functions of x are determined by (Stillman and Wolfe 1977):

$$\frac{\mathrm{d}}{\mathrm{d}x} J_n(x) = \alpha_n(x) J_n(x) + \beta_p(x) J_p(x) + qG(x)\,, \tag{6.264}$$

$$-\frac{\mathrm{d}}{\mathrm{d}x} J_p(x) = \alpha_n(x) J_n(x) + \beta_p(x) J_p(x) + qG(x), \tag{6.265}$$

where q is the electron charge, and $G(x)$ is the photon-induced e–h generation rate at position x. The total current:

$$J = J_n(x) + J_p(x) \tag{6.266}$$

is an x-independent constant.

By integrating (6.264) and (6.265) from 0 to W, analytical relations can be obtained between J and $J_n(x)$, $J_p(x)$. If we assume the carrier generation rate in the space charge region is zero, $G(x) = 0$, and consider separately the hole injection that occurs at $x = W$ and electron injection at $x = 0$, the avalanche gain for

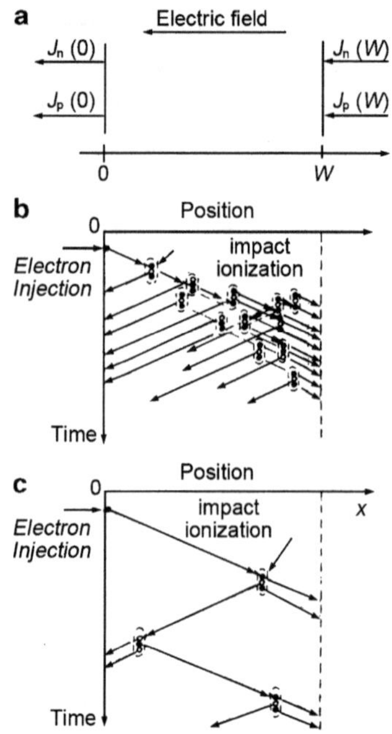

Fig. 6.84 (**a**) The electric field direction in the high-field depletion region, along with the electric current direction for electrons and holes, respectively, (**b**) the avalanche gain process for $\beta_p = 0$, and (**c**) the process for $\alpha_n = \beta_p$

electrons and holes are both obtained (Stillman and Wolfe 1977):

$$\begin{aligned} M_p &= \frac{J}{J_p(W)} = \frac{1}{1 - \int_0^W \beta_p e^{\left[\int_0^W (\alpha_n - \beta_p) dx'\right]} dx} \\ &= \frac{e^{\left[-\int_0^W (\alpha_n - \beta_p) dx\right]}}{1 - \int_0^W \alpha_n \, e^{\left[-\int_0^x (\alpha_n - \beta_p) dx'\right]} dx}, \end{aligned} \tag{6.267}$$

$$\begin{aligned} M_n &= \frac{J}{J_n(0)} = \frac{e^{\left[\int_0^W (\alpha_n - \beta_p) dx\right]}}{1 - \int_0^W \beta_p \, e^{\left[\int_x^W (\alpha_n - \beta_p) dx'\right]} dx} \\ &= \frac{1}{1 - \int_0^W \alpha_n \, e^{\left[-\int_0^x (\alpha_n - \beta_p) dx'\right]} dx}. \end{aligned} \tag{6.268}$$

Clearly, if either the relation for the hole injection occurring at $x = W$, or electron injection at $x = 0$, are such that:

$$\int_0^W \beta_p \, e^{\int_x^W (\alpha_n - \beta_p) dx'} \, dx = 1, \tag{6.269}$$

$$\int_0^W \alpha_n \, e^{-\int_0^W (\alpha_n - \beta_p) dx'} dx = 1 \tag{6.270}$$

then either M_p and M_n will tend to infinity. The corresponding bias voltage is referred to as the avalanche breakdown voltage. Note that the breakdown voltage given by (6.269).6 or (6.270) is identical if $\alpha_n = \beta_p$.

To visualize the influence of the ionization coefficients on the avalanche gain, three cases are next briefly discussed: (1) $\beta_p = 0$, (2) $\beta_p = \alpha_n$, and (3) $\beta_p \neq \alpha_n \neq 0$, while assuming the depletion region is homogenous.

For $\beta_p = 0$, the ionization coefficient for electrons injected at $x = 0$ can be deduced from (6.268):

$$M_n = e^{\int_0^W \alpha_n dx} = e^{\alpha_n W}. \tag{6.271}$$

M_n increases exponentially with $\alpha_n W$. However, in this case, no avalanche breakdown will occur. The current pulse caused by single injected electron gets stronger as the electrons travel through the high-field depletion region, but because the holes that generate no extra pairs tends to zero as the holes move in the reverse direction from right-hand-side to left-hand-side of the high-field depletion region. As a consequence there is a long tail to the current pulse. The evolution of avalanche gain with time is illustrated in Fig. 6.84b.

For $\beta_p = \alpha_n$, however, it is significantly different. From (6.267) and (6.268), it is obvious that:

$$M_n = M_p = \frac{1}{1 - \int_0^W \alpha_n dx} = \frac{1}{1 - \alpha_n W}. \tag{6.272}$$

If $\alpha_n W = 1$, which means that each carrier generates on average one e–h pair during its passage through the high-field depletion region, a real avalanche breakdown occurs. The evolution with time of the avalanche gain under this condition is illustrated in Fig. 6.84c.

In fact, for most semiconductors, both electrons and holes contribute to the ionization process, but normally, $\beta_p \neq \alpha_n$. If α_n and β_p are assumed to be position independent, then the gain for the electron injection at $x = 0$ can be derived from (6.268):

$$M_n = \frac{[1 - (\beta_p/\alpha_n)]\, e^{\alpha_n W[1-(\beta_p/\alpha_n)]}}{1 - (\beta_p/\alpha_n)\, e^{\alpha_n W[1-(\beta_p/\alpha_n)]}}. \tag{6.273}$$

Figure 6.85 illustrates schematically the electron gain as a function of the electric field for different β_p/α_n ratios (Stillman and Wolfe 1977), where $W = 1\ \mu$m, $\alpha_n = 3.36 \times 10^6 e^{-1.75\times 10^6/|E|}$. It is clear that the closer β_p/α_n is to 1, the larger is the change of the avalanche gain caused by a given electric field variation. For $\beta_p/\alpha_n = 0.01$ and a reversed bias of 100 V, 0.5% electric field variation corresponds to a 20% change of the gain, meanwhile for $\beta_p/\alpha_n = 1$, the corresponding gain change is as high as 320%. Such magnitudes of field variations are inevitable for real avalanche photodiodes due to, say doping. Hence for stable operation, it is crucial to select materials with as big a difference between β_p and α_n as possible.

Further analysis suggests that, for $\beta_p \neq \alpha_n$, the avalanche gain also depends on the location where the photon-excited carriers are generated or injected. For $\beta_p > \alpha_n$, holes should be injected near the n-region side, while for $\beta_p < \alpha_n$, electrons near the p-region side of the high-field depletion region are preferred. This guarantees that the electrons/holes pass through the whole depletion region and that ensures an optimum gain is achieved.

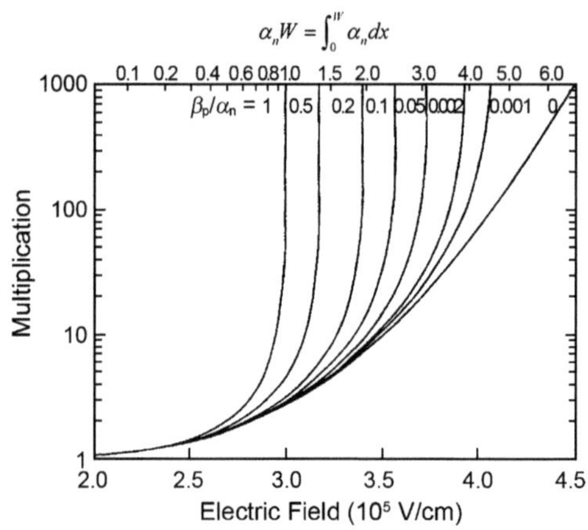

Fig. 6.85 Electron gain as a function of electric field for different β_p/α_n ratios, for $W = 1\ \mu$m, and $\alpha_n = 3.36 \times 10^6 \cdot e^{-1.75\times 10\wedge 6/|E|}$

In fact, the carrier gain is frequency dependent, and its temporal dependence can be described by (6.264a) and (6.264b) (Stillman and Wolfe 1977):

$$\frac{1}{v_n}\frac{\partial J_n(x,t)}{\partial t} = \frac{\partial J_n(x,t)}{\partial x} - \alpha_n J_n(x,t) - \beta_p J_p(x,t) - qG(x,t). \quad (6.274a)$$

$$\frac{1}{v_p}\frac{\partial J_p(x,t)}{\partial t} = \frac{\partial J_p(x,t)}{\partial x} - \alpha_n J_n(x,t) + \beta_p J_p(x,t) + qG(x,t), \quad (6.274b)$$

where v_n and v_p are the drift velocities of electrons and holes, respectively. If the electric field is constant and homogenous, and the photocurrent can be divided into two parts, a DC and an AC part, then:

$$J_n(x,t) = J_{n0}(x) + J_{n1}(x)\,e^{i\omega t},$$
$$J_p(x,t) = J_{p0}(x) + J_{p1}(x)\,e^{i\omega t}. \quad (6.275)$$

The total AC current in an APD also contains a displacement current component, i.e.,

$$J_{ac} = J_{n1} + J_{p1} + \varepsilon\varepsilon_0\left(\frac{\partial E}{\partial t}\right), \quad (6.276)$$

where ε is the relative permittivity, and ε_0 is the dielectric constant in vacuum.

The frequency response of the electron gain for electrons injected at $x = 0$ is determined by:

$$F(\omega) = \frac{|J_{ac}|}{J} = \frac{|J_{ac}|}{M_n J_n(0)}, \quad (6.277)$$

where J is the low-frequency current. For $M_n > \alpha_n/\beta_p$, the gain as a function of frequency is given by:

$$M(\omega) = \frac{M_n}{\sqrt{1+\omega^2 M_n^2 \tau_1^2}}, \qquad \tau_1 \equiv N\frac{\beta_p}{\alpha_n}\tau, \quad (6.278)$$

where τ is the time duration for a carrier passing through the depletion region, and if the saturated velocity of electrons is taken to be equal to that of holes, $v_n = v_p$, so $\tau = W/v_n = W/v_p$. τ_1 is the effective passage time. N is a slowly varying number, $N = 1/3$ if $\beta_p/\alpha_n = 1$, and $N = 2$ if $\beta_p/\alpha_n = 10^{-3}$. Hence, the gain-bandwidth product for the high-frequency region is:

$$M(\omega)\,\omega = \frac{1}{\tau_1} = \frac{1}{N\tau\,(\beta_p/\alpha_n)} = \frac{1}{N\,(W/v_n)\,(\beta_p/\alpha_n)}. \tag{6.279}$$

To achieve a high gain-bandwidth product, the saturated electron drift velocity should be as large as possible, while β_p/α_n and W are as small as possible. It is noteworthy that these parameters are not independent of each other, for Si, a decrease in W results in an increase in β_p/α_n.

Because of symmetry, a similar discussion/results also applies to the case of hole injection at $x = W$, so long as β_p/α_n and M_n are replaced by α_n/β_p and M_p, respectively (Stillman and Wolfe 1977).

Besides the limitation on an APD's device performance by the gain-bandwidth product, the achievable avalanche gain for an APD is also limited due to a space charge effect and a thermal resistance effect.

Melchior and Lynch (1966) have indicated that in the case of a high light intensity irradiation, i.e., the initial total current I_p is much larger than the dark current I_D, the maximum photocurrent gain can be expressed as:

$$(M_{ph})_{max} = \sqrt{\frac{V_B}{n\,I_{ph}\,R}} \equiv \frac{I_{max}}{I_{ph}}; \quad I_{ph} = I_p - I_D \cong I_p, \tag{6.280}$$

where V_B is the reverse break down voltage, I_{ph} is the initial photocurrent, I_p is the initial total current, and I_D is the initial dark current. The dark current is not induced by light irradiation but rather contributed by tunneling or thermal excitation generated e–h pairs. R is an effective resistance that represents contributions from (1) the voltage drop across the load resistance, the contact resistance, as well that from the neutral non-depleted region; (2) the suppression of the electric field cause by space charge in the depletion region; and (3) the reduction of the electron and hole ionization coefficients caused by the thermal resistance. n is a fitting parameter determined from experiment, and is closely related to the electron and hole ionization coefficients and takes on its minimum value when a carrier having high a ionization coefficient is injected into the depletion region. Obviously, the maximum gain is inversely proportional to square root of the initial photocurrent.

If the initial photocurrent is smaller than the initial dark current, the maximum avalanche gain is restricted by the dark current and becomes:

$$(M_{ph})_{max} = \sqrt{\frac{V_B}{n\,I_D\,R}}. \tag{6.281}$$

Therefore, it is crucially important for the dark current to be as low as possible, even if just from the point of view of obtaining a high avalanche gain.

6.6.3 The Basic Structure of an APD

As mentioned earlier, it is necessary to use materials having a large difference between the electron and hole ionization coefficients to ensure a high gain-bandwidth product and low noise avalanche diode. On the other hand, the quantum efficiency of the material should be as high as possible to produce a high detectivity.

The external quantum efficiency of the device is defined as the ratio of the collected number of e–h pairs to the number of photon irradiated onto the light sensitive region of the detector. For most semiconductors, the reflection loss limits the quantum efficiency. Therefore, it is necessary to reduce the reflection loss by a coating technique. If all the photon-induced e–h pairs are collected, and multi-internal reflections are ignored, the external quantum efficiency η can then be expressed as:

$$\eta = (1 - r)(1 - e^{-\alpha x}), \tag{6.282}$$

where r is the front surface reflectivity, α the absorption coefficient, and x is the material thickness a photon travels before it is absorbed.

Figure 6.86 depicts a device structure and its influence on the quantum efficiency (Stillman and Wolfe 1977). p^+ and n^+ regions are employed to form ohmic contacts with the p and n active regions, respectively. Under reverse bias, the width of

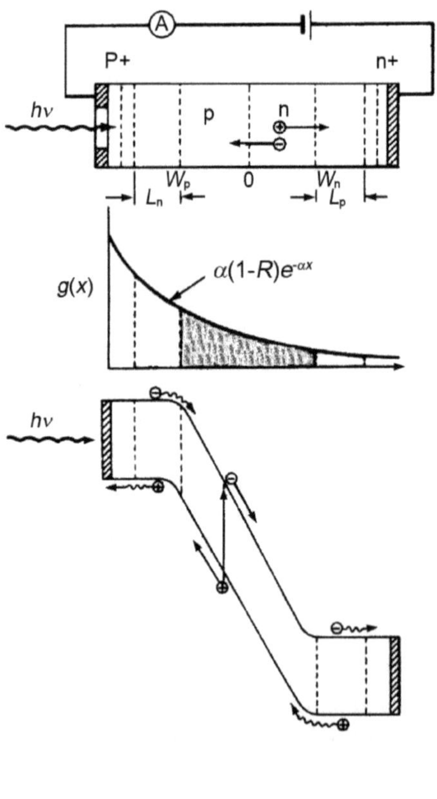

Fig. 6.86 Electron gain as a function of electric field for different β_p/α_n values, where $W = 1\,\mu$m and $\alpha_n = 3.36\times10^6 e^{-1.75\times10^6/|E|}$

the depletion region extends from the p–n junction to W_p and W_n in p and n regions, respectively. The electrons and holes generated in this region are driven by an electric field into the n and p regions, and hence have no chance to recombine before being collected. On the other hand, the minority carriers generated outside the depletion region will diffuse toward it, and some of them will recombine with majority carriers before entering the depletion region, and others enter the depletion region and are finally collected. It can be safely assumed that all the minority carriers located within the diffusion lengths of L_n and $L_{p,}$, but outside the depletion region can be fully collected, and hence, the material thickness contributing to effective absorption is $L_n + W_p + W_n + L_p$.

It is clear that if the material absorption coefficient α is rather high in the concerned wavelength region, W_p and L_n must be as small as possible; while if α is rather low, the thickness of the depletion and diffusion regions must be increased. Nevertheless, too wide a depletion region will increase the carrier transit time through the region significantly, and will in turn result in a negative effect on the device frequency response.

Figure 6.87 plots the photocurrent response of a photodiode to a rectangular light pulse excitation. The light is assumed to be fully absorbed in the depletion region. One component of the photocurrent is the so-called drift current, formed by the carriers generated in the depletion region. Its response is fast, and the rise time is related to the carrier transport time through the depletion region. The other component is the diffusion current, and it produces a tail contributed by the diffusion of minority carriers through the diffusion region into the depletion region. Its response is relatively slow. Hence, to enhance the frequency performance of the device it is crucial to reduce the diffusion component. Because the diffusion length of the minority carriers usually decreases as the carrier density increases, the diffusion current can be depressed by increasing the carrier density in the epilayers. For an avalanche diode, it is possible to reduce diffusion current further by optimizing the device structure design so as to make the diffusion current dominated by the carrier type having a relative low ionization coefficient.

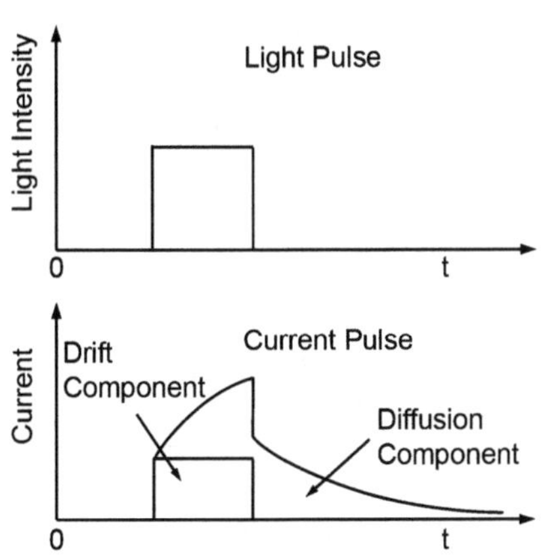

Fig. 6.87 A rectangle light pulse and the induced detector photocurrent response

The device structure also significantly affects the injected carrier type and gain uniformity. For planar structure devices, the electric field at device edge is stronger than that in its center. It is therefore necessary to increase the breakdown voltage at the device edge. For non-planar structured devices, on the other hand, it is necessary to employ a suitable device shape or edge structure to reduce the device surface electric field.

Figure 6.88 illustrates four basic APD structures and their corresponding gain and electric field spatial variation.

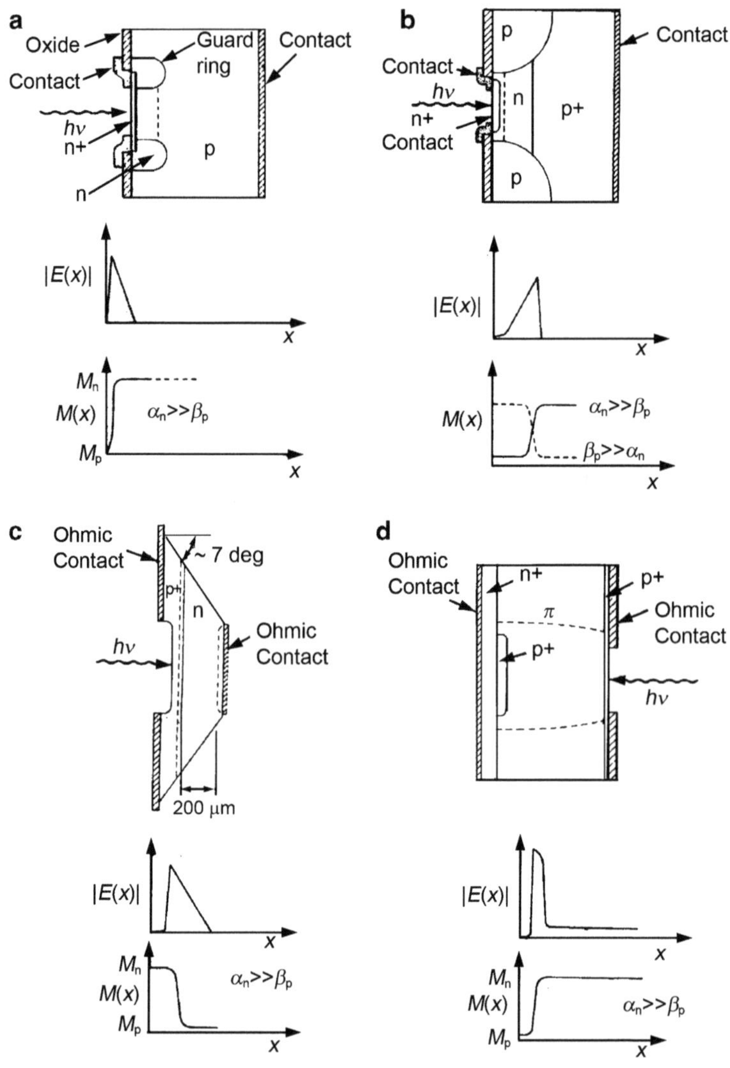

Fig. 6.88 Schematic APD structures: (**a**) a guard-ring, (**b**) an inverted, (**c**) a beveled-mesa, and (**d**) a reach-through

6.6.3.1 Guard-Ring APD

It is easy to use two separate diffusion processes to form an n-type guard-ring and n^+ p active junction in Si. Under a breakdown bias, the thickness of the depletion layer is normally several microns, and is smaller than the depth of the guard ring. This structure is suitable for detecting relatively short wavelength (0.4–0.8 μm) photons. Though the guard ring enhances the device breakdown voltage, it also increases the device capacitance, and hence is disadvantageous for high-speed applications.

6.6.3.2 Inverted APD, Also Known as a Buried APD

This structure has a high-field gain region that is very thin, and is close to the front side of the device. The peak electric field region thickness is only about 5 μm. It is obvious from the gain-position relation that a nearly optimized structure can be achieved by adjusting the doping and thickness of the n region so as to ensure the n region to be just totally depleted at the working voltage for $\beta_p \cong \alpha_n$. However, this structure is not suitable for $\alpha_n \gg \beta_p$ and a strong absorption material.

6.6.3.3 Beveled-Mesa APD

By forming the flank of the APD into a bevel with an obliquity of 6–10°, the lateral electric field can be reduced, and the device can operate under a higher bias voltage. It can be seen from the gain-position relation that this structure is very applicable to materials with $\alpha_n \gg \beta_p$ ($\alpha_n/\beta_p \sim 20$), such as Si. In the long wavelength region, however, the absorption coefficient decreases apparently, leading to a large number of photons being absorbed in the n region, which in turn results in a low gain and a high noise factor. Thus, this structure is not appropriate for long wavelength light detection.

6.6.3.4 Reach-Through APD

In this structure, there is a relatively wide low-field drift region on the front side of the device, and the gain region is very narrow and is located on the backside. Increasing the reverse bias voltage widens the depletion layer, and after reaching a particular threshold voltage, the width of the depletion layer no longer changes with bias voltage. At this point, the electric field in the π region is so high that the carriers will travel with a saturated drift velocity. By fully depleting the π region, the quantum efficiency in the p^+ gain region is frequency independent, and correspondingly, the gain is wavelength independent.

6.6.4 Fundamentals of a Single-Photon Avalanche Diode (SPAD)

If an APD operates under a bias voltage V_A that is higher than its breakdown voltage V_B (the so-called Geiger mode), and is equipped with an avalanche-quenching circuit, it can realize single photon detection, and is therefore called a single-photon avalanche diode (SPAD). The SPAD devices so far reported can be divided into two groups, based on the depletion layer of the p–n junction, which can be thin, typically 1 μm, or thick, from 20 to 150 μm (Cova et al. 1996). Table 6.3 summarizes the main features of two types of Si-based SPADs (Cova et al. 1989; Ghioni and Ripamonti 1991). The Si-based SPAD operates in the near infrared region, while SPADs made of Ge and III–V alloy semiconductors such as InGaAsP can extend their working wavelength to more than 1.6 μm.

Essentially, SPADs are p–n junctions that operate at bias V_A above the breakdown voltage V_B. At this bias, the electric field is so high that a single charge carrier injected into the depletion layer can trigger a self-sustaining avalanche. The current rises swiftly with a nanosecond or subnanosecond rise time to a macroscopic steady level in the milliampere range. If the primary carrier is photo-generated, the leading edge of the avalanche pulse marks the arrival time of the detected photon. The current continues flowing until the avalanche is quenched by lowering the bias voltage to V_B or even smaller. The bias voltage is then restored, to detect another photon. This operation requires a suitable circuit that (1) senses the leading edge of the avalanche current, (2) generates a standard output pulse that is well synchronized to the avalanche rise, (3) quenches the avalanche by lowering the bias to the breakdown voltage, and (4) restores the photodiode voltage to the operating level. This circuit is usually referred to as a quenching circuit. The features of the quenching circuit dramatically affect the operating conditions and therefore, the actual performance of the detector.

In the following, the operating conditions and performance of SPADs will be introduced, followed by a brief discussion of two basic quenching circuits.

Table 6.3 Structure parameters and performance characteristics of two types of Si-based SPADs

	Thick depletion layer	Thin depletion layer
Thickness of depletion layer (μm)	20–150	~1
Breakdown voltage (V)	200–500	10–15
Diameter of active region (μm)	100–500	5–150
Photon detection	50% @ 540–850 nm	~45% @ 500 nm, ~32% @ 630 nm
Efficiency	~3% @ 1,064 nm	~10% @ 830 nm, ~0.1% @ 1,064 nm
Photon time resolution (FWHM)	350–150 ps	~30 ps @ ~10 μm active region diameter, room temperature

6.6.4.1 SPAD Operating Conditions and Performance

The amount by which the bias voltage V_A exceeds the breakdown voltage V_B of the junction is called the excess bias voltage, $V_E = V_A - V_B$. The value of ratio V_E/V_B, determines the detector performance. Since V_B ranges from 10 to 500 V for different SPADs, the V_E values considered are from 1 to about 50 V.

Photon Detection Efficiency

For a photon to be detected, not only must it be absorbed in the detector's active region and generate a primary e–h pair, it is also necessary for the primary e–h pair to succeed in triggering an avalanche. The efficiency of photon detection thus increases with excess bias voltage V_E. In Fig. 6.89, typical values of the detection efficiency are illustrated as a function of excess voltage/wavelength for SPADs with (**a**) a thin junction and (**b**) a thick junction.

Time Resolution

As shown in Fig. 6.90a, the resolution in single photon timing also improves at a higher electric field and hence a higher V_E.

Dark-Count Rate

Even in the absence of illumination, thermal generation effects produce current pulses, and correspondingly, produce dark current-counts. The Poisson distribution

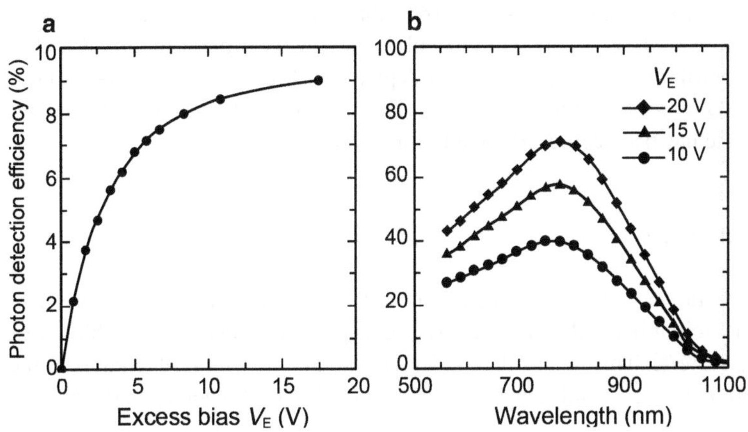

Fig. 6.89 Detection efficiency for (**a**) a thin-junction SPAD as a function of V_E, its detecting wavelength is 830 nm (the width of the p–n junction is 1 μm, $V_B = 16$ V, the active region diameter is 10 μm), and (**b**) a thick-junction SPAD as a function of wavelength (the width of p–n junction is 25 μm, $V_B = 420$ V, the active region diameter is 250 μm)

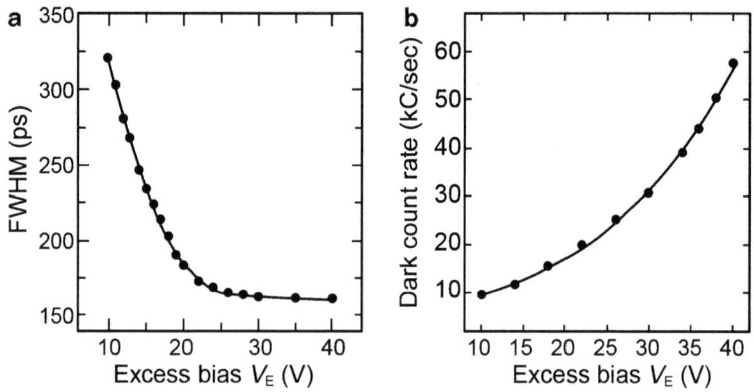

Fig. 6.90 (**a**) FWHM time resolution, and (**b**) dark count rate, as a function of excess voltage V_E for the thick-junction SPAD illustrated in Fig. 6.89b at room temperature

of these dark-count-induced fluctuations represents the internal noise source of the detector. The dark current is determined by material properties and fabrication techniques. As thermal excitation or tunneling effects generate carriers, they can also be multiplied. According to the relation, $g(T) = a_r n_i^2(T)$, where $g(T)$ is the carrier thermal generation rate, a_r is a proportionality factor, n_i is the intrinsic carrier density, the carrier thermal generation rate, and hence the dark current, can be reduced by lowering the temperature. The dark current of SPADs can be expressed as, $I = dQ/dt = \Delta ne/\Delta t$, where the electron charge $e = 1.6 \times 10^{-19} C$, Δn is the number of minority carriers, Δt is a time interval of the order of nanoseconds (ns). The number of electrons passing through a unit sectional area per nanosecond is $\Delta n = I \cdot \Delta t/e = 6.26 \times 10^9 I$. If the dark current is on the order of nanoamperes, Δn is determined to be about six electrons. Obviously, the dark current must be on the order of subnanoamperes to warrant that less than one electron passes through per nanosecond. The dark current pulse contains two parts of initial and second stage contributions (Haitz 1965). As depicted in Fig. 6.90b, the dark count rate of the SPAD increases with excess bias voltage V_E. In practice, a very low dark current count rate has been achieved already for an Si-based SPAD.

Thermal Effects

The breakdown voltage V_B depends strongly on the junction temperature. The thermal coefficient depends on the SPAD device structure and is fairly high, typically around 0.3%/K. This means that under a constant bias V_A, V_E decreases as the temperature increases, leading to a drastic degradation of the device performance.

6.6.4.2 Two Basic Quenching Circuits

Passive-Quenching Circuits

Figure 6.91 shows a simple voltage-mode passive-quenching circuit (PQC) and its equivalent circuit (Cova et al. 1996), where R_L is a high ballast resistor, C_d is the junction capacitance, typically ~1 pF, and C_s is the stray capacitance, typically a few picofarads.

The diode resistance R_d is given by the series combination of the space-charge resistance of the avalanche junction and of the ohmic resistance of the neutral semiconductor region crossed by the current. The R_d value depends on the semiconductor device structure: it is smaller than 500Ω for the types with a wide area and a thick depletion layer, and from a few hundred ohms to several kiloohms for the devices with a small area and a thin junction.

Avalanche triggering corresponds to closing the switch in the diode equivalent circuit. Figure 6.91 shows typical waveforms of diode currents I_d and diode voltages V_d, or of the transient excess voltage $V_{ex} = V_d - V_B$:

$$I_d(t) = \frac{V_d(t) - V_B}{R_d} = \frac{V_{ex(t)}}{R_d}, \tag{6.283}$$

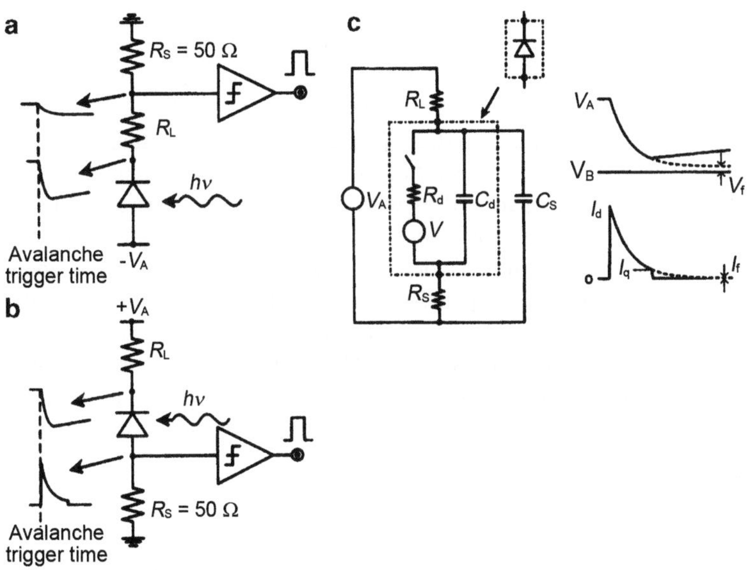

Fig. 6.91 Basic PQCs: (**a**) a voltage-mode output, (**b**) a current-mode output, (**c**) an equivalent circuit of the current-mode output configuration

The avalanche current discharges the capacitances so that V_d and I_d exponentially fall toward the asymptotic steady-state values of V_f and I_f:

$$V_f = V_B + R_d I_f, \quad I_f = \frac{V_A - V_B}{R_d - R_L} \approx \frac{V_E}{R_L}. \tag{6.284}$$

The quenching time constant T_q is set by the total capacitance $C_d + C_s$:

$$T_q = (C_d + C_s) \frac{R_d R_L}{R_d + R_L} \approx (C_d + C_s) R_d, \tag{6.285}$$

for $R_L \gg R_d$.

It can be seen from (6.284) that if I_f is very small, V_f is very close to V_B. The avalanche is self-sustaining above a latching current level $I_q \lesssim 100\,\mu A$ and is self-quenching below it. The I_q value is not sharply defined, however, the behavior of the PQC is the correct one if I_q is set to a value much higher than I_f. As a rule of thumb, I_f should not exceed $20\,\mu A$, that is, the R_L value should be at least $50\,k\Omega/V$ of applied excess bias voltage V_E.

Avalanche quenching corresponds to opening the switch in the diode equivalent circuit. The capacitances are recharged slowly by the electric current in R_L. The diode voltage exponentially recovers toward the bias voltage with a time constant $T_r = (C_d + C_s) R_L$. As the recovery starts, any photons that arrive during the first part of the recovery are almost certainly lost, since the avalanche triggering probability is very low. Subsequently, the arriving photons have a progressively higher probability of triggering an avalanche. Depending on the different connections of R_s with the circuit, the PQC can be either a voltage-mode output or a current-mode output. Figure 6.91a shows a voltage-mode PQC, in which the R_s is in series with the ground lead of the ballast resistor R_L and the output is an attenuated replica of the diode voltage waveform. The voltage-mode PQC is fairly simple and has useful features. Since it produces pulses with longer duration, it is easier to monitor an avalanche pulse sequence on a longer time scale of an oscilloscope. An obvious drawback, however, is that the detector's dead time performance is poorer and the response rate is not fully exploited, because of the intrinsic low-pass filter with time constant T_q that acts on the fast current pulse to produce the voltage waveform. Figure 6.91b illustrates a current-mode PQC, in which the R_s is on the ground lead of the photodiode and the waveform of the output is directly that of the diode current. The current-mode output offers the best performance in high-rate counting and in precision pulse timing and is usually preferred.

Active-Quenching Circuits

In the active-quenching circuits (AQC), a new approach was implemented to simply sense the rise of the avalanche pulse and react back on the SPAD, forcing the quenching and reset transitions into short times by a controlled bias source.

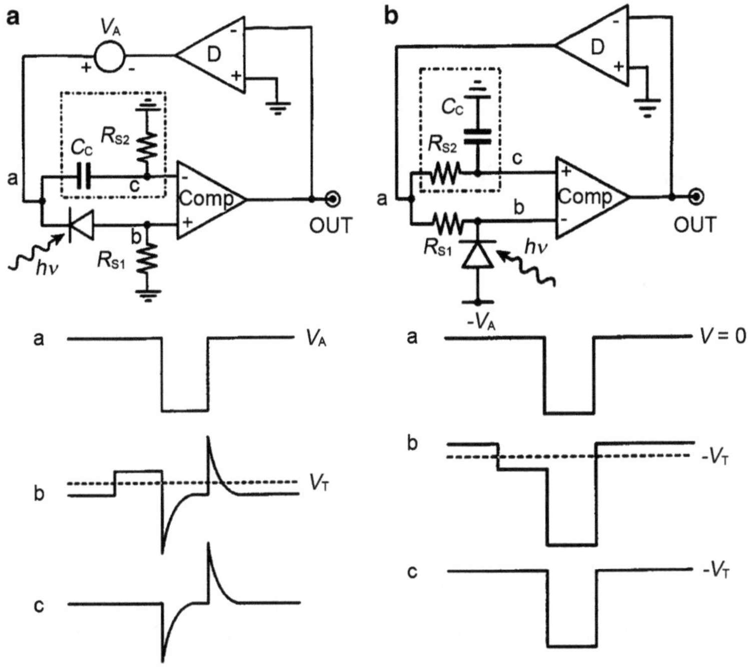

Fig. 6.92 Simplified diagram of two basic AQC configurations with (**a**) opposite quenching and sensing terminals of the SPADs, and (**b**) coincident quenching and sensing terminals of the SPADs. The voltage waveforms drawn correspond to the circuit nodes labeled with the same letter (Cova et al. 1996)

Figure 6.92 depicts simplified diagrams of two basic AQC configurations, one with a quenching terminal opposite the sensing terminal as illustrated in Fig. 6.92a, the other with a coincident quenching and sensing terminal as shown in Fig. 6.92b. In any case, the sensing terminal has a quiescent voltage level at ground potential or not far from it, since it is directly connected to the AQC input. The quenching and reset driver labeled as D in Fig. 6.92 represents the function of a circuit generating a high-voltage pulse when driven from a low-level logic pulse. The amplitude of the quenching pulse should be larger than the excess bias V_E.

With respect to the PQCs, AQCs have two distinct advantages. First, the duration of its avalanche pulse, T_{ac}, is constant, and is set by the quenching delay, given by the sum of the circulation time T_L in the quenching loop, plus rise time T_{aq} of the quenching pulse:

$$T_{ac} = T_c + T_{aq}. \tag{6.286}$$

The smallest obtainable T_{ac} value depends on the required excess bias voltage V_E, since the value of T_{aq} increases with the amplitude of the quenching pulse. For $V_E \leq 1\,\text{V},\ T_{ac} \leq 5\,\text{ns}$, while for $V_E \sim 20\,\text{V},\ T_{ac} \sim 10\,\text{ns}$. Second, the dead time T_{ad} of an AQC is constant, well defined, and can be accurately controlled. Its minimum is given by:

$$(T_{ad})_{min} = 2T_L + T_{aq} + T_{ar}, \tag{6.287}$$

where T_{ar} is the fall time of the quenching pulse. Similar to T_{ac}, $(T_{ad})_{min}$ increases with V_E. For $V_E \leq 1\,V$, $(T_{ad})_{min} \sim 10\,ns$, while for $V_E \sim 20\,V$, $(T_{ad})_{min} \sim 40\,ns$.

6.6.5 *Examples of Single-Photon Infrared Detectors*

The SPAD, which have already proven to be useful as visible and infrared single photon detectors, consist of two types, one is based on Si, Ge, and InGaAs/InP material systems working in the Geiger mode, and the other is still in an exploratory phase and is based on a single electron transistor (SET). An APD device can also be fabricated from the narrow bandgap semiconductor alloy $Hg_{1-x}Cd_xTe$. At a composition, $x \approx 0.7$, the bandgap of $Hg_{1-x}Cd_xTe$ is equal to its spin-orbit split-off energy, so a resonant collision ionization occurs, and the hole ionization coefficient gets large, the device noise gets low, leading to an avalanche gain. Using such a material system, low-noise avalanche diodes with $\lambda = 1.3$ and $1.55\,\mu m$ can be fabricated for optical communication applications. The wavelength can be extended to $2.5\,\mu m$ by optimizing device structures. However, it is argued from a two-dimensional solution of the Poisson equation that the performance of an APD made from HgCdTe cannot be superior to one made from InGaAs/InP (Liu et al. 1992). In what follows, several typical SPAD devices are introduced.

6.6.5.1 Si, and Ge Based SPADs

A Si-based SPAD is one kind of well-studied and well-developed single photon detector. Devices with good performance have already been commercialized for several years (Cova et al. 1996; Lacaita et al. 1988, 1989; Cova et al. 1989; Ghioni and Ripamonti 1991; Ghioni et al. 1988).

Figure 6.93a schematically illustrates an SPAD with a reach-through structure. Its operating wavelength is designed to be 830 nm, for the application of a 1 Gbit/s data communication rate. The whole structure is as follows: on a high resistance p^- type silicon wafer, the high-field region of the active junction is defined by a deep boron diffusion, producing a p region that is about $15\,\mu m$ deep, over which an 5-μm-thick n^+ layer is deposited. The back of the wafer is etched to reduce the active area's thickness to about $35\,\mu m$.

It has been found (Ghioni and Ripamonti 1991) that the time resolution depends on the diameter of the illuminated area, and the t resolution can be improved by shrinking the illuminated area. Figure 6.93b depicts the result obtained with light focused on the center of the device at a spot diameter of $50\,\mu m$. The bias was 40 V above the breakdown and the device was not cooled. The FWHM resolution is about 320 ps.

To enhance device performance (time resolution, etc.), Ghioni et al. (1988) and Lacaita et al. (1989) fabricated SPADs employing a guard-ring APD structure. The active junction was realized by fabricating a shallow n^+ layer ($0.3\,\mu m$) on a

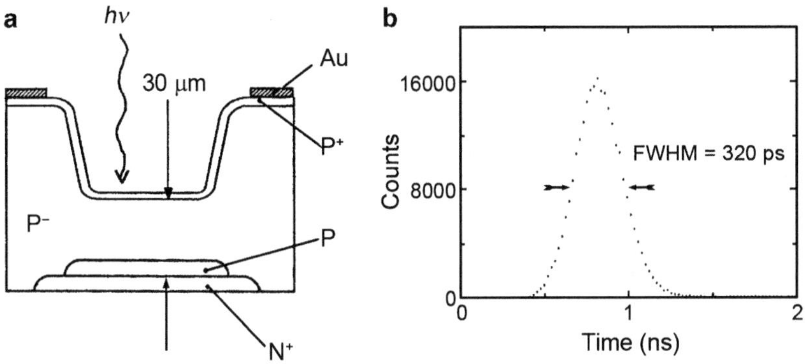

Fig. 6.93 (**a**) A schematic of a reach-through SPADs cross section. (**b**) The time resolution of the device at room temperature for a spot of light located in the center of the device with a diameter 50 μm. The photodiode was operated in a gated mode at 40 V excess bias, with an active quenching circuit (Ghioni and Ripamonti 1991)

0.6 Ω cm p-type substrate, and surrounding it by a 5–8 μm deep-diffusion guard ring to avoid edge breakdown. The breakdown voltage of the device is about 28–33 V. Experimental results indicates that the time resolution of this SPAD is as short as 28 ps at room temperature, and is further reduced to 20 ps if the temperature is set at 208 K (Cova et al. 1989).

Lacaita et al. have found that the time resolution can be improved by increasing the excess bias voltage or by reducing the device temperature (Lacaita et al. 1988).

Recently, Rochas et al. presented a novel idea, a way to fabricate APDs in a standard 0.8 μm silicon complementary metal–oxide–semiconductor (CMOS) process (Rochas et al. 2003). The structure of the integrated APD is a dual junction p^+/deep n-tub/p-substrate, and is sketched in Fig. 6.94a. Its operating voltage is from 2.5 to 50 V. Due to the introduction of a guard ring, the breakdown voltage of the p^+/deep n-tub is about 25 V, while the breakdown voltage of the p-tub/deep n-tub junction is typically 55 V. Therefore, the p^+/deep n-tub junction breaks down first in the planar part of the junction. The second breakdown in the ring appears at a bias voltage roughly 30 V higher. It ensures a proper Geiger mode operation with a potential excess bias voltage of up to 30 V.

For a high-voltage operation, the deep n-tub has a relatively low doping concentration ($\sim 3 \times 10^{16}\,\text{cm}^{-3}$) with the p^+/deep n-tub junction close to the surface at about 500 nm in depth. This doping profile is of the greatest interest for the detection of red and near-IR photons. Extending the depletion region results in increasing the photon collection of those units with high penetration depths. For a Geiger mode operation, it is essential to reduce the thermal generation, i.e., the dark count rate of the diode. The photosensitive area of the SPAD characterized in what follows is then kept deliberately small. The value reported by Rochas et al. is a disk defined by an integrated metal mask to be 7 μm in diameter. Figure 6.94b gives a schematic of the fully integrated single photon detector. A picture of it is shown in Fig. 6.94c.

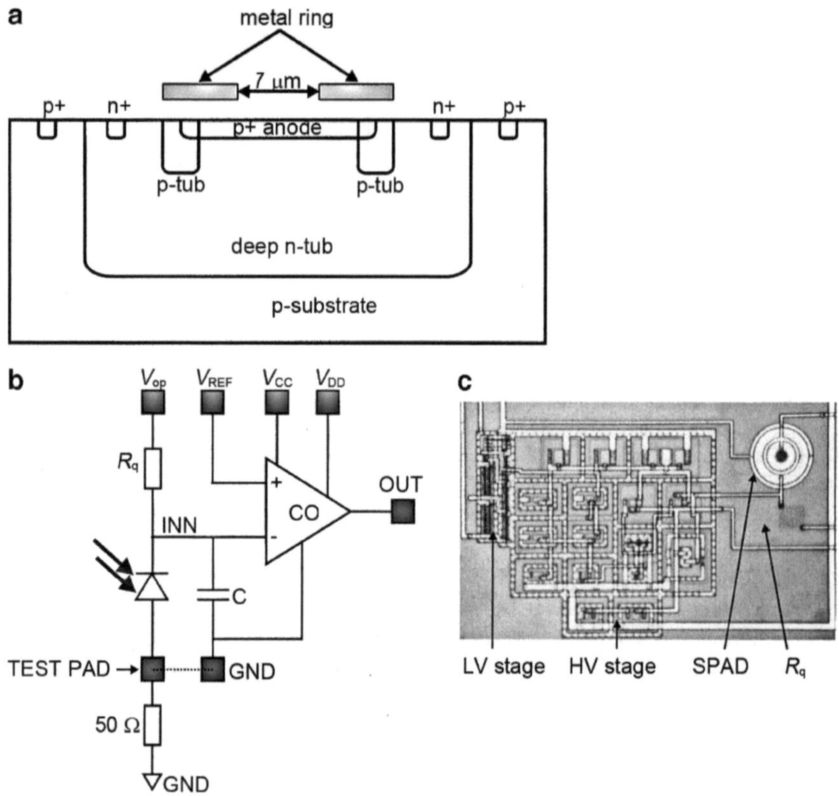

Fig. 6.94 (**a**) The schematic of the APD cross section. The p-tub implantation acts as a guard ring, preventing premature edge breakdown; (**b**) the schematic of the fully integrated passively quenched SPAD; (**c**) a picture of the CMOS high-voltage SPAD. The circular SPAD and the quenching resistor can be seen on the right-hand side. The low-voltage and the high-voltage parts of the comparator can be seen as well (Rochas et al. 2003)

Experimental results indicates (Rochas et al. 2003) that the detection probability increases rapidly with excess bias voltages up to 5 V, and at this value, the detection probability is larger than 20% between 420 nm and 620 nm and still 7% at 750 nm. The single photon detector exhibits a dead time as short as 75 ns. The avalanche current is quickly quenched in less than 3.5 ns. A remarkably short-time resolution has been obtained with values lower than 50 ps for excess bias voltages higher than 5 V. All these values suggest that the CMOS integration opens the way to the fabrication of an extremely compact array with low cost.

To extend the detection wavelength of SPADs above the Si absorption edge (1.1 μm), a large number of studies have been dedicated to Geiger-mode Ge-based APDs (Haecker et al. 1971; Levine and Bethea 1984; Lacaita et al. 1994), and a wavelength as long as 1.6 μm has been achieved. When the detected photon energy is higher than that of a material with an indirect bandgap where phonon-assisted absorption is needed, the detectivity increases as the wavelength decreases. However, for a direct bandgap material once the photon energy becomes much higher than the bandgap energy, the detectivity actually decreases due to an increase of the absorption loss near the surface where surface-assisted recombination occurs (Haecker et al. 1971). By properly selecting operational conditions for Ge-based SPADs, a noise equivalent power of $7.5 \times 10^{-16}\,\mathrm{W/Hz}^{1/2}$ at $1.3\,\mu$m has been obtained at a temperature of 77 K, along with a time resolution of 85 ps, and correspondingly, an equivalent timing bandwidth of 1.8 GHz (Lacaita et al. 1994).

6.6.5.2 III–V Semiconductor SPADs

Because the cut-off wavelength is $1.45\,\mu$m for Ge at 77 K, it is necessary to explore other materials for $1.55\,\mu$m SPAD devices where optical fibers have their minimum loss. It was proven early in 1985 that working in the Geiger mode, the separate absorption and multiplication of InGaAs/InP APDs can detect single photons even at room temperature (Lacaita et al. 1996). Now, commercial InGaAs/InP avalanche photodiodes are available for optical receivers and range finders. As an optical receiver, they generally work with a breakdown voltage bias in amplification region, at a wavelength as long as $1.7\,\mu$m. At 77 K, a carefully selected APD can operate under a bias voltage that is 6 V higher than the breakdown voltage, and can thus be used as an SPAD. A quantum efficiency of 10% is achievable at a wavelength of $1.55\,\mu$m. A detection sensitivity of $3 \times 10^{-16}\,\mathrm{W/Hz}^{1/2}$ has been obtained at a wavelength of $1.3\,\mu$m, together with a time resolution of 200 ps (Lacaita et al. 1996).

In recent years, attempts have been made to extend the wavelength response and to boost the device speed (Liu et al. 1988; Cheng and McGill 1999; Dries et al. 1999; Hirota et al. 2004). Dries et al. have proposed an avalanche photodiode design for wavelengths as long as $2.1\,\mu$m, by combining a 100 period structure consisting of $In_{0.83}Ga_{0.17}As$ quantum wells strain compensated by $In_{0.83}Ga_{0.17}P$ barrier layers with a low noise $In_{0.52}Al_{0.48}As$ gain region (Dries et al. 1999).

This structure is illustrated schematically in Fig. 6.95. First, a 2,500-Å-thick buffer layer of n^+ InP was grown on a (100) n^+-type InP substrate. Next, an undoped, a 2,700-Å-thick $In_{0.52}Al_{0.48}As$ multiplication layer was grown followed by a 500-Å-thick p^+-type InP "field control layer" doped to a concentration of $5\text{–}7 \times 10^{17}\,\mathrm{cm}^{-3}$. This layer is used to separate the high-field $In_{0.52}Al_{0.48}As$ gain region from the low-field 100-period MQW absorption region. A single-quantum-well period consists of six layers as illustrated in the band diagram of Fig. 6.95a. The quantum well is grown systematically using $In_{0.83}Ga_{0.17}P$, InP, $In_{0.83}Ga_{0.17}As_{0.37}P_{0.63}$, and $In_{0.83}Ga_{0.17}As$ with layer thicknesses of 70, 30, 45, and 60 Å, respectively. The entire absorption layer is lightly p^- doped with a Be concentration of $3 \times 10^{15}\,\mathrm{cm}^{-3}$ and has a total thickness of $2.8\,\mu$m, of which $0.6\,\mu$m consists of $In_{0.83}Ga_{0.17}As$ quantum wells. The structure is finally finished with Be-doped ($3 \times 10^{18}\,\mathrm{cm}^{-3}$) p^+-type InP as an electric contact layer covering the MQW absorption region.

A scanning electron micrograph of the epitaxial layers is shown in Fig. 6.95b. In Fig. 6.95c, an Illustration of the completed device is given, with a 100-μm-diameter

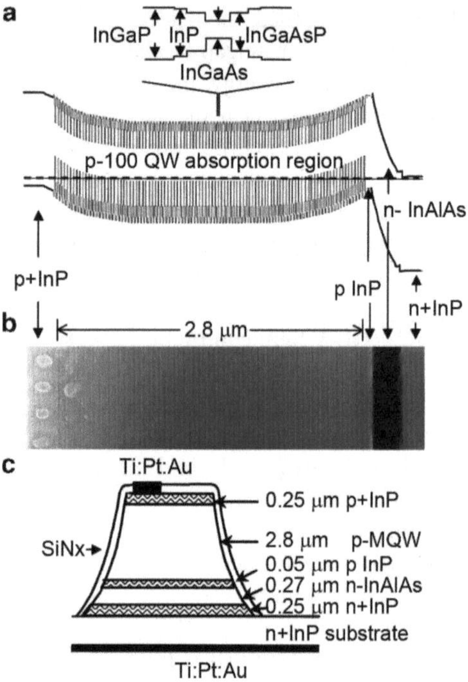

Fig. 6.95 (**a**) Band diagram for a separate absorption, multiplication, and charge layer avalanche photodiode. Above the full band diagram an expanded view of the quantum well absorption region is depicted. The compressive strain of the $In_{0.83}Ga_{0.17}As$ is balanced by the tensile strain of $In_{0.83}Ga_{0.17}P$. (**b**) Scanning electron microscope image of a selectively etched sample. The $In_{0.52}Al_{0.48}As$ gain region is the black band at the right of the device, and the quantum wells are easily observed. (**c**) Illustration of the completed device (Dries et al. 1999)

active region. Experimental results indicate that the peak responsivity of the APD is 45 A/W at 1.54 μm for an optical input power of 106 nW. Given a quantum efficiency of 65%, this corresponds to a photo-carrier multiplication factor of $M = 55$.

The spectral response of this APD is shown in Fig. 6.96 at several multiplication factors. The response extends to 2.0 μm with the exciton transition peak located at $\lambda = 1.93\,\mu$m. The shift in the long wavelength cut-off with increasing M is due to a quantum confined Stark effect. The spectral response is easily tailored by varying both the compositions and the thicknesses of the wells and barrier layers, with a predicted limit to the cut-off wavelength of 2.1 μm (Dries et al. 1999).

With the increasing demand for high-speed optical communications, the need for ultra fast, low-noise, and high gain-bandwidth product photodetectors has become greater than ever. In response to this demand, Hirota et al. have developed an InP-based APD with a p-type neutral absorption layer and refracting-facet structure. This APD shows a maximum DC response of 0.71 A/W at 1.55 μm wavelength, a low breakdown voltage of 16.9 V, and a gain-bandwidth product of 140 GHz. It also exhibits advantages such as its low breakdown voltage (Hirota et al. 2004).

6.6.5.3 Far-Infrared Single-Electron-Transistor SPADs

A far-infrared single-electron transistor (SET) SPAD has been realized by putting a relatively large semiconductor QD (e.g., $700 \times 700\,\text{nm}^2$) in a high magnetic field

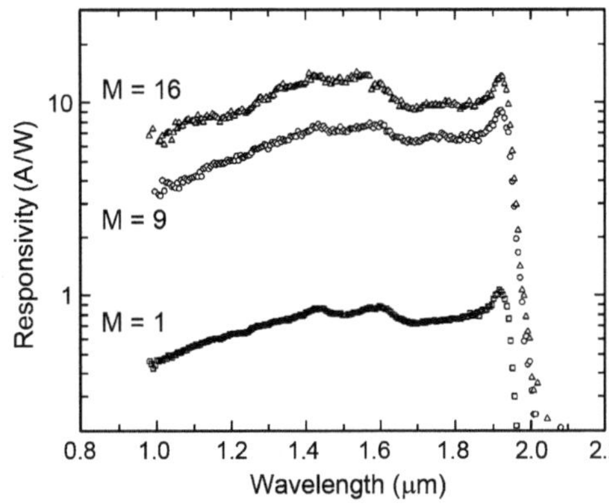

Fig. 6.96 Spectral response of the detector at gains of $M = 1, 9$, and 16. The optical response extends to a wavelength of 2 μm. The theoretical wavelength limit using this strain-compensation technique has been found to be 2.1 μm (Dries et al. 1999)

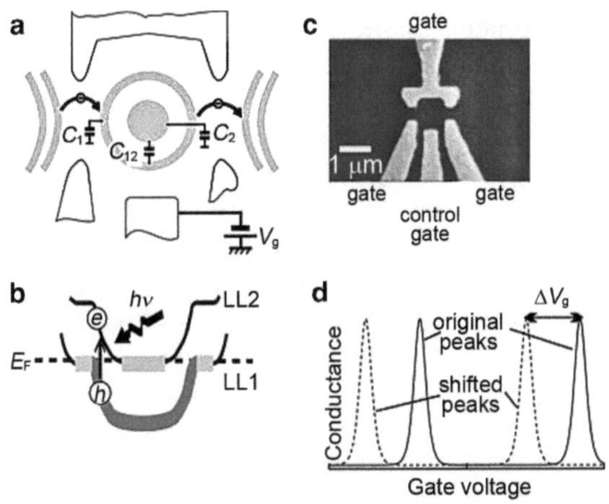

Fig. 6.97 A single-electron transistor with a quantum dot (QD) in a magnetic field. (**a**) The diagram of the QD. The gray regions indicate is metallic inner core and outer ring regions formed by the lowest two Landau levels (LLs). (**b**) The energy spectra of the LLs in the QD. When an e–h pair is excited within the QD by absorption of a far-infrared photon, the excited electron (hole) rapidly falls (climbs up) to the inner core (outer ring) of the QD to polarize the QD. (**c**) Scanning electron micrography of the QD ($700 \times 700\,\mathrm{nm}^2$), formed by laterally confining the 2DEG with negatively biased metal gates. The QD contains about 350 electrons. The QD sample is placed within a mixing chamber of a dilution refrigerator. The magnetic field is applied normal to the plane of the QD with a superconducting solenoid. (**d**) Diagram of Coulomb conductance peaks as a function of the control gate voltage V_g (Komiyama et al. 2000)

at an extremely low temperature (~0.4 K). The operational mechanism of the SET SPAD differs from that of a conventional SPAD. As a direct result, one absorbed photon in the far-infrared region (175–210 μm) leads to a current of 10^6–10^{12} electrons for an SET SPAD. This value is about 10^4 times higher than that of a conventional SPAD.

Figure 6.97 displays the properties of a single-electron transistor with a quantum dot in a magnetic field. The QD is weakly coupled to electron reservoirs through tunnel barriers, as illustrated in Fig. 6.97a and 6.97b. The magnetic field is such that the lowest orbital Landau level, LL1, is filled, while the first excited Landau level, LL2, is occupied with a small number of electrons. At the Fermi level, LL1 and LL2 form two compressible metallic regions, which correspond to an "outer ring" and an "inner core" of the QD.

When a far-infrared (FIR) photon is absorbed by the QD and subjected to a cyclotron resonance, an electron (hole) created in the higher empty LL2 (LL1) will rapidly give up its excess energy to the lattice and fall (climb up) to the inner core (outer ring). As the inner core is now negatively charged by $-e$, the electrochemical potential of the outer ring has decreased by $-\Delta\mu = -eC_2/(C_2C_1 + C_{12}C_1 + C_2C_{12})$, where C_i and C_ij are the capacitances formed between the respective regions. This internal polarization causes the conductance–resonance peaks to shift by $-\Delta V_g \propto -\Delta\mu_1$. This, together with the fact that the excited electron in the inner core has an extremely long lifetime owing to suppression of tunneling, makes single FIR photon detection feasible.

With this structure, Komiyama et al. (2000) have achieved single-photon detection in the wavelength range 175–210 μm restricted by the magnetic-field range (3.4–4.15 T), with a quantum yield, (the ratio of the detected photon number to the incident photon flux on the effective antenna area) of roughly 1%. The effective noise equivalent power reaches the order of $10^{-22}\,\mathrm{W/Hz}^{1/2}$ in the optimum B range, which exceeds other FIR detectors reported in the literature by a factor of more than 10^4. As the intrinsic limit on the speed of the single-electron transistor is greater than 10 GHz, using a high-frequency single-electron transistor will therefore improve the time resolution up into the 10-MHz range.

References

Anderson WW (1977) Tunnel current limitations of narrow bandgap infrared charge coupled devices. Infrared Phys 17:147–164

Baars J, Sorger F (1972) Reststrahlen spectra of HgTe and CdHgTe. Solid State Commun 10: 875–878

Baranov AN, Cuminal Y, Boissier G, Alibert C, Joullie A (1996) Low-threshold laser diodes based on type-II GaInAsSb/GaSbquantum-wells operating at 2.36 μm at room temperature. Electron Lett 24:2279–2280

Benisty H, Sottomayor-Torrès CM, Weisbuch C (1991) Intrinsic mechanism for the poor luminescence properties of quantum-box systems. Phys Rev B 44:10945–10948

Blakemore JS (1962) Semiconductor statistics. Pergamon, Oxford

Bochkarev AE, Dolginov LM, Drakin AE, Eliseev PG, Sverdlov BN (1988) Continuous-wave lasing at room temperature in InGaSbAs/GaAlSbAs injection heterostructures emitting in the spectral range 2.2–2.4 μm. Sov J Quantum Electron 18:1362–1363

Borrello S, Kinch M, LaMont D (1977) Photoconductive HgCdTe detector performance with background variations. Infrared Phys 17:121–125

Broudy RM (1974) NASA Rep CR-132512

Broudy RM, Beck JD (1976) In: Proc Infrared Informat Symp Detector Special Group

Broudy RM, Mazurczyk VJ (1981) Semiconductors and semimetals, vol 18. Academic, New York, p 157

Broudy RM, Mazurczyk VJ, Aldrich NC, Lorenze RV (1975) In: Proc Infrared Informat Symp Detector Special Group

Burstein E (1954) Anomalous Optical Absorption Limit in InSb. Phys Rev 93:632–633

Byungsung O, Choe JW, Francombe MH, Bandara KMSV, Coon DD, Lin YF, Takei WJ (1990) Long-wavelength infrared detection in a Kastalsky-type superlattice structure. Appl Phys Lett 57:503

Capasso F, Tredicucci A, Gmachl C, Sivco DL, Hutchinson AL, Cho AY, Scamarcio G (1999) High-performance superlattice quantum cascade lasers. IEEE J Sel Top Quantum Electron 5:792–807

Chapman RA, Kinch MA, Simmons A, Borrello SR, Morris HB, Wrobel JS, Buss DD (1978) $Hg_{0.7}Cd_{0.3}Te$ charge-coupled device shift registers. Appl Phys Lett 32:434–436

Chen AB, Sher A (1995) Semiconductor alloys. Plenum Press, New York

Cheng XC, McGill TC (1999) Near infrared avalanche photodiodes with bulk $Al_{0.04}Ga_{0.96}Sb$ and GaSb/AlSb superlattice gain layers. J Appl Phys 86:4576

Choi HK, Eglash SJ (1992) High-power multiple-quantum-well GaInAsSb/AlGaAsSb diode lasers emitting at 2.1 μm with low threshold current density. Appl Phys Lett 61:1154

Choi HK, Turner GW (1995) InAsSb/InAlAsSb strained quantum-well diode lasers emitting at 3.9 μm. Appl Phys Lett 67:332

Choi HK, Turner GW (2000) Antimonide-based mid-infrared lasers. In: Helm M (ed) Long wavelength infrared emitters based on quantum wells and superlattices, Chap 7. Gordon and Breach Science, Singapore

Choi HK, Eglash SJ, Connors MK (1993) Single-frequency GaInAsSb/AlGaAsSb quantum-well ridge-waveguide lasers emitting at 2.1 μm. Appl Phys Lett 63:3271

Choi HK, Turner GW, Eglash SJ (1994) High-power GaInAsSb-AlGaAsSb multiple-quantum-well diode lasersemitting at 1.9 μm. IEEE Photon Technol Lett 6:7–9

Choi KK, Levine BF, Malik RJ, Walker J, Bathea CG (1987) Periodic negative conductance by sequential resonant tunneling through an expanding high-field superlattice domain. Phys Rev B 35:4172

Chu JH, Sher A (2007) Physics and properties of narrow gap semi-conductors. Springer-Verlag, New York

Chu JH, Xu SQ, Tang DY (1983) Energy gap versus alloy composition and temperature in $Hg_{1-x}Cd_xTe$. Appl Phys Lett 43:1064

Chu JH, Mi ZY, Tang DY (1991) Intrinsic absorption spectroscopy and related physical quantities of narrow-gap semiconductors $Hg_{1-x}Cd_xTe$. Infrared Phys 32:195–211

Chu JH, Gui YS, Li B, Tang DY (1998) Determination of cut-off wavelength and composition distribution in Hg1-xCdxTe. J Electron Mater 27:718–721

Chu L (2001) PhD thesis. Technische Universität München

Chu L, Arzberger M, Zrenner A, Böhm G, Abstreiter G (1999a) Polarization dependent photocurrent spectroscopy of InAs/GaAs quantum dots. Appl Phys Lett 75:2247

Chu L, Zrenner A, Böhm G, Abstreiter G (1999b) Normal-incident intersubband photocurrent spectroscopy on InAs/GaAs quantum dots. Appl Phys Lett 75:3599

Chu L, Arzberger M, Böhm G, Abstreiter G (1999c) Influence of growth conditions on the photoluminescence of self-assembled InAs/GaAs quantum dots. J Appl Phys 85:2355

Chu L, Zrenner A, Böhm G, Abstreiter G (2000) Lateral intersubband photocurrent spectroscopy on InAs/GaAs quantum dots. Appl Phys Lett 76:1944

Chu L, Zrenner A, Bichler M, Abstreiter G (2001) Quantum-dot infrared photodetector with lateral carrier transport. Appl Phys Lett 79:2249

Colombelli R, Capasso F, Gmachl C, Hutchinson AL, Sivco DL, Tredicucci A, Wanke MC, Sergent AM, Cho AY (2001) Far-infrared surface-plasmon quantum-cascade lasers at 21.5 μm and 24 μm wavelengths. Appl Phys Lett 78:2620

Coon DD, Karunasiri RPG (1984) New mode of IR detection using quantum wells. Appl Phys Lett 45:649

Cova S, Lacaita A, Ghioni M, Ripamonti G, Louis TA (1989) 20-ps timing resolution with single-photon avalanche diodes. Rev Sci Instrum 60:1104

Cova S, Ghioni M, Lacaita A, Samori C, Zappa F (1996) Avalanche photodiodes and quenching circuits for single-photon detection. Appl Opt 35:1956–1976

Dornhaus R, Nimtz G (1976) The properties and applications of the MCT alloy system in narrow-gap semiconductors. In: Höhler G, Niekisc EA (eds) Springer Tracts in Modern Physics, vol 78. Springer, Berlin

Dries JC, Gokhale MR, Forrest SR (1999) A 2.0 μm cutoff wavelength separate absorption, charge, and multiplication layer avalanche photodiode using strain-compensated InGaAs quantum wells. Appl Phys Lett 74:2581

Eisenman WL, Merriam JD, Potter JD (1977) Operational Characteristics of Infrared Photodetectors. In: Willardson RK, Beer AC (eds) Semiconductors and Semimetals, vol 12. Academic, New York

Faist J, Capasso F, Sivco DL, Sirtori C, Hutchinson AL, Cho AY (1994) Quantum cascade laser. Science 264:553

Faist J, Capasso F, Sirtori C, Sivco DL, Baillargeon JN, Hutchinson AL, Chu SNG, Cho AY (1996) High power mid-infrared ($\lambda = 5\mu m$) quantum cascade lasers operating above room temperature. Appl Phys Lett 68:3680

Faist J, Capasso F, Sirtori C, Sivco DL, Hutchinson JN, Cho AL (1997a) Laser action by tuning the oscillator strength. Nature 387:777

Faist J, Gmachl C, Capasso F, Sirtori C, Sivco DL, Baillargeon JN, Cho AY (1997b) Distributed feedback quantum cascade lasers. Appl Phys Lett 70:2670

Faist J, Capasso F, Sivco DL, Hutchinson JN, GeorgeChu SN, Cho AL (1998a) Short wavelength ($\lambda \approx 3.4\,\mu m$) quantum cascade laser based on strained compensated InGaAs/AlInAs. Appl Phys Lett 72:680

Faist J, Sirtori C, Capasso F, Sivco Baillargeon DL, Hutchinson JN, Cho AL (1998b) High-power long-wavelength ($\lambda \approx 11.5\mu m$) quantum cascadelasers operating above room temperature. IEEE Photon Technol Lett 10:1100–1102

Faist J, Muller A, Beck M, Hofstetter D, Blaser S, Oesterle U, Ilegems M (2000) A Quantum Cascade Laser Based on an n-i-p-i Superlattice. IEEE Photon Technol Lett 12:263–265

Ferreira R, Bastard G (1989) Evaluation of some scattering times for electrons in unbiased and biased single- and multiple-quantum-well structures. Phys Rev B 40:1074–1086

Finkman E, Nemirovsky Y (1979) Infrared optical absorption of $Hg_{1-x}Cd_xTe$. J Appl Phys 50:4356

Flatté ME, Grein CH, Ehrenreich H, Miles RH (1995) Theoretical performance limits of 2.1–4.1 μm InAs/InGaSb, HgCdTe, and InGaAsSb lasers. J Appl Phys 78:4552

Garbuzov DZ, Martinelli RU, Lee H, York PK, Menna RJ, Connolly JC, Narayan SY (1996) Ultralow-loss broadened-waveguide high-power 2 μm AlGaAsSb/InGaAsSb/GaSb separate-confinement quantum-well lasers. Appl Phys Lett 69:2006

Garbuzov DZ, Martinelli RU, Lee H, Menna RJ, York PK, DiMarco LA, Harvey MG, Matarese RJ, Narayan SY, Connolly JC (1997) 4 W quasi-continuous-wave output power from 2 μm AlGaAsSb/InGaAsSb single-quantum-well broadened waveguide laser diodes. Appl Phys Lett 70:2931

Gärtner WW (1959) Depletion-Layer Photoeffects in Semiconductors. Phys Rev 116: 84–87

Ghioni M, Ripamonti G (1991) Improving the performance of commercially available Geiger-mode avalanche photodiodes. Rev Sci Instrum 62:163

Ghioni M, Cova S, Lacaita A, Ripamonti G (1988) New silicon epitaxial avalanche diode for single-photon timing atroom temperature. Electron Lett 24:1476–1477

Gianordoli S, Schrenk W, Hvozdara L, Finger N, Strasser G, Gornik E (2000) Strained InGaAs/AlGaAs/GaAs-quantum cascade lasers. Appl Phys Lett 76:3361

Gmachl C, Faist J, Bailargeon JN, Capasso F, Sirtori C, Sivco DL, Chu SNG, Cho AY (1997) Complex-coupled quantum cascade distributed-feedback laser. IEEE Photon Technol Lett 9:1090–1092

Gmachl C, Tredicucci A, Capasso F, Hutchinson AL, Sivco DL, Bailargeon JN, Cho AY (1998a) High-power $\lambda \approx 8\,\mu m$ quantum cascade lasers with near optimum performance. Appl Phys Lett 72:3130

Gmachl C, Capasso F, Tredicucci A, Sivco DL, Hutchinson AL, Chu SNG, Cho AY (1998b) Non-cascaded intersubband injection lasers at $\lambda \approx 7.7\,\mu m$. Appl Phys Lett 73:3830

Gmachl C, Tredicucci A, Sivco DL, Hutchinson AL, Capasso F, Cho AY (1999a) Bidirectional Semiconductor Laser. Science 286:749

Gmachl C, Capasso F, Tredicucci A, Sivco DL, Kohler R, Hutchinson AL, Cho AY (1999b) Dependence of the device performance on the number of stages inquantum-cascade lasers. IEEE J Sel Top Quantum Electron 5:808–816

Gmachl C, Capasso F, Tredicucci A, Sivco DL, Baillargeon JN, Hutchinson AL, Cho AY (2000) High-power, continuous-wave, current-tunable, single-mode quantum-cascade distributed-feedback lasers at λ-5.2 and λ-7.95 μm. Opt Lett 25:230–232

Gmachl C, Capasso F, Sivco DL, Cho AY (2001) Recent progress in quantum cascade lasers and applications. Rep Prog Phys 64:1533–1601

Grein CH, Yong PM, Ehrenreich H (1994) Theoretical performance of InAs/$In_xGa_{1-x}Sb$ superlattice-based midwave infrared lasers. J Appl Phys 76:1940

Gunapala SD, Levine BF, Naresh C (1991) Bound to continuum superlattice miniband long wavelength GaAs/$Al_xGa_{1-x}As$ photoconductors. J Appl Phys 70:305

Haecker W, Groezinger O, Pilkuhn MH (1971) Infrared Photon Counting by Ge Avalanche Diodes. Appl Phys Lett 19:113

Haitz RH (1965) Mechanisms Contributing to the Noise Pulse Rate of Avalanche Diodes. J Appl Phys 36:3123

Hakki BW, Paoli TL (1975) Gain spectra in GaAs double–heterostructure injection lasers. J Appl Phys 46:1299

Hansen GL, Schmit JL, Casselman TN (1982) Energy gap versus alloy composition and temperature in $Hg_{1-x}Cd_xTe$. J Appl Phys 53:7099

Hasnain G, Levine BF, Gunapala S, Naresh C (1990) Large photoconductive gain in quantum well infrared photodetectors. Appl Phys Lett 57:608

Hauser JR (1971) Boundary conditions at p-n junctions. Solid-State Electronics 14:133–139

Helm M (1995) Infrared spectroscopy and transport of electrons in semiconductor superlattices. Semicond Sci Technol 10:557–575

Hirota Y, Ando S, Ishibashi T (2004) High-Speed Avalanche Photodiode with a Neutral Absorption Layer for 1.55 μm Wavelength. Jpn J Appl Phys 43:L375–L377

Hofstetter D, Beck M, Faist J (2002) Quantum-cascade-laser structures as photodetectors. Appl Phys Lett 81:2683

Hooge FN (1969) 1/f Noise is no surface effect. Phys Lett A 29:139–140

Janousek BK, Daugherty MJ, Bloss WL, Rosenbluth ML, O'Loughlin MJ, Kanter H, DeLuccia FJ, Perry LE (1990) High-detectivity GaAs quantum well infrared detectors with peak responsivity at 8.2 μm. J Appl Phys 67:7608

Jelen C, Slivken S, Hoff J, Razeghi M, Brown GJ (1997) Aluminum free GaInP/GaAs quantum well infrared photodetectors for long wavelength detection. Appl Phys Lett 70:360

Kane EO (1961) Theory of Tunneling. J Appl Phys 32:83

Kastalsky A, Duffield T, Allen SJ, Harbison J (1988) Photovoltaic detection of infrared light in a GaAs/AlGaAs superlattice. Appl Phys Lett 52:1320

Kheng K, Ramsteiner M, Schneider H, Ralston J, Fuchs F, Koidl P (1992) Two-color GaAs/(AlGa)As quantum well infrared detector with voltage-tunable spectral sensitivity at 3–5 and 8–12 μm. Appl Phys Lett 61:666

Kim S, Mohseni H, Erdtmann M, Michel E, Jelen C, Razeghi M (1998) Growth and characterization of InGaAs/InGaP quantum dots for midinfrared photoconductive detector. Appl Phys Lett 73:963

Kinch MA (1981) Semiconductors and Semimetals, vol 18. Academic, New York, p 313

Kinch MA, Borrello SR (1975) 0.1 ev HgCdTe photodetectors. Infrared Phys. 15:111–124

Kinch MA, Buss DD (1971) Far Infrared Cyclotron Resonance in HgCdTe. In: Carter D, Bate R (eds) Physics of Semimetals and Narrow Gap Semi-conductors. Pergamon Press, Oxford, p 461

Kinch MA, Brau MJ, Simmons A (1973) Recombination mechanisms in 8–14-μ HgCdTe. J Appl Phys 44:1649

Kinch MA, Borrello SR, Simmons A (1977) 0.1 eV HgCdTe photoconductive detector performance. Infrared Phys 17:127–135

Kinch MA, Chapman RA, Simmons A, Buss DD, Borrello SR (1980) HgCdTe charge-coupled device technology. Infrared Phys 20:1–20

Kingston RH (1978) Detection of Optical and Infrared Radiation. Springer-Verlag, Berlin

Köck A, Gornik E, Abstreiter G, Böhm G, Walther M, Weimann G (1992) Double wavelength selective GaAs/AlGaAs infrared detector device. Appl Phys Lett 60:2011

Komiyama S, Astafiev O, Antonov V, Kutsuwa T, Hirai H (2000) A single-photon detector in the far-infrared range. Nature 403:405

Kosterev AA, Curl RF, Tittel FK, Gmachl C, Capasso F, Sivco DL, Baillargeon JN, Hutchinson AL, Cho AY (1999) Methane concentration and isotopic composition measurements with a mid-infrared quantum-cascade laser. Opt Lett 24:1762–1764

Krishnamurthy S, Chen AB, Sher A, van Schilfgaarde M (1995) Temperature dependence of band gaps in HgCdTe and other semiconductors. J Elec Materials 24:1121–1125

Krishnamurthy S, Berding MA, Robinson H, Sher A (2006) Tunneling in long-wavelength infrared HgCdTe photodiodes. J Elec Materials 35:1399–1402

Kruse PW (1979) In: Infrared and Optical Dectors. Springer-Verlag, Berlin

Kruse PW, McGlauchlin LD, McQuistan RB (1962) Elements of Infrared Technology: Generation, Transmission and Detection. Wiley, New York

Lacaita A, Cova S, Ghioni M (1988) Four-hundred-picosecond single-photon timing with commercially available avalanche photodiodes. Rev Sci Instrum 59:1115

Lacaita A, Ghioni M, Cova S (1989) Double epitaxy improves single-photon avalanche diode performance. Electron Lett 25:841–843

Lacaita A, Francese PA, Zappa F, Cova S (1994) Single-photon detection beyond 1 μm: performance of commercially available germanium photodiodes. Appl Opt 33:6902–6918

Lacaita A, Zappa F, Cova S, Lovati P (1996) Single-photon detection beyond 1 μm: performance of commercially available InGaAs/lnP detectors. Appl Opt 35:2986–2996

Lane B, Wu D, Rybaltowski A, Yi H, Diaz J, Razeghi M (1997) Compressively strained multiple quantum well InAsSb lasers emitting at 3.6 μm grown by metal-organic chemical vapor deposition. Appl Phys Lett 70:443

Lanir M, Shin SH (1980) Characterization of limited-diffusion-volume HgCdTe-CdTe photodiodes. Appl Phys 22:57–60

Lanir M, Vanderwyck AHB, Wang CC (1979a) Ebic characterization of HgCdTe crystals and photo-diodes. J Electron Mater 8:175–189

Lanir M, Wang CC, Vanderwyck AHB (1979b) Backside-illuminated HgCdTe/CdTe photodiodes. Appl Phys Lett 34:50

Lauritzen PO (1968) Noise due to generation and recombination of carriers in p-n junction transition regions. IEEE Trans Electron Devices 15:770–776

Lee SW, Hirakawa K, Shimada Y (1999) Bound-to-continuum intersubband photoconductivity of self-assembled InAs quantum dots in modulation-doped heterostructures. Appl Phys Lett 75:1428

Levine BF (1993) Quantum-well infrared photodetectors. J Appl Phys 74:R1–R81

Levine BF, Bethea CG (1984) Single photon detection at 1.3 μm using a gated avalanche photodiode. Appl Phys Lett 44:553

Levine BF, Choi KK, Bethea CG, Walker J, Malik RJ (1987) New 10 μm infrared detector using intersubband absorption in resonant tunneling GaAlAs superlattices. Appl Phys Lett 50:1092

Levine BF, Bethea CG, Choi KK, Walker J, Malik RJ (1988a) Tunneling lifetime broadening of the quantum well intersubband photoconductivity spectrum. Appl Phys Lett 53:231

Levine BF, Bethea CG, Hasnain G, Walker J, Malik RJ (1988b) High-detectivity $D^* = 1.0 \times 10^{10}\text{cmHz}^{1/2}\text{W}^{-1}$ GaAs/AlGaAs multiquantum well $\lambda = 8.3\,\mu$m infrared detector. Appl Phys Lett 53:296

Levine BF, Hasnain G, Bethea CG, Naresh C (1989) Broadband 8–12 μm high-sensitivity GaAs quantum well infrared photodetector. Appl Phys Lett 54:2704

Levine BF, Bethea CG, Hasnain G, Shen VO, Pelve E, Abbott RR, Hsieh SJ (1990) High sensitivity low dark current 10 μm GaAs quantum well infrared photodetectors. Appl Phys Lett 56:851

Levine BF, Zussman A, Kuo JM, de Jong J (1992a) 19 μm cutoff long-wavelength GaAs/$Al_xGa_{1-x}As$ quantum-well infrared photodetectors. J Appl Phys 71:5130

Levine BF, Zussman A, Gunapala SD, Asom MT, Kuo JM, Hobson WS (1992b) Photoexcited escape probability, optical gain, and noise in quantum well infrared photodetectors. J Appl Phys 72:4429

Levinstein H (1970) Characterization of infrared detectors. In: Willardson RK, Beer AC (eds) Semiconductors and Semimetals, vol. 5. Academic, New York, pp 3–12

Lin CH, Yang RQ, Zhang D, Murry SJ, Pei SS, Allerman AA, Kurtz SR (1997) Type-II interband quantum cascade laser at 3.8 μm. Electron Lett 33:598–599

Liu HC (1992) Photoconductive gain mechanism of quantum-well intersubband infrared detectors. Appl Phys Lett 60:1507

Liu Y, Forrest SR, Ban VS, Woodruff KM, Colosi J, Erikson GC, Lange MJ, Olsen GH (1988) Simple, very low dark current, planar long-wavelength avalanche photodiode. Appl Phys Lett 53:1311

Liu Y, Forrest SR, Loo R, Tangonan G, Yen H (1992) Anomalous multiplication in $Hg_{0.56}Cd_{0.44}Te$ avalanche photodiodes. Appl Phys Lett 61:2878

Long D (1970) On generation-recombination noise in infrared detector materials. Infrared Phys 7:169–170

Long D (1977) Optical and Infrared detectors. In: Keyes RJ (ed) Topics in Applied Physics, vol 19. Springer, New York

Long D, Tredwell TJ, Woodfill JR (1978) In: Proceedings of joint meeting IRIS specialty groups infrared detectors and imaging (U), vol. 1, p 387

Martinet E, Luc F, Rosencher E, Bois Ph, Delaitre S (1992) Electrical tunability of infrared detectors using compositionally asymmetric GaAs/AlGaAs multiquantum wells. Appl Phys Lett 60:895

Mazurczyk VJ, Graney RN, McCullou JB (1974) High-performance, wide bandwidth (Hg,Cd)Te detectors. Opt Eng 13:307–311

Melchior H, Lynch WT (1966) Signal and noise response of high speed germanium avalanche photodiodes. IEEE Trans Electron Devices 13:829–838

Meyer JR, Hoffman CA, Bartoli FJ, Ram-Mohan LR (1995) Type-II quantum-well lasers for the mid-wavelength infrared. Appl Phys Lett 67:757

Mohseni H, Michel E, Sandoen Jan, Razeghi M, Mitchel W, Brown G (1997) Growth and characterization of InAs/GaSb photoconductors for long wavelength infrared range. Appl Phys Lett 71:1403

Moss TS (1954) The Interpretation of the Properties of Indium Antimonide. Proc Phys Soc B 67:775–782

Musienko Y, Reucroft S, Swain J (2000) A simple model of EG&G reverse reach-through APDs. Nucl Instrum Meth A 442:179–186

Nemirovsky Y, Fastaw R, Mayassad M, Unikovsky A (1991) Trapping effects in HgCdTe. J Vac Sci Technol B 9:1829

Namjou K, Cai S, Whittaker EA, Faist J, Gmachl C, Capasso F, Sivco DL, Cho AY (1998) Sensitive absorption spectroscopy with a room-temperature distributed-feedback quantum-cascade laser. Opt Lett 23:219–221

Paldus BA, Spence TG, Zare RN, Oomens J, Harren FJM, Parker DH, Gmachl C, Cappasso F, Sivco DL, Baillargeon JN, Hutchinson AL, Cho AY (1999) Photoacoustic spectroscopy using quantum-cascade lasers. Opt Lett 24:178–180

Paldus BA, Harb CC, Spence TG, Zare RN, Gmachl C, Cappasso F, Sivco DL, Baillargeon JN, Hutchinson AL, Cho AY (2000) Cavity ringdown spectroscopy using mid-infrared quantum-cascade lasers. Opt Lett 25:666–668

Pellegrini PW, Shepherd FD (1983) The evolution of metal silicide schottky-barrier infrared focal plane detectors. Proc of SPIE 409:66–68

Peyton BJ, DiNardo AJ, Kanischak GM, Arams F, Lange R, Sard E (1972) IEEE J Quantum Electron 8:252–263

Phillips JD, Edwall DD, Lee DL, Arias J (2001) Growth of HgCdTe for long-wavelength infrared detectors using automated control from spectroscopic ellipsometry measurements. J Vac Sci Technol B 19:1580–1584

Phillips JD, Edwall DD, Lee DL (2002) Control of very-long-wavelength infrared HgCdTe detector-cutoff wavelength. J Electron Mater 31:664–668

Piotrowski J, Gawron W (1997) Ultimate performance of infrared photodetectors and figure of merit of detector material. Infrared Phys Technol 38:63–68

Price SL, Royd PR (1993) Overview of compositional measurement techniques for HgCdTe with emphasis on IR transmission, energy dispersive X-ray analysis and optical reflectance. Semicond Sci Technol 8:842–859

Pruett GR, Petritz RL (1959) Detectivity and Preamplifier Considerations for Indium Antimonide Photovoltaic Detectors. Proc IRE 47:1524–1529

Reine MB, Sood AK, Tredwell TJ (1981) Photovoltaic infrared detectors. Semiconductors and semimetals, vol 18. Academic, New York, p 201

Rittner ES (1956) Electron processes in photoconductors. In: Breckenridge R, Russell B, Hautz E (eds) Photoconductivity Conference. Wiley, New York

Rochas A, Gani M, Furrer B, Besse PA, Popovic RS, Ribordy G, Gisin N (2003) Single photon detector fabricated in a complementary metal–oxide–semiconductor high-voltage technology. Rev Sci Instrum 74:3263

Rogalski A (2003) Quantum well photoconductors in infrared detector technology. J Appl Phys 93:4355

Rosencher E, Vinter B, Levine B (1992) Intersubband transitions in quantum wells. Plenum, New York

Ryzhii V (1996) The theory of quantum-dot infrared phototransistors. Semicond Sci Technol 11:759–765

Sah CT (1961) Electronic Processes and Excess Currents in Gold-Doped Narrow Silicon Junctions. Phys Rev 123:1594–1612

Sah CT, Noyce RN, Shockley W (1957) Carrier Generation and Recombination in P-N Junctions and P-N Junction Characteristics. Proc IRE 45:1228–1243

Scamarcio G, Capasso F, Faist J, Sirtori C, Sivco DL, Hutchinson AL, Cho AY (1997a) Tunable interminiband infrared emission in superlattice electron transport. Appl Phys Lett 70:1796

Scamarcio G, Capasso F, Sirtori C, Faist J, Hutchinson AL, Sivco DL, Cho AY (1997b) High-Power Infrared (8-Micrometer Wavelength) Superlattice Lasers. Science 276:773

Schmit JL (1970) Intrinsic Carrier Concentration of $Hg_{1-x}Cd_xTe$ as a Function of x and T Using $k \cdot p$ Calculations. J Appl Phys 41:2876

Schmit JL, Stelzer EL (1969) Temperature and Alloy Compositional Dependences of the Energy Gap of $Hg_{1-x}Cd_xTe$. J Appl Phys 40:4865

Shanley JF, Perry LC (1978) Wide bandwidth 10.6 μm (Hg,Cd)Te photodiodes for infrared heterodyne applications. IEEE Int Electron Device Meeting Tech Digest 24:424–429

Sharpe SW, Kelly JF, Hartman JS, Gmachl C, Capasso F, Sivco DL, Baillargeon JN, Cho AY (1998) High-resolution (Doppler-limited) spectroscopy using quantum-cascade distributed-feedback lasers. Opt Lett 23:1396–1398

Shepherd FD (1983) Silicide infrared camera technology. Proc SPIE 395:114–118

Sher A, Berding MA, Van Schilfgaarde M, Chen AB (1991) HgCdTe status review with emphasis on correlations, native defects and diffusion. Semicond Sci Technol 6:C59–C70

Shockley W (1949) The theory of p-n junctions in semiconductors and p-n junction transistors. Bell Syst Tech J 28:435–489

Sirtori C, Capasso F, Faist J, Sivco DL, Hutchinson AL, Cho AY (1995a) Quantum cascade unipolar intersubband light emitting diodes in the 8–13 μm wavelength region. Appl Phys Lett 66:4

Sirtori C, Faist J, Capasso F, Sivco DL, Hutchinson AL, Cho AY (1995b) Quantum cascade laser with plasmon-enhanced waveguide operating at 8.4 μm wavelength. Appl Phys Lett 66:3242

Stier O, Grundmann M, Bimberg D (1999) Electronic and optical properties of strained quantum dots modeled by 8-band k · p theory. Phys Rev B 59:5688

Stillman GE, Wolfe CM (1977) Avalanche Photodiodes. In: Willardson RK, Beer AC (eds) Semiconductors and semimetals, vol 12, Chap 5. Academic, New York, p 291

Sood AK, Marciniec JW, Reine MB (1979a) Moderate temperature detector development. Final Report for NASA contract NAS9–15250

Sood AK, Marciniec JW, Reine MB (1979b) Proceedings of meeting IRIS specialty group infrared detectors (U) vol. 1, p 171

Sood AK, Marciniec JW, Reine MB (1979c) Final Rep US Naval Research Laboratory Contract N00173–78-C-0145

Sze SM (1969) Physics of semiconductor devices. Wiley, New York

Tang DY, Mi ZY (1989) Summy of photoelectric device. Science and Technology. Document, Shanghai

Tang DY, Tong FM (1991) In: Research and Progress of semiconductor devices. Science Press, Beijing, p 1

Tobin SP, Iwasa S, Tredwell TJ (1980) 1/f Noise in (Hg, Cd)Te photodiodes. IEEE Trans Electron Devices 27:43–48

Tredicucci A, Capasso F, Gmachl C, Sivco DL, Hutchinson AL, Cho AY, Faist J, Scamarcio G (1998) High-power inter-miniband lasing in intrinsic superlattices. Appl Phys Lett 72:2388

Tredicucci A, Gmachl C, Capasso F, Sivco DL, Hutchinson AL, Cho AY (1999) Long wavelength superlattice quantum cascade lasers at $\lambda \approx 17\,\mu m$. Appl Phys Lett 74:638

Tredicucci A, Capasso F, Gmachl C, Sivco DL, Hutchinson AL, Chu SNG, Cho AY (2000a) Continuous wave operation of long wavelength ($\lambda \approx 11\,\mu m$) inter-miniband lasers. Electron Lett 36:876

Tredicucci A, Gmachl C, Capasso F, Hutchinson AL, Sivco DL, Cho AY (2000b) Single-mode surface-plasmon laser. Appl Phys Lett 76:2164

Tredwell TJ, Long D (1977) Final Rep NASA Lyndon B Johnson Space Center Contract MAS9–14180, 5S

Van der Wiele F (1976) In: Jespers PG, Van de Wiele F, White MH (eds) Solid stae imaging. Noordhoff, Leyden, p 47

Van der Ziel A (1959) Fluctuation phenomena in semiconductors. Butterworth, London

Van der Ziel A, Chenette ER (1978) Noise in solid state devices. Adv Electron Phys 46:313

Van Vliet KM (1958) Noise in Semiconductors and Photoconductors. Proc IRE 46:1004–1018

Van Vliet KM (1967) Noise Limitations in Solid State Photodetectors. Appl Opt 6:1145–1169

Van Vliet KM (1970) Noise sources in transport equations associated with ambipolar diffusion and Shockley-Read recombination. Solid-State Electronics 13:649–657

Van Vliet KM (1976) Noise and admittance of the generation – Recombination current involving SRH centers in the space-charge region of junction devices. IEEE Trans. Electron Devices 23:1236–1246

Van Vliet KM, Van der Ziel (1977) Physical interpretation of the noise reduction for the generation-Recombination current due to SRH centers in the space-charge layer. IEEE Trans Electron Devices 24:1127–1129

Wang ZG (2001) Nano-semiconductor materials and nano-device research (continued). Semicond Technol 26:18–21

Weiler MH (1981) Nonparabolicity and exciton effects in two-photon absorption in zincblende semiconductors. Solid State Commun 39:937–940

West LC, Eglash SJ (1985) First observation of an extremely large-dipole infrared transition within the conduction band of a GaAs quantum well. Appl Phys Lett 46:1156

Williams RL (1968) Sensitivity limits of 0.1 eV intrinsic photoconductors. Infrared Phys 8: 337–343

Williams RM, Kelly JF, Hartman JS, Sharpe SW, Taubman MS, Hall JL, Capasso F, Gmachl C, Sivco DL, Baillargeon JN, Cho AY (1999) Kilohertz linewidth from frequency-stabilized mid-infrared quantum cascade lasers. Opt Lett 24:1844–1846

Wong JY (1980) Effect of trap tunneling on the performance of long-wavelength Hg1-xCdxTe photodiodes. IEEE Trans Electron Devices 27:48–57

Xu GS, Fang JX (1996) Basic applications studies of HgCdTe materials and devices. In: 95 anniversary collection of the thesis. Shanghai Institute of Technical Physics, p 7

Yang RQ (2000) Novel concepts and structures for infrared lasers. In: Helm M (ed) Long wavelength infrared emitters based on quantum wells and superlattices, Chap 2. Gordon and Breach Science, Singapore

Yang RQ, Lin CH, Murry SJ, Pei SS, Liu HC, Buchanan M, Dupont E (1997) Interband cascade light emitting diodes in the 5–8 μm spectrum region. Appl Phys Lett 70:2013

Youngdale ER, Meyer JR, Hoffman CA, Bartoli FJ, Grein CH, Young PM, Ehrenreich H, Miles RH, Chow DH (1994) Auger lifetime enhancement in InAs-$Ga_{1-x}In_xSb$ superlattices. Appl Phys Lett 64:3160

Yu PY, Cardona M (1999) Fundamentals of semiconductors: Physics and materials properties. Springer, New York

Zegrya GG, Andreev AD (1995) Mechanism of suppression of Auger recombination processes in type-II heterostructures. Appl Phys Lett 67:2681

Zhang D, Yang R, Lin CH, Pei SS, Dupont E, Liu H, Buchanan M (1997) Long-wavelength infrared (~10–15 mm) electroluminescence from Sb-based interband cascade devices. Optics Express 1:97–101

Zhang YH (1995) Continuous wave operation of $InAs/InAs_xSb_{1-x}$ midinfrared lasers. Appl Phys Lett 66:118

Zussman A, Levine BF, Kuo JM, deJong J (1991) Extended long-wavelength $\lambda = 11–15$-μm $GaAs/Al_xGa_{1-x}As$ quantum-well infrared photodetectors. J Appl Phys 70:5101

Appendix

I Various Quantities for $Hg_{1-x}Cd_xTe$

1 *Energy band gap E_g (eV) from (A.1) (Appendix Part II)*

	T					
x	4.2 K	30 K	77 K	100 K	200 K	300 K
0.165	0.0078	0.0175	0.0352	0.0439	0.0816	0.1193
0.170	0.0166	0.0262	0.0436	0.0521	0.0892	0.1263
0.175	0.0255	0.0349	0.0520	0.0604	0.0968	0.1332
0.180	0.0344	0.0436	0.0604	0.0686	0.1044	0.1402
0.185	0.0432	0.0523	0.0688	0.0769	0.1120	0.1471
0.190	0.0521	0.0609	0.0772	0.0851	0.1196	0.1540
0.195	0.0609	0.0696	0.0855	0.0933	0.1271	0.1610
0.200	0.0697	0.0783	0.0939	0.1015	0.1347	0.1679
0.205	0.0785	0.0869	0.1022	0.1097	0.1423	0.1748
0.210	0.0873	0.0956	0.1106	0.1179	0.1498	0.1818
0.215	0.0961	0.1042	0.1189	0.1261	0.1574	0.1887
0.220	0.1049	0.1128	0.1272	0.1343	0.1649	0.1956
0.225	0.1137	0.1214	0.1355	0.1424	0.1725	0.2025
0.230	0.1225	0.1300	0.1438	0.1506	0.1800	0.2094
0.235	0.1312	0.1386	0.1521	0.1588	0.1875	0.2163
0.240	0.1400	0.1472	0.1604	0.1669	0.1950	0.2232
0.245	0.1487	0.1558	0.1687	0.1751	0.2026	0.2301
0.250	0.1574	0.1644	0.1770	0.1832	0.2101	0.2369
0.255	0.1662	0.1729	0.1853	0.1913	0.2176	0.2438
0.260	0.1749	0.1815	0.1936	0.1994	0.2251	0.2507
0.265	0.1836	0.1901	0.2018	0.2076	0.2326	0.2576
0.270	0.1923	0.1986	0.2101	0.2157	0.2401	0.2645
0.275	0.2010	0.2072	0.2183	0.2238	0.2476	0.2713
0.280	0.2097	0.2157	0.2266	0.2319	0.2551	0.2782
0.285	0.2184	0.2242	0.2348	0.2400	0.2625	0.2851

(continued)

(continued)

	T					
x	4.2 K	30 K	77 K	100 K	200 K	300 K
0.290	0.2271	0.2328	0.2431	0.2481	0.2700	0.2919
0.295	0.2358	0.2413	0.2513	0.2562	0.2775	0.2988
0.300	0.2445	0.2498	0.2595	0.2643	0.285	0.3057
0.305	0.2531	0.2583	0.2678	0.2724	0.2925	0.3126
0.310	0.2618	0.2668	0.2760	0.2805	0.2999	0.3194
0.315	0.2705	0.2753	0.2842	0.2885	0.3074	0.3263
0.320	0.2791	0.2838	0.2924	0.2966	0.3149	0.3332
0.325	0.2878	0.2923	0.3006	0.3047	0.3224	0.3400
0.330	0.2964	0.3008	0.3089	0.3128	0.3298	0.3469
0.335	0.3051	0.3093	0.3171	0.3209	0.3373	0.3538
0.340	0.3137	0.3178	0.3253	0.3289	0.3448	0.3607
0.345	0.3224	0.3263	0.3335	0.3370	0.3523	0.3675
0.350	0.331	0.3348	0.3417	0.3451	0.3598	0.3744
0.355	0.3397	0.3433	0.3499	0.3532	0.3672	0.3813
0.360	0.3483	0.3518	0.3581	0.3612	0.3747	0.3882
0.365	0.3570	0.3603	0.3663	0.3693	0.3822	0.3951
0.370	0.3656	0.3688	0.3746	0.3774	0.3897	0.402
0.375	0.3742	0.3773	0.3828	0.3855	0.3972	0.4089
0.380	0.3829	0.3858	0.3910	0.3935	0.4047	0.4158
0.385	0.3915	0.3943	0.3992	0.4016	0.4122	0.4227
0.390	0.4002	0.4027	0.4074	0.4097	0.4197	0.4296
0.395	0.4088	0.4112	0.4157	0.4178	0.4272	0.4366
0.400	0.4175	0.4198	0.4239	0.4259	0.4347	0.4435
0.405	0.4261	0.4283	0.4321	0.434	0.4422	0.4505
0.410	0.4348	0.4368	0.4404	0.4421	0.4498	0.4574
0.415	0.4435	0.4453	0.4486	0.4502	0.4573	0.4644
0.420	0.4521	0.4538	0.4568	0.4583	0.4648	0.4713
0.425	0.4608	0.4623	0.4651	0.4665	0.4724	0.4783
0.430	0.4695	0.4708	0.4734	0.4746	0.4799	0.4853
0.435	0.4781	0.4794	0.4816	0.4827	0.4875	0.4923
0.440	0.4868	0.4879	0.4899	0.4909	0.4951	0.4993
0.445	0.4955	0.4965	0.4982	0.4990	0.5027	0.5063
0.450	0.5042	0.5050	0.5065	0.5072	0.5103	0.5133
0.455	0.5129	0.5136	0.5148	0.5153	0.5179	0.5204
0.460	0.5217	0.5222	0.5231	0.5235	0.5255	0.5274
0.465	0.5304	0.5307	0.5314	0.5317	0.5331	0.5345
0.470	0.5391	0.5393	0.5397	0.5399	0.5407	0.5416
0.475	0.5479	0.5479	0.5480	0.5481	0.5484	0.5486
0.480	0.5566	0.5565	0.5564	0.5563	0.5560	0.5558
0.485	0.5654	0.5651	0.5648	0.5646	0.5637	0.5629
0.490	0.5741	0.5738	0.5731	0.5728	0.5714	0.5700
0.495	0.5829	0.5824	0.5815	0.5811	0.5791	0.5772
0.500	0.5917	0.5911	0.5899	0.5893	0.5868	0.5843
0.505	0.6005	0.5997	0.5983	0.5976	0.5946	0.5915

(continued)

(continued)

	T					
x	4.2 K	30 K	77 K	100 K	200 K	300 K
0.510	0.6093	0.6084	0.6067	0.6059	0.6023	0.5987
0.515	0.6182	0.6171	0.6152	0.6142	0.6101	0.6059
0.520	0.6270	0.6258	0.6236	0.6225	0.6179	0.6132
0.525	0.6359	0.6345	0.6321	0.6309	0.6257	0.6204
0.530	0.6448	0.6433	0.6406	0.6392	0.6335	0.6277
0.535	0.6537	0.652	0.6491	0.6476	0.6413	0.6350
0.540	0.6626	0.6608	0.6576	0.6560	0.6492	0.6423
0.545	0.6715	0.6696	0.6661	0.6644	0.6570	0.6496
0.550	0.6804	0.6784	0.6747	0.6729	0.6649	0.6570
0.555	0.6894	0.6872	0.6832	0.6813	0.6728	0.6644
0.560	0.6984	0.6961	0.6918	0.6898	0.6808	0.6718
0.565	0.7074	0.7049	0.7005	0.6983	0.6887	0.6792
0.570	0.7164	0.7138	0.7091	0.7068	0.6967	0.6867
0.575	0.7254	0.7227	0.7177	0.7153	0.7047	0.6941
0.580	0.7345	0.7316	0.7264	0.7239	0.7127	0.7016
0.585	0.7436	0.7406	0.7351	0.7324	0.7208	0.7092
0.590	0.7527	0.7495	0.7438	0.7410	0.7289	0.7167
0.595	0.7618	0.7585	0.7526	0.7497	0.7370	0.7243
0.600	0.7710	0.7676	0.7613	0.7583	0.7451	0.7319
0.605	0.7801	0.7766	0.7701	0.7670	0.7533	0.7395
0.610	0.7893	0.7857	0.7790	0.7757	0.7614	0.7472
0.615	0.7985	0.7947	0.7878	0.7844	0.7697	0.7549
0.620	0.8078	0.8039	0.7967	0.7932	0.7779	0.7626
0.625	0.8171	0.8130	0.8056	0.8019	0.7862	0.7704
0.630	0.8264	0.8222	0.8145	0.8108	0.7945	0.7782
0.635	0.8357	0.8314	0.8235	0.8196	0.8028	0.7860
0.640	0.8451	0.8406	0.8325	0.8285	0.8112	0.7938
0.645	0.8544	0.8498	0.8415	0.8374	0.8196	0.8017
0.650	0.8639	0.8591	0.8505	0.8463	0.8280	0.8097
0.655	0.8733	0.8684	0.8596	0.8553	0.8364	0.8176
0.660	0.8828	0.8778	0.8687	0.8643	0.8449	0.8256
0.665	0.8923	0.8872	0.8779	0.8733	0.8535	0.8336
0.670	0.9018	0.8966	0.8870	0.8824	0.8620	0.8417
0.675	0.9114	0.9060	0.8962	0.8915	0.8706	0.8498
0.680	0.9210	0.9155	0.9055	0.9006	0.8793	0.8579
0.685	0.9307	0.9250	0.9148	0.9098	0.8879	0.8661
0.690	0.9403	0.9346	0.9241	0.9190	0.8966	0.8743
0.695	0.9501	0.9442	0.9334	0.9282	0.9054	0.8826
0.700	0.9598	0.9538	0.9428	0.9375	0.9142	0.8909

2 Wavelengths corresponding to energy gaps λ_{Eg} (μm)

	T					
x	4.2 K	30 K	77 K	100 K	200 K	300 K
0.165	159.83	70.97	35.26	28.29	15.22	10.41
0.170	74.56	47.37	28.46	23.81	13.92	9.83
0.175	48.64	35.56	23.86	20.56	12.83	9.32
0.180	36.11	28.47	20.55	18.09	11.89	8.86
0.185	28.72	23.74	18.05	16.15	11.09	8.44
0.190	23.84	20.36	16.09	14.59	10.38	8.06
0.195	20.38	17.83	14.52	13.30	9.76	7.71
0.200	17.81	15.86	13.22	12.23	9.22	7.39
0.205	15.81	14.28	12.14	11.32	8.73	7.10
0.210	14.21	12.99	11.23	10.53	8.29	6.83
0.215	12.91	11.91	10.44	9.85	7.89	6.58
0.220	11.83	11.00	9.76	9.25	7.53	6.35
0.225	10.92	10.22	9.16	8.72	7.20	6.13
0.230	10.14	9.55	8.63	8.24	6.90	5.93
0.235	9.46	8.96	8.16	7.82	6.62	5.74
0.240	8.87	8.43	7.74	7.44	6.37	5.56
0.245	8.35	7.97	7.36	7.09	6.13	5.40
0.250	7.89	7.55	7.01	6.78	5.91	5.24
0.255	7.47	7.18	6.70	6.49	5.71	5.09
0.260	7.10	6.84	6.42	6.23	5.52	4.95
0.265	6.76	6.53	6.15	5.98	5.34	4.82
0.270	6.46	6.25	5.91	5.76	5.17	4.70
0.275	6.18	5.99	5.69	5.55	5.02	4.58
0.280	5.92	5.76	5.48	5.35	4.87	4.46
0.285	5.69	5.54	5.29	5.17	4.73	4.36
0.290	5.47	5.33	5.11	5.01	4.60	4.25
0.295	5.27	5.15	4.94	4.85	4.47	4.16
0.300	5.08	4.97	4.78	4.70	4.36	4.06
0.305	4.91	4.81	4.64	4.56	4.25	3.97
0.310	4.74	4.65	4.50	4.43	4.14	3.89
0.315	4.59	4.51	4.37	4.30	4.04	3.81
0.320	4.45	4.38	4.25	4.19	3.94	3.73
0.325	4.32	4.25	4.13	4.08	3.85	3.65
0.330	4.19	4.13	4.02	3.97	3.76	3.58
0.335	4.07	4.01	3.92	3.87	3.68	3.51
0.340	3.96	3.91	3.82	3.78	3.60	3.44
0.345	3.85	3.81	3.72	3.68	3.53	3.38
0.350	3.75	3.71	3.63	3.60	3.45	3.32
0.355	3.66	3.62	3.55	3.52	3.38	3.26
0.360	3.57	3.53	3.47	3.44	3.31	3.20
0.365	3.48	3.45	3.39	3.36	3.25	3.14
0.370	3.40	3.37	3.32	3.29	3.19	3.09
0.375	3.32	3.29	3.24	3.22	3.13	3.04

(continued)

(continued)

x	T					
	4.2 K	30 K	77 K	100 K	200 K	300 K
0.380	3.24	3.22	3.18	3.16	3.07	2.99
0.385	3.17	3.15	3.11	3.09	3.01	2.94
0.390	3.10	3.08	3.05	3.03	2.96	2.89
0.395	3.04	3.02	2.99	2.97	2.91	2.84
0.400	2.97	2.96	2.93	2.92	2.86	2.80
0.405	2.91	2.90	2.87	2.86	2.81	2.76
0.410	2.86	2.84	2.82	2.81	2.76	2.72
0.415	2.80	2.79	2.77	2.76	2.72	2.67
0.420	2.75	2.74	2.72	2.71	2.67	2.63
0.425	2.7	2.69	2.67	2.66	2.63	2.60
0.430	2.65	2.64	2.62	2.62	2.59	2.56
0.435	2.60	2.59	2.58	2.57	2.55	2.52
0.440	2.55	2.55	2.53	2.53	2.51	2.49
0.445	2.51	2.50	2.49	2.49	2.47	2.45
0.450	2.46	2.46	2.45	2.45	2.43	2.42
0.455	2.42	2.42	2.41	2.41	2.40	2.39
0.460	2.38	2.38	2.37	2.37	2.36	2.35
0.465	2.34	2.34	2.34	2.34	2.33	2.32
0.470	2.30	2.30	2.30	2.30	2.30	2.29
0.475	2.27	2.27	2.27	2.27	2.26	2.26
0.480	2.23	2.23	2.23	2.23	2.23	2.23
0.485	2.2	2.20	2.20	2.20	2.20	2.21
0.490	2.16	2.16	2.17	2.17	2.17	2.18
0.495	2.13	2.13	2.14	2.14	2.14	2.15
0.500	2.10	2.10	2.11	2.11	2.12	2.13
0.505	2.07	2.07	2.08	2.08	2.09	2.10
0.510	2.04	2.04	2.05	2.05	2.06	2.07
0.515	2.01	2.01	2.02	2.02	2.04	2.05
0.520	1.98	1.98	1.99	1.99	2.01	2.03
0.525	1.95	1.96	1.96	1.97	1.98	2.00
0.530	1.93	1.93	1.94	1.94	1.96	1.98
0.535	1.90	1.90	1.91	1.92	1.94	1.96
0.540	1.87	1.88	1.89	1.89	1.91	1.93
0.545	1.85	1.85	1.86	1.87	1.89	1.91
0.550	1.83	1.83	1.84	1.85	1.87	1.89
0.555	1.80	1.81	1.82	1.82	1.85	1.87
0.560	1.78	1.78	1.79	1.80	1.82	1.85
0.565	1.76	1.76	1.77	1.78	1.80	1.83
0.570	1.73	1.74	1.75	1.76	1.78	1.81
0.575	1.71	1.72	1.73	1.74	1.76	1.79
0.580	1.69	1.70	1.71	1.72	1.74	1.77
0.585	1.67	1.68	1.69	1.70	1.72	1.75
0.590	1.65	1.66	1.67	1.68	1.70	1.73
0.595	1.63	1.64	1.65	1.66	1.69	1.71
0.600	1.61	1.62	1.63	1.64	1.67	1.70

(continued)

(continued)

x	T					
	4.2 K	30 K	77 K	100 K	200 K	300 K
0.605	1.59	1.60	1.61	1.62	1.65	1.68
0.610	1.57	1.58	1.59	1.60	1.63	1.66
0.615	1.56	1.56	1.58	1.58	1.61	1.65
0.620	1.54	1.54	1.56	1.57	1.60	1.63
0.625	1.52	1.53	1.54	1.55	1.58	1.61
0.630	1.50	1.51	1.52	1.53	1.56	1.60
0.635	1.49	1.49	1.51	1.52	1.55	1.58
0.640	1.47	1.48	1.49	1.50	1.53	1.56
0.645	1.45	1.46	1.48	1.48	1.52	1.55
0.650	1.44	1.45	1.46	1.47	1.50	1.53
0.655	1.42	1.43	1.44	1.45	1.48	1.52
0.660	1.41	1.41	1.43	1.44	1.47	1.50
0.665	1.39	1.40	1.41	1.42	1.46	1.49
0.670	1.38	1.39	1.40	1.41	1.44	1.48
0.675	1.36	1.37	1.39	1.39	1.43	1.46
0.680	1.35	1.36	1.37	1.38	1.41	1.45
0.685	1.33	1.34	1.36	1.37	1.40	1.43
0.690	1.32	1.33	1.34	1.35	1.39	1.42
0.695	1.31	1.32	1.33	1.34	1.37	1.41
0.700	1.29	1.30	1.32	1.32	1.36	1.39

3 *Peak-wavelengths of the photo-conductive response* λ_{peak} *and the cut-off wavelengths* λ_{co} (μm) *for samples with a thickness* $d = 10\,\mu$m *(from (A.2) to (A.3) in Appendix Part II)*

x	T									
	77 K		100 K		213 K		233 K		300 K	
	λ_{peak}	λ_{co}	λ_{peak}	λ_{co}	λ_{peak}	λ_{co}	λ_{peak}	λ_{co}	λ_{peak}	λ_{co}
9.94	18.10	23.94	15.23	20.68	8.74	12.68	8.18	11.91	6.82	
0.185	16.09	20.65	13.80	18.23	8.30	11.80	7.80	11.14	6.57	9.43
0.190	14.48	18.16	12.62	16.29	7.90	11.03	7.45	10.46	6.34	8.96
0.195	13.17	16.20	11.62	14.73	7.53	10.35	7.13	9.86	6.13	8.54
0.200	12.07	14.63	10.77	13.44	7.20	9.76	6.84	9.33	5.93	8.16
0.205	11.14	13.33	10.04	12.36	6.90	9.22	6.57	8.85	5.74	7.81
0.210	10.35	12.25	9.40	11.44	6.62	8.75	6.32	8.41	5.57	7.48
0.215	9.66	11.32	8.83	10.65	6.36	8.32	6.09	8.02	5.40	7.19
0.220	9.06	10.53	8.33	9.96	6.12	7.93	5.88	7.66	5.25	6.91
0.225	8.52	9.84	7.89	9.35	5.90	7.57	5.68	7.34	5.10	6.66
0.230	8.05	9.24	7.49	8.81	5.70	7.25	5.49	7.04	4.96	6.42
0.235	7.63	8.70	7.13	8.33	5.51	6.95	5.32	6.76	4.83	6.20
0.240	7.25	8.23	6.80	7.91	5.33	6.68	5.15	6.5	4.70	6.00

(continued)

(continued)

	T									
	77 K		100 K		213 K		233 K		300 K	
x	λ_{peak}	λ_{co}	λ_{peak}	λ_{co}	λ_{peak}	λ_{co}	λ_{peak}	λ_{co}	λ_{peak}	λ_{co}
0.245	6.90	7.80	6.50	7.52	5.16	6.42	5.00	6.27	4.58	5.81
0.250	6.59	7.42	6.22	7.17	5.00	6.19	4.86	6.05	4.47	5.63
0.255	6.30	7.07	5.97	6.85	4.85	5.97	4.72	5.84	4.36	5.46
0.260	6.04	6.75	5.74	6.56	4.71	5.77	4.59	5.65	4.26	5.30
0.265	5.80	6.46	5.52	6.29	4.58	5.58	4.47	5.47	4.16	5.15
0.270	5.58	6.19	5.32	6.04	4.46	5.40	4.35	5.30	4.07	5.01
0.275	5.37	5.95	5.14	5.81	4.34	5.23	4.24	5.14	3.98	4.87
0.280	5.18	5.72	4.96	5.60	4.23	5.07	4.14	4.99	3.90	4.74
0.285	5.00	5.51	4.80	5.40	4.12	4.93	4.04	4.85	3.81	4.62
0.290	4.83	5.32	4.65	5.22	4.02	4.79	3.94	4.72	3.74	4.51
0.295	4.68	5.14	4.51	5.05	3.92	4.66	3.85	4.59	3.66	4.40
0.300	4.53	4.97	4.37	4.89	3.83	4.53	3.76	4.47	3.59	4.30
0.305	4.39	4.81	4.25	4.74	3.74	4.41	3.68	4.36	3.52	4.20
0.310	4.26	4.66	4.13	4.59	3.66	4.30	3.60	4.25	3.45	4.10
0.315	4.14	4.52	4.02	4.46	3.58	4.19	3.53	4.15	3.39	4.01
0.320	4.03	4.39	3.91	4.33	3.5	4.09	3.45	4.05	3.33	3.92
0.325	3.92	4.26	3.81	4.22	3.43	4.00	3.38	3.96	3.27	3.84
0.330	3.82	4.15	3.71	4.10	3.36	3.90	3.32	3.87	3.21	3.76
0.335	3.72	4.03	3.62	4.00	3.29	3.82	3.25	3.79	3.15	3.69
0.340	3.62	3.93	3.53	3.89	3.23	3.73	3.19	3.70	3.10	3.61
0.345	3.54	3.83	3.45	3.80	3.16	3.65	3.13	3.63	3.05	3.54
0.350	3.45	3.73	3.37	3.71	3.10	3.57	3.07	3.55	3.00	3.47
0.355	3.37	3.64	3.30	3.62	3.05	3.50	3.02	3.48	2.95	3.41
0.360	3.30	3.56	3.23	3.54	2.99	3.43	2.96	3.41	2.90	3.35
0.365	3.22	3.47	3.16	3.46	2.94	3.36	2.91	3.34	2.86	3.29
0.370	3.15	3.40	3.09	3.38	2.88	3.30	2.86	3.28	2.81	3.23
0.375	3.09	3.32	3.03	3.31	2.83	3.23	2.81	3.22	2.77	3.17
0.380	3.02	3.25	2.97	3.24	2.79	3.17	2.77	3.16	2.73	3.12
0.385	2.96	3.18	2.91	3.17	2.74	3.11	2.72	3.10	2.69	3.06
0.390	2.90	3.12	2.85	3.11	2.69	3.06	2.68	3.05	2.65	3.01
0.390	2.84	3.05	2.8	3.05	2.65	3.00	2.64	2.99	2.61	2.96
0.400	2.79	2.99	2.74	2.99	2.61	2.95	2.60	2.94	2.57	2.92
0.405	2.74	2.93	2.69	2.93	2.57	2.90	2.56	2.89	2.54	2.87
0.410	2.69	2.88	2.65	2.87	2.53	2.85	2.52	2.85	2.50	2.82
0.415	2.64	2.82	2.60	2.82	2.49	2.80	2.48	2.80	2.47	2.78
0.420	2.59	2.77	2.55	2.77	2.45	2.76	2.44	2.75	2.44	2.74
0.425	2.54	2.72	2.51	2.72	2.42	2.71	2.41	2.71	2.40	2.70
0.430	2.50	2.67	2.47	2.67	2.38	2.67	2.37	2.67	2.37	2.66
0.435	2.46	2.63	2.43	2.63	2.35	2.63	2.34	2.63	2.34	2.62
0.440	2.42	2.58	2.39	2.58	2.31	2.59	2.31	2.59	2.31	2.58
0.445	2.38	2.54	2.35	2.54	2.28	2.55	2.28	2.55	2.28	2.55
0.450	2.34	2.49	2.31	2.50	2.25	2.51	2.25	2.51	2.25	2.51
0.455	2.30	2.45	2.28	2.46	2.22	2.48	2.22	2.48	2.23	2.48
0.460	2.27	2.41	2.24	2.42	2.19	2.44	2.19	2.44	2.20	2.44

(continued)

(continued)

	T									
	77 K		100 K		213 K		233 K		300 K	
x	λ_{peak}	λ_{co}	λ_{peak}	λ_{co}	λ_{peak}	λ_{co}	λ_{peak}	λ_{co}	λ_{peak}	λ_{co}
0.465	2.23	2.38	2.21	2.38	2.16	2.41	2.16	2.41	2.17	2.41
0.470	2.20	2.34	2.18	2.35	2.13	2.37	2.13	2.37	2.15	2.38
0.475	2.16	2.3	2.15	2.31	2.10	2.34	2.11	2.34	2.12	2.35
0.480	2.13	2.27	2.11	2.28	2.08	2.31	2.08	2.31	2.10	2.32
0.485	2.10	2.24	2.08	2.24	2.05	2.28	2.05	2.28	2.07	2.29
0.490	2.07	2.20	2.06	2.21	2.03	2.25	2.03	2.25	2.05	2.26
0.495	2.04	2.17	2.03	2.18	2.00	2.22	2.00	2.22	2.03	2.23
0.500	2.01	2.14	2.00	2.15	1.98	2.19	1.98	2.19	2.01	2.21
0.505	1.99	2.11	1.97	2.12	1.95	2.16	1.96	2.17	1.98	2.18
0.510	1.96	2.08	1.95	2.09	1.93	2.13	1.94	2.14	1.96	2.15
0.515	1.93	2.05	1.92	2.06	1.91	2.11	1.91	2.11	1.94	2.13
0.520	1.91	2.02	1.90	2.04	1.89	2.08	1.89	2.09	1.92	2.10
0.525	1.88	2.00	1.87	2.01	1.86	2.05	1.87	2.06	1.90	2.08
0.530	1.86	1.97	1.85	1.98	1.84	2.03	1.85	2.04	1.88	2.06
0.535	1.84	1.95	1.83	1.96	1.82	2.01	1.83	2.01	1.86	2.03
0.540	1.81	1.92	1.80	1.93	1.80	1.98	1.81	1.99	1.84	2.01
0.545	1.79	1.90	1.78	1.91	1.78	1.96	1.79	1.97	1.82	1.99
0.550	1.77	1.87	1.76	1.89	1.76	1.94	1.77	1.94	1.81	1.97
0.555	1.75	1.85	1.74	1.86	1.75	1.92	1.75	1.92	1.79	1.95
0.560	1.73	1.83	1.72	1.84	1.73	1.89	1.73	1.90	1.77	1.93
0.565	1.71	1.81	1.70	1.82	1.71	1.87	1.72	1.88	1.75	1.91
0.570	1.69	1.78	1.68	1.80	1.69	1.85	1.70	1.86	1.74	1.89
0.575	1.67	1.76	1.66	1.78	1.67	1.83	1.68	1.84	1.72	1.87
0.580	1.65	1.74	1.64	1.76	1.66	1.81	1.67	1.82	1.70	1.85
0.585	1.63	1.72	1.62	1.74	1.64	1.79	1.65	1.80	1.69	1.83
0.590	1.61	1.70	1.61	1.72	1.62	1.77	1.63	1.78	1.67	1.81
0.595	1.59	1.68	1.59	1.7	1.61	1.76	1.62	1.77	1.66	1.79
0.600	1.58	1.67	1.57	1.68	1.59	1.74	1.60	1.75	1.64	1.78
0.605	1.56	1.65	1.56	1.66	1.58	1.72	1.59	1.73	1.63	1.76
0.610	1.54	1.63	1.54	1.64	1.56	1.70	1.57	1.71	1.61	1.74
0.615	1.53	1.61	1.52	1.63	1.55	1.69	1.56	1.70	1.60	1.72
0.620	1.51	1.60	1.51	1.61	1.53	1.67	1.54	1.68	1.59	1.71
0.625	1.50	1.58	1.49	1.59	1.52	1.65	1.53	1.66	1.57	1.69
0.630	1.48	1.56	1.48	1.58	1.50	1.64	1.51	1.65	1.56	1.68
0.635	1.46	1.55	1.46	1.56	1.49	1.62	1.50	1.63	1.55	1.66
0.640	1.45	1.53	1.45	1.54	1.48	1.61	1.49	1.62	1.53	1.65
0.645	1.44	1.51	1.43	1.53	1.46	1.59	1.47	1.60	1.52	1.63
0.650	1.42	1.50	1.42	1.51	1.45	1.58	1.46	1.59	1.51	1.62
0.655	1.41	1.48	1.41	1.50	1.44	1.56	1.45	1.57	1.49	1.60
0.660	1.39	1.47	1.39	1.48	1.43	1.55	1.44	1.56	1.48	1.59
0.665	1.38	1.46	1.38	1.47	1.41	1.53	1.42	1.54	1.47	1.57
0.670	1.37	1.44	1.37	1.46	1.40	1.52	1.41	1.53	1.46	1.56
0.675	1.36	1.43	1.36	1.44	1.39	1.51	1.40	1.52	1.45	1.55
0.680	1.34	1.42	1.34	1.43	1.38	1.49	1.39	1.50	1.44	1.53

(continued)

(continued)

	T									
	77 K		100 K		213 K		233 K		300 K	
x	λ_{peak}	λ_{co}	λ_{peak}	λ_{co}	λ_{peak}	λ_{co}	λ_{peak}	λ_{co}	λ_{peak}	λ_{co}
0.685	1.33	1.40	1.33	1.42	1.37	1.48	1.38	1.49	1.42	1.52
0.690	1.32	1.39	1.32	1.40	1.36	1.47	1.37	1.48	1.41	1.51
0.695	1.31	1.38	1.31	1.39	1.34	1.45	1.36	1.46	1.40	1.50
0.700	1.29	1.36	1.30	1.38	1.33	1.44	1.34	1.45	1.39	1.48

4 Intrinsic carrier concentrations n_i (cm^{-3}) *(from (A.4) in Appendix Part II)*

	T				
x	30 *K*	77 *K*	100 *K*	200 *K*	300 *K*
0.17	4.28×10^{13}	1.51×10^{15}	3.33×10^{15}	2.30×10^{16}	6.62×10^{16}
0.175	9.68×10^{12}	9.53×10^{14}	2.40×10^{15}	2.00×10^{16}	6.07×10^{16}
0.18	2.06×10^{12}	5.70×10^{14}	1.66×10^{15}	1.71×10^{16}	5.55×10^{16}
0.185	4.30×10^{11}	3.31×10^{14}	1.12×10^{15}	1.46×10^{16}	5.04×10^{16}
0.19	8.86×10^{10}	1.89×10^{14}	7.43×10^{14}	1.23×10^{16}	4.57×10^{16}
0.195	1.81×10^{10}	1.07×10^{14}	4.88×10^{14}	1.03×10^{16}	4.13×10^{16}
0.2	3.67×10^{9}	6.04×10^{13}	3.19×10^{14}	8.63×10^{15}	3.73×10^{16}
0.205	7.39×10^{8}	3.39×10^{13}	2.08×10^{14}	7.19×10^{15}	3.35×10^{16}
0.21	1.48×10^{8}	1.89×10^{13}	1.35×10^{14}	5.97×10^{15}	3.01×10^{16}
0.215	2.97×10^{7}	1.06×10^{13}	8.72×10^{13}	4.94×10^{15}	2.70×10^{16}
0.22	5.93×10^{6}	5.88×10^{12}	5.63×10^{13}	4.08×10^{15}	2.42×10^{16}
0.225	1.18×10^{6}	3.27×10^{12}	3.63×10^{13}	3.37×10^{15}	2.16×10^{16}
0.23		1.82×10^{12}	2.34×10^{13}	2.78×10^{15}	1.93×10^{16}
0.235		1.01×10^{12}	1.50×10^{13}	2.28×10^{15}	1.72×10^{16}
0.24		5.58×10^{11}	9.67×10^{12}	1.88×10^{15}	1.54×10^{16}
0.245		3.09×10^{11}	6.21×10^{12}	1.54×10^{15}	1.37×10^{16}
0.25		1.71×10^{12}	3.99×10^{12}	1.27×10^{15}	1.22×10^{16}
0.255		9.44×10^{10}	2.56×10^{12}	1.04×10^{15}	1.09×10^{16}
0.26		5.21×10^{10}	1.64×10^{12}	8.54×10^{14}	9.65×10^{15}
0.265		2.88×10^{10}	1.05×10^{12}	7.00×10^{14}	8.58×10^{15}
0.27		1.59×10^{10}	6.73×10^{11}	5.74×10^{14}	7.63×10^{15}
0.275		8.74×10^{9}	4.30×10^{11}	4.70×10^{14}	6.77×10^{15}
0.28		4.82×10^{9}	2.75×10^{11}	3.85×10^{14}	6.02×10^{15}
0.285		2.65×10^{9}	1.76×10^{11}	3.15×10^{14}	5.34×10^{15}
0.29		1.46×10^{9}	1.13×10^{11}	2.58×10^{14}	4.74×10^{15}
0.295		8.04×10^{8}	7.19×10^{10}	2.11×10^{14}	4.20×10^{15}
0.3		4.42×10^{8}	4.59×10^{10}	1.73×10^{14}	3.73×10^{15}
0.305		2.43×10^{8}	2.93×10^{10}	1.41×10^{14}	3.31×10^{15}
0.31		1.34×10^{8}	1.87×10^{10}	1.15×10^{14}	2.93×10^{15}
0.315		7.35×10^{7}	1.19×10^{10}	9.42×10^{13}	2.60×10^{15}
0.32		4.04×10^{7}	7.61×10^{9}	7.70×10^{13}	2.30×10^{15}

(continued)

(continued)

	T				
x	30 K	77 K	100 K	200 K	300 K
0.325		2.22×10^7	4.85×10^9	6.29×10^{13}	2.04×10^{15}
0.33		1.22×10^7	3.09×10^9	5.13×10^{13}	1.81×10^{15}
0.335		6.68×10^6	1.97×10^9	4.19×10^{13}	1.60×10^{15}
0.34		3.67×10^6	1.25×10^9	3.42×10^{13}	1.42×10^{15}
0.345		2.01×10^6	7.99×10^8	2.79×10^{13}	1.25×10^{15}
0.35		1.10×10^6	5.09×10^8	2.28×10^{13}	1.11×10^{15}
0.355			3.24×10^8	1.86×10^{13}	9.81×10^{14}
0.36			2.06×10^8	1.51×10^{13}	8.67×10^{14}
0.365			1.31×10^8	1.23×10^{13}	7.67×10^{14}
0.37			8.32×10^7	1.00×10^{13}	6.78×10^{14}
0.375			5.28×10^7	8.19×10^{12}	5.99×10^{14}
0.38			3.35×10^7	6.67×10^{12}	5.30×10^{14}
0.385			2.13×10^7	5.43×10^{12}	4.68×10^{14}
0.39			1.35×10^7	4.42×10^{12}	4.13×10^{14}
0.395			8.57×10^6	3.59×10^{12}	3.65×10^{14}
0.4			5.43×10^6	2.92×10^{12}	3.22×10^{14}
0.405			3.44×10^6	2.38×10^{12}	2.84×10^{14}
0.41			2.18×10^6	1.93×10^{12}	2.51×10^{14}
0.415			1.38×10^6	1.57×10^{12}	2.21×10^{14}
0.42				1.27×10^{12}	1.95×10^{14}
0.425				1.04×10^{12}	1.72×10^{14}
0.43				8.40×10^{11}	1.52×10^{14}
0.435				6.82×10^{11}	1.34×10^{14}
0.44				5.53×10^{11}	1.18×10^{14}
0.443				4.875×10^{11}	1.09×10^{14}

5 Electron effective masses at the bottom of conduction band m_0^*/m_0 (from (A.12) in Appendix Part II)

	T					
x	4.2 K	30 K	77 K	100 K	200 K	300 K
0.170	0.0015	0.0023	0.0038	0.0046	0.0077	0.0107
0.175	0.0023	0.0031	0.0045	0.0053	0.0083	0.0113
0.180	0.0030	0.0038	0.0053	0.0060	0.0089	0.0118
0.185	0.0038	0.0046	0.0060	0.0066	0.0095	0.0124
0.190	0.0045	0.0053	0.0067	0.0073	0.0102	0.0129
0.195	0.0053	0.0060	0.0074	0.0080	0.0108	0.0135
0.200	0.0060	0.0068	0.0081	0.0087	0.0114	0.0140
0.205	0.0068	0.0075	0.0087	0.0094	0.0120	0.0146
0.210	0.0075	0.0082	0.0094	0.0100	0.0126	0.0151
0.215	0.0082	0.0089	0.0101	0.0107	0.0132	0.0156

(continued)

(continued)

	T					
x	4.2 K	30 K	77 K	100 K	200 K	300 K
0.220	0.0090	0.0096	0.0108	0.0113	0.0138	0.0162
0.225	0.0097	0.0103	0.0114	0.0120	0.0144	0.0167
0.230	0.0104	0.0110	0.0121	0.0126	0.0150	0.0172
0.235	0.0111	0.0117	0.0128	0.0133	0.0156	0.0178
0.240	0.0118	0.0124	0.0134	0.0139	0.0161	0.0183
0.245	0.0125	0.0131	0.0141	0.0146	0.0167	0.0188
0.250	0.0132	0.0137	0.0147	0.0152	0.0173	0.0193
0.255	0.0139	0.0144	0.0154	0.0158	0.0179	0.0199
0.260	0.0146	0.0151	0.016	0.0165	0.0184	0.0204
0.265	0.0152	0.0157	0.0167	0.0171	0.0190	0.0209
0.270	0.0159	0.0164	0.0173	0.0177	0.0196	0.0214
0.275	0.0166	0.0171	0.0179	0.0183	0.0202	0.0219
0.280	0.0173	0.0177	0.0186	0.0190	0.0207	0.0224
0.285	0.0179	0.0184	0.0192	0.0196	0.0213	0.0230
0.290	0.0186	0.0190	0.0198	0.0202	0.0218	0.0235
0.295	0.0193	0.0197	0.0204	0.0208	0.0224	0.0240
0.300	0.0199	0.0203	0.0211	0.0214	0.0230	0.0245
0.305	0.0206	0.0210	0.0217	0.0220	0.0235	0.0250
0.310	0.0212	0.0216	0.0223	0.0226	0.0241	0.0255
0.315	0.0219	0.0222	0.0229	0.0232	0.0246	0.0260
0.320	0.0225	0.0229	0.0235	0.0238	0.0252	0.0265
0.325	0.0232	0.0235	0.0241	0.0244	0.0257	0.0270
0.330	0.0238	0.0241	0.0247	0.0250	0.0263	0.0275
0.335	0.0244	0.0248	0.0253	0.0256	0.0268	0.0280
0.340	0.0251	0.0254	0.0259	0.0262	0.0273	0.0285
0.345	0.0257	0.0260	0.0265	0.0268	0.0279	0.0290
0.350	0.0263	0.0266	0.0271	0.0274	0.0284	0.0295
0.355	0.0270	0.0272	0.0277	0.028	0.0290	0.0300
0.360	0.0276	0.0279	0.0283	0.0285	0.0295	0.0305
0.365	0.0282	0.0285	0.0289	0.0291	0.0300	0.0310
0.370	0.0288	0.0291	0.0295	0.0297	0.0306	0.0315
0.375	0.0295	0.0297	0.0301	0.0303	0.0311	0.0319
0.38	0.0301	0.0303	0.0307	0.0309	0.0316	0.0324
0.385	0.0307	0.0309	0.0313	0.0314	0.0322	0.0329
0.390	0.0313	0.0315	0.0318	0.0320	0.0327	0.0334
0.395	0.0319	0.0321	0.0324	0.0326	0.0332	0.0339
0.400	0.0326	0.0327	0.0330	0.0332	0.0338	0.0344
0.405	0.0332	0.0333	0.0336	0.0337	0.0343	0.0349
0.410	0.0338	0.0339	0.0342	0.0343	0.0348	0.0354
0.415	0.0344	0.0345	0.0348	0.0349	0.0354	0.0359
0.420	0.0350	0.0351	0.0353	0.0354	0.0359	0.0363
0.425	0.0356	0.0357	0.0359	0.0360	0.0364	0.0368
0.430	0.0362	0.0363	0.0365	0.0366	0.0369	0.0373
0.435	0.0368	0.0369	0.0371	0.0371	0.0375	0.0378
0.440	0.0374	0.0375	0.0376	0.0377	0.0380	0.0383
0.443	0.0378	0.0379	0.0380	0.0381	0.0383	0.0386

II Some Formulas

1. The energy band gap of $Hg_{1-x}Cd_xTe$:

$$E_g(x, T) = -0.295 + 1.87x - 0.28x^2 + (6 - 14x + 3x^2)10^{-4}T + 0.35x^4 (eV) \tag{A.1}$$

2. Peak-wavelengths of the photo-conductive response λ_{peak} for $Hg_{1-x}Cd_xTe$:

$$\begin{aligned} \lambda_{peak}(\mu m) &= \frac{A(T)}{x - B(T) - C(T)\log(d)}, \\ A(T) &= 0.7 + 2.0 \times 10^{-4}T + 1.66 \times 10^{-6}T^2 \\ B(T) &= 0.162 - 2.8 \times 10^{-4}T - 2.29 \times 10^{-7}T^2 \\ C(T) &= 3.5 \times 10^{-3} - 3.0 \times 10^{-5}T - 5.85 \times 10^{-8}T^2. \end{aligned} \tag{A.2}$$

The thickness, d, is in μm.

3. Cut-off wavelengths $\lambda_{co}(\mu m)$ for $Hg_{1-x}Cd_xTe$:

$$\begin{aligned} \lambda_{co} &= \frac{a(T)}{x - b(T) - c(T)\log(d)}, \\ a(T) &= 0.7 + 6.7 \times 10^{-4}T + 7.28 \times 10^{-8}T^2 \\ b(T) &= 0.162 - 2.6 \times 10^{-4}T - 1.37 \times 10^{-7}T^2 \\ c(T) &= 4.9 \times 10^{-3} + 3.0 \times 10^{-5}T + 3.51 \times 10^{-8}T^2. \end{aligned} \tag{A.3}$$

The thickness, d, is in μm.

4. Intrinsic carrier concentrations for $Hg_{1-x}Cd_xTe$:

$$n_i = (1 + 3.25k_BT/E_g) \cdot 9.56 \cdot 10^{14} E_g^{3/2} T^{3/2} \left[1 + 1.9E_g^{3/4} \exp(E_g/2k_BT)\right]^{-1}$$
$$0.17 < x < 0.443 \text{ between } 4.2\,K < T < 300\,K \tag{A.4}$$

5. Electron mobilities for $Hg_{1-x}Cd_xTe$:

$$\mu_n = [8.75 \cdot 10^{-4}x - 1.044 \cdot 10^{-4}]^{-1} \tag{A.5}$$

6. Dielectric constants for $Hg_{1-x}Cd_xTe$:

$$\varepsilon_\infty = 15.19 - 14.52x + 11.06x^2 - 4.24x^3 \tag{A.6}$$

7. Absorption coefficient for $Hg_{1-x}Cd_xTe$:

$$\alpha = \alpha_0 \exp\left[\frac{\sigma(E - E_0)}{k_BT}\right], (E < E_g) \tag{A.7}$$

$$\alpha = \alpha_g \exp[\beta(E - E_g)]^{1/2}(\mathrm{cm}^{-1}), (E \geq E_g) \tag{A.8}$$

Here,

$$\begin{aligned}
\ln \alpha_0 &= -18.5 + 45.68x \\
E_0 &= -0.355 + 1.77x \\
\sigma / k_B T &= (\ln \alpha_g - \ln \alpha_0)/(E_g - E_0) \\
\alpha_g &= -65 + 1.88T + (8694 - 10.31T)x \\
E_g(x, T) &= -0.295 + 1.87x - 0.28x^2 + (6 - 14x + 3x^3)(10^{-4})T + 0.35x^4 \\
\beta &= -1 + 0.083T + (21 - 0.13T)x
\end{aligned} \tag{A.9}$$

for ($0.165 \leqslant x \leqslant 0.443$), and ($4.2\,\mathrm{K} \leqslant T \leqslant 300\,\mathrm{K}$).
The absorption coefficient α_g at the gap energy for $Hg_{1-x}Cd_xTe$ at 300 K and 77 K, respectively are:

$$\alpha_g = 500 + 5600x, \quad \text{at}300\,\mathrm{K} \tag{A.10}$$

$$\alpha_g = 80 + 7900x, \quad \text{at}77\,\mathrm{K} \tag{A.11}$$

8. Electron effective mass at the bottom of conduction band for $Hg_{1-x}Cd_xTe$:

$$\frac{m_0^*}{m_0} = 0.05966 \frac{E_g(E_g + 1)}{E_g + 0.667} \tag{A.12}$$

Index

MIX
Papier aus verantwortungsvollen Quellen
Paper from responsible sources
FSC® C105338

If you have any concerns about our products,
you can contact us on
ProductSafety@springernature.com

In case Publisher is established outside the EU,
the EU authorized representative is:
Springer Nature Customer Service Center GmbH
Europaplatz 3, 69115 Heidelberg, Germany

Printed by Libri Plureos GmbH
in Hamburg, Germany